The Global Casino

The Global Casino is an introduction to environmental issues which deals both with the workings of the physical environment and with the political, economic and social frameworks in which the issues occur. Using examples from all over the world, the book highlights the underlying causes behind environmental problems, the human actions which have made them issues, and the hopes for solutions. It is a book about the human impact on the environment and the ways in which the natural environment impacts human society.

The sixth edition has been fully revised and updated throughout, with new case studies, figures, and online resources including a complete lecture course for tutors and multiple-choice questions for students. New concepts and topics covered for the first time in this edition include the green economy, the forest transition model, marine microplastic pollution, urban disasters, decommissioning of big dams, and the start of the Anthropocene. Recent international initiatives covered include the Paris Agreement on climate change, the Aichi Biodiversity Targets, and the Sendai Framework for managing disaster risk. New case studies include Morocco's Noor concentrated solar power plant, desert recovery in Kuwait, and river management on the Huang Ho.

Eighteen chapters on key issues follow three initial chapters which outline the background contexts of the physical and human environments and the concept of sustainable development. Each chapter provides historical context for key issues, outlines why they have arisen, and highlights areas of controversy and uncertainty to appraise how issues can be resolved both technically and in political and economic frameworks. Each chapter also contains an updated critical guide to further reading – many of them open access – and websites, as well as discussion points and essay questions. The text can be read in its entirety or individual chapters adopted as standalone reading.

This book is an essential resource for students of the environment, geography, earth sciences and development studies. It provides comprehensive and inspirational coverage of all the major global environmental issues of the day in a style that is clear and critical.

Nick Middleton is a Fellow and Lecturer in Geography at St Anne's College, Oxford University. He also works as an environmental consultant and freelance author, having written more than 300 articles in journals, magazines and newspapers, and 25 books.

The Global Casino
An Introduction to Environmental Issues

Sixth Edition

Nick Middleton

Routledge
Taylor & Francis Group

LONDON AND NEW YORK

Sixth edition published 2019
by Routledge

2 Park Square, Milton Park, Abingdon, Oxon, OX14 4RN
and by Routledge
711 Third Avenue, New York, NY 10017

Routledge is an imprint of the Taylor & Francis Group, an informa business

First edition published by Edward Arnold 1995
Fifth edition published by Routledge 2013

British Library Cataloguing-in-Publication Data
A catalogue record for this book is available from the British Library

Library of Congress Cataloging-in-Publication Data
Names: Middleton, Nick, author.
Title: The global casino : an introduction to environmental issues /
    Nick Middleton.
Description: Sixth edition. | New York : Routledge, 2019. | "[Fifth edition
    published by Routledge 2013]"—T.p. verso. | Includes bibliographical
    references and index.
Identifiers: LCCN 2018018330| ISBN 9781138067844 (hardback : alk. paper) |
    ISBN 9781138067868 (paperback : alk. paper) | ISBN 9781315158402 (eBook)
Subjects: LCSH: Human ecology. | Environmental policy. | Environmentalism.
Classification: LCC GF41 .M525 2019 | DDC 304.2/8—dc23
LC record available at https://lccn.loc.gov/2018018330

ISBN: 978-1-138-06784-4 (hbk)
ISBN: 978-1-138-06786-8 (pbk)
ISBN: 978-1-315-15840-2 (ebk)

Typeset in Minion Pro
by Apex CoVantage, LLC

Visit the eResources: www.routledge.com/9781138067844

Contents

Figures

Tables

Preface

This book is about environmental issues: concerns that have arisen as a result of the human impact on the environment and the ways in which the natural environment affects human society. The book deals with both the workings of the physical environment and the political, economic and social frameworks in which the issues occur. Using examples from all over the world, I aim to highlight the underlying causes behind environmental problems, the human actions which have made them issues, and the hopes for solutions.

Eighteen chapters on key issues follow the three initial chapters that outline the background contexts of the physical and human environments and introduce the concept of sustainable development. The conclusion complements the book's thematic approach by looking at the issues and efforts towards sustainable development in a regional context. The organization of the book allows it to be read in its entirety or dipped into for any particular topic, since each chapter stands on its own. Each chapter sets the issue in a historical context, outlines why the issue has arisen, highlights areas of controversy and uncertainty, and appraises how problems are being, and can be, resolved, both technically and in political and economic frameworks. Information in every chapter of this sixth edition has been expanded and updated to keep pace with the rapid increase in research and understanding of the issues. The chapters are followed by expanded critical guides to further reading on the subjects – including some sources freely available online – guides to relevant sites on the Web and sets of questions that can be used to spark discussion or as essay questions.

I decided on the title *The Global Casino* because there are many parallels between the issues discussed in this book and the workings of a gambling joint. Money and economics underlie many of the 18 issues covered here, which can be thought of as different games in the global casino, separate yet interrelated. Just like a casino, environmental issues involve winners and losers. The casino's chance element and the players' imperfect knowledge of the outcomes of their actions are relevant in that our understanding of how the Earth works is far from perfect. The casino metaphor also works on a socio-economic level, since some individuals and groups of individuals can afford actively to take part in the games while others are less able. Some groups are more responsible for certain issues than others, yet those who have little influence are still affected by the consequences. Different individuals and groups of people also choose to play the different games in different ways, reflecting their cultural, economic and political backgrounds and the information available to them.

The stakes are high: some observers believe that the global scale on which many of the issues occur represents humankind gambling with the very future of the planet itself. Everyone who reads this book has some part to play in the 'Global Casino'. I hope that the information presented here will allow those players to participate with a reasonable knowledge of how the games work, the consequences of losing, and the benefits that can be derived from winning.

Acknowledgements

I am indebted to many people who have helped in a variety of ways during the research and writing of six editions of this book, the first published in 1995. I would particularly like to thank the innumerable undergraduates at St Anne's College, Oxford (plus many at Oriel College for the early editions) who have been exposed to my efforts at communicating the essence of environmental issues. These countless tutorials have provided the foundations for this book. My thanks also go to several of the technical staff at the School of Geography and the Environment, University of Oxford for their patient and efficient help, particularly Ailsa Allen. Thanks also to Sandra Mather of the University of Liverpool who very efficiently drew the new figures for the sixth edition, and two St Anne's students – Kate Bevan and Laura Stone – who carefully put together the lectures that accompany this latest edition. I thank my editors at Routledge, and previously at Arnold/Hodder, for their encouragement and inputs.

The publishers and I are grateful to the following for allowing me to reproduce their photographs: Getty Images (Figures 1.6, 1.13, 1.15, 3.10, 5.8, 6.6, 6.9, 7.6, 9.4, 11.3, 12.2, 12.7, 13.4, 13.6, 13.8, 13.11, 13.12, 14.5, 14.10, 15.1, 16.1, 16.9, 17.2, 17.4, 17.5, 18.6, 18.8, 18.11, 19.2, 19.4, 20.2, 20.5, 21.3, 22.2, 22.9); NASA Earth Observatory (Figures 4.2, 5.7, 6.13, 9.3, 12.5, 13.7, 14.9, 19.5, 21.8, 22.4), and Wellcome Collection. CC BY (Figures 15.5, 16.4, 19.1, 21.7, 21.9). All the other photographs are my own.

Thanks are also due to the following for allowing me to reproduce figures: Cambridge University Press (Figures 2.3, 2.6, 2.9, 14.4, 14.7); Convention on Long-Range Transboundary Air Pollution Steering Body to the Cooperative Programme for Monitoring and Evaluation of the Long-Range Transmission of Air Pollutants in Europe (EMEP) (Figure 12.4); Elsevier (Figures 12.11, 19.3); European Environment Agency (Figures 8.4, 16.10, 17.11, 22.5); Geological Society (Figure 17.3); Global Footprint Network (Figure 22.10); Intergovernmental Panel on Climate Change (Figures 11.1, 11.2, 11.4); International Institute for Land Reclamation and Improvement (Figure 8.12); John Wiley and Sons (Figures 5.3, 9.6); McGill-Queen's University Press (Figure 11.6); Dr Nicholas Howden, University of Bristol (Figure 8.3); Oxford University Press (Figures 12.8, 14.2); Sage Publications (Figure 1.12); Secretariat, Convention on Biological Diversity (Figure 15.3); Swiss Re Institute, sigma No 2/2017 (Figure 21.2); Taylor & Francis (Figures 1.1, 9.1, 13.2, 21.5); The International Tanker Owners Pollution Federation Limited, or ITOPF (Figure 6.12); The Royal Swedish Academy of Sciences (Figure 19.6); UK Environment Agency (Figure 21.14 which contains public sector information licensed under the Open Government Licence v3.0); United Nations (Figures 2.4, 3.5, 4.1, 5.1, 6.2, 18.12); World Bank (Figure 21.11); World Energy Council (Figure 16.8).

Every effort has been made to trace the copyright holders of reproduced material. If, however, there are inadvertent omissions and/or inappropriate attributions these can be rectified in any future editions.

1

The physical environment

The term environment is used in many ways. This book is about issues that arise from the physical environment, which is made up of the living (biotic) and non-living (abiotic) things and conditions that characterize the world around us. While this is the central theme, the main reason for the topicality of the issues covered here is the way in which people interact with the physical environment. Hence, it is pertinent also to refer to the social, economic and political environments to describe those human conditions characteristic of certain places at particular times, and to explain why conflict has arisen between human activity and the natural world. This chapter looks at some of the basic features of the physical environment, while Chapter 2 is concerned with the human factors that affect the ways in which the human race interacts with the physical world.

CLASSIFYING THE NATURAL WORLD

Geography, like other academic disciplines, classifies things in its attempt to understand how they work. The physical environment can be classified in numerous ways, but one of the most commonly used classifications is that which breaks it down into four interrelated spheres: the lithosphere, the atmosphere, the biosphere and the hydrosphere. These four basic elements of the natural world can be further subdivided. The lithosphere, for example, is made up of rocks that are typically classified according to their modes of formation (igneous, metamorphic and sedimentary); these rock types are further subdivided according to the processes that formed them and other factors such as their chemical composition. Similarly, the workings of the atmosphere are manifested at the Earth's surface by a typical distribution of climates; the biosphere is made up of many types of flora and fauna; and the hydrosphere can be subdivided according to its chemical constituents (fresh water and saline, for example), or the condition or phase of the water: solid ice, liquid water or gaseous vapour.

These aspects of the natural world overlap and interact in many different ways. The nature of the soil in a particular place, for example, reflects the underlying rock type, the climatic conditions of the area, the plant and animal matter typical of the region, and the quantity and quality of water available. Suites of characteristics are combined in particular areas called ecosystems. These ecosystems can also be classified in many ways. One approach uses the amount of organic matter or biomass produced per year – the net production – which is simply the solar energy fixed in the biomass minus the energy used in producing it by respiration (see below). The annual net primary production of carbon, a basic component of all living organisms, by major world ecosystem types is shown in Table 1.1. Clear differences are immediately discernible between highly productive ecosystems such as forests, marshes, estuaries and reefs, and less productive places such as deserts, tundras and the open ocean. All of the data are averaged and variability around the mean is perhaps greatest for agricultural ecosystems which, where intensively managed, can reach productivities as high as any natural ecosystem. One of the main reasons for agriculture's low average is the fact that fields are typically bare of vegetation for significant periods between harvest and sowing.

One of the main factors determining productivity is the availability of nutrients, key substances for life on Earth: a lack of nutrients is often put forward to explain the low productivity in the open oceans, for example. Climate is another important factor. Warm, wet climates promote higher productivity than cold, dry ones. Differences in productivity may also go some way towards explaining the general trend of increasing diversity of plant and animal species from the poles to the equatorial regions. Despite many regional exceptions such as mountaintops and deserts, this latitudinal gradient of diversity is a striking characteristic of nature that fossil evidence suggests has been present in all geological epochs. The relationship with productivity is not straightforward, however, and many other hypotheses have been advanced, such as the suggestion that minor disturbances promote diversity by preventing a few species from dominating and excluding others (Connell, 1978).

Table 1.1 Annual net primary production of carbon by major world ecosystem types

Ecosystem type	Mean net primary productivity (g C/m²/year)	Total net primary production (billion tonnes/C/year)
Tropical rain forest	900	15.3
Tropical seasonal forest	675	5.1
Temperate evergreen forest	585	2.9
Temperate deciduous forest	540	3.8
Boreal forest	360	4.3
Woodland and shrubland	270	2.2
Savanna	315	4.7
Temperate grassland	225	2.0
Tundra and alpine	65	0.5
Desert scrub	32	0.6
Rock, ice and sand	1.5	0.04
Agricultural land	290	4.1
Swamp and marsh	1125	2.2
Lake and stream	225	0.6
Total land	324*	48.3
Open ocean	57	18.9
Upwelling zones	225	0.1
Continental shelf	162	4.3
Algal bed and reef	900	0.5
Estuaries	810	1.1
Total oceans	69*	24.9
Total for biosphere	144*	73.2

Note: *The means for land, oceans and biosphere are weighted according to the areas covered by specific ecosystem types.
Source: after Whittaker and Likens (1973).

The relationships between climate and the biosphere are also reflected on the global scale in maps of vegetation and climate, the one reflecting the other. Figure 1.1 shows the world's morphoclimatic regions, which are a combination of both factors. Despite wide internal variations, immense continental areas clearly support distinctive forms of vegetation that are adapted to a broad climatic type. Such great living systems, which also support distinctive animals and to a lesser extent distinctive soils, are called biomes, a concept seldom applied to aquatic zones. Different ecologists produce various lists of biomes and the following eight-fold classification may be considered conservative (Colinvaux, 1993):

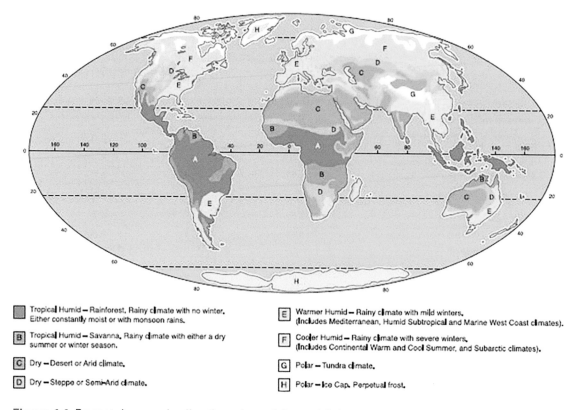

Tropical Humid – Rainforest. Rainy climate with no winter. Either constantly moist or with monsoon rains.

B Tropical Humid – Savanna. Rainy climate with either a dry summer or winter season.

C Dry – Desert or Arid climate.

D Dry – Steppe or Semi-Arid climate.

E Warmer Humid – Rainy climate with mild winters. (Includes Mediterranean, Humid Subtropical and Marine West Coast climates).

F Cooler Humid – Rainy climate with severe winters. (Includes Continental Warm and Cool Summer, and Subarctic climates).

G Polar – Tundra climate.

H Polar – Ice Cap. Perpetual frost.

Figure 1.1 Present-day morphoclimatic regions of the world's land surface (Williams *et al.,* 1993).

1   tundra
2   coniferous forest (also known as boreal forest or taiga)
3   temperate forest
4   tropical rain forest
5   tropical savanna
6   temperate grassland
7   desert
8   maquis (also known as chaparral).

A striking aspect of the tundra biome is the absence of trees. Vegetation consists largely of grasses and other herbs, mosses, lichens and some small woody plants which are adapted to a short summer growing season. The tundra is also notable for receiving relatively little precipitation and being generally poor in nutrients. The cold climate ensures that the rate of biological processes is generally slow and the shallow soils are deeply frozen (permafrost) for all or much of the year, a condition which underlies about 20 per cent of the Earth's land surface. Many animals hibernate or migrate in the colder season, while others such as lemmings live beneath the snow.

The main tundra region is located in the circumpolar lands north of the Arctic Circle, which are bordered to the south by the evergreen, needle-leaved boreal or taiga forests (Figure 1.2). Here, winters are very cold, as in the tundra, but summers are longer. Most of the trees are conifers such as pine, fir and spruce. They are tall and have a narrow, pointy shape which means that the snow tends to slide off their branches, while their needles also shed snow more easily than broad leaves. These adaptations reduce the likelihood of heavy snow breaking branches. Boreal forests are subject to periodic fires, and a burn–regeneration cycle is an important characteristic to which populations of deer, bears and insects, as well as the vegetation, are adapted. Much of the boreal forest is underlain by acid soils.

The temperate forests, by contrast, are typically deciduous, shedding their leaves each year. They are, however, like the boreal forests in that they are found almost exclusively in the northern hemisphere. This biome is characteristic of northern Europe, eastern China and eastern and Midwest USA, with small stands in the southern hemisphere in South America and New Zealand. Tall broadleaf trees dominate, the climate is seasonal but water is always abundant during the growing season, and this biome is less homogeneous than tundra or boreal forest. Amphibians, such as salamanders and frogs, are present, while they are almost totally absent from the higher-latitude biomes.

The tropical rain forest climate has copious rainfall and warm temperatures in all months of the year. The trees are always green, typically broad-leaved,

Figure 1.2 Coniferous forest in Finland, the eastern end of a broad region of boreal or taiga forest that stretches to the Russian Far East.

and most are pollinated by animals (trees in temperate and boreal forests, by contrast, are largely pollinated by wind). Many kinds of vines (lianas) and epiphytes, such as ferns and orchids, are characteristic. Most of the nutrients are stored in the biomass and the soils contain little organic matter. These forests typically display a multi-layered canopy (Figure 1.3), while, at ground level, vegetation is often sparse because of low levels of light. Above all, tropical rain forests are characterized by a large number of species of both plants and animals.

Savanna belts flank the tropical rain forests to the north and south in the African and South American tropics, a biome known as cerrado in Brazil. The trees of tropical savannas are stunted and widely spaced, which allows grass to grow between them. Herds of grazing mammals typify the savanna landscape, along with large carnivores such as lions and other big cats, jackals and hyenas. These mammals, in turn, provide a food source for large scavengers such as vultures. The climate is warm all year, but has a dry season several months long when fires are a common feature. These fires maintain the openness of the savanna ecosystem and are important in mineral cycling.

The greatest expanses of the temperate grassland biome are located in Eurasia (where they are commonly known as steppe; Figure 1.4), North America (prairie) and South America (pampa), with smaller

Figure 1.3 An area of primary tropical rain forest, an evergreen biome with great biodiversity, in Panama, Central America.

expanses in South Africa (veldt). There are certain similarities with savannas in terms of fauna and the occurrence of fire, but unlike savannas, trees are absent in temperate grasslands. The vegetation is dominated by herbaceous (i.e. not woody) plants, of which the most abundant are grasses. The climate in this biome is temperate, seasonal and dry. Typical soils tend to be deep and rich in organic matter.

In many parts of the world, where climates become drier, temperate grasslands fade into the desert biome. Hyper-arid desert supports very little plant life and is characterized by bare rock or sand dunes, but some species of flora and fauna are adapted to the high and variable temperatures – the diurnal temperature range is typically high in deserts – and the general lack of moisture. Some water is usually available via precipitation in one of its forms: most commonly rainfall or dew, but fog is important in some coastal deserts (Figure 1.5). Sporadic, sometimes intense, rain promotes rapid growth of annual plants and animals such as locusts, which otherwise lie dormant for several years as seeds or eggs.

A very distinctive form of vegetation is commonly associated with Mediterranean climates in which summers are hot and dry and winters are cool and moist. It is found around much of the Mediterranean Basin (where it is known as maquis), in California (chaparral), southern Australia (mallee), Chile (mattóral) and South Africa (fynbos). Low evergreen trees (forming woodland) and shrubs (forming scrub) have thick bark and small, hard leaves that make them tolerant to the stresses of climatic extremes and soils that are often low in nutrients. During the arid summer period this biome is frequently exposed to fire, which is important to its development and regeneration.

All these natural biomes have been affected to a greater or lesser extent by human action. Much of the maquis, for example, may represent a landscape where forests have been degraded by people, through cutting, grazing and the use of fire. The human use of fire may also be an important factor in maintaining, and possibly forming, savannas and temperate grasslands. The temperate forests have been severely altered over long histories with high population densities as people have cleared trees for farming and urban development (Figure 1.6). Conversely, biomes considered by many people to be harsh, such as the tundra and deserts, show less human impact. The anthropogenic influence is but one factor that promotes change in terrestrial as well as oceanic and freshwater ecosystems, because the interactions between the four global spheres have never been static. Better understanding of the dynamism of the natural world can be gained through a complementary way of studying the natural environment. Study of the processes that occur in natural cycles also takes us beyond description, to enable explanation.

Figure 1.4 Temperate grassland in central Mongolia is still predominantly used for grazing. In many other parts of the world such grasslands have been ploughed up for cultivation.

Figure 1.5 This strange-looking plant, the welwitschia, is found only in the Namib Desert. Its adaptations to the dry conditions include long roots to take up any moisture in the gravelly soil and the ability to take in moisture from fog through its leaves. The welwitschia's exact position in the plant kingdom is controversial, but it is grouped with the pine trees.

Figure 1.6 The US city of New York, part of one of the world's most extensive areas of urban development. Only a few of the original temperate forest trees survive in parks and gardens. Urban areas are now so widespread that they are often treated as a type of physical environment in their own right (see Chapter 10).

NATURAL CYCLES

A good means of understanding the way the natural world works is through the recognition of cycles of matter in which molecules are formed and reformed by chemical and biological reactions and are manifested as physical changes in the material concerned. The major stores and flows of water in the global hydrological cycle are shown in Figure 1.7. Most of the Earth's water (about 97 per cent) is stored in liquid form in the oceans. Of the 3 per cent fresh water, most is locked as ice in the ice caps and glaciers, and as a liquid in rocks as groundwater. Only a tiny fraction is present at any time in lakes and rivers. Water is continually exchanged between the Earth's surface and the atmosphere – where it can be present in gaseous, liquid or solid form – through evaporation, transpiration from plants and animals, and precipitation. The largest flows are directly between the ocean and the atmosphere. Smaller amounts are exchanged between the land and the atmosphere, with the difference accounted for by flows in rivers and groundwater to the oceans. Fresh water on the land is most directly useful to human society (see Chapters 8 and 9), since water is an essential prerequisite of life, but the oceans and ice caps play a key role in the workings of climate.

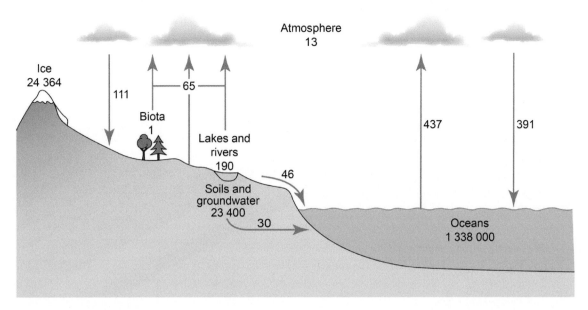

Figure 1.7 Global hydrological cycle showing major stores and flows (data from Oki and Kanae, 2006). The values in stores are in thousand km³, values of flows in thousand km³ per year.

Similar cycles, commonly referred to as biogeochemical cycles, can be identified for other forms of matter. Nutrients such as nitrogen, phosphorus and sulphur are similarly distributed among the four major environmental spheres and are continually cycled between them. Carbon is another key element for life on Earth, and the stores and flows of the carbon cycle are shown in Figure 1.8. The major stores of carbon are the oceans and rocks, particularly carbonate sedimentary rocks such as limestones, and the hydrocarbons (coal, oil and natural gas), plus 'clathrates', or gas hydrates, found mainly in high latitudes and in the oceans along continental margins. Much smaller proportions are present in the atmosphere and biosphere. The length of time carbon spends in particular stores also varies widely. Under natural circumstances, fossil carbon locked in rocks remains in these stores for millions of years. Carbon reaches these stores by the processes of sedimentation and evaporation, and is released from rocks by weathering, vulcanism and sea-floor spreading. In recent times, however, the rate of flow of carbon from some of the lithospheric stores – the hydrocarbons or fossil fuels – has been greatly increased by human action: the burning of fossil fuels, which liberates carbon by oxidation. Hence, a significant new flow of carbon between the lithosphere and the atmosphere has been introduced by human society and the natural atmospheric carbon store is being increased as a consequence (see Chapter 11).

Carbon also reaches the atmosphere through the respiration of plants and animals, which in green plants, blue-green algae and phytoplankton is part of the two-way process of photosynthesis. Photosynthesis is the chemical reaction by which these organisms convert carbon from the atmosphere, with water, to produce complex sugar compounds (which are either stored as organic matter or used by the organism) and oxygen. The reaction is written as follows:

$$6CO_2 + 6H_2O \rightarrow C_6H_{12}O_6 + 6O_2$$

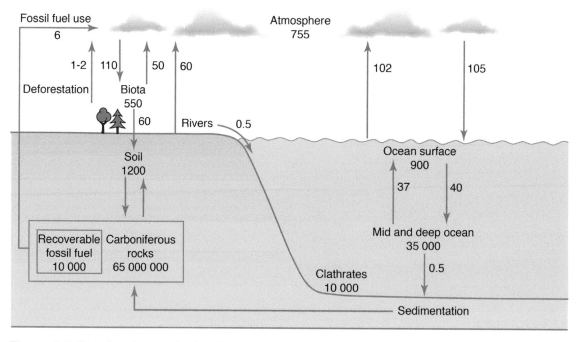

Figure 1.8 Global carbon cycle showing major stores and flows (after Schlesinger, 1991; and Grace, 2004). The values in stores are in units of Pg C, values of flows in Pg C per year. 1 Pg C = 10^{15} g C = 1 billion tonnes of carbon as CO_2.

This equation shows that six molecules of carbon dioxide and six molecules of water yield one molecule of organic matter and six molecules of oxygen. The reaction requires an input of energy from the sun, some of which is stored in chemical form in the organic matter formed.

The process of respiration is written as the opposite of the equation for photosynthesis. It is the process by which the chemical energy in organic matter is liberated by combining it with oxygen to produce carbon dioxide and water. All living things respire to produce energy for growth and the other processes of life. The chemical reaction for respiration is, in fact, exactly the same as that for combustion. Humans, for example, derive energy for their life needs from organic matter by eating (just as other animals do) and also by burning plant matter in a number of forms, such as fuelwood and fossil fuels.

The flow of converted solar energy through living organisms can be traced up a hierarchy of life-forms known as a food chain. Figure 1.9 shows a simple food chain in which solar energy is converted into chemical energy in plants (so-called producers), which are eaten by herbivores (so-called first-order consumers), which, in turn, are eaten by other consumers (primary carnivores), which are themselves eaten by secondary carnivores. An example of such a food chain on land is:

$$\text{grass} \rightarrow \text{cricket} \rightarrow \text{frog} \rightarrow \text{heron}$$

Each stage in the chain is known as a trophic level. In practice, there are usually many, often interlinked, food chains that together form a food web, but the principles are the same. At each trophic level some energy is lost by respiration, through excretory products and when dead organisms decay, so that available

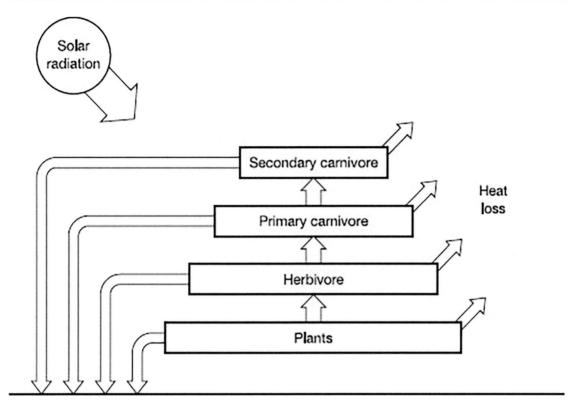

Figure 1.9 Energy flow through a food chain.

energy declines along the food chain away from the plant. In general terms, animals also tend to be bigger at each sequential trophic level, enabling them to eat their prey safely. This model helps us to explain the basic structures of natural communities: with each trophic level, less energy is available to successively larger individuals and thus the number of individuals decreases. Hence, while plants are very numerous because they receive their energy directly from the sun, they can support only successively fewer larger animals. With the exception of humans, predators at the top of food chains – sometimes called 'apex predators' – are therefore always rare.

Food chains, the carbon cycle and the hydrological cycle are all examples of 'systems' in which the individual components are all related to each other. Most of the energy that drives these systems comes from the sun, although energy from the Earth also contributes. All the cycles of energy and matter referred to in this section are affected by human action, deliberately manipulating natural cycles to human advantage. One of the human impacts on the carbon cycle has been mentioned, but humans also affect other cycles. The cycle of minerals in the rock cycle is affected by the construction industry, for example. Human activity affects the hydrological cycle by diverting natural flows: the damming of rivers (see Chapter 9) or extracting groundwater for human use (see Chapter 10). The nitrogen cycle is affected by concentrating nitrogen in particular places such as by spreading fertilizers on fields. Food chains are widely affected: human populations manipulate plants and animals to produce food (see Chapter 13).

However, since all parts of these cycles are interrelated, human intervention in one part of a cycle also affects other parts of the same and other cycles. These knock-on effects are the source of many environmental changes that are undesirable from human society's viewpoint. Our manipulation of the nitrogen cycle by using fertilizers also increases the concentration of nitrogen in rivers and lakes when excess fertilizer is washed away from farmers' fields. This can have deleterious effects on aquatic ecosystems (see Chapters 7 and 8). Excess nitrogen can also enter the atmosphere to become a precursor of acidification (see Chapter 12). One of the effects of acid rain is to accelerate the rate of weathering of some building stones. A better appreciation of these types of changes can be gained by looking at the various scales of time and space through which they occur.

TIMESCALES

Changes in the natural environment occur on a wide range of timescales. Geologists believe that the Earth is about 4600 million years old, while fossil evidence suggests that modern humans (*Homo sapiens*) appeared between 100,000 and 200,000 years before present (BP), developing from the hominids whose earliest remains, found in Africa, date to around 3.75 million BP (Figure 1.10). The very long timescales over which many changes in the natural world take place may seem at first to have little relevance for today's human society other than to have created the world we know. It is difficult for us to appreciate the age of the Earth and the thought that the present distribution of the continents dates from the break-up of the supercontinent Pangaea, which began during the Cretaceous period. Indeed, relative to the forces and changes due to tectonic movements, the human impact on the planet is minor and short-lived. However, such Earth processes do have relevance on the timescale of a human lifetime. Tectonic movements cause volcanic eruptions that can affect human society as natural disasters at the time of the event. Some volcanic eruptions also affect day-to-day human activities on slightly longer timescales, by injecting dust into the atmosphere, which affects climate, for example, and by providing raw materials from which soils are formed. This example also illustrates the fact that the same event may be interpreted as bad from a human viewpoint on one timescale (a volcanic disaster) and good on another timescale (fertile volcanic soils).

The geological period we live in, the Quaternary, constitutes a very minor portion of Earth's history, but it has been a period of great change in the world. Significant variations in global climate have occurred – notably a number of so-called Ice Ages – with associated fluctuations in ice sheets, sea levels and the distribution of plants and animals. We are living during an interglacial period, named the Holocene, which started about 11,500 years ago. This is also an epoch of environmental change, driven increasingly by the actions of modern humans. In fact, humanity's influence has become so great that many authorities now refer to the most recent epoch as the Anthropocene (see Chapter 3).

It is important to realize that the timescale we adopt for the study of natural systems can affect our understanding as well as our perception of them. Many such systems are thought to be in 'dynamic equilibrium', in which the input and output of matter and energy are balanced. This is a state or regime organized around a set of processes that maintain equilibrium by being mutually reinforcing. However, recognition of dynamic equilibrium in natural systems depends upon the timescale over which the system is studied. To take the Earth as a whole, for example, the idea of dynamic equilibrium has been proposed to explain why the temperature of the Earth has remained relatively constant for the past 4 billion years, despite the fact that the sun's heat has increased by about 25 per cent over that period. The Gaia hypothesis suggests that life on the planet has played a key role in regulating the Earth's conditions to keep it amenable to life (Lovelock, 1989). The theory is not without its critics, but even if we accept it, the dynamic equilibrium holds only for the few billion years of the Earth's existence. Astronomers predict that eventually the sun

ERA	PERIOD	EPOCH	START (million years BP)	IMPORTANT EVENTS
CENOZOIC	Quaternary	Holocene	0.01	Early civilisations
		Pleistocene	2.6	First humans
	Tertiary	Pliocene	5	First hominids
		Miocene	23	
		Oligocene	34	
		Eocene	56	
		Paleocene	66	Extinction of dinosaurs
MESOZOIC	Cretaceous		146	Main fragmentation of Pangaea
	Jurassic		200	
	Triassic		251	First birds
PALAEOZOIC	Permian		299	Formation of Pangaea
	Carboniferous		359	
	Devonian		416	
	Silurian		444	First land plants and animals
	Ordovician		488	First vertebrates
	Cambrian		542	
PROTEROZOIC			2500	
ARCHEAN			4600	Formation of Earth

Figure 1.10 Geological timescale classifying the history of the Earth.

will destroy the Earth, so that over a longer timescale, dynamic equilibrium does not apply. This example applies over a very long timescale, and from our perspective the destruction of the Earth by the sun is not imminent, but the principle is relevant to all other natural systems. Adoption of different timescales of analysis dictates which aspects of a system we see and understand because the importance of different factors changes with different timescales. Indeed, even within the lifespan of the Earth, dramatic changes are known to have occurred, such as the progression of glacial and interglacial periods. The longer the timeframe, the coarser the resolution and vice versa. In a simple example, a human being who contracts a cold might feel miserable for a few days, but in terms of that person's lifetime the cold is a very minor influence on their career. This principle is depicted for measurements of mean annual air temperature at Oxford in England in Figure 1.11.

One of the key components of natural cycles and dynamic systems is the operation of feedbacks. Feedbacks may be negative, which tend to dampen down the original effect and thereby maintain dynamic equilibrium, or they may be positive and hence tend to enhance the original effect. An example of negative feedback can be seen in the operation of the global climate system: more solar energy is received at

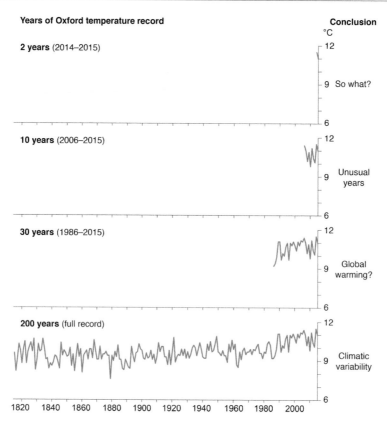

Figure 1.11 Demonstration of how the length of data record (here mean annual air temperature at Oxford, UK) can influence conclusions about environmental variability (using data 1815–2015 at www.geog.ox.ac.uk/research/climate/rms/meanair.html).

the tropics than at the poles, but the movement of the atmosphere and oceans continually redistributes heat over the Earth's surface to redress the imbalance. Positive feedbacks can result in a change from one dynamic equilibrium to another: a change often termed a 'regime shift'. If a forest is cleared by human action, for example, the soil may be eroded to the extent that recolonization by trees is impossible.

This last example illustrates another important aspect of natural cycles: the existence of thresholds. A change in a system may not occur until a threshold is reached: snow will remain on the ground, for example, until the air temperature rises above a threshold at which the snow melts. Crossing a threshold may be a function of the frequency or intensity of the force for change: a palm tree may be able to withstand, or be 'resilient' to, winds up to a certain speed, but will be blown out of the soil by a hurricane-force wind that is above the tree's threshold of resilience. Conversely, thresholds may be reached by the cumulative impacts of numerous small-scale events: regular rainfall inputs of moisture to a slope may build up to a point at which the slope fails, or in the erosion example, the gradual loss of soil reaches the point at which there is not enough soil left for trees to take root and grow.

These two illustrations of how thresholds can be crossed also embody two important ideas on the way change occurs in the physical environment. The hurricane represents a high-magnitude, low-frequency event, for which some use the term 'catastrophe'. An opposing view ascribes change in the environment to small-scale, commonly occurring processes. Of course, most environments are affected by both catastrophic and gradual changes.

To complicate things further, there may be a lag in time between the onset of the force for change and the change itself: the response time of the system. An animal seldom dies immediately upon contracting a fatal disease, for example; its body ceases to function only after a period of time. Likewise in the erosion example, trees are unable to colonize only after a certain amount of soil has been lost. Different parts of our planet have different response times, which can vary from hours to many thousands of years (Table 1.2).

Consideration of feedbacks, thresholds and lags leads to some other characteristics of natural systems – their sensitivity to forces for change, which dictates their ability to maintain or return to an original condition following a disturbance: their 'stability'. A natural system's ability to maintain its original condition with the same functions and processes is dictated by its 'resistance' to disturbance, while the ability to return to an original condition is commonly referred to as 'resilience' (see Box 8.1).

The variability in natural disturbances affecting some environments means that assuming them to be in a more or less stable dynamic equilibrium is not reasonable, however. These are commonly referred to as 'non-equilibrium' environments, of which drylands – deserts and semi-deserts – are a good example. Drylands are highly dynamic and are currently thought of as being in a constant state of change, driven by disturbances such as the variability of fire and insect attack and, perhaps most importantly, the variability of moisture from rainfall. Amounts vary widely, from one intense rainy day in a dry month through seasonal variations to longer periods such as droughts. Many aspects of the physical environment respond accordingly. Perennial plants and small animals respond particularly quickly, so that a different level of their populations can be expected at each particular time. Larger animals respond to such a dynamic environment by moving, sometimes over very large distances, to take advantage of the spatial changes in water and food availability. The dynamism of drylands makes it difficult to assess degradation in these areas (see Chapter 5).

All these considerations on changes in natural ecosystems through time can be assembled into some typical patterns that are illustrated hypothetically in Figure 1.12. The parameters represented on the y axis of the graphs could be a measurement of any physical thing, such as soil organic matter

Table 1.2 Response times of various parts of the Earth system

Component	Response time (range)	Example
Atmosphere	Hours–weeks	Daily heating/cooling – build-up of heatwave
Ocean surface	Days–months	Daily (pm) – seasonal (summer) warming
Vegetation	Hours–decades/ centuries	Sudden leaf loss to frost – tree growth to maturity
Deep ocean	100–1500 years	Time to replace deepest water
Ice sheet	100–10,000 years	Advance/retreat of ice-sheet margins – growth/decay of entire sheet

Source: after Ruddiman (2014: table 1.1).

content, species diversity, carbon dioxide concentration in the atmosphere, or the volume of water flowing along a river channel. Figure 1.12a might represent a mature forest with small variations in biomass with the seasons, and as individual trees grow and die. This constant system could equally be described as stable or as one that is in dynamic equilibrium. It contrasts with the cyclical pattern in Figure 1.12b, which could represent natural cycles of heather burning and regeneration. Figure 1.12c could illustrate natural succession of vegetation in an area with a long-term directional trend. The pattern induced by episodes such as drought, which allow recovery in systems with sufficient resilience (Figure 1.12d), contrasts with a catastrophic disturbance that exceeds resilience, so that the system crosses a threshold resulting in long-term change from one state to another – a regime shift (Figure 1.12e) – such as when soil erosion proceeds to a level where certain types of vegetation can no longer survive in the area.

The graphs shown in Figure 1.12c and Figure 1.12e also illustrate two different forms of change over time. The trend shown in Figure 1.12c is linear, while that shown in Figure 1.12e is non-linear. Non-linear systems typically display abrupt, sometimes unexpected, switches when smooth change is interrupted by a sudden drastic change to a contrasting state. Such regime shifts are often difficult to reverse, thus presenting a substantial challenge to ecosystem management and development goals. In some cases, mismanagement may accelerate or exacerbate regime shifts (see Box 6.1), often because one result of poor management has been a loss of resilience.

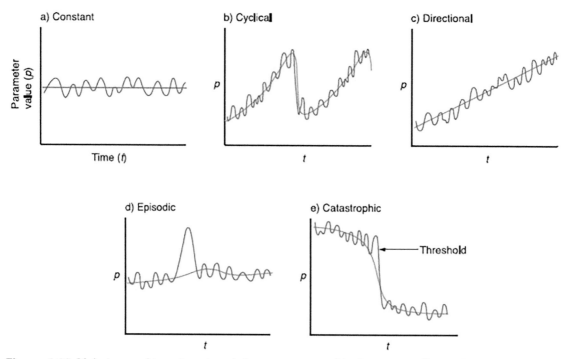

Figure 1.12 Main types of long-term trends in ecosystems, with shorter-term fluctuations superimposed (after Burt, 1994).

SPATIAL SCALES

Just as the choice of timescales is important to our understanding of the natural environment, so too is the spatial scale of analysis. Studies can be undertaken at scales that range from the microscopic – the effects of salt weathering on a sand grain, for example – through an erosion plot measured in square metres, to drainage basin studies that can reach subcontinental scales in the largest cases, to the globe itself. As with time, the resolution of analysis becomes coarser with increasing spatial scale. We draw a line on a world map to divide one biome from another, but on the ground there is usually no line, more a zone of transition, which may itself vary over different timescales.

Similarly, the types of influence that are important differ according to spatial scale. To use another example involving humans, a landslide that results in the loss of a farmer's field may have a significant impact on the farmer's ability to earn a living, but the same landslide has a minimal impact on national food production.

Spatial scale is also important because most environmental phenomena are not evenly distributed over the Earth's surface. The diversity of species, for instance, is very uneven: we think that over half of all species occur in the tropical moist forests, which occupy just 6 per cent of the land surface, and we have recognized numerous biodiversity 'hotspots' (see Chapter 15). There is also the latitudinal gradient of diversity mentioned above. The scale we adopt for an investigation may partly dictate what we study and what we do not study. We can make general statements about the Sahara Desert because it is a certain type of biome, but the Sahara is so large that it inevitably includes a variety of different conditions, some of which may not seem very desert-like. An oasis, where water is abundant, is an obvious example.

Thresholds and feedbacks also have relevance on the spatial scale. Certain areas may be more sensitive to change than others and if a threshold is crossed in these more sensitive areas, wider-scale changes may be triggered. Soil particles entrained by wind erosion from one small part of a field, for example, can initiate erosion over the whole field. On a larger scale, sensitive areas such as the Labrador–Ungava plateau of northern Canada appear to have played a key role in triggering global glaciations during the Quaternary because they were particularly susceptible to ice-sheet growth. A contemporary large-scale example can be seen in the tundra biome (Figure 1.13), which could release large amounts of methane locked in the permafrost if the global climate warms due to human-induced pollution of the atmosphere. Methane is a greenhouse gas, so positive feedback could result, enhancing the warming effect worldwide. Similar concerns surround the West Antarctic ice sheet, which may be inherently unstable and capable of future catastrophic collapse. The volume of ice it contains is equivalent to ~5 m of sea level.

TIME AND SPACE SCALES

The key factors influencing natural events also vary at different combined spatial and temporal scales. Individual waves breaking on a beach constantly modify the beach profile, which is also affected by the daily pattern of tides dictating where on the beach the waves break. Individual storms alter the beach too, as do the types of weather associated with the seasons. However, all these influences are superimposed upon the effects of factors that operate over longer timescales and larger spatial scales, such as sediment supply and the sea level itself (Clayton, 1991).

The range of temporal and spatial scales is illustrated for some biological and climatic processes in Figure 1.14. This emphasizes the fact that various processes in the natural environment (e.g. climatic changes, tornadoes) exist at specific scales, as do its elements (e.g. species, biomes). It is also important to note that no part of the physical environment is a closed self-supporting system; all are a part of larger interacting systems.

Figure 1.13 Tundra in Yukon Territory of northern Canada, part of a biome that may be particularly sensitive to a human-induced warming of global climate and that could create a positive feedback by releasing large amounts of methane, a greenhouse gas (see Chapter 11).

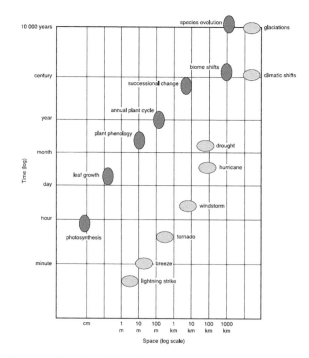

Figure 1.14 The range of temporal and spatial scales at which ecosystem processes exist and operate (after Holling, 1995).

The environmental issues in this book have arisen as a consequence of human activity conflicting with environmental systems. Resolution of such conflicts can only be based on an understanding of how natural systems work. For issues that stem from human impact upon the physical environment, as most do, we need to be able to rank the temporal and spatial scale of human impact in the natural hierarchy of influences on the natural system in question. Inevitably, we tend to focus on scales directly relevant to people, but we should not forget other scales, which may have less direct but no less significant effects. Indeed, successful management of environmental issues relies on the successful identification of appropriate scales and their linkages.

THE STATE OF OUR KNOWLEDGE

We already know a great deal about how the natural world works, but there remains a lot more to learn. We have some good ideas about the sorts of ways natural systems operate, but we remain ignorant of many of the details. Some of the difficulties involved in ascertaining these details include a lack of data and our own short period of residence on the Earth. Direct measurements using instruments are used in the contemporary era to monitor environmental processes. Historical archives, sometimes of direct measurements, otherwise of more anecdotal evidence, can extend these data back over decades and centuries. Good records of high and low water levels for the River Nile at Cairo extend from 641 to 1451 CE, although they are intermittent thereafter until the nineteenth century, and continuous monthly mean temperature and precipitation records have been kept at several European weather stations since the early eighteenth century (e.g. see the 200-year temperature record for Oxford in Figure 1.11). We have a reasonable global coverage of meteorological stations measuring temperature in a systematic manner for the period from 1850 to the present. Other types of written historical evidence date back to ancient Chinese and Mesopotamian civilizations as early as 5000 BP. The further back in time we go, however, the patchier the records become, and in some parts of the world historical records begin only in the last century.

These data gaps for historical time, and for longer time periods of thousands, tens of thousands and millions of years, can be filled in using natural archives. The geological timescale given in Figure 1.10 is based on fossil evidence. Such 'proxy' methods are based on our knowledge of the current interrelationships between the different environmental spheres. Particular plants and animals thrive in particular climatic zones, for example, so that fossils can indicate former climates. The variability of climate during an organism's lifetime can also be inferred in some cases. Study of the width of the annual growth rings of trees (Figure 1.15) gives an insight into specific ecological events that changed the tree's ability to photosynthesize and fix carbon. Essentially similar methods can be used to infer environmental variability from changes in the rate of growth of coral reefs. Other proxy 'palaeoenvironmental indicators' include pollen types found in cores of sediment taken from lake or ocean beds, and the rate of sediment accumulation in such cores can tell us something about erosion rates on the surrounding land. Landforms, too, become fossilized in landscapes to provide clues about past environmental conditions. Examples include glacial and periglacial forms in central and northern Europe, indicating colder conditions during previous glaciations, and fossilized sand dunes in the Orinoco Basin of South America, also dating from periods of high-latitude glaciation, which indicate an environment much drier than that of today.

As with instrumental data and historical archives, natural archives used as proxy variables are patchy in both their spatial and temporal extent. Warm-water coral reefs grow only in tropical waters, ice accumulates only under certain conditions, and not many trees live longer than 1000 years. Even

Figure 1.15 A cut tree trunk showing its annual growth rings containing information about growing conditions during the tree's lifetime.

instrumental data may not be perfectly reliable over long periods of time because methods and instrumentation can change, monitoring sites can be moved, and external factors may alter the nature of the reading. The availability and limitations through time and space of some of the variables used to indicate temperature, a key palaeoenvironmental variable, in the Holocene period are shown in Table 1.3.

Given these gaps in our data and understanding of environmental change through time and across space, academics have explored other kinds of knowledge about how our planet works. One additional source of information about ecosystems and resources is the traditional ecological knowledge (TEK) of indigenous people that a number of research projects have shown can complement science (see Box 11.1). Constant interaction with the physical environment enables many indigenous people to build a knowledge base of the land and develop the sensitivity to recognize critical signs and signals that something unusual is happening. Berkes (2012) points out that systems of TEK build holistic pictures of the environment by considering a large number of variables qualitatively, while science tends to concentrate on a small number of variables quantitatively. Both are important. Quantification has its limits because there is an inverse relationship between the complexity of a system and the degree of precision that can be used meaningfully to describe it. Hence, indigenous ways of knowing show

Table 1.3 Spatial and temporal availability and limitations of instrumental data and some proxy variables for temperature in the Holocene

Variable	Spatial extent	Timescale		
		Interannual	Decades to centuries	Centennial and longer
Instrumental data Contemporary written historical records (annals, diaries, etc.)	Europe from early 1700s, most other coastal regions during nineteenth century. Continental interiors by 1920s, Antarctica by late 1950s	Should be 'perfect' if properly maintained – changes assessable on daily, monthly and seasonal timescales	Site moves, observation time changes and urbanization influences present problems – changing frequency of extremes assessable	As previous, but rates of change to site, instrumentation and urbanization mean absolute levels increasingly difficult to maintain
Proxy indicators	Europe, China, Japan, Korea, eastern North America. Some potential in Middle East, Turkey and South Asia and Latin America (since 1500s)	Depends on function of diary information (freeze dates, harvest dates and amounts, snowlines, etc.). Very difficult to compare with instrumental data	Depends on diary length and observer age. Lower frequencies increasingly likely to be lost due to human lifespan	Only a few indicators are objective and might provide comparable information (e.g. snowlines, rain days)
Tree-ring widths	Trees growing poleward of 30° or at high elevations in regions where cool season suspends growth	Generally dependent upon growing season months. Exact calendar dates determined by cross-dating	Standardization method potentially compromises interpretation on longer timescales	Highly dependent on standardization method. Likely to have lost variability, but difficult to assess
Ice-core melt layers	Coastal Greenland and high-latitude and high-altitude ice caps, where temperatures rise above freezing for a few days each summer	Depends on summer warmth. Unable to distinguish cold years that cause no melt. Rarely compared with instrumental records. Dating depends on layer counting – increasingly difficult with depth	May not respond to full range of temperature variability. Whole layer may melt if too warm; no melt layers if too cold	Increasingly depends on any flow model and layer compaction. Veracity can be assessed using other cores
Coral growth and isotopes	Tropics (between 30° N and S) where shallow seas promote coral growth	Response to annual and seasonal water temperature and salinity. Dating depends on counting. Rarely cross-dated	As coral head grows, low-frequency aspects may be affected by amount of sunlight, water depth, nutrient supply, etc.	Only achieved in a few cases. Veracity can be assessed by comparison with other corals

Source: after Jones et al. (1998: tables 1 and 2).

Table 1.4 Areas where science and traditional ecological knowledge (TEK) can be complementary for population monitoring

Principle	Explanation
Long and short time series	Science is good at collecting data over short time periods over a large area, whereas TEK tends to focus on long time periods, often in small areas, as needed to establish a baseline. Using both together provides more complete information on both time and space scales
Averages and extremes	Much of science collects numerical data, emphasizing statistical analysis of averages. Holders of TEK are very good at observing extreme events, variations, and unusual patterns and remembering them through oral history and social memory
Quantitative and qualitative information	Science demands quantitative data on parts of the system, whereas TEK strives for qualitative understanding of the whole. Understanding of complex systems requires both, so the two are complementary. Qualitative measures can be more rapid and inexpensive, but at the expense of precision
TEK for better hypotheses, science for a better test of mechanisms	TEK provides a short cut to more relevant hypotheses for problem solving but does not usually address mechanisms (the 'why' question). Science has powerful tools for testing the 'why' but can waste time and effort on trivial hypotheses. Using both approaches together benefits from their relative strengths
Objectivity and subjectivity	Science strives to be objective, excluding people and feelings. TEK explicitly includes people, feelings, relationships and sacredness. Science is good at monitoring populations from a distance, but the incorporation of traditional monitoring allows a stronger link between science and community, producing 'science with a heart'

Source: after Moller *et al*. (2004).

us an alternative approach that can complement science. There are several other ways in which the two kinds of knowledge are complementary, as Table 1.4 shows with regard to monitoring wildlife populations.

The realization that TEK can complement scientific approaches to understanding environmental issues that are typically complex, uncertain and multi-scaled is part of a wider trend towards incorporating a range of knowledge and value systems. In consequence, 'stakeholder participation' is increasingly used in environmental studies and decision-making. The quality and effectiveness of consequent policies are thought to be improved in consequence. A participatory approach may also help to build trust and understanding among stakeholders, increasing the chances of them supporting project goals.

It is clear that our understanding of how environments change can be built up only slowly in a patchwork fashion, but the understanding gained from all these lines of evidence can then be used to predict environmental changes, incorporating any human impact, using models that simulate environmental processes. The accuracy of a model can be assessed by comparing its output with any monitored record, records reconstructed from proxy variables, and the understanding provided by TEK, these sources of information allowing us to develop the model as discrepancies are identified. The human impact may still provide further complications, however, because in many instances through prehistory, history and indeed in the present era, it can be difficult to distinguish between purely natural events and those that owe something to human activities (temperature readings at a town that becomes a city are an obvious example because urbanization affects temperature). It is the interrelationships between human activities and natural functions that form the subject matter of this book.

FURTHER READING

Anderson, D., Goudie, A. and Parker, A. 2013 *Global environments through the Quaternary*, 2nd edn. Oxford, Oxford University Press. An assessment of environmental changes during the two to three million years in which people have inhabited the Earth.

Begon, M., Townsend, C. and Harper, J.L. 2005 *Ecology: from individuals to ecosystems*, 4th edn. Oxford, Blackwell. This comprehensive textbook covers all the basics of ecology.

Catalan, J., Pla-Rabés, S., Wolfe, A.P., Smol, J.P., Rühland, K.M., Anderson, N.J., Kopáček, J., Stuchlík, E., Schmidt, R., Koinig, K.A. and Camarero, L. 2013 Global change revealed by palaeolimnological records from remote lakes: a review. *Journal of Paleolimnology* 49: 513–35. Reviewing the value of remote sites for recording broad-scale environmental changes.

Lindenmayer, D.B. and 10 others 2012 Value of long-term ecological studies. *Austral Ecology* 37: 745–57. Their importance to ecology, environmental change, natural resource management and biodiversity conservation.

Nash, K.L. and 10 others 2014 Discontinuities, cross-scale patterns, and the organization of ecosystems. *Ecology* 95: 654–67. An investigation into the importance of scale.

Ruddiman, W.F. 2014 *Earth's climate: past and future*, 3rd edn. New York, Freeman and Company. A readable, systematic expert overview of the Earth's climate system.

WEBSITES

gcmd.gsfc.nasa.gov   a wide range of data on the earth sciences is available through the Global Change Master Directory.

www.fao.org/gtos   the Global Terrestrial Observing System is a programme for observations, modelling and analysis of terrestrial ecosystems to support sustainable development.

www.igbp.net   the International Geosphere-Biosphere Programme's mission was to deliver scientific knowledge to help human societies develop in harmony with the Earth's environment. Although it closed in 2015, the site still holds many resources.

www.inqua.org   the International Union for Quaternary Research oversees scientific research on environmental change during the Quaternary.

www.lternet.edu   the Long Term Ecological Research Network supports research on long-term ecological phenomena in the USA and the Antarctic.

www.pages-igbp.org   PAGES (Past Global Changes) supports research aimed at understanding the Earth's past environment to enable predictions for the future.

POINTS FOR DISCUSSION

■ How and why can we recognize large-scale ecosystem types called biomes?

■ It is a biological fact that predators at the top of food chains are always rare. The only exception is humans and this is why we face so many environmental issues. How far do you agree?

■ What do you understand by the term 'regime shift'? Explain using the concepts of threshold, equilibrium and non-linear system.

■ Prepare a report for a national agency outlining the arguments for and against the funding of long-term environmental monitoring.

■ Explain why studies of the physical environment should be carried out at a range of temporal and spatial scales.

■ Using examples, explain why we use 'proxy' methods in palaeoenvironmental research.

2 *The human environment*

KEY CONCEPTS

tragedy of the commons, debt trap, transnational corporations (TNCs), common property resources, sacred groves, technocentric, ecocentric, traditional ecological knowledge (TEK)

Knowledge of the physical environment only illuminates one-half of any environmental issue, since an appreciation of factors in the human environment is also required before an issue can be fully understood. Relationships between human activity and the natural world have changed greatly in the relatively short time that people have been present on the Earth. A very large increase in human population, along with widespread urbanization associated with advances in technology and related developments of economic, political and social structures, have all combined to make the interaction between humankind and nature very different from the situation just a few thousand years ago. This chapter is concerned with these aspects of humanity, which together provide the human dimensions of environmental issues.

HUMAN PERSPECTIVES ON THE PHYSICAL ENVIRONMENT

Nobody knows what prehistoric human inhabitants thought about the natural world, but fossil and other archaeological evidence can be marshalled to give us some idea of how they interacted with it. The fact that humans have always interacted with the physical environment is obvious, since all living things do so by definition, but the ability of humans to conceptualize has allowed us to formalize our view of these interactions. One such framework, which must have been around in one form or another for as long as people have inhabited the planet, is the idea of resources. We call anything in the natural environment that may be useful to us a natural resource. Aspects of the human environment, such as people and institutions, can also be thought of as resources, while anything that we perceive as detrimental to us is sometimes called a negative resource. Since resources are simply a cultural appraisal of the material world, individual aspects of the environment can vary at different times and in different places between being resources and negative resources. A tiger, for example, can be viewed as a resource (for its skin) or a negative resource (as a dangerous animal). Similarly, different cultures recognize resources in different ways (the Aztecs, for example, were mystified by the Spaniards' insatiable demand for gold, thinking that perhaps they ate it or used it for medical purposes), and some groups see resources where others do not (a grub is a food resource to an Aboriginal Australian, but not to most Europeans). However, resources in the physical environment are not restricted to direct material inputs to society. They also include a wide range of ecosystem services (see Chapter 3) that are not directly consumed, but which are necessary for the maintenance of life and for economies and societies to function (such as water purification and the maintenance of biodiversity). In summary, resources vary in character. The classification shown in Table 2.1 gives some idea as to their characteristics and manageability.

All environmental issues can be seen as the result of a mismatch between extrinsic resources and natural resources: they stem from people deliberately or inadvertently misusing or abusing the natural environment. The reasons for such inappropriate uses are to be found within the nature of human activity.

Table 2.1 A classification of resources

PHYSICAL ENVIRONMENT
Continuous resources, which to all intents and purposes will never run out (e.g. solar energy, wind, tidal energy)
Renewable resources, which can naturally regenerate so long as their capacity for so doing is not irreversibly damaged, perhaps by natural catastrophe or human activity (e.g. plants, animals, clean water, soil)
Non-renewable resources, which are available in specific places and only in finite quantities because although they are renewable, the rate at which they are regenerated is extremely slow on the timescale of the human perspective (e.g. fossil fuels and other minerals, some groundwaters)

HUMAN ENVIRONMENT
Extrinsic resources, which include all aspects of the human species, all of which are renewable (e.g. people, their skills, abilities and institutions)

HUMAN FORCES BEHIND ENVIRONMENTAL ISSUES

The interactions between humankind and the physical environment result from our attempts to satisfy real and perceived needs and wants. The specific actions that cause environmental issues directly are well known. They include such exploits as modifying natural distributions of vegetation and animals, overusing soils, polluting water and air, and living in hazardous areas. At a deeper level, the forces that enable, encourage or compel people to act in inappropriate ways can be traced back to a series of factors that directly or indirectly cause a change in an ecosystem. The term 'driver' is widely used for such forces and the Millennium Ecosystem Assessment (MEA) recognized five main sets of these underlying drivers: demographic; sociopolitical; economic; scientific and technological; and cultural and religious (MEA, 2005). These sets of drivers operate both to promote human impact on the environment and to mitigate such impacts. They occur at all spatial scales, from global to regional and local, and across various timeframes, as well as in differing combinations.

Demographic drivers

Numerous demographic variables have implications for the physical environment, including population size and rate of change over time, a population's structure (e.g. age and gender) and spatial distribution, patterns of migration, and levels of educational achievement. Growth in the global human population is widely recognized as one of the most clear-cut drivers behind increased impact on the environment. The non-linear, exponential rise in human population numbers is evident in Table 2.2. It took some 200,000 years for the population of *Homo sapiens* to reach its first billion, but only just over 100 years to reach its second. The periods to increase by one billion in most recent times have been closer to ten years. The logic is simple and undeniable: more people means a greater need for natural resources, both renewables and non-renewables. As the population grows, more resources are used and more waste is produced.

The relationship between the human population, whose numbers can change, and natural resources, which are essentially fixed, has been an issue for a long time. At the end of the eighteenth century, the Englishman Thomas Malthus suggested that population growth would eventually outstrip food production

Table 2.2 Milestones in world human population

World population (billions)	Year reached	Number of years to increase by one billion
1	1804	
2	1927	123
3	1960	33
4	1974	14
5	1987	13
6	1999	12
7	2011	12
8	2023*	12
9	2038*	15

Source: after UNDESA (2004, 2015).
Note: *estimate.

and lead to famine, conflict and human misery for the poor as a consequence (Malthus, 1798). This Malthusian perspective continues to influence many interpretations of environmental issues. A rapid increase in the population of an area can result in overexploitation of the area's resources, as numerous examples from refugee camps illustrate (see p. 502). However, in many cases the relationship between population numbers and environmental degradation is not straightforward and is complicated by numerous other factors. Indeed, rapid population decline can also lead to environmental degradation if good management practices are neglected because there are not enough people to keep them up. Furthermore, a high population density in itself does not necessarily lead to greater degradation of resources. Political and economic forces also affect the ways in which people use or abuse their resources. Further investigation of the importance of human population numbers is made in Chapter 3.

Sociopolitical drivers

The organization of human society, including economic, political and social structures, forms a diverse set of drivers that influence environmental change. They also underpin the mitigating forces that have been developed by societies to offset some of the damaging aspects of environmental issues. Dramatic transformations in population and technology have been associated with the two most significant changes in the history of humankind – the Agricultural Revolution of the late Neolithic period and the Industrial Revolution of the eighteenth and nineteenth centuries – but these changes have also been reflected in equally marked modifications to the ways that society is organized. The organization of society in turn affects how people make decisions about their interaction with the natural environment. This purposeful collective action – among state, private, and civil society stakeholders – is often called environmental governance. Participation in such governance decisions typically varies between groups within society with different levels of access to power and influence (e.g. women and men, different political parties, ethnic groups). Differences in opinion over environmental matters, such as access to particular resources, can be peaceful or otherwise, on occasion leading to outright conflict (see Chapter 20).

One of the most striking of humankind's social changes is the rise of the city. Although cities have been a feature of human culture for about 5000 years, virtually the entire human species lived a rural existence just 300 years ago. Today, the proportion of the world's population living in cities is more than 50 per cent, and individual urban areas have reached unprecedented sizes. In most cases, urban areas have much higher population densities than rural areas, and the consumption of resources per person in urban areas is greater than that of their rural counterparts. The environmental impacts of cities within urban boundaries are obvious, since cities are a clear illustration of the human ability to transform all the natural spheres (see Chapter 10), but the influence of cities is also felt far beyond their immediate confines. The high level of resources used by city dwellers has been fuelled by extending their resource flows or ecological footprint (see Chapter 3).

Economic drivers

Economic growth and its distribution, globally and within countries, have a great effect on the nature and extent of environmental impacts. Historically, the growth of economies has been fuelled by the use of resources and there are numerous cases of degradation that has occurred in consequence, such as deforestation (Chapter 4), pollution of the atmosphere (Chapters 11 and 12) and the loss of biodiversity (Chapter 15). Such impacts have taken place – and continue to do so – in the places where growth has occurred but also in places where resources have been exploited thanks to important elements of global growth such as colonialism, international trade and the flow of capital. The many processes of globalization are also leading to new interactions between people and economies, helping to make the world an integrated whole.

Although integrated, the world has also become a very unbalanced place in terms of human welfare and environmental quality. There are clear imbalances between richer nations and poorer nations. In general terms, the wealthiest few are disproportionately responsible for environmental issues, but at the other end of the spectrum the poorest are also accused of a responsibility that is greater than their numbers warrant. The factors driving these disproportionate impacts are very different, however. The impact of the wealthy is driven by their intense resource use, many say their overconsumption of resources. The poor, on the other hand, may degrade the environment because they have no other option (Figure 2.1). Economic power is seen as a vital determinant: the wealthy have become wealthy because of their high-intensity resource use, and can afford to continue overconsuming and to live away from the problems this creates. The poor cannot afford to do anything other than overuse and misuse the resources that are immediately available to them, and as a consequence they are often the immediate victims of environmental issues. This difference in economic power is also manifested in political power: the wealthy generally have more influence over environmental governance deci-

Figure 2.1 The health of this argun tree in south-west Morocco is not improved by goats browsing in its canopy, but herders often have few options but to overuse the resources that are immediately available to them.

sions than the poor, although when marginalized people are pushed to the edge of environmental destruction, they may become active in forcing political changes.

Scientific and technological drivers

The development and diffusion of scientific knowledge and technologies can have significant implications for environmental issues. One view sees technological developments as a spur to population growth, agricultural innovations providing more food per unit area and enabling more people to be supported. An opposite perspective sees technological change as a result of human inventiveness reacting to the needs created by more people (Boserüp, 1965). This latter view can be used to suggest that a growing human population is not necessarily bad from an environmental perspective. Any problems created by increasing population will be countered by new innovations that ease the burden on resources by using them more effectively. One illustration of this argument can be drawn from society's use of non-renewable fossil fuel resources. Since fossil fuel supplies will not last forever, and their current use is causing a range of environmental problems, such as global warming and acidification, energy conservation is being promoted and renewable energy supplies are being developed (see Chapter 18).

However, many arguments can be presented to indicate that technological developments are responsible for much environmental degradation. Technology influences demand for natural resources by changing their accessibility and people's ability to afford them, as well as creating new resources (uranium, for example, was not considered a resource until its energy properties were recognized). The Industrial Revolution has been associated

Figure 2.2 A solar-powered kettle in Tibet illustrates a technology devised to relieve pressure on biological resources, since firewood is very scarce on the Tibetan Plateau.

with high population growth and a greatly enhanced level of resource use and misuse; it has promoted urban development, improved transport and the globalization of the economy. But although these developments have undoubtedly increased the scale of human impact on the environment, this is not to say that pre-industrial technologies did not also allow severe impacts (see, for example, prehistorical cases of species extinctions on p. 372).

Virtually every environmental issue can be interpreted as a consequence, either deliberate or inadvertent, of the impact of technology on the natural world. Numerous examples of inadvertent impacts are the result of ignorance: a new technology being tried and tested and found to have undesirable consequences. The use of pesticides and fertilizers has increased food production, but they have also had numerous unintended negative effects on other aspects of the world in which we live (see Chapter 13). Conversely, technology is not environmentally damaging by definition (Figure 2.2), since technological applications can be designed and used as a mitigating force, in many cases in response to a previous undesirable impact.

Cultural and religious drivers

Culture is a very broad term. In one sense, it can be thought of as the values, beliefs and norms shared by a group of people. The group may be identified by its nationality or ethnic origin but also through other ways in which people mix with each other, such as membership of professions and organizations. These associations help to condition individuals' perceptions of the natural world, influence what they consider important, and suggest courses of action that are appropriate and inappropriate.

The extensive characteristics of culture and the overlaps it has with sociopolitical drivers mean that establishing its causal effects is very difficult. However, cultural factors dictate in part the ways in which people perceive certain aspects of the environment, such as plants and animals, perceptions that are subject to change (see, for example, p. 375 on how the perception of the tiger has changed in 100 years). Cultural norms often incorporate taboos – unwritten social rules that regulate human behaviour – on certain actions that affect the environment. Taboos may, at least locally, play a significant role in the conservation of natural resources, species and ecosystems. Good examples are taboos associated with particular areas of habitat, known as sacred groves, that are set aside for religious or cultural purposes. These habitat patches were once widespread throughout India, Africa and Europe. Other types of taboo are associated with specific species, imposing a ban on their destruction or detrimental use, or restricting access to them during a certain time period.

The role of religion in shaping attitudes towards the environment was highlighted in the 1960s when it was argued that the historical roots of the ecological crisis in western industrialized nations could be attributed to Christianity (White, 1967). A materialist, ecologically harmful attitude to nature had been encouraged in the Judaeo-Christian world by a passage in the Book of Genesis in which God gave control and mastery of the Earth to humans. The view has subsequently been discredited but the relationships between religious identification and environmental issues are complex and variable and remain a subject for continuing research.

HUMAN-INDUCED IMBALANCES

Human society has created a set of sociocultural imbalances that is superimposed on geographical patterns of environmental resources on all spatial scales from the global to the individual. These patterns of uneven distribution have been developed by, and are maintained by, the processes and structures of economics, politics and society. These imbalance theories, combined with the human driving force theories outlined above, together make up a diverse set of explanations that have been put forward to explain the human dimension behind environmental issues (Table 2.3).

Ownership and value

Two interrelated theories that explain the underlying causes of imbalance between human activities and the environment stem from differential ownership of certain resources and the values put on them. Some environmental resources are owned by individuals while others are under common ownership. One theory argues that resources under common ownership are prone to overuse and abuse for this very reason – the so-called 'tragedy of the commons' (Hardin, 1968). The example often given to illustrate this principle is that of grazing lands which are commonly owned in pastoral societies. It is in the interest of an individual to graze as many livestock as possible, but if too many individuals all have the same attitude, the grazing lands may be overused and degraded: the rational use of a resource by an individual may not be rational from the viewpoint of a wider society. The principle can also be applied to explain the misuse of other commonly owned resources, such as the pollution of air and water or catching too many fish in the sea.

It is important to note, however, that common ownership does not necessarily lead to overexploitation of resources. In many areas where resources are commonly owned, strong social and cultural rules have evolved to control the use of resources. In situations like this, resource degradation usually occurs because

Table 2.3 Some theories to explain why environmental issues occur

Theory	Explanation
Neo-Malthusian	Demographic pressure leads to overuse and misuse of resources
Ignorance	Ignorance of the workings of nature means that mistakes are made, leading to unintentional consequences
Tragedy of the commons	Overuse or misuse of certain resources occurs because they are commonly owned
Poor valuation	Overuse or misuse of certain resources occurs because they are not properly valued in economic terms
Dependency	Inappropriate resource use by certain groups is encouraged or compelled by the influence of more powerful groups
Exploitation	Overuse and misuse of resources are pursued deliberately by a culture driven by consumerism
Human domination over nature	Environmental issues result from human misapprehension of being above rather than part of nature

Source: after Barrow (1991).

the traditional rules for the control of resource use break down for some reason. Reasons include migration to a new area, changes in ownership rights, and population growth. In examples like overfishing in the open oceans, by contrast, the tragedy of the commons applies because there is no tradition of rules developed to limit exploitation.

A related concept is the undervaluation of certain resources. Air is a good example. To all intents and purposes, air is a commonly owned continuous resource that, in practice, is not given an economic value. The owner of a windmill does not pay for the moving air the windmill harnesses, nor does the owner of a factory who uses the air as a sink for the factory's wastes. Since air has no economic value, it is prone to be overused. A simple economic argument suggests that if an appropriate economic value was put on the resource (see Chapter 3), the workings of the market would ensure that as the resource became scarce so the price would increase. As the value of the resource increased, theory suggests that it would be managed more carefully.

Exploitation and dependency on the global scale

A complex series of economic, political and social processes has resulted in patterns of exploitation and dependency, which are associated with the misuse of resources. Inequalities exist between many different groups of people (see above) and can be identified on several different scales. Three levels are identified by Lonergan (1993): two spatial (international and national), and one temporal (intergenerational).

Some of the main structural inequalities of the global system are shown in Figure 2.3. They have evolved from colonial times to the point, today, where direct political control of empires has been superseded by more subtle control by wealthier countries over poorer ones. The economic dimension is particularly important, since it influences the rate of exploitation of natural resources in particular countries, the relative levels of economic development apparent in different countries, and the power of certain governments to control their own future.

The realities of global inequality in economic terms are stark. One billion people live in unprecedented luxury, while one billion live in destitution. Global income inequality rose steadily over the nineteenth and twentieth centuries although the gap narrowed slightly at the start of the twenty-first century, thanks in part to the Millennium Development Goal of halving the 1990 level of world poverty by 2015. However, the gap remains very large. Whereas the top 20 per cent of the global population, who lived in high-income regions including western Europe, Japan, North America and Australia (collectively regarded as wealthy, developed countries of the 'North'), controlled about 64 per cent of total income in 2007, the bottom 20 per cent had just over 3 per cent (Ortiz and Cummins, 2011). The gap between the richest and the rest was illustrated dramatically in 2016 when it was reported that the world's richest 1 per cent had accumulated more wealth than the rest of the world put together, and just 62 individuals had the same wealth as 3.6 billion people (Hardoon *et al.*, 2016). On an individual level, a child born at the turn of the century in an industrialized country will add more to resource consumption and pollution over his or her lifetime than 30 to 50 children born in developing nations.

The structural aspects of this economic dimension have many facets, including the fact that many poorer countries are in debt to the countries and banks of the rich world (the so-called 'debt trap'), and many less-developed countries rely on a limited number of exports, which are usually primary products such as agricultural goods and minerals. Figure 2.4 illustrates this picture globally with regard to exports, showing that commodity export revenues contribute more than 60 per cent of total goods export earnings in two out of three of all developing countries, half of which are in Africa. This pattern is maintained by biases in the international trade system in which terms of trade and prices for primary exports ('unequal exchange') are set largely by the countries of the North that represent the major markets for these exports.

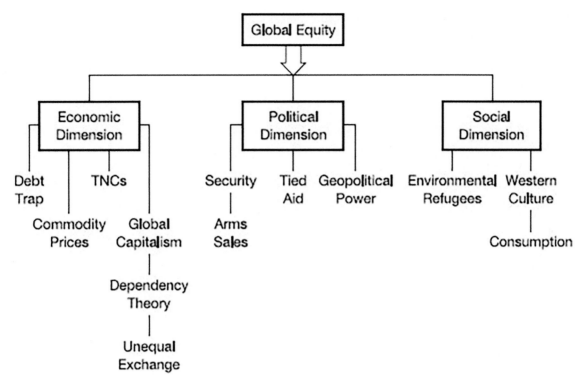

Figure 2.3 Structural inequalities in the global system (after Lonergan, 1993).

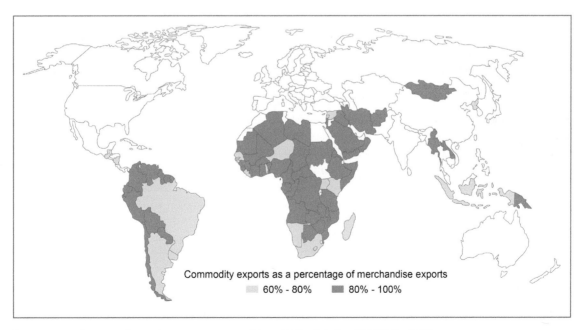

Figure 2.4 Country dependency on commodities, 2012–13 (after UNCTAD, 2015).

Overexploitation of resources in poorer countries often occurs in response to the falling commodity prices that have been typical of recent decades, and the need to service debts.

Transnational corporations (TNCs), the very large majority of which have headquarters in the advanced capitalist regions, particularly Japan, North America and western Europe, also play an important role in the workings of global economics: their corporate strategies have been powerful drivers of the globalization of economic activity. Comparison of the annual turnover of TNCs with the gross national income (GNI) of entire nation-states gives an indication of this role:

> By this yardstick, all of the top 50 global TNCs – including ExxonMobil, General Motors, Ford Motor Company, General Electric, Mitsubishi, IBM, Nestlé, Unilever, BP, Royal Dutch Shell, Siemens and Toyota Motor Corporation – carry more economic clout than many smaller less developed countries (LDCs).
>
> (Knox *et al.*, 2014: 47)

In development terms, investment by TNCs in developing countries has been portrayed as an engine of growth capable of eliminating economic inequality on the one hand, and as a major obstacle to development on the other. Some view such investment as a force capable of dramatically changing productive resources in the economically backward areas of the world, while others see it as a primary cause of underdevelopment because it acts as a major drain of surplus to the advanced capitalist countries (Mackinnon and Cumbers, 2007).

The negative view has been illustrated by comparing the development paths of Japan and Java, which had a similar level of development in the 1830s. While Japan has subsequently developed much further, in spite of a poorer endowment of natural resources, Java has remained underdeveloped, despite a position on main trade routes and a good stock of natural resources, because few of the profits from resource use have been re-invested in the country (Geertz, 1963).

One aspect of these relationships, which some consider to be detrimental to the environment in poorer countries, is the movement of heavily polluting industries from richer countries that have developed high pollution-control standards, to poorer countries where such regulations are less stringent. TNCs may be particularly adept at relocating their production activities in this way if it means saving money otherwise spent on conforming to more stringent environmental requirements. Daly (1993) notes examples of 'maquiladoras', US factories which have located mainly in northern Mexico to take advantage of lower pollution-control standards and labour costs. However, not all assessments of the pollution haven hypothesis have been conclusive. A study by Eskeland and Harrison (2003), who analysed foreign direct investment across industries in Mexico, Venezuela, Morocco and Côte d'Ivoire, found little evidence in support of pollution havens. By contrast, Mani and Wheeler (1998) found clear confirmation of a pollution haven effect, even if they concluded that it has been transient in many countries.

Economic power is closely related to political power, which can be seen in the sale of arms and the giving of aid by richer countries to poorer ones. The workings of global capitalism also have social dimensions, including the spread of western cultural norms and practices (Figure 2.5). The sale of western products can also be seen to reinforce the role of developing nations as suppliers of raw materials, as Grossman (1992) shows for the increasing use of pesticides on Caribbean agricultural plantations where produce is destined for export markets, a trend influenced by foreign aid, among other criteria.

Exploitation and dependency on the national scale

Perhaps the most important underlying aspect of global inequalities is the way in which they combine to maintain a situation where a minority of wealthy nations and most of their inhabitants are able to live

Figure 2.5 Domination of the world's poorer countries by their richer counterparts takes many forms, including what some term 'cultural imperialism', epitomized by this advertisement for Pepsi in Ecuador. Transfer of western technology is often in the form of polluting industries, as shown behind.

an affluent life consuming large quantities of resources at the expense of many poorer nations and their inhabitants. Nevertheless, some countries do move up the economic development scale, and the terms North and South disguise much internal variability. Structural inequalities also exist at the national level, however, and these inequalities similarly have implications for the ways in which people interact with the environment within countries. Some of the aspects of inequality on the national scale, categorized according to their economic, political and social dimensions, are shown in Figure 2.6.

As at the global level, the national scene is often characterized by a small elite group that has more economic, political and social power than the majority of people in more marginal groups. In general terms, this pattern is often manifested in a rural/urban divide: cities are centres of power, although they also typically contain a poorer 'underclass' in both rich and poor countries. Outside the city in the South, most rural people do not have access to economic power. Corruption and centralized control are all too frequent characteristics of political and economic management in developing countries, which often means that local communities lack power over their own resources and how they are managed. These aspects are often tied up with social dimensions such as human rights (Figure 2.7).

Another important social dimension is land tenure; poorer, marginalized groups tend to lack ownership of land and its associated resources. One study of rural villages in India found that about 12 per cent of average household income was derived from natural resources that were open for all to

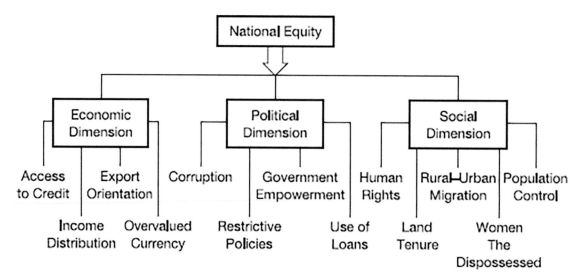

Figure 2.6 Structural inequalities in national systems (Lonergan, 1993).

Figure 2.7 This mural in Buenos Aires recalls the so-called Dirty War of 1976–83, when the military ruled Argentina, a period marked by corruption, economic mismanagement and poor human rights.

use, so-called 'common property resources' (Beck and Ghosh, 2000). When people do not own the resources they use, there is a danger of overexploiting them: the tragedy of the commons. There may also be little incentive to invest in the future productivity of an area if a person's future entitlement to that area is in question. Fertilizing a field to ensure that it retains its productivity means not using resources for immediate benefit: the animal dung used as fertilizer, for example, could have been used as fuel. Instead, those resources are invested for benefits received some time in the future. But if a farmer is not sure that he will still be able to use the field in future years, he may decide not to invest his animal dung.

There are numerous ways in which national inequalities translate into environmental issues. For instance, there is an association between the environmental characteristics of a place and the well-being of its inhabitants: rural people living in chronic poverty (those who remain poor for much or all of their lives) tend to live in areas with low agricultural or natural resource potential. A host of socio-economic, political and historical drivers are also at work to produce a set of typical characteristics associated with geographical concentrations of chronic rural poverty (Table 2.4).

The poor and disadvantaged are frequently both the victims and agents of environmental degradation. The poor, particularly poor women, tend to have access only to the more environmentally sensitive areas and resources. Even where they do have ownership, it is often of low-quality resources. They therefore suffer greatly from productivity declines due to degradation such as deforestation or soil erosion. Their poverty also means that there may be little alternative but to extract what they can from the sparse resources available to them, with degradation often being the result. The high fertility rates typical of poor households put further strains on the stock of natural resources. This set of mutually reinforcing links between poverty and environmental damage, which often operate as a positive feedback, is frequently referred to as the poverty–environment nexus.

Of course, not all poor people 'mine the future' in this way. The poor have used natural resources for many generations, and they can live in harmony with the environment using traditional methods. Problems tend to arise when a poor society is subject to some kind of perturbation or shock, whether natural

Table 2.4 Typical characteristics associated with geographical concentrations of chronic poverty in rural areas

LOW POTENTIAL
Areas with low agricultural or natural resource potential due to combinations of biophysical attributes, including climatic, hydrological, topographical, soils, pests and diseases. (Often crudely equated with drylands and highlands)

REMOTENESS
Areas far from the centres of economic and political activity in terms of physical distance and/or travel time

LESS FAVOURED
Politically disadvantaged areas

WEAKLY INTEGRATED
Areas not well-connected, both physically and in terms of communication and markets

Source: after CPRC (2004).

Figure 2.8 Refugees in northern Thailand who have fled their native lands in neighbouring Myanmar (Burma). A long-running armed struggle between indigenous peoples and the government has been exacerbated in recent years by government-sponsored clearance of tropical forests in northeast Myanmar.

(e.g. floods or droughts) or human-induced (e.g. land tenure changes). Just as in the physical environment, human societies display resistance and resilience to such perturbations (see Chapter 1), but if a shock exceeds a certain threshold, common responses are either to migrate from the affected area or to stay and try to survive.

The impacts can be seen in numerous examples. Political power in the hands of a minority group can be used to force large concentrations of marginal groups into small parts of the national land area, causing environmental degradation through overpopulation. Examples include the homelands system of apartheid South Africa (Meadows and Hoffman, 2002), and numerous similar inequalities exerted by more powerful ethnic groups in many other countries. The exploitation of resources by elite groups can also result in coerced migration of indigenous inhabitants. Such 'environmental refugees' often take refuge in neighbouring countries (Figure 2.8). Political, economic and social forces also combine to make marginal groups more vulnerable to natural hazards and health risks associated with pollution, an issue often referred to as environmental justice.

Intergenerational inequalities

The spatial dimensions of exploitation and dependency are also closely related to the temporal dimension. As wealthy, powerful groups exploit available resources on the global and national scales, the inheritance of future generations is compromised. To take just one example, climate change that takes place due to increases in atmospheric carbon dioxide concentration is thought to be largely irreversible for 1,000 years after emissions stop (Solomon *et al.*, 2009).

Marginalized groups which are compelled to overexploit the resources available to them are doing so at the expense of their own futures. Overusing a renewable resource today leaves that resource degraded for future users. Using a non-renewable resource today means that less of that resource will be available for tomorrow. Mining the future in this way will be offset to some extent, for example by changes in technology which can change our perception of resources, as well as changes in political, economic and demographic factors. Various aspects of intergenerational inequalities are shown in Figure 2.9. Resolving these inequalities across generations is the central aim of 'sustainable development', a concept dealt with in more detail in Chapter 3.

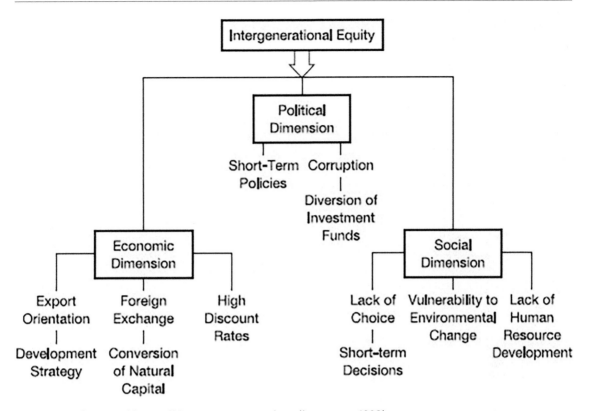

Figure 2.9 Structural inequalities across generations (Lonergan, 1993).

TIME AND SPACE SCALES

The above sections on exploitation and dependency illustrate the fact that, just as in the physical environment, key factors influencing events in the human environment vary at different combined spatial and temporal scales (see Chapter 1). Decisions about environmental governance are taken at all levels, from the individual through the local neighbourhood, city, region, country and continent, up to the global level. None of these decision-makers operates in a closed self-supporting system; all are a part of larger interacting systems. Hence, when surveying the economic and social objectives of an area, for example, it might transpire that local community interests differ from interests at the national level. The costs and benefits of a new large development such as a dam or mine are often quite different at the national scale and at the local level (see Chapters 9 and 19).

Similarly, environmental views and decisions are taken with a wide range of time perspectives. A number of time perspectives have also been illustrated above: contrast the poverty-stricken individual who is compelled to overexploit resources today in order to survive regardless of the longer-term consequences of his actions, with the time taken to develop improved cereal varieties designed to enhance food production. Inevitably, we tend to focus on scales directly relevant to people, but we should not forget other scales that may have less direct but no less significant effects. Indeed, successful management of environmental issues relies on the successful identification of appropriate scales and their linkages in the human environment.

INTEREST IN ENVIRONMENTAL ISSUES

People who derive their livelihoods directly from the physical environment have always been interested in environmental issues. When *Homo sapiens* first emerged as a species in his own right, the availability of wild plants and animals was literally a matter of life and death and an understanding of landscape enabled people to find other basic requirements such as water and shelter. Farmers and other groups directly involved in extracting physical environmental resources continue to rely on their knowledge and understanding of those resources to make a living or simply to survive. Today, environmental issues are also a prominent concern of government and the general public in most countries of the world (Figure 2.10), perhaps more widely so than at any time in the past; but worry over environmental matters among those not directly involved in extracting natural resources is also by no means a recent phenomenon. Observers of the natural world in ancient Mesopotamia, classical Greece, imperial Rome and Mauryan India expressed reservations about the wisdom of human actions that promoted accelerated soil erosion and deforestation. In more recent times, some of the earliest environmental legislation was introduced to protect the quality of natural resources. At the end of the thirteenth century, for example, laws were passed by the Mongolian state to protect trees and animals from overexploitation, and in England atmospheric pollution prompted a royal decree in 1306 forbidding the burning of coal in London. Warning voices were also raised at the damage caused by deforestation and plantation agriculture during the earliest periods of European colonial expansion, on the Canary Islands and Madeira from the early fourteenth century and in the Caribbean after 1560.

Figure 2.10 Public hoarding to promote environmental awareness in Mombassa, Kenya. Concern over environmental issues is increasing in most parts of the world.

Indeed, the origins of today's global concern for the environment can be traced to some of these early European colonial experiences (Grove, 1990). While European colonies were often the sites of environmental degradation, they were also the cradles of modern scientific views on the issues. Colonial administrators and professional scientists working for trading companies in the eighteenth and nineteenth centuries developed western conservationist ethics and environmentally sound management practices. Although such approaches in themselves were not new, since many traditional rural societies had employed similar practices locally for a long time before, this was the first time that a global perspective could be attained. Hence, international travel and observations by naturalist explorers such as Charles Darwin and Alexander von Humboldt allowed them to develop scientific theories that revolutionized western thinking on the way nature was organized and worked.

An ethic of environmental conservation has been a prominent feature of many pre-industrial societies since time immemorial and a critical way this has been achieved has been through religious constraints on the profligate use of common property resources. Hence, so-called sacred groves (see above) have been recognized all over the world, places where a spiritual relationship with the physical environment frequently lends protection to part of a landscape. Of all the great religious traditions, Buddhism is widely recognized as particularly cultivating a conservationist ethos. Traditional nature worship has declined in many regions with the rise of modern industrial societies, but while the Industrial Revolution is rightly associated with an acceleration of resource use, the rise of the city and in many cases the widespread degradation of nature, it also gave rise to its own conservationist ethic. In Britain, the romantic poetry of William Wordsworth was powerful in fostering a love of natural beauty as a counterbalance to industrial society. The works of Henry Thoreau had a similar effect in the USA.

The more recent surge in interest in the natural environment has thrown up a wide range of different approaches to dealing with the issues. The range of contemporary views on environmentalism is illustrated in Table 2.5. This table is not comprehensive, but presents a sample of the diversity of ideas, practices, and organizational cultures embodied in environmental movements.

This plethora of different movements and philosophies can be divided roughly into two camps, those adopting a 'technocentric' approach and those favouring an 'ecocentric' line (O'Riordan and Turner, 1983). Essentially, technocentrism is based on the belief that the way human society interacts with the natural environment only needs minor modifications to deal successfully with environmental issues. Technology and economics will provide the answers to environmental problems, hence existing structures of political power should be maintained, but political, planning and educational institutions should be more responsive to environmental issues. Ecocentrism is rooted in different beliefs. Environmental issues are more the result of problems with our ethical and moral approaches to nature, rather than simply technical difficulties. Fundamental changes in our worldview are needed to solve the issues successfully, along with a redistribution and decentralization of power, and more emphasis on informal economic and social transactions. Among the approaches shown in Table 2.5, the cornucopian school of thought is an extreme example of the technocentric approach, while at the other end of the spectrum deep ecology represents an extreme of the ecocentric position.

Many of the schools of thought outlined in Table 2.5 tend to rely on a scientific approach to environmental issues, certainly when identifying and assessing the importance of an issue, and to some extent in the approaches taken to deal with it. Some movements, however, are critical of science. From the deep ecology viewpoint, environmental problems are the result of crises in society and in science itself (Capra, 1982). Identification of many of today's global environmental concerns was first made more than 100 years ago but most have only gained major political acceptance in recent decades (Table 2.6). The timelags that occur between 'discovery', formulation of policy, and the eventual implementation of policy and management can

Table 2.5 Some contemporary environmental movements and philosophies

Form of environmentalism	Main tenets
Political ecology	Priority given to interconnections between politics and the environment. Includes study of political struggles over natural resources or where outcomes of political struggles are defined by access to natural resources
Free market environmentalism	Emphasis on the free market as the best tool to preserve the environment and promote sustainability. Market-based instruments rather than government intervention can successfully internalize environmental externalities and tackle sources of market failure
Cornucopian (a.k.a. prometheanism, contrarianism, environmental scepticism)	Any environmental issues easily solved by human ingenuity, technological developments and the market economy, but no major problems perceived – the environmental crisis identified by other environmental groups is exaggerated
Social ecology	Views most ecological problems as the result of deep-seated social problems brought on by the products of capitalism. Hence, environmental issues can only be resolved by dealing with problems within society
Ecofeminism	Intellectual and practical movement viewing the domination and oppression of women and nature by men as linked historically, materially, culturally, and ideologically
Deep ecology	Based on nature as having intrinsic value, this philosophy calls for a profound shift in our attitudes and behaviour. It often advocates a strong commitment to direct personal action to protect nature and bring about fundamental societal change
Ecoterrorism	Typified by acts of violence, property damage, or sabotage against individuals and companies in the name of environmentalism. Such direct action is designed to highlight, prevent or interfere with activities deemed harmful to the environment

Source: after Gandy (1996) and Davies (2009).

be explained in several ways. For instance, the many vested interests associated with most environmental issues are highlighted by Likens (2010). Some observers blame the delays in acceptance on scientists themselves, for being poor communicators to a wider audience and for being overcautious in their conclusions (Döös, 1994), but scientists need to gather data and test theories before they can recommend action. Their endeavours are also passed to wider audiences through a multitude of public and private organizations interested in environmental issues, often in association with the news media. Such organizations include government ministries and research bodies as well as many civil society organizations (CSOs) that operate at international, national and local scales. Some would argue that the lag time between identification and acceptance of issues is reason enough to reject a technocentric approach. Indeed, on some occasions political recognition is forced by civil agitation from environmental activists, but science usually plays a role somewhere in the process.

Nonetheless, one important outcome of the current multitude of global environmental issues is a wide acceptance that an exclusively scientific approach, often within a western view of economic development,

Table 2.6 Timetable of major political acceptance of some environmental issues and the dates when they were first identified as potential problems

Environmental issue	Potential problem first identified	Decade of major political acceptance
Air and water pollution	Carson (1962)	1960s
Tropical deforestation	Marsh (1874)	
Acid rain	Smith (1852)	
Stratospheric ozone depletion	Molina and Rowland (1974)	1970s
Carbon dioxide-induced climatic change	Wilson (1858)	
Additional greenhouse gases	Wang *et al.* (1976)	1980s

Source: after Döös (1994).

is not providing all the answers to the challenges facing present and future generations. In other words, science on its own is not up to the task of successfully addressing the challenges of climate change, biodiversity loss, soil erosion, desertification, freshwater shortages, and a host of other social-ecological issues facing humanity. Many now acknowledge that the traditional ecological knowledge (TEK) of indigenous people, accumulated over generations of living in a particular environment, provides valuable additional sources of information about ecosystem conditions and sustainable resource management (see p. 20). Similarly, many natural scientists have begun to acknowledge that an understanding of environmental issues must also include insights from the humanities and social sciences. This idea is based on the simple realization that environmental issues that have in many cases reached crisis proportions are not only issues of the physical environment but also issues of cultural and social environments. These are the systems of representation and the institutional structures through which contemporary societies understand and respond to environmental change. Or, in many cases, fail to understand and respond appropriately, hence contributing to the issue. These aspects of environmental issues have been explored by numerous historians, literary critics and philosophers. In short, the issues we face are very complex, they cut across traditional disciplines and require new, diverse ways of understanding the environment and responding to the ways it changes.

FURTHER READING

Berkes, F. 2012 *Sacred ecology: traditional ecological knowledge and resource management*, 3rd edn. London, Routledge. An assessment of indigenous practices and resource use and how they might complement scientific ecology.

Berry, R.J. 2018 *Environmental attitudes through time*. Cambridge, Cambridge University Press. A historical review and forward look at how people view the environment.

Ehrenfeld, D. 2005 The environmental limits to globalization. *Conservation Biology* 19: 318–26. The environmental problems associated with globalization and their links to socio-economic effects.

Jackson, T. 2009 *Prosperity without growth: economics for a finite planet*. London, Earthscan. A compelling case against continued economic growth in developed nations.

Likens, G.E. 2010 The role of science in decision making: does evidence-based science drive environmental policy? *Frontiers in Ecology and the Environment* 8: e1–e9. Discussion of challenges associated with environmental problems, including the long delay between discovery and relevant policy-making. Open access: onlinelibrary. wiley.com/doi/10.1890/090132/full.

Vlek, C. and Steg, L. 2007 Human behavior and environmental sustainability: problems, driving forces, and research topics. *Journal of Social Issues* 63: 1–19. A social science perspective on links between people and the physical environment.

WEBSITES

www.arcworld.org   the Alliance of Religions and Conservation (ARC) is an international secular organization that aims to assist major world religions to develop environmental programmes.

www.danadeclaration.org   the Dana Declaration calls for a new approach to conservation: one which recognizes the rights and interests of 'mobile' peoples.

wwf.panda.org   one of the largest international environmental NGOs with information and data on a wide range of issues.

www.unfpa.org   the UN Population Fund site.

www.unpo.org   members of the Unrepresented Nations and Peoples Organization (UNPO) include indigenous peoples, occupied nations and minorities who lack representation in major international bodies such as the United Nations.

www.worldbank.org/en/topic/poverty   a World Bank site providing resources and support for people working to alleviate poverty.

POINTS FOR DISCUSSION

- Human society needs physical resources and environmental issues are an inevitable outcome of using them. Do you agree?
- Discuss the premise that all environmental issues are political issues.
- In terms of environmental quality in your local area, are you better or worse off than your parents/grandparents? Explain how you reached your conclusion and speculate about *your* children's future.
- Explain the ways in which the ownership of physical resources is important in determining how they are used.
- How important is religion in shaping attitudes towards the environment?
- Do you agree that the eradication of poverty is necessary before human society can live within the limits imposed by the physical environment?

3

Sustainable development

LEARNING OUTCOMES

At the end of this chapter you should:

- Appreciate that the relationship between environment and society has reached a point whereby we are exceeding the capacity of our planet to support us.
- Understand how ecosystem services are classified.
- Recognize that the Anthropocene is the name for the contemporary age when human action rivals some of the great forces of nature in its impact.
- Appreciate the basic principles of sustainable development.
- Understand that the details of developing sustainably are debatable and decisions will have to be made in the face of uncertainty.
- Appreciate how and why valuing environmental resources economically has emerged as a way to help integrate our use of resources and ecosystem services more sustainably.
- Understand that some believe there is an imbalance between our reliance on economic growth and the importance of development.
- Recognize that population is regarded by many as the key to the whole sustainable development issue.

KEY CONCEPTS

environmental determinism, ecosystem services, Anthropocene, Sustainable Development Goals (SDGs), precautionary principle, maximum sustainable yield, green economy, payments for ecosystem services (PES), IPAT, ecological footprint, carrying capacity

All the environmental issues covered in this book are, by definition, made up of physical and human elements. The approach throughout is to analyse both the physical problems and the human responses to understand why the issues are issues, and how solutions can be found. Although particular issues are apparent at particular spatial scales, all have some manifestation on the global scale. In the case of climatic change brought about by atmospheric pollution, the issue affects the entire planet because the atmosphere is a globally functioning system. In other cases, more localized issues have become so commonplace or affect such a significant fraction of the total global resource that, cumulatively, they occur worldwide. Examples of such cumulative issues include biodiversity loss (through its worldwide distribution of change) and soil erosion (by affecting a large proportion of agricultural land). Since so many environmental issues now are of a global nature, many believe that the time has come for a complete rethink of the way we view these issues, for a change in the philosophy that lies behind the ways in which people interact with the environment, and for a change in the methodology with which the interactions occur. The approach most widely advocated by this rethink is sustainable development, and this chapter looks in more detail at the need for sustainable development and some of the ways in which it might work.

ENVIRONMENT AND SOCIETY

The relationships between people and the environment lie at the heart of geography as an academic discipline. Attempts to explain these relationships have thrown up a number of views that are worth examining. One early outlook on the links between human society and the physical environment saw the one as being determined by the other. So-called 'environmental determinism' regarded the natural environment as the basic factor controlling human development and activities. Hence factors of culture, race and even intelligence were thought to be controlled, perhaps dictated, by the effects of climate, soils and other aspects of the physical world.

The idea was an attractive way of explaining why human behaviour and institutions vary so widely from one place and people to another, and environmental determinism was widely accepted in western geographical thought in the late 1800s and early 1900s. The approach subsequently lost ground because of the way it was used by some as a justification for imperialism, colonial exploitation and racist views. Critics objected, for example, to the belief that bracing climates like that in Britain produced a generally energetic populace, whereas people born and brought up in places with hot, humid climates were inherently more lethargic.

Reactions to the ideas of environmental determinism have thrown up alternative views. Some of its critics have suggested that the environment never determines any particular way of life, but simply offers a number of different possibilities for society to follow. Supporters of this approach have pointed to the fact that different societies have maintained quite different relationships with similar environments.

Others simply reject altogether the idea that human society is reliant on the physical world, preferring to believe that social facts can only be explained by social drivers, an approach variously referred to as cultural, economic or social determinism. But if some aspects of environmental determinism are today considered to be unacceptable or just plain wrong, it is equally foolish to dismiss environmental influences on society entirely.

THE NEED FOR CHANGE

The socio-economic system is just one part of the natural ecosystem in which materials are transformed and energy converted to heat (Figure 3.1). The operation of the socio-economic system is dependent upon the ecosystem as a provider of energy and natural resources, and as a sink for wastes. The ecosystem also provides numerous 'services' by virtue of its processes. These include provisioning, regulating, and cultural services that directly affect people, plus various supporting services that are needed to maintain these other services (Figure 3.2). This ecological, economic perspective emphasizes the fact that resource use and waste disposal take place in the same environment, and that both activities affect the life-support functions of the environment. Hence, the socio-economic system cannot expand indefinitely since it is limited by the finite global biosphere. Until recently, for example, the number of fish that could be sold at market was limited by the number of boats at sea; now it is limited by the number of fish in the sea. Through much of the history of human occupation of the planet, the socio-economic system has been small relative to the biosphere, so that resources were plentiful, the environment's capacity for assimilating wastes was large, and biospheric functions were relatively little affected by human impact. Today the situation is different. As Goodland *et al.* (1993a: 298) put it: 'now the world is no longer "empty" of people and their artefacts, the economic subsystem having become large relative to the biosphere'. As the socio-economic system has grown larger, fuelled by increased 'throughput' of energy and resources, its capacity for disturbing the environment has increased.

Humanity has expended a great deal of effort on modifying ecosystems so that they produce a constant flow of certain preferred ecosystem services such as food, timber and fibre. However, these efforts have often ignored the fact that landscapes simultaneously produce multiple ecosystem services that interrelate in numerous complex and dynamic ways (Bennett *et al.*, 2009). Hence, when people manage ecosystems in an attempt to maximize the production of one ecosystem service, they frequently cause unexpected and unwanted declines in the provision of other ecosystem services, sometimes resulting in regime shifts with sudden losses of other services. Such declines and sudden shifts are problematic because the demand for reliable provision of almost all ecosystem services is increasing (MEA, 2005).

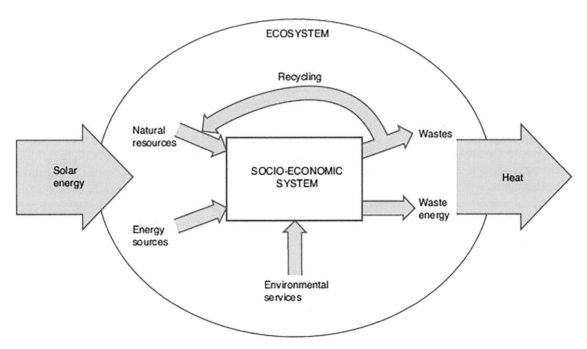

Figure 3.1 The socio-economic system as part of the global ecosystem (after Folke and Jansson, 1992; Daly, 1993).

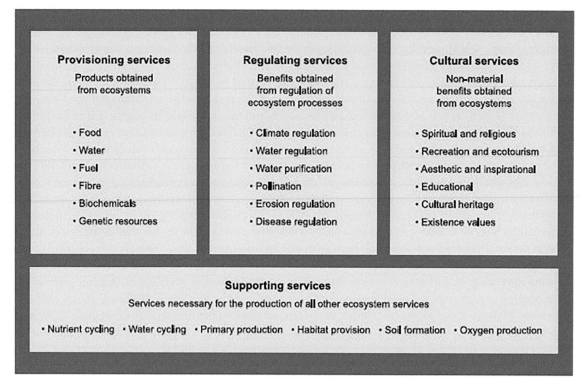

Figure 3.2 A classification of environmental or ecosystem services (after MEA, 2005).

This disruption to the natural environment ultimately feeds back on the operation of society itself, so society's behaviour must conform more closely with that of the total ecosystem because otherwise it may destroy itself. The potential for socio-economic degeneration and eventual collapse to occur when the human system operates too far out of harmony with the environment can be illustrated from both contemporary and historical examples, since although human activity has until recently been within the capacity of the environment on the global scale, breakdown has occurred on more local scales. Overfishing played a key role in the total collapse of Atlantic cod stocks off the coast of Canada in the late twentieth century, with severe social and economic impacts in Newfoundland (see Box 6.1). A lack of harmony between the socio-economic system and the natural ecosystem has also been suggested as a cause of collapse for several ancient cultures. The decline of the Mayan civilization in Central America that began around 900 CE may have been due to environmental degradation as a product of large-scale deforestation, the excessive use of soils and an over-reliance on maize, which failed due to a virus (Figure 3.3). Alternatively, some blame an inability to manage declining water supplies during a period of prolonged drought (Demarest *et al.*, 2004).

A schematic representation of three ways in which human society interacts with the environment is shown in Figure 3.4. The model can be applied on various scales, from the global to the local, and across differing timescales. In cycle A, which is typical of the global economy in historical times, wealth is accumulated largely by degrading the environment. This wealth has brought numerous advances to most parts of the world, which reduce 'stress', a term used here in a wide sense to reflect the general well-being of society. Such advances include sanitation and other facilities, improved health and higher living standards. These improvements have promoted further inappropriate development to continue on cycle A.

Figure 3.3 The ruins at Palenque in southern Mexico, a reminder of the collapse of the Mayan civilization. No one is sure of the true reasons for the downfall of the Mayas, but one theory implicates human-induced environmental degradation as a key factor.

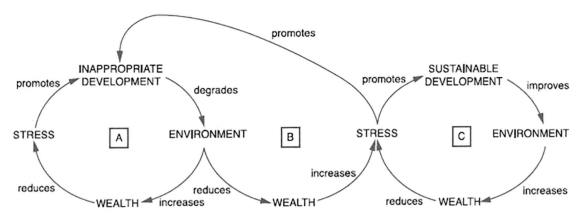

Figure 3.4 Three cycles showing the relationship between modes of development and the environment.

At some point in time, however, cycle A crosses an environmental threshold and the society may enter cycle B. The degraded environment begins to feed back on socio-economic wealth, and stress is increased. Increased stress in cycle B promotes further inappropriate development, particularly when the society in question has limited options as examples of poor, vulnerable groups quoted in this book show (e.g. accelerated soil erosion caused by farmers in Haiti is discussed in Chapter 14), and on the national scale where rapid deforestation in many tropical countries has been pursued in response to the burden of debt and declining commodity prices (see Chapter 4). Continuing on cycle B can ultimately lead to socio-economic collapse, as in the examples of modern fishing industries and ancient civilizations cited above.

Many people believe that, globally, we have now entered cycle B due to the sheer numbers of people on Earth and the resulting scale of human impact on natural systems. There are numerous examples throughout this book which indicate that the productivity of many of the world's renewable resource-producing systems has reached its peak and in some cases is in decline. Human activity is significantly altering several biogeochemical cycles – including the carbon, nitrogen, phosphorus and sulphur cycles – that are fundamental to life on the Earth. We have exploited and controlled nature to such an extent that vast landscapes and entire ecosystems have been fundamentally altered. Roughly 50 per cent of the world's land area has been converted to grazed land or cultivated crops, eradicating more than half of the world's forests in the process. Human activities are likely to be driving the sixth major extinction event in Earth history (MEA, 2005; Steffen et al., 2011). Increased stress is manifested by, among other things, increasing levels of world poverty and fears over the effects of human-induced climatic change.

THE ANTHROPOCENE

The impacts of human actions on the functioning of the Earth system have become so profound and widespread that many observers suggest we have entered a new geological epoch: the 'Anthropocene' (Crutzen and Stoermer, 2000). As the name suggests, the defining feature of this age is the emergence of human action as a critical driver of change in a range of biophysical systems, rivalling some of the great forces of nature in its impact. Other key features of the Anthropocene include a recognition that human influence is occurring at multiple spatial scales, ranging from local to global; that the magnitude and spatial scale of human-induced change may reach thresholds and regional or planetary boundaries that will result in major regime shifts in the Earth system when passed; and that there has been a pronounced surge in human influence and the pace of change since the mid-twentieth century, the so-called Great Acceleration (Steffen et al., 2015).

The term Anthropocene is now widely used to reflect the fact that human action dominates many functions of the planetary system, although the epoch's beginnings remain open to debate (Table 3.1). Some authorities argue for a start at the dawn of agriculture (5,000–10,000 years ago) or even with the extinction of the mammoths and other large mammals (so-called megafauna) towards the end of the Pleistocene epoch 10,000–50,000 years ago, a phenomenon in which prehistoric hunters are implicated (see p. 368). These suggestions effectively imply that the Anthropocene epoch should be merged with the Holocene. Another suggested marker for its start date is the impact that followed the arrival of Europeans in the Americas which led to the rapid deaths of about 50 million indigenous people, a consequent reduction in farming, and regrowth of forests and other vegetation which drew enough carbon dioxide from the atmosphere to produce a marked drop in carbon dioxide (Lewis and Maslin, 2015). Others prefer the early Industrial Revolution (~1800) as a start date for the Anthropocene, or the commencement of the Great Acceleration (~1950), and a beginning some time in the past 50–250 years is most generally agreed.

Table 3.1 Alternative dates proposed for the beginning of the Anthropocene

Year	Main event/marker
1950 CE	Global spread of artificial radionuclides from atmospheric nuclear weapons testing, intensification of fossil fuel use: the Great Acceleration
1800 CE	Early Industrial Revolution, fossil fuel use and global atmospheric change
1610 CE	Dip in atmospheric carbon dioxide related to arrival of Europeans in the Americas and near-cessation of farming: the Orbis Spike
2,000 BP	Anthropogenic soils
5,000–10,000 BP	Early agriculture and global atmospheric change
10,000–50,000 BP	Late Pleistocene megafaunal extinctions

The concept of the Anthropocene has been extensively adopted in academic, technical and public discourses, but the idea has evolved from its beginnings simply as a proposed new geological epoch to become a widely-used metaphor for global change, a novel framework for analyses and a vehicle for views on the relationships between society and nature. The Anthropocene serves to unite physical science perspectives – because our planetary-scale impacts have been identified as changes to physical systems – and the social drivers of environmental change outlined in Chapter 2.

Returning to Figure 3.4, the only appropriate way for humanity to exit cycle B is to enter cycle C. To use a biological metaphor, the socio-economic system needs to be adjusted from its present dominantly parasitic relationship with the environment to a more symbiotic one. The global community does have this option; by adopting sustainable development we can live within the confines of the biosphere, enhance environmental quality and achieve greater wealth by changing the pattern of resource consumption. Adopting cycle C must be a permanent long-term strategy, however, because there is still a danger that reduced stress encourages a reversion to the old ways of cycles A and B.

Examples of societies that have attained a sustainable use of resources can be quoted from both historical and recent times, although the view that all traditionalists were conservationists is somewhat romantic, as indicated by evidence of pre-Hispanic accelerated soil erosion in Mexico (see Chapter 14) and the widespread species extinctions that occurred on tropical Pacific islands prior to the arrival of European settlers (see Chapter 15). Contemporary examples of commercial resource exploitation, in which local communities have had to develop social methods to sustain the resources on which they depend, can also be cited (e.g. Mortimore, 2010). The experience of these societies can be put to good use by other groups seeking to adopt sustainable development.

SUSTAINABLE DEVELOPMENT

Despite the fact that the term sustainable development has become common currency among many groups, it is a confused and sometimes contradictory idea, and there is no widespread agreement as to how it should work in practice (Jabareen, 2008). The phrase is often used interchangeably with sustainability, a way of thinking about how to meet the needs of people and the physical environment simultaneously by enhancing human well-being without undermining the well-being of the planet on which we live. From roots deep in history (Du Pisani, 2006), the concept has developed in the international forum from a document called the *World conservation strategy* (IUCN/UNEP/WWF, 1980), which argued that three priorities should be incorporated into all development programmes:

- maintenance of ecological processes
- sustainable use of resources
- maintenance of genetic diversity.

Sustainable development gained further credence thanks to the World Commission on Environment and Development (also known as the Brundtland Commission after its chair, Gro Harlem Brundtland of Norway) which was formed by the UN in 1983 and reported four years later (WCED, 1987). The Commission emphasized that the integration of economic and ecological systems is all-important if sustainable development is to be achieved, and coined a broad definition for sustainable development which is often quoted: 'development which meets the needs of the present without compromising the ability of future generations to meet their own needs' (WCED, 1987: 43). Although this definition is concise, it is nonetheless open to varying interpretations. What exactly is a *need*, for example, and how can it be defined? Something that is considered a need by one person, or cultural group, may not necessarily be thought of as such by another person or cultural group. Needs may also vary through time, so that it is unlikely (as the definition implies) that those of future generations will be the same as those of present generations. The *ability* of people to meet their needs also changes. Resources are simply a cultural appraisal of the world (see p. 26) and our ability to identify and use them is a function of technology among other things, and technology changes. Likewise, the meaning of *development* can be interpreted in many different ways.

Despite the difficulties in pinning down sustainable development and understanding how it should be applied, it has been adopted by governments, businesses and wider society all over the world. Sustainable development was the central theme of the UN Conference on Environment and Development (UNCED), otherwise known as the Earth Summit, in Rio de Janeiro in 1992 and again at the follow-up Rio+20 Conference in 2012. But although use of the term sustainable development has become increasingly widespread, the fact remains that it is still an ambiguous concept. Perhaps this should not be surprising, since the word 'sustainable' itself is used with different connotations. When we 'sustain' something, we might be supporting a desired state of some kind, or, conversely, we might be enduring an undesired state. These different meanings have allowed the concept to be used in varying, often contradictory ways. The range of views on the matter is illustrated in Table 3.2 where the guiding principles of sustainability as perceived by the dominant economic worldview are contrasted with those put forward by deep ecology. The differences shown in these two perspectives highlight how sustainable development is a political issue as well as a technical one because the way a group goes about dealing with a particular sustainability issue will be greatly influenced by its political beliefs.

Further confusion over the meaning of the term sustainable stems from its use in a number of different contexts, such as ecological sustainability and economic sustainability. A central tenet of ecological sustainability is that human interaction with the natural world should not impair the functioning of natural biological processes. Hence concepts such as 'maximum sustainable yield' have been developed to indicate the quantity of a renewable resource that can be extracted from nature without impairing nature's ability to produce a similar yield at a later date. Economic sustainability tends to give a lower priority to ecosystem functions and resource depletion. It can result in different approaches to the environment depending on the economic objectives of the environmental manager concerned.

Table 3.2 Extremes in the range of views on the guiding principles of sustainability

Dominant economic worldview	Deep ecology worldview
Dominance over nature	Natural harmony: symbiosis
Natural environment is a resource for human use	All nature has intrinsic worth: biospecies equality
Material and economic growth for increasing human populations	Simple material needs: goal of self-realization
Belief in ample reserves of resources	Earth's resources are limited
High technological progress and solutions	Appropriate technology and non-dominating science
Consumerism and growth in consumption	Make do with enough: recycling
National/centralized community	Minority traditions and religious community knowledge

Source: after Gibbon *et al.* (1995); Colby (1989).

These two perspectives have led to a distinction being made between 'strong' and 'weak' sustainability (Neumayer, 2010). Weak sustainability implies that development is sustainable as long as its capacity to generate income for future generations is maintained. A central tenet of this view is an assumption of substitutability between natural capital (natural resources and ecosystem services) and manufactured capital. The focus is on the total capital, but logically this focus can ultimately lead to a situation of near complete environmental devastation, as illustrated by the results of phosphate mining on the Pacific Ocean island state of Nauru (see Chapter 19).

One definite strength of the sustainability idea is that it draws together concerns over economic development, social development and environmental protection, the so-called three pillars of sustainable development. In practice, most would agree on a number of common guiding principles for sustainable development:

- continued support of human life
- continued maintenance of environmental quality and the long-term stock of biological resources
- the right of future generations to resources that are of equal worth to those used today.

The most recent way in which these principles have been applied internationally is through the Sustainable Development Goals (SDGs), a series of guidelines and targets for all countries that were shaped at the Rio+20 Conference. The 17 interconnected SDGs (Figure 3.5) build on the successes of the previous Millennium Development Goals (MDGs), which started a global effort in 2000 to tackle a range of development priorities. The SDGs provide a common plan to meet the urgent environmental, political and economic challenges facing our increasingly crowded planet, including poverty, climate change and conflict.

Figure 3.5 The Sustainable Development Goals (SDGs).

UNCERTAINTY, THE PRECAUTIONARY PRINCIPLE AND ADAPTIVE MANAGEMENT

As Chapters 1 and 2 have shown, our knowledge and understanding of both the physical and human environments are not complete, and one of the greatest challenges of sustainability is society's need to formulate policy and make decisions in the face of this uncertainty. We are uncertain of many aspects of environmental issues, from the local and discrete to the global and complex. These aspects include the shape of future societies, the nature and causes of environmental change and the severity of human impacts. In natural sciences and engineering, uncertainty is commonly understood to be caused by either a lack of information or inherent random variations associated with, for instance, a river system. Uncertainties are also associated with the social system, of course, and ambiguity is a commonly cited third type of uncertainty which can be defined as the degree of confusion that exists among actors in a group (e.g. stakeholders) due to the presence of multiple, valid, and sometimes conflicting, interpretations of issues. These three types of uncertainty are often interrelated: what is known (or not known) about a system is influenced by the framings through which the issues are comprehended (Van den Hoek *et al.*, 2014).

One approach that has been widely adopted to help make decisions about environmental issues in the light of uncertainty is the precautionary principle. The precautionary principle, which developed in Europe, advocates that human responses to environmental issues should err on the side of caution, a sort of 'better safe than sorry' approach to interacting with the environment. It appears in the 1992 Rio Declaration and has been incorporated into the UN Convention on Biological Diversity, the UN Framework Convention on

Climate Change and European Union environmental policy. Like sustainable development itself, the precautionary principle is interpreted in different ways by different people, but Dovers and Handmer (1995) identify some common themes:

- Uncertainty is unavoidable in sustainability issues.
- Uncertainty over the severity of environmental impacts resulting from a development decision or a current human activity should not be an excuse to avoid or delay environmental protection measures.
- The principle recommends we should anticipate and prevent environmental damage rather than simply react after it has occurred.
- The burden of proof should shift from the victim to the developer, so that those proposing an action must show that it will not harm the environment or that whatever practical measures available for preventing damage will be taken.

The basic idea behind the principle has had great popular and scientific appeal, but it is not without its difficulties. For example, the first of the four common themes outlined above may create problems for implementation of the last. The fact that uncertainty is unavoidable could mean that a harmful impact of an action is not known until after the action has taken place. This has happened before, such as with CFCs (see Chapter 11) and nuclear power (see Chapter 18), both technologies that were thought to be totally safe for the environment when first introduced. In situations like these, shifting the burden of proof to the developer of the technology or activity may do little to enhance safety if the harmful effects are only discovered belatedly. Further, shifting the burden of proof may actually make a problem worse by discouraging or delaying the introduction of beneficial technologies.

Perhaps the most severe problem with the precautionary principle lies in the lack of a commonly agreed definition and the associated need for clear practical guidelines on how it should be implemented. In practice, of course, dealing with uncertainty in managing human and natural systems requires constant readjustments to reflect developments in our changing understanding. This is often called adaptive management, a decision-making process under uncertainty that is designed to learn and incorporate new information and thereby improve future decision-making. It should be remembered that these decisions about sustainable development will continue to be informed as much by moral and political issues as by scientific understanding.

VALUING ENVIRONMENTAL RESOURCES ECONOMICALLY

Much research and thinking about sustainable development has focused on modifying economics to better integrate its operation with the workings and capacity of the environment, to use natural resources more efficiently, and to reduce flows of waste and pollution. The full cost of a product, from raw material extraction to eventual disposal as waste, should be reflected in its market price (so-called 'life-cycle analysis', see Chapter 17), although in practice such a 'cradle to grave' approach may prove troublesome for materials such as minerals (see Chapter 19). Many economists contend that a root cause of environmental degradation is the simple fact that many aspects of the environment are not properly valued in economic terms. Commonly owned resources – such as the air, oceans and fisheries – are particularly vulnerable to over-exploitation for this reason, but when such things have proper price tags, it is argued that decisions can be made using cost–benefit analysis. Putting a price on environmental assets and services is one of the central aims of the discipline of environmental economics. This can be done by finding out how much people are willing to pay for an aspect of the environment or how much people would accept in compensation for the loss of an environmental asset. One of the justifications of environmental pricing is the fact that money is the language of government treasuries and big business, and thus it is appropriate to address environmental issues in terms that such influential bodies understand.

There are problems with the approach, however. People's willingness to pay is heavily dependent on their awareness and knowledge of the resource and of the consequences of losing it. Information, when available, is open to manipulation by the media and other interest groups. In instances where the resource is unique in world terms – such as an endemic species (Figure 3.6) or a feature such as the Grand Canyon – who should be asked about willingness to pay? Should it be local people, national groups or an international audience? Our ignorance of how the environment works and the nature of the consequences of environmental change and degradation also presents difficulties. In the case of climatic change due to human-induced atmospheric pollution, for example, all we know for sure is that the atmospheric concentrations of greenhouse gases have been rising and that human activity is largely responsible. However, we do not know exactly how the climate will change nor what effects any changes will have upon human society (see Chapter 11). We can only guesstimate the consequences, so we can only guesstimate the costs. Economists are undeterred by these types of problem: 'Valuation may be imperfect but, invariably, some valuation is better than none' (Pearce, 1993: 5).

Figure 3.6 These dragon's blood trees are endemic to Socotra, an island midway between Yemen and Somalia. The species is unique but not very widely known, so how should it be valued?

The total economic value (TEV) approach is commonly used to compare the diverse benefits and costs associated with ecosystems and measure the ecosystem goods and services used by human beings (Table 3.3). This classification typically divides TEV into two primary categories – use values and non-use values – some of which can be further subdivided. Direct use values correspond broadly to the provisioning and cultural services provided by ecosystems shown in Figure 3.2. These can be consumptive (leaving a reduced quantity of the good available for others to use) or non-consumptive (with no reduction in available quantity). Indirect use values correspond broadly to the regulating and supporting services detailed in Figure 3.2. Option values stem from preserving the option to use ecosystem goods and services at some future time, either by the individual (option value) or by others or heirs (bequest value). These option values may be any of the provisioning, regulating or cultural services shown in Figure 3.2. Non-use values are also often known as existence values because they refer to values given by people to knowing that a resource exists, even if they never use it directly.

Generally speaking, direct use values are the easiest to measure. Measuring indirect use value is often more difficult, but non-use values are more difficult still. Nonetheless, TEV has become a widely used tool for addressing many environmental issues. One example of how it has been harnessed by conservationists is payments for ecosystem services (PES) that are made when valuation shows that indirect uses such as watershed protection provide important benefits (see Box 4.1).

GREEN ECONOMY

Reflecting the environment's value in economic decision-making is an essential element in the idea of the so-called green economy, a concept that has become popular in recent times as a potential remedy to some of the key market and institutional failures associated with conventional economic growth. Although the precise meaning of the term (and related concepts such as 'green growth') has been a matter for debate, in essence a green economy is one that results in improved human well-being and social equity, while significantly reducing environmental risks and ecological scarcities (UNEP, 2011). It is promoted as an effective pathway to advancing the three pillars of sustainable development: economic development, social

Table 3.3 The total economic value (TEV) approach

Total economic value (TEV)			
Use value			Non-use value
Direct use value • Consumptive (e.g. harvesting food or timber products) • Non-consumptive (e.g. bird-watching, water sports)	Indirect use value (e.g. water purification, nutrient cycling)	Option value • Option • Bequest	Existence value

Source: after MEA (2005).

development and environmental protection. The green economy approach requires these three goals to be pursued simultaneously, emphasizing economic development that is resource-efficient, within environmental limits and equitable across society. The green economy, in the context of poverty eradication and sustainable development, was one of two specific themes discussed at the Rio+20 Conference, the other being the institutional framework for sustainable development.

Even when economic values can be assessed, however, differing values potentially derived from a particular area can be of benefit to different sectors of society. For example, the value to the national economy of wilderness areas in Kenya, when conserved as National Parks, has been estimated to be at least an order of magnitude greater per hectare than the land would yield if put to pastoral use (i.e. herding animals). However, management of, say, the Masai Mara Reserve as a game park principally benefits the operators of tourist lodges and tours, while the pastoral Masai derive only 1 per cent or less of the accruing revenue. For the Masai, a more financially worthwhile use of the reserve would be to open it up to pastoralism and/or for poaching ivory and other wildlife products (Holdgate, 1991).

Other difficulties stem from the different ways groups perceive environmental value. Some may consider that certain parts of the environment lie outside the economists' realm, because economists are not actually dealing with values at all, but with prices. Adams (1993) gives the example of a mining company that wanted to exploit a site in Australia's Kakadu National Park, an area sacred in Aboriginal culture. In this case, the application of cost–benefit analysis amounts to immorality: 'The aborigines say "It is sacred." The economist replies "How much?"' (Adams, 1993: 258). It is for these sorts of reasons that some people reject the economic approach to valuing and hence preserving the environment and biological diversity. They argue that economic criteria of value can change and are opportunistic in their practical application (e.g. Ackerman and Heinzerling, 2004). The greed that underlies such systems will ultimately destroy them. It is certainly fair to say that the numerous attempts to put an economic price on aspects of the physical environment have helped to bring their relative value in economic terms to the attention of politicians. However, the true value of the environment in terms of the supporting ecosystem services outlined in Figure 3.2 is so fundamental to the planet and well-being of everything living here as to be priceless.

GROWTH AND DEVELOPMENT

A key issue in the sustainable development debate is the relative roles of economic growth (the quantitative expansion of economies) and development (the qualitative improvement of society). In its first report, the World Commission on Environment and Development (WCED) suggested that sustainability could only be achieved with a five- to tenfold increase in world economic activity in 50 years. This growth would be necessary to meet the basic needs and aspirations of a larger future global population (WCED, 1987). Subsequently, however, the WCED played down the importance of growth (WCED, 1992). This makes sense, because many believe that it has been the pursuit of economic growth, and neglect of its ecological consequences, that have created most of the environmental problems in the first place.

The change in thinking on economic growth has been reflected in the two types of reaction to calls for sustainability that have been made to date: on the one hand, to concentrate on growth as usual, though at a slower rate, and on the other hand, to define sustainable development as 'development without growth in throughput beyond environmental capacity' (Goodland et al., 1993a: 297). The idea of controlling throughput, the flow of environmental matter and energy through the socio-economic system (see Figure 3.1), can be referred back to the ways in which human society interacts with the environment, as shown in Figure 3.4: cycles A and B are based on increased throughput, while in cycle C throughput is controlled to a level within the environment's capacity to support it. Sustainable development – cycle C – means that

Table 3.4 Key rules for the operation of environmental sustainability

OUTPUT RULE
Waste emissions should be within the assimilative capacity of the environment to absorb, without unacceptable degradation of its future waste-absorptive capacity or other important services

INPUT RULE
Renewable resources: Harvest rates of renewables used as inputs should be within the capacity of the environment to regenerate equal replacements
Non-renewable resources: Depletion rates of non-renewables used as inputs should be equal to the rate at which sustained income or renewable substitutes are developed by human invention and investment. Part of the proceeds from using non-renewable resources should be allocated to research in pursuit of sustainable substitutes

Source: after Goodland *et al.* (1993b).

the level of throughput must not exceed the ability of the environment to replace resources and withstand the impacts of wastes (see Table 3.4, which is effectively a set of rules to promote intergenerational equity). This does not necessarily mean that further economic growth is impossible, but it does mean that growth should be achieved by better use of resources and improved environmental management rather than by the traditional method of increased throughput (Costanza *et al.*, 2009), an idea in line with the tenets of the green economy (see above). However, there is also a feeling in some quarters that building a 'post-growth' economy is the more sustainable way forward (Jackson, 2017).

One indication of the degree of change necessary to make this possible is in the ways we measure progress and living standards at the national level. Measurements such as Gross National Income (GNI) and Gross Domestic Product (GDP) are the principal means by which economic progress is judged, and thus form key elements in government policies. But GNI is essentially a measure of throughput, and it has severe limitations with respect to considerations of environmental and natural resources. The calculation of GNI does not take into account any depletion of natural resources or adverse effects of economic activity on the environment, which have feedback costs on such things as health and welfare. Indeed, conventional calculations of GNI frequently regard the degradation of resources as contributing to wealth, so that the destruction of an area of forest, for example, is recorded as an increase in GDP. The need to introduce environmental parameters into national accounting systems is now widely recognized, and adjusted measures of some kind of 'green GNI' are being worked upon to provide a good gauge of national sustainability (Costanza *et al.*, 2009). One such indicator is the Inclusive Wealth Index (IWI), developed to assess changes in a country's productive base (including produced, human, and natural capital) over time and thus provide a measure of both a nation's wealth and the sustainability of its growth (UNU-IHDP and UNEP, 2012).

POPULATION AND TECHNOLOGY

Further investigation of the limiting throughput approach to sustainable development can be conducted by representing human society's impact on the environment by the equation

$$I = P \times A \times T$$

Figure 3.7 The Kombai of New Guinea rely on hunting for their main source of food, using traditional technology – spears and bow and arrows. Their environmental impact is minor, but elsewhere in earlier times small populations of hunter–gatherers are thought to have had considerable environmental impacts.

where I is impact, P is population, A is affluence (measured as the consumption of environmental resources per person) and T is technology (which includes both technologies used, such as cars or bicycles, and also associated political, social and economic conditions, such as subsidies encouraging the use of a particular technology). These three factors can provide the keys to keeping throughput within environmental capacity. This can be done by one of three means (Goodland *et al.*, 1993a):

- limit population
- limit affluence
- improve technology.

This approach to achieving sustainability is attractive if only because of its simplicity. It clearly specifies the key drivers behind environmental change and provides researchers with variables that can be measured in an attempt to understand complex issues. The formula makes it clear that each driver acts in conjunction with the other drivers, so that no one factor can be held singularly responsible for an environmental impact. Changes in one factor are multiplied by the other factors. Hence, for example, if in a particular country P and T remain constant over a period of time while A increases, it would be wrong to argue that impacts are caused by A alone, since P and T, even though they remain unchanged, scale the effects of changes in A.

The equation certainly has applications, but it also contains some flaws, and is based on assumptions that are not always valid. The effects of the driving forces on impacts are assumed to be strictly proportional, but we know that some environmental systems contain non-linear characteristics, including thresholds (see Chapter 1). Hence an ecosystem may be resilient under increasing pressure of human activity until a point is reached at which there is sharp, discontinuous change, such as a regime shift. The relationship also implies that a small human population with limited affluence and rudimentary technology has a small environmental impact, which may indeed be the case in many circumstances (Figure 3.7), but not in all. There is persuasive evidence to suggest that in prehistoric times very small populations of hunter–gatherers may have been responsible for the extinction of several species of large mammal and wide-scale ecological change following the introduction of new fire regimes (see Chapter 15).

Some researchers argue that technological changes are far more significant than population or affluence in determining human-induced environmental change (e.g. Raskin, 1995). Indeed, absolute population numbers, density and growth rates are not always clearly related to degradation (see p. 28). Although greater affluence, as defined above, can equate to greater environmental impact, this is not necessarily so. In many cases, as affluence increases, so does a society's demand for environmental quality, with the consequent adoption of conservation measures and relatively 'clean' technologies. Similarly, population growth rates generally fall, sometimes even to negative values, with rising incomes.

Conversely, there is also a threshold below which decreasing affluence leads to greater degradation rather than less: the mutually reinforcing links of the poverty–environment nexus (see p. 37). Indeed, poverty is a key issue in the sustainability debate, an outcome of inequalities at the global and national scales. As a former executive director of the UN Environment Programme put it: 'The harsh fact remains: conservation is incompatible with absolute poverty' (Tolba, 1990: 10).

The I = PAT equation (sometimes simply referred to as IPAT) also gives little recognition to such factors as beliefs, attitudes, or politics, all important influences on a society's environmental impact (see Chapter 2). Another concern surrounds the geographical scale at which the

Figure 3.8 This severely degraded landscape in Azerbaijan is an illustration of how environmental impacts are effectively exported from one place to another. Oil from the Caspian Sea coast has been exploited commercially in Azerbaijan for more than 100 years. Most of the oil has been exported to both eastern and western Europe, but the impact is clear to see in the region of origin.

equation can be applied. While the formula can be used to sum the human pressures on resources in a closed system, it does not indicate where within the system those pressures will be felt. Some environmental impacts will be local, of course, but others may involve degradation of a globally common resource such as the atmosphere, a sink for wastes such as greenhouse gases, which means that part of the impact is felt by others.

Some of the environmental effects of human activities may be displaced well away from the region of origin. This means that an environmental impact made by the population of one country could be felt in a quite different country, as with some cases of acid rain (see Chapter 12) or pollution of international rivers (see Box 19.1). In addition to such environmental problems obviously crossing borders, trade and other economic relationships between societies may allow harmful effects to be 'exported' to another region in a less literal way. This could be because a poorer society is more willing to incur environmental damage in return for economic gain, or because it is less able to prevent it. The concept of an 'ecological footprint' is one way of accounting for such displaced effects. It is an assessment of the total area required to support the population of a nation, region, city or community – or indeed an individual – and the associated level of resource consumption. This important geographical dimension is illustrated at the global scale by the view that it is the overconsumption of resources in the North that is primarily responsible for most environmental issues (Figure 3.8).

CARRYING CAPACITY

Another facet of the I = PAT formula is its implicit acceptance of the concept of a finite carrying capacity. Carrying capacity is a term with obscure historical origins and the term has been widely used in many different areas of resource management, often meaning different things to different users (Sayre, 2008). One of the most commonly employed meanings is to indicate the point at which human use of an ecosystem can reach a maximum without causing degradation. In other words, carrying capacity is the threshold point of stability. If human activity occurs in a particular area at or below its carrying capacity, such activity can proceed at equilibrium with the environment, but if the carrying capacity of a particular area is exceeded, then the area's resources will be degraded. The important point is that the carrying capacity of an area is defined by the ecology of the area, not by people.

Three simple theoretical views of the relationship between carrying capacity and population are shown in Figure 3.9. Figure 3.9a illustrates a sustainable situation in which population numbers remain more or less stable and within the limits imposed by a fixed environmental carrying capacity. Figure 3.9b, by contrast, illustrates how carrying capacity declines when resources are degraded and human population declines as a consequence. The decline and fall of the Mayan civilization has already been mentioned as an example. Another often-quoted case is thought to have occurred on Easter Island, now known as Rapa Nui, where widespread deforestation probably took place principally for rollers to move the enigmatic statues found all over the island (Figure 3.10). Deforestation resulted in accelerated soil erosion and falling crop yields, leading to food shortages. The inability to erect more statues undermined belief systems and social organization, resulting in armed conflicts over remaining resources. This combination of effects led to a severe decline in human population (Bahn and Flenly, 1992). Subtropical trees and giant palms dominated Rapa Nui's vegetation when it was first colonized by people probably in about the year 1200 (Hunt and Lipo, 2006), but when Dutch explorers landed on the island a few days after Easter Day in 1722, they found a barren landscape with a native society facing dwindling supplies of food and wood. Ever since, Rapa Nui has been considered a textbook example of a culture that doomed itself by destroying its own habitat. Rapa Nui is an extremely remote Pacific landmass, the nearest inhabited land being more than 2000 km distant, so effectively the island was a closed system. A parallel can be drawn with planet Earth isolated in an essentially uninhabitable universe.

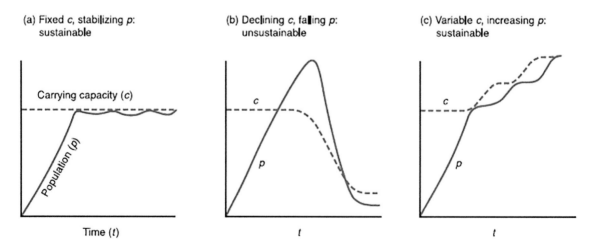

Figure 3.9 Three theoretical variations in carrying capacity and population totals.

This idea of a biologically defined carrying capacity underpins many notions of environmental degradation. It is inherent in Malthusian ideas of overpopulation (see below) and the tragedy of the commons (Figure 3.9b), and is directly related to the idea of maximum sustainable yield (Figure 3.9a). This idea that human society is subject to natural biological laws strikes right at the heart of the sustainable development debate. Some believe that human ingenuity puts us above the laws of nature. The situation in Figure 3.9c is also sustainable, but human population grows because carrying capacity is not fixed; it is actually increased because of new technologies and new resource perceptions (see also below). Meanwhile, some ecologists suggest that the carrying capacities of some natural environments are naturally variable, because the environments are not in equilibrium. Such environments are constantly dynamic, varying in time and space, rendering the idea of carrying capacity meaningless.

Population

Population is regarded by many as the key to the whole sustainable development issue. Some very influential ideas about the relationship between the size of human population and the availability of natural

Figure 3.10 Ancient statues on the deforested Rapa Nui (formerly known as Easter Island) in the Pacific Ocean. Many believe that widespread clearance of trees exceeded the island's carrying capacity, eventually resulting in a collapse of the island's human population.

Figure 3.11 Public information in southern Benin about anti-retroviral treatment for people living with HIV/AIDS. Some view the spread of diseases such as HIV/AIDS as neo-Malthusian checks on population which help to keep numbers within limits set by the planet's resources. The argument is complicated in many areas where people still die of diseases that are curable because they cannot afford the treatment.

resources were developed at the end of the eighteenth century by Thomas Malthus (see Chapter 2). Malthus based his theory on two central principles: that without any checks, human population can grow at a geometric rate; and that even in the most favourable circumstances agricultural production can only increase at an arithmetic rate. Hence, at some point the size of human population will become too large for the food available. Malthus proposed that two sorts of check might act to keep the human population at a manageable size. These were 'preventive checks' such as delaying the age of marriage, which would reduce the fertility rate, and 'positive checks' such as famine, disease and warfare, that would increase the death rate.

Several authors have invoked the Malthusian argument in recent times to suggest that essentially there is a limit to the size of human population on this planet (e.g. Ehrlich and Ehrlich, 2002) and that this limit is imposed by the planet's finite resources (e.g. Meadows *et al.*, 1972, 2004). Examples of Malthus' preventive and positive checks are still plain to see, of course (Figure 3.11). However, needless to say, not everyone agrees with so-called neo-Malthusian scenarios. The opposing school of thought points out that population alone is not a reliable predictor of resource use, because people's use of resources is influenced by economic, environmental and cultural factors.

They see the world's human population not as the passive victims of immutable biological laws of carrying capacity – effectively a form of environmental determinism – but as active creators of new resources thanks to human ingenuity developing technological innovations (Boserüp, 1965: 28). This view is epitomized by the idea that humans are in fact the 'ultimate resource' (Simon, 1981). While population growth may give rise to problems in the short term, these problems become the drivers behind innovations and the creation of new resources. Central to this perspective is the idea that physical resources should not be seen as fixed because they are culturally defined. Hence, the fixed carrying capacity argument does not apply.

The following chapters on individual environmental issues put the theory and concepts outlined in these first three chapters into context by showing how the issues have arisen and how human societies have acted in their attempts to resolve the issues that have been created.

FURTHER READING

Costanza, R., Cumberland, J.H., Daly, H., Goodland, R., Norgaard, R.B., Kubiszewski, I. and Franco, C. 2015 *An introduction to ecological economics*, 2nd edn. Boca Raton, CRC Press. A definitive guide to how the market economy can assist society's attempts to manage the planet.

Jeffers, E., Nogue, S. and Willis, K. 2015 The role of palaeoecological records in assessing ecosystem services. *Quaternary Science Reviews* 112: 17–32. Highlighting how data from natural archives helps our understanding of ecosystem services operating over decadal (or longer) timeframes.

Schröter, M., Zanden, E.H., Oudenhoven, A.P., Remme, R.P., Serna-Chavez, H.M., Groot, R.S. and Opdam, P. 2014 Ecosystem services as a contested concept: a synthesis of critique and counter-arguments. *Conservation Letters* 7: 514–23. A lucid assessment of the pros and cons of the ecosystem services idea.

Steffen, W., Broadgate, W., Deutsch, L., Gaffney, O. and Ludwig, C. 2015 The trajectory of the Anthropocene: the Great Acceleration. *The Anthropocene Review* 2: 81–98. Describing the Anthropocene, including some of the strong equity issues masked by considering global aggregates.

Thiele, L.P. 2016 *Sustainability*, 2nd edn. Cambridge, Polity Press. Focusing on how sustainability problems are defined, by whom and with what solutions, this book looks at the social nature of the issues.

Wiedmann, T. and Barrett, J. 2010 A review of the ecological footprint indicator: perceptions and methods. *Sustainability* 2: 1645–93. A comprehensive review. Open access: www.mdpi.com/2071-1050/2/6/1645.

WEBSITES

sei-international.org   the Stockholm Environment Institute is an independent, international research institute specializing in sustainable development and environment issues at local, national, regional and global levels.

sustainabledevelopment.un.org   the UN's Sustainable Development Knowledge Platform.

www.csin-rcid.ca   a Canadian network that fosters the development and use of sustainability indicators.

www.footprintnetwork.org   the Global Footprint Network aspires to make the ecological footprint as prominent a measure as GDP, fostering a world where all people can live satisfying lives within the planet's ecological capacity.

www.iisd.org   a multimedia resource for environment and development run by the International Institute for Sustainable Development.

www.wbcsd.org   the World Business Council for Sustainable Development site includes case studies of improving environmental performance and online sustainable development courses.

POINTS FOR DISCUSSION

■ Consider what you can do to make your life more sustainable.
■ Which aspect of the Anthropocene is more important to study: its physical science or its social science?
■ Are there contradictions between the sustainable development aims of welfare for all and environmental conservation?
■ Can environmental economics solve all environmental problems?
■ Has our planet's carrying capacity been exceeded?
■ Design a study to test the I = PAT formula in your local area.

4

Tropical deforestation

Forests have been cleared by people for many centuries, to use both the trees themselves and the land on which they stand. Throughout the history of most cultures, deforestation has been one of the first steps away from a hunting–gathering and herding way of life towards sedentary farming and other types of economy. Forests were being cleared in Europe in Mesolithic and Neolithic times, but in central and western parts of the continent an intense phase occurred in the period 1050–1250. Later, when European settlers arrived in North America, for example, deforestation took place over similarly large areas but at a much faster rate: more woodland was cleared in North America in 200 years than in Europe in over 2000 years. Estimates suggest that in pre-agricultural times the world's forest cover was about 5 billion hectares (Mather, 1990). In 2015, the UN Food and Agriculture Organization (FAO) concluded that natural and plantation forests covered just under 4 billion hectares (FAO, 2016a). Although most of this loss has taken place in the temperate latitudes of the northern hemisphere, in recent decades the loss in this zone has been largely halted, and in many countries reversed by planting programmes. Meanwhile, in the tropics, rapidly increasing human populations and improved access to forests have combined in recent times to create an accelerating pace of deforestation, which has become a source of considerable concern both at national and international levels. This chapter concentrates on the humid tropics, while information on forest clearance in the drier parts of these latitudes can be found in Chapter 5.

DEFORESTATION RATES

Despite the high level of interest in deforestation, our knowledge of the rates at which it is currently occurring is far from satisfactory. In part, this is because of the lack of standard definitions of just what a forest is and what deforestation means. A distinction is often made between 'open' and 'closed' forest or woodland, which is sometimes defined according to the percentage of the land area covered by tree crowns. A 20 per cent crown cover is sometimes taken to be the cut-off point for closed forest, for example, while open woodland is defined as land with a crown cover of 5–20 per cent. Measurements of deforestation rates by the FAO define deforestation as the clearing of forestlands for all forms of agriculture and for other land uses such as settlements, other infrastructure and mining. In tropical forests, this entails clearing to reduce tree crown cover to less than 10 per cent, and thus does not include some damaging activities such as selective logging, which can inevitably affect soils, wildlife and its habitat. In this case, the change in the status of the forest is referred to as degradation rather than deforestation.

The collection of data for all countries of the world and their compilation by different experts inevitably cause problems, since different countries and individuals often define the terms in different ways and the quality and availability of data vary between countries. These sorts of difficulties can be considerable. During the course of a recent global assessment of the rate of forest change carried out by the FAO (2010), the number of definitions of forest identified was, extraordinarily, more than 650. Significant differences occur even within a small area such as western Europe. As Mather (2005) points out, if the Swedish definition of forest was used to assess the forested area of Spain, the result would be a forest area that is 5 per cent larger than on the Spanish definition. Using the UK definition would mean an area 8 per cent smaller.

One tool which is being used increasingly to avoid some of the problems of definition and poor data is satellite imagery, which can be evaluated using one set of criteria. But even these data can vary, depending on the satellite sensors used, differing methods of interpretation, and other difficulties such as that of distinguishing between so-called 'primary forest' and 'secondary forest' that is the result of human impact. Selective logging, for example, has gone largely undetected in most satellite-based assessments, but one study that was able to quantify the extent of such thinning of the forested area in the Brazilian Amazon

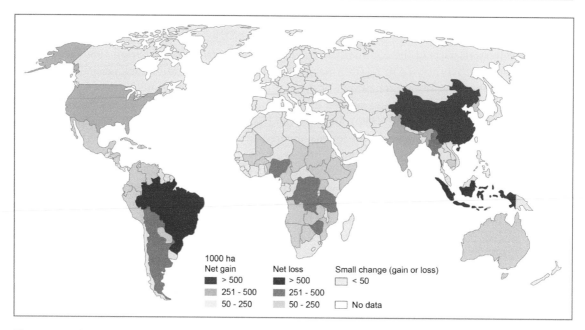

Figure 4.1 Annual change in forest cover by country, 1990–2015 (FAO, 2015).

found that incorporating the area that is selectively logged doubles previous estimates of the total amount of forest degraded by human activities in this area (Asner *et al.*, 2005).

Estimates, from numerous sources, of the annual rate of deforestation of closed forests in the humid tropics (tropical moist forest, which includes two main types: tropical rain forest and tropical monsoon/ seasonal forest) have varied from 11 to 15 million ha for the early 1970s, 6.1 to 7.5 million ha for the late 1970s, and 12.2 to 14.2 million ha for the 1980s (Grainger, 1993b). This author suggests that the uncertainties are primarily the result of lack of attention to remote sensing measurements, and overconfidence in the use of expert judgement. A recent assessment of the rates of change of forest cover by country over 25 years, prepared by the FAO, is shown in Figure 4.1. This map shows that while forest cover in the high latitudes has been generally stable or increasing, many tropical countries have experienced a loss. This survey produced an annual deforestation rate for the tropics of 10.4 million ha in the 1990s, which fell to 6.4 million ha a year from 2010 to 2015. These figures compare favourably with the FAO estimate of the overall annual global loss of all tropical moist forests over the period 1981–90 of 13.1 million ha (FAO, 1995).

CAUSES OF DEFORESTATION

A concise summary of the causes of deforestation in the humid tropics is no easy task, since cutting down trees is the end result of a series of motivations and drivers that are interlinked in numerous ways. Table 4.1 is an attempt to identify the prime direct factors that lie behind deforestation in the major forest regions of the tropics, based on reviews of the large literature on the subject. The importance of these drivers varies from country to country and from region to region, and may change over time. Agriculture, both commercial and subsistence, is frequently associated with forest clearance and degradation in all regions, while mining is less common generally but important in certain areas such as parts of the Congo Basin (see p. 469).

Table 4.1 Key direct drivers of tropical deforestation

Agriculture (commercial)	Clearance for cropland, pasture and tree plantations, usually medium- to large-scale, for international and domestic markets
Agriculture (subsistence)	Includes both permanent and shifting cultivation, usually by smallholders
Wood extraction	Commercial timber, fuelwood and charcoal (domestic and industrial)
Mining	Large-scale industrial, small-scale and artisanal
Infrastructure	Roads, railways, pipelines, electricity transmission, hydroelectric dams
Urban expansion	Settlement growth

Source: after Hosonuma *et al.* (2012); Rudel *et al.* (2009); Tegegne *et al.* (2016).

However, a proper understanding of the deforestation process requires a deeper investigation of the underlying causes behind these drivers. The socio-economic factors that push people to the forest frontier are important. These include poverty, low agricultural productivity and an unequal distribution of land, while the rapid population growth rates that characterize many countries in the tropics also play a part. Access to forests is also significant. Hence, construction of a new road, whether it be by a logging company or as part of a national development scheme, both immediately results in deforestation and also opens the way for settlers who cut down trees for other purposes, such as agriculture. The role of national government is another factor in encouraging certain groups to use the forest resource, through tax incentives to loggers and ranchers, for example, or through large-scale resettlement schemes. On the global scale, international markets for some forest products, such as timber and the produce from agricultural plantations and ranches, must also be considered, an aspect of globalization that has weakened the historically strong relationship between local population growth and forest cover (Rudel *et al.*, 2009; Hosonuma *et al.*, 2012).

Some of the ways in which deforestation in the tropics has changed in recent times are outlined by Rudel (2007) who identifies how a process largely initiated by the state became one driven by private enterprise in the last decades of the twentieth century. During the 1970s, state-run road building and colonization programmes opened up regions for settlement and deforestation throughout the tropics, but by the 1990s these programmes had all but disappeared, to be replaced by private enterprise-driven processes. Since the turn of the twenty-first century, this trend has continued and become more globalized. Leblois *et al.* (2017) identify an increase in agricultural exports from poorer countries with a large amount of remaining forest cover as playing a critical role in driving deforestation.

This transition is reflected in the last 50 years of deforestation in the Amazon Basin in Brazil where government policy was instrumental in initiating deforestation on a large scale in the mid-1970s with a concerted effort to develop the country's tropical frontier. Agricultural expansion was the most important factor responsible for forest clearance at this time, both by smallholders and large-scale commercial agriculturalists, including ranchers producing beef for the domestic market. The Brazilian government's view of the Amazon as an empty land rich in resources spurred programmes of resettlement, particularly in the states of Rondônia and Pará, under the slogan 'people without land in a land without people', along with agricultural expansion programmes and exploitation of mineral and hydroelectric resources.

The 1990s was a period of rapid large-scale expansion of agribusiness – soybean plantations and other mechanized crops – into the Brazilian Amazon, while cattle production intensified, with beef yields increasing fivefold between 1997 and 2007. A sharp growth in the international market for soybeans and beef reflected the globalization of the forces of deforestation. The advance of soybean plantations stimulated further massive government investment in infrastructure such as waterways, railways, and highways (Figure 4.2), which in turn accelerated migration to remote areas and further forest clearing in a process of positive feedback (Fearnside, 2005).

After 2004, when the annual deforestation rate in Brazil's Amazon peaked at over 2.5 million ha, the shift towards export-driven commodity agriculture and timber production continued, but deforestation rates fell (Nepstad *et al.*, 2014). This decline occurred in response to a combination of government policies, supply chain interventions and changes in market conditions. Changes via global supply chains have been

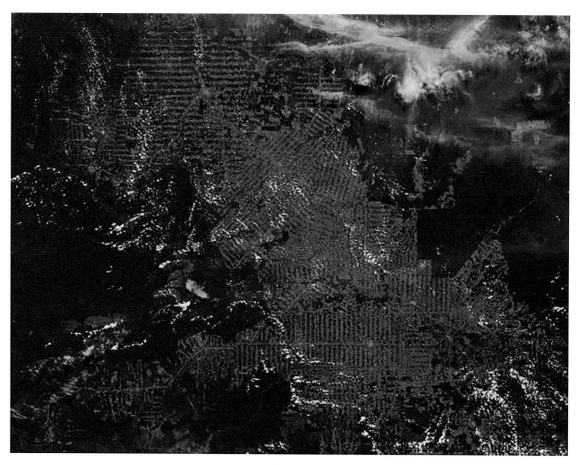

Figure 4.2 The distinctive herringbone pattern of tropical forest clearance by slash-and-burn agriculturists along transport routes in Rondônia, Brazil. Dark areas in this space shuttle photograph are remaining forest. Roads are 4–5 km apart.

particularly influential in Brazil, now a world-leading producer and exporter of beef and soybeans. International concern over tropical deforestation, compounded by high-profile NGO campaigns from organizations such as Greenpeace International, has highlighted the links between forest loss and TNCs that buy agricultural goods from forest-rich regions. In consequence, some retailers and commodity traders have imposed new environmental criteria on their suppliers. Examples of such supply chain interventions include zero-deforestation agreements whereby TNCs have stopped buying produce from farms with recent records of forest clearing. At the next level in the chain, major Brazilian meatpacking companies now avoid purchasing from properties with recent deforestation (Gibbs *et al.*, 2016). However, although Amazon deforestation rates dropped by more than 80 per cent between 2004 and 2014, the Brazilian Amazon still has one of the world's highest absolute rates of deforestation: nearly 0.5 million ha in 2014.

The forces of international economics that are integral to many areas of deforestation have played a central role in West Africa, where in most countries there is hardly any stretch of natural, unmodified vegetation left. A classic example is Côte d'Ivoire. Virtually every study of deforestation in the tropics concluded that Côte d'Ivoire experienced the most rapid forest clearance rates in the world, at around 0.3 million ha per year, in the second half of the twentieth century.

Like most African governments, Côte d'Ivoire has viewed the nation's forests as a source of revenue and foreign exchange. However, given the country's high external debt, and declining international prices received for agricultural export commodities, the country has been faced with little alternative but to exploit its forests heavily. In 1973, logs and wood product exports provided 35 per cent of export earnings, but this figure had fallen to 11 per cent by the end of that decade due to the rapidly declining resource base. In the late 1970s, about 5.5 million m³ of industrial roundwood was extracted annually but this production had fallen below 3 million m³ by 1991. By 1999, Côte d'Ivoire had banned the export of all hardwood with the aim of increasing its income by processing it at home. This resulted in increased log milling and increased manufacture of wood products, but also in lower log prices for forest owners, devaluation of the forest resource and negative impacts on forest management (FAO, 2000). Industrial roundwood production in 2008 was 1.5 million m³.

Deforestation has also been fuelled by a population that grew from 5 million in 1970 to 24 million in 2017, uncontrolled settlement by farmers, and clearance for new coffee and cacao plantations encouraged by government incentives. Monitoring of the country's dense humid forest by satellite shows that forest cover of about 8.4 million ha in 1960 had declined to 2.6 million ha in the 1980s and 1.35 million ha in the 2000s and that the latter periods were characterized by high fragmentation of rain forests in Côte d'Ivoire. Overall, since independence in 1960, the country has lost more than 80 per cent of its forest cover (Koné *et al.*, 2016).

The influence of external factors on national deforestation trends can also be illustrated by the experience of Vietnam, a country that in pre-agricultural times was almost entirely covered in forests but which lost more than a third of its original forest area, much of it during the second half of the last century. Clearance for agriculture took place in the coastal plains and valleys over the last few centuries, and during the French colonial period when large areas in the south were cleared for banana, coffee and rubber plantations, but 45 per cent of the country was still forested in the 1940s. That proportion had fallen to about 25 per cent in the early 1990s as extensive zones were destroyed during the Vietnam War (see Chapter 20) and still greater areas were cleared by a rapidly growing population rebuilding after the war. This trend was reversed when a shift from net deforestation to net reforestation took place during the 1990s, a so-called 'forest transition'. A model to describe this transition from decreasing to expanding forest cover, which has taken place in many developed countries historically, is shown in Figure 4.3.

In Vietnam, the national forest cover reached 38 per cent in 2005 and 42 per cent in 2014 thanks to both natural forest regeneration and plantations, although one-third of the country's forests are degraded and

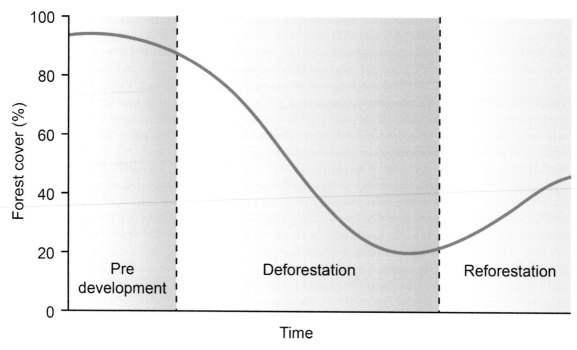

Figure 4.3 Three phases of the forest transition model.

of poor quality (MARD, 2015). Logging was severely restricted in natural forests by a series of policies and Vietnam banned the exports of raw logs. However, it seems that large quantities of illegal logs entered the country, mostly from Cambodia and Laos, so that although Vietnamese forests have been protected and restored, this success has been achieved in part by effectively exporting its deforestation to neighbouring countries (Meyfroidt and Lambin, 2009).

FOREST DISTURBANCE

Deforestation in the tropics, like deforestation anywhere, occurs because people want to use the forests' resources. Hence, tropical deforestation provides human societies with many benefits, such as food, timber and other raw materials, as well as jobs and income to numerous countries, many of them relatively poor. However, the loss of forests may also degrade some of their ecosystem services and much of the concern over the issue of tropical deforestation stems from the perturbation it represents to the forests' role as part of human life-support systems, by regulating local climates, water flow and nutrient cycles, as well as their role as reservoirs of biodiversity and habitats for species. The numerous deleterious environmental consequences resulting from the loss of tropical forests range in their scale of impact from the local to the global. Human clearance is by no means the only form of disturbance that forests experience, however, since the forest landscape is in a state of dynamic equilibrium. Gaps are continually formed in the forest canopy as older trees die, are struck by lightning or are blown over (Figure 4.4). Regrowth occurs continually in such gaps so that small areas are in a state of perpetual flux while the landscape as a whole does not appear to change, prompting the description of many tropical rain forests as a mosaic of patches in different stages of growth (Chambers *et al.*, 2013).

Natural perturbations also occur on larger scales. A tropical cyclone, for example, can cause tremendous damage over huge areas, while volcanic eruptions, earthquakes and associated landslides are frequent in some tectonically active parts of the world such as Indonesia, New Guinea and Central America. Interestingly, some types of disturbance of an intermediate magnitude and frequency – such as frequent flooding – appear to be a driver of high biodiversity in certain tropical forests. The Atlantic rain forest biome in southern Brazil includes a complex mosaic of forest types but flooded areas support greater species diversity than unflooded forest (Marques *et al.*, 2009). Continual disturbances – due to shifting water channels, waterlogging and the unstable soils of flooded forests which may enhance the likelihood of windstorms causing damage to shallow-rooted trees – reduce opportunities for competitive exclusion, meaning that climax is never reached. Through geological time, the global extent of tropical forests has probably also varied with climate and may have been much smaller than its current extent during periods of glacial maxima. Nevertheless, palaeoclimatological and palaeoecological data suggest that the Amazon rain forest has been remarkably resilient through geological time and has been a permanent feature of South America for at least the last 55 million years (Maslin *et al.*, 2005).

The human impact on tropical forests has a long history too. In practice it can be difficult to distinguish in the field between primary forest and secondary forest that is the result of human impact sometimes dating back over long periods. To the casual observer, secondary forests more than 60–80 years old are often indistinguishable from undisturbed primary forests and are, in fact, treated as primary forests in the FAO assessments of tropical forests. Altogether, more than 70 per cent of the world's remaining tropical forest area is thought to be made up of degraded old-growth forests (affected by recurrent fires, road building, selective logging, fragmentation and overhunting) and secondary regrowth forests (FAO, 2010).

Figure 4.4 A fallen tree in Central Africa's Congo forest. This natural form of disturbance is often caused by lightning in the Congo Basin where satellite data indicate an average of 50 lightning strikes per square kilometre a year, higher than anywhere else in the world.

In some cases, archaeological evidence can help us to realize an ancient human impact, as in large areas of tropical forest in Central America that were managed by the Mayan people more than 1000 years ago (Figure 4.5). In other regions, evidence from natural archives is used and these types of palaeoenvironmental investigations indicate that measurable human impacts on equatorial forests can be traced back as far as 3500 years BP in the Amazon and Congo basins, and 8000 BP in South-East Asia (Willis *et al.*, 2004). However, although humans have registered an impact on tropical forests over many centuries, which

Figure 4.5 Secondary tropical forest surrounding ancient Mayan monuments in Guatemala's Tikal National Park. The extent of deforestation during the Mayan period was probably much greater than was once thought.

cumulatively has affected great areas, the intensity, speed and relative permanence of clearance by human populations in the modern era put this form of disturbance into a different category.

CONSEQUENCES OF DEFORESTATION

The consequences of deforestation on hydrology have been widely discussed and an example of the multiple impacts upon aquatic habitats in the rivers of Madagascar is described in Chapter 8 (see Figure 8.1). Deforestation generally leads to an increase in runoff and stream discharge, a causal relationship demonstrated in studies of small watersheds (< 10 km²) throughout the tropics (Sahin and Hall, 1996). Larger river systems are affected in the same way. A 50-year study of the Tocantins River in south-eastern Amazonia showed that widespread deforestation had been largely responsible for a 24 per cent increase in annual mean discharge during a time when precipitation over the basin did not change significantly (Costa *et al.*, 2003).

A particular concern stems from recognition of the forests' role in regulating the flows of rivers and streams from upland catchments, but although there is clearly a need to conserve forest cover in such areas, quantitative evidence of the role deforestation plays in flooding has been difficult to establish. However, an assessment made for developing countries by Bradshaw *et al.* (2007) found clear relationships between flood frequency and natural forest area loss after making adjustments for rainfall, slope and degraded landscape area. Several measures of flood severity (duration, people killed and displaced, and total damage) also showed detectable, though weaker, correlations to natural forest cover and loss.

A less controversial effect is that on sediment loads in rivers due to enhanced runoff and erosion. Another study in South America, this one in the Magdalena River basin in Colombia (Restrepo and Escobar, 2016), found that the river's tributaries experienced increasing trends in sediment load during the period 1980–2010 and that the increases were in close agreement with trends in land use change and deforestation. In many regions, high sediment loads have been the cause of problems, particularly in reservoirs – shortening their useful lives as suppliers of energy and irrigation water, for example, and requiring costly remediation (see Chapter 9).

Soil degradation – including erosion, landslides, compaction and laterization – is a common problem associated with deforestation in any biome, but is often particularly severe in the tropics. Rainfall in the humid tropics is characterized by large annual totals, it is continuous throughout the year and highly erosive. Hence, clearance of protective vegetation cover quickly leads to accelerated soil loss, although the rate will vary according to what land use replaces the forest cover (see Chapter 14). Reviewing work in South-East Asian forests, Sidle *et al.* (2006) conclude that the highest levels of soil loss per unit area impacted generally come from roads and trails, but losses from converted forest lands are high in mountainous areas if sites are heavily disturbed by cultivation and/or when crops or plantations have poor overall ground cover. These authors also note that most of the soil loss studies do not include estimates of landslide erosion, which over the long term may contribute comparable levels of sediment in steep terrain. The probability of landslides does not increase substantially until several years after forest conversion and may then persist indefinitely if weaker-rooted crops or plantations continue.

Soil erosion can also be exacerbated by compaction caused by heavy machinery, trampling from cattle where pasture replaces forest, and exposure to the sun and rain. A compacted soil surface yields greater runoff and often more soil loss as a consequence. Laterization, the formation of a hard, impermeable surface, was once feared to be a very widespread consequence of deforestation in the tropics, but is probably a danger on just 2 per cent of the humid tropical land area (Grainger, 1993a).

The poor nutrient content of many tropical forest soils is a factor that tends to limit the potential of many of the land uses for which forest is cleared. In parts of the Amazon Basin, for example, the available phosphorus and other nutrients in the soil initially peak after the burning of cleared vegetation, giving relatively high yields and consequently high incomes for farmers. But nutrient levels often decline thereafter due to leaching, and are insufficient to maintain pasture growth in less than ten years. Soil nutrient depletion, compaction and invasion by inedible weeds quickly deplete the land's usefulness as pasture. There comes a time when beef cattle ranching is no longer profitable and the land will be abandoned (Figure 4.6).

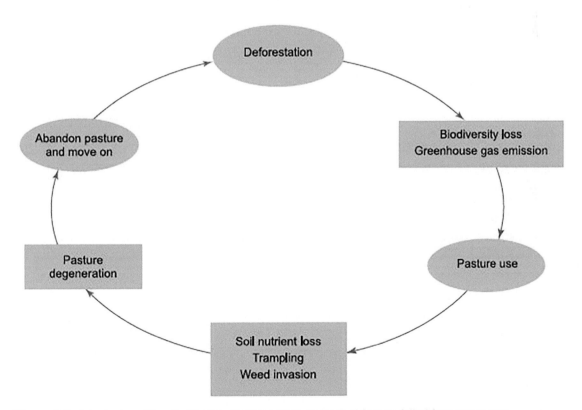

Figure 4.6 A typical positive feedback loop that occurs in tropical forests felled for pasture use.

Similarly, unsustainable shifting cultivation quickly depletes soil nutrients, and, as with the degraded pastures, rapidly diminishing yields from inappropriate land uses encourage settlers to move on and clear further forest areas in a process of positive feedback. Some of the other major constraints for two of the most dominant soil orders found in the Amazon Basin are shown in Table 4.2.

The possible climatic impacts of large-scale forest clearance, resulting from effects on the hydrological and carbon cycles, may occur on local, regional and even global scales. Localized effects of deforestation have been detected in the rainfall records in Central America. In Guatemala and adjacent areas, analysis of data from 266 meteorological stations by Ray *et al.* (2006) showed dry season rainfall to be markedly lower in deforested areas than in equivalent forested zones. In general, the study concluded that dry season deforested habitats have higher daytime temperatures, are less cloudy, and have lower soil moisture than forested habitats in the same ecological zone. The result is hotter and drier air over deforested regions, with lower values of cloud formation and precipitation.

At the regional scale, in a large area like Amazonia, where up to a third of the rainfall is returned to the atmosphere via evapotranspiration, widespread removal of forest could also have a serious impact on the hydrological cycle, with consequent negative effects on rainfall and, hence, continued survival of forest and/or agriculture. Such theories have been tested by simulating future conditions with general circulation models, and the results to date generally agree on a reduction in precipitation and a likely increase in surface temperatures (Lejeune *et al.*, 2015). Several modelling studies also suggest that the relationship between forest and climate in Amazonia may have two stable states and that removal of 30–40 per cent of the forest could cross the threshold between the two, inducing a regime shift by pushing much of the Amazon Basin into a drier climate (Oyama and Nobre, 2003).

The changes brought about by tropical deforestation to surface properties such as albedo and roughness on a large scale could also have knock-on effects in middle and high latitudes via the Hadley and Walker

Table 4.2 Some of the major constraints to agricultural development of two dominant Amazon Basin soils

Constraint	Oxisols	Ultisols
Low plant nutrient levels	✓	✓
Nitrogen losses through leaching	✓	✓
High levels of exchangeable aluminium	✓	✓
Low cation exchange capacity	✓	
Weak retention of bases	✓	
Strong fixation or deficiency of phosphate		
Soil acidity	✓	
Very low calcium content	✓	
Essential to continue to maintain soil organic matter levels		✓
Weak structure in surface layers		✓

Source: after Nortcliff and Gregory (1992).

circulations and Rossby wave propagation, a type of hydroclimatological 'teleconnection' (Avissar and Werth, 2005). Further global effects of large-scale deforestation have been predicted via the carbon cycle. Deforestation releases carbon dioxide into the atmosphere, either by burning or decomposition, contributing to enhanced global warming through the greenhouse effect. Estimates of just how much additional carbon dioxide reaches the atmosphere due to tropical deforestation are uncertain, due to our imprecise knowledge both of deforestation rates and of the amount of carbon locked into a unit area of forest. Although deforestation's contribution to rising atmospheric carbon dioxide concentrations is less than that of fossil fuels, it could contribute up to a third as much again. The possible consequences of climatic change due to the resulting greenhouse warming are discussed in Chapter 11 but feedback effects on tropical forests could be severe. Many climate models project a significant loss of forest cover across Amazonia due to climatic change this century, for example, and in one case a near-permanent El Niño-like state is predicted in the Pacific with growing drought across the Amazon Basin, leaving almost all of Brazilian Amazonia too dry to support rain forest by the year 2100, a phenomenon referred to as Amazon rain forest dieback (Boulton *et al.*, 2017).

One of the most widely discussed consequences of deforestation in the tropics is the loss of plant and animal species. The diversity of life in tropical rain forests relative to other biomes is legendary. Study of a single tree in Peru, for example, yielded no less than 43 ant species belonging to 26 genera, roughly equivalent to the ant diversity of the entire British Isles (Wilson, 1989). Examples of this level of diversity are numerous: one leaf, with an area of 125 cm^2, from a forest in Costa Rica held 49 species of lichens (Lücking and Matzer, 2001), and on just 1 hectare of forest in Amazonian Ecuador, researchers recorded no less than 473 species, 187 genera and 54 families of tree (Valencia *et al.*, 1994), the highest number of tree species found on a tropical rain forest plot of this size, which is more than half of the total number of tree species native to North America.

Although our knowledge of the total number of species on the planet is scant (see Chapter 15), most estimates agree that more than half of them live in the tropical moist forests that cover only 6 per cent of the world's land surface. Many species are highly localized in their distribution and are characterized by intimate links or narrow ecological specialization, resulting in a delicate interdependence between one species and another. Hence they can be destroyed without clearing a very large area. Overall, there is widespread agreement in the scientific community that deforestation in the tropics is a severe threat to the species and genetic diversity of the planet, although the potential magnitude of the loss is unclear.

Not all authorities are so pessimistic, however. Several studies have shown that populations of larger faunal species, such as birds and mammals, are surprisingly resilient in the face of single logging operations. Although the relative abundance of species can change markedly, total species numbers often show comparatively little change in response to such disturbances, perhaps because individuals retreat to untouched forest pockets during logging operations and return after logging has been completed. Putz *et al.* (2012) conducted a meta-analysis based on 109 studies of selective logging of primary tropical forest carried out between 1 and 100 years after a single harvest and found modest impacts on the species richness of birds, mammals, invertebrates and plants. There is also increasing evidence that some species appear to survive in secondary forests (Chazdon *et al.*, 2009), which have received relatively little study in this regard. However, when forest is removed and replaced with a significantly different land use, the impact on biodiversity can be dramatic. Figure 4.7 shows the differences in numbers of bird and butterfly species between primary and logged forest in Malaysia are relatively small, but the two types of plantation shown harbour significantly fewer species. Less versatile and adaptable species, particularly insects, which tend to be very specialized feeders, are still less likely to survive such disturbances. From a purely practical viewpoint, the loss of species, most of which have not been documented, has been likened to destroying a library of potentially useful information having only flicked through the odd book. The types of goods that could be derived from tropical forests include new pharmaceutical products, food and industrial crops, while a diversity of genetic material is of importance to improve existing crops as they become prone to new strains of diseases and pests.

Forest structure and diversity are affected not just by clear-cutting but also by fragmentation (see Chapter 15) and lesser forms of disturbance. Work in central Amazonia suggests that forest fragmentation is having a disproportionately severe effect on large trees, the loss of which will have major impacts on the rain forest ecosystem (Laurance *et al.*, 2000). Large trees are unusually vulnerable in fragmented rain forests for several reasons. They may be especially prone to uprooting and breakage near forest edges, where wind turbulence is increased, because of their tall stature and relatively thick, inflexible trunks. Large, old trees are also particularly susceptible to infestation by lianas (parasitic woody vines that reduce tree survival), which increase markedly near edges. Further, because their crowns are exposed to intense sunlight and evaporation, large tropical trees are sensitive to droughts and so may be vulnerable to increased desiccation near edges.

The loss of large tropical forest trees is alarming because they are crucial sources of fruits, flowers and shelter for animal populations. Their loss is also likely to diminish forest volume and structural complexity,

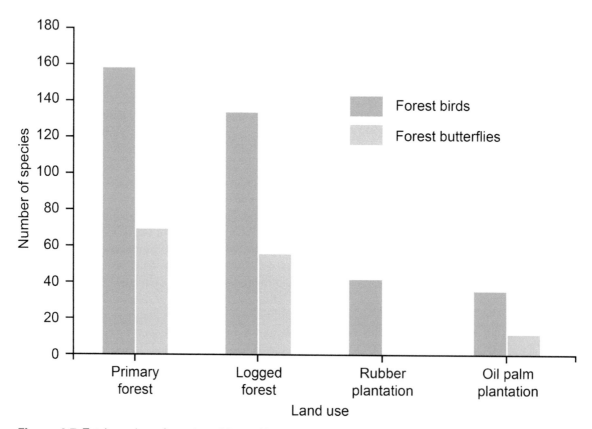

Figure 4.7 Total number of species of forest birds and forest butterflies recorded from different land-use types in southern Peninsular Malaysia and Borneo (after Koh and Wilcove, 2008, from various sources).

promote the proliferation of short-lived pioneer species, and alter biogeochemical cycles affecting evapotranspiration, carbon cycling and greenhouse gas emissions. Equally worrying is the possibility that, because tree mortality is increased in forest fragments and large trees range in age from a century to well over a thousand years old, their populations in fragmented landscapes may never recover.

Fire is another threat to the health of the forest ecosystem affected by logging and clearance. The 1982/83 drought fires that seriously affected over 5 million ha of land in the Indonesian provinces of East Kalimantan and Borneo raged mainly in selectively logged areas where dry combustible material from tree extraction littered the forest floor (Wirawan, 1993). Forest degradation and land clearance were also the root causes of the 1997/98 fire disaster that blanketed nearly 20 million people across South-East Asia in smoke for months, with disastrous consequences for local health (Siegert *et al.*, 2001). The continued use of fire to clear and prepare land on degraded peat, particularly in Kalimantan, has created a non-linear sensitivity to dry conditions and the strong El Niño in 2015 triggered the acceleration of fire across more than 2.6 million ha of Indonesian forests (Field *et al.*, 2016). While the small-scale, low-intensity, low-frequency fires commonly associated with traditional shifting cultivation systems are considered to be a moderate-level disturbance that is likely to stimulate the maintenance of high biodiversity in tropical rain forests, such large-scale fires represent a much greater disturbance from which it will take a very long time to recover.

Another all too common casualty of deforestation in the tropics is indigenous peoples, for

Figure 4.8 The outsiders' perception of indigenous tropical rain forest inhabitants as savages, as this engraving from a seventeenth-century book of travels in Brazil aptly illustrates, still tends to prevail. The all too common consequences for indigenous inhabitants include loss of cultural identity, and suffering due to the introduction of new diseases.

whom death or loss of cultural identity usually follows the arrival of external cultures and the frontier of forest clearance (Figure 4.8). Whether it is deliberate annexation of their territory by the dominant society, or a less formalized erosion of their way of life, the result is often the same. The indigenous Indian population of the Amazon Basin, for example, is thought to have numbered 10 million when Europeans first arrived in South America (Layrisse, 1992), but an estimate by Survival International – an NGO devoted to preventing the annihilation of tribal peoples – puts the total today at some 1 million. To extend the library metaphor with regard to the potential wider uses of biodiversity, the impact of deforestation on tribal peoples is equivalent to killing off the librarians who know a great deal about what the library contains. At a more fundamental level, the plight of indigenous peoples is a simple question of human rights.

TROPICAL FOREST MANAGEMENT

The answers to the problems caused by exploitation of tropical forests are by no means straightforward to formulate or easy to implement. One difficulty that should be appreciated is that most remaining tropical forest areas are remote from national seats of government and that the rule of law in these frontier zones may be less than total. Therefore passing legislation that supposedly limits the drivers of deforestation is often meaningless in practice. In tropical regions, illegal logging rates are thought to be particularly high, with an estimated 50–90 per cent of timber illegally sourced (Nellemann and INTERPOL, 2012). Inevitably, accurate information on the scale of illegal logging is hard to obtain, but estimates of its magnitude from a number of sources in selected countries are shown in Table 4.3. The situation in the lowland forests of the Sunda Shelf in Indonesia is highlighted by Jepson *et al.* (2001) who describe mafia-like logging gangs, backed by army and rebel groups, operating within forests that overlap official park boundaries and ignoring the Indonesian State Forestry Department's sustainable management policy. Government officials who attempt to stop illegal logging face serious intimidation, which includes arson and even murder, while many local communities have signed away their rights to the land they live on to the loggers, both for immediate cash benefit and to avoid retribution from logging gangs.

Preventing this type of systematic tropical forest abuse, a symptom of the underlying drivers of deforestation noted above, means that it is these causes that must be addressed if development is to be directed towards a more sensible and equitable use of the forest resource. Many of the structural changes needed to strike at the root causes of deforestation will require years of endeavour. For the Brazilian Amazon, Fearnside (2005) highlights the importance of slowing population growth and improving employment opportunities both in labour-intensive agriculture and in urban areas outside the Amazon. A range of government subsidies that provide impetus for deforestation should also be reduced or discontinued, including those on agricultural credit and the price of many agricultural products, particularly in areas with high deforestation rates.

Table 4.3 Estimated percentages of illegal logging in selected countries

Country	Estimate % range
Brazil (Amazon)	20–50
Cambodia	90
Cameroon	50–65
Congo, DR	>90
Ghana	34–70
Indonesia	60–80
Myanmar	50
Papua New Guinea	70
Peru	80–90
Vietnam	20–40

Source: after Cashore *et al.* (2016).

While these proposals seem sensible from one perspective, the needs of national economic development cannot be denied. For Brazil to cease development of its largest sovereign natural resource is clearly neither desirable nor feasible. In practical terms, too, the fact that such suggestions often come from developed countries that have previously all but destroyed their own forests in the course of their development can quickly be interpreted as 'ecoimperialism'. There are many parallels between today's developing countries and the past experience of those that have developed further. In Britain, for example, almost the entire land area was covered by forest about 5000 years ago, about the time that human activity began to have an impact on the landscape. By 1900, woodland covered only about 5 per cent of the land surface. As the UK has been through the forest transition, plantations have increased this proportion during the twentieth century to 12 per cent (Forestry Commission, 2003), but most of the lost forests have been converted into farmland, which occupies some 72 per cent of the total land area today, and many wildlife species have become extinct in Britain as a result of this destruction of habitat, which made them more susceptible to hunters (see Table 15.3).

A pragmatic approach to the problems of deforestation cannot seriously obstruct all extension of cultivated lands or all clearance of forest to provide new infrastructure, industrial or urban sites. But this process requires surveys, inventories and evaluation to ensure that each hectare of land is used in the optimum way, to meet both the needs of today and those of future generations. A range of policy initiatives can be used to address the drivers of deforestation and degradation and these policies may be applied at a range of scales, from local to international (Table 4.4). Outsiders can make a positive contribution to controlling the poor use of forest resources (e.g. via global supply chains as illustrated above in Brazil), although direct pressure on governments from well-intentioned individuals abroad can backfire. Improvements in the environmental performance of such players as commercial banks and other investors can be called for, and moves made to tighten their financial approach to reduce the temptation of borrowing countries to run up heavy debts and promote short-sighted land-use policies to pay them back, as in the case of investment in soybean production in southern Brazil. Criticism of the World Bank for its association with wasteful and illegal deforestation, such as the Highway BR-364 in Brazil's Rondônia state, led to the Bank changing its approach towards forests in 1991. The World Bank stopped financing major infrastructure investments that were likely to harm forests and introduced a ban on the financing of commercial logging (World Bank, 2000). Debt-for-nature swaps have also proved to be a successful approach in several countries, making a positive contribution to conservation and developing countries' debt burdens (see Chapter 22).

More recently, the links between tropical deforestation and climate change are set to provide another source of economic incentive to conserve forests via the international markets in carbon created by the Kyoto Protocol and the European Union's emissions trading system (see Chapter 11). International financial flows to tropical forest countries will foster the carbon storage – or sequestration – properties of forests through a programme known as Reduced Emissions from Deforestation and Forest Degradation (REDD) which will effectively pay compensation to developing countries for preserving forests as carbon stores. Over time, the remit of REDD has been enlarged, from its original focus on reducing emissions from deforestation and forest degradation, to give equal emphasis to conservation, sustainable management of forests and enhancement of forest carbon stocks. Now known as REDD+, to reflect these new components, many hope the programme can deliver multiple benefits: simultaneously contributing to climate change mitigation and poverty alleviation, whilst also conserving biodiversity and sustaining vital ecosystem services. The types of intervention strategies that REDD+ funds might be used for include those shown in Table 4.4. However, although the framework for a REDD+ mechanism was finally completed in 2015, after a decade of negotiations, and it is now a prominent mechanism in the

Table 4.4 Examples of intervention options to address deforestation drivers at various scales

Scale of deforestation driver	Possible intervention strategies
LOCAL	
• Agriculture (commercial and subsistence) • Wood extraction • Mining • Infrastructure • Urban expansion	• Encourage farmers to forgo deforestation and invest in more intensive, sustainable production systems • Payments for ecosystem services (e.g. for watershed protection, carbon sequestration) • Establish new protected areas
REGIONAL/NATIONAL	
• National demand for rural products • Infrastructure • Urban expansion	• Suspend access to agricultural credit in areas with highest deforestation rates • Improve law enforcement (e.g. against trade in illegal logging products) • Alter demand (e.g. electrification to reduce demand for fuelwood and charcoal)
INTERNATIONAL	
• International demand for rural products	• Import restrictions • Supply chain interventions (e.g. zero-deforestation agreements, agricultural certification initiatives) • Debt-for-nature swaps

Source: after WWF (2013); Nepstad *et al.* (2014).

Paris Agreement on climate change (see Chapter 11), some are still concerned that it may be ineffective and inefficient (Loft *et al.*, 2016).

Establishing legal protection for forest areas is probably the most common policy used to limit tropical forest degradation and loss. Although legal protection is not necessarily a guarantee of non-disturbance as the example cited above in Indonesia's Sunda Shelf demonstrates, there are some more positive examples. Shah and Baylis (2015) found a range of conservation outcomes in their study of Indonesian protected areas, including Sebangau National Park in Kalimantan and Bantimurung Bulusaraung National Park in Sulawesi, both of which have been very effective in limiting forest cover loss within their park boundaries. In the Brazilian Amazon, more than a third of the forest is legally protected, in indigenous lands and other types of protected areas, and both inhabited and uninhabited reserves appear to be successful in reducing both deforestation and the occurrence of fire, the standard processes heralding conversion of forest to agriculture (Nepstad *et al.*, 2006). That said, protecting the trees is not the same as protecting the entire ecosystem. Harrison (2011) suggests that the ease with which hunters enter many protected areas probably means that most tropical nature reserves may already be considered empty forests, and the consequent disruption of ecological functions caused by the loss of bird and larger mammal species compromises the capacity of these reserves to conserve biodiversity over the long term.

Nonetheless, totally protected areas can never be sufficiently extensive to provide for the conservation of all ecological processes and all species. In this case, the obvious avenues to explore are ways in which a natural forest cover can be used sustainably, to harvest products without degrading the forest. In this case, sustainable forestry means harvesting a forest in a way that provides a regular yield of forest produce without destroying or radically altering the composition and structure of the forest as a whole. In other words, sustainable operations need to mimic natural ecosystem processes.

An obvious product in this respect is timber and some authorities consider that tropical forest that is used for timber production offers the potential not only of conserving the majority of their biodiversity (see above) but also most of their associated ecosystem functions, as well as their carbon, climatic, and soil-hydrological ecosystem services (Edwards *et al.*, 2014). However, attitudes towards the logging industry have become polarized, with some highlighting its role as a primary cause of tropical deforestation and arguing that the international trade in tropical timber should therefore be banned. In practice, however, logging in the tropics almost always involves the removal of selected trees rather than the clear-felling that is much more common in temperate forestry operations. Some damage to other trees occurs even during selective logging, but less damaging techniques are available and if well enforced can greatly reduce smashing and other forms of degradation.

However, most logging in the tropics is probably illegal (see above) and hence likely to be conducted less responsibly, and although logging is rarely a direct cause of total forest clearance, it is often responsible for opening the way to other agents of deforestation, such as agriculturalists. Hence, others argue that despite the poor track record of loggers in the past, they should be encouraged to move away from a mining approach towards timber harvesting through sustainable management. Minimizing damage to the forest, and particularly the forest floor, during timber extraction is important and over the past few decades low-intensity logging practices (as opposed to conventional selective harvesting or clear-cutting) that harvest two to eight trees per hectare have become increasingly common in some parts of the tropics. Directional felling reduces damage to the remaining canopy and facilitates log removal, while pulling out logs by cable reduces damage to the forest floor. Careful planning of extraction tracks can reduce the severity of soil erosion, as can uphill logging, as opposed to downhill logging. The forest is allowed to recover after logging for several years before the next harvest (Fimbel *et al.*, 2001). The cyclical nature and low impact of these harvests on the structure and composition of the forest are designed to minimize negative effects on biodiversity and ecosystem services and to provide a renewable resource of economic value.

Although these low-impact techniques are gaining in popularity, following intense pressure on logging companies from environmentalists, most timber operations are still controlled by entrepreneurs, to whom short-term profits are of prime importance, rather than by foresters, whose duty is to the long-term maintenance of the resources. Ironically, a considerable portion of tropical forest trees felled for timber in many areas is wasted due to inefficient processing using outdated equipment (Figure 4.9). One key policy change that could encourage the wider adoption of low-impact techniques would be to lengthen the periods over which concessions were granted. Typically, concessions are granted for periods of between 5 and 20 years, but if they were increased to at least 60 years, it would be in the interest of the logging companies to protect their forest and produce timber sustainably on perhaps 30- to 70-year rotations. One problem here, however, is that no one can guarantee that there will be a market for tropical timber so far into the future.

Another issue facing those in favour of tropical forest management for timber production is the fact that there are frequently other, more profitable land-use options. Many studies show that the most financially

Figure 4.9 Processing of tropical rainforest logs here at Samreboi in Ghana, where machinery is outdated and inefficient, means wastage rates of 65 per cent for plywood and 70 per cent for planks.

profitable decision is to extract all the economically valuable timber as quickly as possible and then either abandon the area or convert it to soybean fields, cattle ranches, oil palm or pulpwood plantations (Ruslandi *et al.*, 2011). Hence, Putz *et al.* (2012) argue, well-managed tropical forests need to be encouraged as a sort of middle way between deforestation and total forest protection. They suggest that four initiatives can in combination provide sufficient incentives, coupled with adequate enforcement, to promote sustainable forest management.

- assure the legality of forest products through international partnership agreements (e.g. the EU's Forest Law Enforcement, Governance and Trade plan, or FLEGT, and the 2008 Amendment to the Lacey Act in the USA) which will increase market prices and access for legally produced timber while simultaneously promoting more responsible forest management
- strengthen voluntary third-party certification programmes that promote responsible management by securing or even increasing market access and prices for forest products
- harness finance from climate change mitigation programmes such as REDD+ by recognizing the substantial carbon benefits from improvements in tropical forest management

- devolve control over forests to indigenous and other rural communities, clarify their tenure and provide long-term business support for sustainable forest management.

Although logging has received particular attention for sustainable forestry practices, the range of uses for tropical rain forest plants is very wide, as Table 4.5 shows from studies in Mexico. Sustainable production of many of these so-called 'non-timber forest products' can also benefit from the four initiatives outlined above. The top position of medicinal plants indicated in this table has been highlighted by some campaigning groups, claiming that tropical forests contain untold numbers of 'new' drugs just waiting to be developed. However, such claims are debatable. As one recent review concluded: 'The potential of rainforest natural products to become new drugs is still on the horizon' (Jantan *et al.*, 2015: 26). The collection and screening of plants, followed by purification and testing of extracts, are expensive procedures that take many years and they may not be as economical as the computer modelling of molecules and their synthesis. Some biodiversity prospecting programmes collect samples from rain forests and evaluate their medicinal potential, but if and when a marketable new drug emerges from such procedures, the issue of who should share the resulting profits, and in what proportion, is an important one. Efforts are being made to protect the property rights of local communities and to share profits with them under the Convention on Biological Diversity (see p. 386), but the western legal principles upon which such agreements are based may undermine patterns of community ownership and there have been numerous problems of securing

Table 4.5 Useful plants from the tropical rain forests of Mexico

Use	Number of plants	Use	Number of plants
Medicinal	780	Dyes	34
Edible	360	Shade plants	31
Construction	175	Flavour enhancer	24
Timber	102	Gums	20
Fuel	93	Stimulants	16
Honey	84	Fertilizer	15
Forage	73	Hedge plants	15
Domestic use	69	Tannins	12
Crafts	59	Insecticides	11
Poison	52	Perfume	11
Ornamental	51	Nurse crop	10
Tools	51	Latex	10
Ceremonial use	50	Soap	10
Fibres	38		

Source: Toledo *et al.* (1995).

Table 4.6 Some western drugs derived from rain forest plants

Drug	Plant of origin	Use
Quinine	Cinchona tree bark (South America)	Treatment of malaria
Neostigmine	Calabar bean (Africa)	Treatment of glaucoma and to provide blueprint for synthetic insecticides
Turbocurarine	Curare liana (Americas)	Treatment of muscle disorders (e.g. Parkinson's, multiple sclerosis)
Vincristine, vinblastine	Rosy periwinkle (Madagascar)	Treatment of Hodgkin's disease and paediatric leukaemia
Cortisone	Wild yams (Central America)	Active ingredient in birth control pills
Calanolide A	Calophyllumlanigerum (Borneo)	Treatment of HIV

Source: after Jantan *et al.* (2015)

appropriate international patent laws (Straus, 2008). Despite these provisos, however, some of the western drugs that have been developed from tropical forest plants (Table 4.6) indicate the possible scope of future discoveries.

Other non-timber forest products have also been the subject of research. Low-intensity harvesting of fruits, nuts, rubber and other produce should result in a minimum of genetic erosion and other forms of disturbance, and a maximum of conservation. Production and trade of these forest products have been widely encouraged by conservation and development organizations working in tropical rain forests as viable ways of achieving species and ecosystem conservation at the same time as improving local livelihoods. However, such schemes face many challenges (Belcher and Schreckenberg, 2007). Adequate attention needs to be given to appropriate markets and sufficient transport and distribution networks for non-timber products. If the products concerned do not have an international market, such an approach is likely to be less attractive than timber for national governments, because of the hard currency revenues an internationally tradable commodity can command. Where existing small-scale harvesting from tropical forests is already carried out for local markets, the diversity of forest products represents a great strength for the producers, allowing them to respond well to the changing needs and tastes of the market and furnishing them with a good annual income (López-Feldman and Wilen, 2008).

The promotion of non-timber forest products may have greater value when viewed in the wider context of their role in maintaining forest ecosystem services. In some parts of the tropics, forest conservation has been successfully promoted by an economic mechanism designed to turn the ecosystem services of standing forest into a primary means of generating income to support rural populations (Box 4.1).

Some observers are confident that countries can pursue economic development and still maintain a high forest cover given sufficient investment in agriculture, and government dedication to sustainable forest

Box 4.1 Payment for ecosystem services helps forest conservation in Costa Rica

Some of the world's highest rates of deforestation during the 1970s and 1980s were experienced in Costa Rica. Trees were felled for their timber, expansion of the road system and to make way for cattle ranching, activities that were encouraged by government policies. In 1950, forests probably covered nearly 75 per cent of the national territory but this proportion had declined to about 25 per cent by 1995.

Payment for ecosystem services (PES) was formally adopted in an attempt to conserve forests when the Costa Rica Forestry Law 7575 was passed in 1996. The law aims to protect primary forest, allow secondary forest to flourish, and promote forest plantations to meet industrial demands for lumber and paper. These goals are met by paying private landowners who conserve forest, thus helping to safeguard the nation's environmental services in four categories: watershed protection, biodiversity, carbon sequestration, and scenic landscape/tourism. Payments are made under legal contracts agreed when landowners present a sustainable forest management plan.

The rate of deforestation in Costa Rica has certainly slowed considerably in recent decades and the national forest cover had reached about 50 per cent by 2005. The exact contribution made by PES is difficult to establish. In part, these changes can be attributed to Costa Rica's economic transition from an agrarian economy to a service-based one in which the tourism industry became the country's leading sector in earning foreign currency in the mid-1990s. Nonetheless, payments through the PES scheme have certainly lowered the deforestation rate in certain parts of the country. However, the greatest impact of PES on land-use choices is most likely to have been in encouraging forest expansion through natural regeneration and the establishment of plantations.

Source: Daniels *et al.* (2010).

management and to a strong national network of protected areas. Indeed, a cautiously optimistic note is sounded by Meyfroidt and Lambin (2011) when they acknowledge the handful of tropical developing countries that have been through the forest transition (including Vietnam – see above – and Costa Rica – see Box 4.1), a process that should be repeatable elsewhere. Inevitably, some further tropical forest areas will still be lost due to economic problems, international disputes, organizational shortcomings and as policies evolve to make balanced decisions on various alternative land uses. There is a great deal of will, both inside and outside tropical forest countries, to use forests in a sensible manner, but a sustainable future for the world's tropical forests will also depend upon a greater level of international cooperation and equity than we have seen to date.

FURTHER READING

Gaveau, D.L., Sheil, D., Husnayaen, M.A.S., Arjasakusuma, S., Ancrenaz, M., Pacheco, P. and Meijaard, E. 2016 Rapid conversions and avoided deforestation: examining four decades of industrial plantation expansion in Borneo. *Scientific Reports* 6. A long-term examination of oil-palm as a driver of deforestation. Open access: www. nature.com/articles/srep32017.

Ghazoul, J. and Sheil, D. 2010 *Tropical rain forest ecology, diversity, and conservation.* Oxford, Oxford University Press. An excellent introductory text covering biodiversity and forest dynamics, as well as human impact and management.

Laurance, W.F., Kakul, T., Tom, M., Wahya, R. and Laurance, S.G. 2012 Defeating the 'resource curse': key priorities for conserving Papua New Guinea's native forests. *Biological Conservation* 151: 35–40. A range of policy measures designed to improve forest conservation and sustainability.

Lowe, A.J., Dormontt, E.E., Bowie, M.J., Degen, B., Gardner, S., Thomas, D., Clarke, C., Rimbawanto, A., Wiedenhoeft, A., Yin, Y. and Sasaki, N. 2016 Opportunities for improved transparency in the timber trade through scientific verification. *BioScience* 66(11): 990–8. Reviewing scientific methods used to verify global timber supply chains. Open access: https://doi.org/10.1093/biosci/biw129.

Sexton, J.O., Noojipady, P., Song, X.P., Feng, M., Song, D.X., Kim, D.H. and Townshend, J.R. 2016 Conservation policy and the measurement of forests. *Nature Climate Change* 6(2): 192. Outlining the continuing difficulties of assessing the world's forested area.

Temudo, M.P. and Silva, J.M.N. 2011 Agriculture and forest cover changes in post-war Mozambique. *Journal of Land Use Science* doi: 10.1080/1747423X.2011.595834. An assessment of the complex role of shifting cultivation in deforestation in tropical Africa.

WEBSITES

www.fao.org/forestry the FAO site has an extensive forestry section including data and up-to-date reports.

www.globalforestwatch.org Global Forest Watch is an international data and mapping network covering forests inside and outside the tropics.

www.inpe.br the Brazilian Institute for Space Research site includes current deforestation rates in the Brazilian Amazon.

www.itto.int the International Tropical Timber Organization site includes details on the organization and its activities.

www.tropenbos.org the Tropenbos Foundation is dedicated to the conservation and wise use of tropical rain forests and has projects in several countries.

www.un-redd.org the UN-REDD Programme supports national forest conservation programmes in 64 developing countries.

POINTS FOR DISCUSSION

- Since tropical forests are subject to all manner of natural disturbances, why are we so concerned about the human impact?
- Are global supply chain interventions that aim to curb deforestation simply another form of ecoimperialism?
- Deforestation in the tropics is proceeding at a rapid rate. So what? (Answer the question from the viewpoint of: (a) a landless Brazilian peasant; (b) a biotechnology company; and (c) a deep ecologist.)
- To what extent do the arguments in favour of conserving biodiversity also apply to forests in the humid tropics?
- Outline some policy proposals to halt the illegal trade in timber both domestically and internationally.
- How would you set about putting an economic value on an area of tropical forest?

5 Desertification

Human societies have occupied deserts and their margins for thousands of years, and the history of human use of these drylands is punctuated with many examples of productive land being lost to the desert. In some cases the cause has been human, through overuse and mismanagement of dryland resources, while in other cases natural changes in the environment have reduced the suitability of such areas for human habitation. Often a combination of human and natural factors is at work. Salinization and siltation of irrigation schemes in Lower Mesopotamia are thought to have played a significant role in the eventual collapse of the Sumerian civilization 4000 years ago, and similar mismanagement of irrigation water has been implicated in the abandonment of the Khorezm oasis settlements of Uzbekistan dating from the first century CE. The decline of Nabatean towns in the Negev Desert from about 500 CE has been attributed to Muslim–Arab invasions, but a gradual change in climate towards a more arid regime probably played an important role. Invasion and settlement of areas by outsiders have often been highlighted as a reason for the onset of dryland environmental degradation in historical times. The Spanish conquest is cited in South America and a more diverse mix of European settlers was involved in the dramatic years of the Dust Bowl in the US Great Plains in the 1930s.

Desertification became a major global environmental issue in the 1970s after international concern over famine in West Africa focused attention on both drought and desertification as insidious causes. A UN Convention to Combat Desertification (UNCCD) was agreed at the Earth Summit in Rio de Janeiro in 1992 and came into force in 1996. The Convention views desertification as a threat affecting drylands all over the world, in rich and poor countries alike. However, desertification has also become surrounded in controversy, particularly over its nature, extent, causes and effects.

DEFINITION OF DESERTIFICATION

The origins of the word desertification are commonly attributed to a 1949 book by a French forester named Aubreville who worked in West Africa, but the term has been traced back much earlier, to nineteenth-century French colonial North Africa, by Davis (2004). Literally, desertification means the making of a desert, but although the word has been in use for more than 150 years, a universally agreed definition of the term has only recently been reached. A survey of the literature on the subject has revealed more than a hundred different definitions (Glantz and Orlovsky, 1983) and most of these suggest that the extent of deserts is increasing, usually into desert-marginal lands. The idea of a loss of an area's resource potential, through the depletion of soil cover, vegetation cover or certain useful plant species, for example, is included in many definitions. Some suggest that such losses are irreversible over the timescale of a human lifetime.

Most, though not all, authorities agree that desertification occurs in drylands, which can be defined using the boundaries of climatic classifications (Figure 5.1) and make up about 41 per cent of the world's land surface. Arid, semi-arid and dry subhumid zones can be taken together as comprising those dry-lands susceptible to desertification, excluding hyper-arid areas because they offer so few resources to human populations that they are seldom used and can hardly become more desert-like. Recent thinking on desertification refers to it as a process of land degradation in drylands. The concept of degradation is linked to using land resources – which include soil, vegetation and local water resources – in a sustainable way. Land that is being used in an unsustainable manner is being degraded, which implies a reduction of resource potential caused by one or a series of processes acting on the land. The official definition adopted by UNCCD is 'land degradation in arid, semi-arid and dry subhumid areas resulting from various factors, including climatic variations and human activities'. The definition illustrates one of the difficulties of the desertification issue: the term itself covers many different forms of land degradation. These include a reduction of or loss in the biological or economic productivity of rain-fed cropland, irrigated cropland, range, pasture, forest or woodlands.

The relative roles of human and natural factors, such as drought, have long been a subject of debate. Drylands are, by definition, areas that experience a deficiency in water availability on an annual basis, but

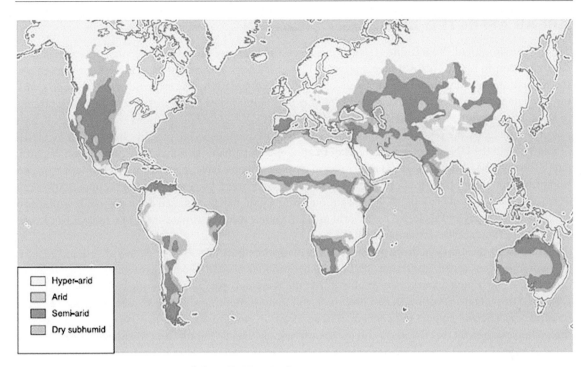

Figure 5.1 The world's drylands (after UNEP, 1992a).

precipitation in drylands is characterized by high variability in both time and space. Rainfall typically falls in just a few rainfall events and its spatial pattern is often highly localized since much comes from convective cells. In addition to this variability on an annual basis, longer-term variations such as droughts occur over periods of years and decades. Dryland ecology is attuned to this variability in available moisture and is therefore highly dynamic. On the large scale, satellite imagery has been used to observe changes in the green vegetation boundary of the Sahel region of Africa of up to 150 km from a dry year to a following wet one (Tucker and Nicholson, 1999).

In practice, it can be difficult to distinguish in the field between the adverse effects of human action and the response of drylands to natural variations in the availability of moisture. A good example is the dramatic rise in the amount of soil lost to wind erosion in parts of the Sahel region during the 1970s and 1980s as recorded by the annual number of dust storm days. At Nouakchott, the capital of Mauritania, dust storms blew on average less than ten days a year during the 1960s, but in the mid-1980s the average had increased to around 80 days per year (Middleton, 1985). This increase in soil loss can be explained both by drought, which characterized the area during the 1970s and 1980s, and by human action, but the relative roles of these factors are difficult to quantify. After the dry period that characterized the 1970s and 1980s, numerous ground- and satellite-based studies have observed a positive trend in vegetation greenness linked to greater rainfall in the Sahel (Brandt *et al.*, 2015). The dust storm frequency at Nouakchott has declined as a result.

AREAS AFFECTED BY DESERTIFICATION

The UN Environment Programme (UNEP) produced the first assessment of the global area at risk from desertification in 1977. This estimate suggested that at least 35 per cent of the Earth's land surface was threatened, an area inhabited by 20 per cent of the world's population. The estimate was based upon what limited data were available at the time and a lot of expert opinion and guesswork. It was a fair first attempt to assess a very complex issue, but subsequent efforts to assess the global proportions of desertification have not greatly reduced the reliance upon estimates. Two subsequent assessments have been made by UNEP (Table 5.1), but the UN agency was accused of treating these estimates as if they were based upon scientific data (Thomas and Middleton, 1994). More recent global estimates have suggested a much smaller proportion of susceptible drylands is affected. The Millennium Ecosystem Assessment figure (Lepers, 2003) came out at about 10 per cent of global drylands, although the regional data sets used did not provide a total coverage of all drylands and included hyper-arid areas. Another recent estimate shown in Table 5.1 indicates 23 per cent of susceptible drylands are affected.

Some people question whether a global assessment of something as poorly understood as desertification is possible, or even useful, given that solutions to the problem should be locally oriented. However, a case can be made for a global assessment if only to put the issue into perspective, to identify specific problem areas, and to generate the political and economic will to do something about it. However, the lack of good scientific evidence to support the concept of desertification continues to dog the issue, resulting in some researchers suggesting that the idea is no longer analytically useful (Behnke and Mortimore, 2016).

Table 5.1 Estimates of types of drylands deemed susceptible to desertification, proportion affected and actual extent

	1977 UNEP	1984 UNEP	1992 UNEP	2009 UN and various sources*
Climatic zones susceptible to desertification	Arid, semi-arid and subhumid	Arid, semi-arid and subhumid	Arid, semi-arid and dry subhumid	Arid, semi-arid and dry subhumid
Total dryland area susceptible to desertification (million km²)	52.8	44.1	51.7	50.8
Proportion of susceptible drylands affected by desertification (%)	75	79	70	23
Total of susceptible drylands affected by desertification (million km²)	39.7	34.8	35.9	11.8

Source: after Thomas and Middleton (1994) and various sources* compiled by Zika and Erb (2009).

'Spreading deserts'

Another area in which many authorities have produced information that can be construed as misleading (Thomas and Middleton, 1994; Behnke and Mortimore, 2016) has been in the numerous statements concerning the rate at which desertification is happening. These statements have been made in the academic literature and in popular and political circles for publicity and fund-raising purposes. Warren and Agnew (1988) quote several examples, including one made in 1986 in which US Vice-President Bush was being urged to give aid to the Sudan because 'desertification was advancing at 9 km per annum' (Warren and Agnew, 1988: 2). More than 20 years later, a UN publication on Sudan referred to the 'estimated 50 to 200 km southward shift of the boundary between semi-desert and desert' that has supposedly occurred since the 1930s (UNEP, 2007: 9). Such pronouncements have become so commonly used that their validity has seldom been questioned, but in reality they are rarely supported by scientific evidence. They also reinforce an image of sand dunes advancing to engulf agricultural land. Although undoubtedly a powerful vision for publicity purposes, and one that is appropriate in some places, the idea is oversimplified and not a generally accurate image of the way desertification works. It is a much more complex set of processes that is more commonly manifested as patches of degradation in the desert fringe.

One of the best-known studies that did set out to measure the rate of desert advance was made in northern Sudan (Lamprey, 1975). The vegetational edge of the desert was mapped by reconnaissance in 1975 and then compared to the boundary drawn from another survey carried out in 1958. The two boundaries indicated that the desert's southern margin had advanced by 90–100 km between the two dates (Figure 5.2), at an average rate of advance of 5.5 km per year over 17 years.

An immediate criticism can be made of the methodology used for the study in that, given the inherent variability of drylands, two 'snapshot' surveys can only be of very limited use in determining significant changes in vegetation, particularly since the 1970s was a period of drought in the Sudan while the 1950s was a decade characterized by above-average rainfall. Reliable conclusions can only be drawn after long-term monitoring that takes into account seasonal and year-to-year fluctuations in vegetation cover. When Lamprey's conclusions were scrutinized by Swedish geographers, they were found to be inaccurate. It transpired that the 1958 desert vegetation line was not based on a ground survey but followed the 75 mm isohyet, and satellite imagery and ground survey data compiled over several years did not confirm Lamprey's findings (Hellden, 1988).

Despite many years of research into desertification since the Lamprey study, there is still a lack of reliable maps of the issue or agreed ways of monitoring desertification at the subnational to global scales, even in the iconic Sahel region (Prince, 2016). Many of the existing maps continue to be based on subjective assessments by experts and are thus difficult to use for comparison with future assessments. One of the consequences is the continuation of unsupported statements about the extent and severity of desertification.

CAUSES OF DESERTIFICATION

The methods of land use that are applied too intensively and, hence, contribute to desertification are widely quoted and apparently well known. They can be classified under the headings of intensive grazing, overcultivation and overexploitation of vegetation. Salinization of irrigated cropland is often viewed as a separate category from these. Although the way in which these inappropriate land uses lead to desertification is also well known in theory, in practice particular areas deemed to be desertified due to a particular cause have often been so described according to subjective assessments rather than after long-term scientific monitoring. Furthermore, although specific land uses have been the subject of most interest from desertification

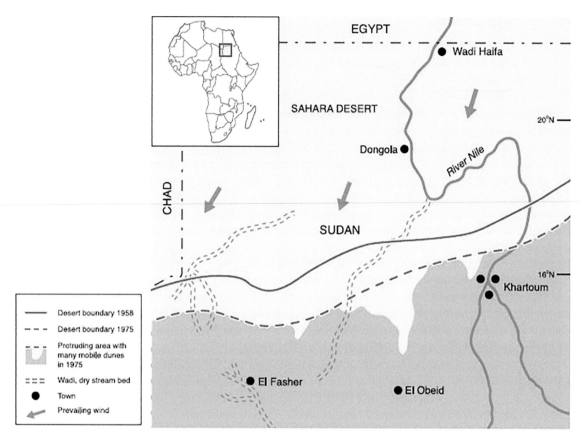

Figure 5.2 Lamprey's 'advancing Sahara', a study that purported to show that the desert was advancing at 5.5 km a year. See text for criticism of these findings.

researchers, it can be argued that permanent solutions to specific problem areas can only be found when the deeper reasons for people misusing resources are identified. Such reasons, many of which are rooted in social, economic and political systems (Table 5.2), enable, encourage or force inappropriate practices to be used. Although an understanding of the physical processes involved in overcultivation, for example, is important, there has generally been too much emphasis on this physical side of the equation, resulting in an over-reliance upon technical fix solutions. It is just as important to understand the underlying driving forces, the amelioration of which is of equal significance in the quest to find long-lasting solutions to desertification problems.

Intensive grazing

The overuse of pastures, caused by allowing too many animals or inappropriate types of animals to graze, can result in a number of environmental impacts. Primary among them are the actual removal of biomass by grazing animals, changes in the types of vegetation growing, and other effects of livestock such as trampling and hence soil compaction, which hampers air and water circulation in the soil, hinders root

Table 5.2 Suggested root causes of land degradation

Natural disasters	Degradation due to biogeophysical causes or 'acts of God'
Population change	Degradation occurs when population growth exceeds environmental thresholds (neo-Malthusian) or decline causes collapse of adequate management
Underdevelopment	Resources exploited to benefit world economy or developed countries, leaving little profit to manage or restore degraded environments
Internationalism	Taxation and other forces interfere with the market, triggering overexploitation
Colonial legacies	Trade links, communications, rural–urban linkages, cash crops and other 'hangovers' from the past promote poor management of resource exploitation
Inappropriate technology and advice	Promotion of wrong strategies and techniques, which result in land degradation
Ignorance	Linked to inappropriate technology and advice: a lack of knowledge leading to degradation
Attitudes	People's or institutions' attitudes blamed for degradation
War and civil unrest	Overuse of resources in national emergencies and concentrations of refugees leading to high population pressure in safe locations

Source: Thomas and Middleton (1994).

penetration and limits seed germination. A common consequence of heavy grazing pressure is a decrease in the vegetation cover, leading to increased soil erosion by water or wind. It is estimated that up to 73 per cent of the world's rangelands may be degraded (Lund, 2007) and overgrazing of rangeland by pastoralists has been the most commonly cited cause of desertification in global drylands for more than 30 years. However, the evidence supporting this link is not always convincing.

The way in which too much grazing can result in bare soils can be illustrated in places where livestock are excluded from certain pastures by fencing, as in the example from north-eastern Kuwait described in Box 5.1. Intensive grazing does not always result in a net loss of vegetation cover, however. In some areas the consequence is a change in the types of vegetation. One way this happens is by the encroachment of unpalatable or noxious shrubs into grazing areas, effectively a regime shift that can occur gradually over time or relatively abruptly. This bush encroachment, which has been recognized in southern Africa since the late nineteenth century, has a number of drivers (O'Connor *et al.*, 2014), but one sequence of events starts with high livestock numbers degrading the grass layer and allowing bushes to take its place. A thick cover of thorny bushes in turn deters grass growth. Livestock find it difficult to enter the thickets, so pasture is lost.

Traditional herders, for whom larger numbers of livestock represent greater personal wealth and social standing, have often been seen as the culprits in these grazing impacts in the developing world. Indeed, traditional communal herding in rangelands has frequently been viewed as a recipe for degradation via the tragedy of the commons. Such arguments have been used as a reason for privatizing access to pastures and encouraging settlement of nomadic pastoralists in many dryland countries of Africa, the Middle East and

Box 5.1 Grazing and vegetation change in north-eastern Kuwait

An interesting sequence of vegetation changes (Figure 5.3) has been observed in north-eastern Kuwait before and after the first Gulf War of 1990–1 as part of the planning process in preparation for ecological rehabilitation at a site proposed for protected area status. In 1988, prior to the war, the area known as Umm Nigga was largely devoid of vegetation, predominantly because of extensive grazing. By 1998, some years after the war ended, much more vegetation had grown as the government had restricted access to Umm Nigga for safety reasons, because of the danger of landmines and unexploded ordinance.

Indeed, the northern portion of Umm Nigga lay within the boundaries of the demilitarized zone (DMZ) adjacent to Iraq, and had been fenced off to prevent public access since 1994. Four years after this, in 1998, plant cover both inside and outside the DMZ was much greater than ten years earlier, but in 1998 the whole Umm Nigga region was declared safe, having been cleared of landmines.

By 2013, well after the return of grazing and other civilian activities, vegetation had decreased in most of the area proposed for protection (except in a zone in the north-west corner where irrigated cultivation began). However, inside the DMZ – which was still fenced off – vegetation cover had continued to increase.

Analysis of the satellite images showed that, inside the DMZ, vegetation cover increased from 0 per cent in 1988 to 40 per cent in 1998, and reached 64 per cent in 2013. This increase in vegetation cover can be attributed very largely to an absence of livestock grazing, allowing seeds stored in the sandy soils to germinate and grow into plants without being eaten. This indicates clearly that intensive grazing in this area had previously resulted in a complete loss of vegetation.

Source: Abdulla *et al.* (2016).

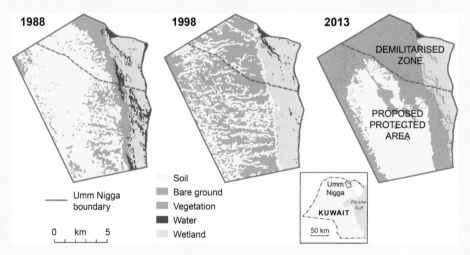

Figure 5.3 Ground surface cover as detected on satellite imagery of Umm Nigga, Kuwait (Abdulla *et al.*, 2016).

Asia. Intensive grazing problems have also been identified on commercial ranches in North America and Australia.

In practice, the reasons for increased grazing pressure on pastures are numerous and complex. They include competition for land as cultivated areas increase, pushing herders into more marginal pastures. In many parts of the Sahel, expanding areas of sorghum and millet cultivation have been primary factors responsible for a major decrease in the availability of range, as Ringrose and Matheson (1992) have identified. The position of herders at the margins of society in the eyes of many central governments has often meant that they are situated at the end of a chain of events that sees the expansion of irrigated land for cash cropping displacing rainfed subsistence cultivators who encroach into traditional grazing grounds, forcing herders into smaller ranges. This situation is described by Janzen (1994) in southern Somalia where expansion of irrigated agriculture on the Jubba and Shabelle rivers has forced small farmers to clear large areas of bushland for cultivation. For nomadic herders, this savanna zone and the river valleys themselves were important grazing lands during the dry season. Resultant intensification of grazing pressure into smaller areas has also been driven by the settling of some nomadic groups, a trend encouraged by government policy in the 1970s, and accelerated more recently by drought.

Although these ideas on intensive grazing problems have been followed by many scientists and policy-makers, recent thinking on the nature of dryland vegetation and ecology has questioned much of this conventional view. Distinguishing between degradation of vegetation and other forms of vegetation change is very difficult, and the importance of intensive grazing as a cause of desertification has probably been exaggerated in many areas.

One commonly quoted situation considered to be typical of desertified areas is the loss of vegetation around wells or boreholes, bare areas that supposedly grow and coalesce. In fact, several studies have failed to show that these 'piospheres' expand over time (e.g. Hanan *et al.*, 1991). Halos of bare, compacted soil 50–100 m in circumference caused by grazing and trampling are undoubtedly characteristic of many rangeland watering holes (Figure 5.4). But such areas are also typified by higher levels of soil nutrients than surrounding areas thanks to regular inputs from dung and urine (Shahriary *et al.*, 2012), which may balance out any negative effects of possible soil loss by erosion. Piospheres are perhaps best interpreted as 'sacrifice zones' in which the loss of the vegetation resource is outweighed by the advantages of a predictable water supply (Perkins and Thomas, 1993).

Other ideas about the effects of grazing on rangeland vegetation are also being substantially revised. Formerly, most claims that a particular area was being 'overgrazed' were based on the idea that the area had a fixed carrying capacity, or theoretical number of livestock a unit area of pasture could support without being degraded. This idea was in turn based on Clements' (1916) model of vegetation succession and ecological stability. But more recent thinking depicts semi-arid ecosystems as seldom, if ever, reaching equilibrium. Rather they are in a state of more or less constant flux, driven by disturbances such as drought, fire and insect attack. Hence, while calculation of a carrying capacity is possible in relatively unchanging environments, it is difficult to apply to non-equilibrium environments like dryland pastures because the number of animals an area can support varies on several timescales: before and after a rainy day, between a wet season and a dry season, and between drought years and wet years. Grazing the number of livestock appropriate for a wet year on the same pasture during a drought could result in degradation, but such disastrous concentrations of livestock rarely occur because animals typically die due to the lack of water first (see section on drought below). Traditional pastoralists have developed many flexible strategies for coping with variations in pasture quality. Mobility is probably the most important of these. Maintaining a herd made up of several different animal species (Figure 5.5) also helps to make the most of the natural dynamism of dryland vegetation.

Figure 5.4 Cattle around a water hole in Niger. Such 'piospheres' are typically devoid of vegetation but the soil has a high nutrient content.

In the light of these changes in our understanding of rangeland ecology, the relative importance of intense grazing and natural stresses is being reassessed. One study of open woodlands in north-west Namibia, an area long thought to have been desertified by local herders overgrazing, found very little evidence of degradation (Sullivan, 1999). This study concluded that previous assessments of the area's ecological health have been based more on perceptions than scientific assessment. These perceptions have been clouded by an adherence to ideas of a fixed carrying capacity based on rangelands as equilibrium systems, and remnants of a colonial ideology that views traditional communal herding as environmentally degrading.

Although views on the sustainability of grazing in rangelands have often become polarized, matching the two principal opposing views of rangeland ecology, recent studies indicate that it is not an either/or issue. Most arid and semi-arid rangelands encompass elements of both equilibrium and non-equilibrium approaches and both grazing and variable rainfall are important drivers of vegetation dynamics on different temporal and spatial scales (Gillson and Hoffman, 2007). Hence, any sustainable approach to rangeland management must take account of temporal variability and spatial heterogeneity (Joly *et al.*, 2012).

Overcultivation

Several facets of overcultivation are commonly quoted. Some stem from intensification of farming, which can result in shorter fallow periods, leading to nutrient depletion and eventually reduced crop yields. Soil

Figure 5.5 Mixed herds are traditionally kept by herders in drylands, as here in Mongolia's Gobi Desert. Different animals eat different types of vegetation and have different susceptibilities to moisture availability.

erosion by wind and water is another result of the overintensive use of soil, resulting from a weakened soil structure and reduced vegetation cover. Monocultures can lead to all these forms of soil degradation, as shown by many years of monitoring of cropland in the semi-arid pampa of Argentina. The results of studies compiled by Wingeyer et al. (2015) indicate that the long-term cultivation of grains (soybeans, maize and wheat) over recent decades has resulted in declining soil quality, in both physical and chemical terms. This includes increased bulk density and decreases in dry aggregate stability, soil organic matter and available nutrients such as phosphorus and nitrogen. The depletion of nutrients means that greater amounts of fertilizers have to be applied to maintain crop yields, while the declines in organic matter and soil stability have meant a greater susceptibility to erosion.

In other cases, erosion has also been caused by the introduction of mechanized agriculture with its large fields and deep ploughing, which further disturbs soil structure and increases its susceptibility to erosive forces. Similar outcomes have also been noted in areas where cultivation has expanded into new zones that are marginal for agricultural use because they are more prone to drought, or are made up of steeper slopes that are more prone to erosion. Some classic examples of overcultivation, their root causes, inappropriate land uses and resulting degradation, are shown in Table 5.3. The wind erosion resulting in the Great Plains and the Virgin Lands is discussed on p. 346.

Table 5.3 Some examples of overcultivation

Area (date)	Root causes	Land use	Environmental response
Argentine pampa (1990–2010)	Market globalization	Extensification of monocultures (particularly soybean)	Wind and water erosion, nutrient depletion, soil organic matter depletion
Niger (1950s and 1960s)	Colonial influence, cash crops	Extensification of groundnut cultivation	Nutrient depletion
Former USSR's Virgin Lands (1950s)	Political ideology, food security	Deep ploughing for cereals	Wind erosion
US Great Plains (1930s)	Pioneer spirit, culture of dominating nature	Deep ploughing for cereals	Wind erosion

Overexploitation of vegetation

Significant areas of forest and woodland occur in many drylands, predominantly in their more humid parts. In developing countries particularly, local communities rely heavily on dryland forest resources for their livelihoods, and the loss and degradation of these landscapes is a growing concern (Nichols et al., 2017). Clearance of forested land is undertaken for a number of reasons, but most commonly to make way for the expansion of grazing or cultivation and/or to provide fuelwood. Such action is a degradation of vegetation resources and also reduces the protection offered to soil by tree cover. Accelerated erosion may result, and over the longer term, soil is deprived of inputs of nutrients and organic matter from decomposing leaf litter. Both of these processes can lead to the degradation of soil structure and fertility. Water tables may also be affected.

In numerous areas, vegetation clearance has a long history. In northern Argentina, the tropical dry forests of the semi-arid Chaco have been exploited for more than 100 years, initially because the wood made excellent railway sleepers, but more recent clearance has been for agricultural expansion, prompted by high grain prices and positive rainfall balances since the 1970s. In recent decades, large areas of forest have been directly replaced by the expansion of soybean cultivation in the Argentine Chaco, north of the pampa grasslands, although cattle ranching has been the most important direct driver of deforestation in the Chaco of neighbouring Bolivia and Paraguay (Fehlenberg et al., 2017). However, Fehlenberg et al. also concluded that deforestation inside the Chaco has been driven by soybean expansion outside the Chaco: pastures being effectively displaced into new areas because rangelands in other regions have been converted into soybean fields.

Elsewhere, however, many of the ways in which people exploit dryland forests – such as harvesting fuelwood and fodder, charcoal making, and grazing – do not lead to a complete clearance but rather a thinning of the forest. Nonetheless, when vegetation is used to a degree that is beyond its natural capability to renew itself, the result is a degraded vegetation cover. An expanding human population is frequently cited as the ultimate driver behind this form of desertification, a neo-Malthusian scenario that may create a positive feedback of continuing degradation when ageing and less resilient forests are more vulnerable to shocks, such as severe droughts and pest invasions.

In the 1970s and 1980s, the environmental problems caused by fuelwood collection were thought to be so acute in Sahelian Africa that a 'fuelwood crisis' was feared. However, this regional environmental disaster did not materialize and in the late 1990s the fuelwood situation in the Sahel was reassessed. The methods used to predict the crisis, comparing current fuelwood consumption with current stocks, annual tree growth and annual population growth, have been criticized by some (e.g. Leach and Mearns, 1988). Poor survey data, the importance of replacement fuels such as crop residues and dung, and the effect of increasing fuelwood prices as local supplies dwindle are highlighted as making the situation more complex than was previously proposed. Overall, it seems that the surveys conducted in the 1970s and 1980s under-estimated the amount of fuelwood available in the Sahel (Ribot, 1999).

Although the predicted fuelwood crisis in the Sahel has not taken place on the scale once feared, environmental degradation has occurred in some areas. Much of this form of desertification is concentrated around urban areas that have expanded very rapidly since the late 1960s as rural migrants fled the effects of drought. Denuded areas susceptible to enhanced erosion have been reported from numerous Sahelian cities, including Khartoum, Dakar, Ouagadougou and Niamey.

However, the complexity of the fuelwood issue can be illustrated by the case of Kano in northern Nigeria where Nichol (1989) found that tree density had increased in the immediate vicinity of the city as the transport of fuelwood by donkey had been displaced by the use of long-distance trucks. A decline in tree density was recorded in the zone 70–250 km from Kano. This view is confirmed in an analysis presented by Mortimore and Turner (2005) who also show diminishing areas of natural woodland since the 1950s in several areas that serve the Kano fuelwood market. However, these authors are not convinced that this trend should necessarily be equated to a degradation problem since 'pollarding selected, mature trees in mixed woodland stimulates regeneration, and may be regarded as a form of rotation; moreover, rates of regeneration have been grossly under-estimated' (ibid.: 584). They also point out that if a 'fuelwood gap' was opening up between demand and supply, the price of fuelwood should increase in consequence, but stable real retail prices of fuelwood in the cities of northern Nigeria suggest the opposite. Fuelwood still accounts for up to two-thirds of energy consumption in Kano (Cline-Cole and Maconachie, 2016). Elsewhere in the Sahel notable positive changes in dryland tree cover have occurred, as Reij *et al.* (2009) report from southern Niger where large-scale reforestation has occurred since the 1980s on an estimated 5 million hectares. The change has been almost entirely through the efforts of local farmers in managing natural tree regeneration on their farmland.

Salinization

Salinization is a form of soil degradation most commonly encountered in dry climatic regions, although it also occurs in more humid environments. Natural, or 'primary', salt-affected soils are widespread in drylands because the potential evaporation rate of water from the soil exceeds the input of water as rainfall, allowing salts to accumulate near the surface as the soil dries. While these primary salt-affected soils are found extensively under natural conditions, salinity problems of particular concern to agriculturalists arise when previously productive soil becomes salinized as a result of poor land management, so-called 'secondary' salinization. Secondary, or human-induced, salt-affected soils cover a smaller area than primary salt-affected soils but secondary salinization represents a more serious problem for human societies because it mainly affects cropland. Major agricultural crops have a low salt tolerance compared to wild salt-tolerant plants (halophytes), so salinization results in a regime shift from high productivity to low productivity and consequently to declines in crop yields. Hence arable land, a scarce and valuable resource in dryland regions, is frequently abandoned when it becomes salinized, due to the very high cost of remediation.

The human activities that lead to desertification through the build-up of salts in soils are well documented and are summarized into five groups in Table 5.4; secondary salinization occurs under a number of circumstances but is most commonly associated with poorly managed irrigation schemes. In some countries, more than half of all irrigated land is affected to some extent by secondary salinization and the process is widely regarded as irrigated agriculture's most significant environmental problem (Ghassemi *et al.*, 1995).

The impact of salinization on crop yields acts indirectly via effects on the soil and through direct effects on the plants themselves (Ashraf *et al.*, 2008). Salt accumulation reduces soil pore space and the ability to hold soil air, moisture and nutrients, resulting in a deterioration of soil structure and a reduction in the soil's suitability as a growing medium. Plant growth is also directly impaired by salinization first of all because salts are toxic to plants, particularly when they are seedlings, and, second, through its effects on

Table 5.4 Major causes and effects of secondary salinization

Cause	Effect
Direct effects of poor irrigation techniques	Four principal causes: water leakage from supply canals, overapplication of water, poor drainage, and insufficient application of water to leach salts away, all contribute to salt build-up in the root zone and/or to creation of perched saline water tables under irrigated fields
Indirect effects of irrigation schemes	River barrages and dams can result in salinization on valley terraces above the dam should a reservoir raise local water tables; irrigation schemes can also increase soil salinity downstream due to an absolute reduction in downstream discharge and/or discharge of saline irrigation wastewater
Effects of vegetation changes	Replacement of native shrubs, trees and grasses with annual grain crops or grassland for grazing reduces evapotranspiration so that rainfall percolates through subsurface sediments to impermeable horizons where it is conducted laterally to lower slope positions, producing extensive areas of salty waterlogged soil; these 'saline seeps' are most common in non-irrigated dryland farming regions
Sea-water incursion	Excessive pumping of coastal freshwater aquifers for irrigation or other uses lowers water table allowing sea water to seep into the aquifer, often resulting in soil salinization; a similar effect can also result from reduced river flows, due to excessive irrigation offtakes or dam building (sometimes exacerbated by drought), allowing enhanced up-channel movement of ocean water
Disposal of saline wastes	Dumping of saline wastes from industrial and municipal sources (e.g. petroleum wells, coal mines, desalination plants) directly onto soils

Source: Middleton and van Lynden (2000).

osmotic pressures. Contact between a solution containing large amounts of dissolved salts and a plant cell causes the cell's lining to shrink due to the osmotic movement of water from the cell to the more concentrated soil solution. As a result, the cell collapses and the plant succumbs.

Salinization also leads to a range of off-site hazards. Drainage from salinized areas often increases the concentration of salts in streams, rivers and wetlands, adversely affecting freshwater biota and in some cases resulting in a loss of biodiversity (see the example of the Aral Sea tragedy on p. 179). Groundwater that has become salinized may no longer be suitable for other human uses (e.g. for drinking), while the capillary rise of salts from such 'aggressive' groundwater into buildings and other structures can cause severe damage to building materials by salt weathering (El-Gohary, 2016).

The economic costs of salinity problems incurred by farmers can be considerable, as Table 5.5 shows for a number of countries. Irrigation schemes in some of the dryland countries of Asia are among the most seriously affected. In Pakistan, where irrigated land supplies more than 90 per cent of agricultural production, salinity problems affect about a quarter of the irrigated area or about 11 per cent of the country's land area (Middleton and van Lynden, 2000). Most of the affected soils are part of the Indus irrigation system, the largest single irrigation system in the world. Salinity and the associated problems of waterlogging are estimated to cost farmers in Pakistan about 25 per cent of their production potential for major crops (Qureshi et al., 2008).

Pakistan's efforts to manage its salinity problems have produced mixed results. Government Salinity Control and Reclamation Projects (SCARPs) were first introduced in the 1950s to address both waterlogging and salinization by sinking tube wells to lower the water table and provide more water for agriculture. These public sector schemes were followed by many private tube wells and in Punjab province the result was a fivefold increase in groundwater abstraction between 1965 and 2002. Surface water irrigation has been supplemented by underground supplies, but the groundwater is often saline and falling water tables are now a major issue. Both supply and demand management are required to offset the problems introduced by a primarily technocentric approach. Qureshi et al. (2010) suggest adoption of water conservation technologies and revision of existing cropping patterns on the demand side, and better regulation of access to groundwater through institutional reforms to improve supply management.

Table 5.5 Economic costs of salinization on irrigated cropland in selected countries

Country/region	Area	Crop	Cost (US$/ha/yr)
Australia	Victoria state	All crops	4550
Central Asia	Aral Sea Basin	All crops	344
India	Haryana state	Rice	313
		Wheat	238
		Cotton	437
		Sugar cane	300

Source: after Qadir et al. (2014).

Poorly managed irrigation schemes are not the only human-induced cause of salinity problems in dry-land environments. Rising water tables can also result when natural vegetation is cleared and replaced with pasture or a crop that requires less moisture for growth. Less evapotranspiration from annual crops or pastures than from native deeper-rooted vegetation increases the amount of water percolating through the soil to recharge aquifers. Rising groundwater mobilizes soluble salts naturally stored deep in soils and these salts can become concentrated by evapotranspiration in the root zone of vegetation. The clearance of native eucalyptus trees from large areas of south-west Western Australia has caused saline groundwater to rise in this way by more than 30 m, leading to the problem of 'saline seep'. Although rainfall has been below the long-term average over most of the region since the mid-1970s, the risks from dryland salinity in Western Australia remain considerable because of the greater impact of vegetation change on groundwater levels (Raper *et al.*, 2014).

Such salinity problems can also be reversed. Salts can be leached from the soil profile using subsurface drainage, by tiles or vertical tube wells, or vegetation that is more salt-tolerant can be introduced. Planting of deep-rooted *Tamarix* on the southern margins of the Taklimakan Desert in China has reduced waterlogging and salinity as well as providing a new source of fuelwood for local villagers (Middleton and Thomas, 1997; see Figure 5.6).

Figure 5.6 Fuelwood harvest from a formerly salinized area on the southern margins of the Taklimakan Desert rehabilitated with *Tamarix*.

Drought

Drought is the principal natural hazard faced by communities living in most drylands and its impacts frequently occur in combination with those of desertification. Three types of drought are commonly recognized: meteorological, agricultural and hydrological. The latter two forms are typically caused by meteorological drought, a period of months to years with below-average precipitation. Agricultural drought occurs when dry soils lead to reduced crop production and pastures, and hydrological drought reflects below-average water levels in lakes, reservoirs, rivers, and aquifers.

Drought is a distinctive natural hazard in several ways. Its onset and end are typically difficult to determine and the impacts of a drought increase slowly, often accumulating over a considerable period, and may linger for years after the drought ends. As a pervasive, chronic hazard, drought impacts are spread over large geographical areas with fuzzy boundaries and sometimes great periods of time, making the quantification of impact (and indeed the provision of relief) far more difficult than for many other natural hazards. Nevertheless, some examples of economic cost estimates indicate the scale of impact:

- Australia's long-lasting Millennium Drought (1997–2010) cost about US$800 million in hydrological ecosystem service losses in the South Australian portion of the Murray–Darling Basin (Banerjee *et al.*, 2013).
- In the US state of California in 2015, the fourth consecutive year of severe drought, the direct agricultural costs of drought were put at about US$1.84 billion (Howitt *et al.*, 2015).
- A severe multi-year drought in north-east Brazil cost the national government nearly US$4.5 billion in emergency response measures over the period 2010–14 (Bastos, 2017).

The impacts in some developing countries can be particularly severe, and chiefly on poorer members of rural societies. Most of the effects of drought on poor rural households occur via its adverse impacts on the quantity and quality of food production. Insufficient water and inadequate grazing can result in huge losses of livestock for herders (Table 5.6), and selling any remaining assets to buffer these losses is undermined as a coping strategy by lower prices, driven by many people selling their possessions at the same time.

Since the mid-twentieth century, global aridity and drought areas have increased substantially, a trend that is particularly marked after the late 1970s, when rapid warming of the atmosphere contributed significantly to global drying. Still greater aridity and widespread droughts are expected in many regions in the coming decades (Park *et al.*, 2018), with clear implications for very large numbers of people in drylands. Droughts cannot be prevented, and the way governments respond to them tends to be reactive. This form of crisis management inevitably treats only the impacts rather than the underlying reasons why some people and some sectors of the economy are vulnerable. Further, providing drought assistance or relief may in effect encourage land managers and others to continue with what are essentially unsustainable land and water management practices. Hence, governments and international bodies are moving towards managing drought risk, in line with the Sendai Framework (see Chapter 21).

Feedbacks

Identifying cause and effect in the desertification issue, and disentangling the relative importance of natural and human-induced environmental changes, are further complicated by several possible feedback mechanisms operating at large scales in dryland areas of the world.

Table 5.6 Effects of some recent droughts on livestock in selected African countries

Drought	Location	Livestock lost
1981–84	Botswana	20% loss of national herd
1982–84	Niger	62% loss of national cattle herd
1991–92	Northern Kenya	70% loss of livestock
1995–97	Southern Ethiopia	78% loss of cattle; 83% loss of shoats
1995–97	Greater Horn of Africa	29% loss of cattle; 25% loss of shoats
1999–2001	Kenya	30% loss of cattle; 30% loss of shoats; 18% loss of camels
2002	Eritrea	10–20% loss of livestock
2004–06	Kenya	70% loss of livestock in some pastoral communities
2008–09	Kenya	50% loss of livestock in some pastoral communities
2010	Niger	80–100% loss of livestock in 20% of pastoral communities
2011	Djibouti	70–80% loss of livestock
2011	Ethiopia	100% loss of livestock in many pastoral communities in Somali region

Source: after Huho et al. (2011).

Changes in albedo over great areas due to a decline in vegetation cover, whether caused by drought or intensive grazing, could have a negative effect on rainfall. Vegetation loss leads to a greater surface albedo, resulting in a cooler land surface, which, in turn, reduces convective activity and hence rainfall. Less rainfall reduces vegetation further, and so on. Some researchers have suggested this feedback helps to explain the major 30-year low-rainfall period in the Sahel – some call it an extended drought, others prefer to speak of a desiccation in climate – dating from the late 1960s, a regime shift from a 'wet Sahel' to a 'dry Sahel' climatic regime that may have been triggered by slow changes in either desertification or sea-surface temperatures (Foley *et al.*, 2003).

The generation of large quantities of atmospheric soil dust from drylands (Figure 5.7) is also likely to have effects on atmospheric processes, including cloud formation and rainfall (Choobari *et al.*, 2014). The drivers behind such wind erosion may also be rooted in the natural operation of processes such as drought or, again, be exacerbated by human activities such as intensive grazing and cultivation. Water droplets in clouds start by forming on small particles such as soil dust but too many dust particles may inhibit precipitation by making many relatively smaller droplets that are not large enough to fall as rain. Another way in which rainfall may be affected is through changes in convective activity brought about by the presence of dust, modifying temperature gradients in the atmosphere. As with the albedo feedback, a possible desertification feedback loop could result from atmospheric dust: suppressed precipitation enhancing drought and thereby the generation of more dust emissions. Proving such feedbacks is difficult, but one attempt (Cook *et al.*, 2009) using a general circulation model showed that the joint effect of reduced vegetation cover and enhanced levels of atmospheric dust probably amplified the impact of drought during the US Dust Bowl in the 1930s (see below).

Figure 5.7 A huge cloud of desert dust billowing out over the Indian Ocean from the Arabian Peninsula in March 2012. Atmospheric dust may have a feedback effect that enhances drought conditions.

Such human-induced environmental changes may have a long history. One suggested explanation for the lack of monsoonal rains in the interior of Australia in the Holocene is that early humans severely modified the landscape through their use of fire (Miller *et al.*, 2005). The theory suggests that regular burning converted the semi-arid zone's mosaic of vegetation types (trees, shrubs and grassland) into the modern desert scrub, thereby weakening biospheric feedbacks that contributed moisture to the atmosphere and resulting in long-term desertification of the continent.

FOOD PRODUCTION

One of the most important reasons for desertification's position as a major global environmental issue is the link between resource degradation and food production. Desertification reduces the productivity of the land and, hence, less food is available for local populations. Malnutrition, starvation and ultimately famine may result. This link was a strong theme in the negotiations leading to the UN Convention to Combat Desertification.

However, the links between soil degradation and soil productivity, and consequently crop yields, are seldom straightforward or clear-cut. Productivity is affected by many different factors, such as the weather, disease, pests, the farming methods used and economic forces. Distinguishing the effects of these various factors is, in practice, very difficult, quite apart from the fact that data on crop productivity are often unreliable and many measurements of soil erosion are also open to question (see Chapter 14). While the theory linking desertification to food production is basically sound, it is often oversimplified and deserves further careful research.

The link between desertification and famine is even more difficult to make. This is not the case in theory, since food shortages stem from reduced harvests and thinner or even dead animals, although such shortages are also caused by factors of the natural environment, particularly drought. In practice, however, famines typically occur in areas that are characterized by a range of other factors such as poverty, civil unrest and war (see Box 13.1). Several authors have emphasized the social causes of severe food shortages (e.g. Sen, 1981). Mass starvation is probably more a function of poor distribution of food, and people's inability to buy what is available, than desertification, although desertification, drought and many other natural hazards may act as a trigger. This tentative conclusion is supported by one of the few studies that specifically aimed to assess the relative importance of these factors, in the Sudanese famine of 1984/85 (Olsson, 1993). Millet and sorghum yields in the provinces of Darfur and Kordofan, the worst affected areas, were reduced to about 20 per cent of pre-drought levels, largely due to climatic factors – the amount and timing of the rains. Desertification, by contrast, was calculated to account for just 10–15 per cent of the variation in crop yields. While drought was undoubtedly a trigger for ensuing famine, the profound causes were different. As fears of crop failures mounted, the price of food in Darfur and Kordofan soared to levels that most rural people were unable to afford. While food was available at the national level, the distribution network was inadequate and famine followed.

UNDERSTANDING DESERTIFICATION

Appropriate solutions to desertification problem areas must be based on an integrated view of human society and the operation of nature in drylands. Variability and resilience are typical of dryland ecology and in many dryland regions people have developed strategies to reduce the risks driven by the unpredictable and generally low levels of rainfall and recurrent droughts. The complex co-evolution of human and ecological systems in drylands is emphasized by the Drylands Development Paradigm (Reynolds et al., 2007), an approach that distinguishes between the 'slow' and 'fast' variables driving change in human and environmental spheres. Slow variables include soil development, geomorphological change, institutional evolution and cultural changes, while fast variables include such things as crop production and household income. Many efforts to address desertification tend to focus on the fast variables that operate at timescales more amenable to consideration by governments, aid donors and dryland populations. However, it is the slow variables that ultimately determine the direction of change over the long term. Examples from two regions illustrate some of the balances and trade-offs between actions and policies designed to address slow and fast variables in drylands.

The US Great Plains

The Dust Bowl of the 1930s on the North American Great Plains is one of the best-known dryland environmental disasters in history. Over a period of 50 years, drought-resistant grasslands were turned into fields of drought-sensitive wheat by a culture set on dominating and exploiting natural resources using ploughs and other machinery developed in western Europe (Worster, 1979). Farmers were spurred on to develop the Great Plains with proclamations such as this from the US Bureau of Soils in 1909:

> The soil is the one indestructible, immutable asset that the nation possesses. It is the one resource that cannot be exhausted; that cannot be used up.

The folly of this statement was exposed in the 1930s, when drought hit the area, as it periodically does, and wind erosion ensued on a huge scale. The most severe dust storms, so-called 'black blizzards', occurred between 1933 and 1938. These huge events carried soil dust northwards as far as Canada and eastwards to New York and out over the Atlantic Ocean. At Amarillo, Texas, at the height of the Dust Bowl period, one month had 23 days with at least ten hours of airborne dust, and one in five storms had zero visibility (Choun, 1936). By 1937, the US Soil Conservation Service estimated that 43 per cent of a 6.5 million ha area at the heart of the Dust Bowl had been seriously damaged by wind erosion. The large-scale environmental degradation, combined with the effects of the Great Depression, ruined the livelihoods of hundreds of thousands of American families.

The lessons learnt from the Dust Bowl years inspired major advances in soil conservation techniques that have been widely applied, but the long-term environmental health of the area remains a subject for debate. Parton et al. (2007) identify two schools of thought: one group sees Great Plains farming, especially dryland grain production, as an ongoing ecological mistake that will lead to another 1930s-type disaster, while another group recognizes the benefits of technical and social innovations, ranging from zero-tillage management (see p. 352) to crop insurance, which they believe have stabilized the agroecosystem on the Plains.

However, soil erosion by wind has not ceased to be a problem and during drought in the 1970s was on a comparable scale to that in the 1930s (Lockeretz, 1978). Investigation of the soil losses of the 1970s illustrates some important underlying causes, indicating that political and economic considerations overshadowed the need for sustainable land management. Large tracts of marginal land had been ploughed for wheat cultivation in the early 1970s, driven by high levels of exports, particularly to the then USSR. To encourage farmers to produce, a federal programme was set up to guarantee them payment according to the area sown, so that the disincentive to plough marginal areas was removed: the farmers were paid whether these areas yielded a crop or not. At the same time, new centre-pivot irrigation technology was widely adopted to water the new cropland, but the use of the rotating irrigation booms required that wind breaks, planted to protect the soil, be removed (McCauley *et al.*, 1981). Wind erosion was again elevated in central and northern parts of the Great Plains over the period 2000 to 2014 (Hand *et al.*, 2017).

Yields of corn and wheat have increased on the Great Plains since the 1940s and the irrigated area of crops had expanded from about 1 million ha in the 1940s to 7 million ha by the 2000s (Parton *et al.*, 2007). However, the increasing use of irrigation has also issued in a new form of desertification on the Great Plains: groundwater depletion. Water pumped from the Ogallala aquifer (Figure 5.8), which stretches across parts of eight Great Plains states from South Dakota to Texas, is essentially a non-renewable resource, since recharge of the aquifer is negligible. However, in the 1960s when water levels in irrigation wells began to decline, farmers did not act to conserve dwindling supplies, but actually increased pumping because those who reduced their consumption still experienced declining water tables as others continued to irrigate, a clear example of the tragedy of the commons. Declining well yields and the rising costs of pumping deeper water are acting to slow aquifer withdrawal rates, and upper limits for groundwater production have been introduced in Texas, but the focus remains on reducing irrigation demand without impacting irrigation

Figure 5.8 Sprinkler irrigation in a field of corn using groundwater from the Ogallala aquifer in Nebraska, USA.

area or crop production levels (Colaizzi *et al.*, 2009). The many years of pumping for irrigation have also contributed to a decline in groundwater quality. Salinity levels in the Ogallala have risen as highly min-eralized groundwater from lower geological formations has welled up into the aquifer (Chaudhuri and Ale, 2014). All in all, despite the booms and busts experienced on the Great Plains throughout the last century, and the adoption of certain technical and social innovations, its farming still cannot be described as sustainable.

The Sahel

Sahelian Africa, where catastrophe in the 1970s and 1980s sparked worldwide interest in desertification, continues to be a focus of global concern. Most inhabitants of this region still rely on a highly variable natural resource base for their livelihoods, through herding and/or rainfed cultivation. Long experience of coping with environmental dynamism has engendered numerous strategies for managing risk and varia-bility (Table 5.7) but many dryland inhabitants remain vulnerable to perturbations, particularly drought, and the longer-term ecological decline of desertification.

The scarcity of water and recurrent droughts long typical of the Sahel were exacerbated by a desiccation of the climate that constituted the clearest and most dramatic example of climate variability measured anywhere in the world to date (Hulme, 2001): averaged over 30-year periods, annual rainfall in the region

Table 5.7 Examples of strategies used in dryland Africa to cope with environmental risk and variability

Activity	Strategies
Cultivation	Shifting between crops, varieties, specializations, in response to rainfall. Sahelian farmers typically cultivate both long- and short-cycle millets, for example, with the aim of spreading risks associated with rainfall timing. Different fields in different parts of the village are planted with different crops to spread risks associated with the characteristic patchiness of rainfall.
Herding	Changing seasonal grazing migrations to take advantage of alternative forage when usual grazing is damaged by drought. Herders' adaptive strategies in East Africa have included accessing tree fodders, selling animals, and intensifying animal health care.
Wild food gathering	Women in particular are repositories of local knowledge on a range of 'famine foods' which might replace conventional food grains.
Income diversification and migration	Farmers harvest natural products, adding value by manufacturing simple objects, labouring and otherwise exploiting multiple (though poorly rewarded) local work opportunities. Migration further afield for temporary, low-paid employment is another option.

Source: after Mortimore *et al.* (2009).

declined by between 20 per cent and 30 per cent between the decades (1930s to 1950s) leading to political independence for Sahelian countries and the decades since (post-1960s). These physical constraints were compounded in many cases by a history of political disadvantage and economic neglect that has resulted in low investment and a general lack of infrastructure, including poor access to education, health facilities and markets. There are also sociocultural issues such as mobility which are important to numerous communities but rarely well incorporated into government policies. The consequence of these drivers, in combination with many others, is stark. Sahelian populations include the poorest, the hungriest, the least healthy and the most marginalized people in the world (Middleton *et al.*, 2011).

The return of predominantly wetter conditions in the Sahel in the mid-1990s (Anyamba *et al.*, 2014) has provided some relief, but the region is still characterized by widespread poverty and the climate-vulnerability of many of its inhabitants. The challenges of poverty cannot be divorced from those of desertification because the degradation of natural resources contributes directly to low levels of well-being via the poverty–environment nexus. Desertification and vulnerability to droughts are important drivers of poverty, and climate-induced crop failures, loss of livestock and food shortages have the greatest impact in poorer households. Hence, environmental sustainability in the Sahel, as elsewhere, goes hand in hand with development. This realization has prompted an approach to dealing with desertification that can build on the resources and capabilities of Sahelian peoples (Mortimore *et al.*, 2009). Elements of this strategy include:

■ improving knowledge to inform sustainable management (e.g. better understanding of Sahelian ecosystem seasonality, variability, and ecosystem services such as water)

- reassessing the economic value of ecosystem services and the prospects they present for developing new market opportunities (e.g. pastoralism is the only viable option for many rural people but it is often undervalued)
- promoting sustainable public and private investments to reverse decades of relative neglect (e.g. investments in rural roads, electricity and education can make important contributions to agricultural growth and poverty reduction)
- improving access to profitable markets (e.g. growing Sahelian cities generate increasing demand for rural produce but access requires better transport and communications)
- supporting institutional changes to strengthen people's rights to natural resources, to better manage risk, and increase resilience in the human-ecological system (e.g. several West African countries have new laws to recognize and regulate access to grazing, thus maintaining the flexibility of mobile pastoralism).

Making the most of dryland opportunities in the Sahel means developing and implementing, through a participatory process, an integrated strategy that will achieve three fundamental aims: enhancing the economic and social well-being of dryland communities, while enabling them to sustain their ecosystem services, as well as strengthening their adaptive capacity to manage environmental (including climate) change.

FURTHER READING

Behnke, R. and Mortimore, M. (eds) 2016 *The end of desertification? Disputing environmental change in the drylands*. Berlin, Springer. A critical collection of views on the desertification concept.

De Vente, J., Reed, M.S., Stringer, L.C., Valente, S. and Newig, J. 2016 How does the context and design of participatory decision making processes affect their outcomes? Evidence from sustainable land management in global drylands. *Ecology and Society* 21(2): 24. Lessons from projects that engage with stakeholders. Open access: www.ecologyandsociety.org/vol14/iss2/art19/.

Kettle, N., Harrington, L. and Harrington, J. 2007 Groundwater depletion and agricultural land use change in the High Plains: a case study from Wichita County, Kansas. *The Professional Geographer* 59: 221–35. Unsustainable use of groundwater from the Ogallala aquifer.

Kimiti, K.S., Western, D., Mbau, J.S. and Wasonga, O.V. 2018 Impacts of long-term land-use changes on herd size and mobility among pastoral households in Amboseli ecosystem, Kenya. *Ecological Processes* 7(1): 4. A study of mobile pastoralism in southern Kenya. Open access: link.springer.com/article/10.1186/s13717-018-0115-y.

Marques, M.J., Schwilch, G., Lauterburg, N., Crittenden, S., Tesfai, M., Stolte, J., Zdruli, P., Zucca, C., Petursdottir, T., Evelpidou, N. and Karkani, A. 2016 Multifaceted impacts of sustainable land management in drylands: a review. *Sustainability* 8(2): 177. A wide-ranging review. Open access: www.mdpi.com/2071-1050/8/2/177/htm.

Middleton, N.J., Stringer, L., Goudie, A. and Thomas, D. 2011 *The forgotten billion: MDG achievement in the drylands*. New York, UNDP-UNCCD. An examination of links between poverty, desertification and drylands. Open access: www.unccd.int/knowledge/docs/Forgotten%20Billion.pdf.

WEBSITES

www.desertknowledge.com.au   Desert Knowledge Australia is an organization that identifies projects that contribute to a social, economic and environmentally sustainable future for Australian drylands.

www.dry-net.org   Drynet is a networking and capacity building initiative involving Civil Society Organizations from across the globe.

www.gndri.net   GNDRI is a network of research institutions committed to dryland research.

www.icpac.net   the Climate Prediction and Applications Centre (ICPAC) for the greater Horn of Africa region has a focus on monitoring, capacity building and disaster risk management.

www.iucn.org/wisp   the World Initiative for Sustainable Pastoralism (WISP) is a global initiative that supports the empowerment of pastoralists to sustainably manage drylands resources.

www2.unccd.int   the UN Convention to Combat Desertification site.

POINTS FOR DISCUSSION

■ Solutions to the world's desertification problems lie not in more technical fixes, but in a greater appreciation of the social side of the issue. How far do you agree?

■ Is desertification inevitable in marginal areas on the edges of deserts?

■ Outline the ways in which salinization occurs in drylands and the problems it poses to human society.

■ How might global climate change affect the desertification issue?

■ Is desertification still a useful concept or has it become redundant?

■ How can land use become more sustainable in (a) the US Great Plains and (b) Sahelian Africa?

6

Oceans

Oceans cover nearly 71 per cent of the Earth's surface and contain 97 per cent of the planet's water. They comprise four main ocean basins – the Arctic Ocean, Atlantic Ocean, Indian Ocean and Pacific Ocean – with a larger number of marginal seas that all together form one single interconnected ocean system. This system plays a fundamental role in the Earth's atmospheric circulation and climate, and marine ecosystems have long provided people with fish and other resources, as well as being a receptacle for wastes. The perception of oceans as being infinitely vast has tended to undermine concern over their increasing exploitation by human society, despite their economic and ecological importance. Public interest tends to focus on certain marine issues, such as the killing of whales and pollution from plastics, while other more pervasive threats – from overfishing, for example – are given less attention. However, consideration of our relatively poor understanding of how oceans work, of the population dynamics of many marine biota and of the effects of certain pollutants, suggests that the arguments for caution in our use and abuse of oceans are strong. In this respect, ownership is an important aspect of many marine environmental issues. Most of the open oceans are common property resources, and the waters that come under national jurisdictions near coasts are subject to influences from other nation-states.

Progress towards managing the numerous issues associated with oceans has generally been slow when compared to many terrestrial environmental issues. This has been for many reasons, not least due to the difficulties associated with managing common property resources and the challenge of access to the enormous areas with no permanent human population. Another key factor has been the perception that the oceans are so vast that the human impact will always be insignificant. Hence, the first comprehensive assessment of the state of our planet's oceans, the World Ocean Assessment, was only published in 2016. However, its message is unambiguous (United Nations, 2016: 40): 'the world has reached the end of the period when human impacts on the sea were minor in relation to the overall scale of the ocean'.

FISHERIES

Global marine fish production is humankind's largest source of either wild or domestic animal protein, and is particularly important in the developing countries. The 1990s saw the end of a 40-year fishing boom. The worldwide catch increased more than four times between 1950 and 1990 but has remained at around the same level since then. In 2014, 82 million tonnes of fish were caught at sea (FAO, 2016b), and many marine biologists suspect that this level is close to the global maximum sustainable catch.

As with terrestrial food crops, we are heavily reliant upon a limited number of the 13,000 or so marine fish species. Ten species made up about 27 per cent of landings in 2013, and many regional stocks show signs of biological degradation. While in the early 1950s, 55 per cent of the world's fish stocks were under-exploited, by 2013 this proportion had declined to just over 10 per cent. The share of fish stocks being overfished (i.e. at biologically unsustainable levels) rose from 10 per cent in 1974 to just over 31 per cent in 2013 (FAO, 2016b).

The rapid development of global fisheries since the 1950s has led to several impacts beyond their overexploitation in certain regions and indeed the collapse of some fish stocks (Box 6.1). Principal among these other impacts are a decline in the biomass of certain predatory species, and the degradation of marine habitats, particularly of the seabed by trawling.

Continuous growth in global catches throughout the 1970s and 1980s was largely based on increased landings of pelagic (free-swimming) species that move in shoals because most demersal (bottom-dwelling) stocks were, and still are, fully fished. This trend was worrying because shoaling pelagic species tend to be at lower trophic levels than demersal species. In a global assessment using ecological modelling of how fish biomass has changed over the last 100 years, Christensen *et al.* (2014) concluded that the biomass of predatory fish in the world oceans declined by two-thirds over the period. In consequence, the biomass of prey fish increased over the last century. Combining these findings – the decrease of high trophic level fish and increase of low trophic level fish – indicates that the trophic structure of marine ecosystems has changed at the global scale.

What has happened globally over the last century has been called 'fishing down marine food webs', in which large, long-lived fishes at or near the top of marine food webs have been initially targeted and then, as these species become depleted, a transition is made to catching smaller, short-lived fishes at lower trophic levels. This results in the size and average trophic level of exploited fish gradually declining, as does the mean trophic level of catches from an ecosystem exploited in this way.

Trawling targets demersal fish but the fishing gear used also has a range of other environmental impacts on the seabed (Eigaard *et al.*, 2015). Short-term impacts include mortality of invertebrates, resuspension and redistribution of sediments, and physical destruction of habitats; long-term impacts may include changes in biodiversity and species composition, and reduction in habitat complexity. The impact of trawling on the seabed has been a topic of scientific research and public attention since the 1880s, although interest has heightened in recent decades. There are considerable variations in the seabed impacts of different fishing gears, and much debate about which fishing methods are most sustainable from an environmental and economic point of view.

Three main management failings have been identified as critical drivers of what many regard as a fisheries crisis:

1   Many of the world's fisheries, and particularly those on the high seas, are still a free-for-all. Free and open access encourages overfishing, as each boat, and each country, tends to catch what it can, leading to a tragedy of the commons situation. Fisheries bodies and agreements are not particularly effective, largely because their members are only weakly committed to cooperating on conserving stocks.

2   Many nations heavily subsidize their fishing fleets, encouraging unprofitable and unsustainable fishing, making overfishing even worse. Reducing or removing fleets would have short- to medium-term economic and social consequences that governments are reluctant to accept (Figure 6.1).

3   Some attempts to conserve fisheries (e.g. introducing closed seasons, or setting limits on the total catch but not on the amount that can be caught by each boat) may unintentionally allow fishing fleets to grow too much. If a fishery is profitable enough, owners will continue to build and operate boats even if they have to be kept in port for part of the year.

Overfishing encouraged by open access and a lack of governance characterized fisheries in the North Sea for much of the twentieth century. As a result, the exploitation of North Sea herring is seen as a classic example of poor fisheries management and the boom and bust consequences of the industrialization of fishing fleets, which led to its virtual extinction as a commercial fishery in the 1970s (Dickey-Collas *et al.*, 2010). Fishing in the North Sea was open-access outside 12-mile national coastal zones and stocks were exploited by fleets from at least 14 different countries until the start of 1977 when all countries around the North Sea extended their so-called Exclusive Economic Zones (EEZ) to 200 miles (322 km) from national coastlines. Abruptly, the North Sea ceased to be a free fishing area and national governments introduced conservation measures within their own areas in response to the collapse in the herring stock. A five-year closure of all fishing between 1977 and 1981 was imposed and generally well respected.

The economic consequences of the closure (bankruptcy for some ship-owners, loss of markets and changes in consumer behaviour) led to an awareness of the need for better management of the North Sea herring to ensure sustainable exploitation. Since 1997, this has been realized through an agreement between the EU and Norway to limit catches so that a core viable population of herring is maintained. This viable population is defined as the total weight of the fish in the stock that are old enough to spawn (the spawning stock biomass, or SSB) of more than 800,000 tonnes. Maintaining this SSB for North Sea herring has meant a management strategy that is both dynamic and iterative (Dickey-Collas, 2016).

Figure 6.1 European fishermen, like these in Portugal, have become subject to increasing restrictions on their activities in an effort to reduce pressure on overexploited fish stocks, but to date such restrictions have not been sufficient to prevent the decline in many stocks.

Identifying the impacts of overfishing can be a complex task for fisheries managers. The location of the world's marine fisheries is governed principally by the distribution of floating plants on which they depend for food, and the most important factor determining this phytoplankton production is the supply of nutrients, which is greatest in areas of upwelling. Climatic circulations can greatly alter the pattern of ocean circulation, however, and hence the pattern of fisheries. Thus, in relatively clear-cut examples of overfishing, natural variability can also play a part, as in the collapse of Atlantic cod off the coast of Newfoundland in the late twentieth century (see Box 6.1). In other cases it can be difficult to distinguish between the impacts of overexploitation and the natural variability of stocks, dependent upon species biology, migratory habits, food availability and natural hydrographic factors. Even with a relatively stable and resilient stock like the North Sea herring, a fishery that has been studied as much as any other comparable marine resource, the relative role played by environmental factors in the 1970s collapse is still not clear (Gröger *et al.*, 2009). The difficulties are greater still with some other shoaling pelagic species such as anchoveta and sardine, the populations of which are notoriously difficult to assess and manage because large interannual variations in recruitment are common, as well as longer-term directional changes lasting for decades or more (e.g. Chavez *et al.*, 2003).

Box 6.1 Collapse of Atlantic cod stocks off the coast of Canada

One of the most infamous examples of total collapse in a fishery occurred off the east coast of the Canadian island of Newfoundland where a moratorium on commercial fishing for Atlantic cod was declared in 1992. There is no doubt that overfishing played a significant part in this forced closure of the fishery after hundreds of years of exploitation. From the late 1950s, catches increased as trawlers began to exploit the deeper part of the fish stock, and landings of Atlantic cod in Newfoundland, which had very rarely exceeded 300,000 tonnes per year, peaked at more than 800,000 tonnes in 1968 (Figure 6.2). Catches of cod thereafter declined very rapidly through the 1970s and 1980s as stocks collapsed to extremely low levels. A small commercial inshore fishery was reintroduced in 1998, but catch rates again declined and the fishery was closed indefinitely in 2003.

Although the catch levels of the 1960s were certainly unsustainable, deleterious environmental conditions also appear to have played a role in the collapse of Newfoundland cod. It is likely that the cod stock did not have an adequate chance to recover from the massive overfishing of the late 1960s before sea temperatures in Newfoundland waters fell below normal for a long, continuous period, an environmental change that also affected the distribution of a number of other fish

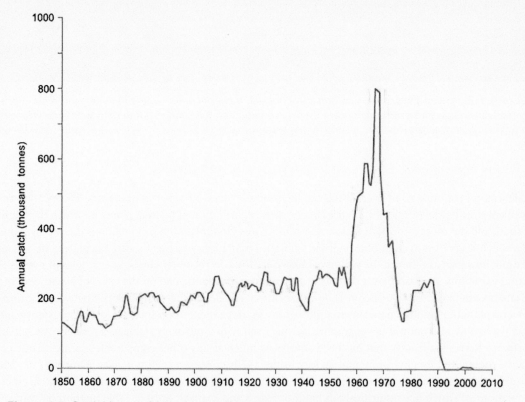

Figure 6.2 Catch history of Atlantic cod off Newfoundland, 1850–2003 (after MEA, 2005).

species in the area. The late 1980s and 1990s was also a time of reduced stocks of capelin, cod's primary food source.

The economic impacts in Newfoundland have been severe and numerous, irrespective of the ultimate drivers behind the fisheries collapse. The 1992 moratorium on cod, and the moratoria of other fish stocks that followed, led to the largest mass layoff of workers in the history of Canada, and the population of Newfoundland fell by 10 per cent in a decade because of shrinking birth rates and substantial outmigration.

Source: Schrank (2005).

Overfishing of large predatory fish species such as cod, and their subsequent collapses in population, can result in the restructuring of entire food webs, so-called trophic cascades that have been documented in several large marine ecosystems. The phenomenon was first identified in the north-west Atlantic where the sharp decline due to overfishing of cod and other predators – including haddock, hake, pollock and plaice – resulted in growing numbers of species that had been preyed upon by cod (small pelagic fish and invertebrate predators, primarily snow crab and shrimp). In turn, their prey (large herbivorous zooplankton) declined substantially. With fewer zooplankton, the phytoplankton they fed on increased. How long it might take for these reciprocal changes in biomass between successive trophic levels to be reversed is a matter for speculation. Some evidence of reversibility has been reported from the waters off Canada's east coast (Frank *et al.*, 2011) and we know from marine ecosystems elsewhere that population numbers of certain species can fluctuate considerably.

One of the best-known examples of dramatically fluctuating population numbers is the Peruvian anchoveta of the south-eastern Pacific. Disruption to population biomass occurs in some years due to the El Niño phenomenon, which intermittently leads to the near total collapse of coastal fisheries. El Niño refers to a distinct warming of the surface sea water in the south-eastern tropical Pacific that occurs around Christmas time along with a rapid change in atmospheric pressure over coastal South America and Indonesia referred to as the Southern Oscillation (the intimate link between El Niño and Southern Oscillation has generated the widely used acronym ENSO). These warmer surface waters occur when the South Pacific trade winds weaken, decreasing the amount of upwelling of water from depth along the coast of South America. This upwelling is very rich in nutrients, hence the abundant fish stocks, so when it cuts out during an El Niño, the number of fish available declines sharply (Figure 6.3).

The impact of this natural variability on the Peruvian anchoveta has also been exacerbated by overfishing. The Peruvian anchoveta was the most important single fishery in the world in the 1960s, providing up to 10 million tonnes of animal protein a year. Annual catches rose from a minor 59,000 tonnes in 1955 to peak in 1970 at just over 13 million tonnes (Figure 6.4). Just prior to 1972, the fishery was so intense that few of the fish caught were more than two years old, and many biologists expressed concern about the dangers of overfishing. The 1972/73 El Niño event caused a reduction in recruitment and subsequent intense exploitation of the remaining stock. Continued heavy fishing after 1972 further reduced the stock. By the late 1970s, annual catches had fallen to less than a million tonnes and directed commercial fishing was abandoned. In 1984, after another intense El Niño event in 1982/83, the landed catch was just 94,000 tonnes. Stocks appear to have collapsed due to the combination of short-term fluctuations caused by El Niño events, and intense fishing pressures (Caviedes and Fik, 1992). The fortuitous coincidence of favourable environmental conditions and controlled fishing has allowed the stock to recover since the historic lows

Figure 6.3 Sea lions off the Pacific coast of South America, part of a marine ecosystem that is intermittently affected by El Niño events. The sharp decline in fish stocks during an El Niño has knock-on effects throughout the food chain, leading to far fewer sea lions.

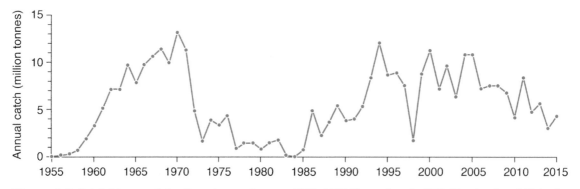

Figure 6.4 Catch history of the Peruvian anchoveta, 1955–2015 (from data in FAO *Yearbooks of Fisheries Statistics, Catchings and Landings*).

of the early 1980s, and landed tonnage was back at nearly 12 million tonnes in 1994. However, the high natural variability of the Peruvian anchoveta was apparent again late in the 1990s when the very intense El Niño event of 1997/98 resulted in a steep decline in total fish caught for 1998. Industrial fishing firms closed fishmeal plants and canneries, and kept many vessels idle, while small-scale artisanal fishermen shifted their efforts to catch tropical species only available during El Niño conditions. Nevertheless, their revenues were limited by drops in market prices and climate-related interruptions to transport on land. By September 1998, unemployment in Chimbote, the largest fishing port in Peru, was so severe that food had to be distributed to thousands of fishermen and their families (Pfaff *et al.*, 1999). Nevertheless, the fishery recovered rapidly after the extreme El Niño event of 1997/98 (Figure 6.4) thanks to a range of coping strategies developed over the years to deal with the effects of climate variability and extreme ENSO events (Schreiber *et al.*, 2011).

Similarly enormous differences between maximum and minimum catches occur for other shoaling pelagic species: the Japanese sardine, for example, appears to be affected by changes in the Kuroshio Current that last for periods of one or more decades. Natural variability similarly contributed to the dramatic collapse of the California sardine fishery, which boomed in the 1930s and 1940s but had almost completely disappeared by 1950. Once thought to be a classic example of the effects of overfishing, these large-scale variations which include decades of almost total absence are now thought to be naturally occurring. It appears that three major Pacific sardine stocks (Japan, California and Chile) fluctuate concurrently with climate–ocean regime shifts that involve large-scale changes in ocean temperatures (McFarlane *et al.* 2000): all three increased in abundance prior to a regime shift in 1947, decreased in abundance until the 1977 regime shift, increased again until 1989 when another regime shift resulted again in decreasing abundance.

The spasmodic population dynamics of the Peruvian anchoveta and Pacific sardine exhibit one of six types of marine fish population dynamics described by Spencer and Collie (1997) and shown in Table 6.1. The patterns observed are consistent with life history traits: stocks with low variability, for example, are generally slow-growing, demersal fish whereas high-variability stocks are typically small, pelagic species. The types of population dynamics shown in Table 6.1 should dictate suitable management policies. While the human impact upon steady-state species is relatively easy to detect and management of their exploitation

Table 6.1 Classification of marine fish population dynamics

Type	Examples	Comments
Steady-state	Pacific cod, Pacific hake	Typically longer-lived demersal species
Low-variation, low-frequency	Pacific Ocean perch (Eastern Bering Sea)	Very long-lived, slow-growing demersal species
Cyclic	Japanese anchovy, Atlantic mackerel, Pacific halibut	Relatively low-frequency cyclic patterns
Irregular	Bay of Fundy scallops, Norwegian herring, North Sea herring	Moderate levels of variation indicating any possible cycles tend to be irregular
High-variation, high-frequency	Dungeness crab, South African pilchard, Pacific herring	Possible cyclic patterns
Spasmodic	California sardine, Japanese sardine, Peruvian anchoveta	Typically small, pelagic species

Source: after Spencer and Collie (1997).

fairly straightforward, irregular and spasmodic stocks are particularly difficult to manage and do not lend themselves to current ideas of sustainable yields on a year-to-year basis. Management of spasmodic stocks may have to alternate between periods of active exploitation and periods of rebuilding, a process that can be facilitated by the existence of alternative fisheries. In some oceans, long-term data indicate a remarkable pattern of synchronized variability in growth and abrupt decline in many of the world's largest fish stocks. As Chavez *et al.* (2003) have highlighted in the Pacific, for example, periods of low sardine abundance have historically been marked by dramatic increases in anchovy populations, and vice versa.

It should be noted that much of the data illustrating these variabilities are for catch landings, and that these are probably only rough indicators of actual population size, but similar patterns are evident where actual biomass assessments have been made. It is possible that these synchronous patterns occur by chance, but it is more likely that they are linked by a climatic mechanism, in the same way that ENSO events in the South Pacific appear to affect weather patterns far removed from the area by so-called 'teleconnections'. It is unlikely that sea temperatures alone could explain the global synchrony in fish population variations, but if a feasible mechanism or mechanisms can be identified (Alheit and Bakun, 2010), it would have important implications for the management and perhaps prediction of global fish supplies.

The Southern Ocean

Ignorance of the workings of marine ecosystems makes management of ocean resources difficult, and this is particularly the case in the Southern Ocean surrounding Antarctica. Historically, the Southern Ocean has been the site of several phases of species depletion caused by overexploitation, illustrating the process of 'fishing down the food web' (Figure 6.5). In the nineteenth century, seals were pushed to the edge of extinction for their fur and oil. They were followed by the great whales in the first two-thirds of the last century (see below), and in the 1970s and 1980s Southern Ocean finfish stocks were heavily exploited, with the marbled Antarctic rock cod and mackerel icefish showing particular signs of decline. Most recently, attention has turned to Antarctic krill, the most abundant living resource of the Southern Ocean. Commercial fishing of the small shrimp-like crustacean began in the early 1970s, but annual catches have not risen above the peak of over half a million tonnes in 1980–81.

The reason for the relatively modest krill catches is the establishment of a limit on harvesting by the Commission for the Conservation of Antarctic Marine Living Resources (CCAMLR), an international body set up in 1981. The CCAMLR aims to manage the sustainable use of Antarctic marine living resources in an attempt to prevent the dangers of overexploitation associated with the common ownership of the Southern Ocean. The devastation of seal and whale populations, and the depletion of some finfish, were clear examples of the tragedy of the commons. Since krill are considered a keystone species, at the centre of the Antarctic ecosystem – providing the major food source for many larger animals such as seals, penguins (Figure 6.6), fish, squid, seabirds and baleen whales – there is a fear that overharvesting of krill might lead to the collapse of the entire ecosystem.

The CCAMLR has a precautionary and ecosystem-based approach to the management of the fishery, but its task is hampered by a lack of knowledge and information on the workings of the Antarctic ecosystem in general, and of krill in particular. Currently the fishery catches about 300,000 tonnes annually, all from the South Atlantic, where the precautionary catch limit is 5.6 million tonnes a year, so it remains one of the ocean's largest known underexploited stocks. However, as Murphy *et al.* (2017) put it: 'Although there is detailed information available on aspects of the krill life-cycle in some regions, there is little quantitative information on the distribution, abundance or basic biology of krill throughout much of the Southern Ocean.'

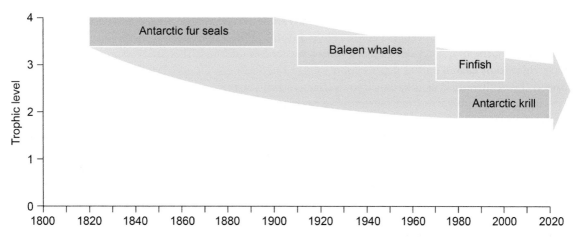

Figure 6.5 Two centuries of biotic resource exploitation in Antarctica illustrating the process of 'fishing down the food web' (after Ainley and Pauly, 2014).

Figure 6.6 Adelie penguins near the Antarctic Peninsula, Antarctica. Like most penguins, the adelie feed primarily on krill.

The abundance of krill appears to fluctuate greatly from one year to the next, but overall their numbers seem to have declined in recent decades as temperatures in the Antarctic have risen since the 1940s and, consequently, the frequency of extensive winter sea-ice has declined. This lower abundance of krill may have significant effects on the marine food web in the Southern Ocean: Trivelpiece *et al.* (2011) have linked the decline in krill abundance to declines in krill-eating penguin populations on the Antarctic Peninsula. The problems of marine resource managers are further exacerbated by a recent explosion in the population of Antarctic fur seals. Evidence gleaned from seal hairs in lake sediment cores taken from one Antarctic island suggests that the current seal population explosion exceeds the range of natural variability (Hodgson and Johnston, 1997), implying that the ecosystem is still responding to the dramatic human impacts of the last few centuries. An important emerging issue is ocean acidification (see below and p. 288): Southern Ocean ecosystems are expected to be rapidly affected because cold waters enhance the solubility of CO_2 and $CaCO_3$. These developments simply underline the importance of understanding the biology and ecology of the entire marine ecosystem before informed decisions can be made about sustainable management.

WHALES

Of all the animals on this planet, none symbolizes the human impact on the environment more powerfully than the whale. There are ten species of so-called 'great whales': the blue, fin, humpback, right, bowhead (or Greenland), Bryde's, sei, minke, grey and sperm whales, and all have been hunted at least to the brink of extinction. Although people have hunted whales for thousands of years, modern commercial whaling for oil, and in time also for meat, is traced back to the eleventh or twelfth centuries when groups of Basque villagers began to hunt right whales from small rowing boats in the Bay of Biscay. Whaling had become big business by the eighteenth and nineteenth centuries, and large whaling fleets from many nations, including England, North America, Japan and the Netherlands, made enormous profits from their sorties.

Commercial whaling focused on the great whales because they are the largest, most profitable species. The history of commercial whaling has seen intensive catches of a species until it has been driven to the brink of extinction, at which point the whalers shifted their attention to a different species in a new area of ocean. The populations of most of the great whale species are now so low as to be effectively extinct from a commercial viewpoint; almost 3 million whales were killed in the twentieth century alone, and several species could ultimately disappear altogether.

The Greenland bowhead whale, for example, was first exploited in 1610 and reduced to near extinction by the mid-nineteenth century. The Arctic bowhead whales, discovered in the mid-nineteenth century and quickly exploited, were almost totally depleted by 1914. As many as 30,000 sperm whales were killed in the Atlantic during peak years in the eighteenth and nineteenth centuries, and by the 1920s the Atlantic sperm whale was practically extinct.

It was early in the twentieth century that whalers began to exploit the previously untouched waters of the Southern Ocean where several species go to feed on krill. The catch rates were further increased with the introduction of factory ships in the 1920s. The humpback was the first species targeted in Antarctic waters, but as its numbers declined, the blue whale became the main quarry. An estimated 250,000 blue whales were quickly killed, leaving around 1000 today. In the second half of the twentieth century it was the turn of the fin and sperm whales. Annual catches of fin whales regularly topped 30,000 in the 1950s. Since the mid-1980s, the scale of whaling has been greatly reduced (see below) and other threats to whales have become more significant (Clapham, 2016). These include ship strikes and entanglement in fishing gear, as well as issues such as ocean noise for which population-level impacts are largely unclear.

Protecting whales

There have been a number of attempts to control the whaling industry over the years. The earliest effort was the Convention for the Regulation of Whaling, drawn up in 1931. But the Convention had little beneficial effect. One of its regulatory instruments, the so-called blue whale unit, effectively lumped all great whales together, so that one unit was made up of successively greater numbers of smaller species: two fin, three humpback, five sei, and so on. Countries were spurred by the profit motive to fulfil their quotas with a few large species rather than a larger number of smaller ones, resulting in a rush on the blue whales. In 1946, this Convention was replaced by the International Convention for the Regulation of Whaling (ICRW), which established the International Whaling Commission (IWC) to discuss and adopt regulations. The IWC has had a chequered history, overseeing as it did the all-time peak in whaling, in the 1960–61 season, when 64,000 whales were caught, but after increasing pressure from outside, the IWC declared a moratorium on all commercial whaling, which entered into force in 1986 and remains in place. Today, the status of the great whales varies widely: some species or populations are recovering from exploitation (e.g. Bryde's whales, minke whales), while a few others remain critically endangered (e.g. North Pacific and North Atlantic right whales).

The whaling moratorium has not been a complete success, with around 1000–2000 caught (mainly Bryde's, sei and minke whales) every year since 1986, as shown in Figure 6.7. Most of these have been caught by a small number of countries – notably Japan, Norway and Iceland – that have defied both the IWC and world opinion. Iceland and Norway lodged official objections to the moratorium and have continued to hunt commercially, while Japan and Iceland have continued to hunt whales for 'scientific purposes', under a controversial exemption clause in the Whaling Convention. Conservationists have argued that effective non-lethal techniques for studying whales have been developed, which do away with the need to butcher the object of the scientists' enquiries. It may come as no surprise to learn that the meat and oil

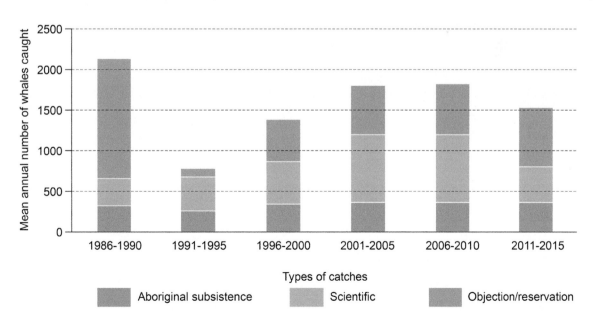

Figure 6.7 Mean annual number of whales caught since the IWC whaling moratorium.

of the whales killed for scientific research is still processed and sold commercially. Some conservationists have also voiced fears for the future viability of the moratorium based upon Japan's powerful position as an aid donor. It is not inconceivable that Japan could use aid as a lever to persuade developing-country IWC members to back its position.

Under IWC regulations some aboriginal subsistence whaling is also permitted in a small number of places (Alaska, Chukotka, Greenland and Bequia), which accounts for a few hundred whales each year. The policy is designed to allow a sustainable harvest of certain whale species by small human populations using traditional methods (Figure 6.8), an approach that effectively employs the I = PAT equation.

Various attempts have been made to dissuade the countries that continue to kill whales commercially and/or scientifically. In 2013, Australia challenged Japan over its research whaling in the International Court of Justice (ICJ), arguing that Japan's authorization of lethal taking of whales for research served no legitimate scientific purpose. The ICJ's judgment, issued a year later, found that the permits being issued by Japan were not for scientific research purposes as provided for in the ICRW. In response, Japan withdrew from the jurisdiction of the ICJ and the Japanese whaling fleet resumed whaling in 2015 despite the ruling (Butler-Stroud, 2016).

Other approaches have also been taken to promote whale conservation further. The Southern Ocean Whale Sanctuary, in which all whaling is banned, was established in 1994, although Japan has a standing

Figure 6.8 An Inuit in Greenland preparing to hunt whales from his kayak with a handheld harpoon and a sealskin float. Some whaling is allowed under IWC rules but only using such traditional techniques.

objection to the sanctuary. Organized whale-watching tours have also been vigorously promoted in recent years as a viable non-consumptive use of these marine mammals. Whale-watching is clearly a more sustainable method of exploiting whales than simply killing them, and the income generated from this form of wildlife tourism can make the organization of these ecotours very profitable in economic terms (Figure 6.9). The promotion of whale-watching has also served a political role in the anti-whaling debate, having been strongly backed by many NGOs seeking to end the practice of whaling (Neves, 2010).

Commercial whale-watching began in the mid-1950s in southern California where migrating grey whales were the attraction. Since then it has become a major tourist industry. In 2008, some 13 million people participated in whale-watching trips in 119 countries and territories, generating total expenditure of US$2.1 billion on food, travel, accommodation and souvenirs (O'Connor *et al.*, 2009).

In the areas of the world where whale-watching is well organized, ships carrying ecotourists abide by strict regulations designed to cause minimal disturbance to the subjects of their searches. Engines must be cut within a certain distance and no vessel is allowed to approach too close. Many of the trips are made on (non-lethal) research ships, with the income from carrying tourists helping to finance the research itself. In some parts of the world, old whaling vessels themselves are used, a development that offsets one of the arguments proposed in favour of continuing the killing of whales: its importance for employment. Nonetheless, not all whale-watching is well organized and repeated exposure to tourism causes disturbances to

Figure 6.9 A whale-watching tour off the coast of British Columbia, Canada. Whale-watching has been demonstrated to be a viable economic alternative to whaling.

the creatures of interest, leading to calls for better regulation and standards for the sustainable management of commercial whale-watching operations (Higham *et al.*, 2016).

Whale-watching tours have even started in Japan, the first setting out from Tokyo for the Ogasawara Islands in 1988. Despite stern opposition from both government officials and whalers, Japan now has one of the world's fastest-growing whale-watching industries. If the trend continues, it will not be long before live whales in Japanese waters are worth more in economic terms than dead ones. It is a trend the IWC hopes to encourage with its backing of whale-watching as a sensible, more humane way of treating these creatures.

POLLUTION

The wide variety of sources, causes and effects of human-induced pollution of the marine environment is indicated in Table 6.2. Pollution can occur because of catastrophic events such as accidents and/or as a chronic problem, due to regular discharges from a sewage outlet for example. Different forms of pollution present challenges over different timescales because of the variable times it takes for pollutants to be assimilated into the marine environment. An indication of the varying persistence of marine pollutants is given in Figure 6.10. Since land-based sources tend to be dominant, the worst effects of these pollutants

Table 6.2 Primary causes and effects of marine pollution

Type	Primary source/cause	Effect
Nutrients	Runoff approximately half sewage, half from upland forestry, farming, other land uses; also nitrogen oxides from power plants and cars particularly	Feed algal blooms in coastal waters. Decomposing algae deplete water of oxygen, killing other marine life. Can spur toxic algal blooms (red tides), releasing toxicants into the water that can kill fish and poison people
Sediments	Runoff from mining, forestry, farming, other land uses; coastal mining and dredging	Cloud water. Impede photosynthesis below surface waters. Clog gills of fish. Smother and bury coastal ecosystems. Carry toxicants and excess nutrients
Pathogens	Sewage; livestock	Contaminate coastal swimming areas and seafood, spreading cholera, typhoid and other diseases
Persistent toxicants (e.g. PCBs, DDT, heavy metals)	Industrial discharge; wastewater from cities; pesticides; seepage from landfills	Poison or cause disease in coastal marine life. Contaminate seafood. Fat-soluble toxicants that bioaccumulate in predators
Oil	46%, runoff from cars, heavy machinery, industry, other land-based sources; 32%, oil tanker operations and other shipping; 13%, accidents at sea; also offshore oil drilling and natural seepage	Low-level contamination can kill larvae and cause disease in marine life. Oil slicks kill marine life, especially in coastal habitats. Tar balls from coagulated oil litter beaches and coastal habitat

Table 6.2 continued

Type	Primary source/cause	Effect
Introduced species	Several thousand species in transit every day in ballast water; also from canals linking bodies of water and fishery enhancement projects	Outcompete native species and reduce marine biological diversity. Introduce new marine diseases. Associated with increased incidence of red tides and other algal blooms
Plastics	Fishing nets; cargo and cruise ships; beach litter; wastes from plastics industry and landfills	Discarded fishing gear continues to catch fish. Other plastic debris entangles marine life or is mistaken for food. Litters beaches and coasts

Source: after Weber (1994: 47, table 3.3); and GESAMP (2009).

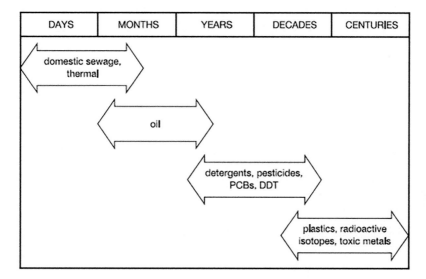

Figure 6.10 Persistence of pollutants in the marine environment (after Meadows and Campbell, 1988).

are concentrated in coastal waters adjacent to areas with large population densities, and in similarly placed regional seas where the mixing of waters is limited. The open ocean, by contrast, is relatively little affected. In this chapter, the focus is on pollution and its effects outside the coastal zone, while coastal pollution is covered in Chapter 7, although in practice the distinction is somewhat arbitrary since coastal pollutants also reach the wider context of the high seas and oceans. Most of the marine issues associated with global warming are covered in Chapter 11.

The Joint Group of Experts on the Scientific Aspects of Marine Environmental Protection (GESAMP), an international group sponsored by the UN, has reviewed the state of our knowledge of pollution in the open oceans (GESAMP, 2009) and concluded that we have a good understanding of the current environmental significance of pollutants on broad temporal and spatial scales, and noted their planet-wide distribution in some cases. Ocean acidification was highlighted as a priority issue. Human-induced increases in the atmospheric inputs of carbon dioxide, and to a lesser extent sulphur dioxide and nitrogen oxide, have generated considerable concern about their roles in ocean acidification. The response of marine ecosystems, coral reefs and fisheries to this acidification is little known but the impacts will be both direct and indirect (see p. 288).

Contamination of the oceans by plastics and other synthetic, non-biodegradable materials from both land- and sea-based sources is another pervasive marine pollution issue of growing concern. Use of plastics globally has been rising almost exponentially since the 1950s and floating plastic is now ubiquitous in global oceans, although highly variable in concentration. Plastic is very durable – it may persist for 200–400 years – and much of it floats in sea water. It is therefore distributed throughout the global marine surface environment, but ocean circulations have created particular concentrations of plastic in each of the five subtropical gyres (Figure 6.11). Once it enters the marine environment, plastic can have a number of consequences (Gregory, 2009). Many marine species – including birds, turtles, seals, whales and dolphins – face the risk of entanglement, ingestion and suffocation, but less frequently recognized problems include hazards to shipping and fisheries. Floating pelagic marine debris has also become a common means by which aggressive alien and invasive species are dispersed. Tiny particles of plastic debris (frequently called microplastics) are so pervasive in marine ecosystems that we find them in seafood and table salt. Borrelle *et al.* (2017) draw a parallel between global plastics production and global carbon emissions, two pollution issues that have progressed at a similar pace in terms of emissions since 1950, but which offer a stark contrast in terms of policy development. They present a strong argument for an international agreement on marine plastic pollution.

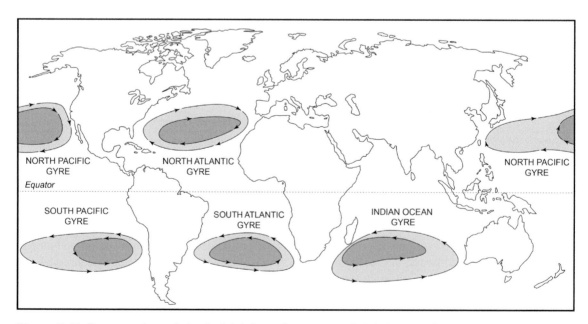

Figure 6.11 Concentrations of plastic debris in surface waters of global oceans (after Cózar *et al.*, 2014).

Pollutants that reach the marine environment from terrestrial sources via atmospheric pathways are also difficult to quantify, not least because they can be transported many thousands of kilometres from source areas before being deposited. The wide dissemination throughout marine ecosystems of persistent organic pollutants, or POPs, has generated increasing international concern because these classes of toxic chemical are long-lived, they accumulate in living tissues, and tend to become more concentrated up the food chain (so-called bioaccumulation). Hence, they pose a health risk to marine biota at higher trophic levels and to human consumers of some sea foods.

POPs are man-made chemicals used in industrial and agricultural applications (Table 6.3) and have been regulated through the Stockholm Convention on Persistent Organic Pollutants since 2004. Although many of these substances have now been banned in numerous parts of the world, POPs from industrialized areas of Europe and North America in particular have contaminated marine environments all over the northern hemisphere, and have been detected in even the most remote and inaccessible habitats on Earth. The findings of Jamieson *et al.* (2017) are particularly alarming: they measured very high levels of POPs in fauna from two of the Pacific's deepest ocean trenches (>10,000 m), indicating that the contaminants had bioaccumulated up the food chain and inferring that these pollutants are pervasive across the world's oceans and to full ocean depth. Analysis of POPs found in the fat or liver of top Arctic predators, such as the polar bear, also demonstrates their bioaccumulation up the marine food web, as Letcher *et al.* (2018) demonstrate in the Canadian Arctic.

Inputs of pollution from shipping, a principal source to the open oceans, arise mainly from operational activities and from deliberate discharges. Oil is the most obvious pollutant involved, but biocides from anti-fouling paints leach into ocean water along shipping routes, wastes are dumped and accidents also contribute.

Oil affects marine life both because of its physical nature (i.e. by physical contamination and smothering), and because of its chemical components (e.g. toxic effects). Plants and animals may also be affected by clean-up operations or indirectly through damage to their habitats. The animals and plants most at risk from physical smothering are those that come into contact with a contaminated sea surface, including marine mammals and reptiles, birds that feed by diving or form flocks on the sea surface, marine life on shorelines, and animals and plants in mariculture facilities. Lethal concentrations of toxic components

Table 6.3 POPs that are widely distributed throughout marine ecosystems

POP	Uses
PCBs (polychlorinated biphenyls)	Used for a variety of industrial applications from the 1950s to the early 1970s
DDTs (dichlorodiphenyltrichloroethane and related chemicals)	Used as an insecticide to increase agricultural production mainly from the 1950s to the early 1970s. Although its application in agriculture is now banned, use of DDT to combat malaria-carrying mosquitoes is still recommended by the World Health Organization, and DDT is used for this purpose in some tropical countries
HCHs (hexachlorocyclohexane isomers)	Organochlorine insecticides used from the 1950s to the 1970s in many countries and as late as the 2000s in some countries

leading to large-scale mortality are relatively rare, localized and short-lived because the most toxic components in oil tend to be those lost rapidly through evaporation when oil is spilt. However, lower concentrations of oil can impair the ability of individual marine organisms to reproduce, grow, feed or perform other functions. Some animals that filter large volumes of sea water to extract food, such as oysters, mussels and clams, are particularly likely to accumulate oil components.

Most oil spills from shipping worldwide result from routine operations such as loading, discharging and bunkering – activities that normally take place in ports or at oil terminals. Most of these spills are small, while accidents tend to give rise to much larger spills, with a fifth involving quantities in excess of 700 tonnes. A number of factors have been responsible for the decline in the number of major spills (>700 tonnes) in recent decades (Jernelöv, 2010). They include improvements to modern tankers – with double hulls lowering spill risk after minor impacts and sectioned tanks meaning leakage will not result in the whole cargo being lost – as well as improvements to navigation brought by the use of GPS (Global Positioning System). The number of major spills over the period 1970–79 averaged about 25 a year, but this average fell below two a year in the second decade of the present century (Figure 6.12). Likewise, the total amounts of oil spilled from tankers in the world's oceans have tended to be lower in recent years. Tankers are not the only source of such major spillages. Blow-outs at oil drilling platforms also contribute. Two particularly large spills have occurred in the Gulf of Mexico in recent decades: an accident at the Ixtoc 1 platform in

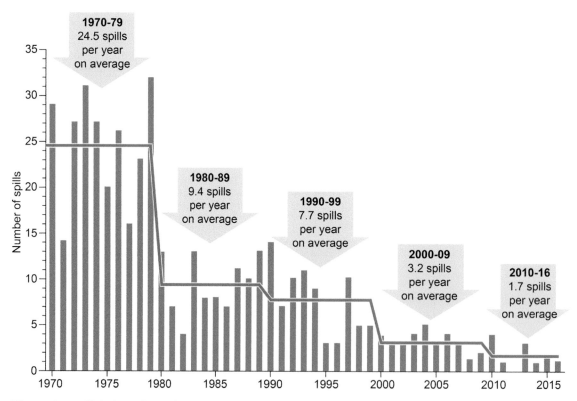

Figure 6.12 Global numbers of oil spills over 700 tonnes, 1970–2016 (www.itopf.com/knowledge-resources/data-statistics/statistics/, accessed December 2017).

1979 released more than 400,000 tonnes of oil, and a similar event at the Deepwater Horizon drilling rig released nearly twice that amount in 2010.

Another contaminant found in many oceans and seas is tributyltin (TBT), which has been widely used in anti-fouling paints applied to ships' hulls. TBT has been linked to problems of reproduction in aquatic creatures. Exposure to TBT has been found to interfere with the hormonal systems of whelks, producing 'gender-bending' effects whereby individuals develop the sexual characteristics of the other sex. The effect has been documented in seas and oceans. The OSPAR Commission (2000) noted impacts on dogwhelks and common whelks in harbours in British and Irish waters, in northern Portugal, north-west Spain, Iceland, Norway, Svalbard and Kattegat.

The application of TBT on small vessels (less than 25 m in length) has been prohibited since 1990 and some biological recovery in the north-east Atlantic has been observed in areas of small-boat use, but the gender-bending effects of TBT occur even at very low concentrations. The OSPAR Commission (2000) found a significant correlation between shipping intensity and TBT levels in marine sediments and the occurrence of gender-bending effects, suggesting that larger ships using TBT anti-fouling paints represent the main source of TBT in the marine environment. The International Maritime Organization's ban on the use of TBT and other similar compounds in anti-fouling treatments on ships longer than 25 m came into force in 2008 but the legacy of historic TBT inputs will continue for some years to come in many areas. Further, TBT replacements used in anti-fouling paints are also intrinsically toxic, suggesting that their effects will also be felt in marine food chains (Turner, 2010).

The International Maritime Organization (IMO) is the United Nations specialized agency with responsibility for the safety and security of shipping and the prevention of marine pollution by ships. It has negotiated a number of international instruments to address state, ship, and port responsibilities, including the London Dumping Convention adopted in 1972, which has largely been responsible for the gradual decrease in the amounts of industrial wastes and sewage sludge dumped at sea since it came into force. However, the oceans are still seen by many as an attractive place to dispose of wastes due to their vast diluting capacity, although the long residence times of some types of waste (Figure 6.10) imply that any contaminants will be removed only very slowly (Jickells *et al.*, 1990). Indeed, the dilute-and-disperse philosophy becomes questionable for persistent pollutants that take very long periods to become harmless (e.g. radionuclides, heavy metals, certain pesticide residues and plastics) and that can be accumulated up the food chain.

The dumping of high- and medium-level radioactive wastes at sea does not occur and the dumping of low-level packaged radioactive waste at sea was voluntarily banned in 1982. During the period 1949–82, the UK and the USA made most use of the 26 sites used in the Atlantic and 21 sites in the Pacific. The total amount dumped was less than 0.1 per cent of the amount that reached the oceans due to atmospheric nuclear weapons testing between 1954 and 1962. This, in turn, is only 1 per cent of the amount found naturally in the oceans. However, the mix of radionuclides involved, their rates of decay and the effects upon biological life are different in each case, and GESAMP (1990: 50) warned that 'dumping cannot be considered safe just because releases of radionuclides are small compared with the natural incidence of radionuclides in the environment'.

A similar statement might be applied to the disposal of low-level liquid waste into the oceans, from nuclear reactors and fuel reprocessing plants, particularly since the discharges are made from point sources. In some parts of the world, these discharges have been made continuously over several decades; the nuclear facilities at Sellafield in north-west England, for example, have been discharging low-level liquid radioactive waste into the Irish Sea since the early 1950s. Such discharges are carefully monitored and limits are set with the aim of maintaining bioaccumulation in organisms well below international standards for various radionuclides. Nevertheless, discharges of nuclear waste to marine and other environments continue to be a source of considerable controversy (see Chapter 18). It is also worth noting that nuclear material for disposal is in many cases derived from military activities, and information on this source of radioactive waste to the oceans, as with other disposal options, is on the whole not available.

REGIONAL SEAS

Numerous regional seas are showing signs of degradation, in many cases with land-based sources of pollution being most prevalent. The difficulties of managing environmental problems in regional seas are similar to those posed by other global 'commons', they are multi-disciplinary and they require international cooperation. In contrast to the relatively well-developed international agreements governing marine sources of pollution, such as the London Dumping Convention, agreements to limit problems that emanate from coastal landmasses tend to be conspicuous by their absence. Some progress has been made in this direction, however, in the form of a series of action plans for regional seas, coordinated by UNEP (Table 6.4).

The Mediterranean Sea was the first regional sea for which such a plan was formulated, in 1975. The Mediterranean Action Plan (MAP) was designed to address pollution in particular, focusing on land-use planning, urban management and pollution prevention, and created a framework for regional consultation

Table 6.4 UNEP Regional Seas Programme

Regional sea area	Year when action plan adopted
Mediterranean	1975
Arabian/Persian Gulf	1979
Caribbean	1981
East Asia	1981
South-east Pacific	1981
Antarctic	1982
Red Sea and Gulf of Aden	1982
West and Central Africa	1984
Eastern Africa	1986
Pacific	1990
Caspian	1992
Baltic	1992
North-west Pacific	1994
South Asia	1995
North-east Atlantic	1998
North-east Pacific	2002
Black Sea	2009

and decision-making which has helped the nation-states involved to define a common approach and take action locally on environmental issues. A pollution monitoring and research programme (MED POL) has also been established, as well as the so-called Blue Plan, which aims to reconcile the varying needs of environment and development. The Horizon 2020 (H2020) initiative to reduce pollution of the Mediterranean by 2020 was endorsed in 2005, but the high level of human impact in the Mediterranean results from multiple drivers, including climatic stress (through increasing temperature and acidification), demersal fishing and ship traffic, in additional to pollution from land (Micheli *et al.*, 2013). Whatever the source of the impacts, the difficulties of organizing successful cooperation between the countries involved – the MAP unites the EU and 21 countries bordering the Mediterranean – have been serious. Indeed, as one study of the Mediterranean concluded, marine governance has an 'adequate legal-institutional structure that exists alongside weak political integration and marked economic differences' (de Vivero and Mateos, 2015: 204). These authors highlight the marked differences in economic development along a north–south divide among Mediterranean countries, with associated limitations in technical and financial capacity, as being the continuing central challenge.

Among many issues raised by the problems of the Mediterranean is that of liability. Most of the pollution entering the sea comes from the northern industrialized countries but the effects are felt, to varying degrees, by all. Should the 'polluter pays' principle be applied? Or should countries pay according to the impact they experience? Other political difficulties arose in the 1990s following the break-up of the former Yugoslavia and continued civil unrest in some of the new Balkan states. Projections for the countries bordering the Mediterranean indicate that the potential threats to the marine environment will continue to increase as one of the main causative factors – the size of the human population – continues to rise. There is no doubt, therefore, that continued successful international cooperation is vital for the management of the Mediterranean, as for other marine environments, and thus such initiatives need to succeed.

The ecosystem of the Black Sea and the adjoining Sea of Azov is another regional sea that has experienced multiple pressures in recent times (Micheli *et al.*, 2013), in this case resulting in a very severe decline in ecosystem health. The outcome has been widespread eutrophication and a fundamental change in the marine ecosystem which has, in turn, had severe repercussions for fisheries and the sea's amenity value (Borysova *et al.*, 2005).

A major part of the Black Sea, one of the world's largest inland seas, became critically eutrophic following a surge in nutrient loads entering the sea over the last few decades of the twentieth century. Much of the increase came from higher loads in rivers, particularly the Danube, reflecting the widespread use of phosphate detergents and agricultural intensification. In the mid-1990s, the overall annual input of nutrients from human activity into the Black Sea amounted to 50,500 tonnes of phosphorus and 647,000 tonnes of nitrogen from riverine and coastal sources such as sewage outlets, with probably another 400,000 tonnes of nitrogen arriving via the atmosphere (Borysova *et al.*, 2005). Large phytoplankton blooms have become increasingly common in the Black Sea as a result (Figure 6.13).

The north-western portions of the sea were particularly affected, exhibiting a regime shift from a diverse ecosystem that supported highly productive fisheries to a eutrophic plankton culture in a marine environment that had become unsuitable for most higher organisms. Increased flows of phosphorus and nitrogen via the Danube do not account for the whole story, however. After the damming of the Danube at the Iron Gates I barrage in 1970 and Iron Gates II in 1983, the dissolved silicate load of the river was reduced by about two-thirds, which probably triggered a dramatic shift in the composition of phytoplankton species in the Black Sea (Humborg *et al.*, 1997). Other rivers in the Black Sea watershed have also been significantly affected by dam construction – Borysova *et al.* (2005) show no fewer than 43 dams on rivers in the Black

Figure 6.13 A satellite image of a vast phytoplankton bloom swirling across the Black Sea in May 2017.

Sea catchment area – and substantial net decreases in silicates reaching the sea from several of these rivers are also likely.

The resulting biogeochemical changes, in combination with intense overfishing, resulted in a trophic cascade affecting species at four trophic levels (Daskalov, 2002). Beginning in the early 1970s, industrial fishing depleted stocks of pelagic predators (bonito, mackerel, bluefish, dolphins), leading to increased abundance of planktivorous fishes, reduced zooplankton and a doubling of phytoplankton. Oil spills, a climate change-induced temperature increase and invasion by an alien jellyfish species introduced through ballast water also contributed to the worsening situation. From the socio-economic perspective, the Black Sea fishing industry, which supported some 2 million fishermen and their dependants, suffered a very sharp decline and the attractions of the sea's beaches have been severely depleted. An assessment of the economic cost of the environmental deterioration to the regional fishery and tourism industries is put at US$500 million annually (Borysova *et al.*, 2005).

Steps towards dealing with the ecological problems began with a legal Convention for the Protection of the Black Sea which entered into force in 1994. Recent years have seen some signs of ecological improvement with decreasing numbers of algal blooms and reduced areas of water depleted in oxygen. The beginnings of system recovery appear to be linked to a reduction in intensive farming practices after the economic collapse of several surrounding socialist republics in the early 1990s, with a consequent reduction in nutrient loading also linked to more recent new sewage treatment plants. Many of the Black Sea's catchment countries are undergoing rapidly changing economic and political conditions, making continued recovery uncertain (Langmead *et al.*, 2010), and there is no doubt that complete rehabilitation will take many years.

FURTHER READING

Baum, J.K. and Worm, B. 2009 Cascading top-down effects of changing oceanic predator abundances. *Journal of Animal Ecology* 78: 699–714. A round-up of evidence for trophic cascades in marine ecosystems.

Chang, S., Stone, J., Demes, K. and Piscitelli, M. 2014 Consequences of oil spills: a review and framework for informing planning. *Ecology and Society* 19(2). An overview of potential oil spill consequences with a focus on tanker accidents. Open access: www.ecologyandsociety.org/vol19/iss2/art26/.

Christensen, V. and Maclean, J. (eds) 2011 *Ecosystem approaches to fisheries: a global perspective*. Cambridge, Cambridge University Press. An accessible overview of ecosystem-based fisheries management.

Dorsey, K. 2014 *Whales and nations: environmental diplomacy on the high seas*. Seattle, University of Washington Press. A study of the twentieth-century whaling industry and the rise of the environmental movement.

Frid, C.L.J. and Caswell, B.A. 2017 *Marine pollution*. Oxford, Oxford University Press. A comprehensive study of marine pollution issues and how they can be managed.

Schreiber, M.A., Ñiquen, M. and Bouchon, M. 2011 Coping strategies to deal with environmental variability and extreme climatic events in the Peruvian anchovy fishery. *Sustainability* 3: 823–46. Lessons from how the Peruvian fishing industry copes with variability. Open access: www.mdpi.com/journal/sustainability.

WEBSITES

fisherysolutionscenter.edf.org   The Fishery Solutions Center provides a collection of research-driven materials on improving fisheries management.

iwc.int   the International Whaling Commission.

www.asoc.org   the Antarctic and Southern Ocean Coalition (ASOC) is the foremost NGO voice for Antarctic environmental protection.

www.itopf.com   the International Tanker Owners Pollution Federation site with information, data and country profiles on marine oil spills.

www.oceansatlas.org   a geographically based information portal on ocean issues.

www.worldoceanassessment.org   site of the first World Ocean Assessment.

POINTS FOR DISCUSSION

- How can we manage the oceans' biological resources when we know so little about them?
- How can we justify efforts to save whales when so many people in the world live in destitution?
- If limited aboriginal subsistence whaling is permitted under IWC regulations, why not a small commercial catch?
- The environmental problems of regional seas like the Mediterranean and the Black Sea are good examples of the tragedy of the commons. How far do you agree?
- For an ocean of your choice, assess which is the more damaging to its ecology: pollution or overfishing.
- Why has the precautionary principle not been applied to exploitation of most of the planet's major fish stocks?

7 *Coastal issues*

LEARNING OUTCOMES

At the end of this chapter you should:

- Understand how the management of physical change along coastlines can adopt two fundamentally different approaches.
- Appreciate how some scientists think that ecological extinctions caused by overfishing precede all other pervasive human impacts on coastal ecosystems.
- Appreciate the wide range of sources, causes and effects of coastal pollution.
- Know the main threats coral reefs face from human activities.
- Understand how ENSO events and global climate change are linked to coral bleaching.
- Know the main threats mangroves face from human activities.
- Appreciate some of the ecosystem services provided by coral reefs and mangroves.

KEY CONCEPTS

erosion, sedimentation, sea-level change, hard engineering, soft engineering, conservative and non-conservative pollutants, no-take areas, coral bleaching, Integrated Coastal Zone Management

---

The coastal zone, which can be defined as the region between the seaward margin of the continental shelf and the inland limit of the coastal plain, is among the Earth's most biologically productive regions and the zone with the greatest human population. It also embraces a wide variety of landscape and ecosystem types, including barrier islands, beaches, deltas, estuaries, mangroves, rocky coasts, salt marshes and seagrass beds. Coral reefs and coastal wetlands are some of the most diverse and productive ecosystems, and 90 per cent of the world's marine fish catch, measured by weight, reproduces in coastal areas (FAO, 2016b). More than 40 per cent of the world's human population is estimated to live within 100 km of a coast, an area that accounts for just 20 per cent of all land area (Martínez *et al.*, 2007), and this coastal zone includes two-thirds of the world's so-called megacities, defined as urban agglomerations with more than 10 million inhabitants (von Glasow *et al.*, 2013). This large and increasing density of human occupation inevitably puts great pressures on coastal resources.

PHYSICAL CHANGES

The configuration of coastlines is constantly changing thanks to the natural processes that operate at a wide range of temporal and spatial scales. Inevitably, all have some effects on human activities in coastal areas. At a large scale, changes in sea level relative to land, due to crustal movements and changes in the overall volume of marine waters, can alter the nature of the coastal zone itself. Some of the knock-on effects for society may be substantial. For example, Nunn (2003) suggests that rising post-glacial sea levels along the coastline of eastern Asia probably played a role in prompting the first human settlement of many Pacific islands. Large numbers of early farmers, particularly in the alluvial plains around the mouths of major rivers such as the Yangtze and the Huang Ho, were displaced by the sea-level rise. These displaced agriculturalists effectively became 'environmental refugees' and had to set out across the Pacific in search of new lands, eventually to become the earliest human settlers on most Pacific islands.

The operation of sedimentation and erosion that is superimposed upon such larger-scale processes also has more localized impacts on the human use of coasts. Sedimentation by the Büyük Menderes River into the Gulf of Miletus on Turkey's Aegean coast over the last 2400 years has today left the ancient port of Haracleia 30 km from the shore (Bird, 1985). Conversely, erosion of Britain's Humberside coastline between Bridlington and Spurn Head has removed a 3 km strip of coastal terrain, with the loss of 30 villages, since Roman times (Figure 7.1). These natural processes which affect the physical make-up of coastlines have also been modified in numerous ways by human populations, particularly in more modern times. In Japan, for example, more than 40 per cent of the country's 32,000 km coastline is influenced by human action, through land reclamation, urban and port facilities or by shoreline protection associated with lines of communication (Koike, 1985).

Attempts to manage physical change along coastlines can be divided into two fundamentally different approaches:

- hard engineering, which involves building physical structures, usually from rocks or concrete, to protect the coast from natural processes, particularly erosion and the risk of flooding
- soft engineering, which makes use of natural systems, such as beaches, salt marshes or mangroves, to help with coastal defences.

Traditional hard engineering structures, such as sea-walls and rock revetments, are designed to absorb the energy of waves in order to reduce or prevent erosion and flooding. Installation of such structures has been successful in protecting local stretches of coastline from physical change but their construction is not without drawbacks. These include the destruction of natural ecosystems and an enhanced danger of erosion along neighbouring parts of the coast. Soft engineering approaches are non-structural and tend to work with nature in protecting a coastline by harnessing natural systems that can absorb marine energy. Whereas hard engineering structures are frequently viewed as expensive, short-term solutions, the soft engineering approach is often advocated for more long-term, sustainable solutions, although of course both approaches come with disadvantages as well as advantages (Table 7.1) and a combined approach may offer the best protection for any particular coastline deemed worthy of protection from physical change.

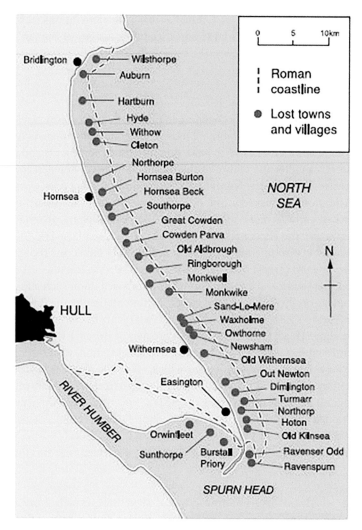

Figure 7.1 Extent of coastal erosion since Roman times in the Humberside region of north-eastern England (after Willson, 1902; Murray, 1994).

Other human activities in the coastal zone can have equally significant, if unintended, effects on its physical properties, not least the removal of coastal materials for construction purposes. In many parts of coastal Indonesia, corals are used for construction, either as building blocks for walls and foundations or crushed and fired to produce lime, an important constituent of cement. The practice was perhaps best known in Bali where coral reefs were broken up and brought ashore to use in the hotel construction boom during the late 1970s and 1980s. At one stage in the 1980s, there were more than 400 coral-burning kilns producing lime on the eastern and southern coasts of the island and the volume of coral mined off the south coast has been estimated at up to 150,000 m³. Without the protection to the coastline provided by the reefs, significant erosion of the beachfront resulted. The Balinese government outlawed coral mining in 1985, and the practice had stopped in most areas by 1990, but millions of dollars have had to be spent

Table 7.1 Some examples of hard and soft engineering coastal protection strategies

Strategy	Description	Advantages	Disadvantages
Hard engineering			
Sea-wall	Large wall of concrete, stone or steel to reflect waves back to sea	Protects cliffs from erosion and offers barrier to flooding	Creates strong backwash eroding beneath the wall, expensive to build and maintain
Gabion	Rocks in metal cages located at cliff base to reduce wave impact	Relatively inexpensive	Limited effectiveness
Groyne	Fence at right angles to beach to trap material transported by longshore drift	Allows beach to build up and protects from flooding	Accelerates erosion on downdrift side and further down coast
Revetment	Slanted structure along cliff to absorb wave energy	Protects cliffs from erosion	Expensive, can create strong wave backwash
Soft engineering			
Beach nourishment	Add sand and shingle to beach	Creates wider beach that slows waves, offering greater protection from erosion and flooding	Can create problems in dredge sites, expensive
Dune regeneration	Creating or restoring dunes by nourishment and/or planting	Creates barrier to wave action and flooding	Barrier to beach access, expensive
Managed retreat	Allowing some previously protected areas to erode and flood by removing protection	Can encourage development of beach or marshland, creating new habitats, inexpensive	Disagreement over which areas so-managed, compensation required for people affected

on artificial measures to protect the eroding beaches, such as sea-walls and breakwaters (Bentley, 1998). Beaches themselves have been mined in many places for building sand, in several areas for heavy minerals such as titanium and zirconium, and on the Namibian coast for diamonds (Defeo *et al.*, 2009).

It is not just activities in the coastal zone itself that have an impact there. River outlets also supply sediment to coasts, so that human modification of river regimes can have a downstream impact. Changes in land use, such as cultivation, can increase sediment loads, for example, while conversely numerous rivers have had their sediment budgets depleted by the construction of dams (see Chapter 9). Many of the world's largest deltas are becoming increasingly vulnerable to flooding due to the effects of sediment being trapped in upstream reservoirs, in combination with sediment compaction due to the removal of oil, gas and water from underlying deltaic sediments, and rising global sea levels (Syvitski *et al.*, 2009). Some of the most vulnerable deltas, many of them densely populated and heavily farmed, are shown in Table 7.2.

A similar threat of marine flooding to London in south-east England, where the dangers from high tides and storm surges are also exacerbated by land subsidence, led to the construction of the Thames Barrier, completed in 1982 (see p. 540). Elsewhere in western Europe, the risk from marine floods has inspired many modifications to coastlines. The storm surge hazards posed to the low-lying coastlines of the North Sea have been combated by a number of hard engineering solutions. In the Netherlands, the Zuider Zee scheme (1927–32) is an early example, which involved a 32-km-long, 19-m-high dam, closing off the vast estuary of the Zuider Zee. In addition to flood protection, closure of the estuary also facilitated land reclamation and improvements to freshwater resource management.

In 1953, when North Sea surge floods drowned 1853 people and seriously damaged 50,000 buildings in the Netherlands, the Delta Project was set in motion to protect the country's south-western coastal area. It provides for the closing off of some of the estuaries discharging into the North Sea – the Rhine, Maas and Scheldt – and three closures have been built: the Veerse Meer, Haringvliet and Grevelingen. The project has effectively shortened the Dutch coastline by 700 km, but its efficiency has compounded the flood risk in coastal parts of the Netherlands via the positive feedback process shown in Figure 7.2. As Filatova *et al.* (2011) explain, the risk of flooding in European water management circles is defined as a function of the probability of a flood event and its potential impact (in terms of monetary damage and human casualties). Improvements to flood defences may decrease the flood risk by decreasing the probability of a flood event, but they also have the effect of attracting more people and economic activity for the very same reason. As the economic value and population density of an area increase, so does the flood risk, because the potential impact increases. The government then comes under more pressure to lower the probability of flooding with further improvements to flood defences.

Table 7.2 Some deltas at particular risk of flooding

Delta	Area < 2 m above sea level (km²)	Twentieth-century reduction in fluvial sediment (%)	Subsurface abstraction of oil, gas or water	Relative sea-level rise (mm/yr)
Chao Phraya, Thailand	1780	85	Major	13–150
Colorado, Mexico	700	100	Major	2–5
Huang Ho, China	3420	90	Major	8–23
Nile, Egypt	9440	98	Major	5
Pearl, China	3720	67	Moderate	7.5
Po, Italy	630	50	Major	4–60
Tone, Japan	410	30	Major	>10
Yangtze, China	7080	70	Major	3–28

Source: after Syvitski et al. (2009).

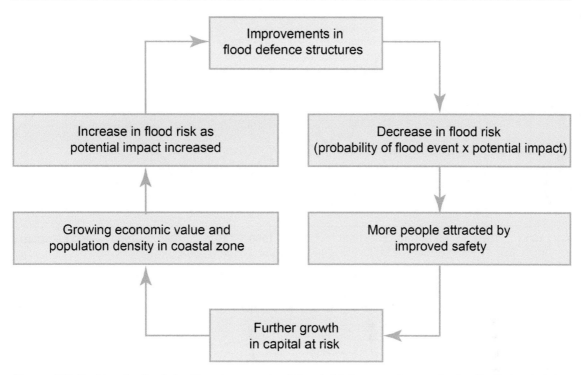

Figure 7.2 Positive feedback fuelling the growth of flood risk in coastal zones in the Netherlands (after Filatova *et al.,* 2011).

Coastlines on inland seas are also susceptible to many of the drivers of change mentioned above, including changes in river discharge. The completion of a dam to facilitate agricultural, industrial and municipal withdrawals across the mouth of the Kora-Bogaz-Gol embayment, an area of high evaporation on the Turkmenistan coast of the Caspian Sea, is thought to have contributed to a decline in sea level in the Caspian of more than 3 m between 1929 and 1977, exposing more than 50,000 km² of former sea bed (Figure 7.3). However, a subsequent rise of nearly 3 m by 1995 suggests that climatic variations influencing fluvial discharge into the Caspian were more important. In fact, large fluctuations in sea level have been identified in the Caspian throughout the Quaternary and although the cause of these changes is much debated, it is most probably related to the fact that the Volga River, which supplies 80 per cent of the inflow of the Caspian Sea, has experienced strong variations in annual discharge due to climatic variability (Kroonenberg *et al.,* 2000). The most recent phase of falling sea level since 1996 is probably driven by increased evaporation rates over the Caspian associated with higher surface air temperature, and this sea level decline is expected to continue under global warming scenarios (Chen *et al.,* 2017a).

HISTORICAL OVERFISHING

Some scientists have traced the degradation of many coastlines back over considerable periods of time and concluded that ecological extinctions caused by overfishing precede all other pervasive human impacts to

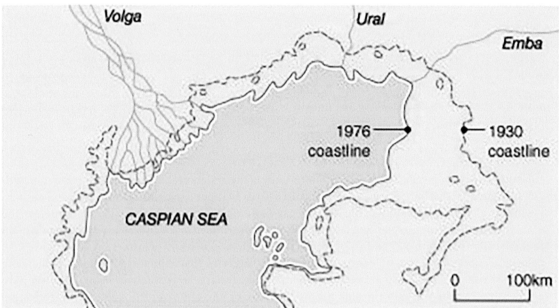

Figure 7.3 Variations in sea level and northern coastline changes of the Caspian Sea (updated after Kroonenberg *et al.*, 2000; Bird, 1985; Ozyavas *et al.*, 2010).

coastal ecosystems (Jackson *et al.*, 2001). Using a range of proxy indicators and historical evidence, the first major human disturbance to all coastal zones studied was found to be overfishing of large marine fauna and shellfish. The effective loss of entire trophic levels has made coastal ecosystems more vulnerable to other human and natural disturbances such as pollution, disease, storms and climate change.

The study examined records from marine sediments dating from about 125,000 years ago, archaeological records from human coastal settlements occupied after about 10,000 years ago, historical records from documents of the first European trade-based colonial expansion in the Americas and South Pacific in the fifteenth century to the present, and ecological studies from the past century to help calibrate the older records. Everywhere, the magnitude of losses in terms of abundance and biomass of large animals, such as whales, manatees, dugongs, monk seals, sea turtles, swordfish, sharks, giant codfish and rays, was found to be enormous. These ecological extinctions caused by overfishing are thought to have led to serious changes in coastal marine ecosystems over many centuries as overfished populations no longer interacted with other species.

Such changes were not always immediate. Timelags of decades to centuries were identified between the onset of overfishing and consequent ecological changes because other species of similar trophic levels took over the ecological roles of the overfished species until they too were overfished or perished as a result of disease epidemics related to overcrowding. The plight of the southern California kelp forests, which began to disappear on a large scale after the 1950s, is shown in Table 7.3. Kelp forests, which represent some of the most productive and diverse habitats on Earth, are usually found in shallow, rocky areas from warm temperate to subarctic regions worldwide, but those off the southern Californian coast began a long process of ecological change dating from the early 1800s when sea otters were hunted to virtual extinction for their fur. The diverse food web of the kelp forest enabled the ecosystem to persist for more than a century as other species took over the sea otters' role in keeping sea urchins, which graze on the kelp, in check. However, widespread loss of the kelp occurred after the 1950s when these other species also became targets for intense human exploitation, enabling a population explosion of sea urchins.

Some recovery of the kelp forests has since been recorded, probably resulting in part from another phase of marine exploitation: a commercial sea urchin fishery that began in the early 1970s. However, the complexity of the issue has also been highlighted by other studies that have questioned the role of overfishing in the ecosystem changes off the coast of southern California. Foster and Schiel (2010) present evidence to indicate that the loss of kelp forest was not as widespread as first thought

Table 7.3 The decline of kelp forests off southern California, USA, due to human impact

Date	Human impact	Ecology
Pre-nineteenth century	Little	Grazing of kelp by sea urchins kept in check mainly by sea otter predation
Early 1800s	Sea otters effectively eliminated for fur trade	Sheephead fish and spiny lobster replace sea otter as main sea urchin predators; abalone compete with sea urchins
Post-1950s	Sheephead, spiny lobster and abalone effectively eliminated	Widespread loss of kelp due to overgrazing by sea urchins
1970s	Commercial sea urchin fishery begins	Some recovery of kelp as sea urchin numbers contained

Source: after Jackson *et al.* (2001).

and that the losses that did occur between 1950 and 1970 were caused primarily by large increases in contaminated sewage discharged into coastal waters, sedimentation from coastal development, and the 1957–59 El Niño event.

POLLUTION

Pollution of the coastal zone comes from activities on the coast, in coastal waters and from inland; in the latter case, usually arriving at the coastal zone via rivers. Pollutants can be divided into non-conservative types that are eventually transformed or assimilated into the biota by physical, chemical, or biological processes such as biodegradation and dissipation, and conservative pollutants that are not biodegradable and are not readily dissipated, and thus tend to accumulate in marine biological systems (Table 7.4; see also Figure 6.10).

The wide range of human activities that contribute pollutants to coastlines, and the pathways by which they reach the coast (see also Table 6.2), are illustrated in an issue that is causing considerable concern: the increasing discharges of nutrients to coastal environments – a problem that is envisaged to constitute a worldwide issue in the next few decades. Human activities have increased nutrient flows to the coast in numerous ways:

- forest clearing
- destruction of riverine swamps and wetlands
- application of large amounts of synthetic fertilizers
- industrial use of large amounts of nitrogen and phosphorus
- addition of phosphates to detergents
- production of large livestock populations
- high-temperature fossil fuel combustion, adding large amounts of nitrogen to acid rain
- expansion of human populations in coastal areas (hence more sewage and storm-water runoff flows).

Table 7.4 Classification of marine pollutants

NON-CONSERVATIVE

- Degradable wastes (including sewage, slurry, wastes from food processing, brewing, distilling, pulp and paper, and chemical industries, oil spillages)
- Fertilizers (from agriculture)
- Dissipating wastes (principally heat from power station and industrial cooling discharges, but also acids, alkalis and cyanide from industries, which can be important very locally)

CONSERVATIVE

- Particulates (including mining wastes and inert plastics)
- Persistent wastes (heavy metals such as mercury, lead, copper and zinc; halogenated hydrocarbons such as DDT and other chlorinated hydrocarbon pesticides, and polychlorinated biphenyls; radionuclides)

Source: after Clark (1992).

The extent of the increase is indicated for a part of the North American seaboard by Nixon (1993), who used a complex mix of historical data to estimate the changes in nutrient inputs to the Narragansett and Mount Hope bays (Figure 7.4). Up to around 1800, the inflow of nutrients from the ocean was five to ten times larger than inputs from the land and the atmosphere, but in the contemporary era about five times more inorganic nitrogen enters Narragansett Bay from the land and atmosphere than from the coastal ocean. The particularly dramatic rise that occurred over a 30-year period from 1880 to 1910 was a response to the proliferation of public water and sewage systems through much of the urban area of the north-eastern USA at this time.

Excessive nutrient flows to coastal environments have been linked to a wide range of adverse environmental impacts that are well summarized by Smith and Schindler (2009). The effects of eutrophication in coastal waters include oxygen depletion and increased phytoplankton blooms, increased incidence of fish kills, reductions in species diversity and reductions in harvestable fish and shellfish. There may also be direct links between eutrophication and the risk of infectious diseases because enhanced nutrient loading can boost the abundance and survival rates of pathogens in aquatic ecosystems. Such diseases include Aspergillus fungi that affect coral reefs and cholera that affects human populations. The probability of cholera epidemics could be influenced via nutrients increasing the abundance of small crustaceans called copepods, which help to spread cholera.

The rise of the eutrophication issue globally is demonstrated by the number of coastal, oxygen-depleted dead zones which has roughly doubled each decade since the 1960s (Diaz and Rosenberg, 2008; see Figure 15.3), and in coming decades this trend may be recognized in part as an early sign of the effects of climatic change. Some scientists fear that the frequency and extent of coastal eutrophication and associated algal blooms may increase with global warming because nitrogen and sulphur stored in soil organic matter will be released into rivers if temperatures rise as predicted. Increased nitrogen losses are particularly likely (Figure 7.5).

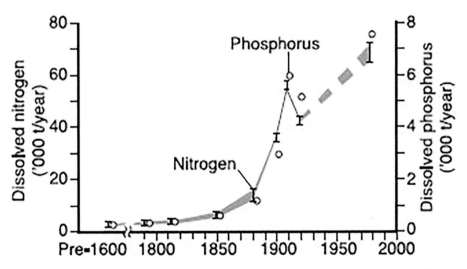

Figure 7.4 Inorganic nitrogen and phosphorus input to Narragansett and Mount Hope bays, north-east USA, pre-seventeenth to late twentieth centuries (after Nixon, 1993).

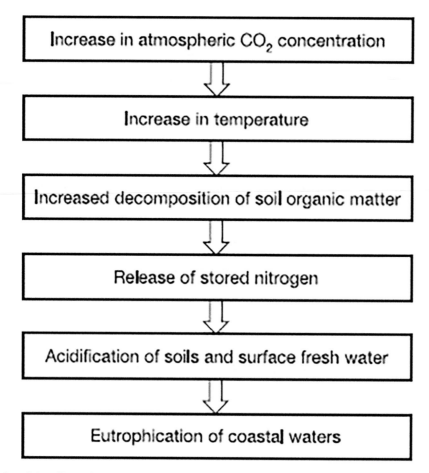

Figure 7.5 Possible effect of global warming on coastal eutrophication through the release of terrestrial nitrates (after Wright and Hauhs, 1991).

All kinds of urban and industrial activities in the coastal zone contribute pollution, the effects of which can become particularly critical in areas where the mixing of coastal water with the open ocean is relatively slow, as in bays, estuaries and enclosed seas (see the account of the Black Sea in Chapter 6). For example, in dryland coastal cities where fresh water is produced by desalination plants, the discharge back into coastal waters of chlorine and several chemical products used in the desalination process may represent a severe pollution problem (Sadhwani et al., 2005).

Petroleum hydrocarbons are a particular hazard to coastal ecology, chiefly at oil terminals but also from accidental spillages. Habitats such as mangroves and marshes are especially vulnerable to oil spills, although wildlife is often the most conspicuous victim. Generally, younger individuals are more sensitive than adults, and crustaceans are more sensitive than fish. The release of 36,000 tonnes of oil into Prince William Sound in Alaska from the grounded supertanker *Exxon Valdez* in March 1989 resulted in the deaths of 36,000 seabirds, 1000 sea otters and 153 bald eagles in the following six months (Maki, 1991).

The loss of these animals from intermediate and top trophic levels generated a cascade of indirect impacts at lower levels that have occurred after timelags of varying lengths. Wildlife has also been affected by the unexpected persistence of toxic oil in subsurface sediments (Bodkin *et al.*, 2012).

Heavy metals and organochlorine compounds have long been recognized as among the most deleterious contaminants to biota in coastal and estuarine waters and a large proportion of these pollutants arrive at coasts in rivers draining areas outside the coastal zone. They have the potential for bioaccumulation in aquatic organisms, sometimes reaching toxic levels and presenting a risk for both wildlife and humans. Since heavy metals and organochlorine compounds cause problems even in very low concentrations, and the measurement of such small quantities is technically difficult, many studies of these coastal pollutants focus on concentrations in marine organisms. Mussels and oysters have been widely used because as filter feeders they bioaccumulate contaminants. In a review of heavy metal pollution in the Arabian Gulf, for instance, Naser (2013) notes that molluscs from waters near outlets from power plants, oil refineries and desalination plants tend to have the highest concentrations.

Several investigations of long-term trends in pollutants have measured concentrations in the feathers of seabirds since old specimens can often be found in museums. Mercury contamination measured in herring gulls and common terns from the German North Sea coast, for example, indicates that the rivers Elbe and Rhine are the major sources of the pollutant. Mercury flows peaked at three times 1880 levels after the Second World War when heavy metals from ignition devices in munitions were released in large quantities (Thompson *et al.*, 1993). Significant declines in recent decades reflect measures to reduce river-borne pollution, however, and the trend is expected to continue. The opposite trend has been detected in ivory gulls, however, a species that has declined by more than 80 per cent since the 1980s in Arctic Canada. Bond *et al.* (2015) used feathers from museum specimens from the Canadian Arctic and western Greenland to assess exposure to mercury over the period from 1877 to 2007 and found that, while there was no significant change in ivory gulls' diet, the feather concentrations of methyl mercury increased by 45 times over the 130-year period. They concluded that continued declines in ivory gull populations are likely given the expected increases in mercury arriving by long-distance atmospheric transport from emission sources in Asia.

HABITAT DESTRUCTION

Coastal habitats, particularly mangroves, salt marshes, seagrass beds and coral reefs, are subject to many types of pressure from human activities. Direct destruction of these habitats for urban, industrial and recreational growth, as well as for aquaculture, combined with overexploitation of their resources and the indirect effects of other activities, such as pollution and sedimentation or erosion, are causing rapid degradation and outright losses in many parts of the world. The sections below detail the situation for two coastal habitats that have come under substantial pressure in recent times.

Coral reefs

Coral reefs are marine ridges or mounds formed as a result of the deposition of calcium carbonate by living organisms, predominantly corals, but also a wide range of other organisms such as coralline algae and shellfish. These reefs occur globally in two distinct marine environments: deep, cold-water (3–14°C) coral reefs, and shallow, warm-water (21–30°C) coral reefs in tropical latitudes.

We know more about the warm-water coral reefs of the tropics than cold-water coral habitats which, although known for centuries to exist, have only recently become the subject of detailed scientific observations. The distribution of cold-water coral reefs is still poorly understood, but current information indicates that they occur in coastal waters, along the edges of continental shelves and around offshore submarine banks and seamounts in almost all of the world's seas and oceans. The largest cold-water reef complex is probably Røst Reef off northern Norway, which was only discovered in 2002. Its size (100 km²) is rather smaller than the largest warm-water reef complex, Australia's Great Barrier Reef, which covers an area of more than 30,000 km² (Wilkinson, 2004).

Despite our relative ignorance of cold-water reefs, we do know that many have already been destroyed or severely damaged. The main threats they face from human activities are outlined by Roberts and Cairns (2014):

- destructive fishing practices, including bottom trawling
- exploration and production of oil and gas
- placement of pipelines and cables
- ocean acidification.

Our understanding of warm-water coral reef ecosystems is much better. These reefs are found in shallow waters (generally <30 m) and are largely restricted to the areas between 30°N and 30°S. They also rank among the most biologically productive and diverse of all natural ecosystems. Warm-water coral reefs cover less than 0.2 per cent of the ocean floor, yet support about a third of all marine species (Fisher et al., 2015). Their high productivity is a function of their efficient biological recycling, high retention of nutrients and their structure, which provides a habitat for a vast array of organisms. From the economic viewpoint, coral reefs protect the coastline from waves and storm surges, prevent erosion, and contribute to the formation of sandy beaches and sheltered harbours. They play a crucial role in fisheries, by providing nutrients and breeding grounds, and provide a source of many raw materials such as coral for jewellery and building materials. Most recently, their potential for tourism has been realized. Globally, some 30 per cent of the world's reefs are of value in the tourism sector, with a total value estimated conservatively at nearly US$36 billion a year (Spalding et al., 2017). Spalding et al. concluded that reef tourism is particularly important in small island economies (Figure 7.6). In each of the Maldives, Palau, Bonaire, the Turks and Caicos Islands, and the British Virgin Islands, coral reefs support over one-third of all tourism value and 10 per cent or more of the entire GDP.

The past 30 years or so have been marked by a growing concern for the future of the world's coral reefs. Reefs have always been subject to natural disturbances such as hurricanes, storms, predators, diseases and fluctuations in sea level, but in recent times they have also come under increasing pressures from human action. Island and coastal societies have long used coral reefs for subsistence purposes, as sources of food and craft materials, but the development of commercial fisheries, rapidly increasing populations still dependent upon a subsistence lifestyle, the growth of coastal ports and urban areas, increases in soil loss due to deforestation and poor land use, and most recently the exponential growth of coastal tourism, have combined to put unprecedented pressures on reefs from both direct and indirect impacts. These localized stresses act in synergy with threats stemming from climate change, including rising water temperatures, more intense solar radiation and ocean acidification (Carpenter et al., 2008).

The wide range of human impacts is indicated in Table 7.5 with examples from Pacific reefs. These local threats can be divided roughly into direct impacts caused by overexploitation of reefs (overcollecting, damaging fishing techniques, and careless or poorly supervised recreational use), and indirect impacts resulting from other activities (siltation following land clearance, damage caused by coastal developments and pollution, and that caused by military activities).

Figure 7.6 Scuba divers swimming near a giant fan coral in the Pacific Ocean near Palau.

In a global assessment published in 2011 (Burke *et al.*, 2011), it was estimated that about 850 million people live within 100 km of a coral reef and are likely to derive some benefits from the ecosystem services that these reefs provide. More than 60 per cent of the world's reefs were considered to be threatened by human activities from one or more local sources such as overfishing (including destructive fishing), coastal development, catchment-based pollution, or marine-based pollution and damage. When these local pressures on coral reefs are combined with thermal stress, which reflects the recent impacts of rising ocean temperatures, some 75 per cent of the world's coral reefs were rated as threatened.

In many cases threats reflect a combination of drivers. The world's greatest concentration of coral reef biodiversity is an area of the Asia-Pacific known as the Coral Triangle, which comprises Indonesia, the Philippines, Malaysia (Sabah), Papua New Guinea, Timor Leste and the Solomon Islands. The Coral Triangle contains 76 per cent and 37 per cent of the world's coral and reef fish species, respectively, but also over 100 million people living in its coastal zones who use this biodiversity to support their livelihoods (Foale *et al.*, 2013). Indonesia and the Philippines have the largest threatened areas of reef, where overfishing and destructive fishing are the main causes of coral decline, followed by pollution from adjacent catchments and coastal development. Managing a sustainable balance between marine biodiversity conservation and the food security of the region's marine-resource dependent people is therefore critical to the Coral Triangle Initiative, a plan devised by the region's governments.

Table 7.5 Human-induced local threats to warm-water coral reefs, with selected examples from the Pacific Ocean

Threat	Example
OVERCOLLECTING	
Fish	Futuna Island, France
Giant clams	Kadavu and islands, Fiji
Pearl oysters	Suwarrow Atoll, Cook Islands
Coral	Vanuatu
FISHING METHODS	
Dynamiting	Philippines
Breakage	Vava'u Group, Tonga ('tu'afeo')
Poison	Uvea Island, France
RECREATIONAL USE	
Tourism	Heron Island, Great Barrier Reef, Australia
Scuba diving	Hong Kong
Anchor damage	Molokini Islat, Hawaii, USA
SILTATION DUE TO EROSION FOLLOWING LAND CLEARANCE	
Fuelwood collection	Upolu Island, Western Samoa
Deforestation	Ishigakishima, Yaeyama-retto, Japan
COASTAL DEVELOPMENT	
Causeway construction	Canton Atoll, Kiribati
Sand mining	Moorea, French Polynesia
Roads and housing	Kenting National Park, Taiwan
Dredging	Johnston Island, Hawaii, USA
POLLUTION	
Oil spillage	Easter Island, Chile (1983)
Pesticide spillage	Nukunonu Atoll, New Zealand (1969)
Urban/industrial	Hong Kong
Thermal	North-western Guam, USA
Sewage	Micronesia
Open-cast mining	New Caledonia, France
Invasive species	Hawaiian islands, USA
MILITARY	
Nuclear testing	Bikini Atoll, Marshall Islands (1946–58)
Conventional bombing	Kwajalein Atoll, Marshall Islands (1944)
Military base expansion	Guam, USA

Source: compiled from sources quoted in IUCN (1988); Nunn (1990); Burke *et al.* (2011).

An illustration of the changing nature of threats to reefs is mapped for Kaneohe Bay, on the east coast of the Hawaiian island of Oahu, in Figure 7.7. Dredging and filling by the military between 1938 and 1950 significantly increased problems of turbidity and sedimentation in most of the bay, and caused direct damage in southern parts where coral was used as fill to create new land. In 1970, while northern reefs were recovering from this impact and colonizing some of the dredged zones, two new sewage outlets caused further reef decline in the southern portion. Closure of these sewage outlets in 1977–78 allowed a rapid recovery in the central part of the bay and a more prolonged recovery in the south. The new sewage outlet outside the bay has not damaged coral, due to the high level of mixing and flushing at the site. Coral reef communities in Hawaii then faced their first recorded large-scale bleaching event (see below) in the summer of 1996 when parts of Kaneohe Bay were severely affected. A second major event in 2014 was similar in duration and intensity but resulted in a much larger area of bleaching and coral mortality (Bahr *et al.*, 2015).

The Great Barrier Reef off the east coast of Australia is one of the world's least threatened, thanks in large part to its status as a marine park. The reef is managed through a system of zoning, and so-called 'no-take' areas, where fishing is prohibited, were expanded in 2004 from 5 per cent of the Great Barrier Reef Marine Park to 33 per cent, in an effort to improve protection. However, parts of the Great Barrier Reef face a pollution threat from land-based sources that have delivered significantly greater levels of contamination over the past 150 years (Table 7.6), with adverse effects on coral and seagrass ecosystems (Kroon *et al.*, 2016). Sediment smothers the living coral and seagrasses, while nutrients inhibit both their growth and reproduction. Herbicides from agricultural runoff also damage seagrasses and other reef organisms by affecting their ability to manufacture food.

Land use in most of the river catchments adjacent to the reef has changed significantly since the 1850s. Grazing has become widespread and large areas of natural vegetation have been cleared to make way for crops. The area used for sugar cultivation in the Great Barrier Reef catchments rose from about 100,000 ha in 1930 to nearly 400,000 ha in the mid-1990s. Pollutants from these catchments, transported to the reef mostly during flood events, have continued to increase and the decline in water quality has a real potential to adversely affect industries such as tourism, and recreational and commercial fishing. Recognition of these dangers has prompted establishment of end-of-river pollution targets for all 26 river catchments adjacent to the Great Barrier Reef, along with a range of policy initiatives designed to reduce land-based pollution (Table 7.7). Nonetheless, the Great Barrier Reef continues to face stress from terrestrial pollution, fishing impacts and climate change, despite being arguably the world's best-managed coral reef system, because, as Brodie and Waterhouse (2012) point out, this is a relative assessment against other reef systems and management regimes and not an absolute claim for effective management.

Indeed, despite being one of the best-protected ecosystems on our planet, the Great Barrier Reef has still experienced a major decline in coral cover in recent decades, with tropical cyclones, coral predation by crown-of-thorns starfish (*Acanthaster planci*), or COTS, and coral bleaching accounting for most of the losses (De'ath *et al.*, 2012). COTS populations exploded at the end of the 1950s and throughout the 1960s in many parts of the Indian and western Pacific oceans, causing widespread damage to reefs. No less than 90 per cent of the fringing reefs of Guam and 14 per cent of the Great Barrier Reef were destroyed, the dead corals being replaced by an algae- and sea urchin-dominated community. Further COTS population explosions have occurred at regular intervals in the Indo-Pacific over subsequent decades. The cause of predator plagues like COTS remains unclear (Hughes *et al.*, 2014) but they are increasingly reported near areas of human activities and two hypotheses have been advanced: the plagues may be initiated – and are certainly exacerbated – by either overfishing of key starfish predators and/or increases in nutrient runoff

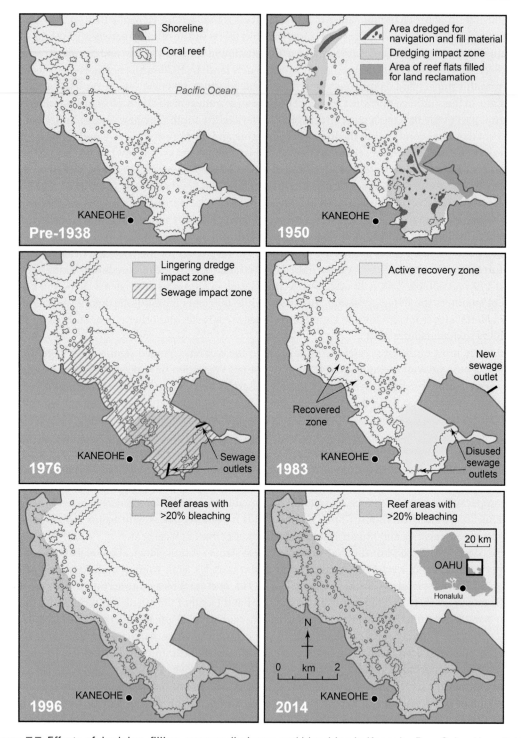

Figure 7.7 Effects of dredging, filling, sewage discharge and bleaching in Kaneohe Bay, Oahu, Hawaii, USA since the 1930s (after Maragos, 1993; Jokiel and Brown, 2004; Bahr *et al.*, 2015).

Table 7.6 Pollution increases to the Great Barrier Reef since c.1850

Pollutant	Increase (%)
Sediment loads	300 to 900
Phosphate discharges	300 to 1500
Total nitrogen load	200 to 400
Pesticide residues	now being detected in tidal sediments

Source: after Great Barrier Reef Marine Park Authority (2001).

Table 7.7 Recent plans to reduce land-based pollution affecting the Great Barrier Reef

Plan	Goal	Timeframe
Reef Plan 2009	Halt and reverse the decline in water quality entering the Great Barrier Reef by 2013	2013
Reef Plan 2013	The quality of water entering the Great Barrier Reef from adjacent catchments has no detrimental impact on the health and resilience of the Reef	2020
Reef 2050 Long-Term Sustainability Plan	Over successive decades the quality of water entering the Great Barrier Reef from broad-scale land use has no detrimental impact on the health and resilience of the Reef	2035

Source: after Kroon *et al.* (2016).

from adjacent coastal areas which favour the planktonic stages of the starfish. It is also possible, however, that the COTS population explosion phenomenon is part of essentially natural cyclical events.

Less uncertainty surrounds the widely reported phenomenon of coral bleaching – where corals lose their colourful symbiotic algae, so exposing their white skeletons – which has resulted in the extensive mortality of reefs in many parts of the world. While bleaching commonly occurs naturally when sea surface temperatures rise during events such as El Niño/Southern Oscillation (ENSO), the rising number of mass coral bleaching episodes around the world since the early 1980s has raised questions about the viability of coral reef ecosystems during a period of rapid climate change. The 1998 coral bleaching event, related to the 1997/98 ENSO, was particularly destructive and global in its impact, resulting in severe damage to an estimated 16 per cent of the world's coral reef area (Wilkinson, 2004). However, suggestions of links to global warming have prompted climate attribution research that has concluded that anthropogenic forcing is likely (>90 per cent chance) to have driven some more recent episodes, including extensive bleaching across the eastern Caribbean in 2005 and the northern Great Barrier Reef in 2016. A growing number of modelling studies indicate that projected ocean warming over the next few decades may make mass coral bleaching a frequent occurrence on reefs worldwide (Donner *et al.*, 2017).

The uncertainties surrounding these two issues of coral reef vitality highlight the general lack of a historical perspective on the environmental parameters affecting reefs and their population dynamics. A heightened realization of the important role of corals in the marine ecosystem, their high biodiversity and their importance to local communities has helped to turn concern at their degradation into action towards managing these key habitats more sustainably. Such management is needed most urgently for those atoll island states whose future is threatened by the altered environmental dynamics of a warmer world (see Chapter 11). In response to global sea-level rise, some reefs appear to 'keep up' with rising water levels, others 'catch up' after a lag period by rapid vertical accretion of coral, while others are terminally affected by initial drowning and effectively 'give up' (Neumann and Macintyre, 1985). Warm-water corals in particular also face a significant threat from the acidification of oceans associated with higher levels of atmospheric carbon dioxide (see p. 288). Although we know little about the degree to which coral species living today may evolve to adapt to the chemical changes, acidification is likely to lower calcification rates in reefs, with possibly dramatic consequences. As one oceanographer puts it: 'It seems highly probable that coral reef ecosystems will cease to occur naturally on Earth' (Tyrrell, 2011: 895).

Mangroves

Mangrove forests are made up of salt-adapted evergreen trees, and are found in intertidal zones of tropical and subtropical latitudes. These diverse and productive wetland ecosystems are under threat from human action in many parts of the world, despite their importance as breeding and feeding grounds for many species of fish and crustaceans, and their role in protecting coasts from erosion, among the many other functions of wetlands in general (see Table 8.5). Estimates of mangrove area in the literature vary considerably, due to differing survey methods and definitions, but many authorities consider mangroves to be among our most endangered natural habitats. Tentative assessments at the global scale suggest that roughly 50 per cent of the worldwide area was lost in the twentieth century and most of that in its last two decades (Valiela *et al.*, 2001). However, comparable multi-year data are not universally available. The Millennium Ecosystem Assessment (2005) put the loss over the period 1980–2000 at about 35 per cent, although this was only for countries with adequate data, representing just over half the total mangrove area. These losses were driven primarily by aquaculture development, deforestation and the diversion of freshwater. Conversion of mangroves to agriculture, urban development and overexploitation for fuelwood or charcoal production are other common drivers of loss, and sea-level rise could be the greatest future threat to these habitats. An overview using globally consistent data from satellite imagery assessed the total area of mangroves in the year 2000 as nearly 138,000 km^2 (Giri *et al.*, 2011), rather less than previously thought, and about 64 per cent of this area is found in just ten countries: those shown in Table 7.8, with the major causes of destruction.

Some of the largest losses along the African coastline have been in the west of the continent. The most frequent cause of mangrove decline in Guinea, Guinea-Bissau and Sierra Leone, formerly some of the most extensive areas of this habitat in West Africa, has been clearance and conversion of the land to rice farming, although exploitation for fuelwood also plays an important part. Clearance for fuelwood has been the major factor behind the almost total disappearance of mangroves in Ghana and Côte d'Ivoire (Sayer *et al.*, 1992) and continues to be the most prominent driver of mangrove loss in Gambia and Senegal (Carney *et al.*, 2014). Nigeria contains the most extensive remaining area of mangrove forests and more than half of these are located in the Niger Delta, although they also declined in the last two decades of the twentieth

Table 7.8 Countries with the greatest mangrove areas and the threats they face

Country	Extent (km²)	Threats
Indonesia	23,324	Shrimp farming (East Java, Sulawesi and Sumatra), logging and conversion to agriculture or salt pans (Java and Sulawesi)
Australia	9780	Minor
Brazil	7676	Timber production, urban development (south-eastern states) and shrimp farming
Nigeria	6540	Urbanization, oil and gas exploitation, dredging and invasive palms
Papua New Guinea	4800	Minor
Malaysia	4726	Agriculture, shrimp farming and urban development
Bangladesh	4370	Minor
Mexico	2992	Agriculture, shrimp farming, urban and tourism development, hurricanes
Myanmar (Burma)	2556	Rice cultivation
Cuba	1634	Minor

Source: Hamilton and Casey (2016); Giri et al. (2011); FAO (2007).

century. James *et al.* (2007) ascribe the loss to urbanization, the dredging of local waterways, activities associated with the oil and gas industries, and the spread of an alien palm species.

In Asia, the clearance of mangrove forests for rice cultivation has been a progressive process over the past century or so in the delta region of the Irrawaddy (or Ayeyarwady), where the harvesting of mangroves for fuel has also intensified, particularly over the past three decades (Hedley *et al.*, 2010). An analysis by Webb *et al.* (2014) using satellite imagery showed that the area in the delta covered by mangroves declined by more than a half between 1978 and 2011, and these authors fear that potentially greater rates of deforestation may occur as Myanmar emerges from a period of international political isolation. The importance of agriculture to the Myanmar economy indicates that further agricultural expansion is likely with increased international investment in combination with insufficient land tenure agreements and poor governance.

By contrast, the area of the world's largest mangrove ecosystem, the forests of the Ganges and Brahmaputra deltas – known as the Sundarbans – has changed little in recent decades (Giri *et al.*, 2011). Parts of the Sundarbans have been protected as a Reserved Forest since 1879. It is a key wildlife habitat for a large number of bird, mammal, reptile and amphibian species, and provides the last stronghold of the Royal Bengal tiger. More than half a million people also depend directly on the Sundarbans for their livelihood. New areas have been planted on offshore islands as part of the management plan, both for fuelwood harvesting and coastal protection against cyclonic storm surges, although the area is under increasing pressure from Bangladesh's growing population.

A mass replanting programme has also been undertaken in post-war Vietnam, successfully re-establishing many thousands of hectares destroyed by spraying with the herbicide Agent Orange (while there has been virtually no natural regrowth since the war). In 2001, these planted mangrove forests made up nearly 80 per cent of the country's total mangrove forest area of 155,000 ha. However, the overall total mangrove area in Vietnam had still declined substantially in the second half of the twentieth century, from nearly 410,000 ha in 1943 (VEPA, 2005). Conversion to agricultural and aquacultural land uses, particularly for shrimp farming, has been the most serious cause in more recent times, but some success in reversing mangrove losses has been achieved in some areas by promoting integrated mangrove–shrimp farming systems. In the Mui Ca Mau, on the southern tip of the Mekong Delta, the mangrove area has increased since the 1990s due to regulations allowing farmers to use land for aquaculture or agriculture if they retain or establish >70 per cent mangrove cover on their farms (Van *et al.*, 2015).

Local populations use mangrove forests for a wide variety of their resources, including wood (for fuel, construction materials and stakes), leaves for fodder, fruits for human consumption, the rich mud that is used as manure, tannin extracted from barks, leaf extracts for medicinal purposes, and captive fisheries of prawns, shrimps and fish (Figure 7.8). The questionable wisdom of destroying this diverse ecosystem is illustrated in economic terms by Sathirathai and Barbier (2001) in their analysis of southern Thailand where many mangroves have been cleared for intensive shrimp aquaculture, which has grown tremendously since the mid-1980s (see Figure 13.13). These authors conclude that although shrimp farms create

Figure 7.8 Captive fishing in the mangroves of the Gulf of Guayaquil, Ecuador. The area has also become the centre of a shrimp-farming industry, one of the country's major sources of foreign exchange, and suffers industrial wastewater pollution from the city of Guayaquil.

great economic benefits for private investors, on balance their value to local communities does not match that of the mangrove forest they replace. Economic returns from shrimp farms are put at about US$200 per hectare, but the value to a local community of harvesting mangrove resources, plus the value to offshore fisheries, is estimated to be between US$1000 and US$1500 per hectare.

Another significant ecosystem service provided by mangroves is coastal protection, a particularly important one in the storm-prone, low-lying coastal zones where they are frequently found (Figure 7.9). This function is thanks largely to the intertwined root systems of dense mangrove forests which dissipate the energy and limit the destructive potential of tsunamis and storm surges from tropical cyclones, offering a level of protection that is superior to anything yet devised by human society (Ostling *et al.*, 2009).

Mangrove loss has also been widespread in Latin America in recent times, and as in some Asian countries much clearance has been to make way for aquacultural production of shrimps for export. The factors affecting this process are often complex, however, as a study of the Gulf of Fonseca on the Pacific coast of Honduras clearly shows (Stonich and DeWalt, 1996). The boom in shrimp aquaculture has been vigorously promoted since the early 1980s by the Honduran government, national and international private investors and several bilateral aid agencies. From the aid agencies' viewpoint, the shrimp export industry was a good way to tackle the country's economic crisis, and a strong emphasis was put on integrating poorer households into the process, mainly through the formation of cooperatives. Unfortunately, small-scale producers have not been profitable and attention has since shifted to larger-scale operations. Nearly 5000 ha of mature mangrove forests were lost along the Gulf of Fonseca coast in the period 1982–92 as the area of shrimp farms increased tenfold. Only about one-fifth of this mangrove loss was attributed to clearance for aquaculture, however, because, among other things, mature mangrove stands are not the most appropriate areas for shrimp farming in this region.

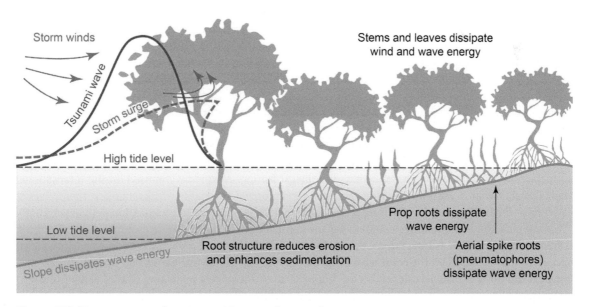

Figure 7.9 How mangrove forests provide coastal protection.

However, a significant portion of the mature mangroves lost was cleared by peasant farmers, and the acid soils and difficulties of managing the shrimp ponds in these areas mean that the probability of failure is high. Further disadvantages for the area's poorer inhabitants have stemmed from the establishment of the larger shrimp operations in the more suitable mudflats. These areas were formerly used by artisanal fishermen when high tides left seasonal ponds rich in fish and shrimp. The transformation of these common property resources into private property, often controlled by foreign investors and national elites, has sparked violent confrontations and grassroots attempts to slow the rapid rise of shrimp aquaculture in the region.

The above catalogue of threats facing mangrove forests will undoubtedly be enhanced in future decades by a range of effects brought about by climatic change (Gilman *et al.*, 2008). A number of climate change components will impact mangroves, including modifications to precipitation and temperature regimes, storm frequency and intensity, and ocean circulation patterns. However, relative sea-level rise is thought to be the most serious driver of predicted future reductions in the area and health of mangroves, as well as other tidal wetlands. In this light, research into the quantity of carbon stored in mangrove forests has been very interesting indeed. The first estimates of above- and below-ground carbon stocks in a mangrove ecosystem in Madagascar carried out by Jones *et al.* (2014) showed that they are amongst the most carbon-rich forests on earth, containing up to 12 times more carbon than undisturbed Amazonian rain forests. These findings provide another compelling argument in favour of mangrove conservation, as a natural means of sequestrating carbon and hence as a critical element in mitigating the impacts of climate change.

MANAGEMENT AND CONSERVATION

Management of the coastal zone is necessary both to limit damaging activities and to protect coastal resources, as well as to restrict development in areas prone to natural hazards such as hurricanes, tsunamis, subsidence and inundation. Effective management strategies depend on many factors. Ownership and governance in the coastal zone are particularly important. Ownership is frequently a complex issue: it is not always clear and can be liable to change, making effective governance more difficult. The break-up of the Soviet Union threw open the question of rights to resources in the Caspian Sea, for example, increasing the number of states round its coastline from two to five, with a commensurate increase in competition for its resources. The Caspian lacks a legal definition. It is either a lake or a sea, but the answer influences how the boundaries are drawn between countries, thereby delineating access to resources (de Mora and Turner, 2004). The five littoral countries (Azerbaijan, Iran, Kazakhstan, Russia and Turkmenistan) agreed to some cooperation on management by signing the Convention for the Protection of the Marine Environment of the Caspian Sea in 2003, but the ownership issue is yet to be resolved. Meanwhile, a lack of adequate management has meant a severe decline in numbers of the Caspian's six commercially valuable sturgeon species, four of which produce 90 per cent of the world's caviar (Figure 7.10). Several sources of pollution, both land-based and offshore, threaten the survival of all fisheries in the Caspian, while overfishing of sturgeon is a particular concern, exacerbated by illegal fishing. A ban was imposed on the international trade in wild caviar in 2006 but it has been difficult to enforce. Criminal networks manifest themselves at all levels of the trade: from the poaching areas to major smuggling operations (van Uhm and Siegel, 2016).

Developing and implementing a strategy for the conservation and sustainable multiple use of coastlines is frequently referred to as Integrated Coastal Zone Management, an approach that aims to balance social and economic demands on the coast with the protection of coastal environments that provide the ecosystem services on which society depends. The importance of adequately managing coastal areas from ecological and economic viewpoints can be illustrated at a regional level by Mauritania's coastal Parc National du Banc d'Arguin (Box 7.1).

Figure 7.10 The Caspian Sea sturgeon, source of most of the world's caviar, is under threat from pollution and overfishing. Inadequate management of these human impacts results partly from a lack of consensus over the ownership of coastal waters.

Box 7.1 International payment for ecosystem services finances coastal conservation and development in Mauritania

The Parc National du Banc d'Arguin is the largest marine protected area in Africa. The park covers an area of 6,450 km^2 along a stretch of coast almost 200 km long in northern Mauritania and includes extensive tidal mudflats, seagrass beds and extremely rich offshore fishing grounds. The site is internationally significant as the most important coastal wetland in Africa through its role as a breeding and wintering ground for millions of waterfowl: it hosts the largest concentration of wintering waders in the world and is the most important breeding area for migrating birds in the Atlantic. It is also a vital spawning ground and nursery for the fish that are Mauritania's main foreign currency earner.

The area became a National Park in 1976, and has subsequently been listed under both the Ramsar and World Heritage Conventions. The objectives of the National Park have evolved over time to include the conservation of fauna and flora alongside local development objectives that focus on maintaining and strengthening the wise use of the mullet fishery by the Imraguen fishermen who reside in the park and use traditional harvesting methods in which they collaborate symbiotically with wild dolphins to catch grey mullet. The number of Imraguen boats has been limited to 100 in order to maintain a sustainable catch, but despite the Imraguen's exclusive fishing rights, the park's waters are under increasing pressure from pirate commercial fishing vessels. These are both small-scale fishers from Mauritania and Senegal, and industrial vessels, often from Europe. In response to these threats, the European Union signed a Fisheries Partnership Agreement with Mauritania in 2006, paying for the rights to fish in Mauritanian waters. Part of this money is specifically set aside to finance the Parc National du Banc d'Arguin, in what is essentially an international payment for ecosystem services because the money helps to conserve the park's role as a nursery ground for commercial species exploited by EU fishing vessels.

Source: Binet *et al.* (2013).

Figure 7.11 The limestone pillars of Halong Bay, a UNESCO World Heritage Site in northern Vietnam, attract visitors from all over the world. Management of the bay's scenic value has to balance other demands on the area's resources from fishing, agriculture and port facilities.

Non-consumptive resource use for tourism is well developed in Vietnam's Halong Bay, a spectacular seascape of some 1600 islands and islets in the Gulf of Tonkin (Figure 7.11) which is also of great biological interest. Halong Bay was inscribed on the UNESCO World Heritage List in 1994. The fact that visitors are attracted by the bay's outstanding scenic beauty should help to ensure its proper protection, although the area in and around Halong is also a major centre for fishing, agriculture and maritime transport.

Sustainable consumptive use of resources is well established in the protected Bangladesh portion of the Sundarbans mangrove forest (Halls, 1997). Wood from the forest reserve provides raw material for the country's only newsprint paper mill, as well as a number of match and board mills and more local uses. The simple management regime involves a 20-year cycle of exploitation. The forest is divided into 20 compartments with each compartment harvested in turn and the cycle repeated at year 21. During the harvest, all trees above a certain diameter are removed as long as such removals do not create any permanent gaps in the forest canopy. In a subsequent operation, all dead and deformed trees are removed. Similar regimes control the use of other mangrove resources (Table 7.9). More than a century of sustainable management has caused very little change to the composition, quality and extent of the Sundarbans forest in Bangladesh.

The number of people actually living in the Sundarbans is relatively small and access by outsiders is strictly controlled, but in other coastal areas management issues are made more complex by large permanent human population densities and the consequent heavy pressures put on resources. The links between coastal problems and pollution sources outside the coastal zone also make integrated management over the entire catchment necessary, given the importance of rivers in contributing to pollution. Several initiatives have attempted to address these issues and improve the environmental quality of Chesapeake Bay, the largest estuary in North America severely affected by coastal eutrophication since the mid-twentieth century. The Chesapeake Bay Program, adopted by the six states and the District of Columbia which together have jurisdiction over the whole catchment, agreed in 1987 to reduce nutrient flows of nitrogen and phosphorus into the bay by 40 per cent by the year 2000. However, as that year approached, it became clear that this goal would not be reached and that numerous interrelated aspects of environmental quality and human activities needed to be addressed in a more comprehensive manner. The Chesapeake 2000 Agreement was reached, and includes more than 100 goals and commitments that together make up one of the most ambitious ecosystem management programmes for any large coastal area (Boesch, 2006). Progress in restoring the Chesapeake

Table 7.9 Sustainable resource management in the Sundarbans forest, Bangladesh

Resource	Uses	Management
Timber	Fuelwood, poles, industrial wood, newsprint	20-year harvesting cycle
Nypa leaf	Roof thatch	3-year harvesting cycle
Honey	Food	Annual collection quotas
Fish	Food	Annual and seasonal catch quotas

Source: after Halls (1997).

Bay ecosystem has certainly been mixed, at least in part because of the many conflicting interests of different communities that rely on the ecosystem services of the catchment.

Many of the management decisions made at any level need to balance conflicting interests and must do so with the cost-effectiveness of particular strategies in mind. These issues are illustrated in the small area of Britain shown in Figure 7.1, where erosion continues to affect existing economic activity. Many farmers along the North Sea coast of Humberside are losing their fields at a rate of up to 2 m/year. The local council is sympathetic to calls for coastal protection schemes, but such projects are costly in financial terms and the priority given to individual farms is likely to be lower than that given to the gas terminals at Easington, which handle 25 per cent of Britain's gas supplies. An added complicating factor is the worry that sediment eroded from the North Sea coast is transported up the Humber estuary and deposited on the river banks. Halting the erosion could therefore increase the flood risk at Hull. The need for an integrated management plan for the coastal zone, which includes a thorough appreciation of marine and terrestrial sediment dynamics, is therefore paramount.

FURTHER READING

Burke, L., Reytar, K., Spalding, M. and Perry, A. 2011 *Reefs at risk revisited.* Washington, DC, World Resources Institute. Open access: www.wri.org/publication/reefs-at-risk-revisited.

Hogarth, P. 2015 *The biology of mangroves and seagrasses*, 3rd edn. Oxford, Oxford University Press. A well-written introductory text.

IOC-UNESCO and UNEP 2016 Large marine ecosystems: status and trends. Nairobi, UNEP. Assessment of 66 relatively large areas of coastal waters. Open access pdf: onesharedocean.org/public_store/publications/lmes-spm.pdf.

Spalding, M.D., McIvor, A.L., Beck, M.W. *et al.* 2014 Coastal ecosystems: a critical element of risk reduction. *Conservation Letters* 7(3): 293–301. A strong argument for the coastal protection benefits of natural ecosystems. Open access: onlinelibrary.wiley.com/doi/10.1111/conl.12074/full.

Swaney, D.P., Humborg, C. and Emeis, K. *et al.* 2012 Five critical questions of scale for the coastal zone. *Estuarine, Coastal and Shelf Science* 96: 9–21. An appraisal of how our understanding and management of coasts depend on drivers operating at multiple scales.

Thomas, N., Lucas, R., Bunting, P., Hardy, A., Rosenqvist, A. and Simard, M. 2017 Distribution and drivers of global mangrove forest change, 1996–2010. *PloS ONE* 12(6): e0179302. Open access: journals.plos.org/plosone/article?id=10.1371/journal.pone.0179302

WEBSITES

www.eucc.nl the European Union for Coastal Conservation is dedicated to the integrity and natural diversity of the coastal heritage, and to ecologically sustainable development.

www.futureearthcoasts.org Future Earth Coasts is a community of organizations, scientists and practitioners devoted to securing sustainable coastal futures in the Anthropocene.

www.gbrmpa.gov.au the Great Barrier Reef Marine Park Authority site.

www.glomis.com site of the global mangrove database and information system.

www.reefbase.org a global database on coral reefs and their resources run by a non-profit research organization.

www.worldfishcenter.org the World Fish Center is an international research organization dedicated to reducing poverty and hunger by improving fisheries and aquaculture.

POINTS FOR DISCUSSION

■ Do you agree that environmental issues in most coastal zones stem largely from the fact that there are simply too many people living near them?

■ Outline situations in which hard engineering structures might be preferable to soft engineering strategies for coastal protection, and vice versa.

■ Explain why overfishing is thought to be the root cause of ecological problems on many coasts.

■ Why might sea levels rise in coming decades, and what problems might this present to coastal environments?

■ Why should we conserve (a) coral reefs, and (b) mangroves?

■ Outline the sort of conflicts you might face in trying to balance social, economic and environmental demands along a particular stretch of coastline you know.

8 Rivers, lakes and wetlands

LEARNING OUTCOMES

At the end of this chapter you should:

- Appreciate the multiplicity of interacting drivers that affect the condition of rivers.
- Understand how the importance of fresh water as a resource has on occasion created zones of conflict.
- Recognize that river restoration is an increasingly important approach to management.
- Know how the Aral Sea has become an iconic example of human-induced environmental degradation.
- Understand the complex recent history of ecological change in Lake Victoria.
- Appreciate the wide variety of reasons for wetland drainage.
- Realize the value of wetlands and their ecosystem services.

KEY CONCEPTS

hydraulic civilization, interbasin water transfer, biochemical oxygen demand (BOD), catchment management plan, cultural eutrophication, Ramsar Convention

---

Most of the world's water is in the oceans, and the largest proportion of fresh waters is frozen in ice caps and glaciers or stored in reserves of groundwater. The water found in rivers, lakes and wetlands together constitutes less than 1 per cent of all the planet's fresh water (see Figure 1.7), yet their relative accessibility has given these sources a disproportionate importance to humanity. Since its earliest inception, human society has seen freshwater bodies as a vital resource, and entire ancient 'hydraulic civilizations' developed on certain rivers several thousand years ago, notably those of the Tigris–Euphrates, the Nile and the Indus, as irrigation agriculture played a formative role in the development of complex social organization. Society is no less reliant on this fundamental natural resource today, both directly and indirectly due to fresh water's critical importance to the functioning of numerous environmental processes, which also provide further resources and ecosystem services on which we rely. Fresh water from lakes, rivers and wetlands is utilized directly for municipal, agricultural, industrial, fisheries, recreational and power generation uses; it forms a convenient medium for transport and a sink for wastes.

Society's use of fresh water, its availability and quality have thrown up numerous issues of controversy, and these issues will become more acute as global water withdrawals continue to increase. Fresh water is a renewable resource, continuously available by virtue of the workings of the hydrological cycle, but this availability is finite in terms of the amount available per unit time in any one region. The amount diverted to service people worldwide has grown both as the human population has increased and as its demand per capita has risen. Globally, total water withdrawn for human uses increased by nearly eight times from about 500 km^3/yr in 1900 to about 4000 km^3/yr in 2010 (Wada *et al.*, 2016). Global data mask regional differences, however, which are essentially functions of climatic influences. All countries suffer from periodic excesses of fresh water in the form of floods, and deficits in the form of droughts, while others also experience perennial water shortages, in some cases due to a complete absence of permanent rivers (e.g. Malta and Saudi Arabia).

Surface freshwater ecosystems are among the most extensively altered on Earth, and an indication of the wide variety of ways in which human activity can adversely affect them is given by noting the effects upon freshwater fish, a biological indicator of ecosystem health (Table 8.1). This chapter deals with surface freshwater ecosystems: rivers, lakes and freshwater wetlands, although the specific issue of big dams is covered in Chapter 9, and some of the effects of particular activities such as deforestation, urbanization and

Table 8.1 Summary of main pressures facing freshwater fish and their habitats in temperate areas

Danger	Effects
Industrial and domestic effluents	Pollution, elimination of stocks, blocking of migratory species
Acid deposition	Elimination of stocks in poorly buffered areas
Land use (farming and forestry)	Eutrophication, acidification, sedimentation
Eutrophication	Algal blooms, deoxygenation, changes in species
Industrial development (including roads)	Sedimentation, obstructions, transfer of species
Warm water discharge	Deoxygenation, temperature gradients
River obstruction (dams)	Blocking of migratory species
Fluctuating water levels (reservoirs)	Loss of habitat, spawning and food supply
Infilling, drainage and canalization	Loss of habitat, shelter, food supply
Water transfer	Transfer of species and disease
Water abstraction	Loss of habitat and spawning grounds, transfer of species
Fish farming	Eutrophication, introductions, diseases, genetic changes

(continued)

Table 8.1 continued

Danger	Effects
Angling and fishery management	Elimination by piscicides, introductions
Commercial fishing	Overfishing, genetic changes
Introduction of new species	Elimination of native species, diseases, parasites
Water recreation	Disturbance, habitat loss

mining on the quality and quantity of freshwater bodies are covered elsewhere (see Chapters 4, 10 and 19). Wetlands found in coastal regions, such as mangroves and estuaries, are covered in Chapter 7, and issues surrounding underground sources of water are covered in Chapters 10 and 13.

RIVERS

Rivers are dynamic features of the physical environment and their dynamism is driven by many factors. The human impact, both direct and indirect, takes many forms. Direct manipulation of the river channel through engineering works, for example, includes channelization, dam construction, diversion and culverting. Some of these activities have very long histories. Deliberate diversion and regulation of the Huang Ho in China, for example, began more than 2000 years ago (Xu, 1993). Water is also commonly extracted from rivers for human use with consequent effects on channel geomorphology and ecology. Such direct and deliberate modifications are complemented by the impacts of land use in the area drained by the river (here referred to as the catchment, but also known as drainage basin, river basin or, in US English, watershed). Important land-use changes include deforestation, afforestation, agriculture and the incidence of fire, with perhaps the most extreme effects produced by construction activity and urbanization. Both the quality and quantity of river water can be altered, sometimes dramatically, at other times more gradually. Variability in rivers, driven by natural as well as human agents of change, inevitably results in dynamism being a characteristic of the resources associated with rivers, a dynamism that on occasion can transform a river into a hazard, when it floods for instance. This changeability can also be a source of conflict among different groups wanting to use them.

Multiple impacts

Many interacting drivers affect the condition of rivers, and drivers in turn interact with river responses to produce a complex set of consequences in any particular catchment. An example of multiple human impacts can be cited from Madagascar, where four of the island's 64 endemic freshwater species are feared extinct and another 38 are endangered due to three main factors: habitat degradation caused by deforestation, overfishing, and interactions with exotic species (Benstead *et al.*, 2000).

The numerous ways in which Madagascar's widespread deforestation has contributed to the degradation of aquatic habitats are shown in Figure 8.1. The effects consequent upon the decline of falling food items

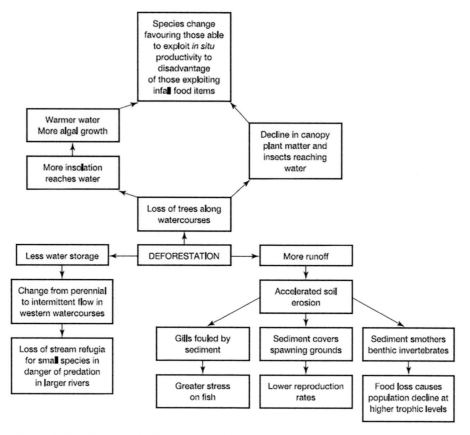

Figure 8.1 Degradation of aquatic habitats due to deforestation in Madagascar (after Benstead *et al.*, 2000).

(plant matter and insects) and the enhanced growth of algae due to warmer waters are frequently dramatic in the short term as fish that feed on falling food are edged out by those able to feed on algae. But these effects need not be irreversible as other tree species colonize river banks to replace primary forest, providing a new source of insects. The changes caused by a more regional decline in water storage and greater runoff are less readily reversed because the forest is usually cleared for crops that have a dampened effect on hydrology.

Problems caused by overfishing and exotic species are also more intractable. Given the rising demand for fish from a rapidly increasing human population and the great logistical difficulties faced in enforcing any sort of environmental regulations, overfishing is likely to remain an issue in Madagascar for a long time to come. Exotics introduced to the island include both aquacultural and ornamental species, and their impact on aquatic ecosystems has been profound. Naturalized exotics have completely replaced native fish in the central highlands of Madagascar and they are widespread in other parts of the island (Stiassny and Raminosoa, 1994).

There are many examples of multiple human impacts on riverine fish biology. Such effects on the River Seine in France date back 2000 years when the common carp became the river's first introduced species, brought by colonizing Roman legionnaires. It was followed in the Middle Ages by other species, including tench and rudd, which escaped from fish-farming ponds kept by noblemen and religious communities. In the late nineteenth century, further biological invaders arrived from rivers further east via canals, followed at the end of the century by deliberate introductions of a number of North American species. In the twentieth century, several native fish disappeared from the Seine as the construction of weirs and locks made it impossible for migratory species to reach their upstream spawning grounds. With the exception of the eel, all of the Seine's migratory species became extinct. The original fish fauna of the Seine probably consisted of about 30 species. Today the river has 46 species, but just 24 of them are native (Boët *et al.*, 1999). The Seine has also been subject to multiple water quality issues resulting from high population density, intensive agriculture and major industries, as well as significant channelization of the river's course, involving straightening of the channel and deepening of the bed, predominantly in the nineteenth century. The effects of the city of Paris are much more widely spread than might be expected, being discernible in water quality 200 km downstream and 300 km upstream in the form of large reservoirs built to protect the city from major floods and droughts (Meybeck, 2002).

Changes to a river over 300 years are reflected in Figure 8.2, which shows the varying rate of fluvial sediment deposition since the late seventeenth century in a river that flows into Chesapeake Bay on the eastern seaboard of North America. The amount of soil eroded from the catchment has altered significantly in response to changing land use in the area (Pasternak *et al.*, 2001). The sedimentation rate before the arrival of European settlers was about 1 mm/yr, but the rate increased by around eight times through early deforestation and agriculture (1750–1820). As deforestation and agriculture intensified (1820–1920), greater erosion followed, and sedimentation rates increased by another three times, peaking in 1850 at about 35 mm/yr. In total, up to 80 per cent of the basin has been deforested. During more recent times (1920 to present), urbanization – protecting soils – and the building of dams – blocking sediment delivery – have combined to reduce erosion and sedimentation by an order of magnitude. The sedimentation rate has been reduced nearly to the background conditions that prevailed in the pre-European settlement era.

River pollution

Sources of water pollution can be traced to all sorts of human activity, including agriculture, irrigation, industry, urbanization and mining. Pollutants may come from a point source, such as a factory or sewage works, or the source may be diffuse, contamination entering the water from runoff or land drainage. Much pollution can be described as chronic, received in the water regularly or continuously, but episodic events also occur, often causing greater damage because of their unpredictability in time and space. Most forms of water pollution can be classified into three categories:

- excess nutrients from sewage and soil erosion
- pathogens from sewage
- heavy metals and synthetic organic compounds from industry, mining and agriculture.

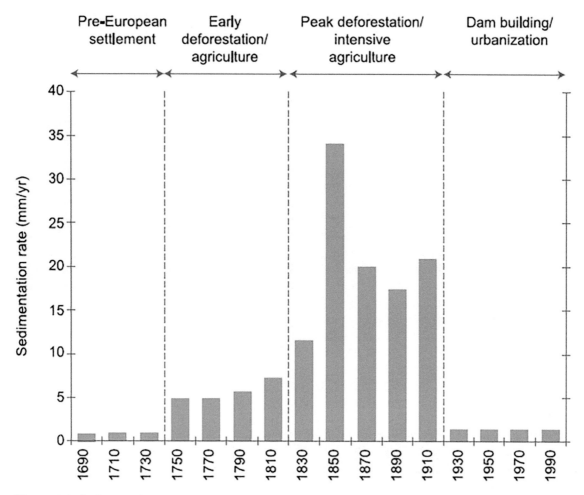

Figure 8.2 Sedimentation rate and historical land use in a catchment that drains into Chesapeake Bay, USA, 1690–1990 (after Pasternak *et al.*, 2001).

Three other forms should also be mentioned:

■ thermal pollution from power generation and industrial plants
■ radioactive substances
■ turbidity problems caused by increased sediment loads or decreased water flow.

Organic liquid wastes can be broken down by bacteria and other micro-organisms in the presence of oxygen, and the burden of organics to be decomposed is measured by the biochemical oxygen demand (BOD). Liquid organic wastes include sewage, many industrial wastes (particularly from industries processing agricultural products) and runoff, which picks up organic wastes from land. As a river's dissolved oxygen decreases, with increasing loads of organic wastes, so fish and aquatic plant life suffer and may eventually die. Heavy volumes of organic wastes can overload a riverine system to the point at which all dissolved oxygen is exhausted.

Pollution of water due to temperature increase, so-called 'thermal pollution', also reduces its dissolved oxygen content. This occurs in two ways. An increase in water temperature decreases the solubility of oxygen, on the one hand, and increases the rate of oxidation, thereby imposing a faster oxygen demand on a smaller content, on the other. Thermal pollution also has a number of more direct adverse effects on river ecology, including a general increase in undesirable forms of algae, and reduced reproduction and growth of some species of fish (Caissie, 2006).

Inorganic liquid wastes become dangerous when not adequately diluted. Even in very small concentrations, however, some heavy metals (such as cadmium, lead and mercury) are particularly dangerous and can bioaccumulate up the food chain, ultimately damaging human health.

Pathogens from human waste spread disease and represent the most widespread contamination of water. Water-related diseases can be classified into those that are water-borne (e.g. diarrhoea, cholera and polio), those that are related to a lack of personal cleanliness (e.g. trachoma and typhoid) and those that are related to water as a habitat for certain disease vectors (e.g. schistosomiasis, malaria and onchocerciasis).

Agriculture is the leading non-point source of pollutants such as sediments, pesticides and nutrients, particularly nitrogen (N) and phosphorus (P). Data on pesticide contamination of rivers are not sufficient to give a clear picture of the situation even in some of the better-monitored parts of the world, but rising levels of usage give reasonable cause for concern, particularly in the developing countries where many pesticides now banned in the developed world are still used (see p. 321). The growing use of synthetic chemical fertilizers, fuelled by increasing demands for food from a growing population, has led to a steady rise in the concentration of N and P in many world rivers and groundwaters in recent times, by about 30 per cent overall between 1970 and 2000 according to estimates made by Seitzinger *et al.* (2010). Although riverine loads increased on all continents over this 30-year period, the greatest increases and largest total loads occurred in southern Asia.

Another important agricultural driver of nitrate that makes its way into rivers is land-use change: the ploughing of permanent grassland leading to reduced levels of soil organic matter and the long-term release of soil nitrogen. The large-scale conversion of grassland to arable farming has had a significant effect on nitrate concentrations in the UK's River Thames which have risen from about 2 mg/l in the first half of the twentieth century to almost 8 mg/l since the 1970s (Howden *et al.*, 2010), and have remained stubbornly high despite Europe-wide interventions to reduce catchment nitrogen inputs since the early 1980s (Figure 8.3).

Riverine pollution from sewage (also another source of nitrogen) is a further acute water quality issue for many of the world's poorer countries where wastewater treatment is inadequate or non-existent. In 2015, the proportion of national population connected to a wastewater treatment plant through a public sewage network was just 12 per cent in Costa Rica, for example (OECD, 2017), although even in some of the richer nations the level of treatment is still far from universal: the equivalent proportions being 58 per cent in Slovenia and 66 per cent in Eire. Sewage effluent can contain many other pollutants, of course. A study of pharmaceuticals in river and tap waters in the Polish capital, Warsaw, highlighted their inadequate elimination in the waste treatment facilities that provide drinking water (Giebułtowicz and Nałęcz-Jawecki, 2014).

In places where the economic power and political will are available, concerted efforts to improve river water quality have scored some notable successes. Improvements have been made to the ecology of Europe's Rhine in recent decades due to a number of interrelated efforts including creation of an International Commission for the Protection of the Rhine (ICPR) in 1950. Improvement in the river's oxygen conditions from a low point in the early 1970s has facilitated a rise in the number of species (Figure 8.4), including the

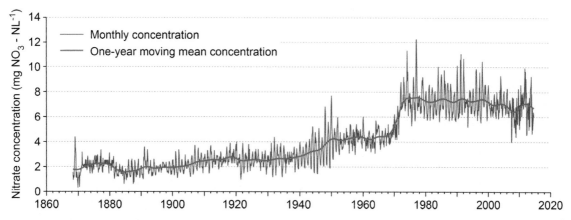

Figure 8.3 Nitrate concentration in the River Thames at Hampton, 1868–2014 (updated after Howden *et al.*, 2010).

return of some that had been considered extinct or severely reduced in numbers. However, many typical indigenous species remain absent, partly because their habitats no longer exist, and because their ecological roles have subsequently been occupied by alien species (Leuven *et al.*, 2009). Another indication of how the Rhine's ecological rehabilitation has not been a smooth process is reflected in the saga of attempts to resolve chloride pollution from mining (see Box 19.1).

In terms of organic wastes, the history of the River Thames provides a clear example of the reversibility of many sorry stories of riverine pollution. The quality of the Thames declined in the nineteenth century as London grew and the flushing water closet became widely used, discharging human excreta into the river. Five cholera epidemics occurred between 1830 and 1871 (see Figure 21.9) and during the long, dry summer of 1858, the so-called 'Year of the Great Stink', Parliament had to be abandoned on some days because of the stench. Such a direct impact on the nation's legislators produced positive action, and conditions had improved by the 1890s with the introduction of sewage treatment plants. During the first half of the twentieth century, however, sewage treatment and storage did not keep pace with the growing population, and dissolved oxygen reached zero at 20 km downstream of London Bridge during many summers. In 1957, scientists at the Natural History Museum declared that the central London section of the Thames was biologically dead.

Following tighter controls on effluent and improved treatment facilities introduced in the latter half of the twentieth century, improvement was gradual, with the mid-1970s generally taken as the time when water quality reached satisfactory levels. In 1974, much publicity accompanied the landing of the first salmon caught in the Thames since 1833, but the river's high levels of nitrate (Figure 8.3) have to date been impossible to reduce.

The ebb and flow of particular water pollution sources in the highly industrialized countries is summarized by Meybeck *et al.* (1989) who outline how the growth of urban areas during the Industrial Revolution created the first wave of serious water pollution, from domestic sources, around the turn of the last century. These have been superseded by industrial pollutants, and during the second half of the twentieth century by nutrient pollution (particularly from agricultural sources) and micro-organisms towards the end of the 1990s. A similar pattern of peaks in serious pollution problems is being experienced in the rapidly industrializing countries, but compressed into the period since 1950, and it is here that some of the most serious

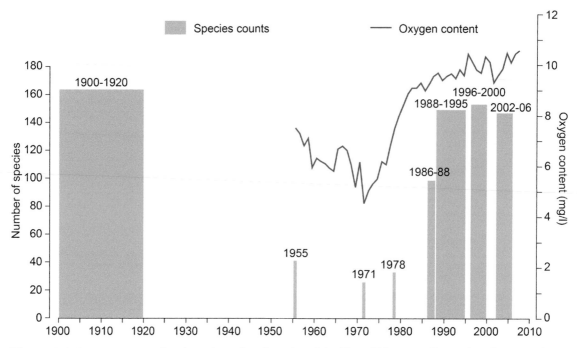

Figure 8.4 Average oxygen levels and species diversity of the River Rhine near Emmerich, Germany (after EEA, 2012).

water pollution problems are being faced today (see p. 228). The other regions where water pollution has reached crisis proportions are in some countries of the former Soviet bloc. Of course, this is not to suggest that all water pollution issues have been resolved in rivers of the global North, as illustrated by the persistently high nitrate levels in the Thames. Indeed, studies of the emerging issue of microplastic pollution have found some of the highest concentrations in river sediments in northern Europe and North America (Hurley *et al.*, 2018).

As with many other pollution issues, it is often the poorer sectors of society that suffer most from the detrimental effects of poor water quality. Nevertheless, most sources of water pollution are well known, and methods for reducing them or their detrimental effects have been devised. Most of today's continuing water pollution problem areas are due to a lack of political will, poor governance both nationally and internationally and/or a lack of funds.

River conflicts

The importance of fresh water as a resource, allied to its uneven geographical distribution in rivers, lakes and aquifers, has inevitably led to political wrangling over the rights of different groups to use water. Some of the oldest rules for water use have been documented in Mesopotamia, dating back as far as 3000 BCE, and it was probably during the time of Islamic rule in Spain that the Tribunal de las Aguas (Water Court) was established on the Plain of Valencia. Set up to adjudicate over water conflicts and problems in the tenth century, it continues to this day, making it one of the oldest active courts in Europe.

On occasion, disagreement over rights to shared water resources can lead to military confrontation, and the notion that water wars may become a leading source of conflict in the twenty-first century has become widespread in some academic and journalistic circles, as well as in political rhetoric (see Chapter 20). However, recent history indicates that armed conflict over international water resources is rare. Wolf *et al.* (2003) found that instances of cooperation over water far outnumbered those leading to conflict when they assessed 50 years of interactions between two or more nations, where water (as a scarce and/or consumable resource) was the driver of events.

This is not to say that water is not frequently the cause of political tension and disputes. International river catchments cover no less than 45 per cent of the world's land surface excluding Antarctica, and a third of these transboundary basins are shared by more than two countries (Giordano and Wolf, 2003), so such tensions are very likely. Common grounds for disagreement over a cross-border river frequently focus on the allocation of rights to volumes of water but may also revolve around issues of quality. Examples include a downstream state's objection to pollution, dam construction, or excessive irrigation by an upstream state, actions that will decrease or degrade the quality of water available downstream. Many of these disputes are settled peacefully by international treaty, but many are not.

Further, not all international treaties are designed to address all players in an international river dispute. A classic example can be quoted from Africa where control of the Nile's waters has long been contested among the ten African states that occupy the river's catchment. Egypt and Sudan reached an international agreement in 1959 governing the volume of Nile water allowed to pass through the Aswan High Dam, but none of the other eight Nile basin countries have agreements over use of the Nile's waters. Given that Egypt and Sudan are the last two states through which the river flows before entering the Mediterranean, similar agreement over water rights with countries upstream is desirable. An important step in that direction was taken in 1999 with the establishment of the Nile Basin Initiative, the first cooperative institution in the basin to include all ten riparian states. However, hammering out the details of a catchment-wide treaty has, to date, proved insurmountable and differences in opinion over rights to Nile water continue to underlie many of the political issues in that part of the world (Cascão, 2009), most recently evident in the downstream reaction to construction of the Grand Ethiopian Renaissance Dam on the Blue Nile River.

Disputes over rivers are not limited exclusively to inter-state conflicts, and tensions frequently emerge at levels other than the nation-state. The Mekong basin illustrates this point, with issues at both international (Figure 8.5) and subnational levels (Sneddon and Fox, 2006). Like many of the world's international river basins, it is simultaneously viewed as an engine of regional economic development, as a crucial basis of local livelihood resources, and also as a vital site for the conservation of biodiversity. Critics of the 1995 Mekong Agreement between four of the catchment's nation-states see the treaty as overly focused on the Mekong's huge hydroelectric potential and capacity to store water for irrigation schemes. Development of this potential is inevitably concentrated at the national level, often with assistance from international development partners such as banks and other governments. These bodies, it is argued, view the Mekong's resources as under-utilized and ripe for development, but the fear is that this stance will marginalize the activities of local resource users who depend on the river for sustenance and livelihoods.

River management

River modification and management are rooted in society's view of rivers both as hazards to be ameliorated and resources to be used. Hence approaches to river management depend upon many different, often

Figure 8.5 The Mekong River basin drains parts of six countries. Hydropower development and the improvement of river navigation, in China particularly, are producing impacts downstream. Long boats using this stretch of the Mekong in Laos are unlikely to be affected but fisheries from the village in the background are expected to be adversely impacted by the ecological changes.

interrelated, river functions, such as navigation, drinking water supply, nature conservation, recreation and flood risk. There are numerous ways in which people have modified rivers as a means of managing these issues: by constructing dams, diverting channels, draining land, building levées, widening, deepening and straightening channels (Table 8.2; see also Chapters 9 and 21). Many of these modifications have very long histories. Downs and Gregory (2004) identify six major chronological phases dating from the earliest classical hydraulic civilizations, although not all phases apply in every geographical region. The beginnings of direct river modification in the UK, for example, are traced back to the first century CE with fish ponds and water mills, and changes to facilitate transportation and to drain land. In England, the medieval period – roughly 950–1250 – was a particularly dynamic era in the history of river engineering, including the digging of canals bypassing difficult stretches of rivers, linking rivers to important production centres, and draining wetlands (Blair, 2007). Severe pollution in the mid-nineteenth century led to an organized system of river management. A marked east–west rainfall gradient, combined with an uneven population distribution (a high proportion live in eastern and south-eastern England where the rainfall is low at 600–700 mm/year), means that many rivers in central and southern England, and many urban rivers throughout the UK, are overexploited. Today, most towns and cities rely on interbasin water transfers, and virtually all

RIVERS, LAKES AND WETLANDS **177**

Table 8.2 Selected methods of river channelization and their US and UK terminologies

Method	US term	UK equivalent
Increase channel capacity by manipulating width and/or depth	Widening/deepening	Resectioning
Increase velocity of flow by steepening gradient	Straightening	Realigning
Raise channel banks to confine floodwaters	Diking	Embanking
Methods to control bank erosion	Bank stabilization	Bank protection
Remove obstructions from a watercourse to decrease resistance and thus increase velocity of flow	Clearing and snagging	Pioneer tree clearance; control of aquatic plants; dredging of sediments; urban clearing

Source: after Brookes (1985).

the major rivers in the UK are regulated directly or indirectly by mainstream impoundments, interbasin transfers, pumped storage reservoirs or groundwater abstractions.

The current phase of river management identified by Downs and Gregory (2004) dates from the late twentieth century, since when conservation and the sustainable use of rivers have become characteristic management objectives. This perspective can be seen as a reaction to the widespread alterations to rivers and their floodplains associated with industrialization, urbanization and agricultural intensification detailed in the previous sections. Attempts to take a properly integrated approach to river management have resulted in a great emphasis on 'catchment management plans', so incorporating management of human activities in the entire basin. However, the importance of making this connection between the channel and its basin is not in itself new. Governmental regulation of timber harvesting along mountain streams so as to maintain channel stability dates back to 806 CE in Japan, for example (Wohl, 2006).

The more precise aims of contemporary sustainable river management in general and catchment management plans in particular depend in part on how 'sustainable' is defined. Nonetheless, four dimensions of sustainable river management are commonly recognized (Gregory *et al.*, 2011):

- the move towards catchment-scale planning (as opposed to reach-scale applications)
- increasing community participation in decision-making
- an ecosystem-based approach
- implementing adaptive management.

In practice, specific objectives can be conflicting, and the relative weight given to each by decision-makers is affected by a wide range of influences that include economic, political and environmental considerations as well as uncertainty in its various forms (see p. 54). River management is no different from any other natural environmental management issue in that it involves compromises.

Objectives also change, for all sorts of reasons. An interesting case in point is the recent history of Cheonggyecheon, a short stream that has flowed through the South Korean capital, Seoul, since its establishment in 1394. In the 1960s, during South Korea's rapid industrialization, Cheonggyecheon was regarded as an obstacle to urban development, concreted over and replaced by an elevated ten-lane highway. However, in the early 2000s the metropolitan government announced it would demolish the ageing highway and restore the stream. After initial public concern at the huge cost, the restoration of Cheonggyecheon is now widely considered to be a great success (Lee and Anderson, 2013), having re-created a waterscape with historical and cultural significance that runs through the heart of the city (Figure 8.6).

In European Union (EU) countries, the EU Water Framework Directive is a powerful driver of river management policies and the directive advocates a 'strong' sustainability platform for management, in which ecosystem protection and the valuation of natural goods and services figure prominently (Newson, 2002). The logic is, of course, simple: the ecosystem health of many degraded rivers must be improved and/ or maintained so that river systems can continue to provide ecosystem goods and services into the future. River restoration is, therefore, an increasingly important approach to management and there are numerous

Figure 8.6 Residents of Seoul enjoy Cheonggyecheon, a short stream that was restored in 2005.

ways in which degradation can be rectified. An example can be cited from Finland where historically extensive damage to migratory salmonid fish stocks occurred due to the dredging of rivers and brooks both to facilitate boat traffic and, increasingly during the last century, for timber-floating. Timber-floating has now almost completely ceased and Finnish water legislation obliges water authorities to make good any damage caused by dredging. The restoration of rapids and their restocking have been increasingly used to rehabilitate degradation caused by dredging (Jutila, 1992).

However, although the transition from a traditional management approach based almost entirely on human demands to a more ecosystem-focused approach is well understood in intellectual terms in Europe and elsewhere, examples of a real shift in practice are relatively few and far between. Common obstacles to the successful implementation of genuinely sustainable river management include fragmented land ownership and ineffective governance (both institutions and organizations) as well as uncertainty stemming from our incomplete understanding of the links between the numerous biophysical processes operating in river catchments.

LAKES

Lacustrine degradation

As with rivers, human use of fresh water from lakes has led to numerous impacts. In some cases, entire lakes have dried up as rivers feeding them have been diverted for other purposes. Owens Lake in California, USA, is a case in point. Levels began to drop in the second half of the nineteenth century due to offtakes for irrigated agriculture, but an accelerated decline in water levels occurred with the construction of a 360-km water export system by the Los Angeles Department of Water and Power during the first 20 years of the last century. The lake, which was 7.6 m deep in 1912, had disappeared by 1930. In its place, the dry bed of lacustrine sediments covering 220 km^2 is a source of frequent dust storms, which are hazardous to local highway and aviation traffic, often exceed California State standards for atmospheric particulates and have increased morbidity among people suffering from emphysema, asthma and chronic bronchitis (Halleaux and Rennó, 2014). From 1930 the Los Angeles system was extended northwards and began to tap streams flowing into the saline Mono Lake in 1941. By 1990, Mono Lake had fallen by 14 m, doubled in salinity, and lost a number of freshwater habitats, such as delta marshes and brackish lagoons that formerly provided lakeside habitats for millions of birds. A repeat of the Owens Lake story was avoided in 1994 when the Los Angeles Department of Water and Power was ordered to ensure protection for Mono Lake and its streams and to effect some restoration of the damaged habitats. Since then, some improvement has been noted in lake level (Figure 8.7) and salinity, although a record drought in California began in 2012. The California Supreme Court ordered that Mono Lake should be returned to a level of 1948 m, a target that will trigger changes in water export rules. It is thought that maintaining this level will be sufficient for a healthy lake ecosystem and acceptable air quality.

Similar offtakes, in this case for agriculture, have had dramatic consequences for the Aral Sea in Central Asia. Diversion of water from the Amu Darya and Syr Darya rivers to irrigate plantations, predominantly growing cotton, has had severe impacts on the inland sea since 1960. Expansion of the irrigated area in the former Soviet region of Central Asia, from 2.9 million hectares in 1950 to about 7.2 million hectares by the late 1980s, was spurred by Moscow's desire to be self-sufficient in cotton. As a result, the annual inflow to the Aral from the two rivers, the source of 90 per cent of its water, had declined by an order of magnitude between the 1960s (about 55 km^3/year) and the 1980s (about 5 km^3/year).

Changes in the level of the Aral Sea have occurred naturally throughout its history, and sea-level variations of up to 36 m are thought to have occurred during the Quaternary due to climatic, geomorphological

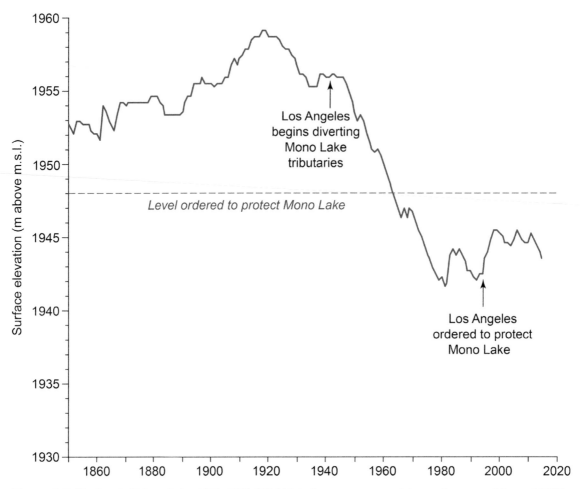

Figure 8.7 The level of Mono Lake, USA, 1850–2016 (data from www.monolake.org/, accessed August 2017).

and tectonic influences (Middleton, 2002). Human use of water resources in the Aral Sea basin dates back more than 3000 years but the unprecedented intensity of water use dating from about 1960 has clearly been responsible for the falling sea levels since then.

In 1960, the Aral Sea was the fourth largest lake in the world, but since that time its surface area has shrunk by 85 per cent (Figure 8.8), it has lost 90 per cent of its volume, and its water level has dropped by more than 25 m. The average water level in the Aral Sea in 1960 was about 53 m above sea level. By 2003, it had receded to about 30 m above sea level, a level last seen during the fourteenth to fifteenth centuries, but in those days for predominantly natural reasons (Boroffka et al., 2006). In some parts, the Aral Sea's remaining waters are several times saltier than sea water in the open ocean.

These dramatic changes have had far-reaching effects, both on-site and off-site. The Aral Sea commercial fishing industry, which landed 40,000 tonnes in the early 1960s, had ceased to function by 1980 (Figure 8.9) as most of its native organisms died out. The delta areas of the Amu Darya and Syr Darya rivers have been transformed due to the lack of water, affecting flora, fauna and soils, while the diversion of river water has

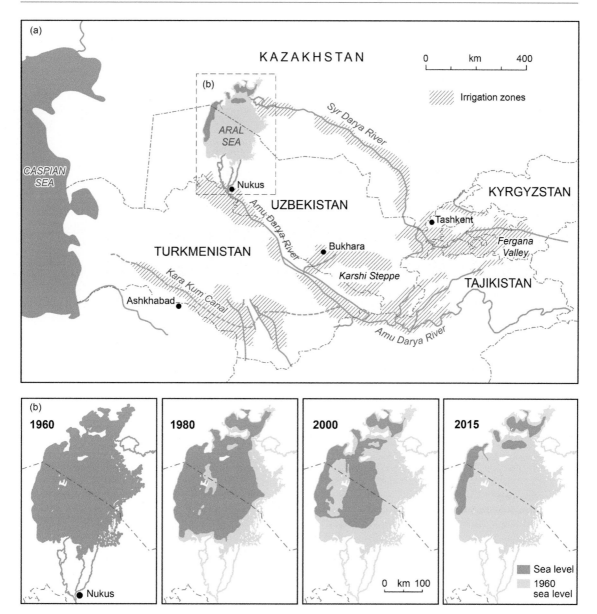

Figure 8.8 (a) Irrigated areas in Central Asia (after Pryde, 1991); and (b) changes in the surface area of the Aral Sea (UNEP, 1992b; Issanova *et al.*, 2015).

also resulted in the widespread lowering of groundwater levels. The receding sea has had local effects on climate, and the exposed sea bed has become a dust bowl from which up to 150 million tonnes of saline material is deposited on surrounding areas each year (Issanova *et al.*, 2015). This dust contaminates agricultural land several hundred kilometres from the sea coast, and is suspected to have adverse effects on human health. The irrigated cropland itself has also been subject to problems of *in situ* salinization and

Figure 8.9 Abandoned trawlers near the former Aral Sea fishing village of Zhalangash, Kazakhstan, victims of perhaps the most extreme example of human-induced environmental degradation in the modern era.

waterlogging due to poor management, with consequent negative effects on crop yields. Drainage water from these schemes is characterized by high salinity and is contaminated by high concentrations of fertilizer and pesticide residues, which have also been linked to poor human health in the region (Crighton *et al.*, 2011).

The declining water levels led to a split in the sea in 1989, creating a small northern lake in Kazakhstan fed by the Syr Darya and a much larger southern lake, divided between Kazakhstan and neighbouring Uzbekistan, fed by the Amu Darya. The southern Large Aral has continued to shrink but the size of the northern Small Aral stabilized in 1996. Since completion in 2005 of a concrete dam blocking a narrow spillway linking the two water bodies, the Small Aral's water content has actually increased for the first time in half a century.

The ecological demise of the Aral Sea and its drainage basin is one of the late twentieth century's foremost examples of human-induced environmental degradation. But it was not the result of ignorance or lack of forethought, since anecdotal or informal projections of disaster were made in some Soviet quarters well before the situation became serious. Seemingly it is the result of an unspecified cost–benefit analysis in which environmental and health impacts were given little consideration (Glantz *et al.*, 1993). The desire to be self-sufficient in textiles, combined with the Soviet belief that human society was capable of complete domination over nature, overrode any fears of the associated environmental implications. It is an example of the dangers of an extreme technocentric approach to the environment and a complete rejection of the precautionary principle.

By contrast, there are numerous examples of lakes being degraded inadvertently, many of which are due to lacustrine sensitivity to pollutants. Some examples of problems caused by pollution, particularly acidification from industrial and mining activity, are covered in Chapters 12 and 19. One of the world's most pervasive water pollution problems, and one that is widespread in all continents, is the eutrophication of standing water bodies (Brönmark and Hansson, 2002). Eutrophication is a natural phenomenon in lakes, brought on by the gradual accumulation of organic material through geological history. But human activity can accelerate the process, so-called 'cultural eutrophication', by enriching surface waters with nutrients, particularly phosphorus and nitrogen. Such increased nutrient concentrations in lakes have been attributed to wastewater discharge, runoff of fertilizers from agricultural land and changes in land use that increase runoff.

Eutrophication is an important water quality issue that causes a range of practical problems. These include the impairment of the following: drinking water quality, fisheries, and water volume or flow. The main causes of these problems include algal blooms, macrophyte and littoral algal growth, altered thermal conditions, turbidity and low dissolved solids. Health problems range from minor skin irritations to bilharzia, schistosomiasis and diarrhoea.

Monitoring of Lac Léman (Lake Geneva) on the border between France and Switzerland indicates how the build-up of nutrient pollutants can be rapid (Figure 8.10). Water quality deteriorated from its relatively clean state during the 1950s as concentrations of phosphorus increased, primarily from point-source

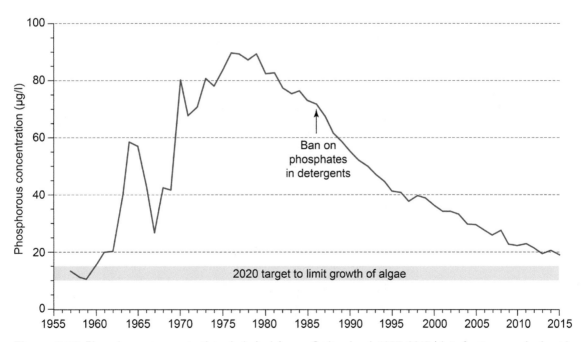

Figure 8.10 Phosphorus concentrations in Lake Léman, Switzerland, 1957–2015 (data from www.cipel.org/, accessed July 2017).

discharges, to reach a critical stage in the late 1970s. In this case the situation has improved following the introduction of better sewage treatment plants and a ban on the use of phosphates in detergents in 1986 (Anneville *et al.*, 2007). Of course, the local anthropogenic impact has also taken place alongside global drivers of environmental change and this became clear in Lake Léman in the mid-1990s. As the water quality management measures were having an effect, major changes appeared in the relative abundance of fisheries landings and it seems that local management to reduce phosphorus loadings played a major role in the recovery of whitefish. At the same time, rising water temperatures due to global warming have increased whitefish larval growth rates and improved whitefish recruitment so that, unexpectedly, climate change and phosphorus reduction have had synergistic effects on the lake's fish population (Anneville *et al.*, 2017).

The introduction of non-native species to lakes is another way in which, often unwittingly, the human impact has caused serious ecological change. The documented history of invasion by aquatic alien species in the North American Great Lakes dates back to the early nineteenth century (Figure 8.11) and includes both deliberate (e.g. via fish stocking or plant cultivation) and accidental releases, most of the latter via the discharge of ballast water from ocean-going vessels. Overall, the North American Great Lakes basin has been invaded by at least 182 non-indigenous species over this period (Ricciardi, 2006), and some of these introductions have had considerable environmental and economic impacts (see p. 395).

Deliberately introducing fish species to lakes to provide employment and additional sources of nutrition was a common component of economic development projects during the 1950s and 1960s. Predatory species such as largemouth, bass and black crappie, introduced into the oligotrophic Lake Atitlan in Gua- temala, succeeded in wiping out many of the smaller fish that had previously been used by indigenous

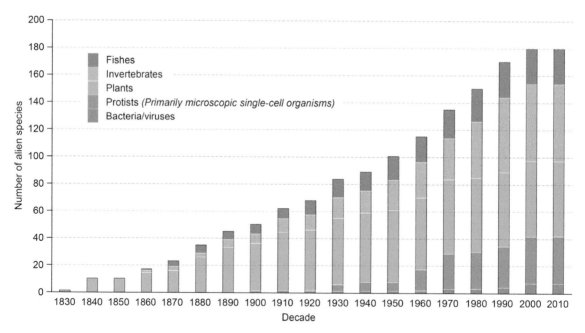

Figure 8.11 Cumulative number of aquatic alien species in the North American Great Lakes by decade (data from sobr.ca/indicator/alien-species-great-lakes/, accessed July 2017).

lakeside people. Similar dramatic effects on the pelagic food web resulted from the introduction of the Amazonian peacock bass to Gatun Lake in the Panama Canal. One serious consequence of this predator's elimination of the lake's smaller native fishes was a dramatic increase in malaria-carrying mosquitoes, since the minnows that had previously occupied the shallower lakeside waters had fed off mosquito larvae (Fernando, 1991). As it happens, there have also been many attempts to control diseases mainly transmitted to humans through mosquitoes (e.g. malaria, chikungunya, yellow fever, zika, and dengue) by introducing non-native fish species to lakes to control mosquito populations, although the efficacy of using these predators for mosquito-population control is controversial and largely unproven (Azevedo-Santos *et al.*, 2016). The relative merits, or otherwise, of ecological changes driven by exotic introductions also depend to a degree on the stance taken towards environmental issues, as illustrated by the case study of Lake Victoria in Box 8.1, an ecosystem that has surprised many biologists with its resilience.

Box 8.1 Ecological change in Lake Victoria

East Africa's Lake Victoria, the largest tropical lake in the world, provides an intriguing example of ecological change driven primarily by exotic introductions. Until the 1970s, the ecology of Lake Victoria was dominated by about 500 species of an endemic fish called cichlids. These fish were extremely diverse ecologically and are thought to have evolved from a few original species in just 15,000 years.

Profound ecological change in Lake Victoria is traced to the early 1960s when the Nile perch, a large predatory species, was introduced to boost the lake's fishing industry. In the first half of the 1980s, the Nile perch suddenly boomed in numbers and cichlids declined dramatically. During the same period, eutrophication of the lake increased strongly and blooms of blue-green algae became common due to a greater influx of nutrients from deforestation, industrialization and other forms of pollution. As many as 200 cichlid species were thought to have been driven to extinction by the combination of predation from the Nile perch and the effects of rising eutrophication (which were probably accelerated in a positive feedback by the loss of some cichlids that fed on algae).

This clear example of a human-induced mass extinction was widely condemned from an ecocentric perspective. However, the Nile perch had generated considerable benefits, expanding an artisanal fishery into a multimillion-dollar export industry for Nile fillets, while improving the incomes and welfare of lakeside communities. Annual fish landings rose from about 40,000 tonnes in the early 1970s to peak at around 550,000 tonnes in the late 1980s. A more technocentric view saw the loss of some cichlid species as the price paid for these social benefits.

In the 1990s, however, further changes were observed in the ecology of Lake Victoria. Catches of Nile perch declined, largely because of overfishing, and the cichlids showed signs of recovery. Evidence emerged indicating that some cichlid species have adapted to the lake's changing conditions and there has been much debate about how many cichlids may have become extinct. Certainly it seems that cichlids are much more resilient than was once thought. Water quality in the lake has stabilized, thanks in part to the construction of new sewage works, although deforestation and erosion are still significant issues that need to be addressed with sustainable land management policies.

Source: Awiti (2011).

Another range of problems common to many lakes are those stemming from the introduction of plant species, particularly waterweeds. The water hyacinth – a native of South America – is a frequent culprit, growing very fast, covering the water surface (so reducing light and oxygen for other plants and for fish), increasing evaporation, blocking waterways and access to fishing grounds, presenting a serious threat to hydroelectrical turbines and providing a suitable habitat for dangerous organisms such as snakes and disease-carrying mosquitoes. The water hyacinth became a major problem in Lake Victoria in the 1990s, although it rapidly declined thanks to control efforts using a weevil and the impact of the cloudy, wet El Niño weather of 1997/98 (Williams *et al.*, 2005).

Finally, in this section, we return to the effects of climate change on lakes. Global warming will affect precipitation and evaporation and consequently lake levels, as well as numerous aspects of lake ecology, as in the Lac Léman example above. Interestingly, surface water temperatures in several lakes have been increasing more quickly than regional air temperatures (Schneider and Hook, 2010). Some of the world's largest lakes are following this trend, which was not expected because these very large masses of water were thought to warm slowly. However, a worldwide synthesis of lake data (O'Reilly *et al.*, 2015) concluded that warming is not occurring everywhere – about 10 per cent of lakes surveyed were cooling – and that surface water warming rates are dependent on combinations of climate and local characteristics. The most rapidly warming lakes are widely distributed geographically, and their warming is associated with interactions among different climatic factors: some ice-free lakes, for example, are experiencing increases in air temperature and solar radiation; some seasonally ice-covered lakes are located in areas where temperature and solar radiation are increasing while cloud cover is diminishing. The lakes with the fastest warming trends are shown in Table 8.3. All are outside the tropics and most are in the northern hemisphere.

Table 8.3 Lakes with fastest warming trends

Lake name (country)	Latitude/ longitude	Years of data	Surface water temperature trend (°C/decade)
Fracksjön (Sweden)	58.15N 12.18E	22	1.35
Superior (Canada/USA)	47.58N 86.59W	25	1.16
Kangaroo Creek Reservoir (Australia)	34.86S 138.79E	13	1.14
Iliamna Lake (USA)	59.47N 155.60W	14	1.03
L223 (Canada)	49.75N 93.75W	14	1.00
Kremenshugskoye (Ukraine)	49.23N 32.79E	23	
Stensjön (Sweden)	58.10N 14.83E	21	0.95
Segozero (Russia)	63.27N 33.71E	20	0.92
Lappajärvi (Finland)	63.27N 23.63E	25	0.89
Rybinskoye (Russia)	58.41N 38.51E	24	0.89

Source: after O'Reilly *et al.* (2015)

Lake management

The restoration and protection of lakes present all sorts of challenges for environmental managers and policy-makers. Because the health of a lake is intimately related to its catchment and to changes humans have made to the lake and surrounding landscape, most of the management plans that exist for lakes are – like those for rivers – based on the principles of a catchment approach (Borre *et al.*, 2001):

- The geographical focus for management includes the lake and its entire catchment.
- Citizen and stakeholder involvement is important throughout the planning and management process.
- Mechanisms are needed to promote cooperation among government bodies and civil society organizations (CSOs) in the catchment.

Creating effective institutional arrangements for implementing this catchment approach is one of the most challenging and important issues facing the world's lakes. The mechanisms used to create them range from cooperative agreements for purposes of policy and planning to national laws and international treaties with full regulatory powers. The management experience of the North American Great Lakes illustrates some of the difficulties involved. A major restoration effort has been in progress throughout the Great Lakes Basin since the US and Canadian governments committed themselves to the Great Lakes Water Quality Agreement in 1972. The agreement adopted an ecosystem approach, which incorporates the political and economic interests of the 'institutional ecosystem' as well as elements of the natural ecosystem, after two previous management strategies had failed to address fully the dangers faced by the lakes. Current efforts are focused on reducing pollution entering the lakes, incorporating the 'critical loads' approach adopted in studies of acidification (see p. 291). Serious pollutants include volatile organic compounds (VOCs), heavy metals, and industrial and agricultural chemicals, which come from both point and non-point sources such as municipal and industrial effluent, rainfall and snowmelt. Special and immediate attention has been given to 43 severely degraded so-called Areas of Concern. Collingwood Harbour, on the Canadian side of Lake Huron, was the first of these sites to be removed from the list of Areas of Concern, in 2000, following successful restoration and protection efforts. The first US site to be formally delisted was Oswego River in 2006, but efforts to revitalize the Great Lakes still require improvements to governance that fully accommodate the complexity of linked social and ecological systems (McLaughlin and Krantzberg, 2011).

The management challenges in some of Africa's lakes have similarities to those just described in terms of unsustainable usage – albeit that pollution tends to be less of an issue – with some additional factors, not least considerable fluctuations in water levels and the high levels of poverty that characterize many of their users. The status of two international African lakes is shown in Table 8.4. Lake Chilwa is shared between Malawi and Mozambique, while the waters of Lake Chad are shared between four countries – Niger, Chad, Cameroon and Nigeria – although declining rainfall and irrigation offtakes have reduced the water available dramatically since the 1960s. A number of management initiatives have been put in place in both lakes, with some success, in efforts to maintain the lakes' ecosystem services and their users' livelihoods.

WETLANDS

The term 'wetland' covers a multitude of different landscape types, located in every major climatic zone. They form the overlap between dry terrestrial ecosystems and inundated aquatic ecosystems such as rivers,

Table 8.4 Environmental issues and adaptive responses in two African lakes

	Lake Chad	Lake Chilwa
Basin population (livelihood support)	37 m (30 m dependent on agriculture and fisheries)	1.5 m (all dependent on agriculture and fisheries)
Lake dynamics	Reduction from 25,000 km² to 500 km² since 1960s	Periodic drying over 10–15-year cycles
Main environmental issues	Reduced river flow; climate variability; habitat and community modification; unsustainable resource exploitation; siltation	Deforestation; wetland degradation; climatic variability; reduced river flows; high population growth; siltation
Institutional framework	Lake Chad Basin Commission	District councils, no basin-wide authority
International treaties	None	Ramsar site; UNESCO Man and Biosphere reserve
Adaptation measures	Efficient irrigation technologies; interbasin water transfer	Afforestation; efficient irrigation technologies; development of small and medium enterprises; commodity value addition

Source: after Kafumbata *et al.* (2014), tables 1 and 2.

lakes or seas. The Ramsar Convention, designed to protect wetlands of international importance (see below), uses a very broad definition, which has been widely accepted:

> Wetlands are areas of marsh, fen, peatland or water, whether natural or artificial, permanent or temporary, with water that is static or flowing, fresh, brackish or salt, including areas of marine water the depth of which at low tide does not exceed six metres.

This definition encompasses coastal and shallow marine areas (including some coral reefs, which are covered in Chapter 7), as well as river courses and temporary lakes or depressions in semi-arid zones. It is only in the past 50 years or so that wetlands have been considered as anything more than worthless wastelands, only fit for drainage, dredging and infilling. But an increasing knowledge of these ecosystems in academic and environmentalist circles has yielded the realization that wetlands are an important component of the global biosphere, and that the alarming rate at which they have been destroyed is a cause for concern. Although estimates vary, wetlands probably currently cover less than 9 per cent of the Earth's land surface (Zedler and Kercher, 2005), and much of the remaining wetland area is degraded to some degree.

Wetlands perform some key natural functions and provide a wide range of ecosystem services (Table 8.5). They form an important link in the hydrological cycle, acting as temporary water stores. This function helps to mitigate river floods downstream, protects coastlines from destructive erosion and recharges aquifers. Chemically, wetlands act like giant water filters, trapping and recycling nutrients and other residues. This role is partly a function of the very high biological productivity of certain wetlands, amongst the highest of any world ecosystem (see Table 1.1). As such, wetlands provide habitats for a wide variety of plants

Table 8.5 Wetland functions

Hydrology	Flood control
	Groundwater recharge/discharge
	Shoreline anchorage and protection
Water quality	Wastewater treatment
	Toxic substances
	Nutrients
Food-chain support/cycling	Primary production
	Decomposition
	Nutrient export
	Nutrient utilization
	Carbon storage
Habitat	Invertebrates
	Fisheries
	Mammals
	Birds
Socio-economic	Consumptive use (e.g. food, fuel)
	Non-consumptive use (e.g. aesthetic, recreational, archaeological)

Source: after Sather and Smith (1984).

and animals. Some wetlands can also act as significant carbon sinks, with important implications for global climate change. All these functions provide benefits to human societies, from the direct resource potential provided by products such as fisheries and fuelwood, to the ecosystem value in terms of hydrology and productivity, up to a value on the global level in terms of the role of wetlands in atmospheric processes and general life-support systems (Zedler and Kercher, 2005).

Wetland destruction

Estimates of the global area of wetlands lost as a result of human activities vary considerably, but it is frequently said to be about half. Records are patchy for most of the world, but a review of 189 reports of change in wetland area by Davidson (2014) concluded that the long-term loss of natural wetlands averages at about 55 per cent but loss may have been as high as 87 per cent since the start of the eighteenth century. Much of this destruction has been in the temperate world, and the prime motivation has been to provide more land to grow food. Losses exceed 50 per cent in Europe and the conterminous US, but more extreme cases include an 80 per cent loss of Pacific Coast estuarine wetlands in Canada and 90 per cent of wetland areas lost in New Zealand (Keddy *et al.*, 2009). The wetlands of Europe have been subject to human modification for more than 1000 years. In the Netherlands, where particularly active reclamation and drainage periods occurred in the seventeenth, nineteenth and twentieth centuries, more than half of the national land area is now made up of reclaimed wetlands (Figure 8.12), and Dutch drainage engineers have hired out their skills to neighbouring countries for more than 300 years.

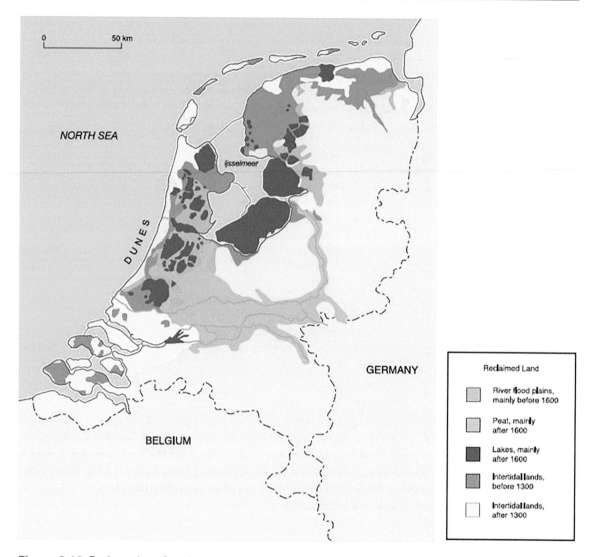

Figure 8.12 Reclamation of wetland types in the Netherlands (after de Jong and Wiggens, 1983).

In the USA, where wetlands are very largely concentrated in the east of the country, with major areas around the Great Lakes, on the lower reaches of the Mississippi River, and on the Gulf of Mexico and Atlantic seaboard south of Chesapeake Bay, the wetland area totalled nearly 90 million ha at the time of European settlement. By 1997, that area had been reduced to an estimated 42.7 million ha (Dahl, 2000). Agriculture has been the major beneficiary from the drainage operations in the USA, while urban and suburban development, dredging and mining account for much of the rest.

Although the rich soils of former wetland areas often provide fertile agricultural lands, conversion to cropland is not always successful. Drainage of the extensive marshlands around the Pripyat River in Belarus during the post-war Soviet period, for example, was largely unsuccessful. The low productivity of reclaimed areas necessitated the application of large amounts of chemical fertilizers, and some former wetland zones also became prone to wildfires (Yatsukhno, 2012).

There have been numerous other reasons for wetland reclamation. On the national scale, in Albania, where about 9 per cent of the country was drained in the period 1946–83, reclamation was as much driven by the fight against malaria as the desire to increase food output, although most of the former marshes were converted to irrigated agriculture (Hall, 2002). Urban sprawl has been a major cause in many areas. Mexico City, for example, was surrounded by five shallow lakes when the Spanish first arrived in 1519, but as the city has expanded since then, all but one small wetland area have been desiccated for their water and land. The remaining area at Xochimilco is threatened by the city's declining water table (Figure 8.13). In South Sudan, the Jonglei Canal Project was designed to utilize the large proportion of the White Nile's discharge that currently feeds the Sudd Swamps. The 360-km artificial channel, planned to bypass and effectively drain the swamps, would provide much-needed additional water to the dryland countries of Sudan and Egypt if it is ever completed – but the project was halted for decades by civil war in the region. Plans to utilize the water resources of wetlands also threaten other sites, such as the inland Okavango delta in semi-arid northern Botswana. The general well-being of the Okavango is also under increasing pressure from cattle encroaching into the lush grazing of the delta, a threat currently held at bay by a 300-km cordon

Figure 8.13 Mexico City's last remaining area of wetland at Xochimilco, which is in danger of drying up due to the continuing need for water from the world's largest city.

fence; but the lure of resources other than water and land has proved to be a serious cause of wetland degradation in many areas.

The fuel resources harboured in wetlands represent perhaps the most important reason in this respect. In many developing countries fuelwood collection is a prime cause of degradation. It is the major threat to wetlands in Central America, as it is in many coastal mangrove swamps in Africa and Asia (see p. 156). In more temperate latitudes, particularly above 45°N, peat is the most important organic fuel of the wetland habitat. Hand-cutting of peat has been practised in rural communities in northern Europe for centuries, and continues today in remote parts of Ireland, Scotland, Scandinavia and Russia, but since the 1950s, peat has been cut from much larger areas with machines. The global production of peat, most of it for fuel but with exploitation for horticultural purposes important in some countries, was 28 million tonnes in 2016, and Finland, Ireland and Sweden were the world's largest producers (USGS, 2016). The Finnish peatlands, covering a total area of some 89,000 km², are some of the most important in Europe and Finland has the highest proportion of wetlands of any country in the world.

The arguments in favour of preserving and restoring peatlands take in all the reasons for wetland preservation in general (Table 8.5), and the importance of peatlands for global-scale stability is increasingly being realized. There is a growing body of evidence to suggest that peatlands play a key role in the initiation of ice ages, through feedbacks with geomorphology, surface albedo, atmospheric moisture and the concentration of greenhouse gases (Franzén and Cropp, 2007). Certainly peatlands are an important and dynamic global carbon pool which has major implications for climate regulation. Indeed, most of the very large carbon pool in peatlands, which occupy about 3 per cent of the world's land surface, has accumulated since the last ice age and it seems that the highest rates of carbon accumulation have occurred during warmer periods of the Holocene. Yu *et al.* (2011) point out that this raises the possibility that peatlands may provide a negative feedback mechanism to future climate change, sequestering more carbon as temperatures rise and thereby causing a cooling effect.

Wetland protection

Realization of the value of wetlands and their ecosystem services has prompted moves to protect this threatened landscape in recent decades, although in many cases conservationists still face a difficult task in protecting wetlands from destructive development projects. A comprehensive national effort towards conservation has been made in the USA, where numerous state and federal laws have a bearing on wetland protection. An attempt was also made to reverse the loss with the Wetlands Reserve Program (WRP), which ran from 1990 to 2014 and sought to protect, restore, and enhance privately owned wetlands. The WRP aimed to improve environmental quality without significantly impacting the agricultural production or the economic well-being of the areas concerned by paying compensation to farmers who took part in the scheme and by allowing certain other money-making activities, such as hunting and fishing, in their wetland areas. This voluntary programme played a significant role in achieving a net gain in wetland area in the USA, up by 12 per cent between 1985 and 2004 (Dahl, 2006).

On the international front, one of the earliest attempts to protect a major world biome was the Ramsar Convention, which aims to protect wetlands of international importance. Formally known as the Convention on Wetlands of International Importance especially as Waterfowl Habitats, it was adopted by 18 countries attending its first conference at Ramsar, Iran, in 1971. The Convention aims to stem the progressive encroachment on, and loss of, wetlands, and represents one of the first attempts to impose external obligations on the land-use decisions of independent states. Each contracting country's main obligation is to designate at least one wetland of international importance for inclusion in a formal list (Figure 8.14). In early 2018, this Ramsar List had 2290 sites, covering 225 million hectares in 169 contracting countries.

Figure 8.14 The Pantanal is a vast wetland covering an area of approximately 230,000 km², straddling the borders between Brazil, Paraguay and Bolivia. Part of the Brazilian Pantanal, in Mato Grosso where this photograph was taken, was designated a Ramsar site in 1993.

The Ramsar Convention has always emphasized that human use on a sustainable basis is entirely compatible with Ramsar listing and wetland conservation in general (Ramsar Convention Secretariat, 2010). The Convention emphasizes that wetlands should be conserved by ensuring their 'wise use', a term that is synonymous with sustainable use (Table 8.6). Wise use is defined as sustainable utilization for the benefit of humankind in a way compatible with the maintenance of the natural properties of the ecosystem. Sustainable utilization is understood as people's use of a wetland so that it may yield the greatest continuous benefit to present generations while maintaining its potential to meet the needs and aspirations of future generations. 'Wise use' may also require strict protection. A good example of more than a century of wise use in the Sundarbans mangrove forest of Bangladesh is shown in Table 7.9.

However, many important wetland sites are still not listed and the Convention lacks any legal powers, relying on persuasion and moral pressure to achieve its aims. It is in the developing countries of the world where most concern over the future of wetlands is focused. The rising resource needs of rapidly growing populations, particularly to expand food production, mean that many unique wetlands remain under considerable pressure. The conservation of wetlands requires that the drivers of wetland loss and degradation, both direct (e.g. pollution, overfishing, water diversion) and indirect (e.g. population growth, economic exploitation), be addressed concurrently. As with rivers and lakes, sensible wetland management can only

Table 8.6 The Ramsar Convention's 'wise use' guidelines

Guideline	Action
Adopt national wetland policies	involving a review of existing laws and institutions designed to deal with wetland matters (either as specific wetland policies or as part of national environmental action plans, national biodiversity strategies, or other national strategic planning)
Develop wetland programmes	inventory, monitoring, research, training, education and public awareness
Take action at wetland sites	involving the development of integrated management plans covering every aspect of the wetlands

Source: after Davis (1993).

be undertaken in these circumstances by incorporating all activities and interest groups present in a particular drainage basin.

The plight of the Azraq Wetland Reserve in Jordan illustrates these points well. This 800 ha mosaic of shallow pools and marshland provides the only permanent water within 12,000 km² of desert and began to be degraded shortly after its designation as a Ramsar List site in 1977. The aquifer feeding the two springs that supplied the wetland with water has been heavily overexploited in recent times, exceeding the annual natural recharge rate of about 35 million m³ (Dottridge and Abu Jaber, 1999) every year since 1983, to supply water to Amman, Zarqa and local irrigation schemes. In 1992, flow from the springs completely ceased and the wetland dried out, allowing slow-burning underground fires to ignite on the peaty soils. Some water has subsequently been restored to Azraq and a management plan has been devised to nurse the area back to health, involving local education and tourist facilities. However, the acute shortage of water in Jordan, where consumption already exceeds renewable freshwater resources, remains a significant threat to the long-term future of the reserve. The Jordanian government has tried to control well expansion and water abstraction by farmers in the Azraq basin but their policies have not always been successful (Al Naber and Molle, 2017).

FURTHER READING

Downs, P.W. and Gregory, K.J. 2004 *River channel management: towards sustainable catchment hydrosystems*. London, Hodder Arnold. A comprehensive coverage of human impacts and river system management techniques often involving the entire drainage basin.

Gilvear, D.J., Greenwood, M.T., Thoms, M.C. and Wood, P.J. (eds) 2016 *River science: research and management for the 21st century*. Oxford, Wiley-Blackwell. Articles from the interface of natural sciences, engineering and sociopolitical sciences.

Kafumbata, D., Jamu, D. and Chiotha, S. 2014 Riparian ecosystem resilience and livelihood strategies under test: lessons from Lake Chilwa in Malawi and other lakes in Africa. *Philosophical Transactions of the Royal Society B: Biological Sciences* 369(1639): 20130052. A review of African lakes and their management challenges. Open access: rstb.royalsocietypublishing.org/content/369/1639/20130052.short.

Lohner, T.W. and Dixon, D.A. 2013 The value of long-term environmental monitoring programs: an Ohio River case study. *Environmental Monitoring and Assessment* 185: 9385–96. Arguments for maintaining long records. Open access: link.springer.com/article/10.1007/s10661-013-3258-4.

Maltby, E. and Barker, T. (eds) 2009 *The wetlands handbook*. Oxford, Wiley-Blackwell. A key text covering the importance of wetlands and how society affects them.

Varis, O., Kummu, M. and Salmivaara, A. 2012 Ten major rivers in monsoon Asia-Pacific: an assessment of vulnerability. *Applied Geography* 32: 441–54. Socio-economic and environmental vulnerability in Asian river basins.

WEBSITES

www.iksr.org site for the International Commission for the Protection of the Rhine with information and data on this major European river.

www.ilec.or.jp site of the International Lake Environment Committee, which promotes sustainable management of the world's lakes and reservoirs; includes news and a large database.

www.internationalrivers.org International Rivers has information, campaigns and publications (e.g. *World Rivers Review*) on many riverine issues.

www.ramsar.org the Ramsar Bureau's website covers reports, news, documents and archives on all aspects of wetlands.

www.wetlands.org Wetlands International, a leading not-for-profit conservation group dedicated to the conservation and restoration of wetlands.

www.worldlakes.org World Lakes is a global network of organizations working for the conservation and sustainable management of lakes.

POINTS FOR DISCUSSION

- Management of freshwater resources is vital for human society but manipulation of the hydrological cycle for our benefit always raises environmental issues. Discuss with specific examples of how problems can be minimized.
- Given that river management has such a long history in some parts of the world, why is river use not yet sustainable?
- What are the arguments for maintaining river water quality?
- If the Aral Sea tragedy is the result of a technocentric approach to using the natural environment, is an ecocentric approach the only way of dealing with the problems?
- Why are not all lakes warming with climate change?
- How and why should wetlands be preserved?

9 *Big dams*

People have constructed dams to manage water resources for at least 5000 years. The first dams were used to control floods and to supply water for irrigation and domestic purposes. Later, the energy of rivers was harnessed behind dams to power primary industries directly, and more recently still, hydroelectricity has been generated using water held in reservoirs, allowing the impoundment and regulation of river flow. The modern era of big dams dates from the 1930s and began in the USA with the construction of the 221-metre

Hoover Dam on the Colorado River. But in the past 50 years or so there has been a marked escalation in the rate and scale of construction of big dams all over the world, made possible by advances in earth-moving and concrete technology. Initially, these dams were for hydroelectricity generation, and subsequently for multiple purposes – primarily irrigation, power, domestic and industrial water supply, and flood control.

Structures above 15 m in height above their foundations are defined as large by the International Commission on Large Dams (ICOLD), and major if they exceed 150 m. About 5000 large dams had been constructed worldwide by 1949, three-quarters of them in industrialized countries, paving the way for the most active phase of large dam construction in the period between 1950 and the mid-1980s. By the end of the twentieth century, there were more than 45,000 large dams in over 140 countries (World Commission on Dams, 2000) and the ICOLD database registered 58,519 dams at the start of 2018. The top five dam-building countries (China, USA, India, Japan and Brazil) account for 73 per cent of all large dams worldwide, and just over 40 per cent of the global total are in China. The pace at which large dams have been built has slowed markedly in recent decades, especially in North America and Europe where most technically attractive sites have already been developed, but another boom in hydropower dam construction is expected in the 2020s to meet growing global energy demands (Zarfl et al., 2015).

Some rivers are more regulated than others, of course. In an assessment of nearly 300 of the world's largest river systems, Nilsson et al. (2005) concluded that just under half were unaffected by dams. However, at the other end of the scale, the flow in some of the world's largest systems is very intensively regulated. Table 9.1 shows those river systems identified in the study where the capacity of existing dams is large enough to hold more than the entire annual average discharge. In terms of number of dams, China's Yangtze River basin is among the most intensively managed, being the site of about 50,000 dams of all sizes, including the Three Gorges Dam, the world's largest by installed capacity (Yang et al., 2011).

In total, the world's major reservoirs were estimated to control about 10 per cent of the total runoff from the land in the 1970s (UNESCO, 1978). That figure had risen to 13.5 per cent by the early 1990s (Postel et al., 1996) and is likely to be more like 15 per cent today. There are no reliable global data for smaller dams, but their numbers are certainly much greater. Indeed, one estimate suggests there may be several million small dams in the USA alone (Renwick et al., 2005). The volume of water trapped worldwide in reservoirs

Table 9.1 Large river systems with high degrees of flow regulation by dams

River system	Region	Degree of flow regulation (%)*
Volta	West Africa	428
Manicougan	Canada	295
Colorado	USA	280
Negro	South America	140
Mae Khlong	Thailand	130
Shatt al Arab (Euphrates–Tigris)	Middle East	124

Note: *A flow regulation of 100 per cent indicates that the system's entire annual average discharge can be held and released by the dams in the river system.

Source: data from Nilsson et al. (2005).

of all sizes is estimated to be no less than five times the global annual river flow (Davies *et al.*, 2000). Such a large-scale redistribution of water is thought to be responsible for a very small but measurable change in the orbital characteristics of the Earth (Chao, 1995).

THE BENEFITS OF BIG DAMS

There is no doubt that many big dam schemes have been successful in achieving their primary objectives and in many respects have made substantial contributions to the sustainable use of the world's resources. Half the world's large dams were built exclusively or primarily for irrigation, and about a third of the world's irrigated cropland relies on dams. The World Commission on Dams (2000) identified seven countries, including such populous nations as India and China, as having more than 50 per cent of arable land irrigated by water supplied from dams. Hydropower currently provides about 16 per cent of the world's total electricity supply and is used in over 150 countries. It represents more than 90 per cent of the total national electricity supply in 24 countries.

Egypt's Aswan High Dam, completed in 1970, provides a more detailed illustration of the benefits of big dam construction. Hydroelectricity generated by the dam is a renewable energy source that is cheap to operate after the initially high capital costs of dam construction, and saves on the purchase of fossil fuels from abroad. For many years after its construction, the Aswan High Dam produced most of the country's electricity and gave numerous villages access to electricity for the first time. Although demand has now surpassed its capacity, the dam still generates about 20 per cent of Egypt's electricity.

The natural discharge of the Nile is subject to wide seasonal variations, with about 80 per cent of the annual total received during the flood season from August to October, and marked high and low flows depending upon climatic conditions in the main catchment area in the Ethiopian highlands. The dam allows management of the flow of the Nile's discharge, evening out the annual flow below the dam (Figure 9.1) and protecting against floods and droughts. Management of the Nile's flow has also had benefits for navigation and tourism, resulting from the stability of water levels in the river's course and navigation channels. Irrigation water for cropland is also provided by the dam's reservoir storage, which has allowed 400,000 ha of cropland to convert from seasonal to perennial irrigation and the expansion of agriculture onto 490,000 ha of new land, a particularly important aspect for a largely hyper-arid country with just 3 per cent of its national area suitable for cultivation (Rashad and Ismail, 2000).

Large dams are often seen as symbols of economic advancement and national prestige for many developing nations, but the huge initial capital outlay needed for construction often means that agendas are set to varying extents by foreign interests. A key element in the financing of Ghana's Akosombo Dam, completed in 1965, was the sale of cheap electricity to the Volta Aluminium Company, a consortium with two US-owned companies that produced aluminium from imported alumina, despite the fact that Ghana has considerable reserves of alumina of its own. The construction of the Cahora Bassa Dam in Mozambique (Figure 9.2) in the 1970s, when the country was still a Portuguese colony, was also largely catering to outside interests. Most of the electricity has always been sold to South African industry, and part of the original reason for flooding the 250-kilometre-long Lake Cahora Bassa was to establish a physical barrier against Frelimo guerrillas seeking independence. The plan for the dam also envisaged the settlement of up to a million white farmers in the region, who, it was thought, would fight to protect their new lands.

In addition, concerns have often been raised about which sectors of society benefit from a new big dam. In the case of Cahora Bassa, poor rural Mozambicans living along the Zambezi River downstream of the dam have been adversely affected (Scodanibbio and Mañez, 2005). The Bayano Hydroelectric Complex in Panama has been criticized for similar reasons. Wali (1988) points out that although Bayano provides 30

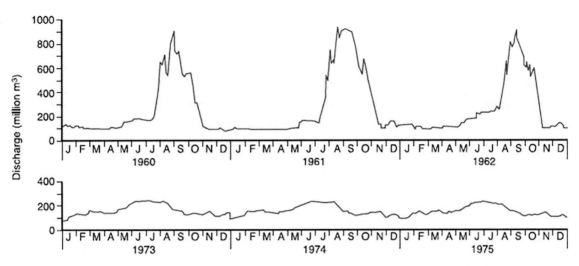

Figure 9.1 Daily discharge regime of the River Nile at Aswan, before and after construction of the High Dam (after Beaumont *et al.*, 1988).

Figure 9.2 The dam on the River Zambezi at Cahora Bassa in northern Mozambique. Just 10 per cent of the dam's electricity-generation capacity would be sufficient for all Mozambique's needs, but the major transmission lines built by the Portuguese serve South Africa, and much of Mozambique is not connected to the dam's supplies.

per cent of the country's electricity, no less than 83 per cent of national production is consumed in Panama City and Colón, so that the dam is reinforcing the concentration of wealth in the urban areas.

Hence, it is clear that the undoubted benefits of big dams are not always gained solely by the country where the dam is located, and that within the country concerned, the demands of urban populations can outweigh those of rural areas. Many of the drawbacks of such structures, however, are borne by the rural people of the country concerned. Despite the success of many big dams in achieving their main economic aims, their construction and associated reservoirs create significant changes in the pre-existing environment, and many of these changes have proved to be detrimental. It is the negative side of environmental impacts that has pushed the issue of big dams to a prominent position in the eyes of environmentalists and many other interest groups.

ENVIRONMENTAL IMPACTS OF BIG DAMS

The environmental impacts of big dams and their associated reservoirs are numerous, and they affect a range of areas including the river channel and its catchment but sometimes beyond. Some idea of the scale of immediate impact is shown in Figure 9.3 in satellite images of the Belo Monte Dam complex on the

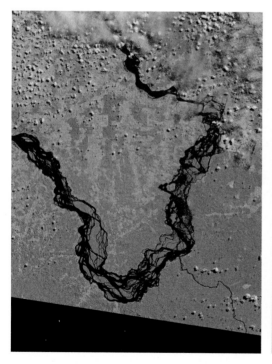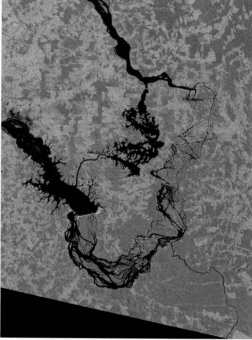

Figure 9.3 Satellite images of part of the Xingu River in Brazil in 2000 (left) and 2017 (right) after construction of two major dams in the Belo Monte Dam complex. The river flows generally northward (up in the images) although in this stretch it dips southward to flow around a 100-kilometre river bend. Creation of two linked reservoirs by 2017 diverted much of the water so that just 20 per cent of the original river now flows around the bend.

Xingu River in northern Brazil, but impacts are associated not just with the dam itself but with all ancillary aspects of a dam project. These include power transmission corridors, pipelines, canals, tunnels, relocation and access roads, borrow and disposal areas, construction camps, and resettlement zones as well as unplanned developments stimulated by the project (e.g. logging or shifting cultivation along access roads).

The temporal aspect of environmental impacts within a certain area is also important. The river basin itself can be thought of as a system that will respond to a major change, such as dam construction, in many different ways and on a variety of timescales. While the creation of a reservoir represents an immediate environmental change, the permanent inundation of an area not previously covered in water, the resulting changes in other aspects of the river basin, such as floral and faunal communities, and soil erosion, will take a longer time to readjust to the new conditions.

The range of environmental impacts consequent upon dam construction, and their effects on human communities, can be considered under three headings that reflect the broad spatial regions associated with any dam project: (1) the dam and its reservoir; (2) the upstream area; and (3) the downstream area.

The dam and its reservoir

The creation of a reservoir results in the loss of resources in the land area inundated. Flooding behind the Balbina Dam north of Manaus, Brazil, has destroyed much of a centre of plant endemism, for example. In some cases the loss of wilderness areas threatened by new dam projects has raised considerable debate, both nationally and internationally. A case in point was the Nam Choan Dam Project on the Kwae Yai River in western Thailand, first proposed in 1982. The suggested reservoir lay largely within the Thung Yai Wildlife Sanctuary, one of the largest remaining relatively undisturbed forest areas in Thailand, containing all six of the nation's endangered mammal species. Debate over the destructive impact of the project resulted in it being shelved indefinitely in 1988 (Boyle, 1998).

Some resources, such as trees for timber or fuelwood, can be taken from the reservoir site prior to inundation, although this is not always economically feasible in remote regions. There are dangers inherent in not removing them, however. Anaerobic decomposition of submerged forests produces hydrogen sulphide, which is toxic to fish and corrodes metal that comes into contact with the water. Corrosion of turbines in Surinam's Brokopondo reservoir has been a serious problem. In a similar vein, decomposition of organic matter by bacteria in reservoirs of the La Grande hydroelectric complex in Quebec, Canada, has released large quantities of mercury by methylation. Mercury has bioaccumulated in reservoir fish tissue to levels often exceeding the Canadian standard for edible fish and in many species the high mercury levels have persisted for decades. In a study of five species taken from reservoirs at the complex between 1978 and 2012 (Bilodeau *et al.*, 2017) marked differences were noted between piscivorous (fish-eating) species and non-piscivorous species (Table 9.2). The differences applied both to the time taken for mercury levels to peak and for mercury levels to return to levels equivalent to those found in fish in surrounding natural lakes. Mercury has also been exported downstream from the reservoirs, causing high concentrations in fish below the dams.

Reservoirs formed by river impoundment typically undergo significant variations in water quality in their first decade or so, before a new ecological balance is reached. Biological production can be high on initial impoundment, due to the release of organically bound elements from flooded vegetation and soils, but declines thereafter. Hence the initial fish yield from Lake Kariba in 1964, its first year of full capacity, was more than 2500 tonnes but by the early 1970s the annual yield had stabilized at around 1000 tonnes (Marshall and Junor, 1981). Such commercial fisheries, often a positive effect of tropical reservoirs, do not always develop, however. The Brokopondo reservoir is a case in point, partly because the most abundant

Table 9.2 Timing of impoundment effects on mercury levels in fishes at La Grande hydroelectric complex

Species	Time to peak concentrations (years)	Time to return to background levels (years)
Piscivorous	9–14	20–31
Non-piscivorous	4–11	10–20

Source: after Bilodeau *et al.* (2017).

large species in Brokopondo is not appreciated as a food fish in Surinam. Indeed, the long-term impact of Brokopondo dam has been a drastic decrease in fish species diversity. Before closure of the dam in 1964 there were 172 species of fish in the Surinam River, but 40 years later Mol *et al.* (2007) found only 41 in the reservoir.

Blooms of toxic cyanobacteria, a type of microscopic algae, may be another facet of the nutrient enrichment that often follows impoundment of new reservoirs particularly in tropical, subtropical and arid regions of the world. These cyanobacterial toxins are potentially dangerous to humans and animals if consumed in sufficient quantities. A range of gastrointestinal and allergenic illnesses can affect people exposed to the toxins in drinking water, food or during swimming, but the most severe case of human poisoning due to cyanobacterial toxins occurred when inadequately treated water from a reservoir was used for patients in a kidney dialysis clinic in the Brazilian city of Curaru in 1996. More than 50 people died due to direct exposure of the toxin to their bloodstream during dialysis (Chorus and Bartram, 1999).

Cultural property may also be lost by the creation of a reservoir – 24 archaeological sites dating from 70–1000 CE were inundated by the Tucuruí Dam reservoir in Brazil, for example – although in some cases such property is deemed important enough to be preserved. Lake Nasser submerged some ancient Egyptian monuments, but major ones – including the temples of Abu Simbel, Kalabsha and Philae – were moved to higher ground prior to flooding (Figure 9.4).

Big dams often necessitate resettlement programmes if there are inhabitants of the area to be inundated, and the numbers of people involved can be very large. Global estimates suggest that 40–80 million people were displaced by reservoirs in the second half of the twentieth century (World Commission on Dams, 2000). Some of the biggest projects in this respect have been in China. The Sanmen Gorge Project on the Huang Ho River involved moving 300,000 people, and the Three Gorges Dam on the Yangtze River involved the displacement of 1.3 million people from 12 cities, 114 townships and more than 1000 villages. Some indication of the trade-off between land lost, people displaced and power generated is given in Table 9.3 for a selection of big dam projects.

Compensation may be offered to the people who are displaced, but in many remote areas inhabitants do not possess formal ownership documents, slowing or preventing legal compensation. For those who have to resettle, helped by government schemes or not, the move can be a traumatic one. The resettlement of 57,000 members of the Tonga tribe from the area of the Kariba Dam on the Zambezi illustrates some of the adverse effects for the people concerned. Obeng (1978) described the culture shock suffered in moving to very different communities and environments. Drawn-out conflicts over land tenure resulted between the new settlers and previous residents, and since the resettlement area was drier than the Tongan homelands,

Figure 9.4 The Temple of Philae, which has been moved to higher ground to avoid inundation by Lake Nasser.

Table 9.3 Hydropower generated per hectare inundated, and number of people displaced for selected big dam projects

Project and country	Approx. rated capacity (MW)	Normal reservoir area (ha)	Kilowatts per hectare	People relocated
Pehuenche (Chile)	500	400	1250	
Guavio (Colombia)	1600	1500	1067	5500
Three Gorges (China)	18,200	110,000	165	130,0000
Itaipu (Brazil and Paraguay)	12,600	135,000	93	8000 families
Sayanogorsk (Russia)	6400	80,000	80	
Churchill Falls (Canada)	5225	66,500	79	
Tarbela (Pakistan)	1750	24,300	72	86,000
Grand Coulee (USA)	2025	32,400	63	

(continued)

Table 9.3 continued

Project and country	Approx. rated capacity (MW)	Normal reservoir area (ha)	Kilowatts per hectare	People relocated
Aswan High Dam (Egypt)	2100	40,000	53	100,000
Lower Sesan II Dam (Cambodia)	400	7500	53	5000
Tucuruí (Brazil)	6480	216,000	30	30,000
Keban (Turkey)	1360	67,500	20	30,000
Batang Ai (Sarawak, Borneo)	92	8500	11	3000
Cahora Bassa (Mozambique)	2075	266,000	8	25,000
BHA (Panama)	150	35,000	4	4000
Kariba (Zimbabwe and Zambia)	1500	510,000	3	50,000
Akosombo (Ghana)	833	848,200	0.9	80,000
Brokopondo (Surinam)	30	150,000	0.2	5000

Sources: Barrow (1981); Dixon *et al.* (1989); Goldsmith and Hildyard (1984); Wali (1988); Goodland (1990); Gleick (1993); McCully (1996).

problems with planting and the timing of harvests were faced. Deprived of fish and riverbank rodents, which traditionally supplemented their cultivated diet, the Tongas faced severe food shortages. When the government sent food aid to relieve the suffering, the food distribution centres became transmission sites for trypanosomiasis, or sleeping sickness.

Development following the construction of big dams can also act as a pull for migrants, bringing associated problems of pressure on local resources. The influx of migrants to the Aswan area led to an increase in population from 280,000 in 1960 to more than 1 million within 30 years, mainly due to the increase in job opportunities (Rashad and Ismail, 2000).

Over the longer term, other effects of reservoir inundation have become evident. The alteration of the environment can have significant impacts on local health conditions. In some cases these can be beneficial. Onchocerciasis, or river blindness, for example – a disease that is particularly common in Africa – is caused by a small worm transmitted by a species of blackfly. The blackflies breed in fast-running, well-oxygenated waters and dam construction can reduce the number of breeding sites by flooding rapids upstream. This has been the case with Ghana's Akosombo and Nigeria's Kainji dams, although the flies may find alternative breeding sites in new tributary streams.

Malaria, conversely, may increase as a result of water impoundment, since the mosquitoes that transmit the disease breed in standing waters. Local malaria incidence increased around Tucuruí, Brazil, and the Srinagarind dam in Thailand, but after reviewing the effects of large dams on the burden of malaria globally, Keiser *et al.* (2005) concluded that an increase is not inevitable. Whether an individual dam project triggers an increase in malaria transmission depends on several factors, including the socio-economic status of local populations and attempts to manage the issue. Controlling mosquito numbers has been carried out with success for several decades by means of water management: fluctuating water levels leave larvae stranded and hence inhibit breeding.

Schistosomiasis, also known as bilharzia – a very debilitating though rarely fatal disease, which is widespread throughout the developing world – is transmitted in a different way: by parasitic larvae that infect a certain aquatic snail species as the intermediate host. The incidence of schistosomiasis was considerably increased by the construction of the Akosombo Dam, with infection rates among 5- to 19-year-olds rising from 15 per cent to 90 per cent within four years of its completion. Similar figures have been reported from other large dams, such as Kariba in Zambia (Steinmann *et al.*, 2006). Exactly how dams increase schistosomiasis is not always clear, but one hypothesis argues that dams block reproduction of the migratory river prawns that eat the snail hosts of schistosomiasis. Evidence to support this theory has been gathered in the Senegal River basin, where prawn populations declined and schistosomiasis increased after completion of the Diama Dam that spans the border between Senegal and Mauritania. In a study of large dams worldwide, Sokolow *et al.* (2017) conclude that nearly 400 million people are at higher risk of schistosomiasis because dams block the migration of snail-eating river prawns.

Other biological consequences of large reservoirs include the rapid spread of waterweeds that cause hazards to navigation and a number of secondary impacts, notably water losses through evapotranspiration. Water-fern appeared in Lake Kariba six months after the dam was closed and after two years had covered 10 per cent of the 420 km^2 lake area. More dramatic still was the spread of water hyacinth on the Brokopondo reservoir, which covered 50 per cent of the lake's surface within two years. The plant was partially brought under control by aerial spraying with herbicide, but this procedure also poisoned many other plants and animals (Pringle *et al.*, 2000).

New reservoirs also have effects on geomorphological and, in some cases, tectonic processes. The trapping of sediment is a particularly important aspect of reservoir impoundment. This siltation has a number of knock-on effects downstream of the dam (see below), but it also seriously affects the useful life of the dam itself. The economic impact of reservoir sedimentation has been calculated by Lahlou (1996) in Morocco, where a dam-building programme launched in 1975 aimed to irrigate 1 million ha of crops by the year 2000. The country's 1995 total of 84 large dams suffered a combined average annual loss of storage due to reservoir siltation of 0.5 per cent per year. This figure translated to annual losses in the year 2000 of 60 mkWh of hydroelectric production, 50 mm^3 of potable water supply, 5000 ha of irrigated cropland and 10,000 jobs.

A wide range of techniques is available to manage reservoir siltation, the cost of which needs to be budgeted for. Kondolf *et al.* (2014) recognize three broad categories:

- methods to route sediment through or around the reservoir
- methods to remove sediments accumulated in the reservoir to regain capacity
- approaches to minimize the amount of sediment arriving into reservoirs from upstream.

Sanmenxia reservoir in China experienced an alarming rate of loss of storage capacity – 50 per cent within just four years of impoundment in 1960 – that prompted a substantial change in the operation of the dam, to achieve a better balance between sediment inflow and outflow (Wang and Hu, 2009). The new regime was designed to reflect the fact that sediment transport in the Huang Ho occurs mostly during the 2–4-month flood season. Reservoir sedimentation has been reduced by releasing silty water during the flood season and storing the clear water that flows at other times of the year. The change in operation of the Sanmenxia dam is thought to have extended its life by 30 years.

Dredging or flushing are other options for managing sedimentation problems. Removing sediment mechanically by dry excavation requires the lowering of a reservoir's water level and is expensive but material dredged can be reused in some cases, as Walter *et al.* (2012) show for the Gallito Ciego reservoir in Peru where sediment has been assessed as an addition to soil on adjacent agricultural fields and as a building material. Tolouie *et al.* (1993) document the case of the Sefid-Rud reservoir, built to supply 250,000 ha of cropland in north-western Iran, which lost over 30 per cent of its storage capacity to sedimentation in the first 17 years after construction. Desiltation successfully restored about 7 per cent of total capacity in seven years (Figure 9.5), but the reservoir had to be emptied during the non-irrigation season to enable sediment flushing. Emptying the reservoir released a highly erosive flow downstream of the dam, and hydroelectricity generation was stopped during the operation.

Local heightening of water tables following reservoir impoundment can have deleterious effects on new irrigation schemes through waterlogging and salinization. Waterlogging is an occasional problem around the Kuban reservoir on the River Kuban, near Krasnodar in southern Russia, when the reservoir is filled above its maximum normal level to aid navigation and benefit rice cultivation. The result has been the ruin of over 100,000 ha of crops, and water damage to 130 communities, including 27,000 homes, 150 km of roads and even the Krasnodar airport (Pryde, 1991). Local changes in groundwater conditions have also affected slope stability, causing landslides around some reservoirs. Water displaced by a landslip at the Vaiont Dam in Italy in 1963 overtopped the dam, killing more than 2000 people in the resulting disaster (Kilburn and Petley, 2003).

The sheer size of some reservoirs can also create new physical processes. Artificial lakes behind dams on the Volga River are so large that storms can produce ocean-like waves that easily erode the fine wind-blown soils lining the shores. When this process undercuts trees, or creates shoals, navigational hazards result (Pryde, 1991).

The triggering of earthquakes by a reservoir was first recognized at Lake Mead in the USA in the 1930s, and the stress changes on crustal rocks induced by huge volumes of water impounded behind major dams have been suspected of inducing seismic activity in many other regions. Nurek Dam on the Vakhish River in central Tajikistan is one of the best-documented examples of a large dam, in this case a 315-metre-high earth dam, causing seismic activity. Filling of the dam, located in a thrust-faulted setting, began in 1967 and increases in water level were mirrored by significant increases in the number of earthquakes during the first decade of the dam's lifetime. Detailed observations at Nurek have demonstrated how even very minor changes in water level of just a few metres can be sufficient to modify the seismicity of the area (McGarr *et al.*, 2002).

Numerous other examples can be cited. In reviewing studies of earthquakes caused by artificial reservoirs, Gupta (2002) identified 95 sites globally where reservoir-triggered seismicity has been reported. The most intense of these events are shown in Table 9.4. The earthquakes at Hsinfengkiang and Koyna in the 1960s caused damage to the dams themselves; Koyna's 1967 earthquake resulted in the deaths of some 200 people and rendered thousands homeless. It was the worst of more than 150 earthquakes of magnitude 4 or greater on the Richter scale at Koyna in 40 years after its reservoir was impounded. An increase in the frequency of reservoir-induced seismic activity at the Three Gorges Dam in China since 2003, when its reservoir was first impounded, has also been accompanied by a rise in the number of landslides and mud-rock flows. Some of these secondary geological hazards have presented severe risks to navigation and significant economic loss. One major landslide in the town of Qianjiangping in 2003 destroyed 346 houses and four factories, resulting in a direct economic loss of US$7 million (Xu *et al.*, 2013). While the local geology and tectonic factors are important in deciding the seismic susceptibility of a site, reservoir characteristics are also significant. The depth of the water column is thought to be among the most important determinants of reservoir-triggered seismicity.

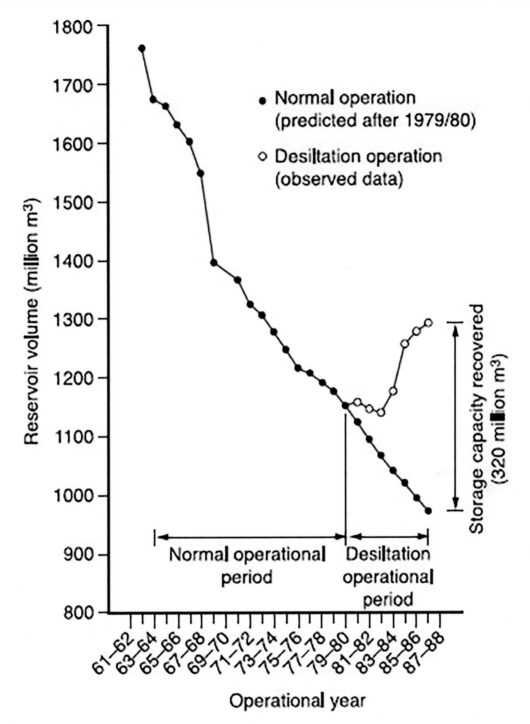

Figure 9.5 Variation of storage capacity in the Sefid-Rud reservoir, Iran, during normal and desiltation operation (after Tolouie et al., 1993).

Table 9.4 Some of the most intense cases of reservoir-triggered seismicity

Dam	Country	Height (m)	Year of reservoir impoundment	Year of largest earthquake
Hsinfengkiang	China	105	1959	1962
Kariba	Zambia–Zimbabwe	128	1958	1963
Koyna	India	103	1962	1967
Kremasta	Greece	160	1965	1966
Aswan	Egypt	111	1964	1981
Benmore	New Zealand	110	1964	1966
Charvak	Uzbekistan	148	1971	1977
Eucumbene	Australia	116	1957	1959
Geheyan	China	151	1993	1997
Hoover	USA	221	1935	1939
Marathon	Greece	67	1929	1938
Oroville	USA	236	1967	1975
Srinagarind	Thailand	140	1977	1983
Warna	India	80	1987	1993

Source: after Gupta (2002).

The creation of new water bodies with large surface areas is thought by many to affect local climate. Thanh and Tam (1990) suggest that Lake Volta has shifted the peak rainfall season in central Ghana from October to July/August, for example, and a study of 92 large dams in North America found influences on local precipitation to be greatest in Mediterranean and semi-arid climates, while for humid climates the influence was least apparent (Degu *et al.*, 2011). Changes in the local temperature regime have been observed at the 45,000 ha Rybinsk reservoir north of Moscow in Russia, where the frost-free period has been extended by 5–15 days per year on average in an area of influence that extends for 10 km around the reservoir's shoreline (D'Yakanov and Reteyum, 1965). Effects on the local temperature regime have also been noted around the 146,000 ha Itaipu Lake, the reservoir formed in 1982 on the border between Brazil and Paraguay. Stivari *et al.* (2005) detect a reduction in the amplitude of the diurnal air temperature cycle around Itaipu, an effect that is linked to the impact on local winds. Evaporation from reservoir surfaces may affect local humidity, and the incidence of fog has been observed to rise in some areas. Indeed, the amount of water evaporated from reservoirs is often considerable, and accounting for water loss through evaporation is a consideration now thought to be worth evaluating when assessing the environmental, social and economic sustainability of a proposed dam or in the evaluation of hydropower as an energy source (Mekonnen and Hoekstra, 2012).

Possible effects on global climate were seldom mentioned in the big dam debate until recently. Indeed, a major argument in favour of big dam construction has been their electricity-generating potential from renewable energy resources without the harmful emissions associated with the burning of fossil fuels.

However, it is now known that hydropower is not a climate-neutral electricity source because reservoirs contribute to climate change by releasing carbon dioxide and methane to the atmosphere from decaying vegetation (Wehrli, 2011). While methane emissions continue slowly after dam closure, carbon dioxide is released in a relatively fast pulse during the reservoir's first decade, but the absolute quantity of greenhouse gas emissions depends on a reservoir's environmental and technical conditions. Estimated maximum emissions may be twice the emissions avoided by refraining from burning fossil fuel according to the IPCC (Schlömer *et al.*, 2014), although on average lifecycle greenhouse gas emissions from hydroelectricity are more than an order of magnitude lower than those of coal and gas.

The upstream area

A variety of upstream impacts can be induced or exacerbated by big dam projects. Some of these, in turn, may impact the dam project itself. Notable in this respect is the improved access to previously remote areas. Deforestation in the watershed above a dam can result in accelerated sedimentation of the reservoir, a problem that is estimated to translate into a loss of irrigated cropland worth nearly US$12 million a year in Iran, for example (Croitoru, 2010). Conversely, afforestation of catchments above dams has been carried out in many areas specifically to limit sediment accumulation in reservoirs. In the UK, for example, many water authorities have bought land in upper catchments to plant new forests.

An increase in forested area upstream of a reservoir is not necessarily sufficient in itself to alter sedimentation rates dramatically, however, as Gellis *et al.* (2006) note at Lago Loíza, impounded in 1953 to supply the Puerto Rican city of San Juan with drinking water. By 1994, the reservoir had lost 47 per cent of its capacity and sedimentation rates during the early part of the reservoir's operation (1953–63) were only slightly higher than the rates during 1964–90 despite an increase in forested area from 7.6 per cent of the basin in 1950 to 20.6 per cent in 1987. The study found that much greater erosion of topsoil was occurring from increasing numbers of construction sites in the basin in the later period as the number of housing units rose by nearly 200 per cent between 1950 and 1990.

The upstream effects of dams on aquatic biota and their habitats have received relatively little attention, but a study of the pallid sturgeon, an endangered large-river species that is endemic to the Missouri and Mississippi rivers in North America, suggests that it is conditions upstream of reservoirs that may have been critical in their decline (Guy *et al.*, 2015). Using a combination of laboratory experiments and field observations at the Fort Peck reservoir in the US state of Montana, the authors concluded that the transition zone between river and reservoir was typically anoxic, or lacking in dissolved oxygen, and that these anoxic stretches of river were responsible for a lack of recruitment in pallid sturgeon. The anoxic condition in the transition zone is a function of a river's lower velocity as it enters the reservoir and the concentration of fine particulate organic material with high microbial respiration which acts to deplete the river's oxygen content.

An interesting example of some of the more general upstream changes brought about by reservoir inundation is provided by Wadi Allaqi in Egypt (Dickinson *et al.*, 1994). Prior to the creation of Lake Nasser by the construction of the Aswan High Dam, Wadi Allaqi was a typical hyper-arid wadi: dry for the most part with unpredictable, short-lived storms providing rare flash floods. Water also percolated underground, typically 30 m below the surface, from the Red Sea Hills to the east. Since 1970, however, both surface and subsurface hydrology have been modified fundamentally by Lake Nasser. The lowest 50 km of the wadi are now permanently inundated and another 40 km are periodically submerged, depending upon the height of the lake. The enhanced availability of water along this 40 km stretch has facilitated a significant change in vegetation, effectively transforming a zone with few resources into one with significant potential for farming or to provide forage for livestock. Wadi Allaqi has been given conservation status by the Egyptian government in order to plan carefully for development of its new resources.

The downstream area

Downstream of a dam, the hydrological regime of a river is modified. Discharge, velocity, water quality, thermal characteristics and seasonal variability are all affected, leading to changes in geomorphology, flora and fauna, both along the river itself and in estuarine and marine environments.

Dams affect the flow of sediment. Globally, the amount trapped in reservoirs is estimated to be more than 100 billion tonnes (Syvitski *et al.*, 2005), and this inevitably means reduced loads in the rivers downstream. The trapping effect of reservoirs along the Huang Ho is depicted in Figure 9.6 which shows annual sediment load measured at Lijin station, 40 km from the river's mouth. Abrupt declines in sediment occurred in 1960, 1968, 1986, and 1999, corresponding to the years the four dams were completed. The overall effect of these dams and a rising demand for the Huang Ho's water has created a scarcity, to the extent that in the early 1990s the river failed to reach the sea on certain days. By 1997 there were 226 'no flow' days, the dry point starting several hundred kilometres inland on some occasions. Since then, the Chinese government has ensured for political reasons that the river always reaches the sea, albeit in small volumes. Hence, the amount of sediment reaching Lijin has been very low since the turn of the current century.

Similar effects on the River Nile have been noted, downstream of the Aswan High Dam, and the lack of silt arriving at the Nile delta has had effects on coastal erosion, and led to salinization through marine intrusion and a decline in the eastern Mediterranean sardine catch (Rashad and Ismail, 2000). To some extent the loss of fisheries of the Nile delta has been offset by a new fishing industry in Lake Nasser, which has provided employment to 6000 fishermen. The salinity problems have been exacerbated by the river's greater salt content downstream of the dam due to the very high evaporation rate over the reservoir, Lake Nasser. Water loss from the reservoir by evaporation is between 10 and 16 billion m^3 every year, equivalent to between 20 and 30 per cent of the Egyptian allocation of water from the Nile Waters Agreement of 1959 (Abd-El Monsef *et al.*, 2015).

The intrusion of sea water into delta areas when river flows have declined due to reservoir impoundment is a common downstream effect. Construction of the Farakka Barrage on the River Ganges in India in 1975 has significantly increased the area affected by salinity problems in Bangladesh as dry-season discharge has declined. This has translated into a cascade of impacts on agriculture, forestry, industry and the supply of domestic drinking water (Mirza, 2004). Ecological shifts brought about by the change in salinity have meant profound changes to spawning and breeding conditions for many species of fish used by local villagers, with consequences for local diets. Agricultural practices have also had to change, as conditions are no longer appropriate for the cultivation of jute, a cash crop. Numerous other aspects of aquatic ecosystems can also be affected by the trapping effect of dams on the flow of nutrients downstream. The downstream ecological effects of dam construction are not limited to rivers, estuaries and coastal areas, as the example of the Black Sea illustrates well (see p. 135).

For communities of fish and other fauna in rivers, lakes and adjacent seas, the building of a dam constitutes a long-lived and complex form of disturbance. By altering the physical–chemical characteristics of river environments a dam typically results in a loss of biodiversity by lowering the diversity of structural features for fish and other aquatic organisms (Poff and Zimmerman, 2010). There are numerous impacts in a variety of habitats, as indicated in Table 9.5 which was compiled from a review of the impacts of dams on Amazonian biodiversity.

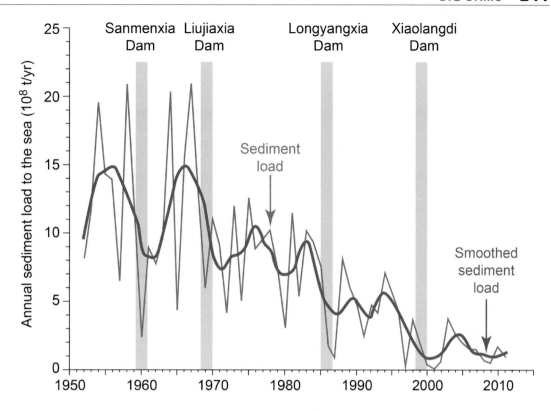

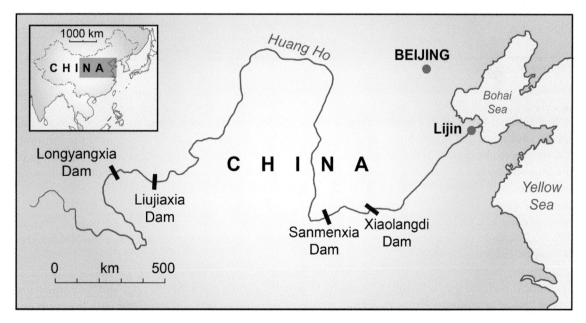

Figure 9.6 Changes in the sediment load of the Huang Ho as measured at Lijin (after Zhao *et al.*, 2013).

Table 9.5 Impacts of dams on biodiversity in Amazonia

Habitat type	Species affected	Extinction process	Scale and magnitude of impact
River channel	Large migratory fish	Migratory disconnectivity	Basin-wide declines or extinctions if species fail to reach headwater breeding grounds
Floodplain	Many fish and aquatic invertebrates	End of seasonal flood cycle	Regional declines and extinctions below dams
	Migratory birds	Loss of ephemeral habitat	Significant loss of staging habitats
Fluvial rocky outcrops	Many endemic fish and invertebrates	Permanent submergence of breeding habitats, food sources and shelters; changes in water quality and velocity	Local to basin-wide declines or extinctions if species are unable to use other habitats
Ephemeral sandy river beaches	Freshwater turtles	Permanent submergence of breeding habitats	Regional declines and extinctions below dams
Fluvial vegetated islands	Microendemic plant communities	Permanent submergence below dams	Regional declines and extinctions below dams
Varzea and *Igapo* forests	Many fish and aquatic invertebrate species	End of seasonal flood cycle including permanent inundation or irregular/ unpredictable floods	Regional declines in reservoir areas and below dams

Source: after Lees *et al.* (2016)

In a study of fish in the Tocantins River, a major tributary of the lower Amazon in Brazil, Lima *et al.* (2018) concluded that reproduction strategies (migration and parental care) and diet were the fish traits mostly affected by closure of the Peixe Angical Dam. Large dams block fish movements that connect populations and enable species to complete their life cycles. This can be particularly disruptive to tropical river fisheries, where many high-value species migrate hundreds of kilometres in response to seasonal flood pulses and have specific habitat requirements, but the impacts are felt in all biomes. The barrier effect of dams in preventing free upstream migration and access to spawning grounds has been evident on salmon and aloses in the River Garonne and its tributaries in south-western France since the Middle Ages (Décamps and Fortuné, 1991).

In the twentieth century, decline in the landed catches of Caspian Sea sturgeon, the source of caviar, is attributable partly to the construction of several large hydroelectric dams on the Volga and the consequent loss of spawning grounds. Although catches recovered to pre-dam levels in the 1980s with the establishment of new sturgeon farms on the Caspian shores, sturgeon stocks have since been seriously reduced again by overfishing, pollution and habitat loss (Khodorevskaya *et al.*, 2009).

Cutting off access to spawning grounds has an impact on the flow of nutrients from ocean environments upstream into riverine environments, as well as vice versa. This is particularly true of anadromous fishes such as salmon that die in a river after spawning. Cederholm *et al.* (1999) outline the essential contributions of nutrients and energy made by Pacific salmon carcasses to both aquatic and adjacent terrestrial ecosystems.

Recognition of the fact that dams obstruct migratory movements and/or decrease the connectivity between specific habitats has led to requirements in many countries for dam-builders to take special measures to protect migratory fishes. A common mitigation practice is the installation of fish passes, but ladders and similar structures do not always produce the desired results. Development of effective passage facilities at hydroelectric dams and other obstructions on rivers was initially directed at high-value migrant adult salmon species in Europe and North America, but it took decades to develop effective facilities. Different species have different needs, preferences and swimming abilities, and even well-designed fish passes will still fail if not positioned correctly (Williams *et al.*, 2012). In a review of large reservoirs as ecological barriers to downstream movements of migratory fish in South America, Pelicice *et al.* (2015) point out that, despite substantial financial investments and engineering efforts, several recent studies have shown that fish passes in South America are largely ineffective because constructed fishways have failed to restore viable populations of migratory fishes.

Of all the downstream effects of impoundment, the most dramatic have occurred on rivers dammed in several places. The Huang Ho has been mentioned in this respect and the Colorado River in North America is another classic example (Box 9.1). A more recent case of large-scale environmental change due to multiple impoundments has occurred in the marshlands of the Tigris–Euphrates delta. Construction of more than 30 large dams on the Tigris and Euphrates in the second half of the twentieth century resulted in a decisive change in the regime and volume of water in the two rivers, considerably reducing water supply and eliminating the floodwaters that fed the wetland ecosystems in the lower basin. The scale of the changes is indicated by the fact that the usable storage capacity of existing dams in the Tigris–Euphrates basin considerably exceeds the annual total discharge of both rivers (see Table 9.1).

Box 9.1 The Colorado River

Demands to serve urbanization, agriculture and hydropower have made the Colorado one of the world's most regulated rivers through the construction of a series of dams in the twentieth century. In consequence, nearly 64 per cent of its runoff is used for irrigation and another 32 per cent is lost to evaporation from its reservoirs. One dramatic change has been in the river's heavy sediment load, which had led the eighteenth-century Spanish explorer Garces to name it 'Colorado' (Spanish for ruddy). Before 1930, the river carried 125–150 million tonnes of suspended sediment each year to its delta in the Gulf of California, but it delivered neither sediment nor water to the sea from 1964, when Glen Canyon Dam was completed, to 1981, when Lake Powell behind Glen Canyon Dam reached capacity for the first time. Since then, river water has reached the Gulf of California in northern Mexico only irregularly, when discharges from dams are allowed. On average, the Colorado now delivers about 100,000 tonnes of sediment to the Gulf of California each year, a load that is three orders of magnitude smaller than the pre-1930 average.

(continued)

The other downstream environmental impacts have been considerable. Several species of fish had evolved to live in the relatively warm, sediment-rich waters that typified the Colorado before construction of its big dams. Colder waters, changes in frequency and timing of major flows, the drop in sediment and other associated changes in riverine habitats brought by the dams have resulted in declines of many of these species. Degraded wetlands and loss of wildlife have also been recorded in the Colorado River delta in Mexico. Although this delta ecosystem has demonstrated considerable resilience, it now relies predominantly on water supplied by agricultural return flows and waste spills that may be further reduced in the future due to climate change and changing land-use practices (e.g. through increased agricultural efficiency and more optimal management of dams) in the USA and Mexico.

Source: Blinn and Poff (2003).

Analysis of satellite images over the lower Tigris–Euphrates indicated that in less than 30 years at least 7600 km² of three primary wetlands (excluding the seasonal and temporary flooded areas) disappeared between 1973 and 2000 (UNEP, 2001). The Central and Al-Hammar marshes in southern Iraq had each been largely desiccated, with more than 90 per cent of their land cover transformed into bare land and salt crusts. While damming of the two rivers in more than 30 places since 1960 has reduced water flow to the entire area, their fate was accelerated by massive drainage works implemented in southern Iraq in the early 1990s following the Second Gulf War in 1991. It is interesting to note that these drainage engineering works in Iraq were to a large extent made physically possible by reduced flow thanks to impoundments further upstream.

The Al-Hawizeh marshes, which straddle the Iran/Iraq border, had been reduced to one-third of their previous area by 2000. Shrinkage continued in the early 2000s and although there was some expansion in the area of Al-Hawizeh from 2011 to 2013 due to increased precipitation, the future of the marsh remains uncertain because of plans for another ten dams upstream in Iraq and Turkey (Daggupati *et al.*, 2017).

The lower Mesopotamian marshlands were once the largest wetland in the Middle East and a biodiversity centre of global importance, as well as being home to an estimated half-million Marsh Arabs, a distinct indigenous people. During the 1990s, most of the Marsh Arab population, whose livelihood has been entirely dependent on the wetland ecosystem for about 5000 years, fled the marshes. Several endemic species of marsh mammals, birds and fish have become extinct or are seriously threatened.

Coastal fisheries in the Persian Gulf, dependent on the marshland habitat for spawning migrations and nursery grounds, have also experienced significant reductions. In the Shatt al-Arab estuary, the decrease in freshwater flows has stimulated sea-water intrusion and disrupted its complex ecology. Although a small proportion of the Marsh Arab population has returned to live within the marshes, most are unlikely ever to go back (Richardson and Hussain, 2006). The destruction of the vast Mesopotamian marshlands has been likened to other human-engineered changes such as the desiccation of the Aral Sea and the deforestation of Amazonia as one of the Earth's major and most thoughtless environmental disasters.

POLITICAL IMPACTS OF BIG DAMS

Big dams are significant politically by their very nature as large endeavours in environmental management. Summoning the political will to embark on such projects, as well as the necessary finance and engineering skills, can take many years. The Three Gorges Project, one of the most ambitious and controversial big dams in recent times, was first proposed in 1919 but 75 years passed before construction began in 1994 (Table 9.6).

Table 9.6 History of the Three Gorges Dam project

Date	Development
1919	Chinese leader Sun Yat Sen proposes construction of a dam at Three Gorges to improve Yangtze River navigation and to make better use of the river's resources
1944	The US Bureau of Reclamation surveys the Yangtze River and drafts a proposal for a dam at Three Gorges
1947	The dam project is suspended by the Chinese government because of high inflation and an economic crisis
1953	Mao Zedong reviews a proposal for constructing reservoirs and requests a dam be built at Three Gorges to control flooding
1954	The Yangtze Valley experiences the second worst floods of the century
1982	Denz Xiaoping pledges to proceed with the Three Gorges Dam project
1984–89	Feasibility studies for the Three Gorges Water Control Project
1992	The People's Congress approves a resolution to proceed with the Three Gorges Dam project
1993	The Three Gorges Project Construction Committee (TGPCC) is formed to represent China's State Council in decision-making
1994	Chinese premier, Li Peng, announces the official launch of construction on the Three Gorges Dam
1996	Two major transportation projects, the Xiling Bridge across the Yangtze and an airport in Yichang, are completed. In May, the US Export-Import Bank withdraws support for the Three Gorges Dam project because of failure to follow the bank's environmental guidelines. In December, the Japanese Export-Import Bank announces financial backing for Japanese companies to participate in Three Gorges
1997	The TGPCC issues corporate bonds to raise construction funds. By September, the first phase of residents to be relocated from the reservoir region are resettled. In October, construction of transportation infrastructure for the project is completed. In November, the Yangtze River is blocked, signalling completion of the first phase of construction on the dam
1998	The Yangtze Valley experiences the worst floods of the century
1998–2006	Phase 2 of the Three Gorges Dam project: construction of the dam, including a flood discharge system and the hydroelectric plant (14 generators go online 2003–06)
1999	4.49 million cubic metres of cement are poured for the foundation of the Yangtze River power plant, a new world record for the construction of a hydropower project
2004–09	Phase 3 of the Three Gorges Dam project: construction on the dam itself completed in 2006 with two five-stage locks and ship lifts. In 2009, another 12 generators are installed, making the Three Gorges Dam fully operational

Source: International Energy Outlook (1998) and media reports.

Increasing public awareness of the environmental and social implications of big dams was a significant driver of the global slow-down in the rate of dam construction over recent decades. Heated debates over the issues have in some cases led to the shelving of construction plans. The Nam Choan Dam Project in Thailand has been mentioned in this respect. Another example is the Myitsone Dam in Myanmar, development of which was suspended in 2011 when civil society organizations opposed the dam, despite a bilateral agreement with China to build it (Chan, 2017). Anti-dam campaigns often drag on for lengthy periods of time, however, and the uncertainty itself can have serious social impacts, as highlighted by Kirchherr *et al.* (2016). These authors studied implementation of Thailand's Kaeng Sue Ten Dam, which has been uncertain for 35 years, and found extreme levels of stress among communities induced by the threat of the project and the fear of compulsory relocation. The study also concluded that the Thai government had withheld public infrastructure investments during the time of uncertainty, further hampering the villages' economic development.

There is little doubt that many of the adverse environmental and social impacts of dams can be greatly reduced by good planning and anticipation, and aid agencies that finance such projects require an Environmental Impact Assessment (EIA) before approval of funding is given. Progress has also been made in the widening scope given to consultation prior to dam construction (Table 9.7). Nevertheless, a sensible operating schedule is also a key factor – many of the problems caused by dams are the result of operators aiming to maximize water use – through releases for hydroelectricity generation and irrigation, for example – to such an extent that other concerns are given too little consideration.

The question of who benefits and who bears the brunt of the environmental costs is also pertinent. They are often different groups of people affected, as numerous examples in this chapter illustrate. Such a divergence can raise serious political issues as in the case of the James Bay Hydroelectric Project on the La Grande River in Quebec, Canada. The indigenous Cree and Inuit communities who had lived in the area probably for thousands of years had not been consulted when it was announced in 1971 that the La Grande River would be dammed, eventually in five places. The Crees and Inuit obtained a court injunction to stop the project in 1973, and although this was quickly overturned, and construction resumed, a political conflict that continues to this day had begun (Desbiens, 2004).

Table 9.7 Broadening of consultation in the design of big dams

Design team	Approximate era
Engineers	Pre-Second World War
Engineers + economists	Post-Second World War
As above + environmentalists	Late 1980s
As above + affected people	Early 1990s
As above + NGOs	Late 1990s
As above + national consensus	Early 2000s

Source: after Goodland *et al.* (1993b).

The increasing sensitivity of the big dam issue, and the need to investigate rigorously the benefits and costs to society of dam projects, prompted the establishment of a World Commission on Dams in 1998 by the World Conservation Union (IUCN) and the World Bank. The Commission was set up to review the environmental, social and economic impacts of large dams, and to develop new guidelines for the industry. It reported in 2000 on its two main goals:

1   to review the development effectiveness of dams, and assess alternatives for water resources and energy development
2   to develop internationally acceptable standards, guidelines and criteria for decision-making in the planning, design, construction, monitoring, operation and decommissioning of dams.

The Commission's framework for decision-making was based on five core values: equity, sustainability, efficiency, participatory decision-making and accountability. It proposed:

■   a rights-and-risks approach as a practical and principled basis for identifying all legitimate stakeholders in negotiating development choices and agreements
■   seven strategic priorities and corresponding policy principles for water and energy resources development – gaining public acceptance, comprehensive options assessment, addressing existing dams, sustaining rivers and livelihoods, recognizing entitlements and sharing benefits, ensuring compliance, and sharing rivers for peace, development and security
■   criteria and guidelines for good practice related to the strategic priorities, ranging from life-cycle and environmental flow assessments to impoverishment risk analysis and integrity pacts.

One approach to the problems associated with big dams involves attempting to restore natural flow regimes and associated ecosystem health and services by modifying dam operations in some way. This strategy is referred to as 're-operation' by Richter and Thomas (2007) who contend that significant environmental benefits can be gained through a careful reassessment of how river water is used and by modifying the uses through more conservative management. However, although re-operation may help to alleviate problems in some areas, big dams are still being built in many parts of the world. Appropriate design (Table 9.8) and management will hence continue to be important for many years to come.

Table 9.8 Characteristics of good and bad dams from an ecosystem standpoint

Good dam	Bad dam
Reservoir with relatively small surface area (often in a narrow gorge)	Reservoir with large surface area
Reservoir that is deep and silts slowly	Relatively shallow reservoir (which sometimes equates to relatively short useful life)
Little loss of wildlife and natural habitat	Considerable flooding of natural habitat and consequent loss of wildlife
Little or no flooding of forests	Submerged forests that decay and create water quality problems

(continued)

Table 9.8 continued

Good dam	Bad dam
River that is relatively small, with little aquatic biodiversity	Large river with much aquatic biodiversity
No problem with large floating aquatic plants	Serious problem with large floating aquatic plants
Many unregulated downstream tributaries	Few or no downstream tributaries
No tropical diseases (i.e. high elevation or mid- to high latitudes)	Location in the lowland tropics or subtropics, conducive to the spread of vector-borne disease

Source: after Ledec *et al.* (1997).

The building of dams on rivers flowing through more than one country brings international political considerations onto the agenda of big dam issues. Such considerations are particularly pertinent in dryland regions where rivers represent a high percentage of water availability to many countries. The main issues at stake here are those of water availability and quality.

In several international river basins, peaceful cooperation over the use of waters has been achieved through international agreement. One such agreement, between the USA and Mexico over use of the Rio Grande, was signed in 1944 and is operated by the International Boundary and Water Commission. This body ensures equal allocation of the annual average flow between the two countries. Similarly, India and Bangladesh signed a water-sharing accord for the Ganges in 1996. It specifies water allocation in normal and dry periods after 20 years of wrangling over the effects of the Farakka Barrage. In Africa, the Nile Waters Agreement between Egypt and Sudan governs the volume of Nile water allowed to pass through the Aswan High Dam, but none of the other eight Nile basin countries has an agreement over use of the Nile's waters (see Chapter 8). This issue has dogged regional politics for many years and has erupted once more following plans to build the Grand Ethiopian Renaissance Dam on the Blue Nile. The Ethiopian government sees the dam as a major driver for national development, while countries downstream are concerned about possible negative impacts (Tawfik, 2016).

Agreements to resolve disputes over water resources have a very long history. The beginnings of international water law can be traced back at least to 2500 BCE, when the two Sumerian city-states of Lagash and Umma reached an agreement to end a dispute over the water resources of a tributary of the River Tigris in the Middle East (Wolf, 1998). Wrangles over water are still a significant potential source of conflict in the Tigris–Euphrates basin due to a lack of agreements in the contemporary era. While there is currently a water surplus in this region, the scale of planned developments raises some concern. Turkey's Southeastern Anatolian Project, a regional development scheme on the headwaters of the two rivers, centres on 22 dams (Kaygusuz, 2010). In early 1990, when filling of the Ataturk Dam reservoir commenced, stemming the flow of the Euphrates, immediate alarm was expressed by Syria and Iraq, despite the fact that governments in both countries had been alerted and discharge before the cut-off had been enhanced in compensation. Syria and Iraq nearly went to war when Syria was filling its Euphrates Dam. Full development of the Southeastern Anatolian Project, expected in about 2020, could reduce the flow of the Euphrates by as much as

60 per cent, which could severely jeopardize Syrian and Iraqi agriculture downstream. The three Tigris–Euphrates riparians have tried to reach agreements over the water use from these two rivers, and the need for such an agreement is increasingly pressing.

Dam removal

A late twentieth-century trend is that of decommissioning dams that no longer serve a useful purpose, are too expensive to maintain safely, or have levels of environmental impact now deemed unacceptable. Another key motivation for dam removal is to rehabilitate rivers by restoring physical and ecological attributes and functions. The desire to restore rivers to their pre-dam ecological state has accelerated in many countries in recent years, particularly in North America. Indeed, in the USA, the decommissioning rate for dams overtook the rate of dam construction in 1998, although most dams removed to date have been small. Decommissioning dams can enable the restoration of fisheries and riverine ecological processes, as experience in North America and Europe (Table 9.9) has shown, but the removal of dams needs to be preceded by careful planning because environmental impacts are also inherent in the removal of such large-scale structures.

These impacts include negative effects on downstream aquatic life due to a sudden flush of the sediments accumulated in the reservoir. The issue is highlighted by Thiebault *et al.* (2017) who found hazardous accumulations of several pharmaceutical products in the sediments collected upstream of a dam near the city of Orleans, France. Evaluation of potential contaminants in sediments has become a routine part of

Table 9.9 Some examples of recent large dam removals

Dam	River	Country	Height (m)	Year completed	Year removed
Bluebird	Ouzel Creek	USA	17	1904 (enlarged 1920)	1990
Glines Canyon	Elwha	USA	64	1927	2011
Condit	White Salmon	USA	38	1913	2011
San Clemente	Carmel	USA	32	1921	2015
Riss East	Four Mile	USA	17	1967	2016
Saint Etienne de Vigan	Allier	France	17	1895	1998
Kemansquillec	Léguer	France	15	1922	2001
Vézin	Sélune	France	36	1920s	2018
Robledo de Chavela	Cofio	Spain	23	1968	2014
Arase	Kumagawa	Japan	25	1955	2018

Source: American Rivers Dam Removal Database and media sources.

pre-removal assessment and purposeful dam removals are engineered to prevent or limit their release (Evans and Wilcox, 2014), in some cases by excavating sediment prior to removal.

There are common responses of rivers to dam removal, but all removals have different starting conditions and are influenced by complex interactions among physical processes, ecological responses and river basin history. In a synthesis of research on dam removals – both large and small – in the USA, Foley *et al.* (2017) observe that the physical response of a river is typically rapid, commonly stabilizing in years rather than decades, but ecological response is much more variable. They also stress that restoring a river to its pre-dammed state may not always be possible.

The restoration of key elements in a river, particularly iconic species such as certain fish, is nevertheless a powerful driver of some dam removal programmes. In France, for example, the 17-metre Saint Etienne de Vigan dam on the Allier, a tributary of the Loire, was removed in 1998 as part of a government effort to restore stocks of Atlantic salmon to the Loire Valley (Van Looy *et al.*, 2014). Removal of the dam, and subsequently three smaller dams, was designed to restore access to salmon spawning grounds. Non-governmental organizations led a campaign to bring back salmon to the Loire Valley, and the French government responded with the Plan Loire Grandeur Nature. Dam removal inevitably becomes highly politicized and at times controversial. In a study of dam removal in Sweden, Jørgensen and Renöfält (2012) highlight some of the contrasting views of such programmes, with groups in favour framing their position around ecosystem services provision (especially for fish) and recreation (fishing), while opponents of dam removal generally frame their position around the potential loss of cultural services (recreation, aesthetics, local heritage) that would ensue from dam removal.

FURTHER READING

Chen, G., Powers, R.P., de Carvalho, L.M. and Mora, B. 2015 Spatiotemporal patterns of tropical deforestation and forest degradation in response to the operation of the Tucuruí hydroelectric dam in the Amazon basin. *Applied Geography* 63: 1–8. A study of how this dam affects surrounding patterns of forest disturbance.

Chen, J., Shi, H., Sivakumar, B. and Peart, M.R. 2016 Population, water, food, energy and dams. *Renewable and Sustainable Energy Reviews* 56: 18–28. This paper examines the impact of global population growth and the increases in consumption of water, food and energy on the development of large dams.

Foley, M.M. and 21 others 2017 Dam removal – listening in. *Water Resources Research* 53: 5229–46. A review of dam removals in the USA. Open access: http://onlinelibrary.wiley.com/doi/10.1002/2017WR020457/full

Hornig, J.F. (ed.) 1999 *Social and environmental impacts of the James Bay Hydroelectric Project*. Montreal, McGill-Queen's University Press. Insights from a variety of disciplines on the controversies surrounding the first mega-scale hydroelectric project in the sub-Arctic, on the La Grande River in Canada.

Yihdego, Y., Khalil, A. and Salem, H.S. 2017 Nile River's Basin dispute: perspectives of the Grand Ethiopian Renaissance Dam (GERD). *Global Journal of Human-Social Science Research* 17(2). Assessment of controversies surrounding what will be Africa's largest dam.

Zhang, Q. and Lou, Z. 2011 The environmental changes and mitigation actions in the Three Gorges Reservoir region, China. *Environmental Science and Policy* 14: 1132–8. Assessment of the world's largest hydroelectric scheme.

WEBSITES

damremoval.eu   Dam Removal Europe is dedicated to the restoration of European rivers.

www.icold-cigb.net   International Commission on Large Dams.

www.internationalrivers.org   International Rivers has information, campaigns and publications (e.g. *World Rivers Review*) on many dam-related issues.

www.narmada.org   site that aims to present the perspective of grassroots people's organizations on the construction of large dams on the River Narmada in Gujarat, India.

www.nwcouncil.org   documents, journals and related information on the management of the Columbia River Basin in North America.

www.therrc.co.uk   the UK-based River Restoration Centre site promotes restoration, habitat enhancement and catchment management and includes a Manual of River Restoration Techniques.

POINTS FOR DISCUSSION

- On the whole, do you consider big dams to be good or bad things?
- Are any of the environmental impacts of big dams irreversible?
- Outline the hazards associated with major reservoirs.
- Is hydroelectricity a sustainable form of power generation?
- What are the political issues associated with big dam development?
- Outline the possible environmental impacts of decommissioning a large dam.

10 *Urban environments*

LEARNING OUTCOMES

At the end of this chapter you should:

- Know that globally most people live in urban environments.
- Recognize the scale of ecological transformation that has occurred in the city-state of Singapore.
- Understand that overuse of groundwater has caused serious rates of subsidence in many cities.
- Know that the residents of most cities face a significant health risk from atmospheric pollution.
- Appreciate the links between environmental problems and social injustice, so that many deprived communities in cities are exposed to disproportionately high levels of environmental risk.
- Understand that self-employed waste pickers are recognized for their valuable contributions to urban sustainability in many cities of the global South.
- Understand some of the ways in which cities can be made more sustainable.

KEY CONCEPTS

megacity, urban sprawl, biotic homogenization, flashy discharge regime, urban stream syndrome, subsidence, urban heat island, smog, environmental justice, post-industrial city, smart city

---

Large numbers of people have lived in close proximity to each other in cities for thousands of years. The first urban cultures began to develop about 5000 years ago in Egypt, Mesopotamia and India, but the size of cities and their geographical distribution expanded dramatically after the Industrial Revolution of the last millennium. This expansion accelerated into the twentieth century which began with 13 per cent of the world's population living in cities and ended with a figure of nearly 47 per cent. Today, for the first time in history, more than 50 per cent of the world's population live in urban areas and this proportion is expected to reach 66 per cent by 2050. In absolute terms, this global urban population was 1 billion in 1960, 2 billion in 1985, and 3 billion in 2002. By 2030 it is expected to near 5 billion (UNDESA, 2006, 2010).

This is not to say that all cities have grown inexorably through the ages. Some of the world's large urban areas have a long history of continuous occupancy and importance, but others have not adjusted successfully to changing circumstances. Baghdad, Cairo and Istanbul (formerly Constantinople) ranked among the world's largest cities at both the beginning and end of the second millennium. Other cities of major importance in 1000 CE have now faded from prominence (e.g. Nishapur, Persia; Córdoba, Spain) or have been abandoned altogether (e.g. Angkor, Khmer Empire).

Nonetheless, growth rates in many of today's major world cities have been unprecedented in recent decades. In 1975, three urban areas – New York–Newark, Tokyo and Mexico City – had a population of more than 10 million people. By the year 2000, the number of these so-called megacities had reached 18, and by 2025 the total is likely to be 29 (Figure 10.1).

Many cities in the industrializing world experienced remarkable growth in the second half of the twentieth century: the populations of Mexico City, São Paulo, Karachi and Seoul growing by more than 800 per cent over that period. The urban area of Mexico City, which is probably the largest conurbation in the developing world, expanded from 27.5 km² in 1900 to cover 1450 km² in 2000. The phenomenal growth of some cities, and the high concentrations of people they represent (the urban density of Mexico City in 2000 was 12,559 people/km²), have created some acute environmental issues both outside and within the city limits.

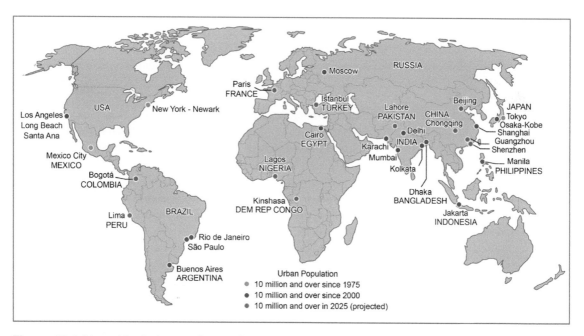

Figure 10.1 Megacities (urban agglomerations with 10 million inhabitants or more) in 1975, 2000 and 2025 (from data in UNDESA, 2006, 2010).

By contrast, in the more prosperous countries, the populations of large cities such as New York and London grew little in the second half of the twentieth century. These urban areas now face different challenges, including generally unplanned incremental development, characterized by a low density mix of land uses on the urban fringe. Such 'urban sprawl' is classically a North American phenomenon, but is also commonly seen throughout Europe where city structure has traditionally been much more compact. Indeed, sprawl is also becoming a feature of the urban periphery in Latin America and Asia too (Douglas, 2008). The development of urban sprawl on all these continents is driven by the same three key factors: a growing population, rising incomes and falling commuting costs.

The growth of cities inevitably occurs at the expense of the natural environment. This modification of nature for urban development takes many forms, affecting the lithosphere, biosphere, hydrosphere and atmosphere. Often deliberate, sometimes inadvertent, these alterations range from the total and dramatic transformation of a landscape to the subtle, partial change in the behaviour of an animal. Overall, cities represent such a dramatic transformation of nature that effectively they can be treated as a type of physical environment in their own right. The scale of ecological change is illustrated in Box 10.1 by the growth of a relatively small and new city: Singapore.

Box 10.1 Transformation of nature in Singapore

In 1819, the island of Singapore was almost entirely covered in primary forest, but most of the trees had been removed by the end of the nineteenth century to make way for the cultivation of cash crops. Since the 1930s, agricultural land has gradually been replaced by the spread of the urban area, which occupied about 50 per cent of the island by the beginning of the twenty-first century, with another 45 per cent of the land area modified in other ways for human use.

Singapore's two centuries of ecological transformation have involved significant losses of the island's original biota. Local extinctions of a wide range of terrestrial and freshwater species have been recorded, but since the first reliable species records were made only in the 1870s, it is likely that many other, unrecorded species have also disappeared.

Less than 200 ha of primary rain forest survives, while a further 1600 ha is covered in tall secondary forest and most of these remaining forest fragments now lie within nature reserves. In 2010, parks and nature reserves constituted 8 per cent of the national land area, which has itself been enlarged through a programme of land reclamation.

A concerted effort to expand the country's territory around its coastline for residential, commercial and industrial use followed independence in 1959, although the history of reclamation in Singapore dates back to the late nineteenth century. In the last 50 years, Singapore's national land area has grown by more than 100 km^2, an increase of nearly 25 per cent (Table 10.1), and this growth continues, with the government aiming to achieve a national land area of 766 km^2 by 2030.

Table 10.1 The increase in Singapore's land area due to reclamation since the 1960s

Year	Total land area (km²)
1966	581.5
1970	586.4
1975	596.8
1980	617.8
1985	620.5
1990	633.0
1995	647.5
2000	682.7
2005	697.9
2010	712.4
2015	719.1

Source: Yearbook of Statistics Singapore, various years.

This programme of land reclamation has occurred at the expense of adjacent coastal ecology, while remaining marine ecosystems, including coral reefs, have been subjected to persistent disturbance in the form of increased sediment loads from the reclamation and dredging activities, as well as shipping, oil-related industries, recreation, and the dumping of urban wastes.

Source: Corlett (1992) ; Brook *et al.* (2003).

Cities represent a completely artificial environment; they absorb vast quantities of resources from surrounding areas and create high concentrations of wastes to be disposed of. In an assessment of the energy and material flows for the world's 27 megacities as of 2010, Kennedy *et al.* (2015) concluded that these urban areas accounted for 9 per cent of global electricity use and 13 per cent of all solid waste produced. The degree to which cities impinge on their hinterlands, their ecological footprint, is indicated by a few examples. About 10 per cent of prime agricultural land has been lost to urbanization in Egypt. The twentieth-century growth of São Paulo was fuelled by the expansion of coffee and sugar-cane plantations in southeast Brazil, helping to make São Paulo the country's most economically developed state with 34 per cent of Brazil's GNP. In consequence, the natural forest cover of São Paulo State has been reduced from 82 per cent in 1860 to no more than 9 per cent in 2010 (Rodrigues *et al.*, 2011). In Iran, the demand for water in Tehran rose during the second half of the twentieth century from 10 million m³ in 1955 to 930 million m³ in 2000, an increase of more than ninety times. This growing demand spurred the construction of a series of dams and canals in the early decades of the last century, to bring water 50 km from the River Karaj to the west, reducing the water available for rural agriculture. By the 1970s, supplies were again running low, so water was diverted more than 75 km from the River Lar to the north-east. Groundwater is tapped for a large proportion of Tehran's needs, with resulting serious rates of subsidence (see Table 10.4 below), and water remains a major constraint on the city's development (Tajrishy and Abrishamchi, 2005).

URBAN BIODIVERSITY

Worldwide, the expansion of cities is a primary driver of habitat loss and species extinction. In a global synthesis of plant extinction rates in urban areas, Hahs *et al.* (2009) found that the legacies of landscape transformations caused by urban development frequently last for hundreds of years. This indicates that modern cities potentially carry a large 'extinction debt' in which the consequences of initial habitat loss are delayed: species that have avoided rapid extinction becoming progressively more vulnerable to the altered dynamics of fragmented landscapes. The impact of urban areas on biodiversity can also extend well beyond a city's immediate boundaries due to its ecological footprint. A study of changes in the abundance of migratory birds concluded that the loss of habitats to urban sprawl in North America appears to be related to declines in the numbers of some birds in the southern hemisphere (Valiela and Martinetto, 2007).

Humans often deliberately influence certain species in urban areas, targeting pests for eradication and introducing domesticated pets, for instance. Indeed, cities have been viewed as homogenizing forces, where some species well-adapted to living in urban areas become common in cities worldwide (McKinney, 2006). Of course, cities have been created to meet the relatively narrow needs of just one species, our own, but others have benefited too. They include creatures that find habitats largely inside buildings, such as the house mouse and Norway rat, and others that use habitats on the outside of buildings (essentially similar to rocky areas) such as the rock dove and peregrine falcon. Urbanization also appears to have evolutionary consequences for some species, adaptation to a city life including changes to breeding, migration and other forms of behaviour that are associated with alterations to their genetic makeup (Partecke and Gwinner, 2007).

Similarly, there is evidence to support the hypothesis that plants in urban areas across the world are being homogenized. In a study of flora in 101 cities worldwide, La Sorte *et al.* (2014) found that although most plants in the cities examined were native and unique to each urban area, there were also distinct homogenizing influences. Prominent among these drivers of biotic homogenization were the effects of invasive species and plants from Europe. Cities have a lengthy history as foci of species introductions, mainly due to the urban role as centres for transport. The global influence of European plants is a function of European colonial history.

SURFACE WATER RESOURCES

Historically, many cities have developed close to bodies of water, particularly rivers, to maximize the associated benefits: an abundant source of fresh water, a ready transport corridor for trade and travel, a direct source of food, and fertile alluvial soils. Hence many great global cities are inextricably linked to their rivers, including Baghdad (Tigris), Buenos Aires (de la Plata), Cairo (Nile), London (Thames), Montreal (Saint Lawrence), Moscow (Moskva), New York (Hudson) and Paris (Seine) to name but a few. Urban development inflicts numerous impacts on water bodies, often involving enormous changes to their form and function by altering hydrology, affecting stream morphology, water quality and the availability of aquatic habitats (Table 10.2). These effects may be abrupt but they can also be continuous and synergistic. A few examples from Asian cities illustrate the huge scale of some impacts on rivers: the development of Tokyo accelerated after the mid-seventeenth-century diversion of the Tone River by >100 km to the east;

Table 10.2 Major impacts of urbanization on waterways

Change in stream characteristics	Effects
Hydrology	Increased magnitude and frequency of severe floods Increased annual volume of surface runoff Increased stream velocities Decreased dry weather base flow
Morphology	Channel widening and downcutting Increased erosion of banks Stream enclosure or channelization
Habitat and ecology	Changes in diversity of aquatic insects Changes in diversity and abundance of fish Destruction of wetlands, riparian buffers and springs
Water quality	Massive sediment pulses Increased pollutant washoff Nutrient enrichment Bacterial contamination Increased organic carbon loads Higher levels of toxics, trace metals, hydrocarbons Increased water temperature Trash/debris jams

Source: after Baer and Pringle (2000).

Shanghai's main river, the Huangpu, is an entirely man-made tributary of the Yangtze River; and Seoul re-created a stretch of the Cheonggyecheon after it was concreted over in the 1960s (see p. 178).

The demand for water, from large cities in particular, also means that urban effects can be far-reaching, as the impact of Los Angeles on certain lakes in California amply demonstrates (see p. 179). In hydrological terms, urban areas are typified by 'flashy' discharge regimes caused by the increased area of impervious surfaces such as tarmac and concrete, and networks of storm drains and sewers. On average, urban watersheds lose 90 per cent of a storm's rainfall to runoff, whereas non-urban forested watersheds retain about 25 per cent of the rainfall (Shang and Wilson, 2009). The timelag between rainfall and peak discharge of a river is also decreased and the peak discharge itself is increased as a result of these impervious surfaces, which has a significant impact on the volume, frequency and timing of floods. In consequence, many urban rivers have been subject to structural measures to reduce the flood hazard, such as channel straightening and channelization, the building of levées, and the stabilization of banks (see Chapters 8 and 21).

A flashy regime is one of a suite of characteristics that combine to produce a consistently observed ecological degradation typical of rivers that flow through cities. Other symptoms of so-called 'urban stream syndrome' include elevated concentrations of nutrients and contaminants, altered channel morphology and stability, and reduced biotic richness, with an increased dominance of tolerant species (Walsh *et al.*, 2005). Poor water quality is one of the most important environmental issues stemming from urban modifications to the hydrological cycle. Runoff from developing urban areas is usually choked with sediment during construction phases, when soil surfaces are stripped of vegetation, and a finished urban zone greatly increases runoff. This drastically modified urban drainage network feeds large amounts of urban waste products – including nutrients, metals, pesticides and organic contaminants – into rivers and ultimately into oceans. Runoff from urbanized surfaces also has significant effects on the thermal regimes of urban rivers.

In ecological terms, Morley and Karr (2002) used measures of a river's invertebrate community to demonstrate an overall decline in stream biological condition as the percentage of urban land cover increased, and there is no doubt that many rivers flowing through cities are heavily contaminated. Pollution is reported to be particularly severe in rapidly industrializing urban areas such as the Pearl River Delta in southern China, one of the most densely urbanized regions in the world and a central hub of Chinese economic growth over recent decades. A wide range of pollutants enter the Pearl River Delta, an area of about 11,300 km^2 with a population of over 52 million living in several major cities (including Guangzhou, Hong Kong and Shenzhen). They include high levels of nutrients, metals, hydrocarbons, PCBs and bacteria originating from domestic sewage and industrial emissions particularly, but also from agriculture and shipping (Qi *et al.*, 2010). Other rivers in industrializing parts of Asia with serious pollution issues include the Buriganga River in Dhaka, Bangladesh (Ahmed *et al.*, 2016); the Citarum River near Jakarta, Indonesia (van der Wulp *et al.*, 2016); and the Yangtze River which flows through numerous cities in China (Floehr *et al.*, 2013). Although the example of the Thames at London (see p. 173) shows how such heavily polluted, near-anaerobic river conditions can be improved, both economic and political resources need to be mobilized before any river can be properly cleaned up, and the process can take a very long time.

The local hydrological impact of the Saudi capital, Riyadh, provides a very contrasting example to the catalogue of riverine ecological problems typically associated with large, rapidly growing cities. Discharge of Riyadh's domestic and industrial wastewater feeds the Riyadh River, which scarcely existed before the 1980s, but now flows throughout the year down what was the sporadic Wadi Hanifa. The water, which is originally derived from desalinated Gulf sea water, is partially treated before being released to flow down the steep-sided wadi and enter open countryside, eventually disappearing 70 km from Riyadh. The new flow has created an attractive valley lined by tamarisk trees and phragmites, which is an important recreational site for Riyadh's 5 million population. Beyond the wadi, significant irrigated agriculture has grown up, drawing on both the river and the groundwater reserves around it (Mahboob *et al.*, 2014).

Indeed, the quantity of water available is another critical environmental issue for many cities, as the situation in Tehran mentioned above indicates. The rising consumption of water from a fast-growing economy and an increasingly urbanized population is having a serious impact on the Huang Ho River, China's second longest (Feng *et al.*, 2012). The Huang Ho basin had a population of some 115 million in 2010 – more than double the 1950 population – with a 39 per cent urbanization rate. The strain on water resources is indicated by the fact that the average urban household in the basin consumes more than twice as much water as a rural household. This calculation uses the idea of 'virtual water', a measure of the total volume of water consumed to create products, and reflects the higher income levels of urban households, allowing them higher consumption rates of all types of water-intensive goods and services, from food to footwear.

Different types of environmental issue are encountered in permafrost areas where surface water and soil moisture are frozen for much, and in some places all, of the year. Frozen rivers and lakes mean that many of the uses such water bodies are commonly put to at more equable latitudes, such as sewage and other waste disposal, are not always available. The low temperatures characteristic of such regions also mean that biological degradation of wastes proceeds at much slower rates than those elsewhere. Hence, the impacts of pollution in permafrost areas tend to be more long-lasting than in other environments.

The nature of the permafrost environment also presents numerous environmental challenges to the construction and operation of settlements – challenges that have been encountered in urban developments associated with the exploitation of hydrocarbons and other resources in Alaska, northern Canada and northern Russia. Disturbance of the permafrost thermal equilibrium during construction can cause the development of thermokarst (irregular, hummocky ground). The heaving and subsidence caused can disrupt building foundations and damage pipelines, roads, railtracks and airstrips. Terrain evaluation prior to development is now an important procedure in the development of these zones, following expensive past mistakes. Four main engineering responses to such problems have been developed: permafrost can be ignored, eliminated or preserved, or structures can be designed to take expected movements into account (French, 2007). Preservation of the thermal equilibrium is achieved in numerous ways, such as by insulating the permafrost with vegetation mats or gravel blankets, and ventilating the underside of structures that generate heat (e.g. buildings and pipelines; see Figure 10.2).

Figure 10.2 Buildings that disturb the thermal equilibrium of permafrost can lead to heaving and subsidence, as seen on the left. One solution is to raise buildings and pipelines up above the ground surface on stilts, so ventilating heat-generating structures (right). Both photographs were taken in Yakutsk, eastern Siberia.

GROUNDWATER

The water needs of urban populations and industry are often supplemented by pumping from groundwater, and pollution of this source is another issue of increasing concern in many large cities. Seepage from the improper use and disposal of heavy metals, synthetic chemicals and other hazardous wastes such as sewage, is the principal origin of groundwater pollution. A serious threat to groundwater quality can occur in parts of urban areas without sewerage systems. One of the most common disposal systems for human excreta in cities in low-income countries is the pit latrine but their use raises concerns over groundwater contamination that may in turn negatively affect human health because most pit latrines are unlined and so inevitably leak bacteria and nitrogen compounds from excreta. A study in an informal urban settlement in Zimbabwe (Zingoni *et al.*, 2005) found that more than two-thirds of boreholes and domestic wells sampled were contaminated with bacteria associated with excreta ('fecal coliforms') and that the highest nitrate concentrations in groundwater were associated with the highest population and pit latrine densities. Another regular source of contamination in snow-belt regions of Asia, Europe and North America is de-icing agents, usually sodium chloride, applied to roads. Salts washed away from urban highways accumulate in soils as well as groundwater (Williams *et al.*, 2000). Contamination of groundwater beneath any city is a serious long-term issue since aquifers do not have the self-cleansing capacity of rivers and, once polluted, are difficult and costly to clean.

A frequent outcome of overusing groundwater is a lowering of water-table levels and consequent ground subsidence, an issue confronted by many urban areas (Table 10.3). Mexico City is one of the most dramatic

Table 10.3 Some examples of subsidence due to groundwater extraction in urban areas

City	Country	Period	Total subsidence (m)	Subsidence rate (mm/yr)
Tehran	Iran	1984–2012	11.7	416
Taipei	Taiwan	1970–2000	2.5	100
Jakarta	Indonesia	1974–2010	4.0	30–100
Osaka	Japan	1935–70	2.9	80
Mexico City	Mexico	1900–99	>8.0	80
Manila	Philippines	1991–2003	>1.0	50–90
Shanghai	China	1921–65	2.6	60
Tokyo	Japan	1895–1970	4.4	59
Bangkok	Thailand	1978–2003	1.0	40
Houston	USA	1906–95	>3.0	34
Tianjin	China	1981–2003	0.6	27
Ravenna	Italy	1897–2002	>1.0	10

Source: Rodolfo and Siringan (2006); Teatini *et al.* (2005); Ovando-Shelley *et al.* (2007); IGES (2006); Abidin *et al.* (2015); Mahmoudpour *et al.* (2016).

examples where use of subterranean aquifers for 100 years and more has caused subsidence in excess of 8 m in some central areas, greatly increasing the flood hazard in the city and threatening the stability of some older buildings, notably the sixteenth-century cathedral. The rate of surface lowering in Mexico City has varied since 1900, accelerating from 30 mm/year up until 1920 to peak at a remarkable 260 mm/year in the early 1950s. Since groundwater pumping was banned in downtown Mexico City in the 1960s, the rate in central parts of the city has slowed to less than 100 mm/year, but water is still drawn from more recent wells sunk on the outskirts of town in the late 1970s and early 1980s and subsidence at some sites near the newer wells has exceeded 300 mm/year (Ovando-Shelley *et al.*, 2007).

Marked episodes of subsidence have also occurred in Japan's capital city, Tokyo, reflecting phases of economic and industrial growth. The problem was first identified in Tokyo's Koto Ward in the southern part of the Kanto Plain in the 1920s as industrial activity grew after the First World War with an associated rapid industrial use of groundwater, but the process came to a halt for some years in the late 1940s following the destruction of industries in the Second World War (Furuno *et al.*, 2015; see Figure 10.3).

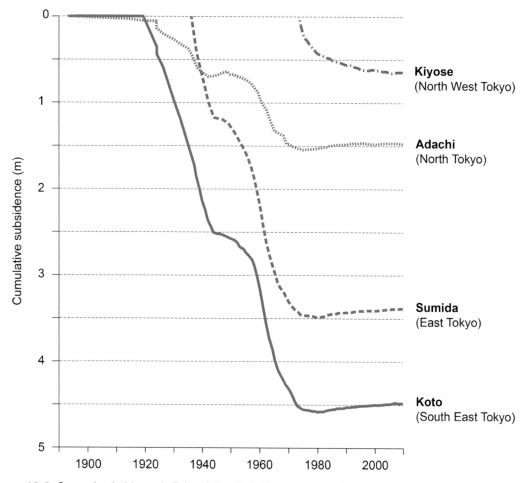

Figure 10.3 Ground subsidence in Tokyo (after Nakajima *et al.* 2012).

Renewed industrial activity during the Korean War accelerated the subsidence process once more, but the trend slowed from the late 1960s with the introduction of pumping regulations, and subsidence has been virtually halted since these regulations were strengthened in 1972, although an intense drought in 1994 saw a fresh intensification of subsidence in areas such as Kiyose on the north-western part of the Kanto Plain. Water supply in Tokyo has now shifted almost completely from groundwater to surface water.

In other coastal cities, depletion of aquifers has created problems of sea-water intrusion. Israel's coastal aquifer, which extends 120 km along the Mediterranean coast, has been heavily exploited for half a century. Overpumping of groundwater in the Tel Aviv urban area depleted groundwater levels to below sea level over an area of 60 km^2 in the 1950s, requiring a programme of freshwater injection along a line of wells parallel to the coast in an attempt to redress the saltwater/freshwater balance. Water was brought by the National Water Carrier from the north of the country to supplement natural recharge by precipitation. Despite these efforts, continued pumping for urban and agricultural use has meant a continued decline in the aquifer's water quality: chloride concentrations increased from 110 mg/litre in 1970 to 190 mg/litre in 1995 and some wells now yield water that neither complies with drinking-water standards nor is suitable for unrestricted agricultural irrigation (Gabbay, 1998). Sivan *et al.* (2005) have shown that sea water moved into the aquifer up to 100 m inland over a few decades.

Rising groundwater levels have become a critical problem for many 'post-industrial' cities as manufacturing industries have given way to service industries, which are much less demanding of water, and legislation has been introduced to control subsidence problems. Leakage from old pipes is an additional common reason for the rise in urban groundwater levels. Such rises affect urban structures that were designed in periods when water levels were depressed, and built without consideration that water levels could later recover. Many European cities currently suffer from this problem, which can result in flood damage to below-ground structures (railway tunnels, parking lots, domestic cellars and entrenched stretches of roads), added costs to any new excavations (because inflow water must be pumped from the excavation site), and a reduced capacity of urban drainage systems to deal with storms. In Barcelona, the period of change from a generally falling to a rising water table occurred in the late 1970s and today water levels have, for the most part, recovered to levels recorded a century ago (Vázquez-Suñé *et al.*, 2005). Groundwater now represents a significant issue for an important part of the city's underground railway system and many buildings built during the period of maximum water level depletion (1950–75). Keeping the underground railway system going requires the pumping of some 15–20 million m^3 of water each year to drain the tunnels.

The rising groundwater problem has also been reported from many Middle Eastern cities where rainfall is commonly low, potential evaporation high, and natural recharge small and sporadic. Inadvertent artificial recharge from leaking potable supplies, sewerage systems and irrigation schemes has caused widespread and costly damage to structures and services, and represents a significant hazard to public health (Bob *et al.*, 2016). A range of measures designed to reduce the impact of rising groundwater prior to and during development are shown in Table 10.4. The cost of implementing mitigation controls during the design phase of a new project is significantly cheaper than costs that would be associated with implementing the controls during or after construction.

THE URBAN ATMOSPHERE

Human activity has had marked effects on the climates of cities. Urban climates tend to be warmer than their hinterlands; they also typically exhibit greater cloudiness and precipitation, more thunderstorms, less snowfall, lower wind speeds and greater frequencies of fog (Oke *et al.*, 2017). One of the clearest, well-documented examples of an anthropogenic modification to climate is the higher air temperatures

Table 10.4 Measures to reduce the impact of rising groundwater beneath urban areas

Strategy	Examples
Preventative actions	Investigation of development sites to identify hazardous areas
	Implementation of suitable building regulations on basement depths, construction materials and methods
	Control water tables by pumping
	Waterproof structures
	Anchor or ballast foundations
	Remove or treat clays that will swell
Water conservation measures	Control of water use and storage
	Leak detection and maintenance of pipes
	Wastewater reuse

Source: after George (1992); Dean and Sholley (2006).

in cities compared to surrounding suburban and rural temperatures, a phenomenon known as the urban heat island. These heat islands develop in areas with a high proportion of non-reflective, water-resistant surfaces and a low proportion of vegetated and moisture-trapping surfaces. Materials used in abundance in cities, such as stone, concrete and asphalt, tend to trap heat at the surface and the relative lack of vegetation reduces heat lost through evapotranspiration (Figure 10.4). The intensity of the urban heat island effect is further enhanced by the addition of heat-trapping pollutants into the urban atmosphere along with heat lost from homes, businesses, factories and transportation. Both types of emission tend to be greater in urban areas than in surrounding regions because cities have higher energy demands as a result of their high population density.

The higher heat-island temperatures experienced in cities have both positive and negative consequences for urban residents. During the winter, cities in cold climates may benefit from the warming effect of their heat islands: higher temperatures can reduce the energy needed for heating and may help melt ice and snow on the roads. In the summertime, by contrast, the same city will probably experience the negative effects of urban heat islands. These include higher energy consumption for air conditioning and a number of health hazards, including elevated atmospheric pollution and heat-related illnesses.

The formation of certain types of urban air pollution, particularly smog, which represents long-term dangers to human health, is accelerated by higher air temperatures. A more episodic health hazard, from heatwaves, is also particularly emphasized in urban areas. High numbers of deaths during particularly hot spells in Chicago in 1995 and in many European cities in 2003 have raised awareness of the serious threat to public health presented by heatwaves (Martiello and Giacchi, 2010). The summer of 2003, which was probably the hottest in Europe since 1500 CE, caused an estimated 20,000–35,000 deaths due to heat stress across the continent. Most of these people were elderly and lived in cities.

Probably the most serious chronic environmental issue pertaining to the urban atmosphere is that of air quality. The principal sources of air pollution in urban areas are derived from the combustion of fossil fuels for domestic heating, for power generation, in motor vehicles, in industrial processes and in the disposal of solid wastes by incineration. These sources emit a variety of pollutants, the most common of which have

Figure 10.4 The combined effects of heat-trapping construction materials, a relative lack of vegetation, heat loss from buildings and their 'canyon effect' tends to make urban areas warmer than their hinterlands. This city is Shanghai in China.

long been sulphur dioxide (SO_2), oxides of nitrogen (NO and NO_2, collectively known as NO_x), carbon monoxide (CO), suspended particulate matter (SPM) and lead (Pb). Ozone (O_3), another 'traditional' air pollutant associated with urban areas and the main constituent of photochemical smog, is not emitted directly by combustion, but is formed photochemically in the lower atmosphere from NO_x and volatile organic compounds (VOCs) in the presence of sunlight. Sources of the VOCs include road traffic, the production and use of organic chemicals such as solvents, and the use of oil and natural gas.

These atmospheric pollutants affect human health (Table 10.5), directly through inhalation, and indirectly through such exposure routes as drinking-water and food contamination. Most traditional air pollutants directly affect respiratory and cardiovascular systems. For example, CO has a high affinity for haemoglobin and is able to displace oxygen in the blood, leading to cardiovascular and neurobehavioural effects. High levels of SO_2 and SPM have been associated with increased mortality, morbidity and impaired pulmonary function, and O_3 is known to affect the respiratory system and irritate the eyes, nose and throat, and to cause headaches. Certain sectors of the population are often at greater risk: the young, the elderly and those weakened by other debilitating ailments, including poor nutrition. In summary, over the long

Table 10.5 Health effects of major air pollutants

Pollutant	Effects
SO_2 (often considered with SPM)	Impaired function of airway and lungs. Increased prevalence of chronic bronchitis and permanent lung damage
SPM	Respiratory problems. Heart disease. High blood pressure. Effects dependent on particle size and concentration – may be no safe level
NO_x	Increases bronchial reactivity, reversible and irreversible lung damage. Sensitizes the lungs to other pollutants. May also affect spleen and liver
CO	Reduces oxygen in the blood, affecting brain, heart and muscle. Developing foetuses very vulnerable
O_3	Inflammation of airway, reduced lung function. Can inhibit immune system response
VOCs	Respiratory problems. Contributes to formation of ground-level ozone
Pb	Affects haemoglobin production, central nervous system and brain function

Source: after Metcalfe and Derwent (2005); Harrison and Hester (2017).

term, numerous studies now indicate that 'inhabiting a relatively polluted city for a prolonged period leads to a shortening of life expectancy' (Maynard, 2004: 10).

Elements of the natural and built environment can also be adversely affected. Sulphur and nitrogen oxides are principal precursors of acid deposition (see Chapter 12), SO_2, NO_2 and O_3 are phytotoxic – O_3, in particular, has been implicated in damage to crops and forests – and damage to buildings, works of art and materials such as nylon and rubber has been attributed to SO_2 and O_3.

In more recent times, these traditional urban air pollutants have been supplemented by a large number of other toxic and carcinogenic chemicals, which are increasingly being detected in the atmospheres of major cities. They include heavy metals (e.g. beryllium, cadmium and mercury), trace organics (e.g. benzene, formaldehyde and vinylchloride), radionuclides (e.g. radon) and fibres (e.g. asbestos). The sources of these pollutants are diverse, including waste incinerators, sewage treatment plants, manufacturing processes, building materials and motor vehicles. Concentrations of these chemicals are generally low, where they are measured, but this occurs at few sites to date.

Monitoring of urban air quality indicates that while many cities in the industrialized countries, and some in newly industrialized countries, have made significant reductions in air pollution in recent decades, rapidly growing urban areas in certain industrializing countries pose serious threats to the millions of people who live in them. The clear distinction in urban air quality between rich and poor countries is indicated in Figure 10.5.

Mexico City provides an example of a megacity in a newly industrialized country that has managed to improve air quality in recent times. In the 1980s, Mexico City had the highest air pollution levels of any major urban area, with WHO guidelines exceeded by a factor of two or more for levels of SO_2, SPM, CO and O_3. Levels of Pb and NO_2 were almost as bad, exceeding WHO limits by up to two times. The city's

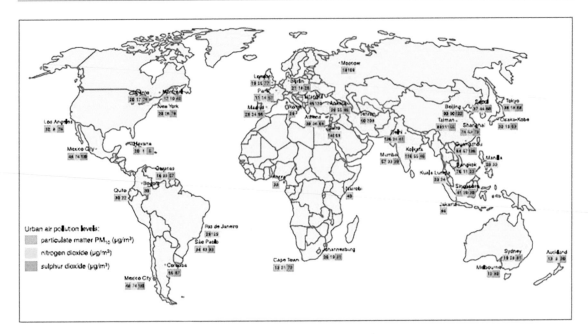

Figure 10.5 Air quality in some major world cities (after UNEP, 2012).

poor air quality is exacerbated by its location, in an elevated mountain-rimmed basin where temperature inversions occur, on average, 20 days per month from November to March, which impairs the dispersion of pollutants (Collins and Scott, 1993). The altitude of Mexico City, 2240 m above sea level, means that the air contains 23 per cent less oxygen than at sea level. Hence, people tend to inhale more air to take in sufficient oxygen, breathing in a larger dose of pollutants at the same time.

Since the 1980s, Mexico City's air quality has improved for several pollutants: downward trends in atmospheric concentrations of SO_2 and CO particularly have been noted (Molina and Molina, 2004). These improvements, which have been brought about despite continued rapid urban growth, have been achieved by a range of measures. They include the use of cleaner fuels, better emission control and technological improvements (e.g. since 1991 it has been obligatory for all new cars to be equipped with catalytic converters), replacement of old industries, and measures to restrict vehicles driving into the city. Mexico City was one of the first cities to implement licence-plate-based driving restrictions, in 1989, a policy that has spurred similar plans in a number of other cities around the world (Table 10.6), although not all have been effective (see p. 409).

Not all efforts to control air pollution in Mexico City have seen immediately positive results. Concern over rising atmospheric Pb levels, which averaged 8 µg/m³ in 1986 (five times the national standard of 1.5 µg/m³), resulted in the national oil company reducing the lead content of gasoline sold in the city in September of that year. The effect on Pb levels was striking: the atmospheric concentration fell below 1.5 µg/m³ by 1993 and has remained below 1.0 µg/m³ since 2000. However, an unexpected side-effect was a dramatic short-term increase in O_3 concentrations, a result of the reaction between atmospheric oxygen and the replacement gasoline additives in ultraviolet sunlight (Ciudad de México, 2010). Ozone carried by dominant winds to the Sierra del Ajusco, to the south-west of the Basin of Mexico, is thought to have significantly reduced the chlorophyll content and growth of pine trees on the mountains (de Bauer and

Table 10.6 Some cities with driving restrictions based on licence-plate numbers

First year of restrictions	City
1982	Athens (Greece)
1989	Mexico City (Mexico)
1990	Santiago (Chile)
1995	São Paulo (Brazil)
1998	Bogotá (Colombia)
2003	La Paz (Bolivia), Manila (Philippines)
2005	San Jose (Costa Rica)
2008	Beijing, Tianjin (China)
2010	Quito (Ecuador)
2014	Paris (France)
2016	Delhi (India), Jakarta (Indonesia), Madrid (Spain)

Source: after Davis (2017).

Hernández Tejeda, 2007). Ozone levels have subsequently been slowly declining since 1992, while concentrations of SPM and NO_2 have also decreased significantly (Figure 10.6).

Many of the rapidly growing industrial cities in Asia continue to suffer from poor air quality. In their survey of atmospheric pollution in the world's megacities, Marlier *et al.* (2016) noted that high and rising SPM levels characterize many Asian cities, where residential energy use is the most significant source of pollutants. Assessing air pollution in three megacities in India (Delhi, Mumbai and Kolkata) over recent decades, Gurjar *et al.* (2016) also highlight increasing trends for NO_x in all three cities, due to higher numbers of vehicles and the high flash point of compressed natural gas engines, which leads to higher NO_x emission, but decreasing trends of SO_2 due to a decrease in the sulphur content of coal and diesel. Attempts to evaluate these high pollution levels in economic terms put the cost to some cities in the hundreds of millions of US dollars. The residents of these cities also suffer the health impacts of poor air quality. Jakarta residents experience particularly high SPM levels and one estimate of the effects on their health suggests that 600,000 asthma attacks, 49,000 emergency room visits and 1400 deaths could be avoided each year if particulate levels were reduced to WHO standards (Ostro, 1994). Nonetheless, the severe pollution conditions observed at several megacities could have been much worse if control measures had not already been introduced. Examples include Beijing, Delhi, Seoul (see p. 303) and Shanghai. The health costs of air pollution are considerable even in relatively clean cities. One estimate of the total economic cost of SPM on health in Singapore put it at 4.3 per cent of national GDP (Quah and Boon, 2003).

The beneficial effects of tighter legislative controls on air quality are indicated by London's mean annual SO_2 concentrations, which fell by an order of magnitude in four decades: from around 300 µg/m³ in the early 1960s to below 30 µg/m³ in the early 2000s (Figure 10.7). The introduction and enforcement of Smoke Control Orders under the 1956 Clean Air Act (amended in 1964 and 1968) are the most important factor responsible for this decline in ambient concentrations. The Act was introduced in response to the infamous London smogs of the 1950s. In the winter of 1952, some 4000 deaths were attributed to one of London's worst

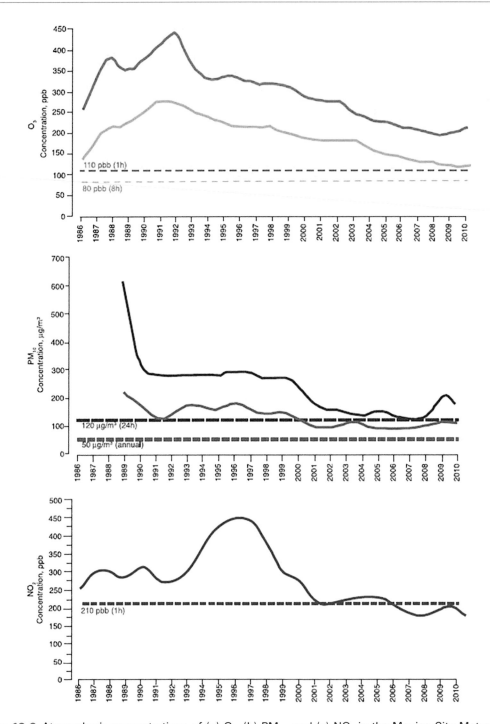

Figure 10.6 Atmospheric concentrations of (a) O_3, (b) PM_{10}, and (c) NO_2 in the Mexico City Metropolitan Area, 1986–2010 (from data in Ciudad de México, 2010).

Note: Dashed lines are national health limits.

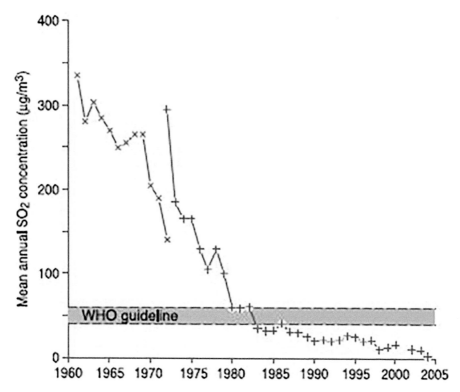

Figure 10.7 Mean annual sulphur dioxide concentrations at two monitoring stations in the City of London, 1961–2004 (data from www.aeat.co.uk/netcen/airqual/, accessed April 2007).

documented smog pollution episodes. In more recent years, air quality in the UK capital has been improved further as a result of implementing a scheme to charge motor vehicles for entering a central London zone on weekdays. Although the charge, introduced in 2003 to reduce traffic congestion, has been successful in its primary aim, it has also led to reductions in air pollution, noise and accidents (Banister, 2008). The congestion charging scheme has been followed up with a system to discourage the most polluting vehicles from entering central London: the introduction in 2008 of a Low Emission Zone (LEZ). The only vehicles allowed to enter the LEZ without paying a daily charge are those that meet certain emissions standards.

Successes have also been recorded in reducing many of the pollutants measured in Tokyo and Seoul (see p. 303) and in some of the US megacities (Parrish *et al.*, 2011). A model for the progression of air pollution problems through time at different levels of development is shown in Figure 10.8. Pollution rises with initial industrial development, to be brought under control through legislation on emissions. Air quality then stabilizes and improves as development proceeds, to be reduced to below acceptable standards by high-technology applications.

However, this is not to suggest that cities in richer, more developed countries of the global North have solved all their air pollution issues. Los Angeles still has a serious O_3 problem, the worst in the USA (see p. 401), and a recent study of SPM in several European countries (Cesaroni *et al.*, 2014) concluded that long-term exposure to particulate matter is associated with heart problems, and that this association persists at levels of exposure below even the current values set by European health authorities.

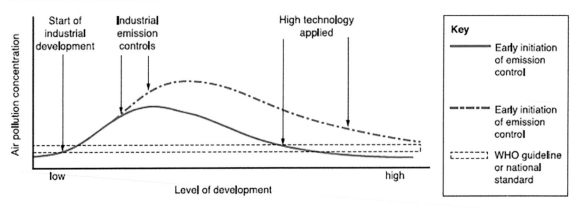

Figure 10.8 Model of air quality evolution with development status (after UNEP/WHO, 1992).

Table 10.7 Some cities with poor household garbage collection facilities

City	Country	Proportion of garbage not collected (%)
Lusaka	Zambia	55
Dhaka	Bangladesh	45
Ghorahi	Nepal	54
Bamako	Mali	43
Moshi	Tanzania	39
Nairobi	Kenya	35
Canete	Peru	27
Bengaluru	India	30
Managua	Nicaragua	18

Source: data from UN Habitat (2010).

GARBAGE

The rapid, and often unauthorized, growth of urban areas has in many cases outpaced the ability of urban authorities to provide adequate facilities, such as the collection of household garbage. Table 10.7 illustrates the scale of the problem in some cities for which reasonable estimates are available. Many other urban areas similarly afflicted are not included due to lack of adequate information. Contrasts are also evident within cities, between more affluent suburbs and poorer areas.

Although the environmental issues associated with garbage do not disappear with its collection (see Chapter 17), uncollected garbage exacerbates many of the environmental hazards covered in this chapter. It can be a serious fire hazard; it attracts pests and disease vectors, creating health hazards; and local disposal by burning or dumping adds to pollution loads and clogs waterways, so increasing the dangers of flooding.

Several animal species, particularly rats, have become adapted to the urban environment by scavenging from urban refuse. Larger species, too, have been drawn to garbage bins and dumps, and are also regarded as pests, such as the urban foxes that inhabit many British cities, and polar bears in Churchill, Manitoba, northern Canada. In Uganda's capital city, Kampala, carnivorous Marabou storks strut the streets like normal citizens, living off garbage and doing a useful job in controlling smaller pests (Figure 10.9). These birds have become so successful that the population of Marabou storks nesting in Kampala is now the largest known breeding colony in Africa, and therefore in the world (Pomeroy *et al.*, 2017).

Some degree of waste recovery occurs in most urban areas. In many cities in developing countries, large numbers of residents are self-employed in the business of garbage recycling (see the example from Dar es Salaam, Tanzania on p. 424). In Brazil, more than 500,000 people survive by collecting and marketing solid waste in large cities, and it is estimated that they reduce the amount of waste that goes into landfills by up to 20 per cent (Fergutz *et al.*, 2011). Increasingly, self-employed waste pickers are being recognized for their valuable contributions to urban sustainability and development, and Dias (2016) documents several cities where urban authorities have acknowledged their contribution and introduced payments accordingly: Bogotá (Colombia), Belo Horizonte (Brazil), and Pune (India).

Figure 10.9 Marabou storks are a common sight in Kampala, Uganda. In the countryside, these scavengers often feed on carrion, and they were probably first attracted to the cityscape by discarded human corpses during the times of Idi Amin and Milton Obote. They now live on garbage and rodents and thus play an important role in removing unhealthy rubbish from the city.

In the Egyptian city of Cairo, the Zabbaleen religious sect has traditionally run a very efficient garbage collection business, scavenging and recycling, feeding edible portions to their domestic livestock and selling inorganic materials to dealers. However, continuation of this community's way of life is jeopardized by the privatization of municipal solid waste services in Cairo and an official policy of moving the Zabbaleen further out of the city, on the grounds that this will turn their neighbourhoods into cleaner and healthier living environments (Fahmi and Sutton, 2010). Elsewhere, metropolitan authorities have long run similar programmes. In Beijing, for example, a state-run recycling scheme has been in operation since the 1950s, and in New York City, Local Law 19, brought into force in 1989, requires all residents, institutions and businesses to separate a variety of materials for collection and recycling.

HAZARDS AND CATASTROPHES

The high concentrations of people and physical infrastructure in cities make them distinctive in several ways with regard to hazards. Where money is available, cities are worth protecting because of the large financial and human investment they represent. Adequate provisions of water supply and sanitation are designed to offset the risks of disease; other infrastructure – such as expensive flood protection schemes – protects against geophysical hazards (see Chapter 21). An idea of the scale of exposure in the world's large port cities to coastal flooding due to sea-level rise and storm surge has been calculated by Hanson *et al.* (2011). This study concludes that the total value of assets exposed across all the port cities was equivalent to around 5 per cent of global GDP in 2005, with the top nine cities ranked in terms of assets exposed being in just three countries: USA (Miami, New York–Newark, New Orleans, Virginia Beach), Japan (Tokyo, Osaka–Kobe, Nagoya) and the Netherlands (Amsterdam, Rotterdam).

Unsurprisingly, the impact of hazards is often marked when they do occur in urban areas because of their great population densities and concentration of buildings, infrastructure and economic activities. This applies equally to natural hazards (see below and Chapter 21) and technological disasters such as industrial accidents and fires. Table 10.8 shows a number of serious disasters that have occurred in cities in recent times. The preponderance of examples from cities in the global South is not coincidental. Some types of urban disaster, such as garbage dump landslips, hardly ever happen in cities of the global North, but even when similar disasters occur in richer cities, the number of casualties tends to be smaller. Further, many urban areas in developing countries are characterized by escalating growth rates and a general lack of finance, which frequently equate to inadequate hazard management. And within these cities, it is usually the poorest sectors of urban society that are most at risk from hazards. Rapid urban growth and rising land prices have used up the most desirable and safest sites in most cities in the developing world, leaving increasingly hazard-prone land for poorer groups. Such hazards include the pervasive dangers of high pollution levels as well as the intensive dangers associated with technological hazards. The accidental discharge from a pesticide production plant in Bhopal, northern India, in 1984, for example, killed 3800 shanty-town dwellers (Dhara and Dhara, 2002). It was primarily caused by inadequate management and lax safety procedures. Many of the other urban disasters shown in Table 10.8 – including those in Cubatão, Mexico City, Metro Manila, Bandung and Dhaka – took place in poorer parts of those cities.

The general quality of the environment in cities is also a function of infrastructure and services, which tend to be less adequate in poorer areas. A study in Buenos Aires of the distribution of infant mortality, a good indicator of the quality of the environment, showed clear correlations with income levels and basic service provision (Arrossi, 1996). Overall infant mortality rates in poorer suburbs of the city were double those in more prosperous areas, and infant mortality rates by 'avoidable' causes (including tetanus, respiratory infections and perinatal jaundice) were up to three times higher in the poorest zones.

Table 10.8 Some examples of urban disasters

City	Date	Disaster	Deaths	Reference
London, UK	December 1952	Five days of dense smog	4000	Bell *et al.* (2004)
Cubatão, Brazil	February 1984	Pipeline gasoline leak and fire	508	de Souza (2000)
Mexico City, Mexico	November 1984	Explosion of LPG storage tanks	>500	de Souza (2000)
Bhopal, India	December 1984	Gas leak from pesticide plant	3800	Dhara and Dhara (2002)
Metro Manila, Philippines	July 2000	Collapse of Payatas Mountain garbage dump	330	Gaillard (2009)
Lagos, Nigeria	January 2002	Explosions at ammunition storage site	>1000	Greene *et al.* (2004)
Bandung, Indonesia	February 2005	Landslide at Leuwigajah garbage dump	143	Lavigne *et al.* (2014)
Dhaka, Bangladesh	June 2010	Fire in multi-storey residential area	124	Mashreky *et al.* (2010)
Tianjin, China	August 2015	Fire and explosions of hazardous chemicals	173	Zhao (2016)
Freetown, Sierra Leone	August 2017	Mudslide after heavy rain at Mount Sugar Loaf, degraded by illegal construction and blasting	>1000	Leone (2017)

High concentrations of poor housing in Third World cities are built on slopes on hillsides prone to sliding (e.g. La Paz, Figure 10.10), or in deep ravines (e.g. Guatemala City), on river banks susceptible to flooding (e.g. Delhi), and on low-lying coastlines prone to marine inundation (e.g. Lagos). A study of Rio de Janeiro's sister city, Niterói, has shown that its poorest housing has been increasingly vulnerable to landslides in recent years (Smyth and Royle, 2000). The causes of landslides during the rainy season are well known: the undercutting of slopes to build houses and the weight of the houses themselves, along with the deforestation of slopes, which causes local water tables to rise and reduces the strength of the regolith. These factors also threaten more prosperous developments, but the occupants of middle- and high-class housing have greater financial resources, which enable them to adopt better construction techniques (including preventative measures such as slope-retention walls and the installation of storm drains) and thus limit the risk. There is little doubt that the economic poverty of Niterói's shanty dwellers is the key to their increased vulnerability: although often aware of slope stability problems, they can do little to abate them.

Even the destruction caused by citywide hazards, such as earthquakes, can be magnified in unstable sites: in Guatemala, 65 per cent of deaths in the capital caused by the 1976 earthquake occurred in the

Figure 10.10 A huge landslide scar marks the highest part of a very steep slope on the outskirts of La Paz, Bolivia. Areas of self-built informal housing that occupy the city's more elevated steeper slopes are under constant threat from such landslips.

badly eroded ravines around the city. In other situations, however, the damage and loss of life caused by earthquakes can be greatest in more built-up parts of the urban environment, when buildings themselves become hazardous if they are not constructed to withstand earth tremors. The 1999 Marmara earthquake in Turkey is an example: most of the 17,000 casualties were city-dwellers living in substandard housing built without proper attention to regulations for earthquake-resistant design to cope with a rapidly rising demand (Özerdem and Barakat, 2000). In the words of Ambraseys and Bilham (2011: 153), 'poor building practices are largely to blame for turning moderate earthquakes into major disasters', contrasting the >200,000 death toll from the 2010 earthquake in Haiti, where most of the casualties were in Port au Prince, with New Zealand's identical magnitude (7) earthquake in 2011 which left large parts of the city of Christchurch badly damaged but killed just 185 people. These authors identify corruption in the construction industry as a major factor, calculating that 83 per cent of all deaths from building collapse in earthquakes over the previous 30 years occurred in countries that are anomalously corrupt. Failure of urban infrastructure following earthquakes is another common cause of damage and loss of life. The most serious earthquake disaster in the USA, in San Francisco in 1906, was largely a function of infrastructural failure. Disruption of gas distribution and service lines caused the outbreak of many fires and interrupted water distribution, so making it difficult to put the fires out.

Table 10.9 Countries with poor drinking-water and sanitation provision in urban areas, 2015

Country	Access to drinking water piped on premises (% of urban population)	Access to basic sanitation (% of urban population)
AFRICA		
Nigeria	3	33
Sierra Leone	11	23
Togo	13	25
Madagascar	16	18
ASIA-PACIFIC		
Marshall Islands	4	84
Afghanistan	31	45
Bangladesh	32	58
Mongolia	33	66

Source: after UNICEF and WHO (2015).

The most critical environmental problems faced in urban areas of the developing world, however, stem from the disease hazards caused by a lack of adequate drinking water and sanitation. Table 10.9 shows some of the countries worst affected. Worldwide, there was a particular effort to improve access to these basic services under one of the Millennium Development Goals (MDGs) and considerable success has been registered in urban areas: well over a billion city-dwellers gained access to improved water sources between 1990 and 2010 and over the same period the proportion of urban residents with access to improved sanitation facilities rose from 76 to 79 per cent (UNICEF and WHO, 2012). Indeed, the MDG drinking-water target, to halve the proportion of the population without sustainable access to safe drinking water between 1990 and 2015, was met in 2010, but this global figure masks some disappointments. Satterthwaite (2016) points out that over 30 countries, most in sub-Saharan Africa, still had more than half of their urban population without water piped on premises in 2015, and that not all the water that is piped to homes is necessarily safe to drink or supplied reliably. Sadly, Satterthwaite also identifies 21 countries, territories and areas that went backwards – i.e. that had a lower proportion of the urban population with water piped on premises in 2015 than in 1990. The cities of sub-Saharan Africa also stand out for their inadequate sanitation provision. There was almost no increase in the proportion of the urban population in sub-Saharan Africa with improved sanitation from 1990 to 2015, and 60 per cent of the region's urban population still lacked improved sanitation in 2015. The actual situation on the ground was probably worse because 57 countries, in Africa and elsewhere, did not assess progress due to a lack of data for 1990, 2015 or both.

Water-borne diseases (e.g. diarrhoea, dysentery, cholera and guinea worm), water-hygiene diseases (e.g. typhoid and trachoma) and water-habitat diseases (e.g. malaria and schistosomiasis) both kill directly and debilitate sufferers to the extent that they die from other causes. Globally, on average, most people without

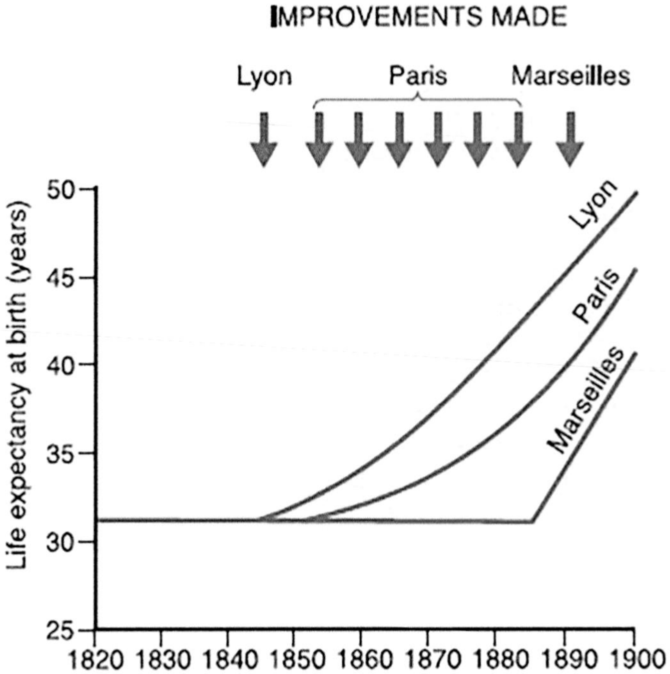

IMPROVEMENTS MADE

Figure 10.11 Rising life expectancy with improvements in water supply and sanitation in three French cities, 1820–1900 (after Preston and Van de Walle, 1978).

an improved drinking-water source live in rural areas and sanitation coverage is much lower in rural than in urban areas. However, as with other forms of urban hazard discussed in this section, it is the less well-off sectors of urban society that are most at risk. Indeed, slum dwellers in developing countries are as badly off if not worse off than their rural relatives. UN-HABITAT (2006) found that the billion people inhabiting the world's slums are more likely to die earlier, and experience more hunger and disease, than those urban residents who do not reside in a slum. Ezeh *et al.* (2017) emphasize that children in slums are especially vulnerable and highlight the global projection that by 2030 about 2 billion people will live in slums, mainly in Africa and Asia.

This dire situation prompted establishment of an MDG to achieve a significant improvement in the lives of at least 100 million slum dwellers by 2020, a target superseded by one of the Sustainable Development Goals that aims to make all cities inclusive, safe, resilient and sustainable. The effects of improvements to water supply and wastewater disposal on life expectancy have been clearly shown in the industrial countries, when services were improved during the nineteenth and twentieth centuries. The trend shown for three major French cities in Figure 10.11 is typical in this respect, with life expectancy increasing from about 32 years in 1850 to about 45 years in 1900, with the timing of changes corresponding closely to improvements in water and sanitation provision.

However, the implementation of improvements in the needy parts of the world's poorer cities is unlikely to adopt similar approaches to those taken historically in the developed countries. Centralized systems built and maintained by subsidized public agencies will not work in today's developing world simply because the numbers of unserved people are too great and the financial costs too high. Community-driven initiatives are much more likely to be used to provide low-cost sanitation, particularly in slum areas. Research has identified four key institutional challenges that community-driven initiatives to improve sanitation in deprived urban settlements typically face:

- a collective action challenge because one person's sanitation problems depend in large part on the sanitation facilities and behaviours of others
- a co-production challenge of working with formal service providers to dispose of sanitary waste safely – neither the state nor the residents can do it alone and collaboration enhances mutual accountability
- an affordability challenge of reconciling the affordable with what is acceptable to both users and local authorities
- a tenure challenge of preventing housing insecurity from undermining residents' willingness to commit to sanitary improvement.

These challenges are not insuperable. McGranahan and Mitlin (2016) highlight two well-documented and relatively successful initiatives – the Orangi Pilot Project (OPP) in Karachi, Pakistan, and an Alliance of Indian partners that began work in Mumbai and spread to other Indian cities. These projects met the challenges primarily through social innovations, but also through the choice and development of sanitation technologies (simplified sewers for OPP and community toilet blocks for the Indian Alliance).

ENVIRONMENTAL JUSTICE IN THE CITY

A theme that has been highlighted in several of the above sections on environmental issues in cities is the existence of particular concentrations of environmental risk in particular parts of the urban landscape and hence the higher exposure of certain sectors of urban society. There is a growing body of research into so-called environmental justice, the idea that links environmental problems and social injustices, much of which has established that many socially deprived, low-income and ethnic minority communities in cities are exposed to disproportionately high levels of environmental risk.

The concept of environmental justice stems largely from the work of community activists and academics in the USA and numerous examples have been cited from that country. To cite just one, Pastor *et al.* (2002) used Geographic Information System analysis to make demographic comparisons among school sites in Los Angeles and found that minority students, especially Latinos, are more likely to attend schools near hazardous facilities and face higher health risks associated with exposure to outdoor air pollution. These types of study are less common on the national scale, but an assessment of air pollution and environmental justice in cities across New Zealand found air pollution levels to be higher in socially deprived areas and neighbourhoods with a high proportion of low-income households (Pearce and Kingham, 2008).

Such spatial inequalities in cities are by no means a uniquely modern phenomenon, however, as demonstrated in a number of historical reviews presented by Massard-Guilbaud and Rodger (2011). Nor are they confined to cities in the global North, of course. Most of the serious urban disasters shown in Table 10.8 occurred in slum areas of the cities concerned and slums are particularly typical of numerous cities in the developing world. These parts of cities are areas of concentrated disadvantage characterized by:

- poverty and social exclusion
- lack of basic services
- substandard housing or illegal and inadequate building structures

- overcrowding and high density
- unhealthy living conditions and hazardous locations
- insecure tenure and irregular or informal settlements

TOWARDS A SUSTAINABLE URBAN ENVIRONMENT

It is probably unreasonable to expect that major cities should be supported by the resources produced in their immediate surrounds, and thus urban areas will always leave a sizeable ecological footprint on their hinterlands and often far beyond. Indeed, studies of modern cities invariably show that their urban ecological footprints are typically two to three orders of magnitude larger than those of the geographical or political areas they occupy and that a city may represent as little as 0.1 per cent of the area of the host ecosystems that sustain it (Rees, 2003). The idea of increasing the sustainability of urban areas receives a great deal of attention from politicians and urban planners but the fact that cities are not isolated from their surrounds means that improvements to urban sustainability should be undertaken as part of wider programmes of sustainable development everywhere.

A theoretical attempt to trace some of the measures needed to improve urban sustainability is shown in Figure 10.12. The unsustainable city (A) consumes energy from non-renewable sources and natural resources of all kinds, producing goods and services as well as all sorts of waste products that are released into the environment with varying degrees of management (e.g. motor vehicle emissions, domestic waste buried in landfills). The city depicted in B has improved its sustainability, drawing energy from renewable sources and consuming fewer natural resources as they are used more efficiently. Wastes have also been reduced by introducing recycling measures. Stronger sustainability is shown in C, where limits have been put on the amounts of inputs and outputs, and better waste management has been introduced in the form of reuse, recovery and recycling, as well as improved prevention (see Chapter 17).

Some of the characteristics of strong sustainability shown in Figure 10.12 have been described elsewhere in this chapter in cities of the developing world where poverty is translated into low resource consumption and great efforts to recycle wastes. Another way in which cities, particularly in the developing world, can improve their sustainability and reduce their ecological footprint is through the promotion of urban agriculture. Lee-Smith (2010) summarizes the importance of agriculture in some African cities, where it also has considerable potential for hunger and poverty alleviation. The proportion of urban households engaged in producing some of their own food ranged between 25 per cent and 50 per cent, helping to improve food security for those households and often contributing to the recycling of nutrients in the city ecosystem, reducing waste and pollution, so contributing to sustainability.

Since most cities grow incrementally, their sustainability is also likely to be developed piecemeal. Improvements to urban sustainability like those outlined in Figure 10.12 are likely to occur relatively slowly, but much faster improvements can be made when new buildings are designed and constructed. Specific steps that can be taken by the architect to reduce the carbon footprint of a new development are shown in Figure 10.13. Much of the additional work on a more conventional design focuses on renewable sources of energy and improving the efficiency of energy use and its conservation (see Chapter 18) which can be combined to produce a carbon-neutral development.

Other opportunities for sustainable urban development present themselves in cities that experience population decline, a trend that has been marked in certain parts of North America, Europe and Japan in particular. A German government-sponsored project, Shrinking Cities, found that globally over the last 50 years or so, more than 450 cities with populations over 100,000 lost at least a tenth of their population, and urban transformations to sustainability may be quite different in shrinking cities as compared with

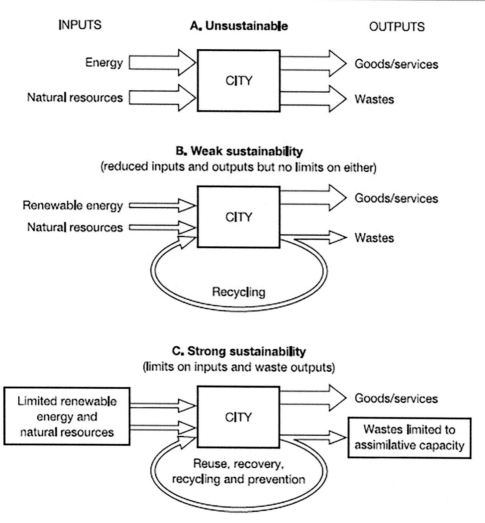

Figure 10.12 Improvements in urban sustainability (after Moffatt, 1999).

growing cities. As Herrmann *et al.* (2016) point out, ecologists working in cities have long recognized the opportunities in urban open spaces, areas that can provide multiple ecosystem services as well as functioning as habitats for a range of species.

While the history of many developed countries' experiences in dealing with urban problems offers numerous indicators as to how the environmental difficulties that plague so many of the developing world's major cities can be solved, the solution to the awesome environmental problems outlined in this chapter appears to depend largely upon available finance and political will. Some of the efforts to combat these problems have been described, but, to conclude, it is useful to examine the experience of Curitiba, a city in south-east Brazil, as a model for other cities where financial resources may be at a premium. Despite a rapidly growing population, rising from 140,000 in 1950 to about 2 million people in 2010, Curitiba has

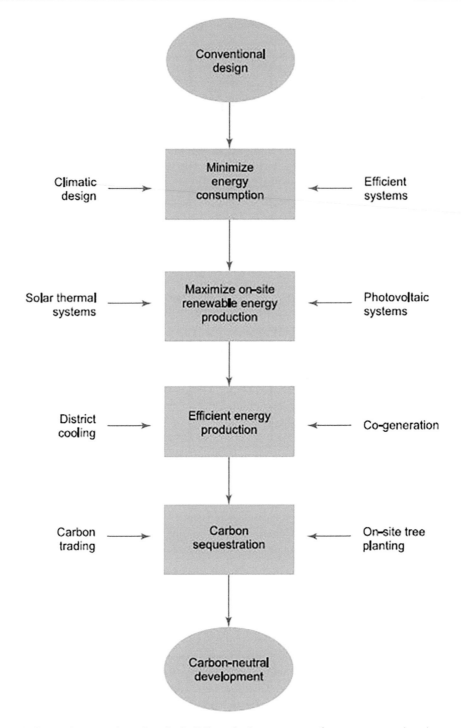

Figure 10.13 Steps that can be taken in building design to move from a conventional approach to a carbon-neutral development.

greatly improved its urban environment with a series of innovative, relatively low-cost transport, land-use and waste disposal measures. Notable achievements have also been made in recreational and flood control facilities, industrial infrastructure, and improving socio-economic safety nets (Schwarz, 2004).

A new public transport system, involving competitively priced bus services using exclusive bus lanes, has been used as a framework for current and future urban development. The system, which carries 2 million passengers per day, has reduced congestion on the roads, and lowered air pollution, as a typical Curitiba private vehicle now uses 30 per cent less fuel than the average for eight comparable Brazilian cities. The city also boasts one of the lowest motor accident rates per vehicle in the country, and its inhabitants enjoy a very low average transport expenditure. Land-use policy has aimed to improve urban conditions, introducing pedestrian precincts and cycleways, concentrating on redeveloping existing sites rather than expanding the urban area, and expanding parks and other green spaces.

Faced with a growing garbage problem, a recycling scheme was introduced in 1989 with education programmes to encourage recycling through municipal collection. Residents were asked to separate garbage before collection, so cutting municipal costs significantly, and more than 70 per cent of households now participate. Effective garbage collection has also meant a significant improvement in the city's squatter settlements (*favelas*). The problem of access to the high-density *favelas* has been tackled by offering bus tickets, food parcels and school notebooks in exchange for household refuse brought to accessible roadsides. The benefits, enjoyed by an estimated 35,000 families, include a marked decrease in litter, rising standards of nutrition and an overall improvement in *favela* quality of life.

Curitiba is widely regarded as Latin America's first 'Smart City' (Saraiva *et al.*, 2018), a term that has gained traction in academia, business and government circles to describe cities that, on the one hand, are increasingly controlled and monitored by ubiquitous computing systems and, on the other, whose economy and governance are driven by innovation, creativity and entrepreneurship. Another key to success in Curitiba has been the city government's promotion of a strong sense of public participation and its emphasis on low-cost programmes. People are encouraged to build their own houses, with city government loans and assistance from city architects, to offset large municipal outlays on housing projects. Old public buses have been converted into mobile schools that tour low-income neighbourhoods. The creativity displayed in Curitiba shows that many of the environmental problems commonly associated with urban areas can be tackled successfully without large financial resources, given strong leadership and guidance to a population that is willing to participate in solving its own difficulties.

FURTHER READING

Andersson, E. 2006 Urban landscapes and sustainable cities. *Ecology and Society* 11(1): 34. A review of how cities can use ecosystem services sustainably. Open access: www.ecologyandsociety.org/vol11/iss1/art34/.

Douglas, I. 2012 *Cities: an environmental history*. London, I.B. Taurus. A global appraisal of how cities have developed.

Herrmann, D.L., Schwarz, K., Shuster, W.D., Berland, A., Chaffin, B.C., Garmestani, A.S. and Hopton, M.E. 2016 Ecology for the shrinking city. *BioScience* 66: 965–73. An overview of how ecology can promote sustainability in shrinking cities. Open access: academic.oup.com/bioscience/article/66/11/965/2754227.

Lapworth, D.J., Nkhuwa, D.C.W., Okotto-Okotto, J., Pedley, S., Stuart, M.E., Tijani, M.N. and Wright, J. 2017 Urban groundwater quality in sub-Saharan Africa: current status and implications for water security and public health. *Hydrogeology Journal* 25: 1093–116. A comprehensive assessment of water quality status of urban groundwater in sub-Saharan Africa. Open access: link.springer.com/article/10.1007/s10040-016-1516-6.

Parrish, D.D., Singh, H.B., Molina, L. and Madronich, S. 2011 Air quality progress in North American megacities: a review. *Atmospheric Environment* 45: 7015–25. A review of recent progress and continuing challenges in Los Angeles, New York and Mexico City.

Perini, K. and Sabbion, P. (eds) 2017 *Urban sustainability and river restoration: green and blue infrastructure.* Chichester, Wiley-Blackwell. Examples of urban river restoration projects in Europe and North America.

WEBSITES

www.gdrc.org/uem   a virtual library on urban environmental management.

sustainablecities.net   Sustainable Cities, based in Vancouver, Canada, has a successful track record of working with cities all over the world to solve urban growth problems.

www.unhabitat.org   the UN's Human Settlements Programme promotes socially and environmentally sustainable towns and cities.

www.urbanecology.org   a US group based in the San Francisco Bay Area dedicated to sustainability in urban environments.

www.urbanecology.org.au   an Australian community-based organization promoting people- and nature-friendly urban settlements.

www.worldurbancampaign.org   The World Urban Campaign (WUC) is an advocacy and partnership platform to raise awareness about positive urban change.

POINTS FOR DISCUSSION

- Are megacities unsustainable by definition?
- Is water the key limiting resource for urban sustainability?
- Will air pollution problems in large cities recede as countries develop?
- Why are some parts of cities more hazardous than others?
- Why are some urban communities exposed to disproportionately high levels of environmental risk?
- Most of the environmental issues associated with urban environments can only be solved with political will and economic resources. Do you agree?

11

Climatic change

LEARNING OUTCOMES

At the end of this chapter you should:

- Understand that climate in the Pleistocene has been characterized by a number of glacial and interglacial periods.
- Appreciate the numerous ways in which human activities inadvertently affect climate.
- Know how our understanding of stratospheric ozone has developed over the last 100 years or so.
- Understand how greenhouse gases are related to global temperature.
- Appreciate how climate change is detected and attributed using climate models.
- Recognize the very wide range of impacts, many of which are already detectable.
- Recognize the uncertainties involved in detecting and attributing impacts.
- Understand the wide range of mitigation and adaptation options available to society.

KEY CONCEPTS

ice age, ozone hole, radiative forcing, Intergovernmental Panel on Climate Change (IPCC), general circulation model (GCM), Montreal Protocol, small island developing states (SIDS), Kyoto Protocol, Paris Agreement, emissions trading, cap and trade, no regrets initiatives, carbon sequestration, Reducing Emissions from Deforestation and Forest Degradation (REDD), Clean Development Mechanism, climate engineering

---

Global climatic change due primarily to increasing atmospheric concentrations of greenhouse gases has dominated the environmental agenda since the mid-1980s and has engendered substantial debate internationally and in many countries. There is a great deal of observational evidence that our climate is changing, and the most likely cause – human activity – has been established with great confidence by a wide range of studies at global and regional scales. There is no doubt that over the past 200 years or so, human action has significantly increased the atmospheric concentrations of several gases that are closely related to atmospheric temperature. It is extremely likely that these increased concentrations, which are set to continue to build up in the near future, are already affecting many climate variables. Predictions of the climatic nature of the Earth into the next century, and the effects of potential climatic changes on other aspects of the natural and human environment, are cautious because our understanding of climate is imperfect, as is our ability to simulate its processes. The implications of climate change for the environment and society will also depend on how humankind responds through changes in technology, economies, lifestyle and policy. Nonetheless, the large majority of scientists, policy-makers and other members of society agree that global climatic change is the greatest environmental challenge facing humankind in the Anthropocene.

QUATERNARY CLIMATIC CHANGE

The Earth's climate has never been static and the human impact on climate has been minor relative to the large-scale perturbations brought about by natural processes. Instrumental or documentary records of climatic parameters are only available for a few thousand years at most, but even on such short timescales some marked variations in climate have been noted (see below). Our knowledge of the Earth's climate over longer time periods is derived from 'proxy' indicators. The first evidence of the existence of former ice ages came from glacial landforms, and buried soils and fossils, which indicated that climate in the Pleistocene (2.6 million to 11,700 years BP) had alternated between cold glacials and warmer interglacials, with interglacial conditions in most mid- and high-latitude regions being similar to those of today. These lines of evidence have subsequently been supplemented with information gleaned from other sources, most notably deep-sea cores.

Explaining these changes in the Earth's climate is not an easy task since there are very many factors that influence our atmosphere and make it dynamic on several different scales. Some important climatic change theories have focused on variations in the amount of solar radiation received by the atmosphere. A composite sequence of these variations due to perturbations in the Earth's orbit around the sun was calculated in the early years of the twentieth century by a Serbian mathematician named Milankovitch. Three types of variation, operating with different periodicities, were identified:

1   orbital eccentricity with a periodicity of about 100,000 years
2   axial tilt or obliquity with a periodicity of about 40,000 years
3   precession of the equinoxes with a periodicity of about 21,000 years.

These periodic variations, which have been confirmed from deep-sea core evidence (Hays *et al.*, 1976), help to explain the occurrence of glacial and interglacial periods on Earth. Over about the last 0.74 million years, for example, eight major glacial–interglacial cycles have occurred, each lasting about 100,000 years, but beginning and ending in quite sudden jumps in temperature. Superimposed on these cyclical fluctuations, there have been more frequent and shorter-term climate changes that are known to have occurred over time periods equal to or even less than a human lifespan (Holmes *et al.*, 2011).

The last glacial is generally thought to have been at a maximum about 20,000 years ago, and since then the Earth's climate has warmed (Figure 11.1a). The most recent phase of the Quaternary is known as the Holocene, which began around 11,700 years ago, and even during this period the Earth's climate has by no means been constant. Indeed, the last 1000 years have seen a relatively warm period during medieval

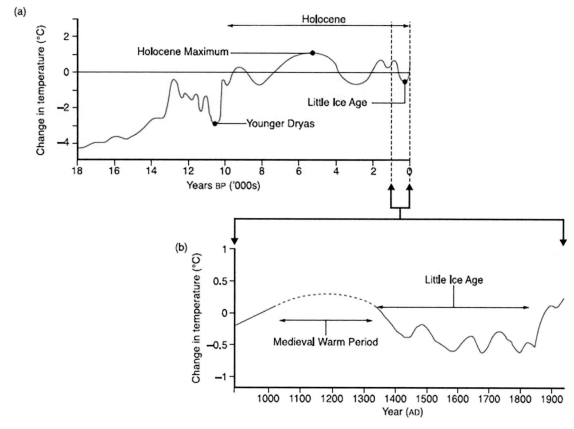

Figure 11.1 Variations in global temperature (after Houghton *et al.*, 1990): (a) over the last 18,000 years; (b) over the last 1000 years.

times and a so-called Little Ice Age during the fifteenth to eighteenth centuries (Figure 11.1b). The end of the Little Ice Age coincided with the era when human activity began to have a significant impact on global temperatures through enhancement of the greenhouse effect (see below).

HUMAN IMPACTS ON THE ATMOSPHERE

Human activities inadvertently affect the workings of the atmosphere in numerous ways, in many cases with possible effects on climatic regimes (Table 11.1). Direct inputs of gases, small particles (aerosols) and heat energy can all affect the operation of climate on various scales. Emissions of aerosols and heat are responsible for local 'heat islands' around urban areas and the creation of photochemical smogs (see Chapter 10), and enhanced inputs of soil aerosols, particularly from agricultural areas in drylands, affect the radiation properties of the atmosphere, possibly resulting in decreased rainfall locally (Cook *et al.*, 2009). On larger scales, gas emissions are responsible for an enhanced global greenhouse effect (see below) and depletion of the concentrated layer of ozone present in the stratosphere.

Table 11.1 Possible mechanisms for inadvertent human-induced climate change

DIRECT ATMOSPHERIC INPUTS

- Gas emissions (carbon dioxide, methane, chlorofluorocarbons, nitrous oxide, krypton-85, water vapour, miscellaneous trace gases)
- Aerosol generation
- Thermal pollution

CHANGES TO LAND SURFACES

- Albedo change (deforestation, afforestation, intensive grazing, dust addition to ice caps)
- Roughness change (deforestation, afforestation, urbanization)
- Extension of irrigation
- Reservoir impoundment

ALTERATIONS TO THE OCEANS

- Current alterations by constricting straits
- Diversion of fresh waters into oceans

Source: after Goudie (2006).

The 'ozone hole'

Stratospheric ozone plays a key role in climatic processes through its capacity to absorb incoming solar ultraviolet radiation. This warms the stratosphere and maintains a steep inversion of temperature between about 15 and 50 km above the Earth's surface, affecting convective processes and circulation in the troposphere below. The history of our understanding of and interaction with stratospheric ozone, including discovery of the 'ozone hole' in 1985, is outlined in Table 11.2. The human-induced depletion of stratospheric ozone over the Antarctic appears to have influenced climate throughout the southern hemisphere, even affecting changes to rainfall in the subtropics (Kang *et al.*, 2011). Much concern has focused on the possible ecological effects of increased ultraviolet radiation reaching the Earth's surface. Modifications to the rate of photosynthesis are likely to have significant impacts on many living organisms, reducing productivity in aquatic life such as plankton, and terrestrial plants (some evidence of deleterious impacts on lacustrine flora and fauna is cited on p. 305). A direct impact on human health is also likely, through increases in the incidence of skin cancers and cataracts.

The issue is no longer confined to the southern hemisphere since stratospheric ozone depletion has now been identified in the Arctic, and at mid-latitudes in both hemispheres. Despite prompt international action to reduce halocarbons containing chlorine and bromine, the compounds mainly responsible for stratospheric ozone depletion, past emissions will continue to cause ozone degradation for decades to come due to the long atmospheric residence time of chlorofluorocarbons, or CFCs. Indeed, ozone destruction over the Arctic in early 2011 was – for the first time in the observational record – comparable to that in the Antarctic ozone hole due to unusually long-lasting cold conditions in the Arctic stratosphere (Manney *et al.*, 2011). However, a chemically-driven increase in polar ozone (or 'healing'), triggered by the phase-out of ozone-depleting chemicals under the Montreal Protocol, appears to have begun over the Antarctic

Table 11.2 Human understanding of, and effects on, stratospheric ozone

Date	Development
1880	Forty years after the discovery of ozone in 1840, its role as a filter to the sun's ultraviolet radiation is recognized, and its concentration in the stratosphere is realized.
Late 1920s	First systematic measurements of the distribution and variability of the ozone layer are made by G.M.B. Dobson, revealing that a maximum occurs during spring at high latitudes.
1950s	Chlorofluorocarbons (CFCs), invented in 1928, come into widespread use, particularly for refrigeration and later in air-conditioning, spray cans, foams and solvents. Hailed as miracle chemicals, CFCs are claimed to be completely harmless to living things.
1970	Concern about the use of supersonic transport planes leads to the discovery of an ozone-destroying catalytic cycle involving nitrogen compounds. Although initial projections of the scale of ozone loss due to these aircraft later prove to be excessive, public awareness regarding the importance and fragility of the ozone layer is raised.
1974	Measurements in the early 1970s reveal a growing abundance of CFCs at ground level, prompting Molina and Rowland (1974) to speculate that CFCs rise to the stratosphere where they break down, producing chlorine, which destroys the ozone layer.
1979	The USA, Canada, Norway and Sweden ban CFC use in nearly all spray cans, and the global use of CFCs slows significantly.
Early 1980s	CFC use increases again, due in part to their use as cleaning agents in the rapidly expanding electronics industry.
1985–86	The 'ozone hole' is discovered in 1985 by scientists from the British Antarctic Survey, who report a deepening depletion in the springtime ozone layer above Antarctica of about 40 per cent. Also in 1985, the Vienna Convention for the Protection of the Ozone Layer is signed. The following year, an Antarctic expedition points to chlorine and bromine compounds as the likely cause.
1987	The UN adopts the Montreal Protocol on Substances that Deplete the Ozone Layer, requiring a freeze on annual use of CFCs by 1990 with a 50 per cent reduction by 2000, and a freeze on annual halon production by 1993.
1988	Depletion of the ozone layer in the northern hemisphere mid-latitudes and in the Arctic is discovered.
1990	Montreal Protocol amended to accelerate emissions reductions, requiring complete phase-out of CFCs, halons (except for essential uses) and other major ozone-depleting substances by 2010.
2016	First signs of healing in the Antarctic ozone layer.
2050	Earliest predicted date for full recovery of damage to the ozone layer.

(Solomon *et al.*, 2016). The timing of the return to pre-1980 values of ozone and ultraviolet radiation cannot yet be predicted precisely, but full recovery of the Antarctic ozone layer is not expected until after 2050.

Other impacts

A number of human modifications to the land surface – particularly vegetation cover and soil characteristics – are suspected of inducing more localized climatic effects. Changing the land cover from trees to grass, for example, reduces leaf area index, increases the reflectivity, or albedo, of the ground surface, decreases its roughness, and alters root distribution and depth. These changes affect the proportion of available water that runs off the ground surface or evaporates, thereby affecting soil moisture and possibly rainfall. Relatively long-lived land cover change can be brought about by deforestation and heavy grazing. Similar changes can also come about through natural processes, however. Drought can alter vegetation cover, and similar effects on albedo have been noted when large amounts of dust particles – from volcanic eruptions or blown from agricultural soils – are deposited on glacier ice.

Changes in surface albedo affect the amount of solar energy absorbed by a surface, and hence the amounts of heat energy the surface subsequently releases. This, in turn, affects such atmospheric processes as convection and rainfall. In simple terms, a greater surface albedo results in a cooler land surface, which reduces convective activity and hence rainfall. Such effects may cause a positive feedback, suggested by some as an explanation for the prolonged drought period in the Sahel after the late 1960s (see p. 106). Warnings that similar feedbacks could operate in areas of widespread tropical deforestation have also been made (Lejeune *et al.*, 2015), and climate model simulations of Amazonia indicate that large-scale conversion of forest to pasture would create a warmer, drier climate. There is also the possibility that deforestation on large scales in the tropics could have impacts on aspects of climate such as precipitation in mid- and high latitudes through hydrometeorological 'teleconnections' (Avissar and Werth, 2005).

Direct human modifications to the hydrological cycle may also have climatic impacts. The rise in the area of irrigated agriculture in many parts of the world (see Chapter 13) and the creation of numerous large water bodies behind dams (see Chapter 9) both modify local albedos, and enhance local evaporation and transpiration rates. Conversion of land use to irrigated agriculture, which aims to modify the distribution of moisture in the root zone, affects the energy balance at the soil surface. In areas where irrigation has been adopted on a large scale this can have demonstrable effects on local temperatures. In parts of the US Great Plains where more than 80 per cent of the land use changed from non-irrigated to irrigated agriculture in the second half of the twentieth century, temperatures during the growing season have dropped by more than 2°C over the same period (Mahmood *et al.*, 2006). The expansion of irrigated agriculture has off-site impacts too. Desiccation of the Aral Sea has occurred as a direct result of irrigation and climate has changed in a band up to 100 km wide along the former shoreline in Kazakhstan and Uzbekistan. The more continental climate is characterized by warmer summers and cooler winters, later spring frosts than previously and earlier autumn frosts, lower humidity and a shorter growing season (Micklin, 2010).

GREENHOUSE TRACE GASES

The most worrying type of human-induced climatic change is that brought about through modifications to the natural atmospheric budgets of so-called greenhouse gases. The atmospheric warming caused by greenhouse gases present in the atmosphere in trace amounts is a natural phenomenon caused by the effect of these gases being mainly transparent to incoming short-wave solar radiation but absorbent of re-radiated long-wave radiation from the Earth's surface. The most important greenhouse gases – carbon dioxide, methane, water vapour and nitrous oxide – all occur naturally in the atmosphere, and without their greenhouse properties the Earth's mean temperature would be about 33°C lower than at present. However, the atmospheric concentrations of some of these gases have increased dramatically over the past 250 years or so due to modifications of biogeochemical cycles by human populations, while new gases with greenhouse properties, notably CFCs, have also been added. Table 11.3 summarizes the main atmospheric gases that influence the operation of the greenhouse effect, many of which also effect the destruction of the stratospheric ozone layer.

The effectiveness of these gases in causing climatic change is assessed in terms of their so-called 'radiative forcing', a measure of how the energy balance in the Earth-atmosphere system is influenced when factors that affect climate are changed. The contributions to radiative forcing from the key greenhouse gases and other important factors are indicated in Figure 11.2 which shows the total forcing relative to the start of the industrial era, which is put at 1750. When radiative forcing from a factor is positive, the energy of the Earth-atmosphere system ultimately increases, leading to warming. A negative radiative forcing, by contrast, means the energy will ultimately decrease, leading to a cooling of the system. As Figure 11.2 indicates, the only increase in natural forcing of any significance since 1750 occurred in solar radiation. Human activities are considered to have had much greater influence and although some are thought to have contributed to global cooling since 1750, these have been outweighed by other actions that have produced a net global warming effect.

Table 11.3 Atmospheric trace gases that are significant to global climatic change

	Carbon dioxide (CO_2)	Methane (CH_4)	Nitrous oxide (N_2O)	Chlorofluorocarbons (CFCs)
Principal natural sources	Respiration, fire, volcanoes	Wetlands	Soils, oceans, estuaries and rivers	None
Principal anthropogenic sources	Fossil fuels, deforestation	Rice culture, cattle, fossil fuels, biomass burning	Fertilizer, land-use conversion	Refrigerants, aerosols, industrial processes
Atmospheric lifetime (years)	100–300	12	121	CFC-11: 45 CFC-12: 100
Pre-1750 tropospheric concentration*	280 ppm	722 ppb	270 ppb	CFC-11: zero CFC-12: zero
Recent (2016) tropospheric concentration*	400 ppm	1834 ppb	328 ppb	CFC-11: 232 ppt CFC-12: 516 ppt

Note: *Concentrations in parts per million (ppm), parts per billion (ppb), or parts per trillion (ppt).

Source: data from cdiac.ess-dive.lbl.gov/pns/current_ghg.html, accessed February 2018.

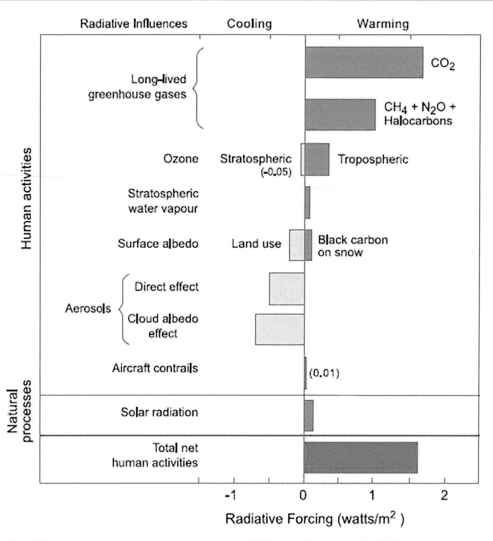

Figure 11.2 Principal causes of global warming since 1750 (after Forster *et al.*, 2007).

Most interest has focused on carbon dioxide as the most important greenhouse gas to have been increased by human action since it has the greatest radiative forcing. Atmospheric concentrations of carbon dioxide have risen by 25 per cent over the last 100 years, with about half of this increase occurring in the past 25 years, and the globally averaged concentration of carbon dioxide in the troposphere reached the symbolic and significant milestone of 400 parts per million for the first time in 2015. The burning of fossil fuels is the most significant source of human additions to atmospheric carbon dioxide (Figure 11.3), while cement manufacture and land-use changes such as deforestation and biomass burning are also important. Global carbon dioxide emissions from fossil fuel combustion and cement manufacture are responsible for more than 75 per cent of the increase in atmospheric carbon dioxide concentration since pre-industrial times. North America and Europe are by far the largest sources of these industrial emissions. Until the

Figure 11.3 Coal-fired power stations like this one at Belchatow, Poland, are major contributors to the human-induced increase in global atmospheric carbon dioxide concentrations.

mid-twentieth century most deforestation also occurred in temperate latitudes, but more recently a greater carbon dioxide contribution has come from deforestation in tropical regions. Estimates of the carbon dioxide due to forest conversion vary because of the problems of assessing deforestation rates in the tropics (see Chapter 4).

The rapid rise in the atmospheric concentration of methane, which is today more than double its pre-industrial concentration, is linked to both industrialization and increases in world food supply. Estimates of the global methane budget indicate that methane emissions associated with human activity (produced by anaerobic bacteria in the standing waters of paddy fields and the guts of grazing livestock by so-called 'enteric fermentation', plus biomass burning and waste) exceed the total natural production, an output dominated by rotting vegetation in wetlands. Like carbon dioxide emissions, industrial sources of methane also contain an important fossil fuel element.

CFCs and other halocarbons are compounds that do not occur naturally. Their development, which dates from the 1930s, was for use as aerosol propellants, foam-blowing agents and refrigerants, and their release into the atmosphere has been inadvertent. Concentrations of these compounds are far lower than those of other greenhouse gases, but the greenhouse-warming properties of CFCs are several thousand times more effective than carbon dioxide. Production of eight halocarbons, including the most abundant CFC-11 and CFC-12, has been severely curtailed by the Montreal Protocol adopted in 1987 (see Table 11.2), and the radiative forcing of stratospheric concentrations of chlorine (the CFC breakdown product that actually destroys ozone) peaked in 2003.

Further indication of the relative importance of human influence on the atmospheric concentrations of greenhouse gases can be gauged from the fact that the contemporary concentrations of the two gases with greatest radiative forcing, carbon dioxide and methane, are both very likely to be much higher than at any time in at least the last 650,000 years. The recent rate of change in concentration is dramatic and unprecedented; increases in carbon dioxide never exceeded 30 ppm in 1000 years, yet in the most recent times carbon dioxide has risen by 30 ppm in less than 20 years (Forster *et al.*, 2007).

GLOBAL WARMING

The theory relating increased atmospheric concentrations of greenhouse gases and global warming is strongly supported by proxy evidence from ice-core data, which show that natural fluctuations in the atmospheric concentrations of greenhouse gases through geological time have oscillated in close harmony with global temperature changes over the past 150,000 years, indicating that the two are almost certainly related (Lorius *et al.*, 1990). Evidence gleaned from a range of other proxy indicators suggests that the twentieth century was the warmest of the last millennium (Jones *et al.*, 2001), and the changes in global mean air temperature measured instrumentally since the mid-nineteenth century are shown in Figure 11.4. Overall, the globally averaged combined land and ocean surface temperature shows a warming of about 0.85°C (0.65 to 1.06) over the period 1880 to 2012, a trend confirmed in multiple independently produced datasets (IPCC, 2014). This recent warming trend partially reflects the end of the Little Ice Age (see Figure 11.1b), but the large majority of researchers think that the trend shown in Figure 11.4 is extremely unlikely to be entirely natural in origin. In large part, it reflects the operation of an enhanced greenhouse effect due to human activities, most importantly pollution of the atmosphere.

Indeed, in recent years, atmospheric scientists have become increasingly confident about the contribution of humans to this contemporary warming. The Intergovernmental Panel on Climate Change (IPCC), an international body of scientists set up especially to look into the issue of global warming, concluded in their first assessment that the observed increase in global temperatures could be due to human influence, but could equally be largely due to natural variability. This first assessment was published in 1990. By the IPCC's fifth assessment, published in 2013, the panel's statements were much more decisive. By this time, the increase in anthropogenic greenhouse gas concentrations plus other human forcings was considered to be extremely likely to have been the major driver of the observed increase in globally averaged temperatures since the mid-twentieth century. Table 11.4 traces the IPCC's increasing confidence in its assessment of the human contribution to contemporary global warming, a confidence based on longer and improved records and our better scientific understanding of the processes involved in global temperature change.

Nevertheless, while the trend in global atmospheric greenhouse gas concentrations has been rising smoothly for many decades, the warming trend over the past 14 or 15 decades has not been continuous through either time or space. Two periods of relatively rapid warming (from 1910 to the 1940s, and again from the mid-1970s to the present) contrast with preceding periods, which were respectively characterized

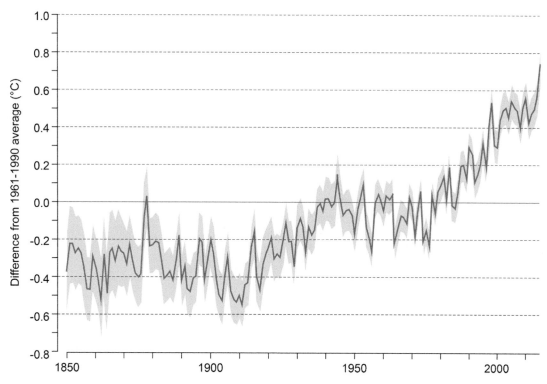

Figure 11.4 Variations in global temperature, 1850–2015 (after Met Office Hadley Centre and CRU).

Table 11.4 The growing confidence in the IPCC's assessment of the human contribution to global warming

IPCC assessment	Attribution of cause
First (1990)	The size of this warming is broadly consistent with prediction of climate models, but it is also of the same magnitude as natural climate variability. Thus the observed increase could be largely due to this natural variability
Second (1995)	The balance of evidence suggests a discernible human influence on the climate
Third (2001)	Most of the observed warming over the past 50 years is likely to have been due to the increase in greenhouse gas concentrations
Fourth (2007)	Most of the observed increase in globally averaged temperatures since the mid-twentieth century is very likely due to the observed increase in anthropogenic greenhouse gas concentrations
Fifth (2013)	It is extremely likely that more than half of the observed increase in global average surface temperature from 1951 to 2010 was caused by the anthropogenic increase in greenhouse gas concentrations and other anthropogenic forcings together

by fairly unchanging (1850s to about 1915) and slightly declining (1940s to 1970s) temperature. However, Easterling and Wehner (2009) demonstrate that decade-long trends with little warming or cooling are to be expected under a sustained long-term warming trend, due to multidecadal internal variability of the ocean atmosphere system.

Spatially, too, global warming has been discontinuous: the two hemispheres have not warmed and cooled in unison, temperature increase is greater at higher northern latitudes, land areas have warmed faster than the oceans and, in contrast to most of the planet, a few parts of the globe have not warmed at all in recent decades. An example here is mainland Antarctica (south of 65°S) where temperatures have remained stable in recent decades, although the Antarctic Peninsula region has experienced rapid warming. Areas that have cooled since 1979 include south-west China and parts of the southern hemisphere oceans, possibly due to changes in atmospheric and oceanic circulation (Trenberth *et al.*, 2007).

Our understanding of the warming effects of enhanced greenhouse gases is further complicated by the observation that highly industrialized areas appear to be warming at a slower rate than less industrial regions. The large emissions of sulphate particles to the atmosphere from industry appear to be retarding warming by reflecting solar radiation back into space (Crutzen, 2006).

Predicting impacts

Most studies of climate change detection (the process of demonstrating that climate has changed in some defined statistical sense) and climate change attribution (the process of establishing the most likely causes for the detected change with some defined level of confidence) use climate models in their work. Climate models are mathematical representations of the processes important in the Earth's climate system which are run on powerful computers. These climate models are also used to predict the conditions of a warmer planet.

There are many different types of model. The most commonly used include General Circulation Models (GCMs) and Earth System Models (ESMs). GCMs depict the climate by simulating physical processes in the atmosphere, ocean, cryosphere and land surface, while ESMs also include some of the biogeochemical cycles such as the carbon cycle. Higher mean temperatures will, of course, affect many other aspects of the climate, such as winds, evapotranspiration, precipitation and clouds, and by simulating the processes in the atmosphere and oceans, in combination with other aspects of the Earth's system represented, and how they respond, the models can be used to predict changes in the distribution of these phenomena through both time and space.

Like all models, climate models are simplifications of the real world and have numerous deficiencies (Washington and Parkinson, 2005). The models are slow to run, costly to use and their results are only approximate. They are, however, being improved all the time. The earliest climate models simulated only the processes in the atmosphere, but adding the effects of the land surface, the oceans, and ice cover – all of which affect atmospheric processes – has increased their sophistication. Prediction of changes in climatic conditions can be made only at certain spatial scales, however. Climate models work by calculating what is happening at a number of specific points at the Earth's surface and in the atmosphere and ocean. Individual models vary, but typical resolutions are between 250 and 600 km horizontally, 10 to 20 vertical layers in the atmosphere and sometimes as many as 30 layers in the oceans.

Climate models are still far from perfect because, among other things, we still do not understand fully all the processes of the climatic system, although we do realize its complexity. Different processes operate on different spatial scales and over varying timescales, and there are numerous feedbacks between different elements of the system, which could enhance or dampen particular environmental responses to change.

Present climate models are still poor at simulating tropical precipitation, ENSO and other tropical variability. They also have significant deficiencies in simulating precipitation frequency and intensity, and the effects of clouds. Indeed, the way in which clouds have been modelled has led to significant uncertainties in climate projections since the IPCC began its assessments. Clouds affect the climate in two fundamental ways: by reflecting incoming solar radiation back to space, which tends to cool the climate, and by trapping outgoing infrared radiation, which tends to warm the climate. As the climate warms, cloud changes might reduce the net incoming energy, offsetting some warming, producing a negative cloud feedback. However, if cloud changes lead to increases in net incoming energy, the change will amplify the initial warming, resulting in a positive cloud feedback. Climate models disagree on the magnitude of the cloud feedback in response to long-term global warming. Their simulations range from a near-zero feedback to a large positive feedback and, despite the importance of the issue, there are few estimates of cloud feedback from real-world observations with which to improve model predictions (Dessler, 2010).

Other possible feedbacks that are poorly represented in many climate models are some of those associated with global vegetation. Terrestrial ecosystems can release or absorb greenhouse gases such as CO_2, methane and nitrous oxide, and they control exchanges of energy and water between the atmosphere and the land surface. An increase in CO_2 may make a further contribution to global warming by inducing a physiological response in vegetation. The way in which plant stomata operate is likely to be affected, with the overall result of reducing transpiration. Carbon dioxide also has a fertilizing effect on plants, so that a CO_2-enriched atmosphere should result in more vegetation on the ground. This structural change in global vegetation will alter the surface albedo and thus produce another feedback on global climate in some way. This is just one of numerous possible climate–ecosystem feedbacks that might amplify or dampen regional and global climate change.

The performance of climate models is tested against our knowledge of present and past climates. Although the models are constantly being improved, they are still far from being perfect representations of the real world. Nevertheless, most people agree on perhaps the most important aspect of global climatic change from the viewpoint of contemporary human societies: the rate of change will be faster than anything we have previously experienced. This being the case, the approximate predictions produced by the models are being used to gain some insight into the nature and conditions of the world we will be inhabiting in the next few generations, since pre-industrial CO_2 levels are expected to double some time in the twenty-first century, with consequent significant changes in very many aspects of the physical environment.

The impacts

Since the atmosphere is intimately linked to the workings of the biosphere, hydrosphere and lithosphere, any changes in climate, both current and future, will have significant effects on all aspects of the natural world in which we live. Research into the possible impacts of global climate change has been one of the central tasks of the IPCC. However, climate and the variability of its elements have always had an impact on human society. The argument that climatic changes, especially those that were abrupt and severe, have been a major cause of great events in human history, particularly the fall of civilizations, was made more than a century ago by Huntingdon (1907). Although Huntingdon's ideas were largely dismissed as being too conjectural and difficult to prove, and his work became too closely associated with the objectionable aspects of environmental determinism (see Chapter 3) and has been largely forgotten, it is interesting to note that climatic explanations for significant historical events have become fashionable once more.

The potential for the sort of disruption to socio-economic systems suggested by Huntingdon has been identified in numerous historical examples studied using palaeoenvironmental evidence (Table 11.5). One

Table 11.5 Some examples of direct climate impacts on human societies inferred from palaeoenvironmental evidence

Location	Period (years BP)	Climate change	Societal impact
W. Asia/Mesopotamia	11,000	cooling	abandonment of hunting–gathering
Libyan Fezzan, Egypt	6000	drying	increasing social complexity and urbanization
W. Asia/Mesopotamia	4200	drying	collapse of Akkadian empire
Indus Valley	4000	drying	decline of Harappan civilization
Peru	1500	drying	decline of Mochica empire (loss of essential irrigation)
Yucatan, Mexico	1200	drought	collapse of Mayan civilization
Mongolia	800	persistent moisture	rise of Mongol empire
Labrador, Canada	650	cooling	migration

Source: after Dearing (2006); deMenocal (2001); Brooks (2006); Pederson *et al.* (2014).

of the most influential of these studies assessed the collapse of Akkad, probably the world's first major empire, in Mesopotamia around 4200 BP (Weiss *et al.*, 1993). The Akkadian empire's demise was linked to the stresses imposed by natural climatic change, largely in the form of increased aridity in this case. Interestingly, other authors have suggested an opposite effect of enhanced aridity, in which the climatic change plays a positive role in the rise of civilizations. Brooks (2006) cites palaeoenvironmental and archaeological data from several locations, including Egypt and elsewhere in the Sahara, to support the idea that increased social complexity occurred as a response to environmental deterioration: people had to retreat to environmental refugia with adequate supplies of surface water.

Although not all authorities agree on the precise role of climate and other environmental variables in past changes to society, few doubt that current and future climatic change will have effects on the world around us and our way of life. The links made by Brooks (2006) between aridity and urbanization in parts of Africa 6000 years ago have been echoed in the contemporary era by Henderson *et al.* (2017). Their study documents strong links between climate and urbanization in sub-Saharan Africa, much of which has experienced a decline in moisture availability over the last 50 years. In parts of the continent with cities that have a strong base of manufacturing for export outside their regions, people have moved into the cities, and this migration, the authors suggest, is a response to the difficulties faced by farmers in a drier climate. In other words, drier conditions increase urbanization. Elsewhere, however, where towns and cities simply service the surrounding agricultural economy, the drying climate has little impact on urbanization because reduced farm incomes reduce demand for urban services.

Shifts in the Earth's climatic zones over several hundred kilometres during the next 50–100 years would affect terrestrial ecosystems, with different species able to adapt or migrate in response to the changes with varying degrees of success. The climate changes recorded in recent decades have already been related to numerous changes among plant and animal communities, some of which are shown in Table 11.6. Shifts

Table 11.6 Evidence of observed ecological changes linked to recent climatic change

Ecology	Location	Observed changes	Climate link
Treeline	Europe, New Zealand, North American Arctic	Advancement towards higher altitudes and latitudes	General warming
Lowland birds	Costa Rica	Extended distribution from lower mountain slopes to higher areas	Dry season mist frequency
Butterflies	North America, Europe	Northward range shifts	Increased temperatures
Wheat	Australia	Increased average yields	More frost-free days
Numerous plant species	Europe	Longer growing season	General warming
Numerous bird species	North America, Europe	Earlier spring migration and breeding	General warming
Caribou	Northern Canada, Alaska	Improved calf survival but reduced health	Warmer air, more insects
Numerous frog species	Central and South American tropics	Decline in numbers due to fungus	Enhanced cloud cover causing cooler days and warmer nights
Bramble Cay melomys	Torres Strait, Australia	Extinction	Sea-level rise

Source: after Walther *et al.* (2002); Hinzman *et al.* (2005); Pounds *et al.* (2006); and Waller *et al.* (2017).

in the distribution, or range, of species poleward and towards higher altitudes have been noted, with some species (e.g. butterflies) responding faster to changes in the location of temperature isotherms than others (e.g. alpine plants). A number of studies have emphasized the sensitivity of mountains, which often hold range-restricted species with limited dispersal abilities (Engler and Guisan, 2009). Some of the other impacts on biodiversity are noted on p. 366.

The possibility that some species are not able to adapt or migrate in time to adjust to environmental changes is illustrated by the disappearance of the Bramble Cay melomys, an Australian rodent that has the dubious distinction of being the first mammalian extinction caused by human-induced climate change (Waller *et al.*, 2017). The extinction is thought to have occurred in a cascade of impacts ultimately driven by sea-level rise: the rodent suffered from a severe reduction in vegetation – its primary food source – on its small, low-lying island habitat, a loss of plant cover that resulted from an increased frequency and intensity of extreme high water levels and storm surges, caused in turn by anthropogenic climate change.

Climatic shifts also affect patterns of agricultural land use, making the production of certain crops and livestock more, or less, suitable in the altered climate, which would not only affect climatic growing conditions but also the hazards from pests and diseases. Increases in minimum temperatures, resulting in fewer frosts, have had an effect on agriculture in the Pampas region of Argentina, although more rainfall has been identified as the greatest climatic influence on the improved yields of four crops during the late twentieth century. The yield increases attributable to changes in climate in the period 1971–99, when compared to the period 1950–70, were 38 per cent for soybean, 18 per cent for maize, 13 per cent for wheat, and 12 per cent for sunflower (Magrin *et al.*, 2005). Similar good news for farmers has been detected in Europe where the growing season has increased by 11 days on average since the early 1960s because of the warming climate. Researchers analysed data from 1959 to 1993 from more than 40 gardens containing genetically identical trees and shrubs in a network that extends roughly between Scandinavia, Macedonia and Ireland. From records for events such as the appearance of leaves or needles, blossoming, and leaves turning colour and falling, they found trends suggesting that over 30 years, on average, the season began about six days earlier and ended about five days later (Menzel and Fabian, 1999). A study using a different methodology – satellite data of vegetation greenness over the period 1982–2011 – found that growing season length increased by 18–24 days per decade in some parts of Europe (Garonna *et al.*, 2014). Other researchers have reported similar shifts in timing and length of the growing season in numerous parts of the world, with most of the evidence indicating a lengthening of the growing season of about 10–20 days in the last few decades, with an earlier onset of the start being the most prominent cause (Linderholm, 2006).

Taking the next step, to assess the impact on food security globally, is fraught with difficulty given the large number of influential factors (see p. 327). The results of one modelling study suggest that global food production should increase with increases in local average temperature over a range of 1 to 3°C, but above this range production will decrease. However, projected changes in the frequency and severity of extreme climate events such as heat stress, droughts and floods are also likely to have significant consequences for food production beyond the impacts due to changes in average conditions, creating the possibility for unexpected changes in yields (Easterling *et al.*, 2007). This theme was pursued by Battisti and Naylor (2009) who used observational data and output from 23 GCMs to show a high probability that growing season temperatures in the tropics and subtropics by the end of the twenty-first century will exceed the most extreme seasonal temperatures recorded from 1900 to 2006. In temperate regions, the hottest seasons on record will represent the future norm in many locations, with potentially disastrous consequences for food production unless farmers adapt to maintain yields by selecting alternative crop varieties, species, and cultivation techniques.

The effects of climate change on food security are just one of the many avenues by which human health may be impacted. Others include extreme weather events, adverse changes in air pollution and the spread of infectious diseases. Infectious disease risks are dynamic and subject to multiple and complex drivers, but one comprehensive review (Watts *et al.*, 2015) highlighted predicted shifts in the geographical distribution of African trypanosomiasis, or sleeping sickness, due to temperature changes, and possible increases in transmission of Hantavirus, West Nile virus and schistosomiasis, or bilharzia, that could be triggered by losses in biodiversity. Overall, this study suggested that the 'implications of climate change for a global population of 9 billion people threaten to undermine the last half century of gains in development and global health' (Watts *et al.*, 2015: 1861).

Many of the ways in which the hydrological cycle responds to global warming stem from the expectation that the water-holding capacity of the atmosphere will increase, roughly exponentially with temperature, and an important anticipated consequence is that wet regions of the planet will become wetter and dry regions drier (Held and Soden, 2006). This theoretical change is being observed in many areas. Global aridity and drought areas have increased substantially since the mid-twentieth century, largely due to widespread drying since the 1970s over Africa, southern Europe, East and South Asia, and eastern Australia, and climate models project increased aridity in these regions (plus the Middle East and most of the Americas) as the twenty-first century progresses (Dai, 2010). Collating projections made by 27 global climate models,

Park *et al.* (2018) found that about 20 to 30 per cent of the world's land surface would become significantly drier if global temperatures reach 2°C above pre-industrial levels but two-thirds of these affected regions could avoid significant drying if warming is limited to a 1.5°C rise. Some of the implications of increasing drought frequency are discussed in Chapter 5 (see p. 105) and the issues are likely to be particularly acute in semi-arid regions and more humid areas where demand or pollution has already created water scarcity (Figure 11.5).

The Mediterranean Basin is a region where a decrease in rainfall and moisture availability has been observed almost everywhere during the twentieth century (Sousa *et al.*, 2011). In north-eastern Spain, an increase in precipitation in winter and summer has been recorded since the 1920s, along with a generally higher number of extreme events separated by longer dry periods (Ramos and Martínez-Casasnovas, 2006). These authors note that the changes in rainfall distribution have had negative effects on the availability of water for crops and have contributed to accelerated erosion in the area. Such changes in the seasonality and intensity of rainfall may also be expected to have an impact on the flood regimes of rivers. Clear patterns of changes in flood timing that can be ascribed to climate effects have been detected over

Figure 11.5 A hoarding urging the residents of Praia, capital of Cabo Verde, to conserve water. Supplies are limited on all of the country's islands due to their arid climate and are likely to become less reliable with the effects of global warming.

the past 50 years across Europe, including later winter floods along parts of the Mediterranean coast and around the North Sea owing to delayed winter storms. Other changes include earlier spring snowmelt floods in north-eastern Europe and earlier winter floods in western parts of the continent caused by earlier soil moisture maxima (Blöschl *et al.*, 2017).

A global flood study that used climate scenarios derived from 21 climate models along with projections of future population found considerable variability in its predictions but one scenario for 2050 predicted that the current 100-year flood would occur at least twice as frequently across 40 per cent of the globe (Arnell and Gosling, 2016). Regional variability was also considerable, but there were consistent increases in flood magnitude across humid tropical Africa, South and East Asia and much of South America, with Asia suffering the most adverse impacts. Most of these types of study address averages rather than specific weather systems associated with particular natural hazards, but atmospheric scientists are now able to assess with some confidence the human impact on specific weather events. A study by Pall *et al.* (2011) was one of the first, showing that increased greenhouse gas emissions substantially increased the risk of flood occurrence during the extensive flooding that occurred in England and Wales in autumn 2000. Interestingly, they were able to access the great computational power needed to generate several thousand climate model simulations by using volunteers among the general public to help on their home computers. Other attribution studies have concluded that other extreme weather events have been made much more likely by anthropogenic climate change (Otto, 2017), including the European heatwave of 2003 and the Russian heatwave of 2010.

High latitudes have long been recognized as areas where global greenhouse warming is likely to be greatest, and surface air temperatures north of 65°N have increased at almost twice the global average rate in the past 100 years (Trenberth *et al.*, 2007). Significant changes are projected in glacial and periglacial processes, affecting glacier ice, ground ice and sea-ice, which, in turn, would affect vegetation, wildlife habitats, and human structures and facilities. A prediction of the geography of changes around the Arctic Circle in northern Canada is shown in Figure 11.6. There is a strong possibility that the Arctic Ocean's ice cover will disappear, facilitating marine transport and exploration for oil, gas and other minerals on the one hand, but also increasing the dangers from icebergs. Year-round shipping is already becoming common in the Canadian Arctic, but vessels servicing mines on winter voyages commonly become stuck in pressured (or ridged) ice for hours to days at a time (Mussells *et al.*, 2017).

Sea-ice constitutes a unique habitat for a number of marine mammals, and some of these are already showing shifts in distribution, compromised body condition and declines in production/abundance in response to sea-ice declines (Kovacs *et al.*, 2010), not least the polar bear (see p. 366). The disappearance of sea-ice has climatic feedback implications because the ice has a high albedo, reflecting a large proportion of sunlight back into space and so providing a cooling effect. The darker areas of open water, which are expanding, absorb greater amounts of solar radiation and increase temperatures. This positive feedback contributes to the increasingly rapid loss of ice. For these reasons, sea-ice has figured prominently in climatic change research (Box 11.1).

The loss of Arctic summer sea-ice is one of several large-scale subsystems of the Earth system that many believe could be approaching a regime shift this century. The phrase 'tipping point' is commonly used to indicate such a system which is close to a threshold, where a small change can have the large, long-term consequences associated with a regime shift (Lenton *et al.*, 2008). Other large-scale elements of the climate system which could pass a tipping point this century (see Figure 22.8) include the Amazon rain forest (through dieback, see p. 77), boreal forests (through dieback, see p. 297), El Niño/Southern Oscillation or ENSO (through a change in amplitude or frequency, see p. 119), the Greenland ice sheet (through melting) and the Indian summer monsoon (through destabilization).

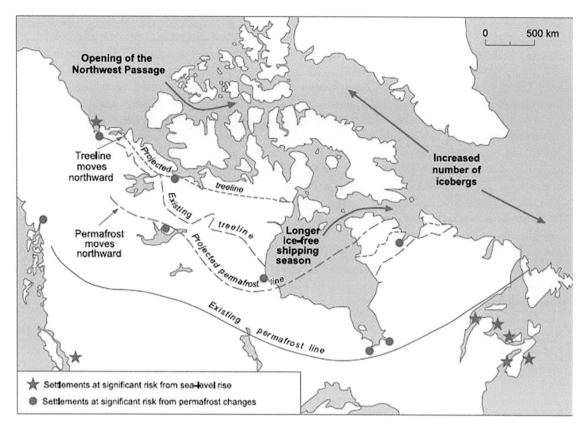

Figure 11.6 Projected changes in northern Canada following climate warming (after Slaymaker and French, 1993; Nelson *et al.*, 2002).

Box 11.1 Sea-ice and climatic change: perspectives from Inuit and scientific communities

Understanding the links between sea-ice and climatic change is important for several reasons. Scientists have focused particularly on how sea-ice regulates energy exchanges between the ocean and the atmosphere and the possible positive feedback mechanisms that may be triggered by interactions between the climate and the cryosphere. This work relies heavily on satellite data and computer modelling to understand the impacts of climatic change on Arctic sea-ice extent, distribution and thickness. Any such changes also have implications for Inuit communities living in the circumpolar Arctic, for whom sea-ice is a vital component of everyday life, as a platform to travel, hunt, fish and/or live on.

Inuit and scientific perspectives on the relationship between sea-ice and climatic change differ in ways that can be complementary. The Inuit focus is on practical consequences of sea-ice change (for their safety, access to marine wildlife, ability to travel and notions of personal/cultural identity). The scientific focus is more on the theoretical consequences (inadequate modelling needs to be improved), although the ultimate goal of the models is to help global society in practical terms.

Inuit traditional ecological knowledge can help climate modellers in several areas, many of which stem from the simple fact that Inuit live permanently in Arctic regions and have accumulated long-term experience and awareness of how sea-ice behaves. Inuit expertise can contribute to western science with local knowledge, providing baseline data and helping to refine scientific models, because computer simulations lack the regional or local-scale aspects that may have some of the most important climatic consequences.

Inuit can also provide scientists with insights into community adaptation. This is important simply because policy-makers who are informed by scientists will find it difficult to propose viable suggestions for adaptation if they do not know the importance of sea-ice to local communities. In turn, scientific tools such as satellite remote sensing and monitoring initiatives could be valuable to Inuit communities in their assessments of rapidly changing ice conditions. Scientific information should be made accessible and understandable to communities and/or local decision-makers to help in the development of their own adaptive strategies.

Source: Laidler (2006).

On land, the northward movement of the permafrost line envisaged in Figure 11.6 has many implications for roads, buildings and pipelines now constructed on permafrost, necessitating reinforcement, and new engineering design and construction techniques. A larger area of thermokarst (irregular, hummocky ground caused by differential ground-ice melt) has been noted already in the Prudhoe Bay Oilfield in Alaska. A sudden increase in the area affected began shortly after 1990, corresponding to a rapid rise in regional summer air temperatures and related permafrost temperatures (Raynolds *et al.*, 2014). A serious feedback aspect of high-latitude permafrost melting is the consequent release of carbon dioxide and methane produced when previously frozen organic matter becomes available for microbial degradation. However, the net outcome of such changes is uncertain given that other trends will offset this release of greenhouse gases at least to some extent. For example, the higher temperatures, longer growing seasons and northward movement of vegetation are all likely to increase the capture and storage of carbon by photosynthesis.

The warming trend documented since the 1940s for the Antarctic Peninsula, among the fastest-warming regions on the planet, has affected the frequency of extensive sea-ice in the Southern Ocean. Cold winters with extensive sea-ice cover occurred on average in four out of five years in the middle of the last century, but have decreased to just one or two years in five since the mid-1970s. One function of this trend appears to be a decline in the abundance of krill in the Southern Ocean, with the knock-on impact of declines in krill-eating penguin populations (Trivelpiece *et al.*, 2011; see also Chapter 6). Warmer air temperatures are also having a predictable effect on glaciers – melting and retreat – in Antarctica and most other parts of the world. Glaciers are receding particularly fast in the Greater Himalayan region (Figure 11.7), popularly known as the Water Tower of Asia because its glacial meltwater feeds Asia's ten largest rivers. The melting

Figure 11.7 Himalayan glacier in retreat in Nepal's Lang Tang national park. Many glaciers in Asia and elsewhere have been retreating over the last 100 years or so, probably due to global warming.

has generated worries about long-term water supplies for hundreds of millions of people in China, India and Nepal (Xu *et al.*, 2009).

Numerous effects of global climatic change will be enhanced when they occur in synergy with other anthropogenic drivers of environmental change. In many semi-arid areas, for example, soil moisture is predicted to be reduced by larger losses to evapotranspiration and decreased summer runoff, and increased rates of soil erosion by wind can be expected as a consequence. Modelling studies of southern Africa by Thomas *et al.* (2005) suggest that during the present century most of the sand dunes in the Kalahari Desert, which have been immobile throughout the Holocene, will become reactivated and mobile due to the combined effects of reduced soil moisture, loss of vegetation and increased wind energy. These impacts are likely to be exacerbated in parts of the region that are under pressure from intensive grazing and cultivation. Other cases of such drivers acting in combination are shown in Table 11.7, although such synergistic effects may also work to ameliorate environmental problems. An assessment of the success of soil conservation measures on the highly erodible loess farmlands of the middle reaches of the Huang Ho River in China found that sediment yields in one of the Huang Ho's tributaries had declined by 74 per cent over the period 1957–89 (Zhao *et al.*, 1992). Just half of this reduction was attributed to improvements in soil and water conservation measures, while the other 50 per cent was a reflection of the shift to a drier climate.

The implications of a warmer world for marine environments are numerous. A basic chemical change that will have far-reaching knock-on effects is the acidification of the oceans, which is occurring because

Table 11.7 Examples of environmental changes affected by synergies between climate change and other human impacts

Environmental change	Climate change impact	Other human impact
Quaternary dune reactivation in the Kalahari	Reduced soil moisture, vegetation loss and increased wind energy	Intensive grazing and agricultural activities
Desiccation of the Aral Sea and associated dust storms	Increased moisture deficit	Excessive irrigation offtake and interbasin water transfers
Coral bleaching on the Great Barrier Reef	Ocean warming	Pollution and physical disturbance
Groundwater depletion in the Great Plains, USA	Increased moisture deficit	Overpumping by centre-pivot irrigation
More frequent flooding of the city of Venice	Sea-level rise	Accelerated subsidence due to the drainage schemes and overpumping from aquifers

some of the CO_2 emitted to the atmosphere by human activities is being absorbed by oceanic waters (see Box 12.1). Another marine issue that has received considerable attention is that of higher sea levels caused both by thermal expansion of the oceans and the added input from melting ice in glaciers, ice caps and ice sheets. As Schneider (1989) suggested, sea-level rise will undoubtedly be the most dramatic and visible effect of global warming in the current 'greenhouse century'.

Given the very large concentration of human population in the coastal zone (see Chapter 7), the consequences for low-lying and subsiding regions are potentially very severe. In their reports, the IPCC has generally assumed a sea-level rise in the twentieth century of 1.5 to 2.0 mm/year, largely on the basis of records from tide gauges (Figure 11.8). More accurate, near-global measurements of sea level from satellite-based sensors indicate that this rate has been accelerating into the twenty-first century (Chen *et al.*, 2017b), a cause for further concern. In the most extreme case, if the Greenland and Antarctic ice sheets were to melt completely, they have the potential to raise the sea level by about 70 m (Alley *et al.*, 2005). Freshwater fluxes from these ice sheets may also affect oceanic circulation, further contributing to climate change effects.

Increased flooding and inundation are the most obvious threats and many low-lying settlements on continental coasts, not least large cities such as Bangkok, Kolkata, Lagos, London, New York, Rio de Janeiro, Shanghai and Tokyo to name but a few, would be affected. The Mediterranean city of Venice is particularly at risk because most of the urban area is only 90 cm above the Northern Adriatic mean sea level (Figure 11.9). The city has also experienced human-induced subsidence in recent decades. While prior to the 1950s the natural subsidence of deltaic coastal plains was compensated for by active sedimentation, the period since the 1950s has been characterized by accelerated subsidence due to the draining of marshes, reduced sedimentation and the overabstraction of water from aquifers. Over the twentieth century relative sea-level rise was 25 cm (about 12 cm of land subsidence and 13 cm of sea-level rise), increasing flood frequency by more than seven times with resulting severe damage to the urban heritage. Projections for relative sea-level rise in the twenty-first century range between 17 and 53 cm by 2100 (Carbognin *et al.*, 2010). Ground

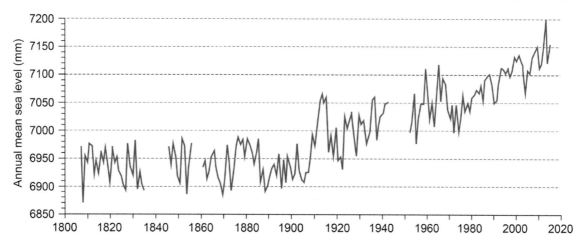

Figure 11.8 Variations in sea level at Brest, France, 1807–2016. This is one of the longest tide gauge records of sea-level change (data from www.psmsl.org/data/obtaining/stations/1.php, accessed February 2018).

Figure 11.9 The Piazza San Marco in Venice is barely above mean sea level and is frequently flooded. Any rise in sea level due to global warming would increase the city's flood frequency.

subsidence has also been a significant issue in Bangkok, Shanghai and Tokyo (see Chapter 10). Densely populated deltas are among the most vulnerable flood-prone regions because of the synergies between a number of drivers (see Table 7.2).

Sea-level rise is arguably the most certain and potentially devastating climate change impact for many small island developing states (SIDS). As Kelman and West (2009) point out in a review of the issues, several SIDS are expected to lose significant proportions of their national territory even under average sea-level-rise scenarios. They include Tuvalu, Tonga, Kiribati, Marshall Islands, Tokelau and the Maldives. Even larger SIDS with much land area well above potential sea-level rise (e.g. Fiji and Samoa) could face severe difficulties since most settlements and infrastructure are in the coastal zone while the hilly, inland regions would experience severe ecological changes in settling all the migrants.

A number of other physical impacts can also be expected for many oceanic islands. As sea levels rise, coastal erosion rates are likely to be affected although the exact nature of these changes will not be straightforward. They will depend on many other factors, including the tectonic history of the island, the rate of sediment supply relative to submergence, the width and growth rate of existing coral reefs and their health, whether islands are anchored to emergent rock platforms, and the presence or absence of natural shore protection such as beachrock or conglomerate outcrops, or vegetation such as mangroves. The range of physical and consequential socio-economic impacts faced by small islands (Betzold, 2015) must be added to the already wide range of natural hazards faced by many SIDS (see Chapter 21), some of which – the meteorological hazards particularly – are likely to be exacerbated by climatic change.

Perhaps the most destructive natural hazard faced by SIDS in low latitudes is the tropical cyclone, and the effect of global warming on cyclone activity has been the subject of considerable debate. Some authors have documented an increase in frequency linked to warmer sea-surface temperatures, such as Holland and Webster (2007) who found that about twice as many hurricanes were forming each year in the Atlantic on average at the beginning of the twenty-first century as 100 years before. However, the IPCC (2012) is more cautious, stating that it has low confidence in any observed long-term (40 years or more) increases in tropical cyclone activity. The IPCC considers that it is too difficult to decide on long-term trends in intensity, frequency or duration because the quality of the observational data is simply not good enough. Most climate models predict future decreases in global tropical cyclone numbers, although with increased intensities for the strongest storms and increased rainfall rates (Walsh *et al.*, 2016). Sea-level rise is also likely to contribute toward increased storm surge risk.

Regardless of the uncertainty over tropical cyclone risks, there are several types of extreme weather event that the IPCC (2012) considers *are* increasingly likely to become more frequent, more widespread and/or more intense in most parts of the world during the twenty-first century. These include heatwaves, droughts and heavy rainfall events. For some types of extreme – notably heatwaves, but also extremes of precipitation – there is now strong evidence linking specific events or an increase in their likelihood to the human impact on our climate (see above). The implications for society are clear.

Responses to global climatic change

The formidable economic, social and political challenges posed to the world's governments and other policy-makers by impending global climatic change are unprecedented. Policy responses can be categorized broadly into those that aim to prevent change and those that accept the changes and focus on adapting to them. In the large literature on climate change, these two aims are usually described as:

- *mitigation* – the reduction of the rate of climatic change via the management of its drivers (the emission of greenhouse gases from fossil fuel combustion, agriculture, land-use changes, cement production, etc.)
- *adaptation* – the process of adjustment to actual or expected climate and its effects, in order to moderate harm or exploit beneficial opportunities.

Mitigation and adaptation complement each other and together can reduce the risks posed by climatic change to society and to the natural environment, although their effects vary over time and space. Mitigation will have global benefits but timelags in the climatic and biophysical systems mean that these will be barely noticeable until around the middle of the twenty-first century. The benefits of adaptation are mainly local to regional in scale but they can be immediate, particularly if they also address vulnerabilities to current climate conditions.

While the issue of climatic change is a truly global one, since all greenhouse gas emissions affect climate regardless of their origin, the costs and benefits of measures to mitigate the effects are likely to be spread unevenly across countries. The issue raises important questions of global equity since, at present, the major proportion of greenhouse gas emissions comes from the industrialized countries, which contain only about 25 per cent of the world's population. Developing world leaders have called for reductions in emissions from the industrialized countries to make more of the planet's capacity for assimilation of greenhouse gases available to those countries that are industrializing now, a plan that should be facilitated by transfers of finance and technology from the North to the South. It has also been pointed out that some of the countries most at risk, the small island states, are effectively subsidizing the economies of industrial countries that are net producers or exporters of CO_2, since more natural CO_2 is fixed by the tropical rain forests, oceans, coral reefs and mangroves of small islands than is emitted locally from these islands (Pernetta, 1992). Given that the impacts of climatic change are also likely to be uneven geographically, with some poorer countries and regions particularly vulnerable (see above), this is clearly a global-scale environmental justice issue. Hence, Table 11.8 indicates some ways of linking the characteristics of climatic change to their implications for social policy and action.

Most countries have accepted the need to make some effort to prevent change, or at least slow its pace, by reducing greenhouse gas emissions. A contribution was made in this respect by the Montreal Protocol (see above) through which governments agreed to reduce consumption and production of substances that deplete the stratospheric ozone layer, many of which also contribute to global warming. Most attention has focused on carbon dioxide, however, an initiative that produced the UN Framework Convention on Climate Change (UNFCCC) which was signed by more than 150 governments at the Earth Summit in 1992 and came into force two years later following the fiftieth country ratification. In effect, signatory states agreed to reduce emissions to earlier levels, in many cases the voluntary goal being a reduction of CO_2 emissions to 1990 levels. The ultimate objective of the UNFCCC is to prevent dangerous anthropogenic interference with the climate system.

An attempt to make agreed reductions legally binding came in 1997 with the Kyoto Protocol, a follow-on to the original treaty, but Kyoto, which only came into effect in 2005, was severely undermined by an inability to enforce targets and timetables. In practice, individual country reductions could be greater or less than those agreed since the Kyoto Protocol also sanctioned the idea of 'emissions trading' between industrialized nations. Hence, if a state's emissions fall below its treaty limit, it can sell credit for its remaining

Table 11.8 Climatic change, environmental justice and the implications for policy

Climate change characteristics	Environmental justice perspective	Implications for action
Greenhouse gas emissions correlate with wealth and growth	Responsibility for climate change lies primarily with richer people in richer countries	Developed countries have an ethical obligation to rapidly provide adaptation support to poor people in poor countries
		Climate mitigation should not constrain energy access for poor people, or the growth paths of poor countries
Climate change impacts differ according to people's power, wealth and the level of dependency on natural resources	The brunt of climate change impacts is borne by poor people in poor countries. They should receive preferential access to adaptation support	There is a compelling need to understand the social dimensions of vulnerability by examining the assets, knowledge, institutions and relationships that different groups have to help them cope with external threats. People can be more or less vulnerable according to age, ethnicity, caste, gender roles, their sources of livelihood, ability to access public support or ability to migrate
	Women will be disproportionately impacted by climate change because social exclusion is strongly gendered in ways that increase vulnerability to climate change for women	
	Indigenous people are among the poorest and most socially excluded people globally. They rely on ecosystems particularly prone to the effects of climate change including polar regions, humid tropics, high mountains, small islands, coastal regions and drylands	An understanding of social difference needs to be translated into guarantees that their enjoyment of fundamental human rights will not be compromised by climate change impacts
Climate change will worsen water stress in many parts of the world through changes in rainfall patterns, glacial and snow melt, and rising salinity in low-lying coastal areas	Poor people will be most severely affected as they have less capacity to extract and store water	Investments in water resources need to take account of the specific needs of poor people, particularly women, and build on local people's knowledge and priorities
	Many women will see an increase in their labour burdens as they have primary responsibility for collecting water in numerous world regions	
Carbon assets (trees, rangelands, etc.) increasingly will be valued for their carbon sequestration properties in the efforts to contain and mitigate climate change	Poor people's rights in carbon assets – whether ownership or use and access rights – are critical to dignity and livelihood	Robust and accountable policy and institutional frameworks must be established to protect poor people's rights in carbon assets and to maximize the income streams they can derive from those assets

Source: after Mearns and Norton (2010).

allotment to another country to help the buyer meet its treaty obligation. An emissions trading scheme was introduced in the EU in 2005 on a similar cap-and-trade basis, with each EU member-state government setting an emission cap or limit on the amount of pollutant that can be emitted for each individual installation covered by the scheme. The idea of emissions trading was pioneered on a large scale by the US Acid Rain Program (see p. 300) and in 2009 the USA's Regional Greenhouse Gas Initiative was initiated.

The first legally binding global climate deal, the Paris Agreement, was made in 2015 under the auspices of the UNFCCC and entered into force in 2016. In the agreement, all countries pledged to work towards limiting global temperature rise to well below 2°C above pre-industrial levels and to strive to limit the increase to 1.5°C. This goal represents the level of climate change that governments agree would prevent dangerous interference with the climate system, while ensuring sustainable food production and economic development.

Practical implementation of the Paris Agreement entails large financial investments to insure against future events that are uncertain. This being the case, the additional beneficial aspects of these and other mitigating policies are widely promoted: these 'no regrets' initiatives are sensible anyway (Table 11.9). For example, planting trees to lock more carbon in the biosphere, so-called 'carbon sequestration', and reducing deforestation have the additional benefits of conserving biodiversity and maintaining environmental services (such activities are frequently referred to under the acronym REDD – Reducing Emissions from Deforestation and Forest Degradation). These initiatives are also, from the global warming perspective, examples of the precautionary principle in operation. Some of these types of project have been established

Table 11.9 Examples of 'no regrets' initiatives to combat global climatic change by mitigating the causes

Initiative	Effect on greenhouse gases	Other benefits
Energy conservation and energy efficiency	Reduced CO_2, methane and N_2O emissions	Conservation of non-renewable resources for current and future generations, reduction of other problems such as acid rain
CFC emission control	Reduced CFC emissions	Reduced stratospheric ozone layer depletion, reduced surface skin cancer and blindness
Tree planting	Increased biosphere carbon sink capacity	Improved microclimate, reduced soil erosion, reduced seasonal peak river flows
Prevent deforestation	Maintain biosphere carbon sink capacity	Biodiversity conservation and maintenance of environmental services
Reduce motor vehicle use	Reduced CO_2 emissions	Improved air quality and congestion
Improve efficiency of N fertilizer use	Reduced N_2O emissions	Improved water quality

under another Kyoto Protocol initiative, the Clean Development Mechanism, which is an arrangement allowing industrialized countries with a greenhouse gas reduction commitment to invest in projects that reduce emissions in developing countries as an alternative to more expensive emission reductions in their own countries.

Nevertheless, these market-based policy solutions to global warming have not been without their critics. As Liverman (2009) points out, a new commodity in carbon reductions has been created and has rapidly become a new form of development investment that some opponents consider to be of questionable value to the poor in the developing world while it has become a new arena for capital investment and speculation.

One of the difficulties governments face in curbing greenhouse gas emissions is the strong link between energy consumption (the principal source of emissions via fossil fuel burning) and economic growth, so that long-term strategies to reduce emissions must uncouple economic growth from growing fossil fuel consumption. Reducing the amount of energy used per unit of GDP will be one element in such a strategy, but there is also a need for a significant shift away from fossil fuels to using more renewable energy sources (see Chapter 18). Many industrialized countries do have experience of economic growth with declines in energy consumption. It occurred during the late 1970s and early 1980s, sparked by the 1970s oil crisis (Goldemberg *et al.*, 1988). Unfortunately, virtually all governments have reinstated their faith in the belief that economic growth must be based on an increase in energy consumption. But the lessons of global warming make it clear that this kind of industrialization, based on an inefficient use of fossil fuel resources, is not a sustainable form of development.

Future emissions of greenhouse gases will be determined by numerous dynamic factors. Some of the key variables are identified in the $I = PAT$ equation (see Chapter 3) which identifies energy-related emissions as a function of population growth, GDP per person, changes in the amount and efficiency of energy consumed, and how much carbon that energy contains. If fossil fuels continue to be used for more than 80 per cent of world energy consumption, as they were in 2015, CO_2 emissions are likely to more than triple by the year 2050, but the phasing in of more renewable energy sources can reduce these levels. Indeed, some optimists believe that a 100 per cent renewable energy system (including transport) is feasible on the national scale using existing technologies (Mathiesen *et al.*, 2011). Interestingly, this study indicates that a complete reliance on renewable energy is possible in Denmark by 2050, but it should also be noted that Denmark is a rare example of a country that has managed to uncouple economic growth from growing fossil fuel consumption: primary energy supply has been kept constant for 35 years, while maintaining stable economic growth.

An alternative approach to reducing the rate of future warming has been suggested by some scientists, in part as a reaction to the slow progress and relative impotence of political solutions. 'Climate engineering' or 'geoengineering' describes a range of proposed techniques which aim to reduce climate warming directly by technical means. They include ideas to reduce the amount of solar radiation received in our atmosphere by injecting large amounts of reflecting particles into the stratosphere or by placing reflecting 'mirrors' beyond the atmosphere. Other proposals focus on accelerating large-scale carbon sequestration by adding very large quantities of lime or iron to the oceans or irrigating huge areas of desert to grow trees. These ideas tend to attract considerable media attention and are controversial among climate scientists (Hulme, 2014). Whether or not they are scientifically viable or technically feasible, a significant argument against these technical fixes is the simple fact that such large-scale manipulation of the physical environment would be likely to have unforeseen impacts, even if they did achieve their primary objectives.

There is also no doubt that warming is still virtually certain to take place in the next few decades because of the increased levels of greenhouse gases already in the atmosphere and timelags in the climate system, and thus a need to prepare for adaptive responses remains. An integral part of this strategy is to continue the funding of scientific research and data collection, which can help to reduce the uncertainty surrounding possible climate change impacts. The strategy will probably also seek to integrate traditional ecological knowledge (TEK) of indigenous peoples with scientific data and analysis of climatic change (Alexander *et al.*, 2011).

These adaptive responses are being undertaken all over the world as individual countries and corporations have developed national climate change strategies. In Europe and elsewhere, adaptation strategies have lagged behind mitigation measures as the primary response to climate change, but this attitude changed after the turn of the century with impacts of climatic change increasingly being observed (Biesbroek *et al.*, 2010). The need for adaptation has been more pressing in more obviously vulnerable countries, such as SIDS comprising low-lying atoll islands. In the Maldives, the government has identified several vulnerable areas and the adaptive measures that could be implemented to reduce climate change impacts (Table 11.10).

Table 11.10 Climate change adaptive measures in the Maldives

Vulnerable area	Adaptation response
Land loss and beach erosion	Coastal protection Population consolidation, i.e. reduction in number of inhabited islands Ban on coral mining
Infrastructure and settlement damage	Protection of international airport Upgrading existing airports Increase elevation in the future
Damage to coral reefs	Reduction of human impacts on coral reefs Assigning protection status for more reefs
Damage to tourism industry	Coastal protection of resort islands Reduce dependency on diving as a primary resort focus Economy diversification
Agriculture and food security	Explore alternative methods of growing fruits, vegetables and other foods Crop production using hydroponic systems
Water resources	Protection of groundwater Increasing rainwater harvesting and storage capacity Use of solar distillation Management of storm water Allocation of groundwater recharge areas in the islands
Lack of capacity to adapt (financial and technical)	Human resource development Institutional strengthening Research and systematic observation Public awareness and education

Source: after MOHA (2001).

The UNFCCC also committed industrialized countries to providing technological and financial resources to help developing countries address climatic change, and hence many aid programmes have now incorporated climate change objectives into their projects. The UNFCCC's financial mechanism is the Global Environment Facility (GEF) which funds projects in both mitigation and adaptation, supporting initiatives in energy efficiency, renewable energy, sustainable urban transport and sustainable management of land use, land-use change and forestry. The GEF also manages two separate adaptation-focused funds, one of which supports technology transfer. There are many similarities between climate change adaptation activities now being supported through international development programmes and the longer-established disaster risk reduction policies and strategies because, as Mercer (2010) puts it, climatic change resulting from human activity can be thought of as a significant long-term global disaster. However, from her work in Papua New Guinea, this author emphasizes that climatic change is one factor among many contributing to vulnerability to hazardous events and hence warns against an over-emphasis on climatic change, which would not adequately address the development concerns of communities.

FURTHER READING

Alexander, C. and 12 others 2011 Linking indigenous and scientific knowledge of climate change. *BioScience* 61: 477–84. An integration of indigenous climate-related narratives and studies from the scientific literature.

Bais, A.F., Lucas, R.M., Bornman, J.F., Williamson, C.E., Sulzberger, B., Austin, A.T., Wilson, S.R., Andrady, A.L., Bernhard, G., McKenzie, R.L. and Aucamp, P.J. 2018 Environmental effects of ozone depletion, UV radiation and interactions with climate change: UNEP Environmental Effects Assessment Panel, update 2017. *Photochemical and Photobiological Sciences* 17(2): 127–79. An authoritative account of stratospheric ozone depletion effects. Open access: pubs.rsc.org/en/content/articlehtml/2018/pp/c7pp90043k.

Hulme, M. 2016 *Weathered: cultures of climate.* London, Sage. Investigation into the idea of climate, setting the scene for climatic change issues.

Liverman, D.M. 2009 Conventions of climate change: constructions of danger and the dispossession of the atmosphere. *Journal of Historical Geography* 35: 279–96. A thoughtful critique of market-based solutions to global warming.

Mearns, R. and Norton, A. 2010 *Social dimensions of climate change: equity and vulnerability in a warming world.* Washington, DC, World Bank. An assessment of the relationship between climate change and key social issues. Open access: openknowledge.worldbank.org/handle/10986/2689.

Skeie, R.B., Fuglestvedt, J., Berntsen, T., Peters, G.P., Andrew, R., Allen, M. and Kallbekken, S. 2017 Perspective has a strong effect on the calculation of historical contributions to global warming. *Environmental Research Letters* 12(2): 024022. Interesting discussion of how much each country has contributed to global warming. Open access: iopscience.iop.org/article/10.1088/1748-9326/aa5b0a.

WEBSITES

ess-dive.lbl.gov   this data archive for Earth and environmental science includes material from the Carbon Dioxide Information and Analysis Center.

ozone.unep.org   the Ozone Secretariat is the Secretariat for the Vienna Convention and the Montreal Protocol.

unfccc.int   site of the UN Framework Convention on Climate Change including reports, data and information on progress towards convention targets.

www.ipcc.ch   information on the Intergovernmental Panel on Climate Change's work assessing the scientific, technical and socio-economic aspects of human-induced climatic change, which includes many open access reports.

www.tiempocyberclimate.org   the Tiempo Climate Cyberlibrary is an information service with publications, data and educational resources on global warming and climate change, with a focus on the developing world.

public.wmo.int   this World Meteorological Organization site has a wide range of information on issues in meteorology, weather and climate.

POINTS FOR DISCUSSION

- Have we done enough to solve the problems associated with the hole in the ozone layer?
- If global climatic change is essentially an environmental justice issue, are those primarily responsible putting enough effort into mitigation?
- Who should pay to reduce global greenhouse gas emissions and how can they be persuaded to do so?
- Outline the pros and cons of tradable emissions permits.
- Is climate engineering the answer to solving the global warming issue?
- How might actions taken to mitigate global climate change affect other environmental issues?

12 *Acidification*

LEARNING OUTCOMES

At the end of this chapter you should:

- Understand the pH scale.
- Appreciate how acidification occurs naturally as part of the Earth's biogeochemical cycles.
- Recognize that ocean acidification is expected to have major implications for marine ecosystems.
- Know that the global geography of maximum sulphur and nitrogen emissions has shifted from Europe and North America to parts of Asia.
- Appreciate the wide range of deleterious effects acid rain has had on ecosystems, human health and the built environment.
- Understand some of the methods used to mitigate the effects of acid rain.
- Appreciate that the pace of environmental recovery varies in different environments.

KEY CONCEPTS

buffering capacity, critical load, chronic acidification, episodic acidification or acid shock, diatom, Oslo Protocol, liming

---

Acidification is a process that occurs naturally in the environment but one that has also been accelerated by a number of human activities. The term is used to describe the loss of nutrient bases (calcium, magnesium and potassium) and their replacement by acidic elements (hydrogen and aluminium), a process that affects all aspects of the natural environment: soils, waters, flora and fauna. Acidification caused by human activity is most commonly associated with atmospheric pollution arising from anthropogenically derived sulphur (S) and nitrogen (N), so-called 'acid rain', a term first used in England about 170 years ago when a Scottish chemist found that local rainfall in Manchester was unusually acidic. Smith (1852) suspected a connection with sulphur dioxide (SO_2) from local coal-burning factories, and commented on the effects the deposits had on buildings and vegetation. The effects of acidification on lake and river ecology in Sweden pushed the issue onto the international agenda in the 1960s and the long-range atmospheric transportation of pollutants from other countries, including Britain, was eventually accepted as a major cause. For much of the nineteenth and twentieth centuries, large areas in Europe and North America were regions of major concern with regard to acid deposition, but parts of Asia have now become seriously affected.

The effects of this deposition on various aspects of aquatic, terrestrial and built environments are still the subject of some controversy, and form a large part of this chapter. However, in more recent times another form of atmospheric pollution has been highlighted as a much wider issue of human-induced acidification: the global decline in pH of the ocean surface due to its absorption of CO_2.

DRIVERS OF ACIDIFICATION

Acidity is commonly expressed using the pH scale, which reflects the concentration of hydrogen ions (H+) in a solution. The scale ranges from 0 to 14 (Figure 12.1), with a value of 7 indicating a neutral solution. Values less than 7 indicate acid solutions (e.g. lemon juice is pH 2.2), while values greater than 7 indicate basic solutions (e.g. baking soda is pH 8.2). The pH scale is logarithmic, so that a solution of pH 4 is ten times more acidic than one with a pH of 5. Precipitation in an unpolluted atmosphere commonly has a pH of 5.6. This slight acidity is due to the ubiquitous presence of carbon dioxide (CO_2) in the atmosphere, which forms a carbonic acid.

The chemistry of the physical environment changes on a range of time and space scales as part of the Earth's biogeochemical cycles. Numerous sources of gases and particulate matter affect the atmosphere and alter pH. Fires, deflation of soil particles and various biological sources all impinge upon atmospheric chemistry and terrestrial ecology in turn, as does the deposition of sea salt during storms in coastal areas. However, emissions from certain volcanoes constitute the world's largest sources of SO_2 and contribute significantly to regional acid loading of local ecosystems.

Similar episodic drivers of acidification can occur in the hydrological cycle, intense events lasting from a few hours to a few weeks which follow periods of heavy rainfall or the early spring snowmelt that release

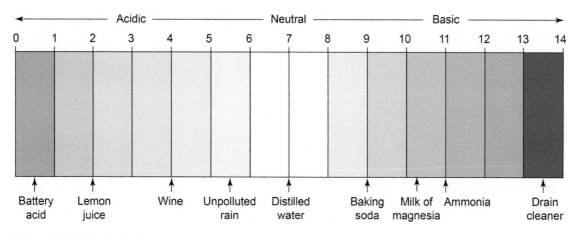

Figure 12.1 The pH scale.

organic acids accumulated over winter. Some freshwater lakes are naturally acidic because they are fed by volcanic or hydrothermal waters or because they have direct contact with massive sulphide deposits. Acid saline groundwater feeds naturally acid saline lakes in Western Australia, some with a pH as low as 1.5 (Bowen and Benison, 2009).

The acidification of soil begins when rocks are first colonized by algae and lichens. Acids produced mainly from the carbon and nitrogen cycles begin to dissolve the rocks and soil minerals to form the soil. Several studies have illustrated the gradual decrease of soil pH with primary succession of vegetation, such as that conducted by Crocker and Major (1955) in Glacier Bay, Alaska. Fine morainic till had a pH of 8.0–8.4 when first exposed by glacier meltback, but had fallen to pH 5.0 beneath alder trees after about 50 years due to leaching by rainwater and modification by developing vegetation. The process of podzolization that occurs in some soils continues the acidification for many years. In the cool, humid conditions of northern Europe, soils have gradually been weathered and leached and organic matter from coniferous forests (acidic needles) has accumulated on and in the surface soil since the last glacial period. The soils have been continuously subjected to acidification processes over the last 8000–10,000 years.

However, this evolution of northern European soils has been increasingly complicated over at least the last 5000 years by the influence of people. The role of humans in accelerating podzolization and the formation of peat bogs has been much debated and explanations have tended to focus on how society has altered vegetation cover. The loss of trees cleared for farming and grazing is thought to have resulted in greater leaching and surface wetness, and invading vegetation – particularly bracken and heather – tended to further produce a more acidic soil than the original forest, so continuing the process.

Land use continues to have an impact. Afforestation in the uplands of Scotland with new conifer plantations (Figure 12.2) in recent decades, for instance, has caused significant declines in soil pH (Grieve, 2001). Forest growth represents a long-term drain on the soil's store of nutrients, which can be represented as a

Figure 12.2 Conifer plantations such as this one in Scotland frequently result in soil acidification.

net uptake of base cations from the soil. The result is acidification of the soil during the life of the trees and if the trees are then felled for their timber the base cations are lost completely from the system. Areas of intensive crop production can also suffer from soil acidification when too much fertilizer, particularly nitrogen, is applied (in excess of crop needs).

The impacts of acid mine drainage on the chemistry of rivers and lakes in mining regions can result in significant changes to pH (see p. 475), but the more widespread anthropogenic sources of acidifying compounds are fuel combustion and agriculture that emit SO_2 and nitrogen oxides (NO and NO_2, which are collectively referred to as NO_x) into the atmosphere. Large quantities of these oxides of sulphur and nitrogen are emitted by the combustion of fossil fuels (coal, oil and natural gas) and industrial processes (principally primary metal production), and it is these gases that are either deposited close to the source, as dry deposition (by gravitational settling and through contact with surfaces), or are converted into sulphuric and nitric acids, and deposited as wet deposition (literally acid rain, snow, sleet, hail, mist and dew) at distances up to thousands of kilometres from the source (Figure 12.3). While these industrial sources are the largest emitters of acid rain pollutants, agriculture makes important contributions in some areas when ammonia from nitrogenous fertilizers and manure escapes to the atmosphere by volatilization.

A combination of fossil fuel use and land-use changes has led to dramatically increasing atmospheric CO_2 concentrations worldwide (see Chapter 11). Some of this CO_2 is absorbed by oceans and reacts with sea water to form carbonic acid, resulting in the acidification of oceans outlined in Box 12.1.

The various impacts of human activity are of course superimposed on natural changes with varying degrees of overlap, which can complicate our understanding of acidification processes in some regions. In a study of upland forest soils in southern Finland, Tamminen and Derome (2005) identified the influence of anthropogenic acid rain deposition since about 1950, the effects of timber harvesting dating from the beginning of the twentieth century and the natural process of podzolization that has been modifying the soil for about the last 10,000 years.

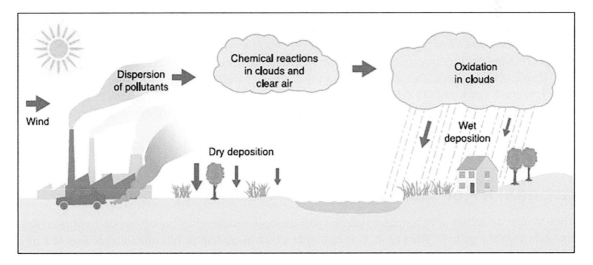

Figure 12.3 Emission, transport, transformation and deposition of pollutants known as acid rain.

Box 12.1 Ocean acidification

Since the beginning of the Industrial Revolution, about 30 per cent of the atmospheric CO_2 emitted by humankind has been absorbed by the oceans. This has had a direct impact on their chemistry, causing a global decline in average surface ocean pH of 0.1 units, from 8.2 to 8.1, a rate that appears to be unprecedented in the geological record. Like global warming, this acidification of the oceans is caused predominantly by greenhouse gas emissions caused by human activities, hence ocean acidification has been referred to as 'the other carbon dioxide problem' (sulphur dioxide and nitrogen oxide are also involved but to a lesser extent).

The response to this rapid acidification by marine ecosystems, including such economically important components as coral reefs and fisheries, is poorly understood. The changes in ocean chemistry will probably affect marine life in three main ways:

■ Many marine organisms use calcium (Ca) and carbonate (CO_3) in sea water to construct their calcium carbonate ($CaCO_3$) shells or skeletons. As pH declines, carbonate becomes less available, making it more difficult for organisms to form skeletal material.
■ Lower pH could affect a variety of physiological processes which could change behaviour and reproduction.
■ Higher concentrations of dissolved CO_2 could alter the ability of primary producers to conduct photosynthesis.

The effects of such changes on marine ecosystems are the subject of much research and speculation. Initial findings suggest that areas of particular concern include coastal margins, deep-sea ecosystems, high-latitude regions and coral reefs in the tropics. A large body of experimental evidence suggests that the distribution of warm-water corals will be dramatically reduced by ocean acidification through lower rates of calcification and hence more fragile skeletons.

Some idea of how marine ecosystems might change with increased acidity can be deduced from areas of the ocean with naturally low pH at present. The seas around parts of southern Italy, for example, have a low pH due to volcanic gas vents that emit CO_2 from the sea floor. Studies here show that the distribution, diversity and nature of marine life change markedly as pH decreases.

It seems highly likely that as pH values in the oceans continue to decline throughout this century, there is great potential for widespread changes to marine ecosystems. This is particularly likely given that the effects of ocean acidification will be felt in combination with the synergistic impacts of other human-induced stresses.

Source: Logan (2010).

GEOGRAPHY OF ACID DEPOSITION

The absolute quantities of atmospheric sulphur and nitrogen produced by human action have risen dramatically over the past 100 years or so, to the point at which human-induced emissions are now of a similar order of magnitude to natural sources. However, the uneven global distribution of industrialization means that the concentration of human-induced acid deposition is highly uneven. Data for the mid-1980s, for example, indicated human-induced sulphur emissions to the atmosphere to be about 60 per cent of the total, but over the industrialized regions of Europe and North America the human-induced flow was thought to be 12 times the natural flow (Bates *et al.*, 1992). Globally, human emissions of sulphur peaked in the 1970s and decreased until 2000, with an increase in recent years due to higher emissions from China, international shipping, and developing countries in general (Smith *et al.*, 2010). The geographical pattern of sulphur emissions has also changed significantly. In 1980, 60 per cent of global emissions were from the North Atlantic basin (the USA, Canada and Europe) but by 2000 this region was thought to make up less than 35 per cent of sulphur emissions worldwide thanks to the collapse of economies in Eastern Europe and concentrated efforts to reduce this form of pollution (see below). The area of maximum emissions shifted towards Asia over this period. Asian sulphur emissions were 26 per cent of the global total in 1980 but had risen to 52 per cent in 2010 (Klimont *et al.*, 2013). This shift in the sulphur emissions pattern, from a centre around the North Atlantic to one dominated by East and South Asia, is expected to continue over the coming decades. Emissions in these regions are dominated by China and India respectively.

Europe and North America are the world's most severely affected large-scale regions because of their long histories of emissions. Maps such as the one of Europe shown in Figure 12.4 give an impression of the scale of the acid rain issue. Over more than a century of sulphur deposition, a broad swath from Wales to

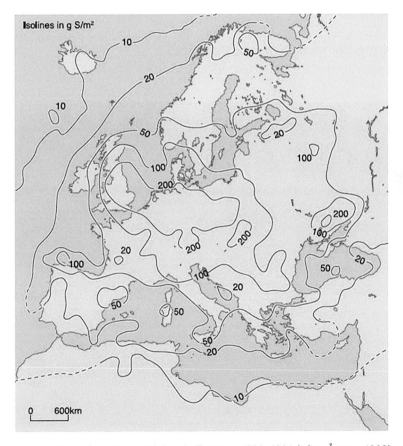

Figure 12.4 Total deposition of oxidized sulphur in Europe, 1880–1991 (after Ågren, 1993).

Poland received more than 200g S/m². However, an increasing body of evidence indicates the presence of smaller acid rain 'hotspots' in many other areas with concentrations of particularly polluting industry. Fioletov *et al.* (2016) highlight the fact that metal-smelting centres constitute some of the largest point sources for SO_2 emissions and that the combined emissions from several smelters in Norilsk, Russia represent one of the largest, if not the largest, anthropogenic source of SO_2 that is clearly seen by satellites. Emissions of both SO_2 and NO_x are also high in many of the world's largest urban areas (see Chapter 10).

East and South Asia have become the world's third and fourth regional-scale acid deposition hotspots, after Europe and North America. Deposition rates of sulphur in parts of China in the early twenty-first century were particularly high: in some areas greater than those seen in central Europe in the early 1980s, when acid deposition in that region was at its peak (Larssen *et al.*, 2006). Since 2007, however, emissions in China have declined by 75 per cent while those in India have increased by 50 per cent, so that India is now surpassing China as the world's largest emitter of anthropogenic SO_2 (Li *et al.*, 2017). Nevertheless, acid rain is not a totally new phenomenon in Asia. Japan has a lengthy history of acid deposition, initial research and policy initiatives being connected to pollution from copper mining in the late nineteenth and early twentieth centuries (Wilkening, 2004).

The movement of acid rain across national political boundaries has long been an important aspect of the issue; indeed, as noted above, the problem first became an international issue when links between acid rain in Sweden were made with polluting industries in other parts of Europe. The issue is also international in East Asia, where Japan has been a net receiver of long-range transboundary air pollutants, particularly from industries in China, since the mid-1980s (Wilkening, 2004). Chinese emissions also contribute to an acidification problem on the Korean peninsula (Figure 12.5).

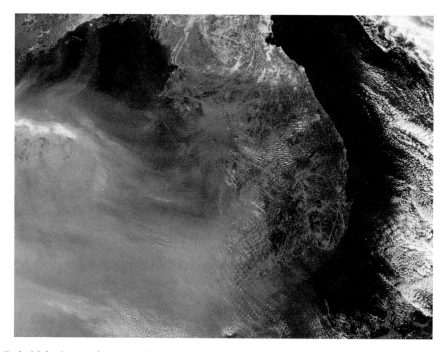

Figure 12.5 A thick plume of smog drifts across the Yellow Sea from China towards Korea in March 2013.

EFFECTS OF ACID RAIN

The actual impact of acid rain on the Earth's surface has been, and continues to be, a subject for some debate. The ecological effects depend on a variety of ecosystem characteristics as shown in Table 12.1. The ability of an ecosystem to neutralize incoming acids, the so-called 'buffering capacity', is particularly important. It is largely determined by the nature of the bedrock: environments on hard, impervious igneous or metamorphic rock, where calcium and magnesium content is low, are most at risk of acidification due to inputs of acid rain. Hence, attention has necessarily focused on areas with low buffering capacity where recent changes in acidification have occurred. Even in these areas, however, different degrees of acidification will have different effects and so the concept of 'critical loads' is now commonly used to determine the susceptibility of a particular environment and to guide air quality management efforts. A critical load is a measure of the amount of acid deposition that an ecosystem can take without suffering harmful effects – in other words, the threshold beyond which damage occurs.

The fact that numerous ecosystems have experienced large decreases in their pH is now well documented. Some of the most convincing long-term evidence comes from work on lake sediments, which analyses the species composition of fossil diatoms (single-cell algae that are sensitive to pH) found in sediment cores. Dramatic increases in acidity over the past 100 years or so have been documented from numerous lakes in North America, northern Britain and Scandinavia.

Linking cause and effect is still by no means straightforward, however, because of the other human impacts that can also affect the pH of soils and water, for example. Long-term diatom data for Swedish lakes indicate the occurrence of a pH increase from about 5.5 to 6.5 dating from the Iron Age (c.2500 BP). This alkalinization was due to extensive burning, agricultural forestry, grazing and other agricultural practices, which increased the base saturation and pH of soils, leading to increased transportation of base cations and nutrients from soils to surface waters. Following this period of alkalinization, the pH of many lakes had fallen to about 4.5 by the beginning of this century and the cessation of many of these practices could account for some of this change (Renberg *et al.*, 1993). Nevertheless, Renberg *et al.* consider acid rain to be more important in producing the recent, unprecedentedly low pHs, and in a similar vein, Patrick *et al.* (1990) looked at the land use and management history around six Scottish lakes over the past 200 years and concluded that acidification can only be explained by acid rain.

These brief examples illustrate some of the varying effects of ecosystem characteristics shown in Table 12.1. Generally, it is interesting to note that in Europe the worst-affected ecosystems are not those where the greatest acid rain deposition has occurred. Scandinavia and northern Britain are on the periphery of maximum deposition shown in Figure 12.4. Many low-alkalinity lakes in the Alps, for example, did not become acidic in the late twentieth century, despite receiving much higher loads of acid rain (Figure 12.6). Buffering capacity is of course one factor that helps to explain this apparent anomaly; another is desert dust transported from the Sahara (Psenner, 1999). This wind-blown material is often rich in calcium and other bases and therefore frequently alkaline, thus increasing an ecosystem's buffering capacity. Saharan dust is commonly deposited over southern and central Europe, but it reaches northern regions much less frequently.

Another facet of the acid rain debate stems from the fact that both sulphur and nitrogen are essential minerals for life, so that additional inputs can, in certain circumstances, be beneficial for organic growth. Moderate inputs of acid rain can stimulate forest growth, for example, in soils where cation nutrients are abundant and sulphur or nitrogen is deficient. A further, arguably positive, aspect of the large quantities of sulphate particles emitted to the atmosphere by human populations is the suggestion that they may act to offset the atmospheric warming associated with the greenhouse effect over some parts of the Earth. This occurs because sulphate particles reflect solar radiation back into space and also act as condensation nuclei,

Table 12.1 Characteristics of ecosystems that affect their risk of acidification

Characteristic	Risk element
Watershed bedrock composition	Rocks such as granite are slow to weather and do not produce neutralizing chemicals. Watersheds that contain these rocks are more vulnerable to acidification
Base nutrient reserves	Higher amounts of elements such as calcium, potassium and magnesium (base nutrients that buffer acids) increase an ecosystem's capacity to neutralize acid inputs
Base nutrient inputs	The input of base nutrients from elsewhere (e.g. wind-blown desert dust rich in calcium) increases an ecosystem's buffering capacity
Soil depth	Shallow soils are often more sensitive to acidification than deep soils
Landscape features	Elevation, edges (e.g. the edge between a forest and field), the presence or absence of vegetation, and the steepness and directional face of a slope all affect acidification
Vegetation type	Different plant species respond in different ways to acid deposition and also contribute differently to ecological processes that regulate acidification
Disturbances	When ecosystems are stressed by disturbances (e.g. insect defoliation, drought, fire) they can become less resilient to acidification
Land-use history	A number of management techniques, such as clear-cutting, can change a forest ecosystem's capacity to neutralize acid inputs

Source: after ESA (2000).

Figure 12.6 Lake Bled, in Slovenia, was not greatly affected by acid rain during the twentieth century, despite being in a part of Europe with a high deposition rate. Inputs of alkaline desert dust blown from the Sahara have further increased its buffering capacity, already high due to surrounding calcareous rocks.

helping clouds to form, making them more reflective and changing their lifetimes. In fact, some scientists have suggested that sulphur be deliberately injected into the stratosphere in an attempt to offset global warming (Crutzen, 2006).

Nevertheless, despite the difficulties and provisos outlined above, there is a large body of evidence that implicates acid rain in numerous deleterious effects upon ecosystems, human health and the built environment.

Aquatic ecosystems

Much of the work on acid rain's impact on aquatic ecosystems has focused on fish, and numerous studies have documented the loss of fish from acidified waters. One survey in Ontario, Canada, identified more than 7000 lakes in an area around Sudbury, a metal mining and smelting centre, that were acidified to pH 6.0, the point at which significant biological damage is expected (Neary *et al.*, 1990). Fish are thought to have been extirpated from Swan Lake, for example, in the 1950s as a result of the lake water acidification. Another study of 1679 lakes in southern Norway showed that brown trout were absent or had only sparse populations in more than half the lakes, and the proportion of lakes with no fish increased with declining pH (Sevaldrud *et al.*, 1980). Thresholds of tolerance to reduced pH can be identified for particular species of fish. A survey of sport-fish populations in Canadian rivers found that Atlantic salmon had been extirpated from seven acidic (pH 4.7) rivers in Nova Scotia that had previously supported the species (Watt *et al.*, 1983). Atlantic salmon were on the decline in other rivers with pH 4.7–5.0, but were stable in rivers with pH 5.0. Different species of fish are affected to varying degrees by reduced pH levels: many cold-water salmonids are generally more sensitive than other species, and among the salmonids, the rainbow trout is the most sensitive.

Fish, like other aspects of the aquatic environment, have been affected both by gradual long-term changes in pH, often referred to as chronic acidification, and more sudden episodic acidification or 'acid shock'. Another factor affecting fish mortality is the increased mobilization of toxic metals by acid deposition. Hydrogen cations introduced into soils by acid rain may exchange with heavy metal cations formerly bound to colloidal particles, thereby releasing the metals into the soil and watercourses. Aluminium is a particular problem and affects fish by obstructing their gills. A gradual loss of fish stocks in many freshwater systems is often more a result of recruitment failure as opposed to outright mortality, as evidenced by higher proportions through time of more mature individuals. The loss of fish from the top of the food chain consequently has knock-on effects at lower trophic levels.

Other aquatic organisms are also affected by acidification. Species higher up the food chain are more likely to be affected indirectly, through changes to their habitats and food sources. The sequence of biological changes caused by acidification monitored in a Canadian lake deliberately dosed with sulphuric acid over an eight-year period is shown in Table 12.2. The crustacean *Mysis relicta* was an early casualty, while several species of fish were seen to decline and eventually disappear, and an increasing growth of filamentous algae was also noted at about pH 5.6.

To sum up, there is no doubt that acid rain has dramatically altered the chemistry of numerous rivers and lakes, particularly in north-western North America and in Europe, chiefly in areas where the buffering capacity of soils is low. These pH changes have had a variety of effects on the ecology of these water bodies. Surface water acidification in East Asia, by contrast, is relatively unusual, despite extremely high deposition rates of sulphur, especially in China and Japan. In part, this is because long-term data on surface water chemistry are limited in the Asian region but the relative lack of evidence for acid rain impacts on aquatic ecosystems is also thought to be due to soils with a high pH or buffering capacity and the generally high alkalinity of inland waters (Duan *et al.*, 2016).

Table 12.2 Sequence of biological changes during experimental acidification of the Ontario Lake 223

Year	pH	Changes observed
1976	6.8	Normal communities for an oligotrophic lake
1977	6.13	Increased chironomid emergence
1978	5.93	Fathead minnow reduced abundance; *Mysis relicta* near to extinction; further increased chironomid emergence
1979	5.64	*Orconectes virilis* recruitment failure; further increased chironomid emergence
1980	5.59	Lake trout recruitment failed; *Orconectes* population decline, parasitism increase, egg attachment failure
1982	5.09	White sucker recruitment failed; pearl dace declined; *Orconectes* near to extinction; chironomid emergence falls back to 1976 levels
1983	5.13	*Orconectes* extinct; mayfly (*Hexagenia sp.*) appears to be extinct

Source: after Howells (1990).

Generally, the impact of acid rain has been assumed to be much reduced in drainage basins covered by soils with a large buffering capacity, often because they tend to be rich in limestone. This assumption has rarely been questioned, but a study of lakes in the US state of New York set out to investigate the acid rain effects in an area with a high buffering capacity and came up with a surprising result (Lajewski *et al.*, 2003). These authors looked at sediments in the Finger Lakes region, an area that received large amounts of acid rain during the twentieth century and that has soils with a high buffering capacity. They found that calcite, a calcium-rich mineral, was detectable in significant quantities in several of the lakes dating from about 1940. Before then, calcite had been virtually absent in the lake sediments for more than 4000 years. It seems that acid rain in the region, perhaps augmented by soil disturbances, increased the weathering of carbonate rocks. This has led to greater flows of base cations into the lakes, thus raising their pH. Unlike regions with low buffering capacity, where acid rain results in the acidification of lakes, acidic deposition across well-buffered terrain may result in the alkalinization of lakes.

Although there are few studies, to date, that investigate acid rain effects on marine ecology, there is some evidence to suggest that increased leaching of trace metals from acidified areas can influence phytoplankton growth in coastal waters (Granéli and Haraldson, 1993). More frequent harmful algal blooms and eutrophication are likely as a result both of this process and of the more direct fertilizing effects of excess atmospheric nitrogen inputs promoting increased phytoplankton growth. The release of atmospheric sulphur and nitrogen by factories and power plants plays a minor role in making the ocean more acidic on a global scale (CO_2 is more significant – see Box 12.1), but the impact is greatly amplified in shallower coastal waters which often have a naturally low pH (Doney *et al.*, 2007).

Terrestrial ecosystems

Acid rain may damage vegetation communities in a number of ways, by the following means:

- increasing soil acidity
- changing nutrient availability

- mobilizing toxic metals
- leaching important soil chemicals
- changing species composition and decomposer micro-organisms in soils.

In China, atmospheric N and S deposition from rapidly expanding industrial areas has probably played a significant role in stimulating soil acidification across northern grasslands (Figure 12.7) where soil pH exhibited a widespread decrease during the 1980s–2000s (Yang *et al.*, 2012). An investigation of major croplands in China reported a similarly significant decline in soil pH over the same period (Guo *et al.*, 2010), but while the latter study acknowledged that acid rain is certainly important regionally in China, it concluded that acidification driven by the overuse of nitrogen fertilizer is at least 10 to 100 times greater than that associated with acid rain. The same essential message of increased acidity on Chinese cropland has been extended to the period 1980–2010 by Zhu *et al.* (2018) who found the highest soil acidification rate occurred in Jiangsu Province due to both high N application rates in chemical fertilizer and high base cation removals by crops and crop residues. A study of soil acidification in a forested catchment in central Japan monitored a rapid decline of soil-surface average pH from 4.5 in 1990 to 3.9 in 2003, an annual rate of decrease that was markedly higher than many previously reported values in Europe and North America (Nakahara *et al.*, 2010). Acidified soil is considered to be one of the most likely causes of the decline of temple and shrine forests in the city of Kyoto (Ito *et al.*, 2011).

Figure 12.7 Keerkante grasslands in north-western China where a decline of soil pH has occurred in recent times. Soil carbonates have buffered the acidification in some areas with implications for the carbon balance in this globally important biome.

Increased nutrient availability is known to reduce species richness in grassland communities, by increasing the productivity of the grassland and the intensity of competition for light, hence favouring a few tall, fast-growing grasses that replace the slower-growing herbs or shrubs. Greater availability of nutrients also changes competition between grasses and heather species in heathlands, and both of these habitats have been affected by sharp declines in plant species diversity in the Netherlands since 1950. Increased soil acidity and higher levels of nitrogen appear largely to blame, both driven largely by ammonia emissions from the country's intensive livestock industry (Roem and Berendse, 2000).

The widespread loss of lichens across many of Europe's industrial areas has been attributed to atmospheric pollution, chiefly by sulphur dioxide, to which nearly all lichens are very sensitive. Particularly severe damage to lichens and mosses has been reported from the most polluted sites, such as in Russia's Kola peninsula, site of two of the world's largest SO_2 and heavy metal emission sources. The moss–lichen layer performs many ecological functions, including biomass production, the regulation of soil temperature and water-holding capacity, and protection of soil from erosive processes. One knock-on effect of the widespread loss of this important layer in Kola is extensive damage to trees and dwarf shrubs caused by frost, as well as extensive soil erosion (Kashulina *et al.*, 1997).

Further east in Arctic Russia, sulphur emissions from the metal-processing facilities at Norilsk have been greater still: in 1988, the amount of SO_2 produced by the Norilsk Metallurgical Combine was equivalent to about two-thirds of emissions from the entire UK. Critical loads for sulphur have been exceeded by a factor of six on the Putorana plateau, site of the world's northernmost open stands of larch trees, causing ecological damage over an area of 400,000 km² (Shahgedanova, 2002). Severe pollution by SO_2 and heavy metals has left the terrain up to 80 km downwind of the smelters completely devoid of plants and the exposed soils have been severely eroded, a landscape for which the term 'anthropogenic desert' has been coined (Figure 12.8). Emissions from the smelter complex at Norilsk have declined in more recent times, along with emissions from other sources of SO_2 in Arctic Russia, and some marginal re-vegetation, primarily by tall shrubs more resilient to the pollutants, has been observed (Nyland *et al.*, 2017). However, Norilsk is still one of the largest point sources of sulphur in the world and comfortably the largest source of SO_2 emissions within the Arctic region.

However, not all cases are so clear-cut. The effect of acid rain on trees and forests has been one aspect of terrestrial ecosystems that has attracted particular interest, and also controversy, since trees are subject to many natural stresses, including diseases, pests and climatic variations such as hard winters and high winds, as well as other forms of pollution (Paoletti *et al.*, 2010). Hence, distinguishing the effects of acid rain from other potential causes of deteriorating tree health is by no means easy, and acid rain pollutants may be just the first link in a chain of degradation as the case of the Kola peninsula described above shows. Indeed, to complicate the issue further, very similar damage to that described for the Kola peninsula is more widespread in neighbouring regions of Finland and Norway, where the cause of moss–lichen layer loss is quite different. Acid rain damage appears to be very limited in Norwegian and Finnish parts of the Barents region, but more widespread loss of the moss–lichen layer is attributed to heavy grazing by reindeer herds.

The symptoms commonly associated with poor tree health – which may be due, at least in part, to acid rain – include thinning of crowns, shedding of leaves and needles, decreased resistance to drought, disease and frost, and direct damage to needles, leaves and bark. A systematic programme monitoring damage to European forests has been carried out since 1986. The programme measures defoliation according to five classes on more than 6000 permanent forest sites across Europe. In the 2014 survey report, about one in every four trees was found to be suffering from abnormal thinning of the crown, having lost more than 25 per cent of their leaves or needles (Sanders *et al.*, 2016). The most severely damaged tree species were

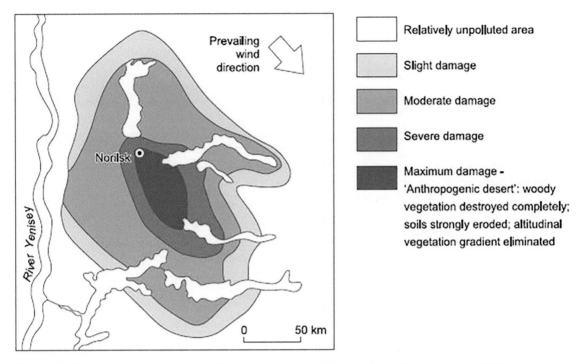

Figure 12.8 Damage to vegetation in the Norilsk region of Siberia (after Shahgedanova, 2002).

deciduous temperate oak species and deciduous tree species in general showed a higher mean defoliation than conifers. Analyses of data series for the period 1992–2014 revealed highly significant increasing trends for defoliation in several species, including ten species of oak and four Mediterranean lowland pine species. Insects were the most frequently occurring directly visible cause of tree damage, followed by abiotic agents such as drought or frost, and fungi. Air pollution was a very minor cause of direct damage, although indirect effects were not assessed and critical loads for acidic deposition were still being exceeded across Europe. Nonetheless, the proportion of plots studied where critical loads were exceeded has been declining, from more than 55 per cent of plots in 1980, and loads for N deposition are expected to be exceeded on 20 per cent of plots by 2020. The survey report concluded that the trend reflects the success of clean air policies (there has been a clear fall in sulphur deposition, though little change in nitrogen) and noted that previous soil acidification is still a burden to forest soils as recovery takes decades.

The complexities of recognizing causes and effects with regard to damaged trees are well illustrated for a number of European forest regions in Table 12.3. Model simulations of vegetation under climatic change indicate further complexities, including the possible large-scale loss of boreal forests to dieback this century. This could occur due to increased water stress, increased peak summer heat stress causing increased mortality, vulnerability to disease and subsequent fire, as well as decreased reproduction rates (Lucht *et al.*, 2006).

Such impacts can have knock-on effects for faunal species that depend upon forests for food and breeding. In the case of birds, some species may face population declines and/or range contractions but others may benefit from the superabundance of dead and decaying standing timber. Some forest-dwelling birds may be affected more directly by the acidification of soils. The reproductive capacity of great tits and other

Table 12.3 Some examples of recent forest decline in Europe and the possible causal sequences

Region	Causes
Central Europe	Dieback of silver fir, a well-documented historical phenomenon that peaked most recently in the 1970s in association with a series of dry summers
Fichtelgebirge (German–Czech border)	Wet deposition of sulphur and nitrogen, SO_2, Ca/Al and Mg/Al ratios in soil affecting tree nutrition via mycorrhizae
Erzgebirge-Silesia (confluence of German–Czech–Polish borders)	SO_2, heavy metals, fluorine, other gaseous pollutants, frost
Inner Bavarian Forest (Germany)	Acid mist, frost shocks, some soil nutrient problems
Athens Basin	O_3 and other oxidant damage to pines
Netherlands	Nitrogen inputs (NH_3 and NO_3 – enhancing sensitivity to other stresses), drought, perhaps O_3

Source: after Freer-Smith (1998).

species has been reduced in European forests as snail populations have declined on soils with falling pH, leading to a calcium deficiency in birds, which, in turn, is manifested in a number of egg-shell defects (Graveland *et al.*, 1994). A similar link between acid deposition and reduced bird abundance has been suggested from research in an acidified forest in central Pennsylvania, USA (Pabian and Brittingham, 2007).

Human health

The effects of acidifying pollutants on human health occur directly through inhalation, and indirectly through such exposure routes as drinking water and food contamination. Most air pollutants directly affect respiratory and cardiovascular systems, and high levels of SO_2 have been associated with increased mortality, morbidity and impaired pulmonary function, although distinguishing between the effects of SO_2 and other pollutants, such as suspended particulate matter, is not always easy. The best-documented effects relate to strong acids in aerosol form (i.e. sulphuric acid, H_2SO_4 and ammonium bisulphate, NH_4HSO_4), and the areas of major concern with respect to inhalation of these aerosols are exacerbation of the effects of asthma and the risk of bronchitis in all exposed persons. Inhalation of highly water-soluble acidic vapours such as SO_2 can also cause breathing difficulties, particularly in asthmatics. Atmospheric concentrations of SO_2 have been shown to have a positive association with daily hospital admissions due to respiratory diseases in the Iranian city of Shiraz, for example, although PM_{10} was similarly related (Gharehchahi *et al.*, 2013).

The release of heavy-metal cations, such as cadmium, nickel, lead, manganese and mercury, from soils into watercourses can eventually reach human populations through contaminated drinking water or fish. Such toxic heavy metals have been associated with a number of health problems including brain damage and bone disorders. Acidified water can also release copper and lead from water-supply systems with similar adverse effects on human health. High levels of lead in drinking water, for example, have been linked to Alzheimer's disease, a common cause of senile dementia.

Materials

Acid deposition has been associated with damage to building materials, works of art and a range of other materials. In all cases, the presence of water is very important, either as a medium for transport and/or in its role in accelerating a vast array of chemical and electrochemical reactions and in creating mechanical stresses through cycles of solution and crystallization.

Acid pollutants can accelerate the corrosion of metals and the erosion of stone. Sulphur compounds are the most deleterious in this respect, although rain made acidic to below a pH of 5.6 by CO_2, SO_2 or NO_x can dissolve limestone. Iron and its alloys, which are widely used in construction, are among the most vulnerable construction materials. The metals are attacked by corrosion, particularly by SO_2. Atmospheric corrosion of iron and steel reached rates of up to 170 mm/year in the most heavily polluted areas during the 1930s, while rates in contemporary Britain for plain carbon steels range from 20 to 100 mm/year (Lloyd and Butlin, 1992). Sulphur corrosion was implicated in the collapse of a steel highway bridge between West Virginia and Ohio in the USA in 1967, a disaster that killed 46 people (Gerhard and Haynie, 1974).

Limestone, marble, and dolomitic and calcareous sandstones are vulnerable to wet and dry acid deposition, mainly of sulphuric and nitric acids, in polluted environments. In the presence of moisture, dry deposition of SO_2 reacts with calcium carbonate to form gypsum, which is soluble and easily washed off the stone surface. Alternatively, layers of gypsum can blister and flake off when subjected to temperature variations. Soluble sulphates of chloride and other salts can also crystallize and expand inside stonework, causing cracking and crumbling of the stone surface. Such processes have caused particularly noticeable damage to statues in many urban and industrial areas and some of the world's most important architectural monuments have been adversely affected, including the Parthenon in Athens and the Taj Mahal in Agra. Limestone monuments outside towns or cities have also been damaged, such as the El Tajin archaeological zone in Veracruz, Mexico (Bravo *et al.*, 2006). A number of research projects have studied the weathering of gravestones by acid deposition, using a micrometer to measure the rate of stone decay around lead lettering that becomes raised above the surface as the stone decays (gravestones come with convenient dates, of course). Mooers *et al.* (2016) present a 120-year record of the spatial and temporal distribution of Carrara marble gravestone decay in the English Midlands, identifying a stark contrast between decay rates in remote rural areas of 0.2 mm/century and a rate of nearly 3.0 mm/century in Birmingham city centre.

Although modern glass is little affected by SO_2, medieval glass, which contains less silica and higher levels of potassium and calcium, can be weakened by hygroscopic sulphates, which remove ions of potassium, calcium and sodium. Paper is affected by pollutants of SO_2 and NO_2 because these are absorbed, making the paper brittle, and fabrics such as cotton and linen respond to SO_2 in a similar way. Damage to materials such as nylon and rubber have also been attributed to SO_2 as well as low-level ozone.

MITIGATING THE EFFECTS OF ACID RAIN

Working on the premise that prevention is better than cure, the most obvious method for mitigating the effects of acid rain is to reduce the emissions of sulphur and nitrogen oxides from polluting sources. There are numerous pathways that can be taken towards achieving this aim. Reducing emissions from power stations can be approached in a number of ways. Energy conservation is the simplest and arguably the most sensible method of reducing emissions from the burning of fossil fuel, and increasing the use of renewable energy sources achieves the same aim (Figure 12.9), although such alternatives also come with

Figure 12.9 Solar-powered water heaters on rooftops in Athens help to reduce acid rain precursors from the burning of fossil fuels in a city where acid rain threatens human health and numerous ancient monuments.

an environmental price (see Chapter 18). Otherwise, there are many existing technologies available that can reduce acid rain pollutant emissions from power stations and other industrial sources. These include:

- fuel desulphurization, which removes sulphur from coal before burning
- fluidized bed technology, which reduces the SO_2 emissions during combustion
- flue gas desulphurization, which involves removing or 'scrubbing' sulphur gases before they are released into the air.

Similar approaches can also be taken to reduce emissions of NO_x from power stations, while catalytic converters and lean-burn engines have been applied to NO_x emissions from motor vehicle engines (see Chapter 16). The end-of-pipe approach has also been implemented to treat ammonia-rich exhaust air from mechanically ventilated animal houses run by the livestock industry, a major contributor to acid deposited on soils in the Netherlands (Melse *et al.*, 2009).

Adoption of such technocentric approaches has frequently been driven by policy. In the case of the Dutch livestock industry, emissions are being reduced in line with increasingly stringent regulations on airborne emissions. A policy instrument that has achieved considerable success in reducing SO_2 emissions in North America is the use of tradable emission permits. Emission trading programmes, also referred to as cap-and-trade policies, establish limits or 'caps' on the allowable levels of total pollution emissions, allocate permits or credits to portions of those caps, and facilitate banking or trading of the permits. The 1990 Clean Air Act Amendments in the USA established the country's Acid Rain Program (ARP), the world's first large-scale cap-and-trade programme for air pollution that began in 1995. The ARP has earned widespread acclaim for its achievements in reducing emissions. With regard to wet sulphate deposition, for example, all areas of the eastern USA have shown significant improvement with an overall 64 per cent reduction in deposition from 1989–91 to 2013–15. Emissions of SO_2 in 2015 were 86 per cent below the 1990 level and annual NO_x emissions in 2015 were 79 per cent below the 1990 level (USEPA, 2016). In 2015, a programme designed to combat air pollution crossing US state boundaries became effective. The Cross-State Air Pollution Rule (CSAPR) requires certain states in the eastern half of the USA to improve air quality by reducing power plant emissions that cross state lines and contribute to pollution in downwind states.

Political aspects of emissions reduction

The history of emissions reduction in Europe and North America provides an interesting insight into how voluntary international environmental protocols have been used to address the issues of long-range transboundary air pollution. The acid rain issue has been the subject of several international environmental agreements, starting with the Convention on Long-Range Transboundary Air Pollution (CLRTAP), which was formally signed by 35 states (33 European countries, the USA and Canada) in 1979. Three follow-up protocols have regulated sulphur emissions:

- the 1985 Helsinki Protocol on the Reduction of Sulphur Emissions or their Transboundary Fluxes by at least 30 per cent
- the 1994 Oslo Protocol on Further Reduction of Sulphur Emissions
- the 1999 Gothenburg Protocol to Abate Acidification, Eutrophication and Ground-Level Ozone.

Targets set under these protocols were progressively more ambitious and specific as the science supporting the policies became stronger. For example, the United Kingdom did not sign the 1985 Helsinki Protocol, being unwilling to take the issue of transboundary transportation of acid rain pollutants seriously enough to take preventative action. While other European countries, particularly those convinced that their ecology was being adversely affected by pollution from their neighbours, were clamouring for concerted action, the UK's response was to call for more research (Dudley, 1986). The UK government's logic was contrary to the precautionary principle but was based on the fact that action to reduce emissions would cost a great deal of money, and therefore they had to be convinced that the money was being spent correctly. Evidence that was convincing enough for some countries was not convincing enough for others. Other countries that refused to join the 30% Club, such as Spain and Poland, also did so primarily for economic reasons. In 1986, however, the UK did concede the need to reduce SO_2 emissions and began to introduce flue gas desulphurization equipment to a number of large British power stations.

The UK did then sign the 1994 Oslo Protocol, which marked a significant milestone in international pollution control since it was the first time that different targets were set for each signatory country. While the Helsinki Protocol mandated uniform emissions cuts of 30 per cent from 1980 levels, the Oslo Protocol mandated country-specific emissions cuts reflecting goals based on the critical loads approach, setting maximum permissible emissions based upon the ability of the environment to withstand pollution. The goals of the Oslo Protocol have been reinforced and extended since 1999 by the Gothenburg Protocol which aims to mitigate a broader range of pollution issues related to acidification, eutrophication and ground-level ozone. Negotiations in Gothenburg were conducted on the basis of scientific assessments of pollution effects and abatement options, including the economic costs of controlling emissions of the various pollutants.

Although some authors have questioned the effect of international environmental protocols on sulphur emissions (Aakvik and Tjøtta, 2011), the success of several industrialized countries in reducing emissions over the period 1980–2010 is clear in Table 12.4. In percentage terms, sulphur dioxide emissions from Canada declined by 70 per cent over the 30-year period; from the UK by 92 per cent. These decreases have been achieved in many cases through a combination of pollution control strategies, energy conservation and fuel switching. Emissions in other industrialized countries, in Asia, for example, have similarly declined, and the effect on atmospheric concentrations in many cities in the more developed countries is illustrated by the trends shown for Tokyo and Seoul in Figure 12.10. However, the trends in some other Asian countries have been in the opposite direction. Data compiled by Smith *et al.* (2010) show that sulphur emissions in China and India roughly tripled over a slightly shorter period (1980–2005). China was easily the world's largest SO_2 emitter in 2007 when its emissions peaked at 36.6 million tonnes (Li *et al.*, 2017), although since then emissions in China have declined substantially while those in India have increased by 50 per cent (see above). China's success in reducing its sulphur emissions is due to stricter emission control measures implemented by a programme to install flue gas desulphurization devices on power plants and the phasing out of small power generation units (Zhao *et al.*, 2009).

Recent trends for nitrogen oxides show more gentle declines in Europe and North America and increased emissions in China and India. Agriculture and motor vehicles are additional major sources of NO_x and larger emissions probably reflect the intensification of food production and increases in national vehicle fleets as well as the boom in energy production from coal. NO_x emissions from power plants are still

Table 12.4 Changes in national emissions of sulphur dioxide and nitrogen oxides in selected countries, 1980–2010

Country	Annual emissions (thousand tonnes)			
	1980	1990	2000	2010
SULPHUR DIOXIDE				
USA	23,500	20,940	14,830	15,170
Canada	4640	3202	2320	1370
Poland	4100	3210	1560	970
UK	4880	3710	1230	410
NITROGEN OXIDES*				
USA	22,120	23,160	21,550	17,690
Canada	1960	2480	2460	1370
UK	2580	2890	1790	1110
Poland	1230	1280	840	870

Note: *Emissions for nitrogen oxides are given as nitrogen dioxide equivalents.
Source: CLRTAP emission database at www.ceip.at/, accessed July 2012.

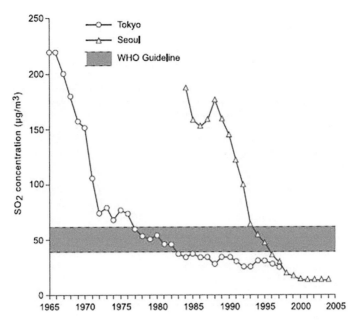

Figure 12.10 Decline of mean annual sulphur dioxide concentrations at Tokyo and Seoul city centres (extended after Dhakal, 2004).

rising in China due to the relatively poor performance of the commonly used abatement techniques such as low-NO_x burners. In India, NO_x emissions from power plants increased by at least 70 per cent during 1996–2010. As in China, this rise was dominated by coal-fired power plants, NO_x emissions from which are not regulated in India (Lu and Streets, 2012).

Environmental recovery

Even given the actual and planned reduction in emissions of sulphur and nitrogen oxides, the response of affected environments is not necessarily known or guaranteed. In the case of building stone weathering, rates appear to respond fairly promptly. Micro-erosion meter measurements have been taken on the balustrade of St Paul's Cathedral in London since 1980, a period over which atmospheric SO_2 concentrations fell from a daily average of 80 ppb (in the early 1980s) to less than 3 ppb by the late 2000s. Thirty years of measurements have shown that erosion rates on the balustrade declined from 0.049 mm/yr to 0.035 mm/yr over the same period. Interestingly, the rate has changed little since the 1990s and rates of erosion now approach those found on natural limestone surfaces (Inkpen et al., 2012; see Figure 10.7).

However, in many instances there is likely to be a timelag between emissions reduction and a detectable reduction in environmental effects. In ecosystems, it is thought that recovery from acidic deposition is probably a complex, two-phase process in which chemical recovery precedes biological recovery. Work in the English Lake District, for instance, has shown improvements in freshwater ecology following two decades of declining acid emissions. Tipping et al. (2002) found acid-sensitive stoneflies that had been absent in surveys conducted during the 1960s and 1970s had returned by 1999. Diatoms found in lake sediments had also responded to the increasing pH in some of the lakes surveyed, though not in all. A similarly complex story can be told in Ontario, Canada, where the combined output of SO$_2$ and toxic metal emissions from the Sudbury metal smelters has fallen by about 95 per cent since the 1960s and the concentrations of copper and nickel in nearby lakes have declined in consequence, while pH has improved considerably (Table 12.5). Biological recovery in these acidified lakes has lagged behind chemical recovery. Using algae as indicators, Swan Lake started to show signs of biological recovery in 1984 although progress was slowed by a drought-induced re-acidification event in 1988. Recovery at Tilton Lake has been much slower (Tropea *et al.*, 2010). Since 1990, Daisy Lake has been invaded by a wide variety of fish species from a downstream lake, which in turn has allowed fish-feeding birds to resume reproduction, but recovery has been much slower on the lake's shoreline due to continued acid- and metal-contaminated drainage water from its catchment areas (Gunn *et al.*, 2016).

Generally the time for biological recovery is better defined for aquatic than terrestrial ecosystems. Evidence from the Hubbard Brook Experimental Forest, a long-term research site in the White Mountains of New Hampshire in the north-eastern USA, suggests that for acid-impacted aquatic ecosystems, stream macroinvertebrate and lake zooplankton populations would recover in three to ten years after favourable chemical conditions are re-established. Fish populations would follow within a decade following the return of the zooplankton on which they depend. However, a distinction should be made between waters that are chronically acidic and those subject to episodic acidification. In upland regions of the north-eastern USA, the former is probably less widespread than the latter (Lawrence, 2002). One study concluded that 10 per cent of lakes in the Adirondack region of New York were chronically acidified but that an additional 31 per cent of lakes in the area continue to be susceptible to acid shock.

Such susceptibility to episodic acidification is related to the state of surrounding terrestrial ecosystems, particularly the decreased neutralization capacity of soils. Soils heavily leached by acid rain, in some cases losing as much as half their calcium and magnesium contents, may take many decades to recover (Kaiser, 1996). Trees probably respond positively to more favourable atmospheric and soil conditions over a period of decades (Driscoll *et al.*, 2001). There is widespread evidence that lichens have returned relatively quickly to areas where they had disappeared due to historically high SO$_2$ concentrations, as shown by Ahn *et al.* (2011) in the urban area of Seoul and by Pescott *et al.* (2015) in many parts of the UK, where reductions in air pollution are also associated with increases in species of moths that feed on lichen.

However, in the case of acidified lakes, some researchers believe that it might take more than 100 years before fresh waters regain something close to their pre-acidification species composition with maintained functions. This view emphasizes the continued need for complementary local measures such as appropriately managed land use and the continued implementation of liming programmes designed to increase

Table 12.5 Average pH in three lakes within 20 km of Sudbury, Ontario

Lake	1970s	1980	1985	1990	1995	2000	2005	2010	2015
Daisy	4.1 (1969)	–	4.5	4.9	5.3	6.2	6.4	6.6	6.8
Swan	3.9	4.7	5.2	5.1	5.6	5.5	5.8	–	–
Tilton	–	5.0	5.6	5.8	6.2	6.7	6.7	–	–

Source: Keller *et al.* (2004); Tropea *et al.* (2010); Gunn *et al.* (2016).

water pH. Liming is carried out in several European countries and in North America. Large-scale freshwater liming programmes were initiated for lakes and rivers in Norway and Sweden in the early 1980s with a primary initial goal of restoring waters identified as being of special value for fishing. These programmes continue today and have been chemically successful, reducing acidity in streams and lakes. Positive effects on fish include increased diversity and in some cases increased abundance or improved reproductive success. Sometimes, it has been necessary to re-stock water bodies to replace lost populations, as in the Otra River in southern Norway, where salmon catches resumed in the 1990s after three decades (Figure 12.11). Biological improvements have not been reported everywhere, however, with a number of studies in Europe and North America reporting mixed results on the biological responses to liming (Clair and Hindar, 2005). Such programmes are probably best viewed as offering partial remediation of acidification impacts, rather than complete long-term recovery of ecosystem structure and function.

Recovery of lacustrine ecosystems is also likely to be slowed further by the combined effects of acid rain and other major environmental issues. The Otra River, for instance, is also impacted by hydroelectric dams and increasingly by climate change, with up to 20 per cent more precipitation predicted for this region of Norway by the year 2100 (Wright *et al.*, 2017). In their study of lakes in the Sudbury area, Tropea *et al.* (2010) concluded that biological recovery was being impeded by multiple stressors such as residual metal contamination, climate warming and an environmental legacy associated with the loss of soils. Another study of lakes in Ontario related increased rates of death and disease among fish and aquatic plants to both global warming and the depletion of the ozone layer (Schindler *et al.*, 1996). The lakes' levels of dissolved organic carbon were found to have fallen due to both climate warming and acidification. Since carbon absorbs ultraviolet radiation, which has increased due to stratospheric ozone depletion, ultraviolet radiation has penetrated lake waters to much greater depths, adversely affecting the health of aquatic organisms. No fewer than 140,000 lakes in North America are estimated to have carbon levels low enough to be at risk from such deep ultraviolet penetration. The deleterious effects can also be compounded by another aspect of climate, when sulphur compounds stored in lake sediments are oxidized in response to receding lake levels during drought periods (Yan *et al.*, 1996).

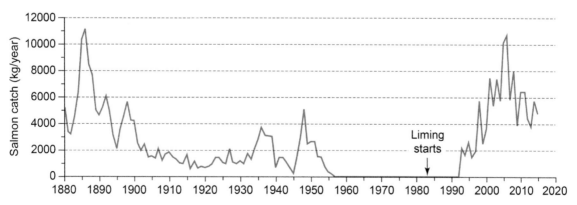

Figure 12.11 Salmon catch in the Otra River, southern Norway, 1880–2015 (after Wright *et al.*, 2017, with data from Statistics Norway).

FURTHER READING

Carn, S.A., Fioletov, V.E., McLinden, C.A., Li, C. and Krotkov, N.A. 2017 A decade of global volcanic SO_2 emissions measured from space. *Scientific Reports* 7. Monitoring of emissions that are difficult to assess.

Duan, L., Yu, Q., Zhang, Q., Wang, Z., Pan, Y., Larssen, T., Tang, J. and Mulder, J. 2016 Acid deposition in Asia: emissions, deposition, and ecosystem effects. *Atmospheric Environment* 146: 55–69. The state of acidification in Asia. Open access: www.sciencedirect.com/science/article/pii/S1352231016305374.

Hédl, R., Petrík, P. and Boublík, K. 2011 Long-term patterns in soil acidification due to pollution in forests of the Eastern Sudetes Mountains. *Environmental Pollution* 159: 2586–93. A study that stretches from before to after the period of major industrial pollution in the Czech Republic.

Lawrence, G.B., Burns, D.A., and Riva-Murray, K. 2016 A new look at liming as an approach to accelerate recovery from acidic deposition effects. *Science of the Total Environment* 562: 35–46. A reappraisal of liming as a management option with an emphasis on North America.

Rice, K.C. and Herman, J.S. 2012 Acidification of Earth: an assessment across mechanisms and scales. *Applied Geochemistry* 27: 1–14. A comprehensive review of relevant human activities.

Sullivan, T.J. and 10 others 2018 Air pollution success stories in the United States: the value of long-term observations. *Environmental Science and Policy* 84: 69–73. This paper has a major focus on acidification. Open access: www.sciencedirect.com/science/article/pii/S1462901117312352.

WEBSITES

hubbardbrook.org   the Hubbard Brook Ecosystem Study carries out long-term research into acid rain in the USA and has a wide range of educational resources.

ocean-acidification.net   a huge source of information on ocean acidification for scientists, policy-makers and the public.

www.acap.asia   the Asia Center for Air Pollution Research focuses on acid deposition in East Asia.

www.emep.int   the site of the programme for monitoring and evaluation of the long-range transmission of air pollutants in Europe gives access to numerous publications and data.

www.epa.gov/airmarkets   the US Environmental Protection Agency's site contains reports, data and maps of deposition in the USA.

www.ngu.no/Kola   the Kola Ecogeochemistry Project site has details on this historical survey in the Barents region, which includes two of the world's largest SO2 emission sources.

POINTS FOR DISCUSSION

- Since acidification occurs naturally in the physical environment why are we so concerned about human-induced acidification?
- How has the concept of thresholds in the physical environment been used to help resolve the acid rain issue?
- Why are some of the worst-affected ecosystems not those with greatest acid deposition?
- Examine the reasons for the timelag between reductions in acid rain emissions and biological recovery.
- Assess the links between acidification and global warming.
- Why is it so difficult to attribute forest decline directly to any one cause?

13

Food production

LEARNING OUTCOMES

At the end of this chapter you should:

- Understand how the beginnings of agriculture were linked to the process of domestication.
- Appreciate the global implications of the Green Revolution and the Livestock Revolution.
- Understand some of the environmental implications of intensive farming.
- Appreciate the controversies surrounding the use of modern biotechnology in food production.
- Understand the range of factors affecting food security.
- Know that waste and over-eating make large contributions to food system losses worldwide.
- Understand some of the environmental impacts associated with aquaculture.

KEY CONCEPTS

domestication, Green Revolution, Livestock Revolution, organochlorine pesticides, integrated pest management, biotechnology, GM food, terminator technology, agri-environment schemes, food security, organic farming, Blue Revolution

---

The origins of agriculture, which includes both crop and livestock production, are traced back 10,000 years to the start of the Holocene. Before that, the genus *Homo* had lived for more than 2 million years by gathering plants and hunting wild animals for food, and *Homo sapiens* continued this practice for 100,000 years. For the past 5000 years, virtually the entire world's population has been reliant on farmers and herders who consciously manipulate natural plants and animals to provide food. As the global population has grown, the area used for food production has expanded and production techniques have become increasingly sophisticated. Both the extensification and the intensification of food production have thrown up numerous environmental issues. Many of these are covered in this chapter, but other environmental impacts of agriculture are dealt with elsewhere. These include issues associated with the clearance of forests for farming (Chapter 4), the environmental impacts of heavy grazing in drylands (Chapter 5), soil erosion on agricultural land (Chapter 14) and some of the impacts associated with irrigation (Chapters 5, 8 and 9). Most of the world's food supply is produced on land, but fish farming has grown in importance in recent years and is detailed in this chapter. Nevertheless, the majority of fish are still hunted, and environmental issues surrounding this form of food production are covered in Chapter 6.

AGRICULTURAL CHANGE

A few societies still continue to hunt and gather their food with little attempt at agricultural management (Figure 13.1) but these peoples do so at low population densities. Agricultural management enables more calories of food to be obtained from a given area of land over a more predictable period of time than the simple collecting of wild foods. Whether agriculture was invented in response to rising human populations or its development drove human population growth is a matter for debate (Gignoux *et al.*, 2011). However, it is likely that social, economic, technological and environmental factors all contributed in varying

Figure 13.1 A Biaka woman out foraging for fruits, roots and leaves in the Congo forest of central Africa. The Biaka, or pygmies, are one of the few societies that continue to rely on hunting and gathering for their food.

combinations to the early transition from hunters and gatherers to herders and farmers, and this 'Neolithic Revolution' was associated with the emergence of urban civilizations on some of the world's great rivers (the Nile, Tigris–Euphrates, Indus and Huang Ho) between 3500 and 5500 years ago, and in MesoAmerica and the Central Andes around 3000 years ago.

Part and parcel of the start of agriculture was the process of domestication, the deliberate selection and breeding of wild plants and animals that led to morphological or genetic change. By cultivating plants, for example, farmers chose particular seeds that offered the best results and by planting more of them increased their chances of survival. The dominance of particular characteristics was thus deliberately increased. In crops valued for their edible seed, such as wheat, a shift towards non-shattering occurred quickly, and hence the natural mechanism for seed dispersal was lost and the plant became dependent on humans for survival. The origins of domestication, which can be seen as one of the most important human impacts on the natural environment, have been traced to diverse parts of the world using archaeological evidence (Figure 13.2). The process of selection and breeding continues today in a more sophisticated manner using such techniques as tissue culture and genetic engineering (see below).

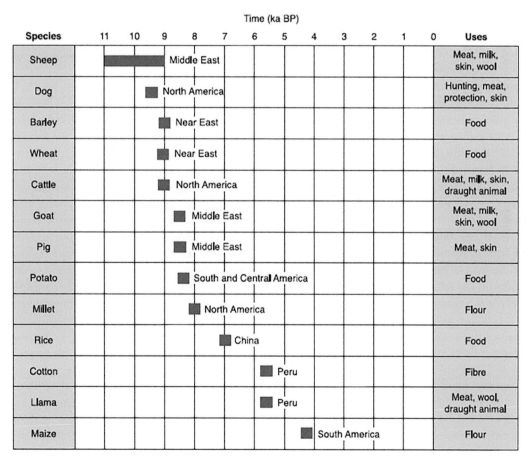

Figure 13.2 Origins of domesticated plants and animals (after Williams *et al.*, 1993).

Irrespective of the original causes of the emergence of agriculture, there is no doubt that the ability to provide more food from a unit area of land per unit time has been associated with the exponential growth in the world's human population. A global population thought to have been around 5 million 10,000 years ago increased by a factor of 20 to 100 million in 5000 years (May, 1978). The 500 million mark was passed about 300 years ago and by 1804, 1000 million (1 billion) people inhabited the planet. By this time, the exponential rate of population growth was clear and world population had surpassed 4 billion by 1974, 6 billion by 1999 and is projected to reach 8 billion by 2023 (see Table 2.2).

Until the early 1980s, world agricultural production had kept pace with population growth chiefly through expansion of the cultivated area. Hence, as population grew exponentially, cropland area expanded dramatically. Indeed, more land was converted to cropland in the 30 years after 1950 than in the 150 years between 1700 and 1850 (MEA, 2005). The expansion of cropland globally stopped in the 1980s but this period of stability ended in the early 2000s and since 2002 crop production area has increased at nearly 10 million hectares per year and 60 per cent of this expansion is due to increased production of rice, wheat and maize, the major cereals (Grassini *et al.*, 2013).

Today, after 10,000 years of agriculture, cultivated systems – areas where at least 30 per cent of the landscape is in croplands, shifting cultivation, confined livestock production or freshwater aquaculture – now cover one-quarter of the Earth's land surface. By clearing natural vegetation and replacing it usually with fewer domesticated species, or by grazing livestock, agriculturalists have been responsible for some of the most widespread changes wrought by humans on the natural environment during the Anthropocene. Forests and grasslands of the temperate latitudes have been the most severely altered, but all the world's major biomes, with the exception of the polar ice caps, have been affected (Figure 13.3).

Continued increases in food production since the 1950s – to keep up with population growth, which has continued at unprecedented rates – have been achieved chiefly by increasing crop yields. The average cereal yield in Britain, for example, has risen steadily from about 2 tonnes per hectare in the mid-1930s to more than 7 tonnes per hectare in the 2000s, when yield increases stalled (Robinson and Sutherland, 2002). In fact, the rising trend in yields has a long history in many parts of the world and has been the result of numerous improvements in agricultural efficiency. The tonnage of wheat harvested from an average hectare of English cropland in the 2010s was a third more than in the 1980s, four times that in the 1930s, seven times that in 1800 and 13 times that of 1300. The increases were achieved first by the replacement of open-field farming by mixed farming, and then by the replacement of mixed farming by increasingly mechanized farming with an increasing use of chemical pesticides, including insecticides, herbicides, fungicides, and bactericides (Grigg, 1992). One of the outcomes of using more machinery on arable land has been the removal of hedgerows to enlarge field sizes, which in England is akin to a reversion to the large open fields of the medieval period as illustrated in Table 13.1. The average field size in arable Cambridgeshire has more than doubled since 1945 (Robinson and Sutherland, 2002), but modern cultural attachment to hedgerows with their flora and fauna, which became established only in the eighteenth century, caused strong reactions and the biodiversity value of maintaining hedgerows has also been realized in more recent times (Staley *et al.*, 2013). In consequence, contemporary agricultural policy in Britain increasingly recognizes the importance of hedgerows for sustainable agriculture and the conservation of rural biodiversity.

Worldwide, it is the adoption of modern chemical farming techniques, both in the developed countries and in parts of the developing world during the so-called Green Revolution of the 1960s and 1970s, which enabled the continued increase in yields. Indeed, Hertel *et al.* (2014) demonstrated that the substantial

Figure 13.3 A landscape transformed by agriculture on the Portuguese North Atlantic island of Faial in the Azores.

Table 13.1 Changes in the length of hedgerows in three parishes in Huntingdonshire, England, since the fourteenth century

Date	Length of hedges (km)	Date	Length of hedges (km)
Pre-1364	32	1780	93
1364	40	1850	122
1500	45	1946	114
1550	52	1963	74
1680	74	1965	32

Source: after Moore *et al.* (1967).

Table 13.2 Energy efficiency of some agricultural production systems

Agricultural system	Energy input (kcal/ha/year)	Energy production (kcal/ha/year)	Energy efficiency (production/input)
Hunter–gatherer	2685	10,500	3.9
Pastoralism (Africa)	5150	49,500	9.6
Peasant farming (Mexico)	675,700	6,843,000	10.1
Estate crop production (Mexico)	979,400	3,331,230	3.4
Estate crop production (India)	2,837,760	2,709,300	0.9
Maize (USA)	1,173,204	3,306,744	2.8
Spinach (USA)	12,800,000	2,900,000	0.2

Source: after Pimental (1984).

increase in crop yields associated with the Green Revolution led to more than a 200 per cent increase in crop production with just an 11 per cent expansion in the global cropland area over the period between 1961 and 2006. Large fields and mechanization have gone hand in hand with several other key components of these techniques:

■ application of increasing amounts of synthetic fertilizers to supply plant nutrients (nitrogen, phosphorus and potassium)
■ expansion and improvement of irrigation systems
■ protection of crops from diseases, pests and weeds with synthetic chemicals
■ development of new varieties of crop plants that are more responsive to fertilizers, more resistant to pests and with a higher proportion of edible material.

These modern farming techniques require greater inputs of energy than more traditional approaches. As Table 13.2 shows, a highly intensive spinach farm in the USA not only needs nearly 19 times more energy per hectare than a Mexican peasant to produce its crop, but also does so 50 times less efficiently.

LIVESTOCK INDUSTRY

The environmental impacts that have come with modern intensive farming will form the focus of much of this chapter, but parallel intensification has occurred over the past 50 years or so in the livestock industry, particularly in the industrialized world where livestock products provide 30 per cent of the calorie intake, in contrast to the less-developed countries where they provide less than 10 per cent (Grigg, 1993). Factory-style farms are commonplace as pigs and poultry have been moved from the farmyard to indoor feeding facilities (Figure 13.4), and cows are fattened with special feeds. Abundant feed grain, drugs to prevent

Figure 13.4 This organic accredited poultry farm in Elstorf, Germany, is typical of the factory-style indoor feeding that is common in many parts of the world.

disease, and improved breeds have enabled animals to be raised faster and leaner as meat production for consumers rich enough to eat it regularly has become the focus of livestock production, taking over from the multiple benefits of manure, draught power, milk and eggs. These large-scale, intensive operations, in which animals are raised in confinement, already account for three-quarters of the world's poultry supply, 40 per cent of its pork, and more than two-thirds of all eggs (Bruinsma, 2003). Production and consumption of meat, milk and eggs continue to increase particularly quickly in much of the developing world, driven by growth in population, income and urbanization, prompting talk of a 'Livestock Revolution' (Delgado, 2003).

The intensification of modern livestock farming has raised a number of environmental issues, not least of which are the concerns over animal welfare, which have engendered a minor shift back towards 'free range' animal products in some western countries. An increasingly serious issue is that of pollution from farm wastes: principally water pollution by nitrates and through eutrophication; air pollution, particularly ammonia and greenhouse gas emissions; and soil pollution arising from nutrient accumulation (Martinez *et al.*, 2009). While the traditional mixed farm operated as a closed system, with manure being spread on cropland to provide valuable nutrients, specialization in the livestock industry has meant a trend towards concentration in areas distant from centres of arable production. Leakage of manure, with a high biochemical oxygen demand, from storage tanks has become an increasingly serious point-source pollutant, depleting the dissolved oxygen in groundwater, wells and rivers (see Chapter 8). Nitrogen in manure is also lost to the atmosphere as ammonia, contributing to acid deposition (see Chapter 12).

In contrast to intensified livestock farming, extensive grazing of livestock is still practised in numerous parts of the world, many of them in poorer and drier areas, and globally livestock uses about 80 per cent of all agricultural land combined (Herrero *et al.*, 2015). Desert-marginal rangelands support about 50 per cent of the world's livestock (Allen-Diaz *et al.*, 1996) and pastoralism remains the only viable livelihood for many rural dryland populations. Nonetheless, in some circumstances the pressure of grazing too many livestock has raised the issue of overgrazing which has figured largely in the desertification debate (see Chapter 5).

CROPLAND

Fertilizers

Manure and other materials have long been used as natural fertilizers, and are still applied to cropland in many parts of the world, but the rise of modern farming techniques has been underpinned by a rapid increase in the use of manufactured chemical, or synthetic, fertilizers to supply plant nutrients. Specially manufactured chemical fertilizers date back to the mid-nineteenth century when a factory producing superphosphates by dissolving bones in sulphuric acid was opened in the English county of Kent, but it was the advent of the Haber-Bosch process (the industrial fixation of N_2 into ammonia, NH_3) in 1909 that really boosted the production of nitrogenous fertilizers. Their use has become globalized in the last 50 years or so. World production of synthetic nitrogenous fertilizers in 1920 was 150,000 tonnes nitrogen content and had risen to 3.7 million tonnes in 1950. By 1975, production had increased more than ten times to 42 million tonnes, and it continued to rise to 85 million tonnes in 2000 when manufacture absorbed about 1 per cent of world energy consumption (Dawson and Hilton, 2011). It is now estimated that about half of the global human population is dependent on the increased yields of agricultural crops thanks to the use of nitrogenous fertilizers (Fowler *et al.*, 2013). Similarly rapid increases have occurred in the quantities manufactured of the other two main fertilizers: phosphates and potash. This rise in supply has been a key component of the increase in average world cereal grain yields from 1.13 t/ha in 1948–52 to 3.1 t/ha in 1999, and the maintenance of cereal production per head at the present level to 2030 will require a further doubling of chemical nitrogen consumption (Gilland, 1993, 2002).

The recent increase in synthetic fertilizer use in the large majority of countries, in combination with several other human activities such as fossil fuel burning and the modification of natural ecosystems through land clearance and wetland drainage, has led to unprecedented effects on the nitrogen cycle. Human activities now contribute about 210 million tonnes to the global supply of biologically available or 'fixed' nitrogen, a similar amount to that made available through natural processes, although there is still considerable uncertainty in some of these estimates (Table 13.3). The rising trend in the use of nitrogenous and other fertilizers has fuelled concerns over the environmental effects of nutrients that are not taken up by crops. In practice, fertilizers are lost from a field because of excessive or poorly timed applications. For wheat, rice and maize – which together account for about 50 per cent of current nitrogenous fertilizer use globally – the nitrogen use efficiency is typically less than 40 per cent. In other words, most applied fertilizer is either washed out of the root zone or is lost to the atmosphere by denitrification before it is assimilated into the crop biomass (Canfield *et al.*, 2010). These lost nutrients cause a cascade of environmental impacts and human health problems:

Table 13.3 Estimates of global sources of fixed nitrogen

Source	Release of fixed nitrogen (million tonnes/year)	Uncertainty of estimate
ANTHROPOGENIC		
Fertilizer (Haber-Bosch process)	120	+/- 10%
Agriculture (nitrogen-fixing crops and grazing)	60	+/- 30%
Combustion (fossil fuels and biomass burning)	30	+/- 10%
Total: human sources	210	
NATURAL		
Marine ecosystems	140	+/- 50%
Terrestrial ecosystems	58	+/- 50%
Lightning	5	+/- 50%
Total: natural sources	203	

Source: after Fowler *et al.* (2013).

- Excess nitrate and phosphate are lost by runoff and leaching, and enter rivers, lakes, groundwater and coastal waters where they can cause eutrophication and, in the case of nitrate, may be associated with 'blue baby syndrome', reproductive risks and stomach cancer.
- Excess nitrate can be converted by denitrification into nitrous oxide, which is released into the atmosphere (nitrous oxide is a greenhouse gas and also contributes to stratospheric ozone depletion).
- Excess nitrogen can be converted by volatilization into ammonia, which is released into the atmosphere where it can contribute to the effects of acid rain (see Chapter 12).
- Excess nitrogen has been associated with declines in forest health, changes in species composition in many ecosystems and losses of biodiversity in temperate grasslands.
- Excess nitrogen and phosphorus can indirectly affect the abundance of infectious and non-infectious pathogens, sometimes leading to disease epidemics. Important infectious diseases that may be affected include malaria, West Nile virus, cholera and schistosomiasis.

Even the expected result of fertilizer use – a rise in crop yields – can have unexpected environmental repercussions. A loss of coastal vegetation and adverse changes in soil properties in parts of the Canadian Arctic in recent decades have been identified as an off-site impact of the increased use of nitrogen fertilizers in the USA. The link is the lesser snow goose, which breeds in the eastern and central Canadian Arctic and sub-Arctic, and winters in the southern USA and northern Mexico. Goose numbers increased by 57 per cent annually from the late 1960s to the mid-1990s, largely because of increased survival in response to a US agricultural food subsidy. The rise in numbers complements the increased use of fertilizers and a corresponding rise in yields of rice, corn and wheat along the flightpaths and on the wintering grounds (Abraham *et al.*, 2005). The vegetation loss and soil changes have occurred due to many more geese foraging in their sub-Arctic migration areas and Arctic breeding colonies. The authors of this study suggest that the introduction of liberal hunting regulations may reduce the snow goose population size in the near term, but the revegetation of Canadian coastal ecosystems will take decades to achieve.

Irrigation

Since the beginning of crop cultivation, irrigation has been used to improve yields by compensating for low precipitation. Irrigation probably first began in the eighth millennium BP when two river-basin civilizations, on the Tigris–Euphrates of Lower Mesopotamia and on the Nile in Egypt, first began to sustain settled agriculture. In modern times, irrigation is often one of the key tools used to intensify agriculture, and the global area of irrigated cropland increased more than fivefold in the twentieth century (Figure 13.5); it stood at more than 300 million hectares in 2015. Its importance is summed up by Neumann *et al.* (2011) who point out that 24 per cent of all harvested cropland is under irrigation and more than 40 per cent of the global cereal production takes place on irrigated land. In many dryland countries, the national area under cultivation would be considerably smaller without irrigation, and in the extreme case of Egypt it would be almost zero (Figure 13.6).

Unfortunately, however, poorly managed irrigation systems can lead to a wide range of environmental problems. Excessive applications of surface water can result in rising water tables, which cause salinization and waterlogging. These processes reduce crop yields on irrigation schemes all over the world (see Table 5.5), and can accelerate the weathering of buildings, as Akiner *et al.* (1992) have documented for numerous examples of ancient Islamic architecture in Uzbekistan. Salinity problems can also result from the overpumping of groundwater in coastal locations, with freshwater aquifers becoming contaminated by

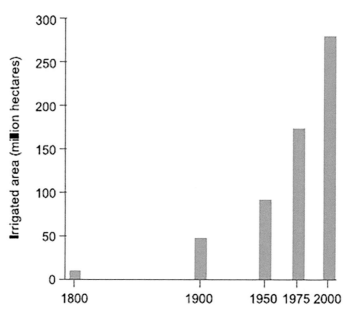

Figure 13.5 Rise in the global area of irrigated cropland, 1800–2000 (from data in FAO, 1993; Gleick, 1993; FAOstat, 2000).

Figure 13.6 Cropland irrigated with water from the River Nile in Egypt. Irrigation first began on the Nile 6000 years ago, and agriculture in modern Egypt is no less dependent upon the river.

sea-water intrusion, as Zekri (2008) describes in the Batinah coastal area of Oman, site of about half of all the country's cropland.

Overpumping of groundwater resources can rapidly deplete supplies, which may threaten the long-term viability of the irrigation schemes themselves. Many examples involving centre-pivot irrigation technology can be quoted, such as the depletion of the Ogallala aquifer beneath the Great Plains in south-west USA (see p. 109) and the rapid drop in groundwater levels in Saudi Arabia that occurred with the brisk expansion of irrigated agriculture in the late twentieth century (Figure 13.7). In the Wadi As-Sirhan area in the north of the country, Othman *et al.* (2018) demonstrate that overpumping has resulted in land subsidence and the development of sink holes and fissures, as well as an increase in the frequency of shallow earthquakes.

Some of the largest land-subsidence rates recorded are in the San Joaquin Valley, California, USA, one of the most intensively farmed agricultural areas in the world. Total subsidence reached 9 m in some areas of the valley by 1970 after nearly half a century of groundwater extraction. Rates of land subsidence tended to decline in the later decades of the twentieth century following the importation of surface water to the valley in the late 1960s. Hence, the average rate of surface lowering at benchmark S661 in the Los Banos-Kettleman City area of the valley slowed to 46 mm/yr over the period 1971–95, having averaged 325 mm/yr between

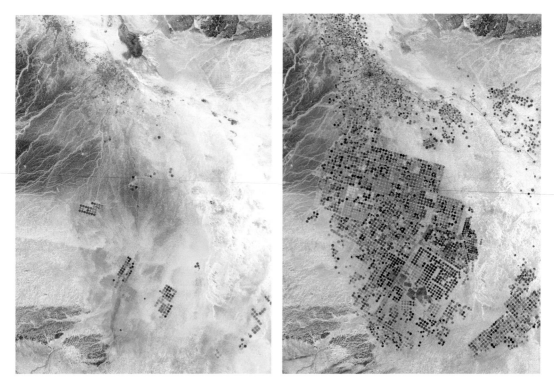

Figure 13.7 The dramatic increase in use of centre-pivot systems to irrigate crops with groundwater in the Wadi As-Sirhan Basin in Saudi Arabia can be seen by comparing these two satellite images taken in 1991 and 2012.

1943 and 1971. The total cumulative land subsidence at this point over the entire 52-year period 1943–95 was 10.2 m (Basagaoglu *et al.*, 1999). About half of the San Joaquin Valley's 1.5 million ha of land under irrigation has experienced subsidence (Zektser, 2000).

Importing water to the San Joaquin Valley eased the subsidence problem to some extent but may have simply transferred the water-related issues elsewhere. Experiences of managing groundwater overpumping in Mexico and India demonstrate that overpumping can be reduced by better pricing of electric power. Raising the price of electricity can increase the cost of irrigation water, encouraging farmers to use groundwater more efficiently (Scott and Shah, 2004).

An increased incidence of water-associated diseases is another problem often associated with the introduction or extension of irrigation schemes. Sharma (1996) documented how in India during the 1990s irrigation projects improved breeding sites for the dominant malaria-carrying mosquito, leading to the rapid spread of what became known as 'irrigation malaria' among a population of some 200 million people (see also p. 204). The effect has been documented in irrigation schemes all over the world, particularly in dryland areas where standing water is less common prior to the development of irrigation. A study by Bett *et al.* (2017) in eastern Kenya concluded that the introduction of irrigation increases the risk of several mosquito-borne infections besides malaria (Rift Valley fever, West Nile virus and Dengue fever), although irrigation also provides a protective effect against some other diseases that thrive in areas with high live-stock population densities.

Other detrimental health effects can stem from the excessive use of pesticides on irrigation schemes, as exemplified by Central Asian cotton plantations (see below), an example that also illustrates the scale of potential off-site environmental effects on the Aral Sea (see p. 179). The Aral Sea disaster is an extreme example of how agricultural expansion and intensification have altered the quantity and quality of water flows resulting in a catastrophic ecosystem regime shift. Gordon *et al.* (2008) emphasize the urgent need to develop new ways of anticipating and managing such non-linear environmental changes as global demands for agriculture and water continue to grow.

Pesticides

Losing a portion of a field's crop to pests has been a problem for farmers since people first started cultivating soils. Worldwide, an estimated 67,000 different pest species attack agricultural crops and about 35 per cent of crop production is lost to them each year. Insects cause an estimated 13 per cent of the loss, plant pathogens 12 per cent, and weeds 10 per cent (Pimental, 1991). Hence, pests represent a serious problem for farmers and for human society in general.

The use of synthetic chemical sprays to control agricultural pests and diseases dates back to the 1860s when they were used to control Colorado beetle, which damaged potato crops in the USA, and vine mildew in France. Pesticides (which include insecticides, herbicides, fungicides and bactericides) entered general use in many of the world's agricultural areas along with synthetic fertilizers in the 1950s, following the discovery of DDT in Switzerland in 1939 and 2,4-D in the USA during the Second World War. This development has been a response to an age-old problem but also an attempt to limit the escalation of the pest problems that came about as a result of changing cultivation practices, such as shortened fallow periods, narrow rotation and the replacement of mixed cropping by large-scale monocultures of genetically uniform varieties. Indeed, many scientists have long suspected that the spread of industrial agriculture and the concomitant loss of natural habitat areas have resulted in greater insect pest pressure on crops, which in turn has led to an increased use of insecticides in a process of positive feedback. The causal chain of events suggests that conversion of diverse natural plant assemblages to monocultures reduces the abundance and diversity of natural enemies of crop pests, hence making the use of synthetic chemical pesticides more likely. Further, increases in the size, density and connectivity of monocultural crop patches facilitate the movement and establishment of crop pests, leading to higher pest pressure, again increasing the likelihood of pesticide use. Evidence to support this theory has come from a range of cropping systems in the US Midwest where Meehan *et al.* (2011) found that the proportion of harvested cropland treated with insecticides increased with the proportion of cropland, and decreased with the proportion of semi-natural habitat, in a study of more than 500 counties in seven states.

Worldwide, annual sales of pesticides grew from around US$7 billion in the early 1970s to nearly US$40 billion in 2007 (Tolba and El-Kholy, 1992; Grube *et al.*, 2011). Most are used in the industrialized countries, but while consumption in these countries has levelled off in recent years, sales to developing countries continue to grow. Most pesticides are used in agriculture, with about 10 per cent used in public health campaigns such as the fight against mosquitoes, which transmit malaria.

One unfortunate aspect of the increasing use of pesticides is the development of resistance to the chemicals. While most individuals of a pest population die on exposure to a pesticide, a few individuals survive by virtue of their genetic makeup and pass on this resistance to future generations. Increasing resistance is often countered by stronger doses of chemicals or more frequent applications, at increasing cost to the farmer, which also speeds up the trend towards resistance, so encouraging the development of new chemicals. Elimination of one pest can also result in the rapid growth of secondary pest populations.

Pesticides are designed to kill living things, whether they be insects, weeds or plant diseases, but any amount of chemical that misses its intended target can affect numerous other aspects of the environment. In practice, the proportion may be large (Pimentel, 2005) and is greatest when pesticides are applied by aircraft (50 to 75 per cent not reaching the target; Figure 13.8), but is still considerable with ground application equipment (10 to 35 per cent). The remainder becomes a contaminant. The resultant effects on non-target plants, animals, soil and water organisms, and the pollution of land, air and water, which can also affect human populations, have caused considerable concern.

Natural predators of the problem pest may be destroyed, for example, thereby increasing the need for more pesticide applications. Other creatures useful to the farmer, including pollinators such as bees, may be affected. Bees are an interesting case in point because their numbers have been in decline in numerous parts of the world. There are many potential drivers of these declines including habitat loss, disease and parasites, but the intensification of farming and associated increased use of agrochemicals is widely considered to be one of the major factors. Some insecticides influence the cognitive abilities of bees, impairing their performance and ultimately reducing the viability of bee colonies. The extensive use of insecticides

Figure 13.8 Aerial crop spraying with pesticide in New South Wales, Australia.

also appears to have adverse effects on the health of both honey bees and wild bumblebees, making them more susceptible to parasite infections and viral diseases (Sánchez-Bayo *et al.*, 2016).

On the national scale, the excessive use of pesticides has widespread effects. An assessment in the USA demonstrated the extensive occurrence of pesticide residues in rivers and groundwater, with concentrations in many streams at levels that may have effects on aquatic life and fish-eating wildlife (Gilliom, 2007). The pesticides detected included some that had not been used in the country for decades (organochlorine compounds such as DDT), indicating the persistent nature of some compounds. The dangers posed to human health by high concentrations of residues are illustrated in Central Asia, where Glantz *et al.* (1993) suggest that pesticide applications have historically been on average up to ten times the levels used elsewhere in the former USSR or the USA. These high levels have been related to deteriorating human health via contaminated surface water and groundwater in regions of Kazakhstan and Uzbekistan dominated by cotton monoculture (Crighton *et al.*, 2011), and impacts upon human health have been exacerbated by the relative dearth of medical and health facilities in the Aral Sea basin.

More direct exposure to dangerous levels of pesticides occurs in numerous ways. Agricultural workers are particularly at risk if not adequately protected during pesticide applications, as is often the case on plantations in the developing world. In one tragic incident in the 1970s, 1500 male banana-plantation workers in Costa Rica became sterile after repeated contact with dibromochloropropane (Thrupp, 1991). A number of cases have also been documented in which seeds treated with pesticides have been eaten rather than planted. Wheat seeds treated with mercury-based fungicides and distributed as food in Iraq caused the hospitalization of 370 people in 1960 and 6530 in 1972, with 459 deaths in the latter case (Bakir *et al.*, 1973). Bakir *et al.* note similar cases in Guatemala in the early 1960s and Pakistan in 1969. One estimate suggests that as many as 26 million people worldwide may be affected by pesticide poisonings each year, with 220,000 deaths (Richter, 2002).

Another pathway by which pesticides can become dangerous to human health is via the accumulation of some compounds up the food chain, to reach damaging levels in higher species that may be eaten by people. DDT, a particularly persistent compound, is the classic example of a long-lasting pesticide residue that bioaccumulates in this way (Table 13.4). High concentrations of DDT have caused considerable increases in mortality among seabirds and birds of prey (see p. 367), and DDT has also been reported at hazardous concentrations in human mothers' milk. Dangerous levels of DDT and other organochlorine pesticides have been measured in top predator tigerfish from rivers flowing through Kruger National Park in South Africa, and Gerber *et al.* (2016) report that a significant human cancer risk is associated with the consumption of fish from these rivers.

Realization of the dangerous effects of many pesticides has encouraged some governments to introduce programmes to reduce their use in agriculture. Some of the countries most heavily dependent upon synthetic chemicals, such as Denmark, the Netherlands and Sweden, have been the most active in this regard (Gianessi *et al.*, 2009). These moves follow the introduction of complete bans on the use of some of the more toxic and persistent pesticides in many developed countries. Long-lasting organochlorine insecticides, such as DDT, aldrin and dieldrin, and herbicides such as 2,4-D and 2,4,5-T, are among the pesticides now rarely used in the more developed countries, although these compounds are still common inputs to agriculture in many developing countries (Kesavachandran *et al.*, 2009).

In some cases, however, such bans have not been without their critics. The use of dieldrin in locust-control programmes is a good example. It was successfully used in Africa throughout the 1960s and 1970s by spraying strips of land near locust hatching areas. Although dieldrin breaks down quickly in tropical sunlight, the strips remained lethal to hoppers for six weeks. Concern over dieldrin's lethal effects on birds, its persistence in the food chain, and its toxicity to people, led to a ban on its use under the 2001 Stockholm

Table 13.4 Bioaccumulation of DDT residues up a salt marsh food web in Long Island, USA

Organism	DDT residues (parts per million)
Water	0.00005
Plankton	0.04
Silverside minnow	0.23
Pickerel (predatory fish)	1.33
Needlefish (predatory fish)	2.07
Heron (feeds on small animals)	3.57
Herring gull (scavenger)	6.00
Fish hawk (osprey) egg	13.80
Merganser (fish-eating duck)	22.80
Cormorant (feeds on larger fish)	26.40

Source: after Woodwell et al. (1967).

Convention on Persistent Organic Pollutants (POPs). A much shorter-lived replacement, fenitrothion, was advocated, but fenitrothion's three-day active period means that spraying has to be delayed until mature locusts are swarming, and then they must be sprayed directly. This approach requires precise monitoring and timing. Swarms also need to be sprayed much more frequently and in larger volumes, making the environmental advantages over dieldrin marginal. Using fenitrothion also takes more work, more spray and more money to stop a locust swarm (Skaf, 1988).

Alternatives to pesticides

Despite the widespread use of pesticides, about one-third of total crop production is still lost to insects, weeds and plant diseases, a similar proportion to that lost in pre-synthetic chemical times (Pimental, 1991). Although this percentage has been maintained during a period of rapidly increasing crop production, the success of chemicals in controlling pests has not been as great as was once hoped. Alternatives to synthetic chemicals have long been used and, given the problems of resistance and contamination associated with chemical use, these alternatives are being given more attention by food producers all over the world.

There are five main alternative approaches to chemical pest control:

- environmental control
- genetic and sterile male techniques
- behavioural control
- resistance breeding
- biological control.

A combination of these approaches is often employed as a form of integrated pest management (IPM), which acknowledges that both crops and pests are part of a dynamic agricultural ecosystem, and aims to limit pest outbreaks by enhancing the effects of natural biological checks on pest populations and only resorting to chemicals in extreme cases.

Environmental control measures incorporate all human alterations to the environment, which range from simple measures, such as digging up the egg pods of pests and the planting of trap crops to lure pests away from the main crop, to larger-scale efforts such as the draining of wetlands that harbour pests. The special breeding for release of large numbers of genetically altered, often sterile, individuals has been used in a number of cases, the aim being to decimate pest populations by preventing reproduction. The technique was pioneered in the control of the screw-worm fly, a pest affecting cattle. Behavioural control techniques that aim to control pests by use of sex pheromones and by attractant or repellent chemicals are in the early stages of development, while the selected breeding of organisms resistant to pests has long been practised and is now a more precise science thanks to modern biotechnology (see below).

The most widely practised and successful method of biological control, which uses living organisms as pest control agents, has been the introduction and establishment of appropriate species to provide permanent control of a pest. An early large-scale example was in the 1880s when the vedalia beetle and a parasitic fly were successfully introduced in California, USA to control cottony-cushion scale, a tiny insect affecting citrus orchards. In a review of the field, Barratt *et al.* (2018) point out that the use of biocontrols declined rapidly from the mid-1940s with the growth of the synthetic pesticide industry but returned in the 1960s following publication of the book *Silent Spring* (Carson, 1962), which criticized the widespread use of agricultural pesticides because of their deleterious environmental impacts.

Methods of biological control now face renewed criticism because of possible impacts on non-target species. A particularly damaging example occurred on the French Polynesian island of Moorea, where the giant African snail was introduced in the nineteenth century by colonial governors with a yearning for snail soup. However, the snail spread across the island and caused widespread destruction to crops, so a predaceous snail, the Ferussac, was brought in to control the giant African snail in 1977. Unfortunately, the Ferussac found native snail species to be more palatable targets and, by 1987, seven species endemic to Moorea had become extinct in the wild, although six survive in captivity. The same tragic story of an ill-judged biological control programme on the other islands of French Polynesia has reduced 61 endemic tree snail species to just five (Coote and Loève, 2003).

BIOTECHNOLOGY

Although the term biotechnology is a recent one, people have manipulated the genetic makeup of crops and livestock for thousands of years by selecting and breeding according to their needs – the process of domestication. Similarly, the production of bread, cheese, alcohol and yoghurt are age-old examples of the deliberate manipulation of organisms or their components to the benefit of society and are, therefore, examples of biotechnology. Some of the recent steps in the cross-breeding of wheat strains are shown in Figure 13.9. Higher yields have been attained by cross-breeding with varieties that had more ears per stem, but the extra weight often caused the stem to break, making it impossible to harvest. Cross-breeding with short-stemmed varieties solved this problem. Diseases and pests are a constant problem which can be combated with pesticides and further cross-breeding.

For the last five millennia, virtually the entire world's population has been reliant on farmers and herders who consciously manipulate natural plants and animals to provide food. This means that the crops we grow for food and the animals we raise to eat are radically different from those that existed in the 'natural' state. The same even applies to our pets and the plants in our parks and gardens. Thousands of years of selective breeding have boosted the yields of crops, the milk production of cows, the amount of meat on cattle, and

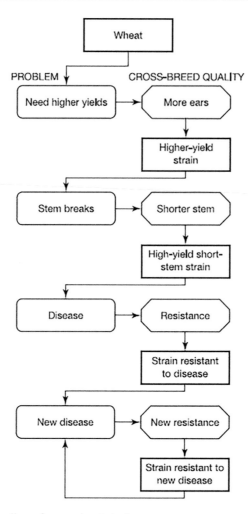

Figure 13.9 Steps in the breeding of new wheat strains.

the colours and sizes of our flowers and dogs. Virtually every living thing that humankind exploits has been genetically modified in a major way (Figure 13.10).

Such procedures can now be conducted much more precisely using genetic engineering, which transfers useful DNA material from one organism into the genetic makeup of another organism. Today we use the term 'genetically modified', or GM, to refer to food produced from plants or animals that have had their genes changed by scientists in the laboratory rather than farmers in the field. In many respects, genetic engineering is no different in principle from conventional breeding, except that it is faster. However, genetic engineering can go further than conventional breeding. Scientists can now select a single gene for a single characteristic and transfer that stretch of DNA from one organism to another. The implanted genes may come from other species and can cross between the animal and plant kingdoms. For example, inserting genes for scorpion toxin or spider venom into maize and other food crops can act as a 'natural pesticide' to deter insects and birds from feeding on them.

Figure 13.10 All herds of domesticated livestock, like these dairy cows near Lismore in Eire, are the result of innumerable generations of selective breeding.

The scope for modifying the properties of commercial food sources is very wide. Among the many applications, DNA can be transferred to increase resistance to pests and diseases, to improve tolerance of herbicides used for weed control, to modify the timing of fruit ripening, to enhance nutritional value, and to increase tolerance of altitude or temperature regimes. The most successful traits to date have been those aimed at the farmer (e.g. herbicide tolerance for soybean, oilseed rape, cotton and maize; insect resistance for cotton and maize). Traits affecting the quality or the nutritional value of the product have been more difficult to develop and market.

Once a new variety of plant has been created using recombinant DNA techniques, it can be mass-produced using cell or tissue culture, another important biotechnology technique. The approach enables a complete plant to be regenerated in the laboratory from a single cell exposed to nutrients and hormones, and is commonly used for clonal propagation, the mass production of clones or genetic duplicates that can then be planted like conventional seedlings.

DNA-based methods of modifying the genetic makeup of crops and livestock have not replaced 'conventional' forms of plant or animal breeding; the latter are still widely used alone and in conjunction with DNA-based techniques. However, the new biotechnology methods used in agriculture have had numerous consequences (Herdt, 2006). These include widespread commercial applications in a limited number of countries, significant economic contributions to farmers, continuing controversy over their environmental impacts, a proliferation of regulations (covering the technology and property rights), a wide range of changing public reactions, and a relatively small contribution to increasing food production, nutrition, or farm incomes in less-developed countries.

Several potential environmental impacts have raised concern. The notion that introducing a single gene (that releases a toxin, for example, to endow resistance to a pest) to a plant will produce just a single result (resistance to the pest) is naïve since any human-induced change is bound to have other knock-on effects. Nevertheless, built-in resistance to a pest produces benefits in terms of reducing synthetic chemical pesticide use, of course. However, such GM crops can be toxic to non-target species, including 'friendly' species such as beetles, bees and butterflies. Another difficulty of permanently inducing resistance to a pest by genetic modification is that no control over the expression of the new gene is yet possible. Hence, the pest population will be exposed to the toxin continually and eventually the pest will develop a resistance to the toxin. Further, genetically altered plants or micro-organisms might have a competitive advantage that could disturb natural ecology if they were released into the wild. The potential effects could echo those caused by unintended releases of exotic insects and other organisms into environments where they have proliferated to the detriment of many other species, as the above example of the Ferussac snail in Moorea illustrates (see also p. 323). Many feel that the current pace at which GM foods are being introduced is too fast. Adoption of the precautionary principle is urged, running indoor laboratory trials first, only gradually moving toward unrestricted outdoor release once the trials have proved satisfactory.

The risk of GM genes escaping into the wild could be reduced by incorporating 'terminator technology'. These are genes that are expressed in embryos at a very late stage of development allowing crops to develop normally but killing them when they are mature, so effectively making the harvested seed sterile. This modification helps to ensure that there are no offspring, although it does not eliminate the risk completely. Terminator technology, however, has raised considerable protest over its ethical implications (Yusuf, 2010). This type of genetic manipulation would effectively place the company owning the GM seed in a very powerful position: the farmer has to buy a new batch of seeds from the company each year. Many fear that this would stifle competition, and operate against the common interests of crop producers and consumers. The feared impact would be particularly great in developing countries. Here the age-old practice of seed saving from one year to the next, which enables hundreds of millions of poor farmers to subsist, would effectively cease.

The new techniques also highlight the need to maintain the reservoir of genetic diversity in the wild, to provide new genetic material that could be of use to humankind for food production and in other applications such as pharmaceuticals. The dangers of our increasing reliance on a small number of food crops – 75 per cent of food comes from just 12 crops (Groombridge and Jenkins, 2000) – are enhanced by their limited genetic diversity, a situation exacerbated by crop cloning, as examples of major crop failures through history illustrate (Table 13.5). The constant need for new genetic material, much of it found in wild relatives of domesticated crops, can be shown by just one example: Plucknett and Smith (1986) demonstrated that a new variety of sugar cane is required in Hawaii about every ten years to adapt to pests and to maintain yields. While realizing the need to protect and maintain genetic diversity in the wild, regulating the use of such material, thereby enabling compensation to be paid to the countries of origin, is another pertinent issue, particularly for developing countries. It is an aspect of the biodiversity debate that the Convention on Biological Diversity specifically recognizes and aims to tackle.

Public attitudes to biotechnology and GM foods have been divided in many parts of the world. On the one hand the potential benefits of improving agricultural productivity and alleviating hunger are highlighted, on the other are views reflecting deeper concern with a technocentric approach to managing the environment. The debate over GM foods is one of the biggest issues facing contemporary agriculture and

Table 13.5 Crop failures attributed to genetic uniformity

Date	Location	Crop	Cause and result
900	Central America	Maize	Anthropologists speculate that collapse of Mayan civilization may have been due to a maize virus
1846	Ireland	Potato	Potato blight led to famine in which 1 million people died and 1.5 million emigrated
Late 1800s	Sri Lanka	Coffee	Fungus wiped out homogeneous coffee plantations on the island
1943	India	Rice	Brown spot disease, aggravated by typhoon, destroyed crop, starting 'Great Bengal Famine'
1960s	USA	Wheat	Stripe rust reached epidemic proportions in Pacific north-west
1974–77	Indonesia	Rice	Grassy stunt virus destroyed over 3 million tonnes of rice (the virus plagued South and South-East Asian rice production from the 1960s to the late 1970s)
1984	USA	Citrus	Bacterial disease caused 135 nurseries in Florida to destroy 18 million trees

Source: after WCMC (1992: 428, table 27.22).

is often characterized as a clash between pro-science and anti-science forces (Linnhoff *et al.*, 2017). It is a debate that is likely to continue for some time to come.

FOOD SECURITY

An adequate and regular supply of food is needed by everyone to grow, develop, remain active and sustain good health. Hence, people who do not have enough to eat on a regular basis may become undernourished and can ultimately die of starvation. When access to nutritious, safe foods is limited or uncertain, we talk of 'food insecurity'.

The security of food supplies can be studied at all scales, from the international and national levels to more localized scales such as the household. Food security is dependent upon both supply and demand. Adequate national or local availability of food is a necessary condition for household food security, but widespread famines in the Sahel and East Africa in the 1970s revealed that adequate national and international supplies do not necessarily prevent extensive food insecurity. The importance of the demand side in the food security equation was highlighted in the work of Sen (1981) and his entitlement theory. He identified two important situations in which access to food by some sectors of society is diminished. Examples of these are a failure of crops due to drought in subsistence agriculture, and a sudden change in the market, such as a rapid increase in prices. Both scenarios can result in food insecurity for some groups of people.

At the country level, two types of food insecurity are often identified: 'chronic' and 'transitory'. Chronic food insecurity refers to situations where, on average, food availability is below the level required. Root causes of this chronic condition include economic poverty and a poor endowment of natural resources such as soils. The short-term decline in food due to drought, fluctuations in income or prices is referred to as transitory food insecurity.

Two countries that have experienced successive food crises in recent times are Ethiopia and Eritrea. In both cases, their chronic food insecurity is punctuated by occasional incidents of widespread hunger and famine which are characterized as having been triggered by drought, with rural overpopulation and land degradation serving as underlying causes. Warfare has also been identified by White (2005) as an important contributory factor in both countries, escalating food insecurity in numerous ways including the loss of access to arable land and pastures, disruption of trade and access to markets and relief supplies, and the diversion of resources to the war effort. The combination of conflict and climate-related shocks to the food supply system is typical of many countries identified as being particularly food insecure in Box 13.1.

Box 13.1 Food security, conflict and climate

The majority of people who face chronic food insecurity and malnourishment live in countries affected by conflict. In many cases, the wide-reaching effects of conflict are exacerbated by climate shocks, and in 2016 much of the planet suffered adverse weather conditions due to the strong El Niño of 2015–16. The food crises of 2016 were widespread and severe, in the case of Yemen affecting the entire national population (Table 13.6).

Nonetheless, many of these cases of conflict and food insecurity had been long-running by 2016. Serious conflict erupted in Syria in 2011 and in Yemen in early 2015. Insecurity had been rife in many parts of the Democratic Republic of Congo for over 20 years and Somalia for 25 years. Afghanistan had endured protracted conflict for almost 35 years. Such long-running conflicts seriously undermine development and society's coping mechanisms so that when adverse weather conditions occur, the impact on food security is particularly clear.

Table 13.6 Countries with significant food-insecure populations, conflict and climate-related shocks in 2016

Country	Food-insecure population (millions)	Main climatic driver of food insecurity
Yemen	14.1	Flooding, heavy rains and tropical cyclones
Afghanistan	8.5	Floods, landslides in winter; drought in Ghor province
Syria	7.0	Drought in Aleppo, Idlib and Homs
Congo, DR	5.9	El Niño
South Sudan	4.9	Drought and floods
Sudan	4.4	El Niño
Somalia	2.9	El Niño-related drought
Burundi	2.3	El Niño
Central African Republic	2.0	Localized floods
Iraq	1.5	Drought

Source: FSIN (2017).

Food security has been improved in numerous parts of the world thanks to many of the techniques detailed in this chapter. However, although at the global scale we produce sufficient food to feed the entire world's population, the number of chronically undernourished people in the world was estimated to be 815 million in 2016 (FAO, IFAD, UNICEF, WFP and WHO, 2017). One of the UN's Sustainable Development Goals (SDGs) adopted in 2014 is to end hunger, achieve food security and improved nutrition, and promote sustainable agriculture worldwide. This involves ambitious targets to end hunger and all forms of malnutrition by 2030.

SUSTAINABLE AGRICULTURE

Although global food production has increased significantly over the past half-century or so, one of the most important challenges facing society is how to feed an expected population of some 9 billion by about 2050. To meet the expected demand for food without significant increases in prices, it has been estimated that we need to produce 70–100 per cent more food, in light of the growing impacts of climate change and concerns over energy security (Godfray *et al.*, 2010). Given the numerous environmental impacts associated with agriculture outlined in this and other chapters, it is equally important that the necessary increase in production is achieved in as sustainable a manner as possible.

However, perhaps a more immediate priority for enhancing the availability of food should be the reduction of waste in current production systems. Incredibly, approximately 30–50 per cent of all food produced is wasted, although the drivers are complex and differ considerably between rich and poor countries as well as between rich and poor population groups (Pelletier *et al.*, 2011). In the developing world, losses are typically caused by a lack of infrastructure for storage, processing and distribution, whereas in the developed world the bulk of food waste occurs in the home. The relatively low price of food, high expectations of food cosmetic standards, and the increasing disconnection between consumers and how food is produced are all contributory factors. Hall *et al.* (2009) indicate that 40 per cent of the available food in the USA is wasted.

Another interesting perspective on how food is consumed is the finding by Alexander *et al.* (2017) that over-eating is at least as large a contributor to food system losses worldwide as consumer food waste. This study points out that globally just over a third of all people are overweight, while about 12 per cent of people were undernourished between 2010 and 2012. Since the analysis conducted was done at the global level, it averaged out the wide range of nutritional consumptions between individuals. Hence, the losses associated with over-eating were probably underestimated because overconsumption of food is partially offset by people who do not have enough to eat (Figure 13.11).

Improving the sustainability of agriculture should focus on the production of crops and livestock using methods that are more closely based on the functioning of natural systems and the use of ecological processes, as opposed to a large-scale manipulation of nature and the high inputs of energy indicated in Table 13.2. Organic farming, which aims to produce food with minimal harm to ecosystems, animals or humans, is one approach that is becoming increasingly popular in parts of the western world as a sustainable alternative to intensive farming practices. Critics argue that organic agriculture tends to produce lower yields and thus needs more land to produce the same amount of food as intensive farming, so undermining the environmental benefits. A major review (Seufert *et al.*, 2012) confirmed that organic yields are typically lower than conventional yields, but found that under certain conditions (i.e. good management practices, particular crop types and growing conditions), organic systems can nearly match conventional yields. These authors concluded that organic agriculture can be an important tool in sustainable food production while urging better understanding of the factors limiting organic yields, as well as assessments of the many social, environmental and economic benefits of organic farming systems.

Figure 13.11 More than a third of all people are overweight and over-eating is thought to be as large a contributor to food system losses worldwide as consumer food waste.

An assessment of the impacts on biodiversity of organic farming, relative to conventional agriculture, carried out by Hole *et al.* (2005) identified a wide range of organisms – including birds, mammals, invertebrates and flora – that benefit from organic management through increases in abundance and/or species richness. Three broad management practices common to organic farming were highlighted as being particularly beneficial for farmland wildlife. These are the prohibition/reduced use of synthetic chemical pesticides and inorganic fertilizers, sympathetic management of non-cropped habitats, and the preservation of mixed farming. However, the review also concluded that it is unclear whether an organic approach provides greater benefits to biodiversity than carefully targeted prescriptions applied to relatively small areas of cropped and/or non-cropped habitats within conventional agriculture (so-called 'agri-environment schemes'). It is also worth noting that most of the review's conclusions applied to cultivated areas because our knowledge of the impacts of organic farming in pastoral and upland agriculture is limited.

A wider perspective on the sustainability of food production can be evaluated using life-cycle assessment (LCA). By assessing the environmental impacts of a particular product or farming system, LCA can reveal where in the life cycle the impacts are greatest and, perhaps, how they can be reduced (Roy *et al.*, 2009). Much popular discourse on the sustainability of food has focused on transportation, assuming that small-scale, local food systems use less energy than those dependent on global distribution networks, an assumption based on the simple notion of 'food miles', a weighted-average distance travelled for a food product. While there is an undeniable logic to this approach, LCA indicates that transport is a minor element in the overall food system energy use in industrialized countries. A study of the energy used in the US food system by Cuellar and Webber (2010) found that transport accounted for just 3 per cent, while the largest share of the total energy used was in the household. Travel to purchase food, storage in freezers/refrigerators, meal preparation and clean-up accounted for 28 per cent of energy use in the US food system.

A further argument to suggest that the home is where some of the greatest improvements can be achieved in food production sustainability is based on the types of food we eat. Godfray *et al.* (2010) point out that the conversion efficiency of plant into animal matter is about 10 per cent, making a strong case for vegetarianism. In simple terms, more people could be supported from the same amount of land if they did not eat meat. The argument against meat is bolstered by the findings of Clark and Tilman (2017) who looked at a range of environmental impacts associated with different foods and concluded that ruminant meat (beef, goat and lamb/mutton) had impacts 20–100 times those of plants, while milk, eggs, pork, poultry and seafood had impacts 2–25 times higher than plants per kilocalorie of food produced. However, the rapidly increasing demand for meat and dairy products in developing countries, the so-called Livestock Revolution (see above), is a major challenge to the global food system and the argument that all meat consumption is bad is too simplistic. Although many animals are fed on cereals that could feed people directly, a very substantial proportion of livestock is grass-fed and much of this grassland is unsuitable for conversion to arable land.

AQUACULTURE

A notable recent development in global food production is the rise in importance of aquaculture, which is dominated by fish farming but also includes the production of other aquatic organisms such as seaweeds, frogs and turtles. Between 1987 and 1997, the global production of farmed fish and shellfish more than doubled in weight and value, by which time farmed fish accounted for more than a quarter of all fish directly consumed by people (Naylor *et al.*, 2000), a proportion that had reached 44 per cent by 2014. Asia accounts for about 90 per cent of world aquacultural output (Figure 13.12), and China is by far the world's largest producer. Most of the Chinese production comprises freshwater species such as carp and tilapia, and globally aquacultural output is split about 60/40 between inland and marine production.

Fish farming has a long history in parts of East and South Asia, dating back to 5000 BP in China, and has been practised in Europe since before medieval times. At its simplest level, fish are raised in ponds, where they find their own food, from eggs collected in the wild, and are harvested when grown to a certain size. More sophisticated approaches involve a series of ponds or cages for different levels of development, including special nursery ponds and/or hatcheries. Provision of feed and control of natural predators, competitors and diseases are also common practice. Particular advances in fish-farming techniques have occurred in the past 50 years: selective breeding of stock is now widely practised in more advanced farms, reproduction is controlled by hormonal injection and disease by antibiotics. Such intensive management has been used mainly in more developed countries for species that command a high market price. The spectacular development of the fish farming industry in Norway (dominated by Atlantic salmon and rainbow trout) is an important example. Initiated in the early 1970s, fish farming has grown to become one of the country's largest export industries by economic value, with production passing 1.3 million tonnes in 2016, 95 per cent of which is exported (Statistics Norway: www.ssb.no/en/fiskeoppdrett/). Such intensive aquaculture is, however, associated with a number of adverse impacts on the environment. These include the destruction of natural habitats, displacement of native fish species, control of wildlife that might prey on the farmed species, the spread of disease, and local contamination of waters, particularly by organic

Figure 13.12 A sea bream fish farm off the island of Shikoku, Japan.

particulate wastes from feed and faeces which can cause eutrophication and can smother sea- or lake-bed organisms.

Viral diseases represent a serious problem in Norwegian salmon farms. Disease often leads to substantial economic losses, and the possible spread of viruses from salmon farms to wild fish is a major public concern. After the Norwegian government established a set of environmental goals for sustainability in the aquaculture industry, an annual risk assessment of Norwegian salmon farming has been conducted since 2010. The threat of viral diseases and farmed salmon lice to wild salmonid populations is one of the central issues, along with the possible impact of escaped farmed salmon on wild populations, and the local and regional impact of nutrients and organic load. After a few years of these risk assessments, a review by Taranger et al. (2014) concluded that there was an urgent need for better knowledge about such impacts, to implement improved monitoring programmes for the most important hazards and also to improve procedures for risk assessments.

Disease has also been an issue in Asia. The rapid growth in aquacultural production of shrimp and prawns since the mid-1980s in Thailand, for instance, has been punctuated by periodic declines due to disease outbreaks (Figure 13.13). The most severe crash in production dates from 2012 when acute hepatopancreatic necrosis disease (AHPND) spread to Thailand after beginning in China in 2009 and reaching Vietnam in 2010 and Malaysia in 2011. Thai production fell from an all-time high of about 600,000 tonnes in 2011 to less than half that amount by 2014 due to a combination of shrimp mortality from AHPND and a decline in pond stocking as farmers feared their stocks would succumb to the disease (Thitamadee et al., 2016).

The conversion of natural habitats for aquaculture farming has been a primary cause of mangrove destruction in many coastal areas of the tropics, including Thailand where shrimp and prawn production has largely been at the expense of mangroves (see p. 158). The loss of habitat has a feedback effect on marine production. It has been estimated that for every kilogram of shrimp farmed in Thai shrimp ponds developed in mangroves, 400 g of fish and shrimp are lost from capture fisheries (Naylor et al., 2000).

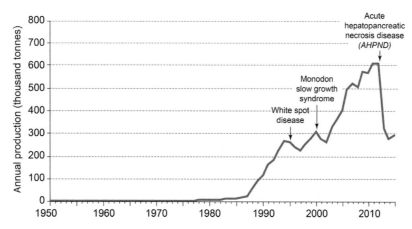

Figure 13.13 Production of shrimp and prawns in Thailand, 1950–2015 (FAO Fishery Statistics), with start dates of disease outbreaks.

Salinity problems due to sea-water intrusion have been encountered in fish-farm areas on Taiwan's western coastal plains, where the extraction of groundwater from the aquifer to supply the fishponds has also caused land subsidence. The subsiding plain has left fish farms susceptible to greater coastal erosion and storm damage (Lin, 1996). Another consequence of the excessive pumping of water from the aquifer in this area is thought to be a more dangerous form of groundwater poisoning. In the Yun-Lin region, Liu *et al.* (2003) have shown that the area of groundwater overpumping is coincident not only with areas of high sea-water salinization, but also with arsenic pollution. The arsenic is released into the groundwater by excess dissolved oxygen introduced by overpumping.

Some predictions suggest that global production from aquaculture will continue to rise significantly, which has engendered talk of a 'Blue Revolution' to rival the agricultural Green Revolution. In essence, this would amount to a major shift from capture to farming or husbandry in water, which would parallel that which occurred when hunting and gathering gave way to agriculture on land. Some observers believe the recent rise in aquacultural output is relieving pressure on ocean fisheries, but the farming of carnivorous fish species actually has the opposite effect because it is based on large inputs of wild fish for feed. Eel, salmon and trout produced in intensive aquacultural systems are fed two to five times more fish protein, in the form of fish meal and fish oil, than their actual weight (Tacon and Metian, 2008).

Cheaper and more ecologically sound management practices have long been established in China, however, where carp are reared alongside pigs with reciprocal recycling of nutrients between land and water. Other species, such as tilapia, a prolific breeder that can also be fed mainly on wastes, are widely farmed in South Asia due in part to the traditional dietary importance of fish. Here, and in many other parts of the world, aquaculture has the additional benefit of helping to deal with problems of waste disposal. The biggest of many such systems which are based on sewage is in the Indian city of Kolkata, where the large-scale use of sewage for fish culture began in the East Kolkata Wetlands in the 1930s (Nandeesha, 2002).

In the contemporary era the Kolkata sewage-fed ponds, called 'bheries', produce about 7000 tonnes of fish annually for the local market. Water and sewage are channelled into shallow lakes covering about 4000 ha, and carp and tilapia are introduced after an initial bloom of algae. Thereafter, the lakes are fed with additional sewage on a monthly basis. Health concerns over fish raised in this manner can be eliminated by retaining sewage in settling ponds for 20 days before introducing it into the fish ponds, or by moving fish to clean-water ponds before harvesting. Some bheries are managed as individual operations, but most are worker cooperatives. Despite the importance of this fishery to sewage treatment and pollution control, income generation and employment for the poor, the Kolkata lakes have been gradually depleted in the face of increasing demands for housing land. Li *et al.* (2016) used satellite images to demonstrate that the wetland area decreased by nearly 18 per cent between 1972 and 2011, although designation in 2002 of the 12,500 ha East Kolkata Wetlands as one of India's Ramsar sites should help to protect this unique example of a multiple-use wetland.

FURTHER READING

Alexander, P., Brown, C., Arneth, A., Finnigan, J., Moran, D. and Rounsevell, M.D. 2017 Losses, inefficiencies and waste in the global food system. *Agricultural Systems* 153: 190–200. An assessment of losses at every stage in the food system. Open access: www.sciencedirect.com/science/article/pii/S0308521X16302384.

Barzman, M., Bàrberi, P., Birch, A.N., Boonekamp, P., Dachbrodt-Saaydeh, S., Graf, B., Hommel, B., Jensen, J.E., Kiss, J., Kudsk, P. and Lamichhane, J.R. 2015 Eight principles of integrated pest management. *Agronomy for Sustainable Development* 35: 1199–215. Principles set out by the EU for sustainable farm management. Open access: link.springer.com/article/10.1007/s13593-015-0327-9.

Battye, W., Aneja, V.P. and Schlesinger, W.H. 2017 Is nitrogen the next carbon? *Earth's Future* 5: 894–904. Reviewing the environmental impacts of synthetic nitrogen fertilizers. Open access: onlinelibrary.wiley.com/doi/10.1002/2017EF000592/full.

Gordon, L.J., Peterson, G.D. and Bennett, E.M. 2008 Agricultural modifications of hydrological flows create ecological surprises. *Trends in Ecology and Evolution* 23: 211–19. A cautionary look at the wider implications of agriculture.

Li, X., Li, J., Wang, Y., Fu, L., Fu, Y., Li, B and Jiao, B. 2011 Aquaculture industry in China: current state, challenges, and outlook. *Reviews in Fisheries Science* 19: 187–200. An overview of the world's largest producer and consumer of aquacultural products.

Murphy, D. 2011 *Plants, biotechnology and agriculture.* Wallingford, CAB International. A good assessment of how plants function and how people manipulate them.

WEBSITES

www.cgiar.org   CGIAR, an international organization that contributes to food security and poverty eradication in developing countries.

ddsindia.com   the Deccan Development Society is a grassroots organization working with women in villages in India on their autonomy over production, seed stocks and many other aspects of food.

www.fao.org   the UN Food and Agriculture Organization site has a very wide range of information, data and reports on all aspects of food production.

www.icgeb.trieste.it   the site of the International Centre for Genetic Engineering and Biotechnology, which has a focus on the developing world, has a wide range of information.

pan-international.org   the Pesticide Action Network is a network of non-governmental organizations (NGOs), institutions and individuals working to replace hazardous pesticides with ecologically sound alternatives.

www.was.org   the World Aquaculture Society site.

POINTS FOR DISCUSSION

- Explain the pros and cons of the shift from hunting and gathering to *agriculture.*
- Why should we be concerned about the human impact on the global nitrogen cycle?
- Is biotechnology the answer to problems of world food supply?
- Before we try to increase food production still further, we should address the problems of waste. Discuss.
- Will all farmers be practising sustainable agriculture by 2050?
- Outline the probable benefits and possible problems that might arise from a 'Blue Revolution'.

14

Soil erosion

Soil erosion is a natural geomorphological process that occurs on most of the world's land surface, with the principal exception of those areas on which soil eroded from elsewhere is deposited. This movement of soil is usually slow, only infrequently occurring in dramatic events, so perception of its seriousness, measurement of its progress and persuasion of the need to take ameliorative action are not always regarded as a priority. A fundamental distinction is made between natural or geological soil erosion – a phenomenon that forms an integral pathway in numerous biogeochemical cycles – and rates that are accelerated by human activity, and it is accelerated erosion that has received most scientific attention. Such attention is not new, however. Ancient Greek and Roman observers of the natural world commented on the effects of human activities such as agriculture and deforestation on soil loss.

Soils provide many ecosystem services, and are important to human communities in several ways:

- as a medium in which crops, forests and other plants grow
- for their filtering, buffering and transformation activity, between the atmosphere, groundwater and plant cover, servicing the environment and people by protecting food chains and drinking-water reserves
- as biological habitats and reserves of genes and seeds
- by serving as a spatial base for society's structures and their development (e.g. the construction of buildings and dumping of refuse)
- as a source of raw materials (e.g. clay, sand and gravel for construction), and also as a reserve of water and energy.

Hence the loss of soil from an area by erosion, the most widespread form of soil degradation (a term which also includes such processes as acidification, salinization and compaction), is frequently regarded as an important environmental issue. This importance is all the more pressing when we consider the fact that the world's human population is increasing, along with its use of resources, and we know that natural rates of soil formation are slow, essentially making soil a finite resource.

SOIL EROSION DRIVERS

A hierarchy of groups of drivers, both natural and societal, all contribute to the occurrence of erosion and the rate at which it progresses (Figure 14.1). At the immediate level, soil is principally eroded by the forces of water or wind acting on the soil surface, and on steep slopes by mass movement, although other agents can also contribute. The actions of animals and humans are important among these other agents. Grazing animals, for instance, can cause erosion by trampling on bare soil, and 'tillage erosion', the redistribution of soil by farmers preparing for cultivation, is a widespread human force.

Water erosion occurs both by the action of raindrops, which detach soil particles on impact and move them by splashing, and by runoff, which transports material either in sheet flow or in concentrated flows, which form rills or gullies. Wind moves soil particles by one of three processes depending on the size and mass of the particles: the largest particles move along the surface by the process of surface creep; sand-sized particles usually bounce along within a few metres of the surface by saltation; and finer dust particles are transported high above the surface in suspension. Mass movement occurs principally by landsliding and various forms of flow, depending upon the amount of moisture in the soil profile.

When and where soil erosion occurs is determined by the mutual interaction of the erosivity of the eroding agent and the erodibility of the soil surface. These variables of erosivity and erodibility change through time and space, at varying rates and differing scales, so that the relationship between the variables is in a constant state of flux. The various factors affecting erosivity and erodibility in relation to water erosion are shown in Figure 14.2. Most of the numerous human activities that affect the erosion system do so by altering the erodibility of the soil surface (see below).

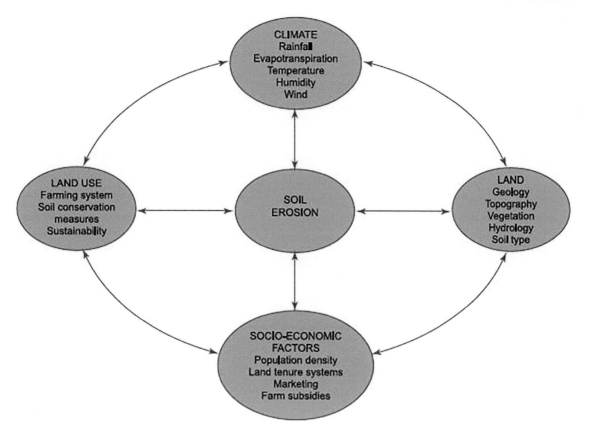

Figure 14.1 Factors that cause soil erosion and their interactions (after Lal and Stewart, 1990).

MEASURING SOIL EROSION

Soil erosion is commonly measured using one of a number of techniques, as shown in Table 14.1. These approaches do not all measure the same processes, however. Some are designed to monitor the redistribution of soil particles over distances of a few millimetres, centimetres or metres on a field surface. Others measure soil loss on a larger scale, such as from a whole field, or the quantity of soil that leaves an entire catchment in a river. An equivalent variation in temporal scale is implicit in Table 14.1: processes operating over small spatial scales (e.g. raindrop impact) occur in fractions of seconds, while catchment scale processes are usually monitored over much longer timescales (i.e. seasons, years, decades or geological timescales).

Measurements of the rate of soil loss from an area are calculated from the amount of sediment deposited in a splash cup, tray, in lakes or on the sea bed or in soil profiles. Erosion rate is usually expressed in terms of weight (t/ha/yr) or volume (m³/ha/yr) of soil lost. One important aspect of the wide range of techniques used to monitor soil erosion, and the fact that they do not all measure the same processes, is the difficulty

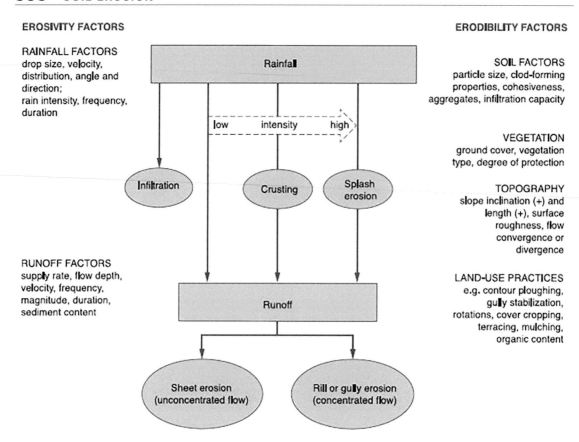

Figure 14.2 Main factors affecting types of soil erosion by water. Generally, erosion will be reduced if the value of an erosivity factor is reduced and/or the value of an erodibility factor is increased, with the exception of factors shown with a (+), where the reverse is the case (after Cooke and Doornkamp, 1990).

Table 14.1 Soil erosion measurement techniques over a range of spatial scales

Research technique	Spatial scale	Processes measured
Splash cup	mm_2	Rain splash dominant; overland flow/deposition limited. No gullies, stream bank erosion or mass movements
Laboratory tray	cm_2	Rain splash dominant; overland flow/deposition limited. No gullies, stream bank erosion or mass movements
Field plot	m_2	Rain splash and overland flow; some deposition. Some gullying and mass movements possible. No stream bank erosion

Research technique	Spatial scale	Processes measured
Field	ha	Rain splash, overland flow and deposition. Gullying and mass movements possible. No stream bank erosion
Sub-catchment	ha–km$_2$	Rain splash, overland flow and deposition. Gullying possible. Some stream bank erosion
Catchment/landscape	km$_2$	Rain splash, overland flow and deposition. Some gullying and mass movement possible. Stream bank erosion

Source: after Verheijen *et al.* (2009).

of comparing rates measured at different time and space scales. It is important to realize that the rates measured are typically scale-dependent. Characteristically, the larger the area under consideration, the lower will be the rate obtained. Hence, results of soil loss and sediment delivery obtained at one spatial scale cannot and should not be extrapolated to another scale (De Vente and Poesen, 2005).

Many older measurements of erosion rates are based on studies of experimental plots, usually of a standard size (22 x 1.8 m), on which controlled experiments are undertaken by varying such factors as slope angle, crop type and management practice. This approach suffers from the serious drawback that actual landscapes have much more complex topography than standard plots. As Boardman (2006: 76) puts it, 'It is clear that using standard experimental plots will not give estimates of what is happening in the landscape as a whole as regards erosion rates.' Certain types of erosion form, such as rills and gullies, may not develop at all on the small scale of a plot. Some measurements on real fields have therefore concentrated on these important forms, with field workers repeatedly visiting an area to measure the volume of rills and gullies, an exercise that can be supplemented by aerial photography.

Measurement of suspended sediment loads in rivers is another commonly used approach, which gives a rough estimate of current erosion on slopes and fields in the catchment area. Not all eroded soil reaches the river, however, and there have been many attempts to estimate 'sediment delivery ratios'. The technique is further hampered by the fact that a proportion of a river's sediment load is eroded from the river's own banks, although methods of 'sediment fingerprinting' have been developed to assess the contributions of different sediment sources (D'Haen *et al.*, 2012).

A technique used to estimate both erosion and deposition measures the radionuclide caesium-137 present in a soil profile, which can be compared to amounts in nearby undisturbed profiles. The caesium-137, most of which has been released by atmospheric nuclear weapons testing in the 1950s and 1960s, was deposited globally (plus some in Europe and Western Asia by the 1986 Chernobyl disaster) and adsorbed by clay minerals in the soil. Once adsorbed, chemical or biological removal of caesium-137 from soil particles is insignificant and it is assumed that only physical processes moving soil particles are involved in caesium-137 transport. Hence, caesium-137 has been used as a tracer to provide average rates of soil movement over the medium term (30–50 years) and patterns of erosion and deposition at particular sites by water and, to a lesser extent, by wind. Such work has focused primarily on soil loss from agricultural land but the technique has also been used to assess erosion rates on footpaths (Rodway-Dyer and Walling, 2010). Other radioisotopes have also been investigated for soil erosion research because caesium-137 has a half-life of 30 years and will become increasingly difficult to measure in coming decades. Plutonium-239, which has a half-life of 24,110 years, is one of the possible alternatives (Lal *et al.*, 2017).

Field measurements and experiments have also provided data from which general relationships between the factors affecting soil erosion have been derived, and these relationships have been incorporated into models used to predict erosion. The most widely used is the Universal Soil Loss Equation (USLE), developed in the USA to estimate soil erosion by raindrop impact and surface runoff from US fields east of the Rocky Mountains under particular crops and management systems. The USLE is calculated as:

$$E = R \times K \times L \times S \times C \times P$$

where E is mean annual soil loss, R is a measure of rainfall erosivity, K is a measure of soil erodibility, L is the slope length, S is the slope steepness, C is an index of crop type, and P is a measure of any conservation practices adopted on the field.

The USLE has been revised and modified since its first publication in 1965 and can be adapted for use outside the temperate plains of North America, although it has often been used inappropriately, without such adaptations. An equivalent wind erosion equation has also been developed and both of these relatively simple models have been further refined to produce more complex computer-run models such as EPIC (Erosion-Productivity Impact Calculator), one of the few erosion models that can be used to predict erosion by both water and wind.

Despite the availability of numerous measurement and prediction techniques, the fact remains that there is virtually no reliable soil erosion data for most of the world's land surface. An attempt to overcome this deficiency was made with the Global Assessment of Human-Induced Soil Degradation (GLASOD), which employed more than 250 local soil experts around the world to give their opinion on soil degradation problems to supplement what data are available. The project followed a strict methodology by dividing the world's land surface into mapping units corresponding to physiographic zones, and for each unit an assessment was made of water and wind erosion as well as chemical and physical degradation processes. The degree of degradation, its extent and causes were assessed to produce an overall estimate of degradation severity in each unit. Asia and Europe were considered worst-affected by water erosion, each with just over 10 per cent of land area affected. Africa and Asia were most affected by wind, each with just over 5 per cent of land area affected (Deichmann and Eklundh, 1991). Nonetheless, the scarcity of hard evidence continues to hamper scientific delimitation of areas of soil erosion concern and there have been calls for serious erosion hotspots to be identified (e.g. Boardman, 2006).

IMPACTS OF EROSION

The movement of soil and other sediments by erosive forces has a large number of environmental impacts that can affect farmers and various other sectors of society. Many of these effects are consequent upon natural erosion, but are exacerbated in areas where rates are accelerated by human activity. The scale of human impact on global erosion systems is such that human society is now considered to be more effective at moving sediment than the sum of all other natural processes operating on the surface of the planet by an order of magnitude, with agriculture and construction being the most widespread and effective human drivers of soil loss (Wilkinson, 2005).

Table 14.2 Some environmental hazards to human populations caused by wind erosion and dust storms

Entrainment	Transport	Deposition
Soil loss	Sand-blasting of crops	Salt deposition and groundwater salinization
Nutrient, seed and fertilizer loss	Radio communication problems	Reduction of reservoir storage capacity
Crop root exposure	Microwave attenuation	Drinking-water contamination
Undermining structures	Transport disruption	Burial of structures
	Local climatic effects	Crop growth problems
	Air pollution	Machinery problems
Respiratory problems and eye infections	Reduction of solar power potential	
	Disease transmission (human)	Electrical insulator failure
	Disease transmission (plants & animals)	Disruption to power supplies

Source: Middleton (2017).

The environmental effects associated with erosion occur due to the three fundamental processes of entrainment, transport and deposition. This threefold division is used in Table 14.2 to illustrate the hazards posed to human populations by wind erosion, while the following sections are divided into on-site and off-site effects.

On-site effects

The loss of topsoil changes the physical and chemical nature of an area. Deformation of the terrain due to the uneven displacement of soil can result in rills, gullies, mass movements, hollows, hummocks or dunes. For the farmer, such physical changes can present problems for the use of machinery, and in extreme cases such as gullying, the absolute loss of cultivable land (Figure 14.3). Deposition of soil within a field may also result in burial of plants and seedlings; loss of soil may expose roots; and sand-blasting by wind-eroded material can both damage plants and break down soil clods, impoverishing soil structure and rendering soil more erodible. Splash erosion can cause compaction and crusting of the soil surface, both of which may hinder germination and the establishment of seedlings, while exposure of hardpans and duricrusts presents a barrier to root penetration.

Erosion has implications for long-term soil productivity through a number of processes. The top layer of the soil profile, the A horizon, is where most biological activity takes place and where most organic material is located. Hence depletion of the A horizon preferentially removes organic material, soil nutrients,

Figure 14.3 A 15-metre-deep gully in central Tunisia.

including fertilizers and even seeds, and can reduce the capacity of the soil for holding water and nutrients. The subsoil, or B horizon, is much less useful for plant growth.

There is no doubt that soil erosion negatively affects crop yields although relating erosion rates to losses in productivity is not always straightforward, not least because the relationship is highly variable depending upon soil type and crop type, and because crop yields are affected by numerous other factors. These include the amount and timing of rainfall, temperature, pests, diseases and weeds. However, many experiments have shown that as erosion proceeds, crop yields do decline. Figure 14.4 illustrates this for two staple crops in south-west Nigeria for a range of slope angles on the same soil type. The relationship can be both linear and non-linear, depending on the slope angle involved.

In economic terms, Berry *et al.* (2003) summarize a number of studies on farmland in Africa and Asia to show that the cost of soil erosion is typically 3–7 per cent of agricultural GDP. Putting such figures in context, Cohen *et al.* (2006) calculate that in Kenya national erosion losses are equal in magnitude to national electricity production or agricultural exports. Yield declines can, of course, be compensated for by adding fertilizers, assuming the farmer can afford to do so. However, in some cases severe erosion may result in a farmer having to abandon the field. Bakker *et al.* (2005) found this effect has occurred during the past century on the Greek island of Lesvos, where large areas used to cultivate cereals have been turned over to extensive grazing. High rates of erosion were shown to be an important driver of the abandonment and reallocation of land uses from arable to pasture.

Off-site effects

The off-site effects of eroded soil are caused by its transport and deposition. Material carried in strong winds can cause substantial damage to structures such as telegraph poles, fences and larger structures by

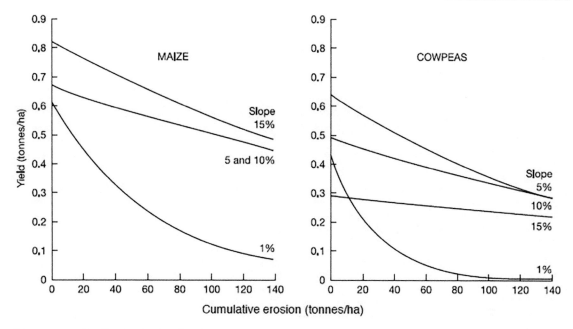

Figure 14.4 Decline in yields of maize and cowpeas with cumulative loss of soil in south-west Nigeria (Lal, 1993).

abrasion. Material transported in dust storms, which can affect very large areas, severely reduces visibility (Figure 14.5), causing a hazard to transport, and adversely affects radio and satellite communications. The transport of soil dust contaminated by organisms or toxic chemicals spreads human and plant diseases. Inhalation of fine particles can aggravate human diseases such as bronchitis and emphysema, and a wide variety of infectious diseases have also been associated with dust exposure. Elderly people, young children, and individuals with heart problems are at the greatest risk from health effects of dust storms (Schweitzer *et al.*, 2018). An increase in the incidence of meningococcal meningitis in the Sahel zone and Horn of Africa has been associated with airborne desert dust. The annual meningitis epidemics in West Africa, which affect up to 200,000 people between February and May, are closely related in their timing to the Harmattan season when Saharan dust is blown across the region (Sultan *et al.*, 2005). The nutrients attached to soil particles transported by water erosion can cause the eutrophication of water bodies (see Chapters 7 and 8). Other off-site impacts of erosion include damage to infrastructure caused by muddy floods (see Box 14.1) and associated public health issues.

The deposition of eroded material can also cause considerable problems for human society. The flood hazard can be increased due to river-bed infilling, and the siltation of reservoirs, harbours and lakes presents hazards to transport and loss of storage capacity in reservoirs. The economic costs thus imposed are often considerable. Dredging of harbours and channels used for navigation in the USA costs more than US$250 million a year, a price tag that reaches up to US$5 per tonne of soil deposited in some areas (Hansen *et al.*, 2002). Deposition of sediment reaching the coastline can adversely affect marine environments used by local populations, including coral reefs and shellfish beds.

Despite the numerous negative aspects of sediment deposition for human society, the positive side of the equation should also be highlighted. Relatively flat, often low-lying areas of deposition provide good, fertile

Figure 14.5 A dense dust haze obscures the harbour bridge in Sydney, Australia, in September 2009. A large part of eastern Australia was affected by this dust storm that deposited up to 7 million tonnes of topsoil off the east coast.

land for agriculture and many other human activities. Floodplains are an obvious example, as are valley bottoms that receive material transported from valley slopes. Wind-deposited dust, or loess, also provides fertile agricultural land – the loess plateau of northern China is the country's most productive wheat-growing area, for example – although it requires careful management because it remains highly erodible. Similarly, evidence is growing that dust deposition may make an important contribution to ecosystem nutrient budgets on land, in lakes and rivers and at sea. Because desert dust can be transported many thousands of kilometres through the atmosphere, these contributions can be far removed from source areas. For example, Saharan dust blown across West Africa in the Harmattan wind is thought to provide key nutrients to the humid tropical rain forests of coastal Ghana, some 2000 km away, but also to the Amazon rain forest at least another 5000 km across the Atlantic Ocean (McTainsh and Strong, 2007).

ACCELERATED EROSION

Although some of the adverse effects of soil erosion for human society outlined above may occur due to natural rates, by far the most pressing problems are found in areas where accelerated rates occur. Natural rates vary enormously depending on such factors as climate, vegetation, soils, bedrock and landforms. Information on natural erosion rates is used as a yardstick against which to measure the degree to which human action exacerbates natural processes and the sort of rates that soil conservation measures might aim to achieve (see below). Rates of soil loss also need to be compared to rates of soil formation, another area where our knowledge is limited. Natural rates of soil formation depend on the weathering of rocks, the deposition of sediment and the accumulation of organic material from plants and animals. They are generally taken to be less than 1 t/ha/year, and for many practical purposes acceptable target rates for soil conservation, or tolerable soil erosion rates, are commonly set at several times that level, depending on local conditions. Historically, notions of tolerable soil erosion have stemmed largely from the impact of erosion and soil loss on agriculture, but more recently the impact of erosion on other ecosystem services has also been appreciated, and Verheijen *et al.* (2009) propose an upper limit of 1.4 t/ha/year for conditions prevalent in Europe. In some countries, lower limits have been established. Verheijen *et al.* note that the threshold tolerated in Switzerland is 1 t/ha/year for most soils and double that in Norway. The Swedish Environmental Protection Board considers a soil loss of 0.1–0.2 t/ha/year as a recommended limit for preventative measures to be applied on arable land (Alström and Åkerman, 1992).

The two most common effects of human activity that lead to accelerated erosion are modifications to, or removal of, vegetation, and destabilization of natural surfaces. Such actions have a variety of motives: vegetation may be cleared for agriculture, fuel, fodder or construction; vegetation may be modified by cropping practices or deforestation for timber; land may be disturbed by ploughing, off-road vehicle use, military manoeuvres, construction, mining or trampling by animals. Other processes of erosion also follow such disturbances: effects on slope stability and resulting mass movement problems can be caused by highway construction. Transport routes can cause accelerated landsliding by increasing disturbing forces acting on a slope, both during construction when cuts and excavations remove lateral or underlying support (Figure 14.6),

Figure 14.6 A landslide blocking a road in the Khasi Hills in north-east India. Very heavy rainfall in this area, where average annual totals approach 12,000 mm in some parts, causes frequent landslides during the monsoon.

and through earth stresses caused by passing vehicles. Other human activities that can increase the chance of slope failure do so by decreasing the resistance of materials that make up slopes. This can occur if the water content is increased, as happens when local water tables are artificially increased by reservoir impoundment, for example.

The initial impact of certain activities may be reduced when a new land use is established, however. Many construction activities initially cause marked increases in soil loss, but erosion rates can be reduced to below those recorded under natural conditions when a soil surface is covered by concrete or tarmac. Conversely, soil erosion problems may become displaced by the effects of construction, with water flows from drainage systems causing accelerated soil loss where they enter the natural environment.

Another human impact that may result in temporarily accelerated erosion rates occurs through the use of fire as a management tool. The burning of bush and grasses is a long-established and widely used management technique in the savannas of Africa, used to encourage tender green shoots from perennial grasses for livestock to feed on and to release phosphorus and other nutrients for use by crops. The exposed soil is particularly susceptible to erosion if a new vegetation cover does not become established before an erosive event. Numerous studies have found that severe fires result in an increase in runoff and sediment yields, often by several orders of magnitude. This has been attributed to several factors, including the loss of protection from surface cover, enhanced soil water repellence, and soil sealing by sediment particles and/or by particles of ash (Larsen *et al.*, 2009).

There is widespread agreement that the prime causes of accelerated soil erosion are deforestation and agriculture. Deforestation removes the protection from raindrop impact offered to soil by the tree canopy, and reduces the high permeability humus cover of forest floors, a permeability that is enhanced by the many macropores produced by tree roots. Cultivation also removes the natural vegetation cover from the soil, which is particularly susceptible to erosion when bare after harvests and during the planting stage. Some crops, such as maize and vines, usually leave large portions of the ground unprotected by vegetation even when the plants mature, which can result in particularly high rates of soil loss (Table 14.3). Furthermore, mechanical disturbance and compaction of the soil by ploughing and tilling can enhance its erodibility.

An illustration of the dramatic effects on soil loss instigated by deforestation is shown in Figure 14.7 which is derived from analysis of sediment and pollen in a lake bed in an area cleared of woodland in the mid-nineteenth century in Michigan, USA. The initial response to clearance was a sharp rise in erosion by 30 to 80 times the rates derived for the pre-settlement period. Fluctuating rates characterized a period of about 30 years as a new steady state was reached under farmland, which is nonetheless some ten times greater than under undisturbed woodland. Similar patterns of accelerated soil loss by water erosion are typical of areas deforested in the humid tropics, as outlined for South-East Asian forests in Chapter 4 (see p. 75).

Some of the most significant cases of accelerated erosion by wind have occurred in dry grassland areas used for grain cultivation. Probably the most infamous case of wind erosion in the western world came after widespread ploughing of the grasslands of the US Great Plains, which created the Dust Bowl of the 1930s (see p. 108), but similar environmental mishaps occurred in the former USSR in the 1950s and in a copycat exercise in the Mongolian steppes in the 1960s. An area nearly the size of Spain – some 45 million hectares of steppe grassland, so-called Virgin Lands – was put to the plough in northern Kazakhstan, western Siberia and eastern Russia between 1954 and 1963 (Figure 14.8). Deep ploughing was employed, removing the stubble from the previous year's crop to allow planting earlier in the year, so reducing the chance of losses to early snows at harvest time. Land was also used more intensively, doing away with alternate years when soil was traditionally left fallow under grass. Wind erosion soon began to take its toll (Table 14.4), coming to a head in the early 1960s when drought hit the region.

Table 14.3 Some examples of mean annual erosion rates measured in European vineyards

Country (region)	Measurement scale	Average erosion rate (t/ha/year)
Italy (Sicily)	m² – ha	102.2
Italy (Emilia-Romagna)	m² – ha	20.7
Slovakia (Vràble)	m² – ha	14.6
France (Burgundy)	ha	12.0
France (Burgundy)	m² – ha	8.3
Slovenia (Maribor)	m²	1.0

Source: after Prosdocimi et al. (2016); and Novara et al. (2011).

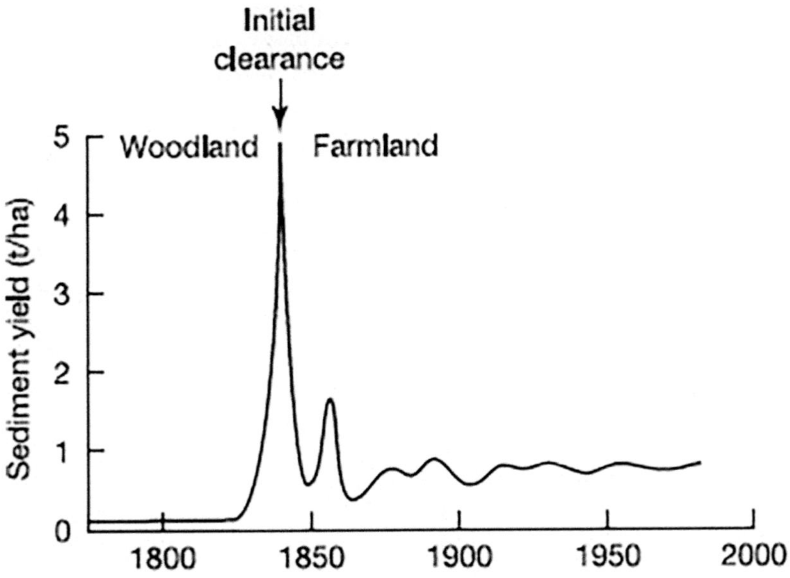

Figure 14.7 Historical reconstruction of sediment yield at Frains Lake, Michigan, USA (Davis, 1976).

The introduction, from outside, of agricultural techniques that are inappropriate to local conditions is a widely cited driver of accelerated erosion. The US Dust Bowl was driven by settlers using machinery developed in western Europe, and soil degradation in many parts of Latin America has similarly been blamed on management approaches brought by Spanish and Portuguese colonists in combination with an intensification of land use. However, evidence from lake sediment cores in central Mexico suggests caution over the ease with which invading Iberians are blamed for unbalancing traditional, supposedly harmonious, systems of soil use. Analysis of the cores indicates that pre-Hispanic agriculture in the basins of Patzcuaro and Zacapu was not as conservationist in practice as was previously thought. Several periods of accelerated

Figure 14.8 A monument to the expansion of wheat cultivation on the steppes of Kazakhstan, part of the Soviet Union's Virgin Lands programme of the 1950s which has been plagued by accelerated wind erosion.

Table 14.4 Effect of the Virgin Lands Scheme on the frequency of dust storms in the Omsk region of western Siberia

Station	Mean annual number of dust storm days		Increase
	1936–50	1951–62	
Omsk, steppe	7	16	× 2.3
Isil'-Kul'	8	15	× 1.9
Pokrov-Irtyshsk	4	22	× 5.5
Poltavka	9	12	× 1.3
Cherlak	6	19	× 3.2
Mean	6.8	16.8	× 2.8

Source: after Sapozhnikova (1973).

erosion have been identified that occurred before the arrival of the Spanish and were of comparative magnitude to those during colonial times. This suggests that the introduction of the plough did not have a greater impact on soil erosion than traditional methods (O'Hara and Metcalf, 2004).

Spanish and Portuguese colonists also introduced exotic animals – sheep, goats and cattle – to graze the pastures of Latin America and their different grazing strategies may have had a significant impact on soil and vegetation. This is because native domesticated camelids – alpacas and llamas – are low-impact grazers since they lack hooves that compact the soil. The development of severely eroded terrain riven by steep-sided gullies, so-called 'badlands', in the southern Bolivian Andes is thought to be partly associated with the introduction of goats and cattle by the Spaniards 500 years ago. Indeed, it seems that the area's grazing history has altered its sensitivity to erosion, to such an extent that contemporary erosion rates are high despite current low grazing intensity (Coppus *et al.*, 2003).

In Madagascar extensive erosion on hillslopes is commonly associated with contemporary large-scale deforestation and subsistence agriculture (Figure 14.9). The erosion is represented by widespread gullies known as 'lavaka' from the Malagasy word for hole but these features have received relatively little attention from scientists and their origins are not necessarily straightforward (Zavada *et al.*, 2009). Only a small proportion of lavakas can unequivocally be linked to human activities, particularly deforestation and

Figure 14.9 Sediments from severe soil erosion in north-western Madagascar colour the Betsiboka River.

controlled burning of grassland. Others may have formed naturally in response to Madagascar's climatic regime and continuing tectonic uplift of the central highlands.

Although serious soil erosion is often associated with the humid tropics and semi-arid areas, it does give cause for concern in more temperate regions. The intensification of agriculture in the UK, for example, has brought erosion problems to the fore, with about a third of arable land in England and Wales classified as being at moderate to high risk of erosion (Evans, 2005). Sandy and peaty soils in parts of the Vale of York, the Fens, Breckland and the Midlands suffer from deflation during dry periods, but water is the most widespread agent of soil loss. The central reason for the increase in erosion in the past 30 years or so has been a shift in the sowing season, from spring to autumn, for the main cereal crops of wheat and barley. The change has come about because autumn-sown 'winter cereals' produce higher yields, but the shift in sowing season leaves winter cereal fields exposed in the wettest months – October and November – in many parts of the country, since arable fields are at risk from erosion until about 30 per cent of the ground is covered by the growing crop. Several other aspects of more intensive arable farming have also contributed to the enhanced erosion rates, including the expansion of arable crops on to steeper slopes made possible by more powerful tractors, and the creation of larger fields by removal of walls, hedges, grass banks and strips. The larger fields have longer slopes and larger catchment areas, allowing greater volumes of water to be generated on slopes and in valley bottoms. Driving these changes are a range of socio-economic and political forces (see below). Muddy runoff from these fields causes considerable off-site damage to property, roads and freshwater systems, particularly in southern England. These muddy floods are also common in agricultural areas of central Belgium and northern France where they more typically occur in summer, associated with heavy thunderstorms on fields planted with maize, potatoes and sugar beet (Evrard *et al.*, 2010).

SOIL CONSERVATION

Since soil is such a vital natural resource, using it sustainably involves employing methods to reduce accelerated erosion. Sustainable land management incorporates many techniques for protecting cultivated soils from erosion. Table 14.5 shows examples employed in sub-Saharan Africa, many of them with long histories of use. Overall, soil conservation systems can be classified into three groups:

1 agronomic measures, which manipulate vegetation to minimize erosion by protection of the soil surface
2 soil management techniques, which focus on ways of preparing the soil to promote good vegetative growth and improve soil structure in order to increase resistance to erosion
3 mechanical methods, which manipulate the surface topography to control the flow of water or wind.

Maintaining a sufficient vegetative cover on a soil is sometimes referred to as the 'cardinal rule' for erosion control. One of the oldest of agronomic measures designed to reduce soil erosion is to rotate the location of crops, so-called shifting cultivation, in which an area of forest is cleared and cultivated for a year or two and then allowed to revert to scrub or secondary forest. An essentially similar method involves rotating crops grown in rows with cover crops such as grasses or legumes grown on the same field every other year. Mulching – the practice of leaving some residual crop material, such as leaves, stalks and roots, on or near the surface – is another widely used agronomic technique. It is successful in reducing erosion and in reducing the loss of water from fields by decreasing evaporation. But crop residues also provide a good habitat for insects and weeds so that the method often requires higher chemical inputs of pesticides

Table 14.5 Some examples of soil and water conservation systems from sub-Saharan Africa

Country (region)	Average annual ppt (mm)	Technique(s)	Major crops
Burkina Faso (Mossi)	400–700	Stone lines, stone terraces, planting pits	Sorghum, millet
Cape Verde (São Antão)	400–1200	Dry stone terraces	Sugar cane, sweet potatoes, maize
Chad (Ouddal)	250–650	Earth bunds	Sorghum, millet
Mali (Djenne-Sofara)	400	Pitting systems	Sorghum, millet
Niger (Ader Doutchi Maggia)	300–500	Stone lines, planting pits	Sorghum, millet
Rwanda (north-west)	1100–1400	Terraces	Beans, bananas, sorghum
Somalia (Hiiraan)	150–300	Earth bunds	Sorghum, cowpeas
South Africa (Free State Province)	550	Basin and mulch	Maize and sunflower
Uganda (Kabale District)	750	Trash lines from crop residues	Maize, millet and sorghum

Source: after Critchley *et al.* (1994); and Biazin *et al.* (2012).

and herbicides where available, which is expensive and increases hazards of off-field pollution and the killing of non-target species.

Mulching can also be conducted using vegetable matter from elsewhere. Grass is widely used on the central plateau of Burkina Faso just before the rainy season from February to May where soils are liable to crust formation, which produces considerable runoff during heavy rainstorms (Slingerland and Masdewel, 1996). The mulch is not only an effective method for soil and water conservation; the grass helps to fertilize the soil by decomposition, and attracts termites that burrow into the soil, breaking up the crust and increasing soil porosity and permeability. Although the termites eat the grass, if sufficient is used, the mulch still protects the soil from rainfall and reduces runoff.

Not all examples of agronomic measures have used local species for controlling erosion and a cautionary tale can be cited in this respect from the US southern states where a perennial vine called kudzu was introduced from East Asia in 1876 and widely planted for erosion control (Forseth and Innis, 2004). Kudzu is popularly known as 'the vine that ate the south' due to its rapid growth rate and tendency to smother whatever it grows on, including trees, utility poles and houses (Figure 14.10). Re-designated as an invasive plant in 1950, kudzu now covers an estimated 3 million ha of the USA and represents a substantial threat to native ecosystems and, ironically, agricultural land.

Most soil management techniques are concerned with different methods of soil tillage, an essential management technique that provides a suitable seed bed for plant growth and helps to control weeds. However, different types of soil respond in different ways to tillage operations and a range of methods have been

Figure 14.10 Kudzu vines covering trees and telegraph poles along a road in Kentucky, USA.

developed to reduce erosion effects. Strip or zone tillage leaves protective strips of untilled land between seed rows, requiring weed control on the protective strips. In recent decades, the advent of modern herbicides has permitted development of other tillage methods designed to leave varying degrees of vegetative matter from the previous crop still on the soil surface to provide protection against erosive forces. Minimum tillage incorporates the idea of stubble mulching mentioned above, while zero-tillage leaves most of the soil covered with plant residues. Zero-tillage technology allows the farmer to lay seed in the ground at the required depth with minimal disturbance to soil structure, the specially designed machinery eliminating the need for ploughing and minimizing the tillage required for planting. In addition to being very effective at controlling erosion, these forms of tillage have effectively revolutionized agricultural systems in many parts of the world because they have allowed individual farmers to manage greater amounts of land with reduced energy, labour and machinery inputs (Triplett and Dick, 2008). They also improve the efficiency of water and fertilizer use so that many crops produce higher yields. Their adoption became widespread in the USA in the 1980s and subsequently in Australia, Canada and parts of South America.

Zero-tillage practices were widely adopted in the pampa region of Argentina in the early 1990s, particularly on soybean farms. They were promoted by an innovative partnership involving farmers, researchers, extension workers, and private companies and are credited with improving soil fertility by reversing decades of degradation. The associated reduction in costs for farmers in Argentina plus increases in gross

Figure 14.11 Terracing is an ancient soil and water conservation technique typically employed on steep slopes like here in the Himalaya of central Nepal.

income through improved production and productivity were estimated to be more than US$16 billion between 1991 and 2008 (Trigo *et al.*, 2009).

Mechanical methods, which are normally used in conjunction with agronomic measures, include such techniques as the building of terraces and the creation of protective barriers against wind, such as fences, windbreaks and shelter belts. Terracing is a very ancient soil conservation technique, dating back 5000 years in Palestine and Yemen, for example (Wei *et al.*, 2016). Terraces are commonly built on steep slopes and effectively transform them by creating a series of horizontal soil strips along the slope contours (Figure 14.11). They intercept runoff, reducing its flow to a non-erosive velocity, thus also acting to conserve water. If well maintained, terraces are very effective, but they are costly to construct and their physical dimensions act as a constraint on the use of mechanized agriculture.

Check dams are constructed in a similar way to control gully erosion. The dam impedes the flow of water generated by a rain storm, causing the deposition of sediment – which helps to build up a soil – and encouraging the ponded water to soak into this alluvial fill. A small check dam might then support a single fruit tree, for example. Larger dams may allow a cereal crop to be planted in a small field. In Tunisia this type of small check dam and its associated terraced area is called a 'jessr' (plural: jessour). Jessour, which cover an area of about 4000 km² in the south of the country, allow crops to be grown in an environment that

Table 14.6 Effect of low-cost soil conservation techniques on erosion and crop yields

Method	Decrease in erosion (%)	Increase in yield (%)
Mulching	73–98	7–188
Contour cultivation	50–86	6–66
Grass contour hedges	40–70	38–73

Source: after Doolette and Smyle (1990).

is otherwise too arid for agriculture. The most common crops are olive trees and drought-resistant annual grains with a short growing period such as wheat and barley (Schiettecatte *et al.*, 2005).

Windbreaks perform the same sort of function against wind erosion: they reduce wind velocity, thereby lowering its erosivity, and encourage the deposition of material that is already entrained. Fences or walls placed at right angles to erosive winds serve this purpose, or windbreaks may be created from living plants such as trees or bushes, in which case they are known as shelter belts. Numerous other benefits to crops can also often be associated with the establishment of windbreaks. These include increased soil and air temperatures, reduced pest and disease problems, and an extended growing season in sheltered areas. The planting of shelter belts has a long history in many parts of the world. In Russia, the additional importance of living shelter belts for carbon sequestration has recently been demonstrated by Chendev *et al.* (2015): the trees improve soil quality by enhancing soil organic matter content while providing a sink for atmospheric carbon in tree biomass and soil organic matter.

The beneficial effects of some of these erosion control techniques for soil retention and crop yields are shown in Table 14.6, in which the range of percentage figures is derived from a review of more than 200 case studies.

Implementation of soil conservation measures

Given that we have a good understanding of the factors affecting accelerated soil erosion, the problems caused by the process, and the methods for its control, the question 'Why is soil erosion still a problem?' is an obvious one to ask. Innumerable soil conservation projects have been implemented, particularly in the poorer areas of the world, but often with disappointing results. In many cases, farmers simply do not adopt agricultural innovations such as the soil conservation techniques outlined in the previous section. There are numerous explanations for this non-adoption, but there appear to be few if any universal answers that regularly explain the phenomenon (Knowler and Bradshaw, 2007). An analysis of the reasons for the failure of such projects made by Hudson (1991) found that the poor design of projects was a fundamental flaw. He concluded that errors are mainly a function of incorrect assumptions made at the design stage, by both donor agencies and host governments. The main donor errors are:

- over-optimism, including overestimating the effect of new practices
- overestimating the rate of adoption of new practices
- overestimation of the ability of the host country to provide back-up facilities

- underestimation of the time required to mobilize staff and materials for the project
- frequently, a quite unreal estimate of the economic benefits.

Some of the main problems arising from the assumptions of the host governments are:

- overestimating their capacity to provide counterpart staff and the funds for the recurrent costs arising from the project
- a tendency to underestimate the problems of coordination among different ministries or departments
- a tendency to overestimate the strength of the national research base and its ability to contribute to the project.

The importance of engaging farmers as active players in efforts to conserve soil, rather than passive adopters of technology, is highlighted in projects that focus on a society's 'social capital' as an effective way of achieving greater impacts. In the broadest sense, social capital refers to the interconnectedness among individuals in a society and considers relationships as a type of asset, and there is some evidence to suggest that social capital can both create incentives and remove barriers to the adoption of soil conservation practices. Government support for social capital has been identified as an element in the success of Landcare in Australia, a grassroots movement dedicated to managing environmental issues in local communities (Sobels *et al.*, 2001).

Other observers of the soil erosion issue believe that attention must also be focused on the central reasons for people misusing soils, which lie at a deeper level. These underlying driving forces are social, economic and political in nature, and include such factors as population growth, unequal distribution of resources, land tenure, terms of trade and subsidies, colonial attitudes and legacies, and class struggles. These factors limit the options available to poorer sectors of society who may be forced into degrading their soil resource simply in order to survive.

This approach to understanding the political ecology of degradation might be based on the 'Chain of explanation of soil erosion' proposed by Blaikie (1989), which analyses the causes of erosion from the land user at one end of the spectrum all the way up to the nature of international relations at the other. A typical example is a farmer who is forced to cultivate steep slopes because his original, more productive fields have been permanently flooded by a hydroelectric power project financed by international donors. Where does the ultimate blame lie? The farmer? Local or national government? The governments donating the foreign aid? There is no straightforward answer. This political ecology approach applies equally to erosion issues that impact richer countries, as illustrated by the example of muddy floods in southern England detailed in Box 14.1.

In other situations, erosion may result because it is not perceived as a serious problem since it often proceeds at relatively imperceptible rates, or because its effects are masked by fertilizers. Allied to this problem, the benefits of soil conservation on the farm may be long-term, whereas farmers may have a short-term view of the relative importance of maintaining their income or repaying their debts. The costs involved in implementing soil conservation measures may also be high and there is an ethical and practical question of who should pay: the individual farmer or society as a whole? Often it is the off-site impacts of soil erosion that cause more immediate concern if a farmer is found liable for damage caused by sedimentation of material eroded from his fields. In a UK study of the cost-effectiveness of erosion control measures based on an ecosystem services approach, Posthumus *et al.* (2015) concluded that many measures result in negative financial and economic returns, which explains why UK farmers are generally reluctant to implement erosion control measures without compensation.

Box 14.1 Managing muddy floods in southern England

A complex set of drivers, allied to the involvement of numerous stakeholders and institutions of governance, has hampered the successful management of muddy floods in southern England. The impacts of muddy flooding are serious both on-site and off-site. Loss of soil from a farmer's arable field involves gullying and damage to crops and the muddy runoff causes damage to property, roads, rivers and lakes.

Successful management of muddy floods is complicated by the large number of stakeholder groups involved and their varied expectations. The stakeholders include farmers and home-owners, local government and commercial insurance companies. Typically the financial costs of a muddy flood are borne by these groups. Individual farmers commonly bear the costs of on-site damage while off-site damage is paid for by local councils (e.g. damage to public infrastructure such as roads) and insurance companies (e.g. damage to private properties). This relatively simple allotment of costs may be further complicated, however, if legal action is taken against farmers, which sometimes happens.

Managing muddy floods is also complicated because it is an issue that involves overlap between policies in several different areas: agriculture, environment, land-use planning, and water management. In England, the institutions involved include departments of local, national and European government. For example, the European Common Agricultural Policy (CAP) encouraged conversion of chalk landscapes in southern England from grass and spring cereals to winter cereals, a land-use change that led directly to the muddy erosion problems. CAP reforms in the 1990s introduced agri-environmental schemes that pay farmers to implement environmentally friendly techniques, but these incentives encourage farmers to tackle the symptoms of the problems (e.g. runoff, erosion), rather than the causes themselves (i.e. intensive agriculture).

In England, local government authorities have no legal responsibility to protect people from muddy floods, but in practice they often help to organize emergency defence measures such as ditches, dams and pipes. Farmers frequently feel responsible for erosion on their land and implement their own initiatives to alleviate muddy flooding, such as the building of retention dams to protect their own or others' property from floods.

Muddy flood management can either focus on soil and water conservation or on property protection, but not all stakeholders necessarily agree on the best approach. These and other issues need to be overcome if an integrated scheme to manage erosion and muddy floods is to be created.

Source: Evrard *et al.* (2010).

Another interesting stance on the issue of different assessments of erosion is presented by Warren *et al.* (2003). These authors contrast a scientific view of erosion in south-western Niger with that of local villagers practising subsistence agriculture. Scientific measurement using caesium-137 found the erosion rate to be over 30 t/ha/yr but local farmers showed relatively little commitment to preventing the problem. The study concluded that erosion may be unavoidable in the face of multiple risks, given the current constraints of labour and the need to produce a crop every year under conditions of variable and unpredictable rainfall. Despite the scientific evidence to support the imposition of a rigorous system of soil conservation, it was felt that such outside interference might threaten the cohesion of the very community it would be designed to assist. This example highlights the apparent mismatch that can occur between a scientific perspective and that taken by local farmers. As a result, a participatory approach to soil erosion issues is probably best (Barrios *et al.*, 2006), recognizing that local farmer knowledge offers many insights into the sustainable management of soils and that the integration of local and technical knowledge systems helps extension workers and scientists work more closely with farmers.

FURTHER READING

Arnalds, A. 2005 Approaches to landcare: a century of soil conservation in Iceland. *Land Degradation and Development* 16: 113–25. Interesting lessons from probably the world's oldest soil conservation service.

Chendev, Y.G., Sauer, T.J., Ramirez, G.H. and Burras, C.L. 2015 History of east European chernozem soil degradation; protection and restoration by tree windbreaks in the Russian steppe. *Sustainability* 7: 705–24. An interesting study of wind erosion control. Open access: www.mdpi.com/2071-1050/7/1/705/htm.

Fifield, L.K., Wasson, R.J., Pillans, B. and Stone, J.O.H. 2010 The longevity of hillslope soil in SE and NW Australia. Catena 81: 32–42. A study showing that modern rates of erosion are markedly higher than either soil formation rates or natural erosion rates.

Gomiero, T. 2016 Soil degradation, land scarcity and food security: reviewing a complex challenge. Sustainability 8: 281. An overview of the links between soil and food. Open access: www.mdpi.com/2071-1050/8/3/281/htm.

Middleton, N. and Kang, U. 2017 Sand and dust storms: impact mitigation. Sustainability 9: 1053. A review of methods employed to reduce wind erosion impacts. Open access: www.mdpi.com/2071-1050/9/6/1053/htm.

Panagos, P., Imeson, A., Meusburger, K., Borrelli, P., Poesen, J. and Alewell, C. 2016 Soil conservation in Europe: wish or reality? Land Degradation and Development 27(6): 1547–51. The importance of good policies to promote soil conservation. Open access: onlinelibrary.wiley.com/doi/10.1002/ldr.2538/full.

WEBSITES

globalsoilbiodiversity.org   the Global Soil Biodiversity Initiative promotes the translation of expert knowledge on soil biodiversity into environmental policy and sustainable land management.

www.ieca.org   the International Erosion Control Association is the world's oldest and largest association devoted to the problems caused by erosion and sediment.

www.isric.org   the International Soil Reference and Information Centre contains information on international research programmes on all aspects of soil degradation.

www.soils.org   the Soil Science Society of America site has news and educational resources.

www.swcs.org   the Soil and Water Conservation Society site includes news, conferences and publications.

www.wocat.net   the World Overview of Conservation Approaches and Technologies (WOCAT) promotes the documentation, sharing and use of knowledge to support soil and water conservation.

POINTS FOR DISCUSSION

- Assess the role of soil in the classification of ecosystem services outlined in Chapter 3.
- Should we be concerned about soil erosion? If so, why?
- Why are not all erosion measurement techniques comparable?
- Design a research programme to determine whether the implications of soil erosion are more serious on-site or off-site.
- Outline the main ways in which people influence erosion rates.
- There are many techniques to prevent soil erosion, so why are there still areas where soil erosion is a problem?

Biodiversity loss

LEARNING OUTCOMES

At the end of this chapter you should:

- Appreciate that our understanding of biodiversity is still insufficient in many respects.
- Recognize that we are widely thought to be responsible for the Earth's current sixth mass extinction event.
- Know the five direct drivers of biodiversity loss.
- Understand that many species are at risk from more than one threat and that threats can change through time.
- Recognize that island species are especially vulnerable.
- Understand that human perceptions of biodiversity can change.
- Appreciate that the establishment of protected areas is the leading method of biodiversity conservation.
- Understand that species diversity is geographically uneven and biodiversity hotspots have been identified where large numbers of endemic species are heavily threatened.

KEY CONCEPTS

extinction, mass extinction event, endemism, K-strategist, r-strategist, keystone species, extinction debt, biological invasion, protected area, Convention on International Trade in Endangered Species (CITES), *ex situ* conservation, hotspot, Aichi Biodiversity Targets

---

Biodiversity, a term that refers to the number, variety and variability of living organisms, has become a much-debated environmental issue in recent times. Although we live on a naturally dynamic planet, where species are prone to extinction and ecosystems are always subject to change, biodiversity has become an issue of major concern because of the unprecedented rate at which human action is causing its loss in the Anthropocene. The very large and rapid increase in human population over recent centuries has been accompanied by an unprecedented scale of modification and conversion of ecosystems for agriculture and other human activities. At the same time, we have documented increasing numbers of cases where species have been driven to extinction by human activities.

Although the exact definition of biodiversity (or biological biodiversity) is the subject of considerable discussion, it is commonly defined in terms of genes, species and ecosystems, corresponding to three fundamental levels of biological organization. Some authorities also include a separate human element in their definitions: cultural diversity. The Convention on Biological Diversity defines biodiversity as 'the variability among living organisms from all sources including, *inter alia*, terrestrial, marine and other aquatic ecosystems and the ecological complexes of which they are a part; this includes diversity within species, between species and of ecosystems'. Genetic diversity includes the variation between individuals and between populations within a species. Species diversity refers to the different types of animals, plants and other life-forms within a region. Ecosystem diversity means the variety of habitats found in an area. Much of this chapter is focused at the species level because this is where most research is concentrated. Some aspects of genetic and ecosystem diversity are dealt with in more detail elsewhere (see Chapters 4 and 13).

UNDERSTANDING BIODIVERSITY

The 158 states that signed the Convention on Biological Diversity at the UN Conference on Environment and Development in Rio de Janeiro in 1992 agreed that there was a general lack of information on and knowledge of biodiversity, and that there was an urgent need to develop scientific, technical and institutional capacities to provide the basic understanding on which to plan and implement appropriate measures. Progress has been made since then, of course, but understanding is still insufficient on many aspects of biodiversity. Although scientists have been systematically counting and classifying other living organisms for two and a half centuries, we remain remarkably ignorant of the most basic of information concerning the living things with which we share the Earth. While the numbers of species in some groups of organisms are relatively well known, our knowledge of others is extremely imprecise. The number of bird species, for example, is close to 10,000 (Sibley and Monroe, 1990), but estimates of the number of insect species vary widely. Expert estimates of the total number of all species range between 2 million and 100 million. About 1.5 million species have been scientifically catalogued to date, and a reasonable estimate of the total number of species on the planet is 2–8 million (Costello *et al.*, 2013).

The geographical pattern of biodiversity is not even. Uncertainties about *where* species are may be more limiting than not knowing *how many* species there are (Pimm *et al.*, 2014). Most estimates agree that more than half of all species live in the tropical moist forests that cover just 6 per cent of the world's land surface. The factors that influence the distribution of biodiversity are numerous, and probably change in their importance at different scales of both space and time. Table 15.1 is an attempt to summarize some of the most significant environmental variables.

Our knowledge of the world's biomes and individual ecosystems is also unsatisfactory. Indeed, although several systems of classification have been developed, no single measure of ecological community diversity can be uniformly applied to all ecosystems. We are also still trying to understand the ways in which biological diversity affects the functioning of ecosystems and the provision of ecosystem services. Although genetic diversity is the ultimate basis for evolution and for the adaptation of populations to their environment, it is even less well understood. We know very little about the genetics of most living organisms. The few exceptions are for a handful of species identified as having direct importance to certain forms of economic activity, such as agriculture (see p. 324) and human health.

Table 15.1 A hierarchical framework for processes influencing species diversity

Spatial scale	Dominant environmental variables	Temporal scale
Local: within communities, within habitat patches	Fine-scale biotic and abiotic interactions (e.g. habitat structure, disturbance by fires, storms)	~1–100 years
Landscape: between communities; turnover of species within a landscape	Soils, altitude, peninsula effect	~100–1000 years
Regional: large geographical areas within continents	Radiation budget and water availability, area, latitude	the last 10 000 years (i.e. since end of last glacial)
Continental: differences in species lineages and richness across continents	Aridification events, Quaternary glacial/interglacial cycles, mountain-building episodes (e.g. Tertiary uplift of the Andes)	the last 1–10 million years
Global: differences reflected in the biogeographical realms (e.g. distribution of mammal families between continents)	Continental plate movements, sea-level change	the last 10–100 million years

Source: after Willis and Whittaker (2002).

Given the poor state of our knowledge, it is likely that many species that we have never known about have become extinct in recent times. This is probably the case even for groups of organisms that are well documented, such as birds (Lees and Pimm, 2015). Among those organisms that have been described, problems emerge in documenting their disappearance. Just because no member of a species has been documented for some time does not necessarily mean that it has become extinct: absence of evidence is not evidence of absence. A classic example is the 'rediscovery' of the coelacanth (*Latimeria chalumnae*), a deep-sea fish long thought to have gone extinct at the end of the Cretaceous period (Figure 15.1).

Despite these and other difficulties, estimates of the rate at which species are becoming extinct, mostly due to human action, indicate that the rate has been growing exponentially since about the seventeenth century. Many current and projected estimates of species loss are based upon the rate at which habitats are being destroyed, degraded and fragmented – the most serious threats to species diversity – coupled with biogeographical assumptions relating to numbers of species and area of habitat. We should be aware, however, that estimates for habitat loss in tropical forest areas, the most diverse ecosystems, are themselves subject to wide variations (see p. 67). Another, in some ways more effective, strategy has been to analyse extinction in groups where the size of the species pool is quite well known. Examples include North American birds, tropical palms and Australian mammals.

While some earlier projections of global extinction rates suggested that 20–50 per cent of species would be lost by the end of the twentieth century (Myers, 1979; Ehrlich and Ehrlich, 1981), these now seem exaggerated. Nonetheless, more contemporary estimates are still alarming. Globally, current rates of extinction are thought to be about 1000 times the background rate of extinction (Pimm *et al.*, 2014) and humans have

Figure 15.1 The coelacanth was once known only from fossils and was thought to have gone extinct with the dinosaurs 65 million years ago, until the first living coelacanth was discovered in 1938.

probably increased the extinction rate for all species over the past few hundred years by as much as three orders of magnitude (MEA, 2005).

THREATS TO BIODIVERSITY

Ecosystems change and species become extinct under natural circumstances. We know that the Earth's climate is dynamic over a variety of timescales, and plant and animal communities have to adapt to these changes or run the risk of extinction. Species may also become extinct due to a range of other natural circumstances, such as random catastrophic events, or through competition with other species, by disease or predation. Indeed, as one biologist puts it: 'Extinction is a fundamental part of nature – more than 99 per cent of all species that ever lived are now extinct' (Jablonski, 2004: 589).

Studies of the fossil record, an excellent natural archive of extinctions, show us that long geological periods when the rate of species extinction was fairly uniform have apparently been punctuated by catastrophic episodes of mass extinction, during which global biodiversity was sharply reduced. In the last 570 million years of Earth's history, five 'mass extinction events' have occurred, each thought to have removed more than 60 per cent of marine species. The most severe was during the late Permian period some 245 million years ago, and the most recent was at the end of the Cretaceous period 65 million years ago when the dinosaurs and several other families of species were wiped out.

Currently, however, there is widespread fear that another – a sixth – mass extinction event is occurring, one in which the Earth's human population is playing the key role. Although we are not sure of the current rate of species extinction, five principal pressures on biodiversity or direct drivers of biodiversity loss have been identified (MEA, 2005):

- habitat loss and degradation
- climate change
- excessive nutrient load and other forms of pollution
- overexploitation and unsustainable use
- invasive alien species.

THREATENED SPECIES

Some species are particularly at risk from the threat of extinction simply because they are only found in a narrow geographical range, or they occupy only one or a few specialized habitats, or they are only found in small populations. Species that occur in only one location are known as 'endemic'. Other factors may also affect the degree of risk faced by certain species. These include:

- low rates of population increase
- large body size (hence requiring a large range, more food and making the species more easily hunted by humans)
- poor dispersal ability
- need for a stable environment
- need to migrate between different habitats
- perceived to be dangerous by humans.

Combinations of some of these characteristics are found in species known as K-strategists, and it is these species that are generally more likely to become extinct because they tend to live in stable habitats, delay reproduction to an advanced age and produce only a few, large offspring. By contrast, species that produce many offspring at an earlier age and have the ability to react quickly to changes in their environment, are known as r-strategists, and it is their speedier turnover and flexibility that make r-strategists less likely to experience extinction.

The wider ecological implications of the loss of a certain species vary between species. Another important aspect of the extinction issue is the fact that certain keystone species may be important in determining the ability of a large number of other species to persist. Hence, the loss of a certain keystone species could potentially result in a cascade of extinctions. Such fears have been expressed over tropical insects, many of which have highly specialized feeding requirements.

Species known to be at risk are documented, according to the severity of threat they face and the imminence of their extinction, by the International Union for Conservation of Nature and Natural Resources (IUCN, also known as the World Conservation Union) on so-called Red Lists (Table 15.2). The general term 'threatened' is used to refer to a species of fauna or flora considered to belong to any of the categories shown in Table 15.2. A Red List of Ecosystems is also under development by IUCN, with an aim to evaluate all of the world's ecosystems by 2025.

Table 15.2 IUCN Red List categories

Extinct	No reasonable doubt that the last individual has died
Extinct in the wild	Only known to survive in cultivation, in captivity or naturalized well outside past range
Critically endangered	Extremely high risk of extinction in the wild in immediate future
Endangered	High risk of extinction in the wild in near future
Vulnerable	High risk of extinction in the wild in medium-term future
Near threatened	Close to qualifying for critically endangered, endangered or vulnerable
Least concern	Not near threatened
Data deficient	Data insufficient to categorize but listing highlights need for research, perhaps acknowledging the need for classification
Not evaluated	Not assessed against the criteria

THREATS TO FLORA AND FAUNA

Most of the factors currently threatening species of both fauna and flora are induced or influenced by human action. Such actions may be deliberate, as in the case of destruction by hunting, or inadvertent, as in the case of destruction or modification of habitats in order to use the land for other purposes. In practice, many species are at risk from more than one threat. For example, non-human primates, our closest biological relatives, are most at risk from extensive habitat loss but other important threats are bushmeat hunting, the illegal trade of primates as pets and primate body parts, climate change and infectious diseases caught from humans. Estrada *et al.* (2017) point out that these pressures often act in synergy, exacerbating primate population declines. Some 60 per cent of primate species are now threatened with extinction and about 75 per cent have declining populations. Threats can also change through time. The decline of the New Zealand mistletoe (*Trilepidea adamsii*) began as its habitat was reduced by deforestation, first by the Maoris and at an accelerating rate by British settlers in the late nineteenth century. The population was further reduced by collectors, and the decline of bird populations that were responsible for seed dispersal, due to forest clearance. The final specimens, which disappeared from North Island in 1954, may have been eaten by brush-tailed possums deliberately introduced from Australia during the 1860s to establish a fur trade (Norton, 1991).

The detailed nature of threats to specific flora and fauna is frequently complex and variable. Individual drivers of extinction may be obscured because of a substantial timelag between cause and effect, dubbed the 'extinction debt'. In other cases, extinction may result from a suite of pervasive secondary processes and synergistic feedbacks that have a greater total effect than the sum of individual effects alone (Brook *et al.*, 2008). The loss of habitat by deforestation, for example, can cause some extinctions directly by removing all individuals over a short time period, but it can also be indirectly responsible for lagged extinctions in several ways. These include making the hunting of large mammals easier, facilitating invasions, eliminating prey and generally altering biophysical conditions. The combination of hunting and habitat destruction by deforestation has put paid to many original animal species in Britain over the centuries (Table 15.3).

Table 15.3 Dates when the last wild member of selected animal species was killed in Britain

Animal	Year
Wolverine (*Gulo gulo*)	8000 BP
Elk (*Alces alces*)	3400 BP
Lynx (*Lynx lynx*)	200 BP
Brown bear (*Ursus arctos*)	500
Beaver (*Castor fiber*)	1100s
Wild swine (*Sus scrofa*)	1260
Wolf (*Canis lupus*)	1743
Great auk (*Pinguinus impennis*)	1844
Goshawk (*Accipiter gentilis*)	1850
White-tailed sea eagle (*Haliaetus albicilla*)	1918
Muskrat (*Ondatra zibethicus*)	1935

Source: from information in Rackham (1986); Peters and Lovejoy (1990); Yalden (1999).

Habitat loss and degradation

The destruction of habitats is widely regarded to be the most severe threat to biological diversity, while fragmentation and degradation, which are often precursors of outright destruction, also present significant cause for concern. A summary of this chief driver of biodiversity loss worldwide is provided by Boakes *et al.* (2010) who point out that over the last 10,000 years vegetation across about half of the Earth's ice-free land surface has been cleared or become otherwise dominated by human activity and that a further 10 million km² of natural habitat is predicted to be lost to cultivation or grazing by 2050. Habitat destruction and degradation are implicated in the decline of over 85 per cent of the world's threatened mammals, birds and amphibians, and the IUCN Red List assessments show agriculture and unsustainable forest management to be the greatest cause of these losses. As with many forms of human impact on the environment, the rate of habitat loss has accelerated in more recent times. A quarter of the world's forest and half its grassland have been converted to agriculture since 1700.

The destruction of tropical rain forests, coral reefs, wetlands and mangroves, documented elsewhere in this book, is particularly serious given the high biodiversity of these habitats. Hence, the large majority of current human-induced extinctions are occurring in the world's tropical rain forest areas, and insects are the order of species most at risk. Nevertheless, other habitats have suffered similarly severe destruction. The global area of steppes, savannas and grasslands declined from about 32 million km² in 1700 to 18–27 million km² in 1990 (Lambin *et al.*, 2003).

There are numerous examples of species that have been driven to the edge of extinction because of the loss of their habitat. Internationally, one of the best-known threatened species, the giant panda (*Ailuropoda melanoleuca*), has suffered from progressive human encroachment into its habitat. Once found throughout much of China's high-altitude regions and beyond, the species is now confined to 24 sites near Chengdu and Xian (Figure 15.2). The panda's heavy reliance upon its specialized bamboo diet, which occasionally requires forays into lowland regions during natural bamboo die-offs, puts the normally shy creature in direct conflict with encroaching human populations.

These factors, combined with the extreme difficulties encountered by attempts at captive breeding, look set to add the panda to a long list of species whose ultimate fate can be traced to destruction of their habitat by humans; this is despite huge efforts to conserve the species. A network of panda reserves, designed to protect 60 per cent of the animal's present range, has been established in China, but not all of these protected areas have been successful in their prime objective. Research on the 200,000 ha Wolong Nature Reserve has shown that panda habitat is still being destroyed faster inside one of the world's most high-profile protected nature reserves than in adjacent areas that are not protected. Further, the rates of destruction

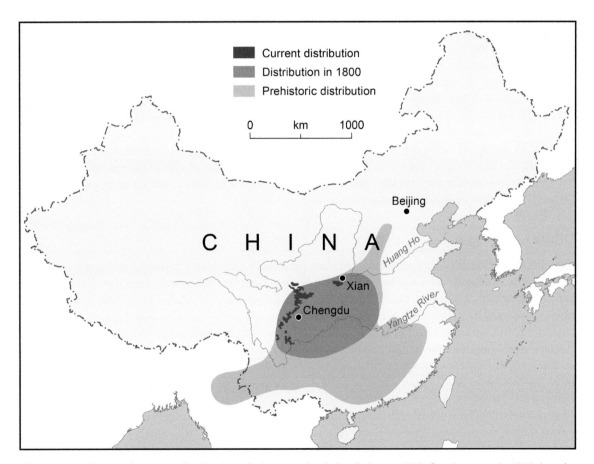

Figure 15.2 Past and present distribution of giant pandas (after Roberts, 1988; Servheen et al., 1999; Loucks *et al.*, 2001).

were higher after the reserve was established in 1975 than before (Liu *et al.*, 2001). It appears that the Wolong Nature Reserve has been a victim of its own success. Towns and settlements have thrived in and around Wolong since the reserve was established and began attracting tourists to the area. The upturn in the local economy, fuelled by tourism, has been a major factor behind an increase in resident human population in the area of 70 per cent since the reserve was created. The people are cutting more trees for fuel and other uses, destroying prime panda habitat.

Fragmentation of habitats not only reduces the area of habitat but also affects normal dispersion and colonization processes, and reduces areas for foraging. Fragmentation of the panda's habitat is not just a threat to the number of bamboo species remaining. Small, isolated populations of giant pandas also face a risk of inbreeding (Hu *et al.*, 2017). This could lead to reduced resistance to disease, less adaptability to environmental change, and a decrease in reproductive rates. Fragmentation also increases the ratio of edge to total habitat area, hence increasing dangers from outside disturbances such as the effects of hazards (e.g. fire and diseases) and invasion by competitors and/or predators. Research in the Brazilian Amazon on dung beetles shows that they are significantly affected by fragmentation, with small forest fragments having fewer species of beetles, lower population densities for each species and smaller-sized beetles than undisturbed forest areas. Since dung beetles are a keystone species in the forest ecosystem, through their role in burying dung and carrion as a food source for their larvae – thus facilitating the rapid recycling of nutrients and the germination of seeds defecated by fruit-eating animals, and reducing vertebrate disease levels by killing parasites that live in the dung – the effects on community interactions and ecosystem processes are probably very widespread (Barragán *et al.*, 2011).

The prevalence of dams (see Chapter 9) has put rivers among the most intensively fragmented ecosystems on Earth, and their fragmentation at specific points is significantly different from the fragmentation of terrestrial habitats. The effects of dams on fish species have been documented all over the world, population declines being due primarily to the change from a riverine to a lacustrine environment and the isolation of migratory fishes from their spawning and feeding grounds. Genetic change and extinction can result if critical habitats are lost or become inaccessible, as Lees *et al.* (2016) document for a range of habitats affected by large dams in the Amazon Basin (see Table 9.5). Rivers also act as pathways for plant dispersal which can be equally disrupted by dams.

Climate change

The effects of climate change on biodiversity are projected to become a progressively more significant threat in the coming decades but they have already been detected, numerous ecological changes having been attributed to changes in climate during the twentieth century (see Table 11.6). Warming temperatures, more frequent extreme weather events and changing patterns of rainfall and drought are expected to continue to have significant impacts on biodiversity both directly and indirectly. The warming climate is considered a profound threat to the survival of polar bears (*Ursus maritimus*) because of its direct effect on the sea-ice of the circumpolar Arctic. Loss of sea-ice, the primary habitat of polar bears as a platform from which to hunt seals, negatively affects the long-term survival of this specialized marine carnivore (Stirling and Derocher, 2012). One forecast suggests that about two-thirds of the world's polar bears could be extirpated by 2050. Other Arctic marine mammals that are also closely dependent on sea-ice include walrus (*Odobenus rosmarus*) and a number of whale species. Reviewing the evidence of sea-ice conditions, Laidre *et al.* (2015) found that major changes had already occurred to the duration of the summer period of reduced ice cover between 1979 and 2013. In all Arctic regions except the Bering Sea, the duration of the summer period increased by 5–10 weeks over this timespan. In the Barents Sea, the summer period increased by more than 20 weeks. However, assessing the impacts of these changes on Arctic marine mammals is very difficult because there are few studies long enough for reliable trend analysis.

An indirect impact of warming is via changes in the dynamics of diseases which have already affected numerous species of amphibian (Pounds *et al.*, 2006). Pounds *et al.* identified large-scale warming of sea-surface and air temperatures as a key factor in the extinction of the Monteverde harlequin frog (*Atelopus sp.*) and the golden toad (*Bufo periglenes*) from the mountains of Costa Rica, along with the disappearance of numerous other amphibians endemic to the American tropics. They concluded that the demise of these species occurred due to a fungus that has thrived in the higher temperatures.

One obvious response of many species to changing patterns of climate is to migrate, but the loss and fragmentation of habitats make this difficult in many parts of the world. Synergistic effects are also likely because the processes of climatic change and habitat loss are happening concurrently. A global systematic review of the evidence (Mantyka-Pringle *et al.*, 2012) indicates that areas with high temperatures where average rainfall has decreased over time add to the negative effects of habitat loss on species density and/or diversity, a finding that has important implications for conservation management strategies.

Pollution

Human-induced modifications to habitats can also lead to species loss in a more pervasive manner through stress or subtle effects on ecological processes. Pollution in its various forms is probably the most significant issue in this respect. Examples of such effects are referred to in many of the chapters in this book; hence, just a few illustrations will be given here.

The detrimental effects of agrochemicals have been widely documented (see Chapter 13), with some of the most deleterious being observed in creatures at the top of the food chain where some materials bioaccumulate. Raptors are typical in this respect, suffering a steep decline in population numbers in the latter half of the twentieth century in Europe and North America with the rise in use of organochlorine pesticides – particularly DDT, which had the effect of thinning egg shells, leading to breakage in the nest, and alterations in breeding behaviour. Use of the highly toxic dieldrin and aldrin in cereal seed dressings and sheep dip, which occurred from 1956 in the UK, also led to widespread declines through increased adult mortality. However, the phasing out of organochlorine pesticides in many western countries has resulted in bird of prey populations rising once more. In the UK, the population of peregrine falcons (*Falco peregrinus*), numbering around 850 pairs in the 1930s, had fallen to 360 pairs by 1962, but had risen to 1050 pairs in 1991, 1426 pairs in 2002 and 1505 in 2014 (Banks *et al.*, 2010; Hayhow *et al.*, 2017).

Numerous forms of water pollution can have detrimental effects on aquatic flora and fauna. Acidification of lakes and streams, for example, has affected waterbirds and invertebrates (see Chapter 12). An example of a species extinction due to pollution is the harelip sucker (*Lagochila lacera*), discovered in the Chickamauga River in Tennessee, USA, in the late nineteenth century and named as a new genus and species. The splitmouth was very common in the region and found in the Tennessee, Cumberland, White and Ohio river drainages to Lake Erie. The species was last seen in the Auglaize River, Ohio, in 1893. Its extinction – the first North American freshwater fish species known to die out entirely due to human action – appears to have been caused by continuous silting and pollution of its habitat (Maitland, 1991). Rivers carrying heavy nutrient loads can severely degrade water quality in coastal waters, creating 'dead zones' where decomposing algae use up oxygen in the water to leave large areas virtually devoid of marine life. The number of these zones has increased dramatically in the past 100 years or so (Figure 15.3).

Air pollution can also have insidious effects on plants and animals. Atmospheric acidification has been linked to numerous impacts on trees and other floral species (see Chapter 12); the widespread

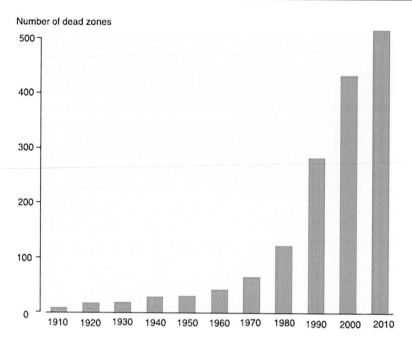

Number of dead zones

Figure 15.3 Increase in marine 'dead zones' globally, 1910–2010 (after Secretariat, Convention on Biological Diversity, 2010).

impoverishment of lichens, which are very sensitive to air pollution, across large parts of Europe and North America has been linked to sulphur dioxide and, to a lesser extent, to photochemical smogs (Purvis, 2010). Atmospheric pollutants from heavy industries such as smelting can be devastating to surrounding ecology (see Chapter 19).

Overexploitation

Humans have long been implicated in the extinction of species through overexploitation, particularly by hunting for food. Many believe that such effects can be traced back to the Stone Age and to the late Pleistocene, with the loss of large mammals such as the woolly rhinoceros, woolly mammoth and sabre-toothed tiger being intimately linked to overhunting, although the role of abrupt and substantial climatic change is also implicated in the extinction of such creatures (Lorenzen *et al.*, 2011). Larger creatures have been suffering from overexploitation ever since. In Roman times, hunters and trappers extirpated the elephant, rhino and zebra in Africa north of the Sahara, and lions from Thessaly, Asia Minor and parts of Syria. A similar fate befell tigers in Hercynia and northern Persia. In many instances, the loss of a species has knock-on effects for other organisms. The death of the last dodo (*Raphus cucullatus*), killed by seafarers on Mauritius in 1681, has meant that another of the island's endemic species, the tambalocque tree (*Sideroxylon sessilisiorum*), has been unable to reproduce for the past 300 years because the fruit-eating bird prepared the fruit in its gizzard for germination (Temple, 1977), an extinction debt of more than three centuries.

The advent of the firearm, a dramatic improvement in hunting technology, greatly enhanced humans' ability to exterminate creatures in large numbers, resulting in the decimation of the buffalo (Figure 15.4)

Figure 15.4 Wild buffalo, or bison, were found in large herds over a very wide area in North America before 1800 but their numbers were decimated by hunters with modern firearms in the nineteenth century. Today, small herds are maintained in just a few areas, including here in Alberta, Canada.

and the extinction of the passenger pigeon in North America, for example. The North American passenger pigeon (*Ectopistes migratorius*) is widely believed to have been the most abundant bird ever to have inhabited the planet. Conservative estimates suggest that its numbers may have been around 10,000 million in the first half of the nineteenth century. They lived in huge, very dense flocks, some containing more than 2000 million birds, which darkened the skies for up to three days on flying past. They were killed in very large numbers for their meat, and this intense exploitation, coupled with destruction of breeding habitats, led to a precipitous decline in numbers. The last passenger pigeon seen in the wild was shot in 1900 and the last known individual died a lonely death in Cincinnati Zoo in 1914. In just a few decades, the planet's most abundant bird species had been rendered extinct (Blockstein, 2002).

In some parts of the world, migrating birds continue to be killed in huge numbers by hunters. Some of the worst culprits are the European Mediterranean countries such as Italy and Malta (Figure 15.5). For rare species or those with small breeding populations this illegal hunting could have a significant impact on their long-term persistence in Europe and in some cases globally (Raine *et al.*, 2016). Overzealous hunting is also still a very serious threat to numerous large mammals, particularly for those whose products fetch high prices in local and international markets. The population of African elephant (*Loxodonta africana*), for example, was decimated by poachers in many areas during the 1970s and 1980s, before a ban

Figure 15.5 Hide for shooting migratory birds in rural Umbria, Italy. Italian hunters are thought to kill as many as 3 million birds a year.

on international trade in ivory was introduced in 1989 (although even this ban has been controversial – see below). Similarly, the market for whale products fuelled the hunting of oceanic cetacean species, many of which currently have indeterminate chances of survival (see p. 124).

The market for certain floral species is also an underlying factor behind their overexploitation. Cacti and orchids are particularly at risk from collectors and many species of tree have been dramatically reduced in their range by selective logging. International trade in wild plants increased markedly in North America and Europe after the Second World War as the demand for exotic plants rose in line with increased disposable incomes and more affordable greenhouses (Lavorgna *et al.*, 2017). The market's profitability also boosted the trade of endangered species, and some rare species have declined as a result, such as *Ariocarpus* (a genus of Mexican cacti) which became virtually extinct in the wild by the end of the 1970s.

Despite the increasing importance of international markets in driving the decline of many species of fauna and flora, this is not to discount more localized causes of overexploitation. In a review of hunting impacts on vertebrate populations in the tropical forests of South-East Asia, Harrison *et al.* (2016) highlight evidence from multiple sites that indicates animal populations have declined precipitously across the region since 1980, and that many species are now extirpated from substantial portions of their former ranges. Hunting is by far the greatest immediate threat to the survival of most of the region's endangered vertebrates. Much of the hunting is for meat, and the sale of wild meat is largely a local issue, most being consumed in villages, rural towns, and nearby cities.

Invasive alien species

The increased trade, travel and transport of goods associated with globalization have brought many great benefits to large numbers of people but they have also facilitated the spread of numerous biological invasions by alien species. The ecological impacts of such invasive species are wide-ranging and several examples are given in Chapter 16 (see p. 392) and in the following section. In some cases invading species out-compete and displace native species; in others they have significant impacts on the structure and functioning of ecosystems, by disrupting pollination, for instance, or altering energy and nutrient flows. Other impacts occur via predation, parasitism, disease, or hybridization. The overall effect in many places has been dubbed 'biotic homogenization', the process by which disparate regions become more similar in their species composition over time. This loss of regional distinctiveness of biotas worldwide is a relatively new aspect of the biodiversity crisis (Olden, 2006) that is thought to rival the loss of species through extinction in its seriousness, although there have also been some dissenting voices from scientists who consider the threat from invasive species may have been exaggerated (Russell and Blackburn, 2017).

Island species

Although islands make up only a minor proportion of the world's land surface area, they feature prominently in any account of threatened flora and fauna. Island species are considered to be especially vulnerable for a number of reasons. Many are endemic species that have evolved in isolation and are thus particularly susceptible to introduced competitors, diseases and predators. The physical size of many islands also means that human activities can rapidly degrade large portions of their ecosystems and can have a great impact on relatively small island populations.

The importance of islands for the conservation of global plant diversity is summed up by Caujapé-Castells *et al.* (2010) who point out that although islands make up only 5 per cent of the Earth's land surface, they are home to about a quarter of all known extant vascular endemic plant species. Further, of some 80 documented plant extinctions in the last 400 years, about 50 were island species. For island avifauna (birds) the likelihood of being threatened with extinction is 40 times greater than that for continental bird populations. In consequence, oceanic islands contain more than a third of all globally threatened bird species (Trevino *et al.*, 2007).

For obvious reasons, it is often the largest species on islands that first suffer from direct human action. On Madagascar, nine genera and at least 17 species of giant lemurs have become extinct, along with several species of flightless bird and giant tortoise, since humans arrived there about 2300 years ago. There is good evidence to suggest that people were intimately involved in these extinctions (Burney *et al.*, 2004), although several forms of human impact have been implicated and climate is another suspected driver of the changes (Table 15.4). Some form of synergistic combination of several drivers is also a possibility, and the same negative synergies of resource overuse, fire-mediated vegetation change, and biological invasion continue to endanger several smaller lemur species and a number of other organisms in Madagascar today.

Severe ecological damage on many islands dates from the arrival of explorers and colonizers from the European maritime powers, for whom oceanic islands had strategic importance. Goats were introduced to

Table 15.4 Hypotheses proposed to explain extinctions in late Holocene Madagascar

Hypothesis	Rate of change	Pattern of change	Process of change
Great Fire	Rapid	Simultaneous throughout	Landscape transformed by humans introducing fire – extinctions follow due to forest loss
Great Drought	Slow	Southern region only	Climatic desiccation, extinctions due to spread of semi-arid conditions
Blitzkrieg (overhunting)	Rapid	Wave across island	First-contact overkill of naïve fauna, beginning at coastal settlements and moving inland
Biological Invasion	Moderately rapid	Multiple waves across island	Introduced cattle and other herbivores disrupt natural vegetation and compete with native herbivores
Hypervirulent Disease	Very rapid	Panzootic disease pattern	Unknown pathogen(s) introduced by humans, spreading quickly; lethal to wide array of mammals, and possibly some birds and reptiles
Synergy	Very slow	Mosaic	Full array of human impacts played a role, but to differing extents in various regions; factors interacting to multiply effects; some amplification by background climate change

Source: after Burney et al. (2004).

the South Atlantic island of St Helena in 1513, and within 75 years, vast herds roamed the island. St Helena's endemic plants had evolved without large grazing animals to contend with and so had few defences against them. Botanists first reached St Helena in the early 1800s, so we can only guess at the damage done by then. Today, 50 endemic species are known, of which seven are extinct, but some estimates put the original number of endemics at more than a hundred. St Helena remains an island with severely threatened endemic flora: seven of its endemic plant species are endangered and nine are critically endangered (Caujapé-Castells et al., 2010).

European colonizers are by no means the only culprits, however. Many studies of bird bones from archaeological sites on islands in the tropical Pacific indicate that Polynesians, the first human colonists, were responsible for the large-scale loss of numerous species by hunting, habitat modification and the introduction of predators (Steadman, 2006). Polynesians are thought to have exterminated more than 2000 bird species (some 20 per cent of the world total) using just Stone Age technology. Damage to the wildlife of the Hawaiian islands, where almost 100 per cent of the native insects are endemic, as well as 98 per cent of the birds, 93 per cent of the flowering plants, and 65 per cent of the ferns, began long before the arrival of Captain Cook, the first European, in 1778. Polynesians, who had settled the island by 750 CE, cleared extensive lowland areas for cultivation and introduced rats, dogs, pigs and jungle fowl. In some parts of Hawaii, there is evidence to suggest that habitat was lost not by direct clearance but as a result of rats introduced by Polynesian colonizers. Athens (2009) hypothesized that rat numbers increased exponentially in the absence of significant predators or competitors, feeding on a largely endemic vegetation that had evolved in the absence of mammalian predators. It seems that the native lowland forest on the 'Ewa Plain in Oahu declined very quickly, eaten by rats after colonization of Oahu but before the Polynesians settled the plain. Palaeontological evidence suggests that at least half of the known total of 83 species of Hawaiian birds became extinct in the pre-European period (Olson and James, 1984). However, further introductions and disturbance of the natural habitat since 1778 have resulted in the loss of at least 23 bird species and 177 species of native plants.

In some cases, the introduction of one exotic species to an island may result in the local extinction of numerous native species. An example here is the brown tree snake (*Boiga irregularis*) which has been introduced onto a number of Pacific islands with devastating effects on endemic birds. The brown tree snake is a slender, climbing snake that typically grows to 1–2 m in length. It is a nocturnal predator that hunts for food at all levels within forests. It was accidentally introduced to the island of Guam, probably from New Guinea, in the 1940s. Before the brown tree snake's arrival, Guam was effectively snakeless, harbouring only the harmless, wormlike blind snake. Consequently, Guam's wildlife evolved in the absence of snake predators. Without behaviours developed to protect them against snakes, many of the island's species became easy prey for the new arrival.

As the alien snake became established on the island during the 1950s, it started to decimate Guam's native birdlife. The number of brown tree snakes began to grow exponentially and bird populations declined in response. Nine of the 11 species of native forest-dwelling birds have been extirpated from Guam. Three of these were endemic species and are now extinct (Table 15.5). Several other bird species not shown in Table 15.5 continue to exist only in very small numbers, and their future on Guam is perilous. In total, 22 of Guam's 25 resident bird species, including 17 of 18 native species, have been severely affected by snakes. Other wildlife has also fallen prey to the voracious predator. As Table 15.5 indicates, two of Guam's three bat species have disappeared. Similarly, of the island's 12 native lizards, six species survive and only three are now common on the island (Wiles *et al.*, 2003). All of these creatures played a crucial role in the workings of Guam's ecosystems, as pollinators, insect predators and by dispersing seeds. Indeed, the potential for predator invasions to cause indirect, pervasive cascades of knock-on impacts is highlighted on Guam by Rogers *et al.* (2017). They looked at two fleshy-fruited tree species and found that the loss of many of the island's fruit-eating birds has resulted in a severe decline in the reproduction rate of the trees because the birds provided the critical service of dispersing seeds.

Similarly severe repercussions for endemic species, in this case of snails, have occurred on the South Pacific island of Moorea following the deliberate introduction of the predaceous Ferussac snail (*Euglandina rosea*). The Ferussac snail was brought to the island in an attempt at biological control of a previously introduced species – the giant African snail (*Achatina fulica*) – when it attained pest status (see p. 323).

In other cases the exotic species introduced may represent a threat even after the people responsible for their introduction have gone. This is the case with the endangered short-tailed albatross (*Diomedea albatrus*), one of many albatross species facing extinction, whose chicks and eggs are easy prey for feral cats and rats on Torishima island, one of the Japanese Izu Islands, its main nesting site (Collar *et al.*, 1994). Once seen throughout the North Pacific Ocean, the short-tailed albatross has suffered a long history of decline. In the late nineteenth century its numbers began to fall because its long white wing and tail feathers became popular in the manufacture of pen plumes and its downy body feathers to stuff feather beds. Although a ban on the collection of short-tailed albatross feathers was instituted in 1906, it was not very effective. Indeed, illegal feather collection continued until the 1930s. Collection only declined then because the species was no longer economically significant, its numbers having been reduced so drastically. Once

Table 15.5 Bird and bat species lost from Guam

Species	Endemic to Guam	Last confirmed record in the wild
Birds		
Nightingale reed-warbler (*Acrocephalus luscinia*)		1969
Brown booby (*Sula leucogaster*)		1979
White-tailed tropicbird (*Phaethon lepturus*)		1982
Guam flycatcher (*Myiagra freycineti*)	●	1984
Rufous fantail (*Rhipidura rufifrons*)	●	1984
Bridled white-eye (*Zosterops conspicillatus*)		1984
Mariana fruit-dove (*Ptilinopus roseicapilla*)		1985
Micronesian honeyeater (*Myzomela rubratra*)		1986
White-throated ground-dove (*Gallicolumba xanthonura*)		1986
Guam rail (*Gallirallus owstoni*)	●	1987
Micronesian kingfisher (*Todiramphus cinnamominus*)		1988
Chestnut munia (*Lonchura atricapilla*)		1994
Bats		
Little marianas fruit bat (*Pteropus tokudae*)		1968
Pacific sheath-tailed bat (*Embollonura semicaudata*)		1972

Source: modified after Wiles *et al.* (2003).

thought to number around 5 million individuals, the species was mistakenly declared extinct in 1949 until ten individuals were discovered on Torishima in 1951.

Introduced species on Torishima are not the only current threat the short-tailed albatross faces. People no longer live on Torishima but this is because it is an active volcano. In 1902, an eruption killed all 125 of the island's human occupants. A repeat eruption, if it occurred during the albatross breeding season, would wipe out most of the world population that nests on the ash-covered slopes. The short-tailed albatross is also threatened by longline fishing while at sea.

CONSERVATION EFFORTS

The two ends of the spectrum of arguments in favour of conserving biodiversity are rooted in morals and pragmatism. Some believe that the destruction of any living organism is morally unacceptable, and that people who share the planet with plants and animals have no right to exterminate other species. More technocentric arguments point out that the extermination of other species is not in the interests of humankind because biodiversity is useful to us through the full range of ecosystem services it provides. It appears that the impact of biodiversity loss on ecosystem functioning is comparable in magnitude to other major drivers of global change (Tilman *et al.*, 2012), with many significant consequences for human societies. There are also potential provisioning services of biodiversity not yet realized. The importance of maintaining genetic diversity in the wild because it might one day prove useful for agricultural crops is outlined on p. 326.

Conversely, however, certain species are considered detrimental to human activities and are therefore the subject of deliberate attempts to control or exterminate their populations. Species regarded as pests are obvious examples here. While deliberate efforts to make an organism extinct are not common, the smallpox virus (although not strictly a species) is a case in point. Subject to a successful international effort to eradicate the disease by vaccination, the virus remains only in laboratory stocks in Russia and the USA. Periodic recommendations to destroy these stocks have been countered by preservation arguments because they can be used for research purposes (Arita and Francis, 2014).

Human perceptions of biodiversity are always subject to change. A century ago, the tiger was considered to be a pest by authorities in many parts of Asia and hunters claimed thousands (Figure 15.6). Today the official view is quite the opposite (see below). Similarly, for a long time many societies considered wetlands to be worthless and so these habitats have been altered in many parts of the world to make the land more

Figure 15.6 A tiger-shooting party of Europeans on elephants in India some time in the nineteenth century.

obviously useful for human activities. It is only relatively recently that the inherent values of undisturbed wetlands and the ecosystem services they provide have been widely recognized (see p. 192).

Nonetheless, efforts to preserve aspects of biodiversity considered to be worthy of protection have a long history. Some of the earliest examples involve species given spiritual or religious significance, such as the South Asian river dolphin (*Platanista gangetia*) of the sacred Ganges River which figures in Hindu religious tracts dating back thousands of years. The dolphin became one of the world's first protected species, given special status under the reign of Emperor Ashoka, one of India's most famous rulers, some 2300 years ago.

All conservation efforts require a certain degree of knowledge about the biota to be protected, of course. We cannot protect species that have not been identified, unless they become unintentional beneficiaries of efforts designed for other species. Even basic information about many species is not always available, as outlined earlier in this chapter, a problem that can undermine conservation efforts. Counting the number of surviving individuals in a known species can be difficult. Indeed, a direct head count is often effectively impossible, so indirect sampling methods have to be used.

In the case of the giant panda, that icon of the conservation movement, assessing their numbers is difficult despite the fact that the individuals are physically large, and thus in theory relatively easy to see. Obtaining accurate data on panda populations is hard because individuals are elusive, wary and not easy to observe in the wild. One traditional method involves analysing panda droppings and differentiating individual pandas from the size of pieces of bamboo bitten off and found in the faeces, hence allowing an estimate of the number of pandas present. However, this method is now thought to underestimate the number of individuals. Survey methods that analyse panda DNA in the droppings directly are able to distinguish between individuals much more reliably. The first use of these techniques in the Wanglang Nature Reserve indicated that the number of surviving giant pandas was no less than double that previously estimated (Zhan *et al.*, 2006).

There are numerous ways of approaching biodiversity conservation and the dominant methods used have changed considerably over the last 100 years or so. An overriding focus on protected areas has more recently been complemented with attempts to regulate trade in endangered species and their products, integrated conservation and development projects, and the introduction of conservation contracts in which financial compensation and rewards are offered to people conserving environmental services. At about the same time, the dominance of national governments as enforcers of conservation management has given way to more multi-stakeholder forms of governance in which non-governmental and international organizations play roles, and a variety of market-based and negotiation approaches are being used to achieve conservation goals (Swallow *et al.*, 2007). Of course any particular conservation effort will need to focus on the relevant drivers of biodiversity loss, which may involve the need to manage multiple threatening processes simultaneously over lengthy periods (Brook *et al.*, 2008).

Habitat protection

Since the destruction, modification and fragmentation of habitats are the primary threats to biodiversity, the protection of habitats from the causes of their destruction is the most obvious conservation method. Currently, protected areas cover about 15.4 per cent of the Earth's land area and 3.4 per cent of the global ocean area. On land, the proportion varies widely at the national level (Table 15.6). While some countries give some form of protection to more than one-third of their land area (e.g. Bhutan, Bulgaria), others protect less than 1 per cent of their national space (e.g. Yemen). In practice, some types of habitat are better protected than others, largely due to the fact that the actual designation of sites is often based more upon socio-economic and political factors than on conservation ideals. Biomes such as mixed mountain systems and island systems, which are typically not intensively used by humankind, are better represented than systems such as temperate grasslands and lakes. It comes as no surprise to learn, therefore, that the world's largest protected area is the Northeast Greenland National Park, which covers 972,000 km^2 of Arctic desert and ice cap that has a very small human population. Also in the top five largest protected areas are the Rub al Khali Wildlife Management Area in Saudi Arabia, 640,000 km^2 of sandy desert, and the Chang Tang Nature Reserve, 298,000 km^2 of the Tibetan Plateau, a huge area of alpine and desert steppe at elevations of 4800 m and higher (Figure 15.7). Like Greenland, the Rub al Khali and Chang Tang are very sparsely populated by people.

Table 15.6 Protected areas (terrestrial) in selected countries

Continent/country	Area protected (thousand km^2)	Area protected (% land area)
ASIA (excl. Middle East)		
Bhutan	19.2	48
Japan	109.9	29
Uzbekistan	15.2	3
EUROPE		
Bulgaria	38.5	35
Sweden	66.9	15
UK	70.5	29
MIDDLE EAST AND N. AFRICA		
Algeria	174.2	8
Israel	4.2	20
Yemen	3.5	<1
SUB-SAHARAN AFRICA		
Botswana	169.4	29
Ethiopia	209.8	19
Liberia	3.9	4
NORTH AMERICA		
Canada	964.2	10
Mexico	285.0	15
USA	1280.5	14
CENTRAL AMERICA AND CARIBBEAN		
Costa Rica	14.3	28
Cuba	18.5	17
Jamaica	1.8	16

(continued)

Table 15.6 continued

Continent/country	Area protected (thousand km²)	Area protected (% land area)
SOUTH AMERICA		
Chile	139.9	18
Surinam	21.4	15
Uruguay	6.2	3
OCEANIA		
Australia	1487.7	19
Fiji	1.0	5
New Zealand	87.8	33

Source: UNEP-WCMC (2018) Protected Area Profiles from the World Database of Protected Areas at www.protectedplanet.net, accessed January 2018.

Figure 15.7 The Chang Tang – a high-altitude desert area in Central Asia that offers few resources for human occupants – is the site of one of the world's largest nature reserves.

The largest marine protected areas are Marae Moana (almost 2 million km^2) surrounding the Cook Islands in the Pacific Ocean and the Ross Sea Region Marine Protected Area in the Southern Ocean (1.6 million km^2), but generally marine habitats are less well protected than those on land. Also as on land, marine protected area coverage is uneven. Reserves are often absent where threats to biodiversity are greatest, such as fishing grounds and oil and gas leases. Beyond the 322-kilometre (200-nautical mile) limits of national jurisdiction, the Exclusive Economic Zones, less than 1 per cent of open waters are protected, compared with 8 per cent of continental shelves (Spalding *et al.*, 2013).

The types and quality of protection and management in these areas also vary greatly. A lack of political and financial support limits the success of many protected areas. The need to maintain levels of protection is well illustrated by the Operation Tiger reserves set up in several Asian countries in the early 1970s to protect the Asian tiger from local people faced with the loss of their domesticated livestock, and in some cases fellow villagers. Threats from the tiger have been enhanced in many areas due to the loss of its forest habitat to agricultural lands.

While Operation Tiger has long been hailed as a conservation success story, the story's most recent chapter has not been a happy one. Further loss of habitat and continued poaching have resulted in dramatic changes to tiger populations. Tigers now occupy just 7 per cent of their historical range (Dinerstein *et al.*, 2007), an expanse that once stretched from the Caspian Sea to the island of Bali in Indonesia. Even in India, widely considered to be the stronghold for wild tigers, conservationists were shocked to discover in 2005 that intense poaching had eliminated all tigers from supposedly well-protected sanctuaries such as the Sariska Tiger Reserve, and had depleted populations in other tiger reserves such as Bandhavgarh and Rhanthambore.

Conflict between the aims of conservation and those of local people is a key issue in the threatened species debate since without local involvement in both the design and management of protected areas, adequate protection can only be achieved if the park agency has the authority and ability to enforce regulations. This is not necessarily desirable and often unattainable, and this realization represents a major shift in practical conservation philosophy, away from a 'wilderness conservation approach' that excludes people from protected areas towards integrating local communities into conservation efforts. This greater recognition of local human (especially indigenous) populations is part of a wider shift in the global approach to protected area establishment and management that involves less emphasis on biological diversity and more on nature conservation in a broader sense, along with an attempt to ensure full and effective participation from all stakeholders involved (Dudley *et al.*, 2014).

Even in those areas that receive maximum protection, however, active conservation management is usually needed because often it is not sufficient simply to cordon off an area and leave it be. Such management is not always an easy task, however. A ban on hunting, for example, is theoretically easy to introduce, and grazing goats, sheep, pigs and even rabbits are relatively easy to control and even eliminate, but a threat from an introduced plant is much more intractable.

Effective management is also dependent upon a sufficient understanding of the ecology of the organisms in question, which is not always the case. Although the gradual decline in British colonies of the large blue butterfly (*Maculinea arion*), a dry grassland species, was largely due to the fact that 50 of its former 91 sites were ploughed or otherwise fundamentally changed in the period 1800–1970 (Figure 15.8a), its eventual extinction from Britain in 1979 was due to ignorance. Conservationists were aware of the large blue's highly specialized life cycle: eggs are laid on *Thymus praecox*, on which the larvae feed briefly before being adopted and raised by *Myrmica* spp. of ant. But the disappearance of the large blue from another 41 sites, including four nature reserves where *Thymus* and *Myrmica* remained abundant, proved enigmatic. The realization that only one species of *Myrmica* ant could act as host, and that regular heavy grazing was necessary to maintain appropriate soil surface temperatures came too late to save the large blue.

A similar pattern of events occurred with the heath fritillary (*Mellicta athalia*), a species that has declined as its woodland clearing habitat has disappeared and because specially designated nature reserves were not adequately managed (Figure 15.8b). The heath fritillary is still one of the rarest butterflies in Britain and is now found only in a few parts of Cornwall/Devon, Kent/Essex and Exmoor. However, a clearer understanding of the makeup of its narrow niche and how it can be managed may yet save it from extinction (Hodgson *et al.*, 2009). Likewise, the lessons learned over the loss of the large blue have been put to good use in the re-introduction of a similar subspecies from Sweden. Following a re-establishment trial on the edge of Dartmoor in 1983, it has since been re-introduced and subsequently spread to more than 30 other sites, in the Cotswolds in Gloucestershire, and the Polden Hills and Mendip Hills in Somerset (Thomas *et al.*, 2009). The UK now supports some of the world's largest known populations of the large blue and the UK population's genetic diversity is similar to that of the original Swedish population (Andersen *et al.*, 2014), an important measure for its long-term viability and its evolutionary potential to react to, for example, climate changes.

Bans on hunting and trade

Legislative bans on threatening activities have been widely implemented to safeguard certain species. One example is the ban on the catching and selling of coelacanths (*Latimeria chalumnae*) introduced in the

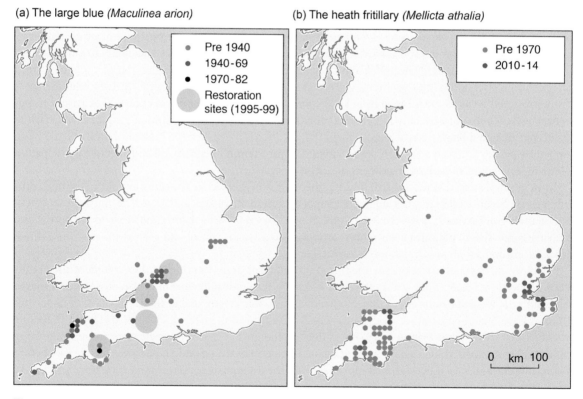

Figure 15.8 Decline in the distribution of butterflies in southern Britain: (a) the large blue (*Maculinea arion*); (b) the heath fritillary (*Mellicta athalia*) (after Heath *et al.*, 1984; Asher *et al.*, 2001).

Indian Ocean island state of Comoros as part of a conservation plan to protect the small population of this fish, first found alive in 1938, which previously had been thought to have become extinct 65 million years ago (Ribbink and Roberts, 2006).

However, a ban on hunting and trade is, in practice, like any legislation, only as effective as the ability of states to uphold it. If enforcement is weak, and incentives are large enough, poachers will continue to operate in spite of the law. The seemingly intractable problem of African elephant poaching during the 1980s inspired drastic conservation measures, with some countries implementing shoot-on-sight policies to deal with poachers in protected areas. But the placement of the African elephant on Appendix I of the Convention on International Trade in Endangered Species (CITES) in 1989 dramatically reduced the world price of ivory, resulting in a downturn in poaching of the African elephant. Nonetheless, the listing on Appendix I was not welcomed by all conservationists. In the southern African countries of Botswana, Namibia, South Africa and Zimbabwe, well-managed elephant herds are large and expanding, to the extent where they have to be culled to offset the risk of degradation in their ranges. The ban meant that ivory from the culled animals could not be sold to maximum profit. As a result, a decision was taken in 1997 to allow some limited trade in African elephant ivory, resulting in two sales of legal ivory from southern African countries, in 1999 and 2008. The decision has been controversial, with some conservationists and animal welfare organizations arguing that the sales result in higher demand for ivory, leading in turn to a rise in elephant poaching (Stiles, 2009).

Trade bans are not simple instruments, and the case of the Old World fruit bats (or flying foxes) illustrates some other associated difficulties. The flying foxes constitute a single family (*Pteropodidae*) with 41 genera and 161 species, which play an important ecological role among Pacific islands as pollinators and seed dispensers for hundreds of plant species, many of which have important economic and subsistence value to humans (Fujita and Tuttle, 1991). One of the central threats to the fruit bats is from hunters. Indigenous people on Guam have eaten them for at least 2500 years, but Guam's bat population was decimated in the years after the Second World War with the proliferation of firearms used in hunting. Many other South Pacific islands' fruit bat populations have subsequently suffered declines as exporters have supplied the Guam market. Listing on CITES Appendix I has had some success in curbing the threat to some of the species, but problems have arisen because of the lack of wildlife inspectors to uphold the law and the fact that several other islands in the region are, like Guam, effectively part of the USA, and therefore the trade is considered domestic rather than international (Sheeline, 1993).

Another challenge to those who attempt to enforce trade bans is the rise in popularity and accessibility of the internet where trading online offers convenience, anonymity and a wider market. Sellers openly advertise illegal wildlife, and reports indicate that the flourishing trade of prohibited plants and animals online is a cause of conservation concern for a broad range of species (Lavorgna, 2014).

Some conservationists argue that bans can only be a short-term measure to protect certain species. In the case of the South American vicuña a strict ban has allowed numbers to recover and subsequent sustainable harvesting of the vicuña's very valuable fleece (Box 15.1). In other cases, many feel that long-lasting protection can only be envisaged if the people who demand certain products can be persuaded against them. Concerted publicity campaigns by animal rights groups have had considerable effect in this respect in certain European countries where ornamental furs are concerned. The fact that if a market for products persists, the threat it represents to species does not disappear, can be illustrated by the case of the saiga antelope (*Saiga tatarica*), which inhabits the dry steppe grasslands and semi-arid deserts of Central Asia. Hunting of the saiga during the nineteenth century in Russia (Figure 15.9) pushed the species to the brink of extinction before being outlawed in the 1920s. Subsequently, controlled cropping allowed the saiga to recolonize most of its original range, but deterioration in law enforcement after the collapse of the Soviet

Box 15.1 Vicuña conservation and the luxury fashion market

The story of the vicuña, a species of New World camel native to the Andes in South America, shows how wildlife conservation can be balanced against demand for the products of biodiversity if the products can be harvested sustainably. Wild vicuña were hunted to the brink of extinction for their highly-prized fleeces, the fibres being woven into luxury fashion items. By 1960, the vicuña population had dropped from some 2 million a few hundred years ago to an estimated 10,000 individuals, but strict conservation regulation using CITES and a Vicuña Convention has helped to rebuild populations.

This success resulted in a shift in international policy from strict preservation (CITES Appendix I) to sustainable use (CITES Appendix II), allowing trade in fibre shorn from live animals. Most vicuña management programmes are based on the idea of community-based wildlife management, which aims to maintain wildlife habitats and preserve species while simultaneously improving the social and economic well-being of local communities. The approach is rooted in the poverty–environment nexus, the mutually reinforcing links which effectively mean that poverty reduction and environmental protection are complementary goals. Hence, vicuña fibre is produced mainly by extremely poor rural Andean communities who capture, shear and release vicuña in the wild because commercial farm management is virtually impossible. The capture and release system mimics the ancient Inca method of harvesting vicuña fleece.

The vicuña is usually considered a 'success story' in the conservation world, but its contribution to poverty alleviation is less clear-cut. Sales of vicuña fibre do make a contribution towards improving local livelihoods in the Andes but these poor communities receive a very small proportion of the profits relative to traders and international textile companies. Vicuñas compete with domestic animals for grazing, resulting in conflict over the use of rangelands. This problem is particularly acute because many poor communities do not have legal rights to the rangelands they use. A greater economic gain from harvesting their fibre would provide vicuña with better protection, as would proper legal protection for communal rights to use both land and wildlife in remote parts of the Andes.

Source: Lichtenstein (2010).

Union in the early 1990s resulted in an upsurge in the illegal trade in saiga horn destined for medicinal markets, particularly in China. The result has been not only a catastrophic fall in numbers but also a severe distortion in the saiga population's sex ratio because only the males have horns. In consequence, there are serious fears of a reproductive collapse (Kühl *et al.*, 2009).

Off-site conservation practices

While the conservation of species is best achieved by their maintenance in the wild, through protected area programmes and legislative measures, other practices may be necessary for species whose populations are too small to be viable in the wild or are not located in protected areas. Maintenance of species in artificial conditions under human supervision is a strategy known as off-site, or *ex situ*, conservation. The off-site

Figure 15.9 Saiga antelopes fleeing a hunter in Siberia. This engraving was made in the late 1800s when the saiga was widely hunted for its meat, hide and horn. Having recovered from the brink of extinction thanks to a hunting ban introduced in the 1920s, the saiga became threatened again by a resumption in illegal trade in the 1990s.

approach includes game farms, zoos, aquaria and captive breeding programmes for animals – although in reality few of these are actively involved in the conservation of endangered species – while plants are maintained in botanical gardens, arboreta and seed banks.

In practice, off-site techniques complement on-site approaches in a number of ways, such as by providing individuals for research, the results of which can be fed back into management techniques in the wild. Perhaps the ideal way in which the maintenance of captive individuals of species can help in the biodiversity issue is by providing individuals that can be re-introduced into the wild.

Such programmes are expensive, logistically demanding, and require long-term management and monitoring to assure and assess their success. Plant re-introductions have relatively low rates of long-term success (Godefroid *et al.*, 2011). Re-introduction of animals is also a difficult task, but one notable success has been the re-introduction of the Arabian oryx (*Oryx leucoryx*) to Oman, a species that became extinct in the wild in 1972. Individuals kept in captivity in the Middle East and elsewhere were released to the wild in Oman in batches throughout the 1980s and are now established, although the scheme's success led to a rise in poaching in the 1990s. A fenced reserve and anti-poaching patrols have helped to secure the future of the species in the wild, and an equivalent re-introduction programme has also been established in neighbouring Saudi Arabia (Islam *et al.*, 2011). A similar initiative began in Mongolia in 1992 for the Przewalski horse (*Equus ferus przewalskii*), which became extinct in the wild in the 1960s. Individuals were taken to Asia from zoos and game parks in Canada, the Netherlands and Russia and in 1997 the first harem group was released into the wild in the Dzungarian Gobi desert. By 2009 the population was considered entirely free-ranging although two-thirds of them perished in the very harsh winter of 2010 (Kaczensky *et al.*, 2011). However, given the long-term commitment and substantial funding required for such operations, their contribution to the maintenance of species diversity can only be limited. Perhaps the greatest value of these programmes is in their symbolic and educational importance (Figure 15.10).

Hotspots

Given the large numbers of species thought to be at risk of extinction it is unlikely that the conservation efforts outlined above can be successfully applied to save them all, if only for lack of sufficient funding. This being the case, some conservationists have used the fact that species diversity is geographically uneven to concentrate efforts on particularly endangered areas with high biodiversity. Myers *et al.* (2000) identified

Figure 15.10 Feeding an Aldabra giant tortoise (*Geochelone gigantea*) on the Indian Ocean island of Changuu. This rare species, endangered by poachers supplying the illegal exotic pet trade, is protected here in a sanctuary that is open to visitors. Income from tourists helps to finance the programme, which includes educational visits from local schoolchildren.

Figure 15.11 Cacti on an island in the Salar de Uyuni in southern Bolivia, part of the Tropical Andes biodiversity hotspot, an area that contains about a sixth of all plant life in less than 1 per cent of the world's land area.

25 biodiversity 'hotspots', where exceptional concentrations of endemic species are undergoing exceptional loss of habitat, as priority areas for conservation efforts. As many as 44 per cent of all species of vascular plants and 35 per cent of all species in four vertebrate groups are confined to these 25 hotspots, which comprise just 1.4 per cent of the Earth's land surface (Figure 15.11). The risks of continued human impact in these hotspots are emphasized by the fact that human population growth rates in these areas are higher than the global average (Cincotta *et al.*, 2000).

For a long time it was thought that most marine species were less likely than terrestrial species to become extinct as a consequence of human activities because of their vast geographical ranges in the oceans. But similar hotspots have also been identified for coral reefs (Roberts *et al.*, 2002). The ten richest centres of coral reef endemism cover 15.8 per cent of the world's coral reefs (0.012 per cent of the oceans) but include about half of the restricted-range species. Eight of the ten coral reef hotspots are also adjacent to a terrestrial hotspot. Similar spatial concentrations of species richness have also been identified for marine mammals (Pompa *et al.*, 2011). Worryingly, six global hotspots of marine biodiversity (based on global distributions of fish, marine mammals and seabirds) are also areas most affected by both climate change and industrial fishing (Ramírez *et al.*, 2017).

CONVENTION ON BIOLOGICAL DIVERSITY

Increasing realization of the adverse human impact on biodiversity in recent decades has given rise to a number of international efforts to curb the worst effects. Some of the international agreements have focused on specific habitats, such as the Ramsar Convention for wetlands (see p. 192), others on particular threats such as international trade (e.g. CITES). The culmination of these efforts was the Convention on Biological Diversity, which came into force in 1993. It is designed to ensure the conservation of biological resources and their sustainable use, and to promote a fair and equitable sharing of the benefits arising from genetic resources. In the latter respect, this means regulating biotechnology firms, access to and ownership of genetic material, and ensuring that compensation is paid to developing countries for extraction of their genetic materials and to the holders of associated traditional ecological knowledge (TEK).

The Convention requires countries to develop and implement strategies for the sustainable use and protection of biodiversity. These strategies include the traditional approaches to maintaining biodiversity outlined above. However, scientists agree that it is impossible to shield all genes, species and ecosystems from human influence. The aim must be towards adapting all human activities so that they can take place in ways that minimize adverse impacts on the planet's biodiversity. Targets have been set to reduce the rate of biodiversity loss, including the largely unsuccessful 2010 Biodiversity Target, now replaced by 20 Aichi Biodiversity Targets based on a set of strategic goals to be achieved by 2020 (Hagerman and Pelai, 2016). The Aichi Targets were designed to be SMART (specific, measurable, ambitious, realistic, and time-bound) but midway global assessments indicated that the majority of targets are unlikely to be met.

FURTHER READING

Aiyadurai, A. 2016 'Tigers are Our Brothers': understanding human–nature relations in the Mishmi Hills, Northeast India. *Conservation and Society* 14: 305–16. A study of alternative ways of understanding tigers. Open access: www.conservationandsociety.org/text.asp?2016/14/4/305/197614.

Heywood, V.H. 2017 Plant conservation in the Anthropocene – challenges and future prospects. *Plant Diversity.* This paper addresses the key issues underlying our failure to meet agreed plant conservation targets. Open access: www.sciencedirect.com/science/article/pii/S2468265917300847.

Lees, A.C. and Pimm, S.L. 2015 Species, extinct before we know them? *Current Biology* 25(5): R177–80. A cautionary tale that focuses on Brazil's coastal forests. Open access: www.sciencedirect.com/science/article/pii/S0960982214016182.

Lyons, S.K., Smith, F.A. and Brown, J.H. 2004 Of mice, mastodons and men: human-mediated extinctions on four continents. *Evolutionary Ecology Research* 6: 339–58. A good assessment of the hypotheses proposed to explain late Pleistocene megafaunal extinctions.

Skandrani, Z., Lepetz, S. and Prévot-Julliard, A.C. 2014 Nuisance species: beyond the ecological perspective. *Ecological Processes* 3(1): 3. A thought-provoking analysis of how pigeons have been viewed by society in France since the seventh century. Open access: ecologicalprocesses.springeropen.com/articles/10.1186/2192-1709-3-3.

Sodhi, N.S. and Ehrlich, P.R. (eds) 2010 *Conservation biology for all.* Oxford, Oxford University Press. Open access: www.mongabay.com/conservation-biology-for-all.html.

WEBSITES

www.cbd.int   official site for the Convention on Biological Diversity.

www.cites.org   official site for the Convention on International Trade in Endangered Species of Wild Fauna and Flora.

www.conservation.org   Conservation International is a major NGO that promotes biodiversity conservation.

www.gbif.org   the Global Biodiversity Information Facility is an international initiative that encourages free and open access to biodiversity data on the internet.

POINTS FOR DISCUSSION

■ Where would you place the sixth mass extinction event among other major human impacts on the environment in the Anthropocene?

■ Is the loss of biodiversity simply the price we pay for progress?

■ Are there any circumstances in which you would feel the human-induced extinction of a species is morally justified?

■ Since the destruction, modification and fragmentation of habitats is the primary threat to biodiversity, why don't we protect more habitats?

■ Outline the main difficulties in upholding legal bans on hunting and trading threatened species.

■ Is it fair to say that we cannot preserve all biological diversity, so all we have to do is decide which bits are most important?

16 *Transport*

LEARNING OUTCOMES

At the end of this chapter you should:

- Understand that mobility has many major environmental effects.
- Know that all forms of transport consume resources in their creation, construction and operation.
- Recognize that new transport routes also spawn environmental impacts indirectly by improving access to resources.
- Appreciate that the movement of people has throughout history assisted numerous biological invasions, both deliberate and inadvertent, often with great ecological significance.
- Recognize that transport is a major source of air pollution.
- Appreciate that the use of lead in fuel has been phased out in virtually all countries of the world.
- Understand the lengthy history of biofuel use in Brazil.
- Recognize that road, air and rail transport are major sources of noise in both urban and rural areas.
- Recognize the wide variety of approaches to traffic management strategy.

KEY CONCEPTS

transported landscape, frontier expansion, contrail, smog, catalytic converter, biofuel, zero-emission vehicle (ZEV), *Nefos*

Movement is a basic element of the daily rhythm of life in every society, a fundamental human activity and need. Transport modes make such movement easier, whether it be a trip to the shop, or the movement of raw materials or goods from one place to another. From the earliest times, transport has been an integral part of the evolution of civilizations, and rapid transport and good communications using modern transport networks of road, rail and air are essential elements of all advanced economies, while global economic integration also relies upon efficient maritime transport. Development of most parts of the less prosperous economies is likewise intimately associated with transport facilities because positive feedbacks link transport and the economy, the growth of one frequently stimulating growth in the other.

Transport is becoming faster and more efficient, and there is no better indicator of this trend than the relentless growth of average distance travelled per person per day. Such mobility comes with an environmental price tag, and the effects of transport on the environment are wide-ranging. Some of the major issues are summarized in Table 16.1; they include air pollution and noise from road and air traffic, and marine pollution from shipping (see Chapter 6). The transport sector of economies is also a major consumer of resources, including energy, minerals and land.

IMPACTS ON LAND

All forms of transport consume land resources, whether for nodes such as airports, railway stations and ports, or for the route corridors of roads, railways or canals. New transport developments are welcomed because they are designed to improve accessibility, but they are sometimes opposed by certain groups because they alter the nature of the landscape in which they are placed. Concern for the impacts of new transport routes on the countryside is not a recent phenomenon. Canal and railway companies in eighteenth- and nineteenth-century Britain faced opposition from landowners, artists and writers on environmental grounds in the same way that motorway construction is opposed today by individuals and pressure groups intent on preserving the countryside. In many countries, however, desire by governments to improve communications often overrides other concerns, even to the detriment of areas supposedly protected from damage by developments. New motorways are often built in response to congestion problems as more cars are used (Figure 16.1), but unfortunately for those who decry the

Table 16.1 Major environmental effects of transport

LAND
- Use and degradation of land and its associated ecosystems
- Excavation and use of minerals (e.g. gravels) for road construction
- Generation of solid waste as vehicles are withdrawn from use

BIOSPHERIC
- Introduction of immigrant species to new environments
- Barriers to migration of species

ATMOSPHERIC
- Emissions of greenhouse gases, particulates, fuel and fuel additives
- Noise and vibrations

HYDROLOGICAL
- Contamination of surface and groundwater from surface runoff, and spillages of petrol and oil and transported substances
- Modifications of hydrological regimes during construction of roads, ports, canals and airports

Source: modified after OECD (1991a).

Figure 16.1 A highway in China's capital, Beijing, where the number of cars has increased very rapidly in recent years, a trend that has spawned more infrastructure. The city has six ring roads.

loss of landscapes to motorways, more roads can also create more traffic, spurring the demand for still further expansion of road networks.

Transport developments not only despoil landscapes, but also consume resources in their construction and operation. There are numerous historical examples of the impacts new routes have upon local resources. The introduction of wood-burning steamships to West African rivers, for example, which occurred in the 1830s on the River Niger, heralded a significant pulse of deforestation along river banks. Similarly, when rail travel began with the line from Dakar to Saint Louis in Senegal in 1885, timber was felled on a large scale to clear the way for tracks, to make sleepers, to build bridges and to fuel steam engines. In many countries today, road building is the largest consumer of crushed stone from quarries, which leaves permanent scars on the landscape. A single kilometre of two-lane highway can consume more than 15,000 tonnes of aggregates, for example.

Impacts upon the landscape can, in turn, present hazards to the transport routes themselves, as well as other land uses in the vicinity. In permafrost areas, the importance of maintaining the thermal regime of ground ice during and after construction of routeways was learnt the hard way during the construction of the Alaska Highway in the early 1940s. Even minor disturbances to the thermal equilibrium caused permafrost to thaw, presenting problems of heaving and subsidence due to frost action, and the creation of impassable mires. Soil erosion and problems of slope failure are other hazards the highway engineer has to confront. Areas receiving high rainfalls are among the most susceptible, and slopes that are lacking in vegetation are often the most likely to fail. The major stability problems faced on steep slopes along the Kuala Lumpur–Ipoh highway in peninsular Malaysia are shown in Table 16.2. Serious soil erosion problems can also occur when vehicles are used off roads. Such off-road driving was found to be a prime cause of surface destabilization and consequent soil erosion by wind in a desert area near the city of Las Vegas in the south-west USA (Goossens and Buck, 2011), and the atmospheric dust presents a range of health hazards to city residents.

New transport routes also spawn environmental impacts indirectly through their *raison d'être* of improving access to places and resources, thereby encouraging mining, industrial and other land uses. In Russia, for example, construction of the Trans-Siberian railway in the late nineteenth century opened up the vast mineral resources of Siberia, resulting in numerous sites of environmental degradation (Figure 16.2). Economic development in many other parts of the world has also been driven by improved access to new natural resources in what is regarded as the classic model of frontier expansion, which continues in the contemporary era (Barbier, 2012). In regions where land-use regulations and the rule of law are limited, the impacts of new roads and other transport infrastructure can be exacerbated. In the Brazilian Amazon, for example, it is estimated that for every kilometre of legal road there are nearly three kilometres of illegal roads (Barber et al., 2014). These roads are reported to facilitate a variety of illegal activities, such as timber theft, poaching, illicit drug production and illegal gold mining.

Agriculturalists too often follow in the wake of new transport routes, to convert new land to food production. Cultivation only began in Australia, for example, with the arrival of European settlers, and in modern times it is generally agreed that agriculturalists are a primary driver of tropical forest loss, their movement often closely following transport routes. In Brazil, large-scale investment in transport infrastructure has encouraged migration to remote areas and consequent forest loss in several states (see Chapter 4). Similar frontier agricultural expansion is occurring in many developing economies at the expense of other natural habitats.

Table 16.2 Gradient and proportion of slopes (%) with major slope stability problems on the Kuala Lumpur–Ipoh highway, Malaysia

Stability problem	Slope gradient		
	33–45 degrees	63 degrees	83 degrees
Slope failure	8	19	0
Slumps	21	31	54
Gullies	11	8	0

Source: modified after Bayfield et al. (1992: 80, table 2).

Figure 16.2 Heavy industry on the Trans-Siberian railway just outside Irkutsk, Russia.

IMPACTS ON THE BIOSPHERE

Transport both facilitates and hinders movements of flora and fauna, and opens up new routes for species dispersal. The movement of people has assisted numerous biological invasions of great ecological significance, both deliberate and inadvertent (e.g. Simberloff, 2009), and competition from introduced species remains a major threat to the continued existence of native species on the global scale (see Chapter 15). Recognition of the considerable disturbances that can be caused by this form of biological pollution prompted the inclusion of a call to prevent the introduction of alien species in the Convention on Biological Diversity, although implementing such an objective is no easy task.

History is punctuated with many examples of plants and animals deliberately transferred between regions for subsistence and commercial purposes. Many of the first settlers on tropical Pacific islands are thought to have left areas we now know as New Guinea and the Solomon Islands on their ocean-going rafts stocked with pigs, chickens, yams, taro and coconuts. These food sources both kept them going on long voyages at sea and could also be established in new lands, so that the settlers effectively recreated familiar environments with their 'transported landscapes' (Nunn, 2003).

Once human colonizers were established in new territories, they frequently transported newly-discovered useful local plants back to the colonizers' original territories. New crops such as potatoes, maize, tomatoes

Figure 16.3 A carefully managed facility outside Riyadh in Saudi Arabia where plants brought from desert areas all over the world are cultivated to ascertain their suitability for use on the Arabian peninsula.

and tobacco from the Americas were introduced and cultivated in western Europe, for example, and several of today's staple African food crops (e.g. maize and cassava) were brought to the continent from Latin America by outsiders. Indeed, one estimate suggests that introduced crop and livestock species account for no less than 98 per cent of food produced today in the USA (Pimental *et al.*, 2014). The introduction of alien species continues today and can bring great benefits if managed properly (Figure 16.3), but there are many cases where introduced species have become such successful colonizers that they have given cause for concern.

The golden apple snail is one such example, introduced from South America to South-East Asia in 1980 to be cultivated as a high-protein food source for local eating and as potential gourmet export item. These intended uses were not commercially successful, and the snails escaped or were released into rice paddies, feeding on rice seedlings and causing significant crop damage in many South-East Asian countries. The golden apple snail reproduces rapidly and acts as a vector of diseases, which makes it a very serious pest to agriculture, the wider environment, and human health. It has also been introduced and become established in parts of North America, Europe and some Pacific islands. These snails are in the Global Invasive Species Database as one of the 100 worst invasive alien species in the world (www.iucngisd.org/), and the cost of attempts to manage the problem globally is put at US$1200 million annually (de Brito and Joshi, 2016). Several attempts have been made to quantify the economic impact of all invasive alien species at a national

THE ISTHMUS OF SUEZ MARITIME CANAL: THE CUTTING NEAR CHALOUF.—see preceding page

Figure 16.4 Engineering work on the Suez Canal, an engraving made in c.1864. Since its completion in 1869, the Suez Canal has enabled the exchange of many fish and other marine species between the Red Sea and the eastern Mediterranean.

level, yielding total annual costs in the billions of US dollars for Canada, China, Great Britain and the USA (Williams *et al.*, 2010).

There are also numerous cases of unintentional dispersal of, and subsequent colonization by, species that have 'hitched a ride' on various forms of transport. Modes of transport include automobiles (e.g. as seeds stuck in mud on tyres), aircraft (e.g. diseases/pests on food carried by passengers or on cargo) and ships (e.g. species attached to hulls). Ships have been a particularly strong influence on aquatic ecosystems, with numerous marine and freshwater species becoming harmful invaders after dispersal in or on ships (Keller *et al.*, 2011), many arriving in ballast water or as a result of canal building (Figure 16.4). Following these introductions, native fish species often disappear or are greatly reduced due to competition or predation.

A classic example is the invasion of the North American Great Lakes by two species from the Atlantic, the alewife and sea lamprey, following completion of the Erie Canal (now the New York State Canal) and the Welland Canal during the nineteenth century (Figure 16.5). As a result, the once common Atlantic salmon, lake charr, lake trout and lake herring have been severely depleted by competition for food from the alewife and predation by the sea lamprey. Since several of these greatly reduced native species are of commercial importance, the Great Lakes Fishery Commission currently spends about US$25 million per year in lamprey control, through the use of barriers, traps, periodic applications of a toxicant in their spawning areas, and release of sterile males (Rasmussen *et al.*, 2011). The barriers and toxicants have some negative effects on non-target species, but these are considered acceptable by fishery managers in return for protecting highly valued fishes.

More recently, the Great Lakes have been invaded by another species, the zebra mussel: a small, striped native of the Caspian Sea, which probably arrived in the ballast tanks of a European tanker. Within two years of their first appearance in 1988, zebra mussel densities had reached 70,000 individuals per square metre in parts of Lake Erie, choking out native mussels in the process. There has been a subsequent spread of the invader so that all five of the Great Lakes are infested, as well as numerous other inland lakes and rivers. The zebra mussel can now be found in the waters of 23 US states and two Canadian provinces (Ontario and Quebec) with the potential to cause enormous damage to drinking-water treatment plants, electric power generation facilities, dams, boats and fisheries, as well as devastation to endemic aquatic communities (Connelly *et al.*, 2007). A round-up of economic impact studies of zebra mussels on ecosystem services in the Great Lakes by Pejchar and Mooney (2009) put the net costs to aquaculture at about US$32 million

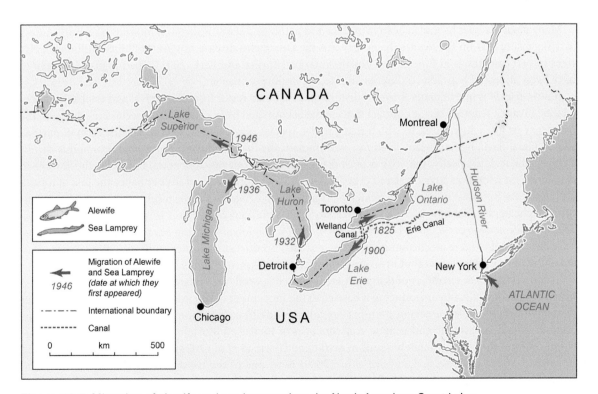

Figure 16.5 Migration of alewife and sea lamprey into the North American Great Lakes.

per year and the annual cost to a range of water-dependent facilities at US$11 million. One reaction to the zebra mussel invasion was the introduction in 1990 of mandatory controls on the release of ballast water from commercial ships in the Great Lakes in an attempt to prevent any future biological invasions.

Similar examples can be quoted for land transport. Evidence from Gabon indicates how the little red fire ant has penetrated the tropical forests of central Africa some 60 times faster along logging roads than through undisturbed forest. The ants appear to be responsible for causing mortality or blindness in native species such as primates, leopards and many invertebrates (Walsh *et al.*, 2004).

The dispersion of diseases is a particularly notable aspect of the transport revolution, with rapid journey times sharply reducing the natural checks upon the spread of diseases. The point can be illustrated by the case of smallpox:

> In Columbus's time, crossing the Atlantic was slow compared to the progression of smallpox. Since all carriers of the virus manifest symptoms of the disease, most of the infected travellers would have either become sick and died or recovered before reaching the New World. As a result, smallpox probably did not reach the Americas until several decades after Columbus's voyage.
>
> (Levins *et al.*, 1994: 57)

Today the situation is very different; travel to almost anywhere in the world can be accomplished in a few days at most, less than the average incubation time of many disease pathogens. Modern rapid transportation can turn a local disease into a worldwide pandemic. A good example is the outbreak of cholera in South America in 1991, thought to have been introduced from China in the ballast water of a freighter that docked at a port in Peru (see p. 531).

Many plant diseases have also been introduced to native populations through modern transport. A devastating example from the twentieth century was the Dutch elm disease fungus that killed millions of elm trees in two pandemics in Europe and North America (Brasier and Kirk, 2001). The fungus is thought to have been dispersed between continents on infected imports of timber.

A new disease is one of many forms of disturbance that may result from the arrival and establishment of a biotic invader. Another serious impact that can be encouraged by invasive plant species is a change in the frequency or intensity of fires, to which many native species may not be adapted. The invasion of parts of the US Great Basin by European cheatgrass has dramatically increased the frequency of fires in this shrub-steppe habitat. Historically, the return period for fire was about 60–100 years, but since the European cheatgrass invasion fires have occurred every 3–5 years. The more frequent fires enhance the loss of topsoil and nutrients, but most importantly retard the regeneration of native shrubs resulting in virtual monocultures of cheatgrass that have reduced biodiversity across millions of hectares in western North America. This invasive issue is expected to become more serious with global warming because higher temperatures have been found to increase cheatgrass biomass and seed production (Blumenthal *et al.*, 2016). Some other examples of ecosystem impacts of invasive plant species are shown in Table 16.3.

In contrast to these examples of transport facilitating dispersal, the construction of new routes can also have the opposite effect by introducing a barrier to the movement of organisms. This barrier effect, either through mortality during crossing attempts (so-called roadkill) or behavioural avoidance, fragments populations. The demographic and genetic consequences of increased mortality and isolation lead to population declines that may ultimately result in extinction (Forman *et al.*, 2003). Such effects may be marked in tropical forests because, as Laurance *et al.* (2009) note, one of the most striking features of forests in the tropics is the high proportion of species that tend to avoid even narrow (<30 m wide) clearings or forest edges. Avoidance of forest edges and clearings is believed to occur for several reasons, including

Table 16.3 Some examples of invasive plant species that have altered ecosystem properties

Ecosystem impact	Plant species	Location
Excessive light use (shading)	blackberry, raspberry	Galapagos Islands
Excessive water use	black wattle	South Africa
Excessive soil-carbon use	Monterey pine	Ecuador (páramo grasslands)
Fire promotion	European cheatgrass	Great Basin, USA
Fire suppression	catclaw mimosa, giant sensitive plant	Australia (mesic forest)
Salt accumulation	common iceplant	Mexico; California, USA

Source: after Henderson *et al.* (2006).

adaptations for flying in dense, cluttered environments and requirements for dark, humid microclimates or specialized food resources.

A study in Norway assessed the impact upon reindeer of the construction of roads and power lines associated with hydroelectric development (Nellemann *et al.*, 2003). These authors found substantial reductions in use by reindeer of areas within 2.5–5 km of roads and a substantial increase in reindeer abundance in the few remaining undisturbed areas in both winter and summer, including in the insect harassment period. In an example involving railways, Ito *et al.* (2017) tracked the movements of Mongolian gazelle and the Asiatic wild ass on both sides of the Ulaanbaatar–Beijing railway in Mongolia's Gobi-steppe grassland, and found that most of the tracked animals never crossed the railway. This barrier across migration routes prevents gazelles from reaching their traditional food-rich winter grazing areas, potentially increasing their winter mortality due to starvation.

The numerous forms of pollution derived from transport sources also have impacts on the surrounding biosphere. De-icing salt, usually sodium chloride, applied to roads during winter in mid-latitude, high-latitude and other snow-belt regions can have deleterious effects on nearby vegetation, as Zítková *et al.* (2018) show in the Czech Republic. These researchers examined the impacts on Norway spruce and found that the health condition of the trees declined with increasing concentrations of sodium and chlorine ions measured in their needles. In North America, de-icing salts have elevated chloride levels in many lakes, streams and wetlands to such an extent that concentrations frequently exceed the chronic and acute thresholds established for the protection of freshwater biota. Using laboratory experimentation, Hintz *et al.* (2017) concluded that the salinization can trigger a trophic cascade of negative impacts in freshwater communities.

Noise has been shown to affect some wildlife, particularly those species that incorporate sound into their basic behaviour, such as birds. Disturbance from noise was thought to be a possible cause of the difficulty male willow warblers had in attracting and keeping a mate along roadsides in the Netherlands due to the distortion of their song by traffic (Reijnen and Foppen, 1994). Emissions of nitrogen oxides from passing motor vehicles on roads in southern England were found to have caused greater growth in heathland plant species up to 200 m away from the road and consequent changes in species composition (Angold, 1997). The effects of pollutants can be complex and synergistic. A study in Switzerland found that physiological stress in plants caused by road pollutants made them more susceptible to attack by pests, with greater infestations of aphids on trees along roadsides (Braun and Fluckiger, 1984). These types of ecological effect associated with roads may be very widespread in countries with dense transport networks. Riitters and Wickham (2003) estimated that approximately 83 per cent of the land area in the continental USA is within slightly more than 1 km of any road, and only 3 per cent of the area slightly more than 5 km away.

ATMOSPHERIC IMPACTS

The atmospheric impacts of transport routes and their vehicles include microclimatological effects and the effects of pollution. Microclimate and vegetation-edge effects were studied by Young and Mitchell (1994) in a broadleaf forest in North Island, New Zealand. They found that more than 100 years after road construction microclimatic edge effects were discernible about 50 m into the forests, influencing processes such as germination and the early establishment of young trees. However, the pollution problems associated with transport are better known. Transport routes act as corridors for the dispersion of pollutants from vehicles with numerous consequent effects. These pollutants include noise (see below), light, particulate matter such as sand and dust, and a number of gases. Transport is a major source of air pollution due to its heavy dependence upon the combustion of fossil fuels, either in vehicles or at power stations. The major pollutants involved are:

- carbon dioxide
- carbon monoxide
- nitrogen oxides
- hydrocarbons
- sulphur oxides
- lead
- suspended particulate matter.

These pollutants have a wide variety of environmental impacts, being detrimental to human health (see Chapter 10) and a significant disruption to flora and fauna (see above). However, the pollution issue of greatest concern in recent years is the transport sector's role as a major and increasing source of greenhouse gases, principally carbon dioxide.

Pollution emissions from aircraft are unique in that they are largely released into the sensitive upper atmosphere at 8–12 km above the Earth's surface. At these altitudes, the emissions are longer-lived than at ground level and are more effective at causing chemical and aerosol effects (Lee *et al.*, 2009). The emission of gases and particles from aircraft directly into the upper troposphere and lower stratosphere has an impact on atmospheric composition. These gases and particles alter the concentration of atmospheric greenhouse gases, including carbon dioxide (CO_2), ozone (O_3), and methane (CH_4); trigger formation of condensation trails (contrails); and may increase cirrus cloudiness, all of which contribute to climate change (Dessens *et al.*, 2014).

Emissions of nitrogen oxides and water vapour during cruise mode are perhaps the most critical in this respect. In the troposphere, nitrogen oxides contribute to ozone formation, in turn contributing to smogs at ground level, while as a greenhouse gas, nitrogen oxides emitted directly to the troposphere may remain resident there as much as a hundred times longer than those released from terrestrial sources. Although estimates are difficult, some suggest that up to 30 per cent of aircraft emissions occur while cruising in the stratosphere, and in this region nitrogen oxides deplete ozone at a level that some researchers regard to be equivalent to the depletion caused by CFCs. Sulphur dioxide emitted by air traffic may also contribute to stratospheric ozone destruction.

Water vapour released during stratospheric flight is another area of some concern. Since the natural water vapour content of the stratosphere is low, the effect of contrail additions can be marked, leading to a higher frequency of cirrus clouds. This impact of contrails on cloud formation affects the radiation balance of the atmosphere both by trapping outgoing longwave radiation emitted by the Earth and atmosphere (so-called positive radiative forcing) and by reflecting incoming solar radiation (negative radiative forcing). On average, the longwave effect dominates and the net result of contrails is one of atmospheric warming. Night flights appear to be more responsible for this warming effect than daytime air traffic, and winter flights similarly account for a larger proportion than summer flights (Stuber *et al.*, 2006). Long residence times, thin air and low temperatures mean that contrail cirrus formed due to water vapour emissions at flight altitudes represent the largest single contribution to global warming associated with aviation (Kärcher, 2016). The volume of aviation passenger traffic grew strongly, at a rate of around 5 per cent per year, over the period 1992–2015, despite world-changing events such as the First Gulf War, the World Trade Center attack and outbreaks of Severe Acute Respiratory Syndrome (SARS), and such strong growth rates look set to continue (Lee *et al.*, 2009), making these environmental impacts the subject of increasing concern in years to come.

Road transport is the largest source of pollution in the transport industry thanks to the heavy reliance on motor vehicles in most countries. Concern over the environmental effects of motor vehicle pollution has prompted some concerted action to reduce certain pollutants. Lead is a particular example, due to its association with reduced mental development in infants and children, inhibition of haemoglobin synthesis in red blood cells in bone marrow, and impairment of liver and kidney function. Indeed, lead is toxic even at very low levels: no threshold blood concentration has yet been identified below which no adverse health effects occur (Schwartz, 1994). Besides its direct health risk through inhalation, lead can also enter the food chain via accumulation in soil and contamination of drinking water.

Lead was first added to fuel in the 1920s to increase the output of power from the internal combustion engine. Lead increases the octane rating of the fuel and reduces engine 'knock', a measure of how sensitive the fuel/air mixture is to engine pre-ignition. Reducing knock tendency (raising the octane rating) meant that engines could be designed for more power. Lead's use in fuel as an anti-knock agent has been reduced considerably in many countries in recent decades, with consequent reductions in concentrations in human blood. The reduced lead content of fuel was also partly a response to the realization that pollution by this highly toxic metal had dramatically increased at great distances from areas where motor vehicles are used. The concentration of lead in Greenland ice and snow increased by about 200 times between 5500 years BP and the mid-1960s, but has decreased significantly since then (Boutron *et al.*, 1991). In many countries, lead-free petrol has been widely introduced, with its adoption by motorists encouraged by lower taxes. The year from which only unleaded petrol was sold in a selection of countries is shown in Table 16.4. By 2017, just three countries still sold leaded petrol: Algeria, Iraq and Yemen.

Japan was one of the first countries to start reducing the lead content of its fuel, in 1970, in response to concerns about the widespread contamination of Tokyo's atmosphere. By the early 1980s, only 1–2 per cent of petrol sold in Japan contained lead, and no leaded fuel has been produced or used in Japan since 1986. A similar approach in North America took rather longer: about 20 years. Leaded gasoline was completely phased out by 1990 in Canada, and by 1996 in the USA.

This 'phase-in, phase-out' approach required the maintenance of a dual distribution system – for both leaded and unleaded fuel – for many years, which was very expensive. More recently, other countries have been able to take more immediate action. El Salvador, for example, went from over 90 per cent leaded

Table 16.4 Year from which all petrol sold has been unleaded, selected countries

Country	Year
Japan	1986
Canada	1990
Guatemala	1991
Brazil	1993
Austria	1993
Sweden	1994
Slovakia	1995
El Salvador	1996
USA	1996
South Korea	1996
Mexico	1998
UK	1999
Russia	2000
Vietnam	2001
Kyrgyzstan	2002
Uruguay	2003
Nigeria	2004
Burkina Faso	2005
Cuba	2006
Morocco	2009
Tajikistan	2010

Source: updated from Lovei (1998).

petrol in 1995 to 100 per cent unleaded petrol in 1996. In China, the strategy began in 1997 and conversion to unleaded petrol occurred city by city, starting with Beijing, and province by province, a process that took two years before the entire country was practically lead-free.

Many countries in sub-Saharan Africa also banned leaded petrol only relatively recently following an agreement reached in 2001 by 17 states to introduce a ban in their respective countries by 2005 (Figure 16.6). This relatively late ban in Africa and elsewhere reflected a number of factors, among them the average age of vehicles. There was concern over the effects of unleaded fuel on vehicles designed to run on leaded petrol, which focused on engine valves, some of which may wear out more rapidly when run on unleaded fuel. Today, most of these worries have been largely dispelled, except in some older, pre-1980 engines, for which other fuel additives are available to substitute the lubrication function of lead. In fact it is now thought that the economic savings on maintenance of vehicles using unleaded fuel significantly outweigh any potential negative side-effects, such as the increased wear on valves. The combustion of

Figure 16.6 A rural petrol stand in Burkina Faso, where the sale of leaded fuel was discontinued in 2005. The youngest generation benefits most from the phasing out of leaded petrol since children and infants are worst affected by lead poisoning.

unleaded fuel, compared with leaded fuel, extends the life of spark plugs from 6000 miles to more than 50,000 miles, extends oil change intervals by a factor of two to four, and results in less engine and exhaust system corrosion.

Some of the most serious air pollution issues associated with motor vehicles occur in large urban areas (see Chapter 10) and the US city of Los Angeles has a long history of pollution management dating from the first episodes of photochemical smog in the city in the early 1940s. In the mid-1950s, for example, California established the USA's first state agency to control motor vehicle emissions, and the state has frequently been a national pioneer in pollution control. Elimination of the use of lead in gasoline (petrol), for example, occurred in California in 1992, four years before elimination in the USA as a whole. In addition to the lowering of atmospheric lead levels, ozone concentrations have also fallen in southern California (Figure 16.7). The federal health standard for ozone has been tightened, but the number of days per year when the 1997 level was exceeded was still over 50 for most years after the turn of the century, while the 2015 standard was exceeded on more than 100 days a year for most of the 2000s. Motor vehicles in Los Angeles have been running on a specially reformulated petrol, designed to reduce the emission of VOCs, since the beginning of the 1990s. Reformulated petrol was introduced to southern California under the 1990 Clean Air Act and it is part of the reason for the reduction in ozone pollution over Los Angeles.

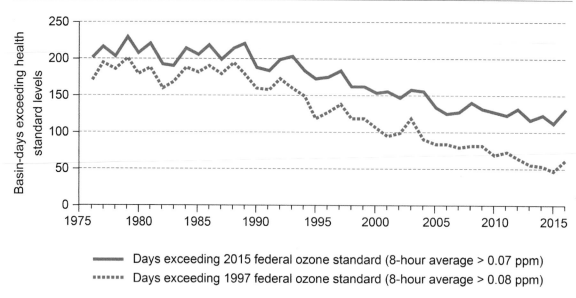

Figure 16.7 Number of days when atmospheric levels of ozone exceeded federal standards in southern California, USA, 1976–2016 (data from www.aqmd.gov/, accessed January 2018).

Much of the success in reducing pollutants from motor vehicles in Los Angeles and elsewhere in the western world is attributable to the increasingly widespread use of catalytic converters, which were first developed in the USA in the 1950s and 1960s, and fitted to US cars from the mid-1970s. The catalytic converter is a device fitted to vehicles, which removes certain pollutants by a chemical reaction. The pollutants targeted in vehicle exhausts are unburnt hydrocarbons, carbon monoxide and nitrogen oxides, which are converted by platinum, palladium and rhodium into carbon dioxide, water vapour and nitrogen. The ability of the three-way catalytic converter to reduce by more than 80 per cent the output from the internal combustion engine of these three pollutants has led to its widespread adoption. Catalytic converters had penetrated the entire passenger car market (i.e. they were used on every car on the road) in the USA by the early 1990s, with similar levels of market penetration in Japan and the European Union (EU) for new cars a few years later. Since these systems are only effective with lead-free petrol, the introduction of catalytic converters has contributed to the reduction in lead pollution from motor vehicles.

The use of catalytic converters, and other emission-reducing technologies, such as lean-burn engines, electronic fuel injection, computer-controlled spark ignition and exhaust recirculation systems, can only go so far in reducing polluting emissions. A technical fix to the vehicle engine system or an additive to fuel cannot solve the global emission problem. Indeed, recent experience indicates that these efforts do not keep pace with increasing traffic. It seems that we may have to abandon crude oil as a primary energy for transportation within four decades, due to dwindling supplies (see Chapter 18). Even if this estimate is pessimistic, the end of crude oil is undoubtedly a foreseeable event, and thus provides humankind with the opportunity 'to redesign the complete energy supply and conversion chain from mining and refining to the combustion in the vehicle engines in such a way that it can meet the global ecological criteria' (Svidén, 1993: 145).

Some examples of switches to less-polluting fuels have already occurred. Diesel has some advantages over petrol. Diesel engines use around 30 per cent less fuel, hence they emit less CO_2; they also emit fewer hydrocarbons and fewer nitrous oxides. Conversely, diesel engines produce larger quantities of sulphur dioxide, particulate matter, noise and smoke. Other alternative fuels are also in use in various parts of the world. One example is liquefied petroleum gas (LPG) which, like all natural gas derivatives, contains less carbon than gasoline and hence produces less CO_2, carbon monoxide and hydrocarbons, as well as less nitrous oxides, when burnt.

Other alternatives already in use include methanol and ethanol, which when burnt produce lower quantities of CO_2 and nitrous oxides than petrol. Methanol and ethanol also have the advantage that they can be produced renewably from biomass, examples of so-called biofuels (see Chapter 18). Biofuels have a particularly long history of use in Brazil, where ethanol fermented from sugar-cane juice has been used in fuel since the 1940s thanks to considerable policy support from the government (Box 16.1). Another fuel that is a strong candidate as a cleaner replacement for those derived from crude oil is hydrogen, sometimes referred to as the ultimate clean combustible fuel since its combustion emits only water vapour and small quantities of nitrogen oxides. Indeed, hydrogen has already been used as a fuel for manned spaceflights, but its viability for more widespread use is in the long term, given present difficulties in production, storage and distribution (Moriarty and Honnery, 2009).

Electrically driven vehicles also present some hope for future development when the electricity is generated from renewables, particularly given that the energy efficiency of the various types of electric vehicle is much higher than that of combustion engines. A major challenge with electric road vehicles is the batteries, which are heavy, expensive, and of uncertain durability, factors that mean at present electric vehicles are small in size and are typically driven for relatively short trips. Developing the technology for such vehicles has been encouraged by legislation in some cases and a leader in this respect is the US state of California, where a statewide mandate was introduced in 1990 requiring that 2 per cent of all passenger vehicles sold in the state by 1998 be emission-free (so-called Zero Emission Vehicles, or ZEVs), increasing to 10 per cent in 2003. Although the pace of change has not been as fast as hoped, several other US states joined California by adopting a ZEV mandate in 2013.

Nevertheless, such technological developments to reduce pollution would still leave many of the undesirable aspects of society's heavy dependence on cars, such as congestion and sprawl, unresolved (Figure 16.9). Hence, a wider perspective on the issue from the policy viewpoint is needed, which should include significant efforts to change behaviour in addition to those aimed at changing technology (see below).

NOISE

Road, air and rail transport is a major source of noise in both urban and rural areas. Transport noise affects people to varying degrees, ranging from mild annoyance and minor interruptions to everyday activities, to mental and physical damage. Acute exposure to noise is believed to cause changes in blood pressure and heart rate and the release of stress hormones; persistent exposure to noise is thought to increase the risk of cardiovascular disorders, including the risk of strokes (Sørensen et al., 2011), and in younger members of society, noise enhances the detrimental impact of air pollution on children's lung function (Franklin and Fruin, 2017). As such, noise is a significant aspect of the transport sector's environmental impact.

It is difficult to arrive at a definition of what level of sound constitutes a problem or a nuisance, and hence when a certain sound becomes unacceptable, but guidelines are available. Transport noise is usually measured using the logarithmic decibel (dB) scale. At 80 dB, people standing next to each other would need to shout to be heard, and at 90 dB – a typical level for a heavy lorry passing at 7 m – temporary loss

Box 16.1 Biofuels in Brazil

Brazil has one of the world's oldest biofuels programmes, dominated by the production of ethanol for fuel since the early twentieth century. A law passed in 1941 made the addition of ethanol to gasoline mandatory and domestic use of ethanol was accelerated during the global oil crisis of the 1970s because international oil prices were high (with the additional purpose of protecting the sugar industry from falling international sugar prices). The government supported research into alcohol-fuelled vehicles and subsidized prices at the pumps. Since then, the mandate for blending ethanol with conventional petrol has ranged from E10 (10 per cent ethanol) to E25, depending on sugar prices and ethanol supply prospects. The ethanol sector received a new boost in 2003 with the introduction of flex-fuel vehicles capable of running on any mix of ethanol and petrol, and these flex-fuel vehicles now dominate new vehicle registrations (Figure 16.8)

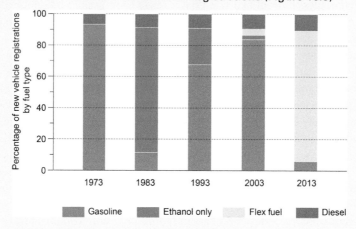

Figure 16.8 New vehicle registrations in Brazil by fuel type (after WEC, 2017).

Converting sugar cane to ethanol is very efficient, as the process is fuelled by the electricity and heat generated by burning the crop's crushed residue, known as 'bagasse'. This makes sugar-cane ethanol particularly sustainable, and provides life-cycle savings in greenhouse gas emissions of between 70 and 100 per cent compared with conventional petrol. These savings are much higher than for most other biofuels produced from starch and vegetable oil (e.g. comparative emissions savings for corn ethanol are between 20 and 50 per cent).

The Brazilian government hails its biofuels programme as a shining example of 'clean' and renewable energy that reduces greenhouse gas emissions, substitutes fossil fuels, creates jobs and spurs rural development without undermining food production. However, some criticisms have arisen over human rights and environmental justice due to poor working conditions on sugar-cane plantations and displacement of smallholders as well as environmental impacts in certain parts of the country. These issues have been addressed by the sugar-cane ethanol industry with voluntary governance and certification mechanisms. The Brazilian government has also forbidden sugar-cane expansion into sensitive ecosystems including the Amazon rain forest and the Pantanal wetlands.

Source: de Andrade and Miccolis (2011).

Figure 16.9 Severe traffic jams on the roads of Manila surround an overhead train that helps to relieve the congestion.

in acuteness of hearing would be felt for a few minutes. The 120 dB noise typically heard at 200 m from a jet aircraft on take-off is approaching the level at which physical pain, and ultimately deafness, are experienced. Generally, a continuous noise tends to be regarded as less of a nuisance than individual events, and the time of day when a noise is heard is also a factor – night-time noise being less acceptable than that during daylight hours. In the EU countries, two important indicators linked to harmful effects are set in the Environmental Noise Directive:

- 55 dB: the day, evening, and night-level indicator designed to assess annoyance
- 50 dB: the night-level indicator designed to assess sleep disturbance.

Road traffic noise, both inside and outside urban areas, is the dominant source affecting human exposure above the annoyance level of 55 dB in Europe. More people are affected in cities than in rural areas, and the percentage of urban residents exposed to road traffic noise that disturbs their sleep in a selection of European capital cities is shown in Figure 16.10. Another city with high levels of noise pollution from road traffic is the capital of Greece, where nearly one in four residents of the Municipality of Athens (the nucleus of the metropolitan area) live under levels of traffic noise greater than 72 dB. These excessively loud noise levels are recorded in the majority of residential neighbourhoods because, as Sapountzaki and Chalkias

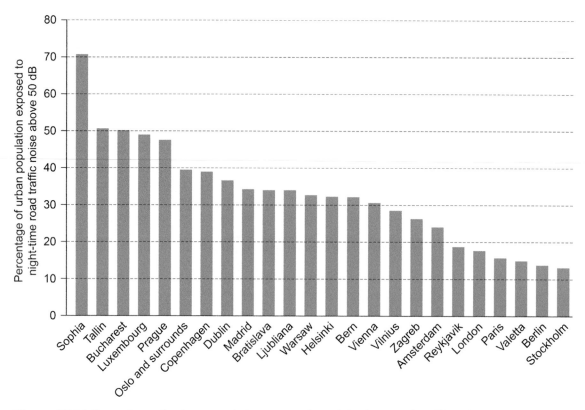

Figure 16.10 Percentage of urban population exposed to night-time road traffic noise above 50 dB in selected European capital cities (after EEA, 2017)

(2v005: 123) put it, 'noise pollution from road traffic in Athens is not concentrated at specific spots or along ribbon developments; it sprawls everywhere'.

These authors highlight a range of geographical, socio-economic and cultural factors that have combined to produce such a noisy city. These include:

- lack of greenery to formulate quiet areas
- extensive use of motorcycles, which are cheap to run and handy for avoiding traffic congestion
- the hot, dry microclimate which, combined with the dense arrangement of buildings, favours the use of large windows that are left wide open in summer for ventilation purposes
- extensive use of horns owing to traffic chaos and the cultural obsession of many drivers with traffic priority
- overnight recreational activities entailing high traffic loads during day and night
- culture of tolerance of excessive noise levels.

Mitigation of noise problems can be approached on four fronts:

1 Reduction of noise at source – through vehicle design, traffic management, tunnelling and noise-abatement procedures for aircraft.

2   Measures to control noise along its transmission path – mainly by barriers such as fences and embankments, and the use of buildings as noise barriers.
3   Measures to protect the observer from noise at the point of hearing – such as through building design, ensuring smaller windows on the noisiest façades and incorporating double glazing and acoustic insulation.
4   Land-use planning and zoning – effective on the larger scale by zoning noisy traffic paths away from residential and working areas.

The best results for noise abatement are derived from a combination of all these approaches. This is the stance taken to mitigate noise around US airports by the US Federal Aviation Administration which combines source reduction (quieter aircraft), land-use planning and management, noise-abatement operational procedures, and operating restrictions. This balanced approach has had considerable effect according to Waitz *et al.* (2005) who show that about 70 million people were exposed to a day–night noise level of 55 dB from commercial aircraft noise in the USA in 1975 but that the total had dropped below 10 million by 2000. Technological improvements probably represent the most important driver of this fall in the numbers of people affected by noise at airports, a decline that is mirrored elsewhere in the world. As Suau-Sanchez *et al.* (2011) point out, aircraft coming off the production line today are about 75 per cent quieter than in the 1970s.

TRANSPORT POLICY

The future environmental impacts of transport, as with many other environmental issues, will depend largely on policy-makers. Passenger travel by public transport and bicycles, for example, both more environmentally sound methods of travel than the car, depend upon appropriate emphasis being given to such modes. Yet ownership and use of personal cars have increased rapidly in developed and developing nations in recent times, despite the negative aspects of these increasing levels of motorization. Road transport is a hotly debated issue, perhaps, as Howey (2012: 28) suggests, 'because of our peculiar relationship with cars – to many people they represent a lifestyle choice, a statement about power, social status and self-esteem, not simply a means to get from A to B'. The relentless rise of the car also reflects policy and planning decisions that feed back on society and change lifestyles: decisions that reinforce dependence upon cars increase car use. A similar attitude has led in most countries to the great increase in road haulage of freight at the expense of rail, for example. Projections indicate there will be 2 billion motor vehicles on the roads by 2030, most of them still powered by fossil fuels (Gross, 2016).

Technological innovations can only go so far in reducing the environmentally damaging aspects of transport, and will need to be complemented with attempts to change behaviour alongside better planning that can reduce the need for travel that is costly in environmental terms. More practical urban and suburban land-use patterns can help to reduce car trips, both by shortening them and by reducing the need to travel by car. Public transport, cycling and walking need to be encouraged and provided for as alternatives (Figure 16.11).

The current obsession with the car, which characterizes many governments' attitude to transport, might change if the true costs of the car and other transport modes are calculated. External costs, which a particular mode of transport imposes on society as a whole, are difficult to quantify but they include traffic congestion, road accidents, pollution, dependence on imported oil, solid waste, loss of cropland and natural habitats, and climatic change. Life-cycle assessment methods have been applied to transport since the late 1990s and have provided assessments of such full environmental costs, also including impacts due to

Figure 16.11 Use of electrified trams like this one in Riga, Latvia, reduces the number of car journeys, relieving congestion and cutting down on urban pollution.

vehicle production and scrapping, and the extraction and production of fuel (Chester and Horvath, 2009). However, even if these life-cycle factors are included in the costs of transport (internalized), they have little effect on transport volumes, in part because rising levels of income reduce the effectiveness of higher prices.

An illustration of the difficulties involved in changing people's behaviour, even when the costs of personal mobility are well known to be high, can be made in the case of Athens, Greece. The city's air quality deteriorated significantly after the 1970s to become one of the worst in Europe, and road traffic is the principal source of pollutants that include ozone and nitrogen dioxide from petrol-driven vehicles, and black smoke particles emitted by diesel engines. Other common atmospheric pollutants produced by burning fossil fuels are also present, including carbon monoxide and sulphur dioxide. Athenians call their pollution the *Nefos* (literally meaning cloud); *Nefos* pollutants have negative effects on public health (Samoli *et al.*, 2011), damage the city's numerous ancient monuments (Kambezidis and Kalliampakos, 2012), and probably retard the growth of trees in nearby areas (Saitanis *et al.*, 2003).

Motor vehicles have been one of the main targets of local authority action to combat the *Nefos* and in 1982 Athens was the first major world city to introduce a scheme designed to reduce the number of private cars in the city centre based on licence plate numbers (see Table 10.6). Vehicles with licence plates ending in an even number were only allowed access on even-numbered dates, while odd-numbered dates were reserved for cars with odd-numbered licence plates. Many Athenians reacted by buying a second vehicle, with a licence plate allowing them in on the days when their first car was banned. These second cars tended to be older and more polluting, so air quality actually got worse. A year later, the authorities modified the scheme, and warned that the regulations would be changed again if people defied the spirit of the programme. In 1985, the impact on vehicle numbers was assessed. Although there were 22 per cent fewer private cars in the city centre, atmospheric pollution showed little change. It transpired that many people were travelling into town by taxi instead. Their numbers had increased by 26 per cent. It was decided to include taxis in the ban but the taxi drivers went on strike and deliberately staged traffic jams until their exemption was restored (Elsom, 1996).

In 1995, all motor vehicles were banned from 2.5 km² of the city centre for a three-month experimental period, but the experiment was not made permanent. During smog alerts, various levels of restrictions on vehicle access are still implemented, while factories are required to reduce or stop production. Vehicle emission tests have also been introduced, but with limited success. Improvements to air quality in Athens have been observed since the 1990s, despite a significant increase in the number of vehicles, thanks to these and other measures (Progiou and Ziomas, 2011): concentrations of carbon monoxide (CO) and nitrogen oxides (NO_x) have declined steadily since 1990, probably as a result of the increasing number of motor vehicles fitted with three-way catalytic converters under EU law. Exhaust particulates (PM10) also declined by around 25 per cent over nine years of study.

Many observers believe that most of the environmental issues that make transport unsustainable can only be properly addressed by focusing on ways in which behaviour can be changed, to reduce the demand for the activity that produces the problem in the first place (Taylor, 2007). As the Athens example shows, such changes can be very difficult to make, but moderation of the environmental impacts of transport will come about ultimately from government, through regulation, legislation, and other ways in which transport modes are encouraged and discouraged. A wide array of approaches to traffic management have been trialled, particularly in major cities (Table 16.5).

Policies that attempt to change behaviour are also likely to be important in attempts to 'decarbonize' transport, a strategy that many consider to be essential given the growing transport-related emissions of CO_2. The priority is summed up by Banister *et al.* (2011: 247) when they state that 'The recent global exponential growth in transport is unsustainable and must end unless the transport sector can decarbonize.' Although these authors are not optimistic about the prospects for transport's decarbonization, they present a wide range of possible policy measures, emphasizing a radically holistic approach that should go beyond business-as-usual approaches to reducing greenhouse gas emissions based on technological optimism and investment in green infrastructures. Their suggestions include putting a price tag on CO_2 consumption, reducing travel demand, and unconventional instruments not normally considered in transport policy such as banning commercial advertisements for high-CO_2 vehicles or types of travel.

In many European countries, a start has been made in this direction with the EU's Renewable Energy Directive which requires 10 per cent of all energy used in transport to come from renewable sources by 2020, a target that will probably be delivered almost entirely through the use of biofuels. The urgency of totally decarbonizing transport is highlighted by Skinner *et al.* (2010) who assess the policy options for achieving this goal in the EU by 2050.

Table 16.5 Some approaches to traffic management strategy

Approach	Examples
Operating restrictions and pricing	Low/zero emission zones (LEZ/ZEZ): pricing or restrictions based on emissions status of vehicles
	Vehicle operating and access restrictions: by license plate, time-of-day, or route
	Parking management: supply and pricing strategies
Lane management	High occupancy vehicle (HOV), high occupancy toll (HOT), and eco-lanes
	Bus and/or truck lanes
Speed management	Variable speed limits
	Speed control devices: traffic calming such as humps, chicanes, micro-roundabouts
Traffic flow control	Electronic toll collection
	Traffic signal timing: signal coordination, adaptive signal systems, transit signal priority
Trip reduction strategies	Shared-ride programmes
	Employer programmes for trip reduction: flex-time, telework
	Pedestrian and bicycle facilities: roadway and trip-end facilities

Source: after Bigazzi and Rouleau (2017).

FURTHER READING

Borda-de-Água, L., Barrientos, R., Beja, P. and Pereira, H. (eds) 2017 *Railway ecology*. Cham, Springer. A book giving an overview of the impacts of railways on biodiversity. Open access: link.springer.com/book/10.1007/978-3-319-57496-7.

Chester, M.V. and Horvath, A. 2009 Environmental assessment of passenger transportation should include infrastructure and supply chains. *Environmental Research Letters* 4: 024008. Analysis shows how air pollutant emissions are much greater when the entire vehicle life cycle is accounted for.

Fahrig, L. and Rytwinski, T. 2009 Effects of roads on animal abundance: an empirical review and synthesis. Ecology and Society 14(1): 21. Part of a special feature on Effects of Roads and Traffic on Wildlife Populations and Landscape Function. Open access: www.ecologyandsociety.org/issues/view.php?sf=41.

Goodenough, A.E. 2010 Are the ecological impacts of alien species misrepresented? A review of the 'native good, alien bad' philosophy. *Community Ecology* 11: 13–21. A critical review.

Hirota, K. 2010 Comparative studies on vehicle related policies for air pollution reduction in ten Asian countries. *Sustainability* 2: 145–62. A review of attempts to improve urban air quality. Open access: www.mdpi.com/2071-1050/2/1/145.

Leao, S., Ong, K.L. and Krezel, A. 2014 2Loud? Community mapping of exposure to traffic noise with mobile phones. *Environmental Monitoring and Assessment* 186(10): 6193–206. This paper presents a mobile phone application developed to monitor, assess and map exposure to traffic noise.

WEBSITES

aqicn.org/here   worldwide real-time air quality data for many cities.

greenercars.org   a guide to 'green' cars and trucks.

www.aqmd.gov   southern California's Air Quality Management Plan site has news, pollution data and information on transport programmes.

www.issg.org   the Invasive Species Specialist Group site contains a wealth of information on problems, management solutions, etc.

www.sustrans.org.uk   a UK charity promoting travel by foot, bicycle or public transport.

www.transportenvironment.org   Transport & Environment is a leading NGO dealing in smarter and greener transport policies at the EU level.

POINTS FOR DISCUSSION

- Is it possible to prevent biological invasions?
- Are Zero Emission Vehicles a realistic possibility?
- Is air pollution just an unfortunate side-effect of our need to move?
- If you were asked to propose measures to make transport more sustainable in your area, what would you suggest?
- Keep a personal transport diary for a week, recording distances and mode of travel. How many trips could have been made by walking or cycling?
- Assess the ways in which transport is related to other environmental issues covered in this book.

17

Waste management

Wastes are produced by all living things – excreted by an organism or thrown away by society because they are no longer useful and if kept may be detrimental. But the makeup of waste means that once discarded by one body, its constituents may become useful to another. In the natural world, a waste produced by an animal, for example, is just one stage in the continual cycle of matter and energy that characterizes the workings of the planet: an animal that urinates is disposing of waste products its body does not need, but the water and nutrients can become resources for other organisms. Similarly, sewage from a town can be used by bacteria that break it down in a river, for example, or an old shirt discarded by one person might be worn by another, often poorer, individual. The term 'waste' is therefore a label determined by ecology, economics and/or culture.

Much of the waste produced by human society consists of natural material and energy, although we have also created products not found in the natural world, such as CFCs and plastics, which can become wastes. The issues surrounding wastes stem from problems of disposal. People's use of resources, and hence production of wastes, have accelerated in the period since the Industrial Revolution, and this fact, combined with the growing number of people on the planet, has created increasing volumes of waste that need to be disposed of. This disposal requires careful management since too much waste disposed of in a certain place at a particular time can contaminate or pollute the environment, creating a hazard to the health, safety or welfare of living things. Examples of waste products and their impacts on the environment are found throughout this book. They include agricultural and food waste (Chapter 13), wastes produced by energy production (Chapter 18) and mining (Chapter 19), and pollution of the atmosphere (Chapters 11, 12 and 16), aquatic environments (Chapters 6, 7 and 8) and urban environments (Chapter 10).

TYPES OF WASTE

Waste can take many different forms: solid, liquid, gas, or energy in the form of heat or noise. The problems associated with its disposal can stem from a range of other properties, including its chemical makeup, whether it is organic or inorganic, the length of time taken for it to be broken down into less hazardous constituents, and its reactivity with other substances – so-called synergy. The sources of waste are also diverse, stemming from virtually every human activity, including:

- mining and construction
- fuel combustion
- industrial processes
- domestic and institutional activities
- agriculture
- military activities.

Collection of data on waste has had a relatively low priority in many countries, and the methods for their compilation vary widely. Statistics that are available are often given as weights, but the large differences between quantity and quality mean that these figures can only give a general indication of the nature of the waste disposal problem.

Since some wastes are inherently more dangerous than others, categories such as 'special', 'controlled' and 'hazardous' are often identified and such wastes dealt with in a more careful manner. In the UK, where the disposal of different types of waste is regulated by different laws, so-called special wastes, for example, are those deemed dangerous to life and are subject to regulations designed to track their movement from production to safe disposal, or from 'cradle to grave'. Although there is no universal agreement over what constitutes hazardous wastes, they include substances that are toxic to humans, plants or animals, are flammable, corrosive, or explosive, or have high chemical reactivity. Such hazardous substances include acids and alkalis, heavy metals, oils, solvents, pesticides, PCBs (polychlorinated biphenyls) and various hospital wastes.

DISPOSAL OF WASTE

All wastes are disposed of into the environment, but some enter the environment in a more controlled manner than others (Figure 17.1). Some wastes are emitted directly from the source without treatment, others are collected and sometimes treated before disposal. Wastes produced from the combustion of fuel by motor vehicles, for example, are emitted directly into the atmosphere, while domestic sewage wastes are often collected by municipal authorities and disposed into specific locations such as a river or ocean. Three of the compartments into which wastes are emitted – air, rivers and oceans – are usually publicly owned and this common ownership has facilitated unregulated emissions of wastes. In many countries, however, the degradation of these common property resources by wastes has spawned numerous controls on their emission.

This chapter is concerned with wastes that are actively managed and where a wide range of management options are available, as Table 17.1 indicates for hazardous wastes. The following sections will look more closely at two of the most commonly used options: landfill and incineration.

Landfill

Dumping of waste is by far the most commonly used method of disposal in most countries. For example, mining and quarrying wastes commonly remain on the land surface as tailings and spoil heaps, or occasionally are dumped at sea (see Chapter 19). A significant quantity of other wastes is buried in landfill sites, and in many countries it is the most commonly used management method for disposing of municipal wastes (Figure 17.2) as well as being a frequently used method for hazardous waste disposal. However, in some parts of the world the quantities of material being disposed of in landfill are declining. About 64 per

Figure 17.1 The plastic bag is a potent symbol of the throw-away society. Produced in very large numbers, it is designed to be durable, lightweight and cheap. However, its convenience comes at a high environmental price in pollution terms, which has led to the development of biodegradable alternatives.

Table 17.1 Treatment and disposal technologies for hazardous wastes

General approach	Specific technology
Physical/chemical	Neutralization
	Precipitation/separation
	Detoxification (chemical)
Biological	Aerobic reactor
	Anaerobic reactor
	Soil culture
Incineration	High temperature
	Medium temperature
	Co-incineration
Immobilization	Chemical fixation
	Encapsulation
	Stabilization
	Solidification
Dumping	Landfill
	Deep underground
	Marine
Recycling	Gravity separation
	Filtration
	Distillation
	Solvent extraction
	Chemical regeneration

Source: after Tolba and El-Kholy (1992).

cent of municipal waste generated in OECD countries was destined for landfill in 1995 and this proportion is expected to reach 50 per cent by 2020 as recycling rates improve (Zacarias-Farah and Geyer-Allely, 2003).

If properly designed and managed, landfill sites may do little harm to the environment and can eventually provide a surface for such land uses as playing fields or reforestation, though not usually for heavy structures such as housing. The two main hazards associated with landfills are leakage of toxic leachates, which can contaminate surface and groundwater, and leaving refuse open to the air, which can allow infestation by rats or fermentation bacteria, which can generate methane and cause a fire hazard. Each day's addition must be covered with soil to prevent infestation and fermentation, although in some cases methane generation is actively encouraged and collected for use as a renewable fuel to generate electricity or heat (Themelis and Ulloa, 2007). Four typical landfill designs are shown in Figure 17.3. If the site is located on impermeable strata, problems from leaching to groundwater are not faced (Figure 17.3a), but when sited on an aquifer, various approaches can be taken to prevent leaching, such as an impermeable landfill

Figure 17.2 Trucks delivering waste for burial in landfill, the most common method of waste disposal in many countries.

lining of plastic or rubber (Figure 17.3b) or by careful control of the local groundwater table (Figure 17.3c). In these designs the aim of the landfill is to concentrate, isolate and contain wastes to minimize the hazard they represent, but in areas where leachate is not expected to be produced in harmful quantities, a dilute and disperse philosophy may be adopted, allowing seepage into the aquifer (Figure 17.3d).

Designs like that shown in Figure 17.3d have been used all too commonly in the past for disposal of hazardous wastes, and leakage from such old sites is causing increasing concern in many countries. A classic example of the dangers and costs of such sites is at Love Canal, a disposal site in the city of Niagara Falls, north-eastern USA, where between 1942 and 1953 a chemical company disposed of more than 20,000 tonnes of chemical wastes. Not long after the landfill was sealed, the site was redeveloped. A school was constructed right at the edge of the filled canal and many houses were built in the surrounding area. Problems with odours and residues, first reported at the site during the 1960s, came to a head more than 20 years after the redevelopment. Heavy rains in the winter of 1975 and spring of 1976 caused land subsidence and created pooling of surface waters, which were heavily contaminated with numerous toxic chemicals from the dump. Infiltration of these waters into nearby buildings caused a public outcry over the possible health

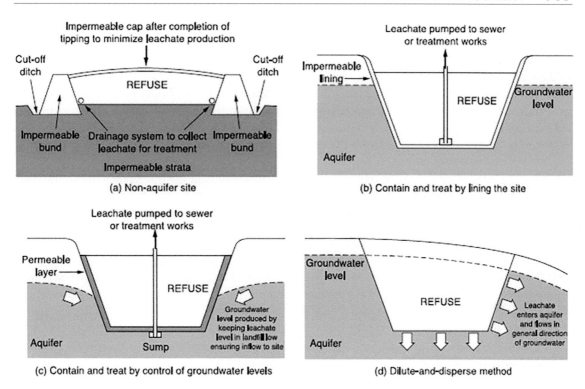

Figure 17.3 Four landfill designs (after Swinnerton, 1984).

hazards, and after a startling increase in skin rashes, miscarriages and birth defects, a state of emergency was declared at Love Canal by President Carter in 1978, the first executive order to address a hazardous waste problem in the USA. Residents from 239 houses were relocated in that year, followed by another 710 families in 1980. Subsequently, the school and many houses were razed to the ground. Clean-up of the site and relocation costs amounted to about US$250 million, but the long-term health implications for former residents are still being assessed (Gensburg *et al.*, 2009).

The Love Canal incident played a major role in inducing the US Congress to create in 1980 the so-called Superfund, a programme run by the US Environmental Protection Agency (USEPA) to provide financial support for clean-ups of such environmental disasters. The USEPA also created a National Priorities List of hazardous waste sites that needed remediation. Subsequent to the Love Canal incident, many of the world's industrialized countries have also begun a slow and costly process of identifying and cleaning up potentially dangerous old landfill sites. The costs of clean-ups in the USA are shared between the current and previous landowners, those who have dumped waste at the site and those who have transported it there, with some money also provided by the Superfund. Legal arguments over who should pay are proving to be a considerable delaying factor in the process, however.

Further evidence of the need to tackle sites containing hazardous wastes comes from continuing worries over the possible health effects among people living near them. One study of landfills containing hazardous industrial waste in five European countries showed a strong association between mothers living within 3 km of a site and a significantly raised risk that their babies would be born with birth defects (Dolk *et al.*, 1998). Deformities and illnesses such as spina bifida, cardiac septa (hole in the heart), and malformations of the arteries and veins were noted in the analysis of children born to mothers who lived near 21 landfills in Belgium, Denmark, France, Italy and the UK. Investigation of the pathways by which people living near such sites may be exposed to toxic waste will help to establish one way or the other whether the link between hazardous waste and birth deformities is a real one, but the study also showed a fairly consistent decrease in the risk of birth defects with increasing distance away from the sites, suggesting the strong possibility of a causal relationship.

Although today's landfill sites are subject to much stricter controls in many countries than in the past, the future of burial in landfills is uncertain, given the pressure on available space and opposition from local residents to the creation of new sites. However, the decline of waste disposal at sea, which has been banned for many waste forms (see Chapter 6), is likely to increase the pressure for more sites, although some of the additional burden will undoubtedly be absorbed into the incineration option. One way of making landfill a more sustainable option for waste management is to view it as a treatment process that can allow eventual re-assimilation of the site into the surrounding environment (Westlake, 1997), and the reclamation of resources buried in landfill by so-called landfill mining may be part of the remediation process (Hogland *et al.*, 2004). The location, design and operation of landfills must be appropriate to local conditions, and reduce risks associated with such sites to acceptable levels. The dangers of not so doing were illustrated at Love Canal.

Incineration

The controlled burning of waste at a high temperature, designed to attain its complete combustion, is increasingly used by many countries to reduce the bulk of wastes and to break down hazardous compounds, so rendering them less dangerous. Incineration can also be combined with energy recovery (Figure 17.4). In Denmark, for example, municipal solid waste is an important fuel in the national energy system: about 20 per cent of the district heating and about 4 per cent of the electricity was produced from waste in 2007 (Fruergaard *et al.*, 2010).

Incineration has proved to be a good disposal method for worn tyres, which are otherwise problematic to get rid of. Landfill sites do not welcome them since tyres represent a fire hazard and tend not to stay buried. In practice, many are simply stockpiled, at the rate of hundreds of millions each year globally. The high energy content of tyres makes them ideally suited for energy recovery through combustion, however, allowing waste tyres to be used as supplemental or alternative fuel in various industries such as cement and paper mills (Gieré *et al.*, 2004; see also Box 17.1).

Incinerators do, however, release harmful emissions into the atmosphere, such as particulate matter, heavy metals and trace organics (in particular, highly toxic dioxins), as well as numerous gases (NO_x, CO, CO_2, SO_2). Such emissions can be limited with a range of approaches, such as improved combustion techniques, sorting of wastes prior to incineration, and by fitting pollution control devices. Site selection for incinerators, as for landfills, is another controversial issue for the waste disposal industry. While people may recognize that such facilities are necessary, the 'not in my backyard' syndrome is a powerful force in determining where such facilities are located. In practice, many such locally undesirable land uses, or LULUs, are sited in poor and minority communities. A national study of French towns (Laurian, 2008) concluded that the issue was clearly one of environmental justice, finding that towns with high proportions of immigrants tend to host more hazardous sites (these included industrial and nuclear sites as well as incinerators and other waste management facilities). However, although at first such studies may appear to

Figure 17.4 A waste incineration plant that also generates electricity in Zürich, Switzerland.

reflect discrimination in siting procedures, research in the USA suggests that the situation may not always be so clear-cut. In some examples, LULUs appear to change community dynamics by driving down property values, resulting in a higher proportion of African-Americans and the poor around such sites, so that the dynamics of the housing market are more to blame than any discrimination in the choice of LULU site (Been, 1994).

Strong public feelings have often been foremost in pushing for more stringent emission standards for incinerators. In many European countries, these standards are now stricter than those applied to other energy sources, and some observers fear that the use of incinerators may be discouraged as a consequence. In practice, the total environmental impact of incinerators should be assessed and compared to similar measures for both alternative energy production and alternative waste treatment facilities (Astrup *et al.*, 2015). Interestingly, public outcry against using the incineration option for municipal waste disposal in Delhi, India has come from two rather different viewpoints (Demaria and Schindler, 2016). Middle-class residents oppose waste-to-energy incinerators because of their deleterious impact on air quality, while the city's informal waste-pickers fear that the authorities' decision to embrace waste incineration threatens their access to waste and hence their livelihoods.

Part of the total environmental impact of waste incinerators stems from the need to dispose of the ash generated by incineration, a final waste product that is usually buried in landfill sites. Pressure on available space for landfills, combined with the pollution dangers of heavy metals and dioxins remaining in the ash, has prompted research into the possible reuse of incinerator ash. One form of treatment developed in Japan is to convert incinerated waste ash into melted slag, which is then artificially crystallized to form stones (Nishida *et al.*, 2001). Crystallization gives the stones a hardness equal to natural stones, decomposes most of the dioxins in the process, and successfully contains hazardous heavy metals. The stones can then be used in civil engineering works as aggregates and building materials.

The construction industry is also one of the main users of the ash produced in Swedish combustion plants, incorporating it into the construction of roads and surfaces in landfills, but also as ballast or filler in concrete. However, since non-coal ashes in Sweden tend to be rich in calcium, they are also used in forestry to balance the pH of acidic soils, and have been shown to improve growth rates of spruce trees (Ribbing, 2007).

International movement of hazardous waste

As controls on the disposal of hazardous wastes have tightened in developed countries, there has been movement of both operations and wastes themselves to areas where legislation is poorly enforced or less stringent. During the 1980s, it became increasingly apparent that such trade in hazardous wastes was on the increase, both between industrialized and less-developed countries, and also between western and eastern Europe. International concern over this trend led ultimately to the Basel Convention on the Control of Transboundary Movements of Hazardous Wastes and their Disposal, which came into force in 1992. The Convention is based on a series of guiding principles originally adopted by the OECD countries, three of which are particularly important (OECD, 1991a):

■ The principle of non-discrimination – OECD members will apply the same controls on transfrontier movements of hazardous wastes involving non-member states as those applied to movements between member states.
■ The principle of prior informed consent – movements of waste will not be allowed without the consent of the appropriate authorities in the importing country.
■ The principle of adequacy of disposal facilities – movements of waste will only be permitted when wastes are directed to adequate disposal facilities in the importing country.

The Basel Convention also bans exports to countries that have not signed and ratified the treaty, and incorporates requirements to control the generation of hazardous wastes and obligations to manage them within the country of origin unless there is no capacity to do so. In practice, however, much hazardous waste is still traded internationally despite the Convention. A case of growing concern is the international trade in waste from all types of electrical or electronic equipment (e-waste), which is produced and traded in very great quantities. Largely, though by no means exclusively, the trade is from wealthy countries to low- and middle-income countries where recovery of valuable constituents (e.g. iron, aluminium, copper, gold, silver) is frequently unregulated (Figure 17.5). Safety concerns arise because e-waste also contains hazardous materials such as lead, mercury, chromium and flame retardants. Adverse health effects are related to contamination of the air, soil and water for people working and living at or near informal e-waste processing sites (Heacock *et al.*, 2016).

Following continuing concerns that the Basel Convention has not been strong enough, a number of signatory countries agreed in 1994 to add an amendment that actually bans altogether the export of hazardous wastes from OECD to non-OECD countries. However, this amendment, known as the Basel Ban, had not entered into force by early 2018 because it is not supported by a group of developed countries

Figure 17.5 Dismantling e-waste for recycling in West Bengal, India.

responsible for generating a large proportion of the world's hazardous waste, nor by a number of significant countries in the developing world.

Andrews (2009) argues that a complete prohibition on hazardous waste trade between North and South is both misguided and futile given the difficulties of enforcement, and that the Basel Convention is currently failing to prevent developing nations from being used as a dumping ground for the industrialized world's hazardous waste. He concludes that the Convention's institutions and procedures need reform and that the cornerstone of these reforms should be a strengthening of the procedure for prior informed consent. An independent body should be created to approve any shipment of hazardous waste and avoid abuse of the process by incompetent, corrupt or desperate operators on both sides of the North/South divide.

REUSE, RECOVERY, RECYCLING AND PREVENTION

Although landfill and incineration are currently the most frequently used methods of waste disposal, they are generally considered to be low down on the list of possible techniques devised from an environmental impact perspective, as the widely applied hierarchy of treatment techniques for solid wastes indicates (Table 17.2). Reusing a product, as opposed to discarding it, obviously makes environmental sense, and the advantages of waste recovery and recycling have also long been recognized. As for reuse, these advantages include a reduction of resource consumption, and a curb on the cost and inadequacies of disposal. In practice, waste products are reused when it is economically viable to do so and viability is assessed for a variety of motives. The history of industrial development is punctuated with examples of waste products being reappraised and turned into valuable resources. In the early nineteenth century, for example, Britain's fledgling chemical industry on Merseyside was using the Leblanc soda process to produce alkalis by treating salt with sulphuric acid, producing highly corrosive hydrogen chloride as a waste product. Realization that this pollutant was a lost resource, combined with fears of legal action from local landowners over the effects on surrounding vegetation, spurred industrialists to convert the hydrogen chloride to chlorine, which was used to make bleaching powder (Elkington and Burke, 1987). Another example from the nineteenth century concerns a resource that, today, most people take for granted. In the 1870s, when Nicolaus Otto found that a mix of flammable gas and air could be spark-ignited within the cylinder of a piston machine to generate movement, the flammable gas used was obtained from gasification of a waste product that came from the petroleum refinery process that produced paraffin for lamps. The waste product was called petrol (Svidén, 1993).

The perception shift that underlies these and many other examples highlights the fact that something that is considered a waste today can become a resource tomorrow. As Dijkema *et al.* (2000) put it, a substance or object is qualified as waste when it is not used to its full potential. This type of thinking has prompted formal identification of industrial ecology as a discipline: studying energy and material flows through industrial systems, using wastes from one process as inputs into others. When these sorts of links are established, a logical next step is to encourage traditionally separate industries to locate near one another and benefit from regional synergies or 'industrial symbiosis', as the example from Australia in Box 17.1 illustrates.

Reuse, recovery and recycling

There is a relationship between affluence and waste. In general terms, richer societies, and richer members of society, produce more waste than their less affluent counterparts. Indeed, realization of the economic value of certain wastes promotes reuse, recovery and recycling. In many developing countries, informal scavenging of municipal wastes from city dumps is widely practised (see Chapter 10) and the materials recovered are put to a variety of uses (Figure 17.7). It provides employment to individuals who otherwise would have none, particularly important in those countries where social security systems are inadequate

Table 17.2 The waste hierarchy: priorities for treatment

CLEANER TECHNOLOGY	1.	Prevention
RECOVERY	2.	Reuse
	3.	Recycling
	4.	Main use as fuel or other means of generating energy
DISPOSAL	5.	Incineration or composting
	6.	Landfill

Box 17.1 Industrial symbiosis in the Australian minerals industry

Gladstone Industrial Area in Queensland provides a good example of the mutual benefits derived from industrial symbiosis or regional synergy as a means to reduce disposal of wastes. Gladstone is a heavy industrial region dominated by one sector: alumina and aluminium and its supplier of energy, a coal-fired power station. Its other major processing industries are cement, chemicals and oil shale while the area also has a large multi-cargo port. The Gladstone Industrial Area has five major waste synergies, as depicted in Figure 17.6. All of these synergies have benefits measured in economic, environmental and local community terms.

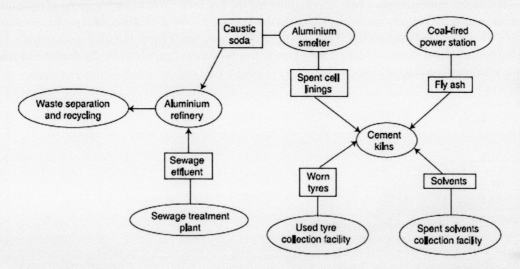

Figure 17.6 Industrial symbiosis in the Gladstone Industrial Area, Queensland, Australia (after Van Beers *et al.*, 2007).

■ Alternative fuels for cement production. The cement plant substitutes several alternative fuels for coal to power its kilns: worn tyres which had previously been sent to landfill; solvent-based fuels prepared from hazardous wastes; and spent cell linings from the local aluminium smelter.

■ Wastewater reuse. The alumina refinery uses partially treated sewage effluent for mud-washing, one of the processes involved in refining alumina from bauxite ore.

■ Waste separation and recycling. The alumina refinery opened a waste sorting facility to reduce its use of landfill which has now totally ceased. After its first year of operation, the recycling rate (mostly metal, cardboard and wood) had passed 90 per cent.

■ Fly ash reuse. Fly ash from the Gladstone power station is used to improve cement quality, also reducing use of conventional raw materials.

■ Caustic recovery. Caustic soda, a by-product of aluminium smelting, is used at the nearby alumina refinery.

Source: Van Beers *et al.* (2007).

or non-existent. These people are usually among the poorest sectors of society and often form discrete social groups or belong to minorities. Examples include the Zabbaleen in Egypt; Mexico's Pepenadores, Catroneros and Buscabotes; Basuriegos, Cartoneros, Traperos and Chatarreros in Colombia; Chamberos in Ecuador; and Cirujas in Argentina. One estimate suggests that in Asian and Latin American cities up to 2 per cent of the population earns a livelihood by scavenging, recovering materials to sell for reuse or recycling, as well as a range of items for their own consumption (Wilson *et al.*, 2006). However, in individual cities with a combination of particularly rapid growth, poor governance and a dearth of employment opportunities, the proportion can be significantly greater. In the Democratic Republic of Congo, an estimated 45 per cent of the urban residents in Kinshasa have turned to solid waste as a source of livelihood and income generation. These informal waste-pickers also provide an essential service in a city where less than 30 per cent of the solid waste generated is collected by formal waste management teams (Simatele and Etambakonga, 2015).

Studies of this informal sector have also highlighted the important role of such scavenged resources as inputs to small-scale manufacturing industries. In the Tanzanian capital of Dar es Salaam, these industries provide consumer goods that are much cheaper than if they were made from imported raw materials. Low-cost buckets, charcoal stoves and lamps are all made from scavenged metals from city dumps and sold to people who live in squatter settlements in the city (Yhdego, 1991).

Figure 17.7 A novel reuse for soft drinks cans in Namibia.

In today's more developed countries such informal waste recovery was also common more than 100 years ago, but more formalized recovery and recycling schemes have now sprung up. These schemes are also most advanced for metals and some other materials such as paper and glass, and collection points for such solid wastes have become a familiar sight in many cities in recent years. Some of the benefits of these schemes in environmental terms are shown in Table 17.3.

Among the critical factors affecting the quantities of material that are recycled are the capacity of the recycling plant and equipment, and the size of the market for recycled materials. In some cases, these factors have been influenced by legislation. In extreme cases, some types of packaging are banned altogether. In Denmark, for example, beer and soft drinks were simply not allowed to be sold in cans for many years (although the ban was repealed and replaced with a deposit and return system in 2002). More commonly used measures include laws designating a certain level of use of recycled materials and the imposition of a tax on products that do not incorporate a certain amount of recycled material. The Canadian city of Toronto, for example, has a local law which states that daily newspapers must contain at least 50 per cent recycled fibre or the publishers will not be allowed to have vending boxes on the city's streets. In the European Union (EU) a raft of directives have been introduced to encourage and impose higher recovery and recycling rates as part of a drive to foster a 'green economy' (see also below).

Denmark was one of the first countries to introduce a comprehensive waste taxation scheme to promote reuse and recycling as part of a waste management policy developed in the 1980s (Andersen, 1998). The policy was in response to the country's serious waste disposal problem: per capita generation of waste was among the highest in Europe, Denmark was running out of landfill space and there was widespread public concern about air pollution from incinerators. The tax is levied on most of the household and industrial waste delivered to Danish landfills and incinerators and the actual amount paid has increased every year since its introduction in 1987. The tax is designed to reduce the overall quantity of waste, to ease the burden on the country's limited landfills and to lower dioxin emissions from incineration. It has also been an integral part of Denmark's various national waste management plans, the first of which was introduced in 1989 with the aim of achieving a 54 per cent recycling rate for all waste by 1996. No taxes are levied on waste that is reused or recycled.

The Danish waste-reduction policy, in which the taxation system has played an important role, has registered considerable success. In the ten years from 1987 to 1996, Denmark achieved a reduction of 26 per

Table 17.3 Environmental benefits of substituting secondary materials for virgin resources

Environmental benefits	Aluminium	Steel	Paper	Glass
Reduction (%) of energy use	90–97	47–74	23–74	4–32
air pollution	95	85	74	20
water pollution	97	76	35	–
mining wastes	–	97	–	80
water use	–	40	58	50

Source: after Bartone (1990).

cent in the quantity of waste delivered to landfills and incinerators, and reached an overall recycling rate of 61 per cent. By 2011, the national recycling rate reached 61 per cent with 29 per cent going to incineration (which generates considerable amounts of energy – see above), and 6 per cent of waste disposed of in landfill (Andersen and Larsen, 2012). The ultimate aim is to transform Denmark into a zero waste society.

Taxes on waste are used to increase the economic incentives to recycle, but such intervention is not always necessary. In economic terms, the large energy savings gained from recycling of most metals make the practice worthwhile anyway. In the case of aluminium, the conversion of alumina to aluminium by electrolysis – the Hall–Héroult process – accounts for 80 per cent of the energy used in the entire production process, a stage that is bypassed when aluminium is recycled (Figure 17.8). These and other benefits have encouraged the increasing use of recycling in those industries that use metals.

Recycling also allows considerable energy savings in the glass industry. Over half of the energy consumption in the glass production process is used for melting and the melting point for broken or waste glass is lower than that of the mineral raw materials (mainly silica sand, soda ash and limestone). Adding recycled glass thus reduces energy use and the associated emissions such as carbon dioxide, given that most furnaces use natural gas or fuel oil. As a general rule, 10 per cent extra waste glass results in a reduction of furnace energy consumption by between 2.5 and 3 per cent (Beerkens and van Limpt, 2001). For many countries, the average recovery rate for glass has risen in recent decades, and in a number of European countries recycled material makes up more than 80 per cent of all glass used (Figure 17.9).

An industrial process that has a long history of waste product recycling is the steel industry. Steel slags from the Basic-Bessemer or Thomas process have been used as a phosphate fertilizer since 1880. The hardness and structural stability of blast furnace and steel slags have also enabled their use in the preparation of materials such as ceramic glass, silica gel, ceramic tiles and bricks as well as aggregates for road construction and as armour-stones in hydraulic engineering works, to stabilize and refill river beds and banks scoured by erosion (Motz and Geiseler, 2001). These uses help to relieve the pressure on natural aggregates like gravel, sand and processed rocks, the excavation of which leaves scars on the landscape. Steel converter slag has also been used more recently to treat wastewater. The process utilizes the slag's iron oxide, or magnetite component, which can adsorb nickel, a heavy metal, from wastewater (Ortiz et al., 2001).

Wastewater is commonly reused or recycled in many countries, and is widely recognized as a significant, growing and reliable water source that is particularly important in drylands where water is a scarce resource. Indeed, as Bakir (2001) points out, wastewater production is the only potential water source that will increase as the population grows and the demand on fresh water increases. The use of treated or untreated wastewater in landscaping and agriculture is common in many countries of the Middle East and North Africa, including the United Arab Emirates, Oman, Bahrain, Egypt, Yemen, Syria and Tunisia. Water resources management strategies in several countries, such as Jordan, sensibly consider wastewater as a part of the water budget.

The reuse of wastewater occurs in many ways, including industrial processes (particularly cooling, as in power stations), fire-fighting, aquaculture and aquifer recharge. Domestic use (e.g. toilet flushing and garden watering) is also commonplace and advanced treated recycled water is seen as an increasingly likely option to augment drinking-water supplies, particularly in cities (Rodriguez et al., 2009). The long experience of Windhoek in Namibia shows that this is a viable option. Situated near one of the world's driest deserts and prone to recurrent droughts, the city has used reclaimed water for drinking since 1969 (Lahnsteiner et al., 2018). During half a century of operation, the safety of the scheme has been verified by epidemiological studies and no health problems reported. Only domestic sewage is used for potable reclamation in Windhoek, which is feasible technically but has also required information and education campaigns in

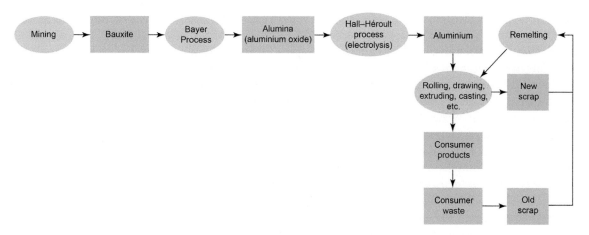

Figure 17.8 Stages in the production of aluminium goods.

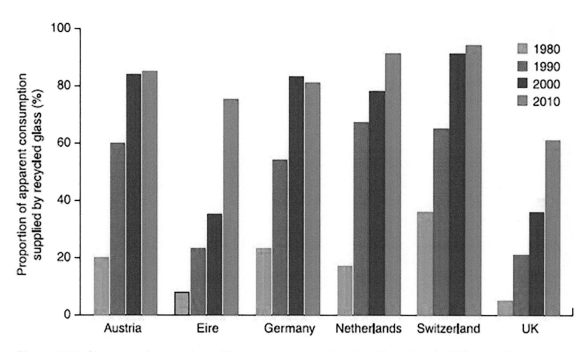

Figure 17.9 Glass recycling in selected European countries (updated from data in OECD, 2005).

order to make the water acceptable to consumers. Perhaps the largest challenge to a wider reuse of recycled water to augment drinking-water supplies in other parts of the world will be the issue of public perception.

Agricultural irrigation is one of the most widespread destinations for wastewater and this practice has a very long history dating back to the ancient Egyptians, the Mesopotamians and Indus valley societies. The water itself is valuable of course, but domestic wastewater is also typically high in key plant nutrients (phosphorus and nitrogen) derived from sewage. In most cases the irrigated fields are located in or around the urban areas where the wastewater is generated. In Mexico, for instance, wastewater from almost all cities that have a sewerage system is used in this way to irrigate about 150,000 ha of crops nationally. The largest such zone is in the Mezquital Valley where wastewater and storm runoff channelled out of Mexico City considerably increase yields (Table 17.4) and farmers pay a premium to use the wastewater (Duran-Alvarez and Jimenez-Cisneros, 2016). However, these increased yields come at an environmental price, since none of Mexico City's wastewater receives any conventional sewage treatment before it is used in this way. One characteristic of sewage is its high heavy metal content, a function of sewerage systems mixing industrial and household wastes, and long-term applications of wastewater in the Mezquital Valley have led to an accumulation of heavy metals in soils and their uptake into the crops themselves. There is also evidence of an increase in parasitic infections among agricultural workers and their families due to their exposure to raw sewage (Siebe and Cifuentes, 1995).

Over 80 per cent of all wastewater is discharged without treatment, but the proportion varies by country (WWAP, 2017). On average, high-income countries treat about 70 per cent of the municipal and industrial wastewater they generate and this figure drops to just 8 per cent in low-income countries. A global pledge to halve the proportion of untreated wastewater and substantially increase recycling and safe reuse is part of one of the Sustainable Development Goals. One of the dangers associated with reusing water on sewage-irrigated crops is the possibility of transmitting infectious diseases to the general public. An outbreak of cholera in the capital of Israel, Jerusalem, in 1970 was thought to be caused by consumption of vegetables irrigated with wastewater. The risks of this kind of disease transmission can be limited by simple measures, however. Irrigation with wastewater during planting is less hazardous than during the growing cycle, and certain crops (e.g. fruit and vegetables) are more likely to carry disease than others. Israel has progressively tightened its wastewater irrigation regulations in response to the 1970 cholera outbreak and by 2010 was using more than 70 per cent of the country's sewage for irrigation (amounting to half of all irrigation water) through both large-scale treatment and aquifer recharge and storage schemes producing high-quality effluent for irrigation, as well as many small-scale projects (Scheierling et al., 2011).

In many countries where sewage is treated, the sludge that remains after treatment is also reused as a fertilizer due to its high nutrient content. About half the sewage sludge produced in the UK, and one-third of that in France and Germany, is reused in this way, but in these and other countries, further use is limited by

Table 17.4 Comparative crop yields (t/ha) with different irrigation water sources in the Mezquital Valley, central Mexico

Crop	Groundwater irrigation	Untreated wastewater irrigation
Alfalfa	70	120
Barley	2	4
Chili	7	12
Maize	2	5
Tomato	18	35

Source: after Duran-Alvarez and Jimenez-Cisneros (2016).

the high heavy metal content. Hence, sewage sludge not used as fertilizer is disposed of by ocean dumping, dumping in landfill and municipal garbage dumps, or by incineration. Indeed, continued concern over the pollution content of sewage sludge, combined with a rising demand for organic and quality-assured food products, prompted Switzerland to end its use as a fertilizer in 2005.

As population and industrialization continue to increase, however, and the treatment of water and sewage becomes more widely used in order to reduce the pollution impacts from untreated sewage outlets, so more sludge will need to be disposed of. Switzerland has incinerated all of its sewage sludge since 2005, but a more innovative approach to this increasing dilemma of disposal can turn a problem into a benefit by realizing the value of the heavy metals themselves. One estimate of the potential yield of some of the high-value metals in sewage sludge suggests that global production of palladium from sewage sludge could be of the same order of magnitude as that from mining production (Table 17.5). The future disposal of sewage sludge should concentrate on the control of pollutants at source and the extraction of metals (Lottermoser and Morteani, 1993), an approach that would yield numerous benefits, including:

■ more widespread use of sludge as a fertilizer
■ revenue from metals extracted
■ conservation of geological metal resources
■ saving on expensive waste repository space
■ prevention of environmental impacts on terrestrial and marine ecosystems.

Waste prevention: cleaner production and extended producer responsibility

There is no doubt that the best way to manage waste is to prevent it at source wherever this is possible. The argument that prevention is better than cure is put by UNEP:

> When end-of-pipe pollution controls are added to industrial systems, less immediate damage occurs. But these solutions come at increasing monetary costs to both society and industry and have not always proven to be optimal from an environmental aspect. End-of-pipe controls are also reactive

Table 17.5 Estimated worldwide annual production of noble metals from sewage sludge compared with current production from geological resources

Metal	Current production (t/year)	Production from sewage sludge (t/year)
Gold	1600	100
Platinum	100	8
Palladium	100	80

Source: after Lottermoser and Morteani (1993).

and selective. Cleaner production, on the other hand, is a comprehensive, preventative approach to environmental protection.

(UNEP IE/PAC, 1993: 1)

Cleaner production is achieved by examining all phases of a product's life cycle from raw material extraction to its ultimate disposal – so-called 'life-cycle assessment' or LCA – and reducing the wastefulness of any particular phase. LCA involves identifying and quantifying the environmental loads involved in each stage of a product's life cycle (e.g. the energy and raw materials consumed, the emissions and wastes generated) and assessing the options available for reducing these impacts (Figure 17.10). A number of companies have attempted to develop 'closed-loop' processing cycles in which waste products are completely recycled and never enter the environment to become pollutants. In this way, material use is optimized, waste is minimized, and the overall environmental impact of production and consumption is significantly reduced.

The prioritization of preventative measures over end-of-pipe approaches, which has occurred at the same time as the rise of LCA, also reflects a shift from the 'command-and-control' approach to environmental policy-making to a less prescriptive, goal-oriented approach. Many of the resulting environmental policies and laws that have been developed in the past two decades or so are based on the principles of 'extended

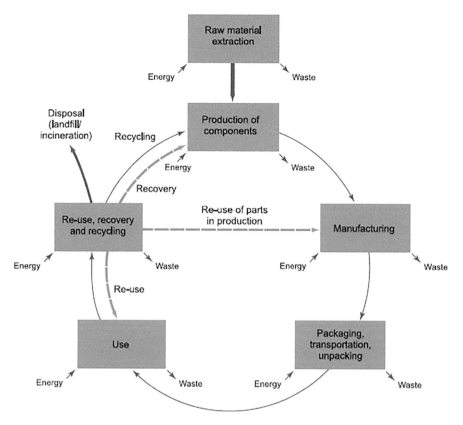

Figure 17.10 Stages in the life cycle of a product, with inputs and outputs.

producer responsibility': increasingly, countries in Europe and Asia have passed laws that require the manufacturer of a product to take responsibility for environmental impacts throughout the whole product life cycle, from resource extraction to recycling, reuse and disposal. The extended producer responsibility concept can be seen as a method of integrating sustainable development principles into international trade based on the polluter pays principle, recognizable in international environmental law (Kibert, 2004). It has been implemented in a variety of ways, including administrative, economic and informative instruments as indicated in Table 17.6.

One consequence of extended producer responsibility is the increasingly common requirement for a manufacturer to take back its product at its 'end-of-life' and dispose of it safely and sustainably. This has been implemented for relatively simple wastes such as packaging but also for much more complex disposal issues such as motor vehicles (Walls, 2006). Germany's Green Dot packaging ordinance, created in 1991 in response to the country's severe landfill shortage, was the first large-scale extended producer responsibility programme that assigns the full collection and recycling costs to producers. Under the law, producers of all kinds of packaged products are required to either individually take back their packaging (including plastic, metal, glass and paper) or join the Duales System Deutschland, a producer responsibility organization. Material-specific recycling targets of 60–75 per cent had been routinely met by 2000 and the volume of material used in packaging has dropped dramatically since the programme began.

Table 17.6 Examples of approaches to extended producer responsibility (EPR)

EPR approach	Types of tools	Examples
Product take-back programmes	Mandatory take-back	Packaging (Germany)
	Voluntary or negotiated take-back	Packaging (Netherlands, Norway)
Regulatory approaches	Minimum product standards	Electronic and electrical equipment, batteries
Prohibitions of certain hazardous materials or products	Disposal bans	Electronic and electrical equipment in landfills (Switzerland)
Cadmium in batteries (Sweden)	Mandated recycling	Packaging (Germany, Sweden, Austria)
Voluntary industry practices	Voluntary codes of practice	Transport packaging (Denmark)
	Leasing, labelling	Photocopiers, vehicles
Economic instruments	Deposit-refund schemes	Beverage packaging (South Korea, Canada, USA)
	Advance recycling fees	Electronic and electrical equipment (Switzerland, Sweden)
	Fees on disposal	Electronic and electrical equipment (Japan)

Source: after OECD (2001).

Table 17.7 Estimates of number of end-of-life vehicles (ELVs) in selected countries

Country	Number of ELVs (million units/year)
Australia	0.5
Brazil	1.0
Canada	1.2
China	3.5
France	1.6
Germany	0.5
Italy	1.6
Japan	3.0
Spain	0.8
South Korea	0.7
UK	1.2
USA	12.0
Worldwide	40.2

Source: after Sakai *et al.* (2014).

The disposal of old motor vehicles has become an increasing focus of extended producer responsibility legislation. A law for recycling end-of-life vehicles (ELVs) came into effect in Japan in 2005, and similar laws have been introduced in the EU, Korea, China and Taiwan, while in the USA, ELV recycling is managed under existing laws on environmental protection. The scale of the ELV issue in certain countries is indicated in Table 17.7. Under the EU directive, car manufacturers have become responsible for disposing of ELVs in the EU. The manufacturers are responsible for all or a significant part of the costs of take-back, reuse and recycling of vehicle components, while the national governments are obliged to provide authorized treatment facilities for the ELVs (Smink, 2007). Similar take-back schemes are in place for other goods such as computers, mobile phones, refrigerators and air conditioners.

WASTE MANAGEMENT IN THE GREEN ECONOMY

Improvements in waste management – reuse, recovery, recycling and waste prevention – are all critical advances towards minimizing the flows of materials and their wastes into and out of societies. Cutting waste as close as possible to zero therefore optimizes material resource flows, making our use of resources more efficient in what can be called a circular economy. Keeping already extracted resources in use should reduce dependence on raw materials, thereby reducing environmental pressures in those places where they are extracted and processed. However, closed-loop processing cycles alone are not sufficient to prevent further impacts on the environment and human health and well-being. Some argue that the circular economy approach needs to go beyond waste management, and facilitate a transition to a green economy by re-thinking the ways we produce, consume and dispose of products (EEA, 2015).

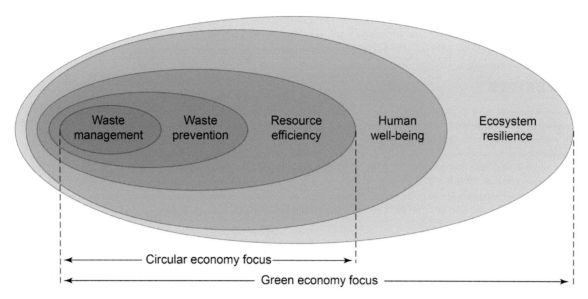

Figure 17.11 The green economy as an integrating framework for policies relating to material use (source: EEA, 2015).

An attempt to show how a focus on the green economy can go further than the circular economy is depicted in Figure 17.11. The green economy approach extends the emphasis beyond waste and material resources to the more general management of water, energy, land and biodiversity, in accordance with objectives of ecosystem resilience and human well-being. The green economy also addresses wider economic and social issues, such as competitiveness and environmental justice concerns surrounding exposure to pollution and access to green spaces.

FURTHER READING

Heacock, M., Kelly, C.B., Asante, K.A., Birnbaum, L.S., Bergman, Å.L., Bruné, M.N., Buka, I., Carpenter, D.O., Chen, A., Huo, X. and Kamel, M. 2016 E-waste and harm to vulnerable populations: a growing global problem. *Environmental Health Perspectives* 124(5): 550. A review focusing on the harmful health effects. Open access: www.ncbi.nlm.nih.gov/pmc/articles/PMC4858409/.

Merrild, H., Larsen, A.W. and Christensen, T.H. 2012 Assessing recycling versus incineration of key materials in municipal waste: the importance of efficient energy recovery and transport distances. *Waste Management* 32: 1009–18. A demonstration that for certain wastes incineration with energy recovery has less environmental impact than recycling.

Nahman, A. 2010 Extended producer responsibility for packaging waste in South Africa: current approaches and lessons learned. *Resources, Conservation and Recycling* 54: 155–62. An examination of various approaches to implementing EPR for packaging waste in South Africa.

Rodriguez, C., Buynder, P.V., Lugg, R., Blair, P., Devine, B., Cook, A. and Weinstein, P. 2009 Indirect potable reuse: a sustainable water supply alternative. *International Journal of Environmental Research and Public Health* 6: 1174–209. A review of water recycling for drinking purposes. Open access: www.mdpi.com/journal/ijerph.

Schneider, P., Anh, L.H., Wagner, J., Reichenbach, J. and Hebner, A. 2017 Solid waste management in Ho Chi Minh City, Vietnam: moving towards a circular economy? *Sustainability* 9(2): 286. An overview of waste management in an Asian megacity and the options for recycling. Open access: www.mdpi.com/2071-1050/9/2/286.

WWAP (United Nations World Water Assessment Programme) 2017 The United Nations World Water Development Report 2017. *Wastewater: the untapped resource*. Paris, UNESCO. Opportunities from improved wastewater management. Open access: unesdoc.unesco.org/images/0024/002471/247153e.pdf.

WEBSITES

www.basel.int   site for the Basel Convention on hazardous wastes.

www.epa.gov/superfund   the US Environmental Protection Agency's Superfund site with details of Love Canal and many other hazardous waste sites.

www.iswa.org   the International Solid Waste Association covers all aspects of solid waste management.

www.pro-e.org   information on countries affected by the EU directive on packaging waste.

www.unep.fr/scp   UNEP's Sustainable Consumption and Production Branch works to promote sustainable resource management in a life-cycle perspective.

zwia.org   the Zero Waste International Alliance promotes positive alternatives to landfill and incineration.

POINTS FOR DISCUSSION

- Are all waste products simply resources we haven't properly appreciated yet?
- How would you set about trying to prove a causal relationship between hazardous landfill sites and deteriorating human health?
- Is a zero waste society really possible?
- Is extended producer responsibility the answer to environmental problems caused by wastes?
- Assess all the various forms of waste generated every month by your household and what proportion is reused or recycled.
- Dilute and disperse or containment?

18

Energy production

Energy has long been a basic requirement of human societies. Archaeological evidence from caves in Africa suggests that fire was used by early humans, the hominids, as long ago as 1.5 million years BP, and fire has since fuelled the technologies on which civilized societies have been built. Today, regular supplies of energy drive the motors, appliances, cities, industries and transport on which the lifestyle of most of the

Earth's human population depends. Through time, the ability of human cultures to access energy, and the amounts of energy used, are often seen as indicators of society's level of development and resource use. Energy use's attendant environmental issues have long been with us, however. One of the first examples of environmental legislation in England occurred during the reign of Edward I when coal burnt for industrial and domestic purposes caused so great a smoke nuisance that the nobility, strongly backed by London residents, were able to obtain a royal proclamation forbidding coal burning in 1306 (although it proved impossible to enforce). Since then, the generation of energy and the use of fuels in multifarious ways to drive the industries and machinery of modern societies have facilitated an unprecedented level of both deliberate and inadvertent impacts on the global environment. The production, transportation, conversion and use of energy, particularly that derived from fossil fuels, are responsible for some of the world's most serious environmental challenges. These include global climatic change (see Chapter 11), acid rain (see Chapter 12) and pollution from motor vehicles (see Chapters 10 and 16), but human society's harnessing and use of energy lie behind virtually every one of the environmental issues found in this book. This chapter will focus on energy sources, specifically on the possible future of those not based on fossil fuels.

ENERGY SOURCES

Ultimately, most of the energy we use today comes from the sun. This radiant energy is converted by photosynthesis into chemical energy, which makes plant life and ultimately all animal life possible. It is also used by human society both as biomass and as fossil fuels, which were themselves once living plant matter. Solar radiation is also used directly by human society and indirectly by harnessing the movement of wind and rivers as part of the hydrological cycle. Three other ultimate energy sources are also commonly used: the processes of cosmic evolution preceding the origin of the solar system (nuclear power); the forces of lunar motion (tidal power); and energy from the Earth's core (geothermal power).

These sources of energy are often categorized into renewable and non-renewable resources, based on a timescale of human lifetimes. Hence, although fossil fuels are renewable over geological time, the rate at which we are using them effectively means that they are finite. Estimates of just how much oil, gas and coal remain for society to use are constantly being revised – and are to an extent a function of technology, price and the rate at which we use them – but they are ultimately non-renewable. Assessments for the end of 2016 suggest that we have about another 50 years' worth of both oil and natural gas at current production rates, and about 150 years for coal (BP, 2017). While exhaustion of these energy sources is not imminent, the finite nature of fossil fuels, combined with increasing awareness of the environmental impacts of their use, has spurred interest in the conservation of energy and the development of more sustainable forms of production from renewable resources. For policy-makers, the major driver of efforts towards renewables, as well as energy-efficiency conservation measures, is the desire to lower carbon dioxide emissions (see Chapter 11).

ENERGY EFFICIENCY AND CONSERVATION

Perhaps the most obvious approach to reducing the environmental impacts associated with energy use is to make energy production more efficient and to reduce the amounts of energy that are wasted during its use. Better energy conservation can be promoted in many sectors of society and the adoption of energy-saving technology and behaviour is usually encouraged by a fairly rapid return in terms of reduced energy costs.

The International Energy Agency (IEA) has developed a set of 25 energy-efficiency policy recommendations (Table 18.1) for seven priority areas across a typical national economy. In industry, the conservation

Table 18.1 The IEA's 25 energy-efficiency recommendations

Across sectors

1. Energy-efficiency data collection and indicators
2. Strategies and action plans
3. Competitive energy markets, with appropriate regulation
4. Private investment in energy efficiency
5. Monitoring, enforcement and evaluation of policies and measures

Buildings

6. Mandatory building codes and minimum energy performance requirements
7. Aiming for net zero energy consumption buildings
8. Improving energy efficiency of existing buildings
9. Building energy labels and certificates
10. Energy performance of building components and systems

Appliances and equipment

11. Mandatory energy performance standards and labels for appliances and equipment
12. Test standards and measurement protocols for appliances and equipment
13. Market transformation policies for appliances and equipment

Lighting

14. Phase-out of inefficient lighting products and systems
15. Energy-efficient lighting systems

Transport

16. Mandatory vehicle fuel-efficiency standards
17. Measures to improve vehicle fuel efficiency
18. Fuel-efficient non-engine components
19. Improving operational efficiency through eco-driving and other measures
20. Improve transport system efficiency

Industry

21. Energy management in industry
22. High-efficiency industrial equipment and systems
23. Energy-efficiency services for small and medium-sized enterprises
24. Complementary policies to support industrial energy efficiency

Energy utilities

25. Energy utilities and end-use energy efficiency

Source: IEA (2011).

of energy is one of the basic aims of campaigns for cleaner production (see p. 430), while in the domestic and commercial sectors, substantial savings can be made with relatively simple building design and technologies. In developed countries of the global North, where residential and commercial buildings account for 20–40 per cent of total energy consumed (Pérez-Lombard *et al.*, 2008), energy is mainly used for space heating and cooling (in residential buildings) and lighting (in commercial buildings). At the national level, some of the opportunities for reducing energy consumption in US homes are outlined by Pimentel *et al.* (2009). These authors point out that an estimated 25 per cent of residential heating and cooling energy is lost each year directly through windows, and that nationwide about 17 per cent of all energy is used for lighting in buildings. Replacing all incandescent bulbs, which convert only 10 per cent of energy received into light (the rest is converted to heat), with compact fluorescent lights in US homes would cut electrical demand for lighting by about 25 per cent. The most energy-consuming appliance in the average US home is the hot water heater, but its consumption can be cut by half by implementing a few simple changes, including the installation of a water-conserving showerhead, limiting showers to five minutes, and reducing water temperature.

Figure 18.1 Energy-efficiency label used on household appliances in EU countries.

Many countries have introduced regulations to encourage and/or impose lower energy consumption. A popular measure is the introduction of mandatory energy-efficiency labelling for new domestic electrical appliances such as refrigerators and washing machines (Figure 18.1), an effort to inform consumers and so speed up the diffusion of more energy-efficient equipment. Another common measure is minimum energy-efficiency standards which have been introduced for new appliances, new cars and new buildings. The minimum standards approach has been employed in numerous countries effectively to ban the purchase and use of less efficient technologies such as incandescent light bulbs.

The introduction of these measures in many countries has produced a positive trend in the sales of more energy-efficient appliances, but the consequent energy savings have been eroded by a rapid increase in the use of electrical and electronic equipment (entertainment, office equipment, communication/internet, white appliances with embedded electronics), along with the proliferation of gadgets with electronic controls, and which are typically connected to the electricity supply all the time. Research for an EU project on standby energy use – electricity used by appliances and equipment while they are switched off or not performing their primary function – indicates the large potential energy savings (De Almeida *et al.*, 2011). Across the EU, standby electricity consumption represents about 11 per cent of the total annual electricity consumption per household, and globally standby power use is roughly responsible for 1 per cent of total carbon dioxide emissions.

For homes, offices and factories, policy measures may at some point combine to result in the zero energy building, a concept that has received increasing attention in recent years (Marszal *et al.*, 2011). A number of prototypes have been constructed, each building designed to generate all of the energy it requires through a combination of energy efficiency and renewable energy generation technologies. Almost all of them rely on a combination of:

- high-efficiency lighting and appliances
- renewable energy sources (e.g. photovoltaic cells, solar water heaters and geothermal heat pumps)
- passive solar design techniques (e.g. passive solar heating, insulation, controlled windows, shading, interior space planning, and landscaping).

Another area with substantial room for improvement in efficiency is in the generation of electricity. In many developing countries, power plants that run on fossil fuels use five or six units of fuel to make one unit of electricity. In more developed countries, the ratio is about three to one but can reach two to one in combined-cycle gas turbines. This ratio can be improved further by the cogeneration of heat and power (CHP). Much of the energy wasted in conventional electricity generation is in the form of hot water so using this water in some way makes economic and energy-efficient sense. In Denmark this hot water is increasingly used in district heating systems for houses, factories and offices. While in a country such as the UK most buildings are supplied with cold water that is heated by individual boilers, some Danish buildings have received hot water from district heating boilers since the 1920s. Providing this hot water from a CHP plant is more efficient still: reducing fuel consumption by about 30 per cent (Figure 18.2). Production of electricity and district heating in Denmark is among the most efficient in the world: about two-thirds of

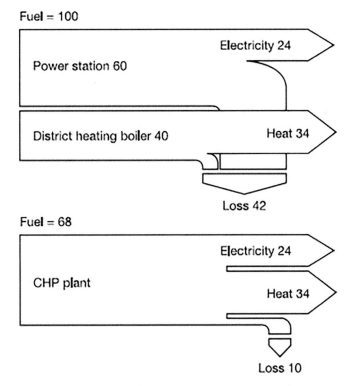

Figure 18.2 Energy efficiency with combined production of power and heat (distributed through a district heating system). With cogeneration, fuel consumption is reduced by about 30 per cent.

Danish-produced electricity is cogenerated with heat, and district heating provides warmth to about 60 per cent of Danish households (Chittum and Østergaard, 2014). More buildings are still being connected to CHP heating, and existing district heating plants are being converted to CHP plants run on natural gas and organic waste biomass (see below).

In many countries of the developing world, electricity is not widely used and most of the population relies on household solid fuels (biomass fuels such as fuelwood, charcoal, dung, or crop residues as well as coal). Globally this equates to some 3 billion people. Energy lost by burning these fuels on open fires can be reduced by using closed stoves, while converting fuelwood to charcoal is more energy-efficient still. Using improved solid fuel stoves also brings other benefits: they are safer, produce less household air pollution, and reduce environmental impacts that are often associated with fuelwood collection. However, encouraging people to use solid fuel stoves has not been straightforward. A review of the challenges in Asia, Africa and Latin America conducted by Rehfuess *et al.* (2014) identified 31 factors influencing uptake and concluded that the context was critical. Among the important factors were the need to offer technologies that meet household needs and save fuel, user training and support, effective financing, and appropriate government action to encourage use.

RENEWABLE ENERGY

The current use of renewable energy sources is most significant in the developing countries where biomass fuels represent a major energy source. Hydropower provides the main source of electricity for some countries, such as Austria, Brazil, Canada, Norway and Switzerland. Overall, of course, fossil fuels currently make the largest contributions to global energy supply (Figure 18.3).

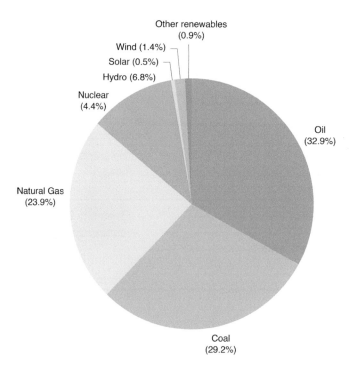

Figure 18.3 Global total primary energy consumption, 2015, excluding traditional biomass (WEC, 2016).

Historically, the cost of renewable energy supplies (other than biomass fuels) has not been competitive with fossil fuels in many high-income countries, largely because of the lack of finance and effort put into their development, and the subsidies given to energy production from oil, coal and natural gas. In more recent times, the renewable energy sector has been the target of numerous national policy interventions incentivizing the use of the various renewables, and these policies, most driven by the desire to reduce greenhouse gas emissions, are having a positive effect on the development and cost-effectiveness of renewables. Renewables also have a number of other advantages in addition to the environmental benefits most have relative to the burning of fossil fuels. One such advantage is that most renewable energy equipment is small, allowing much faster construction than conventional technologies. While most large conventional facilities are constructed in the field, most renewable energy equipment is constructed in factories where manufacturing techniques can facilitate cost reduction. The small scale of equipment also makes for a short time between design and operation, so that modifications stemming from field tests can quickly be incorporated into new designs. Hence, many generations of technology can be introduced in a short period. Another facet of small-scale equipment is its suitability for independent, stand-alone units that do not need to be connected to the existing power grid.

There are, however, some difficulties with certain renewable energy sources. Large-scale renewable energy systems, like their conventional equivalents, can face difficulties of social acceptance, particularly by communities in the direct vicinity of renewable power plants. Local public resistance against infrastructure projects of any kind is often referred to as the 'Not In My Back Yard' (NIMBY) syndrome, and NIMBYism has been an obstacle to the development of many renewable energy sources in Europe and North America (Devine-Wright, 2014). On the more technical side, the variable nature of wind and solar power sources is an issue and tidal power, while predictable, is intermittent. This can be resolved by appropriate use or storage of energy in many cases, but not all. It is unlikely that solar power will ever contribute a large proportion of national energy requirements in high-latitude areas because of the lack of solar intensity and the long, dark winters. It is also important to note that not all renewable sources are appropriate for all types of energy need: wind and wave energy will not fuel a motor vehicle directly, for example, although electricity so generated may supply battery-powered vehicles. The following sections will look at the main forms of renewable energy to assess their future potential to contribute to world energy needs.

Hydropower

The mechanical power of moving water is an energy resource with a long history, used by the ancient Greeks to turn water wheels to grind wheat into flour. The first hydroelectric power plant was installed in north-east England in 1870 and since then hydropower has become the only renewable resource used on a large scale today for electricity generation. At the beginning of 2016, hydropower supplied 71 per cent of all renewable electricity globally, equating to just over 16 per cent of the world total (WEC, 2016). The top six countries in terms of hydroelectricity generated are shown in Table 18.2. Dams and their reservoirs can also provide a range of other services beyond electricity supply, but the main constraints on further development are the associated social and environmental impacts outlined in Chapter 9, although many of these issues can be mitigated by improved planning with the inclusion of public participation at an early stage. It should also be noted that reservoirs used for hydroelectricity generation are likely to emit significant quantities of atmospheric pollutants (see p. 209).

Although much uncertainty surrounds the estimates of potential hydroelectric power that could be generated, there is agreement that, to date, only a relatively small percentage of this resource has been developed. The proportions of undeveloped potential range from 47 per cent in Europe and North America to

Table 18.2 Largest generators of hydroelectricity, 2016

Country	Total capacity (GW)
China	305
Brazil	97
USA	80
Canada	79
Russia	48
India	47
World total	1,096

Source: after REN21 (2017): table R5.

92 per cent in Africa (Kumar *et al.*, 2011), and the potential for small-scale hydroschemes, designed to serve local needs, is thought to be particularly great. In most cases, small-scale hydro is 'run-of-river', with no dam or water storage infrastructure, and is therefore considered to be one of the most cost-effective and environmentally benign energy technologies (Sachdev *et al.*, 2015).

Conversely, however, the potential for hydroelectric power generation will be affected by changes to the hydrological cycle associated with global climate change. A modelling study by Turner *et al.* (2017) identified several clearly impacted hotspots, the most prominent of which encompasses the Mediterranean countries in southern Europe, North Africa and the Middle East. Countries in the Balkans were the most vulnerable, some projected to lose up to 20 per cent of national electricity generation by the 2080s under a high emissions scenario. On the other hand, a number of countries in Scandinavia and Central Asia were projected to generate a significantly greater amount of hydroelectricity (up to 15 per cent in some cases) without investing in new power generation facilities.

Wind energy

The wind's energy has been used by sailing ships and windmills to power human activities for thousands of years, and wind pumps are still a common feature of rural landscapes today, particularly in dry regions where they are used to pump groundwater to the surface (Figure 18.4). Modern wind turbines for electricity generation are a comparatively recent phenomenon, which some suggest have the potential to supply as much as 20 per cent of global electricity demand by 2050 if ambitious efforts are made to reduce greenhouse gas emissions (Wiser *et al.*, 2011).

The large-scale generation of wind-derived electricity became viable in the 1970s due to technical advances and government support, first in Denmark at a relatively small scale, followed by developments at a much larger scale in the US state of California during the 1980s, and in Denmark, Germany and Spain in the following decade. Denmark has continued to put a great deal of effort into developing its wind power potential, encouraged by investment incentives and enthusiastic public support. The achievements have been considerable: turbines in the Danish countryside and offshore generated 42 per cent of its electricity consumption in 2015, up from 3 per cent in 1992. This proportion is planned to increase to 50 per cent of

Figure 18.4 Wind power is widely used in rural areas, particularly to pump groundwater to the surface in regions where surface water is in short supply, as shown here on the Mediterranean island of Malta.

electricity consumption by 2020 (Hvelplund *et al.*, 2017) and a large part of this will come from offshore wind farms in the North Sea and the Baltic Sea. Following establishment of the world's first offshore wind farm in 1991 at Vindeby (now decommissioned), Denmark has rapidly developed its offshore generating capacity in the twenty-first century (Table 18.3). The turbines at the Horns Rev II farm in the North Sea and at Rødsand II in the Baltic are four times as big as previous models, with blades 60 metres in diameter on 55-metre towers. Long-term development of the technology means that Danish offshore wind power has become as cheap to produce as building new coal-fired power stations and comparable in cost to gas-fired stations, without the disadvantage of CO_2 emissions.

There are several environmental impacts associated with wind farms. Noise was a serious concern surrounding earlier generations of turbines, but modern designs make little sound above the rush of the wind. Problems for birds include deaths by collision and modification of their territories. Most wind farm impact studies have focused on large bird species such as eagles and other raptors but small bird populations are also impacted (Farfán *et al.*, 2017). The land requirement of large wind farms has been another issue of concern, although land between individual turbines can still be used for other activities such as farming and ranching. Perhaps the most serious environmental issue associated with wind power in many areas is based on aesthetic grounds, many people objecting to how wind turbines and related infrastructure fit into the surrounding landscape. Turbines are often particularly visible because they are sited at high elevations to capture the strongest and most consistent winds and although aesthetic concerns cannot be fully mitigated,

Table 18.3 Danish offshore wind farms

Name	Start-up year	Total capacity (MW)
Tunø Knob	1995	5
Middelgrunden	2001	40
Horns Rev I	2002	160
Samsø	2003	23
Rønland	2003	17
Frederikshavn	2003	8
Nysted/Rødsand I	2003	165
Horns Rev II	2009	209
Sprogø	2009	21
Avedøre Holme	2009/10	11
Rødsand II	2010	207
Anholt	2012	400
Nearshore (2 projects)	2019?	350
Horns Rev III	2020?	400
Kriegers Flak	2021?	600

Source: DEA (2017).

an assessment of visual impacts is often required as part of the siting process. Other recommendations to minimize visual intrusion include turbines of similar size and shape, the use of light-coloured paints, and ensuring that all blades rotate in the same direction (Wiser *et al.*, 2011).

Many of these drawbacks associated with wind farms may be avoided by constructing the turbines off-shore, an approach that has great potential along much of the western European coastline. In an assessment of the potential impacts on the marine environment, Petersen and Malm (2006) point out that effects during the construction phase should be negligible if appropriate care is taken to avoid areas containing rare habitats or species. Similarly, disturbances associated with windmill operation – noise, vibrations and electromagnetic fields – are considered to be of minor importance to the marine environment, although the impact on birds can be expected to be essentially similar to those on land. These authors highlight effects below the water surface as having potentially the greatest impact, particularly through the creation of artificial reefs on windmill bases. This so-called 'reef effect' will inevitably alter submarine sediment flow and is likely to have a significant impact on local species composition and biological structure, which may be negative. On the positive side, Hammar *et al.* (2016) point to evidence that offshore wind farms can generate increased biodiversity and abundance for many groups of species which, alongside the indirect impacts of reduced fishing inside the wind farm, can be valuable from a conservationist viewpoint. These authors conclude that offshore wind farms can be at least as effective as existing marine protected areas in terms of creating refuges for fish, marine mammals and their habitats.

Solar power

The sun's energy is the most abundant of all energy resources on our planet and in the very long term – beyond this century – solar power in its various forms is expected to be able to meet the bulk of the world's energy needs. Solar energy also offers significant potential over the shorter term and although not all countries are equally endowed with solar energy, it can make a significant contribution to the energy needs of almost every country (Arvizu *et al.*, 2011). Converting solar energy for human use is done with a range of technologies that can deliver heat, cooling, natural lighting, electricity, and several types of fuel. Solar energy can be used in three main ways:

- passively by designing buildings to optimize the sun's light and heat energy
- actively to heat water and building space
- to generate electricity.

Generating electricity from the sun's energy is achieved by using two principal methods:

- Solar energy is converted directly into electricity in a device called a photovoltaic (PV) cell.
- Solar thermal energy is used by concentrating the sun's rays with mirrors to heat a liquid, solid or gas which is then converted to electricity via a heat engine and generator. Frequently referred to as concentrated solar power (CSP).

PV power is produced when individual light particles – photons – absorbed in a semiconductor such as silicon create an electric current. PV electricity is thus created with no pollution and no noise. PV systems also need minimal maintenance and no water, and have been particularly successful in remote areas such as deserts (Figure 18.5). The clear skies and large number of sunshine hours typical of deserts make these environments particularly suitable for solar power generation but unfortunately they also have drawbacks. High temperatures can adversely affect PV performance (Diarra and Akuffo, 2002), and the deposition of dust on PV cells and CSP mirrors reduces power output, typically by 20 to 50 per cent (Sarver *et al.*, 2013).

PV systems can operate on any scale, from portable modules for remote communications and instrumentation to huge power plants covering millions of square metres, enabling them to be positioned near to the electricity users and reducing the need for transmission systems. Indeed, PV cells can be integrated into buildings, usually on the roof, to provide electricity directly to residential and commercial premises. The cost to the consumer of electricity from PV arrays has fallen rapidly in recent decades, so that they are cost-effective even at high latitudes and in relatively cloudy regions, as well as in all sorts of remote areas. Their environmental impacts are generally low, even when used on a large scale, relative to other energy systems, including other renewables (Hernandez *et al.*, 2014).

The other way in which solar radiation is used directly to generate electricity is the solar thermal electric generator, which uses mirrors to track the sun and focus its rays to heat usually a fluid in a pipe. The heated fluid is then used to create steam that drives a turbine generator, or directly to produce steam and hot water for industrial and domestic purposes. An advantage of such CSP systems is that they are essentially similar to conventional fossil-fuel-generated electricity – both use a hot fluid – and hence benefit from much current expertise on power generation. Several designs have been developed, ranging in scale from an individual parabolic mirror that focuses the sun's rays onto its receiver, to a field of sun-tracking mirrors, which all reflect solar energy to a receiver mounted on a central tower. Generating electricity by solar thermal means has a relatively small environmental impact, although water usage may be a concern and the mirrors

Figure 18.5 Solar panels outside a nomadic herder's tents in the Gobi Desert, Mongolia.

can cause some bird deaths through collisions and burns. As with PV systems, the land requirements for centralized systems are perhaps the most significant issue that may cause public concern (Box 18.1).

One country with considerable potential for expanding solar power production is India, where almost one-fifth of the population is currently without access to electricity. The government has set an ambitious target of achieving 100 GW of installed solar capacity by 2022, made up of 40 GW of rooftop solar and 60 GW of medium- and large-scale grid-connected power plants. This plan would make a significant contribution to India's ambitious climate pledge to lower its carbon intensity by a third from its 2005 level by 2030. In a critical review of the solar plan, Hairat and Ghosh (2017) see an over-reliance on imported PV systems and advocate a better balance between PV and CSP, as well as a greater emphasis on decentralized solar systems in rural parts of the country where electrification can contribute significantly to economic development.

Biomass

Plant matter, or biomass, is a form of energy that has been used by people for heating, lighting and cooking since the discovery of fire. In many developing countries, fuelwood is still the main source of energy (Figure 18.7), and approximately 3 billion people, about half the world's population, rely on traditional biomass for cooking and heating (UNDP and WHO, 2009). Although in some parts of the developing

Box 18.1 Community acceptance of a large-scale solar power project in Morocco

Morocco has limited domestic fossil fuel reserves and imports almost all of its primary energy, but aims to generate over 40 per cent of electricity from renewable sources by 2020. The country has a predominantly desert climate and solar power will make an important contribution. The first site to be developed for the large-scale collection of solar power is the Noor Ouarzazate Concentrated Solar Power (CSP) Project, commissioned in 2016 (Figure 18.6). It is expected to be the world's largest solar power plant at a cost of some US$9 billion.

Early surveys conducted to assess local community acceptance of Noor Ouarzazate have been positive, particularly because solar power is perceived to be environmentally friendly, with a notable lack of concern about aesthetic disruption of the landscape, and little NIMBYism. However, researchers warned that local knowledge about the project was very low and that expectations regarding job creation and electricity prices were high. They advised those responsible for the project to keep these issues in mind to avoid disappointing the local population in the future.

Figure 18.6 Part of the Noor CSP plant near Ouarzazate in southern Morocco.

Source: Hanger *et al.* (2016).

world fuelwood collection is carried on unsustainably, with environmental impacts (see p. 156), biomass has great potential to be used as a renewable fuel in many different applications.

Biomass fuels come in a variety of forms, both unprocessed (wood, straw, dung, vegetable matter and agricultural wastes) and processed (e.g. charcoal, methane from biogas plants and landfills, logging waste and sawdust, and alcohol produced by fermentation). These fuels were widely used in the more developed

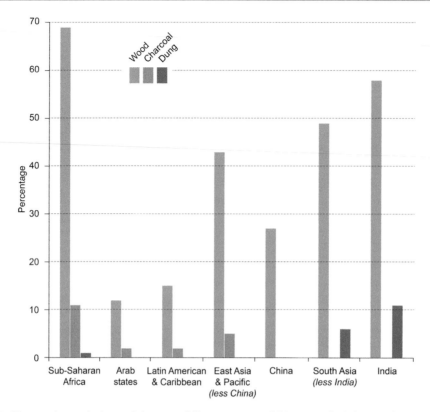

Figure 18.7 Share of population relying on different types of biomass fuel for cooking by developing regions, 2007 (%) (after UNDP and WHO (2009): figure 8).

countries before the large-scale switch to fossil fuels that came with industrialization, but biomass also has a more recent history of use in the industrialized countries. Industrial and agricultural residues have been used for many years to provide electricity in conventional steam-turbine power generators. In some countries where biomass contributes a fairly high percentage of national energy consumption, including Austria and Denmark, wood logs and woody waste from the pulp and paper industries are a significant component. Some countries, notably Brazil, substitute fuel derived from crops for petrol used in motor vehicles (see Chapter 16), while on the local scale, municipal wastes are burnt in incinerators, and landfills are tapped for emissions of gaseous energy (see Chapter 17).

Sustainably managed, the use of biomass fuels produces no net emissions of carbon dioxide since the amount released into the atmosphere by burning will be taken up by growing plants. Although more effort can be made to use residues from other activities, and from the harvesting of forests, dedicated energy plantations are probably the largest potential source of biomass. Appropriate energy crops include oil seed rape for conversion into biodiesel, and sugar cane and sugar beet for ethanol. Indeed, biofuels have been promoted in recent times by government policies particularly in the European Union, the USA and Brazil (see Box 16.1), to reduce dependence on fossil fuel imports, to increase energy supply security and to reduce greenhouse gas emissions. Biofuel plantations have expanded rapidly in many countries of the South in consequence, although they have also sparked controversy and have been blamed for food shortages and higher food prices (Michalopoulos *et al.*, 2011). Such concerns can be eased if the land used for energy plantations is deforested or otherwise degraded in some way, the planting with energy crops representing an environmental improvement, but there are many factors that determine the sustainability of energy plantations, including their social impact. Nonetheless, biomass has the potential to make a very significant contribution to world energy needs. Estimates made by Beringer *et al.* (2011) assumed a range of sustainability requirements to safeguard food production, biodiversity and terrestrial carbon storage and concluded that a combination of all biomass sources might still provide 15–25 per cent of the world's commercial energy demand in the twenty-first century.

Tidal power

Energy in the oceans, which cover two-thirds of the planet's surface, is found in the form of tides, waves, ocean currents, temperature differences, salt gradients and marine biomass. However, although the total amounts available are large, only a small fraction is likely to be utilized in the foreseeable future, both because ocean energy is spread diffusely over a wide area and because much of the available energy is distant from centres of consumption. To date, only tidal power has been seriously investigated and technically proven. It is one of the oldest forms of energy exploited by human society – tide mills were operated on the coasts of Britain, France and Spain before 1100 CE. Tides ebb and flow because of the gravitational interaction between the sun, the moon and Earth and the centrifugal force of the Earth–moon system. Their movements are therefore highly predictable, which gives tidal power a distinct advantage over many other renewable energy forms. Tidal energy is harnessed in much the same way as that of hydropower, by building a barrage across a suitable estuary. But whereas a hydroelectric dam depends on water flowing in one direction, a tidal barrage must allow the bay behind it to alternately fill and empty with the tides. As water flows out of an estuary from high tide to low tide, it passes through turbines which generate electricity.

The amount of energy available from a site depends upon the range of the tides and the area of the enclosed bay. The number of sites suitable for modern commercial-scale energy production is limited, however, and only a few of these have been developed (Table 18.4). The world's first operational tidal power plant is associated with a barrage across the La Rance estuary on the Brittany coastline of northern France. This 0.8-kilometre-long barrage, which was built in the early 1960s, also serves as a highway bridge linking the settlements of St Malo and Dinard. La Rance has an average tidal range of 8 m and the barrage's 24 turbines produce enough electricity for a city of 300,000 people. La Rance was superseded as the world's largest operational tidal power station in 2011 when the plant at Sihwa was inaugurated in South Korea.

The potential for tidal energy at the head of the Bay of Fundy, on the Atlantic coast of Canada, where the average tidal range of 10.8 m is among the largest in the world, has long been recognized. In 1984 a plant was commissioned at Annapolis across a small inlet on the Bay of Fundy's east coast but, despite its large tidal energy potential, Canada has concentrated upon the development of its substantial conventional hydropower reserves.

Other locations where tidal energy could viably be harnessed include the Severn Estuary in Britain where the average tidal range is 8.8 m. Indeed, most of the potential for harnessing tidal energy in Europe lies on the British coast, where the exploitation of all practical sites could yield up to 20 per cent of the electricity needs of England and Wales. The Severn and the Mersey estuaries are among the UK's best sites, but the environmental effects of such schemes have featured highly in the debate over whether or not to go ahead with barrages across these estuaries. Despite the environmental advantages of tidally generated electricity, including the lack of pollutants generated and the protection barrages offer to coastlines against storm-surge tides, such schemes inevitably modify the hydrodynamics of their estuaries. Alterations to the tidal range, currents and the intertidal area within the barrage would affect several other environmental

Table 18.4 Operational tidal energy plants

Location	Mean tidal range (m)	Installed capacity (MW)	Start-up year
La Rance (France)	8.0	240.0	1966
Kislaya (Russia)	2.4	0.4	1968
Jiangxia (China)	7.1	3.2	1980
Annapolis (Canada)	6.4	17.8	1984
Strangford Lough (UK)	3.6	1.2	2008
Sihwa Lake (S. Korea)	5.6	254.0	2011

Source: after Bahaj (2011).

parameters, including sediment movement and water quality, which would in turn have an impact on the food chain, ultimately affecting bird and fish populations.

Geothermal energy

Geothermal energy is tapped from the heat which flows outwards from the Earth's interior energy, both from the core as it cools and from the decay of long-lived radioactive materials in rocks. On the global scale, it is most accessible along the boundaries between the Earth's crustal plates, but many geothermal possibilities also exist away from these areas of concentration.

Hot waters from the Earth have been used therapeutically and for their mineral salts since Roman times, when geothermal spring water was also used to heat bath houses. This energy has been exploited commercially, to provide heat, mechanical power and electricity, since the early 1900s. In 1930, for example, large-scale geothermal heating systems for buildings were built in Iceland, and similar operations subsequently were constructed in France, Italy, New Zealand and the USA.

The location of Iceland on the Mid-Atlantic Ridge, the boundary between the North American and Eurasian plates, makes it one of the most tectonically active places on Earth, with a large number of volcanoes and hot springs. Geothermal resources are therefore particularly abundant in Iceland (Figure 18.8), where they provide nearly 70 per cent of the country's primary energy supply (Ragnarsson, 2015). In 2015, about 90 per cent of all energy used for house heating came from geothermal sources, mostly via district heating systems. Other uses include the heating of swimming pools, greenhouses and fish farms, as well as a number of industrial applications. Hot water that has been used to heat houses is also reused to melt snow and ice on pavements and in parking spaces. Electricity generation from geothermal energy has increased significantly in recent years and in 2015 Iceland generated 29 per cent of its total electricity needs in this way. It was one of just seven countries that derived at least 10 per cent of national electricity production from high-temperature geothermal resources (Bertani, 2016). The others were Costa Rica, El Salvador, Kenya, New Zealand, Nicaragua and the Philippines.

Most prevailing geothermal systems are hydrothermal, tapping aquifers of hot water a few kilometres below the surface, while other approaches are largely in the research stage. One such alternative uses energy from hot, dry rocks which is exploited by fracturing them at depth and pumping water down from the

Figure 18.8 Bathers in the Blue Lagoon, south-western Iceland. The warm waters are fed by the water output of the geothermal power station in the background.

surface to be heated and then returned to the surface. The drawback of this technique is that the injection of fluids into underground rock formations is capable of inducing earthquakes. In a review of injection-induced earthquakes, Ellsworth (2013) highlights the example of Basel, Switzerland, where the injection of water under high pressure into impermeable basement rocks beneath the city as part of an enhanced hot dry rock geothermal project initiated several small earthquakes. The project had to be abandoned because of the seismic activity, resulting in protracted legal action over compensation for damages.

Overall, the environmental impacts associated with the use of geothermal energy can be divided into temporary ones due to drilling and exploration, and more permanent issues resulting from well maintenance and power plant operation, although Bayer *et al.* (2013) emphasize that general assessments are hard to make because many aspects are controlled by geological and other local factors and are, therefore, site-specific. Power plants take up land and may attract objections on aesthetic grounds. Air pollution is the greatest concern for most geothermal plants, since gases released during power production are not easily re-injected back into the thermal reservoir, although geothermal plants generally produce fewer gaseous emissions than their fossil fuel equivalents. Other environmental problems include:

- solid waste and residual water disposal
- emissions to soil and water
- geomorphological hazards (earthquakes, subsidence, landslips)

- water use
- noise.

Geothermal energy could play a key future role for some countries where potential is high and energy demands are currently low (e.g. Djibouti, El Salvador, Kenya, and the Philippines). Indeed, Bertani (2016) identifies 40 countries, located mostly in Africa, Central/South America and the Pacific, which can be 100 per cent geothermal-powered. On the global scale, Bertani suggests that by 2050 it is realistic to foresee 8 per cent of total world electricity production generated from geothermal sources, serving 17 per cent of the world population.

The future for renewable energy

A clear North–South divide is recognizable when it comes to assessing national energy budgets. On the one hand, most industrialized countries have fuelled economic growth for 100 years and more with power generated principally from fossil fuels and are currently looking to decrease their reliance on these conventional, non-renewable resources. Conversely, inhabitants of many poorer developing nations continue to be reliant predominantly upon fuelwood as their principal energy source (Figure 18.9), while their national

Figure 18.9 Yaks laden with firewood on the Nepal–Tibet border. Nearly half the human population, mostly in the poorer parts of the world, rely on biomass, a renewable fuel, for their daily energy needs.

governments aspire to a process of industrialization similar to that followed by their richer counterparts, although the fossil fuel route to industrialization is becoming less favoured.

In the countries of the North, however, the contribution of renewables varies considerably, as Table 18.5 indicates for electricity consumption in Europe. Some countries with favourable natural environments already rely predominantly upon renewables. Mountainous terrain has encouraged hydropower to become the foremost supplier of electricity in Austria, Norway and Sweden. Geothermal energy and hydropower are important in Iceland, while the renewable contribution in Finland – a country with a large forested area – is due mainly to electricity production from woody biomass as well as hydropower.

Natural endowments are not enough in themselves, of course. Political will is also necessary. Denmark's rise in its renewable contribution is largely due to the recent efforts to develop wind power. The less impressive contribution of renewables in the UK is despite the fact that the country has great potential for electricity generation from wind (Sinden, 2007).

It is likely in the foreseeable future that energy demand will continue to rise as economic growth proceeds, despite the feeling in many quarters that truly sustainable development should aim to uncouple economic growth from rising energy use (see p. 280). Increasing the efficiency with which energy is used can help to offset this rise, but is unlikely to absorb all the additional needs. Furthermore, the environmental impacts of conventional energy sources, notwithstanding their eventual depletion as non-renewable resources, will continue to play a role in deciding to what degree renewable energy sources contribute. In this regard, the primary concern of governments in recent times has been how the consumption of fossil fuels has accounted for the majority of global anthropogenic greenhouse gas emissions, and it is the potential of renewables to mitigate climatic change that will continue to be the most powerful driver behind their continued development.

Many now talk of a 'global energy transition' from fossil fuels as the dominant global energy source towards renewable energy as the leading provider. In 2015, renewables contributed an estimated 19 per cent of total final energy consumption globally; and in 2017 renewables made up nearly 25 per cent of global electricity production (REN21, 2017). Martinot (2016) highlights the fact that more than 160 countries have set themselves future targets for renewable energy use, many of them ambitious. In 2007, the EU set an energy consumption target of 20 per cent from renewables by 2020, part of a 20-20-20 target with a 20 per cent reduction in EU greenhouse gas emissions from 1990 levels and a 20 per cent improvement in energy efficiency. For 2030, the EU target is 27 per cent, but individual countries have been more ambitious: Denmark aims to attain 100 per cent of its electricity from renewables by 2035, Germany 80 per cent by 2050. Driven by these targets, numerous future energy scenarios predict sizeable shares of electric power being contributed by renewables globally and in specific regions, with many projections indicating a 40–80 per cent share of total electricity generation by 2050. Renewable fuels processed from crops are also increasingly popular for motor vehicles (see p. 404).

The need for government support, particularly in the research and development phase, but also later in the energy marketplace, is critical to achieving these targets, but even where the appropriate political decisions are made to move towards renewable energy sources, other difficulties may still arise because of the environmental effects of renewables. Some of these issues are being faced in New Zealand, where an already high level of electricity generation from renewables is being expanded further. Hydroelectricity generation, currently the country's largest source of renewable energy, could probably be increased by up to 50 per cent, but there are several barriers to realizing this potential (Kelly, 2011). Most untapped big rivers are in conservation areas and the distances between remote rivers and areas of consumption would require substantial transmission infrastructure, including an undersea high voltage cable between the country's

Table 18.5 Renewable electricity as a percentage of gross electricity consumption in selected European countries, 2000–15

Country	Contribution of renewables to gross electricity consumption (%)			
	2000	2005	2010	2015
Austria	72	62	66	70
Belgium	2	2	7	15
Denmark	16	25	33	51
Finland	29	27	28	32
France	15	14	15	19
Germany	7	10	18	31
Hungary	1	4	7	7
Iceland	100	95	92	93
Netherlands	4	6	10	11
Norway*	112	97	98	106
Slovenia	31	29	32	33
Sweden	55	51	56	66
UK	3	4	7	22

Note: *Norway's share is >100% because some renewable electricity generated domestically is exported to other countries.

Source: updated after EEA (2005) from Eurostat.

two main islands because most of the remaining potential is in South Island and most of the consumers are in North Island.

Another interesting case is Sweden, where electricity production has been essentially carbon dioxide-free since the 1980s, a status which most other countries in Europe aspire to reach by 2050. For decades, hydro and nuclear have provided the large majority of Sweden's electricity, but discussion over reducing the reliance on nuclear power has been driven by environmental and safety concerns. The strength of public feeling over the nuclear issue grew during the 1970s, to a level in 1976 where the ruling Social Democrats, who had governed the country for 44 years and were responsible for Sweden's nuclear energy programme, lost their majority in parliament. In a pioneering move based on a 1980 referendum decision, Sweden shut its first nuclear reactor in 1999 as part of a planned complete withdrawal of nuclear power, but public feelings on what should replace nuclear energy were almost as strong (Löfstedt, 1998). Biomass is the most viable alternative, because solar power is not sufficient due to Sweden's geographical position at a high latitude; wind power's potential has been limited by public opposition and slow technical progress; and further development of hydroelectricity is constrained by the 1987 Natural Resources Law, which protects Sweden's remaining free-flowing rivers. But an increase in the use of biomass fuels also attracted opposition from ornithologists and one environmental organization, who argued that it would adversely affect biodiversity. The controversy was eased but not eradicated in 2010 when the Swedish parliament voted to repeal the policy to phase out nuclear power (see below) as part of a programme which stipulates that renewable sources should supply all energy produced by 2040.

The debate in Sweden emphasizes the fact that all energy sources entail some form of environmental impact. Indeed, as the output increases from particular renewable sources, these impacts are likely to increase. In fact in some cases this scaling up may introduce novel effects. Extracting the maximum level of wind power globally, for instance, may result in significant climatic effects similar in magnitude to those associated with a doubling of atmospheric CO_2 (Miller *et al.*, 2011). A sustainable energy policy can only aim to minimize these impacts because they will not disappear.

NUCLEAR POWER

When energy generated by nuclear fission was first developed for civil uses in the 1950s, having grown out of a small number of national nuclear weapons programmes, it was heralded as cheap, clean and safe. The subsequent image of nuclear power has changed considerably since those times, and today it is one of the most controversial forms of energy from both the economic and environmental perspectives. By 2017, nuclear reactors were operating in 31 countries and nuclear power stations have been supplying over 10 per cent of world electricity production since the late 1980s, although the percentage has been declining for the last 20 years (Figure 18.10). For some countries the proportion is much higher: four European states derived more than 50 per cent of their electricity in this way in 2016 (France 76 per cent, Ukraine 57 per cent, Slovakia 56 per cent, Hungary 53 per cent). Ultimately, nuclear power is a non-renewable form of energy production since uranium, the fuel used, is a mineral product. Nevertheless, current reserves are

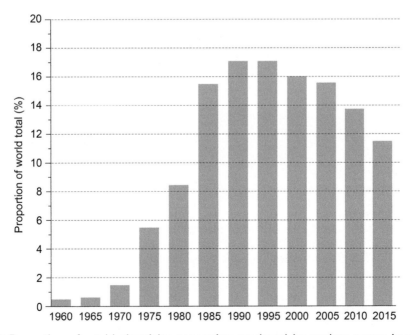

Figure 18.10 Proportion of world electricity generation produced by nuclear power in selected years, 1960–2015.

thought to be enough to cover global needs for about 100 years (WEC, 2016) and advances in technology have meant that increasing amounts of uranium can be reprocessed and used again because spent nuclear fuel when it leaves the reactor still contains some 95 per cent of its original energy content. Reprocessing and recycling of unspent uranium and the plutonium generated during its residence in the reactor can therefore extend the lifetime of nuclear power still further, but in the shorter term the future of the industry will be dependent largely upon the associated environmental concerns.

These concerns are based on the fact that nuclear material is radioactive and hence can be highly dangerous to living things, and that for some radioactive elements this radioactivity is very long-lasting. The potential for these materials to be released into the environment in dangerous concentrations can be traced through the nuclear fuel cycle – from mining, conversion, preparation of reactor fuel, and the management of wastes – and can arise again with the decommissioning of old nuclear power plants. Considerable research has been focused on the dangers of increased mortality rates from cancers around operational nuclear plants. In the 1980s and 1990s, increased incidences of childhood leukaemia were reported near several nuclear facilities in the UK but the cause or causes were not conclusively linked to nuclear emissions. Since then, researchers in many countries have investigated the risk of leukaemia and other cancers in children living in the vicinity of nuclear power plants, producing mixed results (Kühni and Spycher, 2014): some studies observing such associations, others concluding that no causative effect was detectable. Such research around nuclear installations will undoubtedly continue, but greater concerns abound over the disposal of nuclear wastes and the risk of major accidents.

Nuclear waste

Nuclear waste is commonly categorized into three classes (Table 18.6). In most of the countries that produce nuclear waste, low-level material makes up the large majority of waste by volume, but accounts for a minor proportion of the radioactivity in waste. High-level waste, by contrast, is volumetrically small but accounts for the largest share of radioactivity. Liquid low-level wastes have long been disposed of by adopting a 'dilute and disperse' philosophy. The British nuclear power complex at Sellafield on the Cumbrian

Table 18.6 Categories of nuclear waste

LOW-LEVEL WASTE
Liquid and solids lightly contaminated by short-lived radionuclides (e.g. discarded protective clothing, contaminated building materials, uranium mine tailings)

INTERMEDIATE-LEVEL WASTE
Materials contaminated by long-lived radionuclides such as plutonium and transuranic elements, most of which arise from the processes of energy production and reprocessing (e.g. fuel cladding, control rods, liquids used to store spent fuel before reprocessing)

HIGH-LEVEL WASTE
Materials contaminated by highly active radionuclides with long half-lives, which may cause significant increases in the temperature of wastes (e.g. non-reprocessed spent fuel, liquid wastes produced during reprocessing of spent fuel)

coast, for example, has been discharging such low-level liquid waste into the Irish Sea since the early 1950s. The disposal of solid wastes, by contrast, has aimed to contain the material and then dispose of it. Until 1982, some low-level and selected intermediate-level wastes were contained in drums, which were dumped at sea, but this practice has been suspended and burial is now the preferred option. For low-level and short-lived, intermediate-level wastes, shallow burial is practised in such countries as the UK and the USA, while in France, surface storage is the preferred disposal option. In Sweden, however, a submarine repository has been constructed for this short-lived material below the bed of the Baltic Sea.

Overall, the volumes of radioactive waste are very small when compared to other waste products generated by human activity. Although estimates are difficult to come by, in part because of the political sensitivity that surrounds the nuclear power industry, the total volume of waste for long-term disposal produced in the UK between the 1950s and the 1980s is put at about one million cubic metres – a heap of waste packages less than the height of a house and 500 m square (Chapman and Hooper, 2012). Substantially larger volumes can be expected from decommissioned reactors, but despite the amounts involved, disposal of high-level waste, which will remain dangerous for many thousands of years, is one of the key issues facing the nuclear industry. High-level waste is heat-generating and needs to be cooled and then managed safely for a period that is far longer than we believe human civilization has existed to date. Although the nuclear industry is convinced that burial of high-level and long-lasting, intermediate-level waste material in suitable geological formations, at least several hundred metres deep, is the best method, opponents of the nuclear industry argue that this is not an acceptable solution. They advocate surface storage until a better management method can be devised. Public sensitivity over the issue of nuclear waste disposal is uniquely great, and although the technical aspects of high-level waste disposal have been solved to the satisfaction of most in the nuclear industry, the political aspects remain to be resolved.

Nuclear accidents

Concern at the potentially huge damaging effects of an accident at a nuclear power facility is the other major environmental issue surrounding nuclear power. Despite the stringent checks and safety measures employed at such facilities, accidents at nuclear reactors do happen. The International Atomic Energy Agency (IAEA) has developed an International Nuclear Events Scale (INES) which classifies such events on seven levels: Levels 1–3 are called 'incidents' and Levels 4–7 'accidents'. The scale is exponential, indicating that the severity of an event is about ten times greater for each increase in level on the scale. Table 18.7 gives details on the effects with examples.

Two Level 7 accidents have occurred to date: at Chernobyl in Ukraine (then part of the USSR) in April 1986, and at the Fukushima nuclear power plant in Japan following an earthquake and tsunami in March 2011. The event at the Chernobyl nuclear power plant is the most serious accident to have occurred at a nuclear power station, and it has haunted the world's nuclear industry since. Chernobyl involved the release of ten times more radioactivity than Fukushima and, after comparing the two environmental disasters, Steinhauser *et al.* (2014) concluded that, in almost every respect, the consequences of the Chernobyl accident clearly exceeded those of the Fukushima accident.

Radionuclides from the explosion at Chernobyl were dispersed throughout the northern hemisphere in trace amounts, with particular 'hotspots' in areas where rainfall washed radioactive material from clouds: in parts of the European Soviet republics, Austria, Bulgaria, Finland, Germany, Norway, Romania, Sweden, Switzerland, the UK and Yugoslavia. Most concern has focused on the biomedical dangers presented to humans by the radionuclide deposition. Initial fears over iodine-131 led to the destruction of fruit and vegetables grown in the open air and milk from cows grazing on contaminated grassland, but since iodine-131

Table 18.7 Nuclear power plant accidents on the International Nuclear Events Scale (INES)

INES level	Effects on people and environment	Examples
7 Major Accident	Major release of radioactive material with widespread health and environmental effects requiring implementation of planned and extended countermeasures	Chernobyl (USSR – Ukraine) 1986 – External release of a significant fraction of reactor core inventory. Widespread health and environmental effects Fukushima (Japan) 2011 – Reactor shut down after Sendai earthquake and tsunami; failure of emergency cooling caused an explosion. Widespread environmental effects
6 Serious Accident	Significant release of radioactive material likely to require implementation of planned countermeasures	Kyshtym (USSR – Russia) 1957 – Significant release of radioactive material to the environment from explosion of high activity waste tank
5 Accident with Wider Consequences	Limited release of radioactive material likely to require implementation of some planned countermeasures. Several deaths from radiation	Windscale (UK) 1957 – Release of radioactive material to the environment following fire in a reactor core Three Mile Island (USA) 1979 – Severe damage to reactor core
4 Accident with Local Consequences	Minor release of radioactive material unlikely to result in implementation of planned countermeasures other than local food controls. At least one death from radiation	Tokaimura (Japan) 1999 – Fatal overexposure of workers following a criticality event
3 Serious Incident	Exposure in excess of ten times the statutory annual limit for workers. Non-lethal deterministic health effect (e.g. burns) from radiation	Sellafield (UK) 2005 – Release of large quantity of radioactive material, contained within the installation
2 Incident	Exposure of a member of the public in excess of 10 mSv. Exposure of a worker in excess of the statutory annual limits	Atucha (Argentina) 2005 – Overexposure of a worker at power reactor exceeding annual limit Cadarache (France) 1993 – Spread of contamination to an area not expected by design
1 Anomaly	Overexposure of a member of the public in excess of statutory annual limits	—

Source: IAEA.

has a half-life of only eight days, attention soon shifted to caesium-134 and -137, the latter with a half-life of 30 years. Caesium accumulates up the food chain from the soil through vegetation to contaminate meat, necessitating special measures to restrict the movement and sale for consumption of livestock in several regions, some more than 1000 km from the accident in Scandinavia and Britain. Other dangerous radionuclides involved include strontium-90 (half-life, 29 years) and plutonium-239 (half-life, 24,000 years).

Although the health risks are thought to be small, restrictions on foodstuffs from parts of Europe and the former Soviet Union will need to be maintained for at least 50 years after the incident. The consumption of forest berries, fungi and fish, which contribute significantly to people's radiation exposure, will remain particularly dangerous in the more contaminated parts of Ukraine and Belarus (Beresford *et al.*, 2016).

The main overall biological effects of contamination from the Chernobyl catastrophe are shown in Table 18.8. Different ecosystems absorb radionuclides in different ways, and forests were some of the most contaminated. Foliage absorbed radionuclides in rainfall along with carbon-14 and tritium from the air by photosynthesis. When foliage fell in the autumn, most of these contaminants were transferred to the ground, concentrating in leaf litter and the thin upper soil layers. Coniferous forests kept radionuclides in their canopy longer, and such trees were consequently felled and disposed of in some heavily affected areas.

There is generally considered to be a direct relationship between the levels of soil and plant contamination, and a direct relationship between the level of radioactive contamination of agricultural lands and contamination levels in animals, notably in meat and milk products. Concern over these effects led to the official designation of contaminated territories that account for no less than 23 per cent of the surface area of Belarus, 5 per cent of Ukraine and 1.5 per cent of the Russian Federation. The population of these territories is around 6 million people (UNDP/UNICEF, 2002). Since the Chernobyl accident, about 350,000 people have been relocated away from the most severely contaminated areas (Figure 18.11).

The human health effects of the Chernobyl disaster are the subject of investigation and monitoring at both national and international levels. Although most of the long-term health effects of Chernobyl will be statistically undetectable, because they will be spread through a population of hundreds of millions over several decades, some evidence has emerged. The occurrence of thyroid cancer in children has increased

Table 18.8 Main biological effects of the Chernobyl catastrophe

Level	Short-term response	Long-term response
Biosphere	Global and local changes in radionuclide accumulation and dispersion	Global and local disturbances in genetic and phenetic structure of the biosphere
Ecosystem	Changes in ecosystem diversity, stability and development patterns	Changes in co-evolution process and ecosystem succession
Human population	Changes in birth and mortality rates	Changes in mutation rate, natural selection intensity and adaptive reaction
Human individual	Disturbances in physiology and behaviour	Changes in probability of cancer, hereditary abnormalities and other diseases

Source: after Savchenko (1991).

Figure 18.11 Beds in a deserted nursery school in Pripyat, the town built to house workers for the Chernobyl nuclear power plant. Pripyat was abandoned in 1986 following the Chernobyl accident.

markedly in areas close to Chernobyl, mainly due to the accumulation in the thyroid gland of iodine-131 ingested in food and inhaled from the initial radioactive cloud. The disease is normally rare in children but in the ten years to 2002 more than 4000 cases of thyroid cancer were diagnosed in Belarus, Russia and Ukraine among those who were children and adolescents at the time of the accident, most of which can be attributed to radiation exposure from the accident (Chernobyl Forum, 2005). Previous experience of the irradiation of the thyroid gland indicates that cases related to the exposure will continue to occur for at least 50 years after exposure.

Expected increases in the incidence of leukaemia have been less marked in the most heavily contaminated areas. The Chernobyl Forum (2005) concluded that there was no convincing evidence that the incidence of leukaemia had increased in children or those exposed *in utero*, but acknowledged that it may be too early to evaluate the full radiological impact of the accident. Several studies have reported on the long-term psychological impacts of the disaster, some suggesting that the mental health effects were the most significant public health consequence of Chernobyl (Bromet *et al.*, 2011).

An early estimate of the direct and indirect economic costs of the Chernobyl accident was calculated as at least US$15 billion, 90 per cent of which would be in countries of the former USSR (Tolba, 1992). Disruption of agricultural activities, relocation of people, and consequent psychological stress were the main

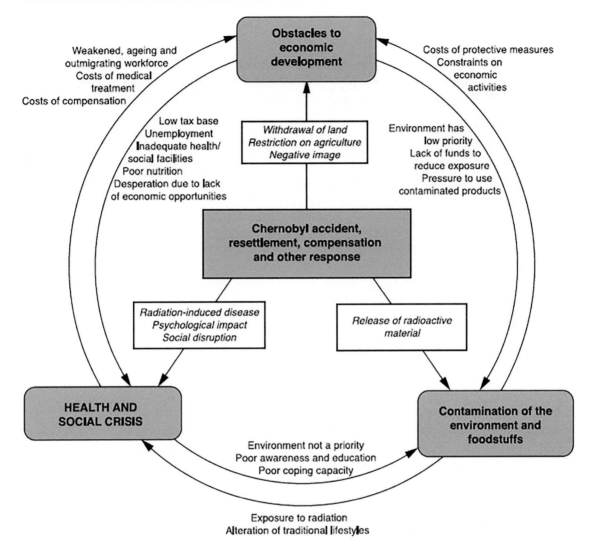

Figure 18.12 The downward spiral in communities affected by Chernobyl (after UNDP/UNICEF, 2002).

immediate consequences of the disaster. More recent assessments of the costs are much greater. The government of Belarus estimated that losses over the 30 years following the accident would amount to US$235 billion, while the Ukrainian government estimated the loss as US$148 billion over the period from 1986 to 2000 (UNDP/UNICEF, 2002). These two countries have had to introduce a special Chernobyl emergency tax to help pay for the world's most severe nuclear disaster. The effects of the Chernobyl accident and the additional economic problems faced in the post-Soviet era have combined to produce a downward spiral of health and well-being in the communities affected (Figure 18.12).

The future for nuclear power

Concern over the environmental effects associated with nuclear power generation means that the nuclear industry is conducted under some of the strictest controls and regulations of any major industry, and certainly more stringent than for other forms of power generation. However, assessing the risks of nuclear accidents or environmental damage from unsafe waste disposal is hampered by a lack of data. In one sense this can be seen as a good thing (i.e. there have been few accidents), but there is widespread suspicion that many, mostly small incidents at nuclear power plants are simply not reported publicly (Ha-Duong and Journé, 2014). One recent statistical analysis of 216 nuclear energy accidents and incidents by Wheatley *et al.* (2016), using a dataset twice as large as the previous best available, reached some startling conclusions. According to this analysis, with 388 reactors in operation worldwide, there is a 50 per cent probability of a Fukushima-like event every 60–150 years, and a Three Mile Island event (see Table 18.7) every 10–20 years. This study also concluded that in economic terms the average cost of nuclear power plant accidents and incidents per year is around the cost of the construction of a new plant.

Despite significant reforms following past disasters, the risks associated with nuclear power generation continue to rate highly in the public consciousness, and the lack of public confidence stemming from these concerns is perhaps the major problem facing the nuclear power industry today. The issues have consequently become subject to a high level of political, media and public relations activity. For some countries, perception of the risks associated with nuclear power has resulted in national programmes being stopped, scaled down or subjected to a concerted effort to phase them out altogether, as happened in Sweden in recent decades (see above). In Germany over the last 50 years, energy policy has swung between strong enthusiasm for nuclear energy and deep scepticism, with the most significant energy policy changes occurring in response to nuclear accidents (Renn and Marshall, 2016). The Fukushima disaster provided a distinct thrust towards the launch of Germany's policy of *Energiewende* ('energy transformation') in which nuclear power will be phased out by 2022.

Abandonment of nuclear power was reversed in Sweden, however, when the issue of climate change was put ahead of nuclear decommissioning. Indeed, the rise in concerns over global warming and its links to power plants based on fossil fuels has seen a considerable rejuvenation of interest in nuclear energy in recent years. Fossil fuel plants contribute significantly to emissions of greenhouse gases, an environmental issue to which nuclear power plants do not contribute. While many of the technologies for renewable sources of power continue to be developed, yet are currently unable to provide power on a sufficiently large scale for modern society, carbon-neutral nuclear energy becomes increasingly attractive in some eyes. Lovelock (2006), for example, argues passionately that nuclear electricity generation is not only a secure, safe and reliable source of energy but also the only way to counter the environmental dangers of global climate change. As with renewable energy, the future use of nuclear power will be a political decision in which environmental aspects will undoubtedly play a significant role.

FURTHER READING

Khandelwal, M., Hill, M.E., Greenough, P., Anthony, J., Quill, M., Linderman, M. and Udaykumar, H.S. 2017 Why have improved cook-stove initiatives in India failed? *World Development* 92: 13–27. A study of efforts over many decades.

Kyne, D. and Harris, J.T. 2015 A longitudinal study of human exposure to potential nuclear power plant risk. *International Journal of Disaster Risk Science* 6: 399–414. A study in environmental justice showing that minority groups in the USA are exposed to the highest levels of risk. Open access: link.springer.com/article/10.1007/s13753-015-0075-0.

Martini, D.Z., Sanches, I.D.A., Galdos, M.V., da Silva, C.R.U. and Dalla-Nora, E.L. 2018 Land availability for sugarcane derived jet-biofuels in São Paulo–Brazil. *Land Use Policy* 70: 256–62. Assessing how the aviation industry can replace fossil fuels with jet-biofuels grown in a Brazilian state.

Patterson, W. 2007 *Keeping the lights on: towards sustainable electricity*. London, Earthscan. A readable account arguing that electricity is about infrastructure and should be perceived as a commodity.

Peake, S. (ed.) 2017 *Renewable energy: power for a sustainable future*, 4th edn. Oxford, Oxford University Press. An introduction to the economic, social, environmental and policy issues surrounding all the main renewables.

Yablokov, A.V., Nesterenko, V.B. and Nesterenko, A.V. 2009 *Chernobyl: consequences of the catastrophe for people and nature*. New York, New York Academy of Sciences. A comprehensive assessment of the Chernobyl accident, its long-term effects and consequences.

WEBSITES

cleancookstoves.org   the Global Alliance for Clean Cookstoves promotes adoption of clean cooking solutions and fuels.

www.iaea.org   the International Atomic Energy Agency site covers many aspects of atomic energy, including up-to-date data and reports.

www.irena.org   the International Renewable Energy Agency promotes the adoption and sustainable use of all forms of renewable energy.

www.reeep.org   the Renewable Energy and Energy Efficiency Partnership is a global public–private partnership that promotes renewable energy and energy efficiency in developing countries and emerging markets.

www.renewableenergy.com   an industry hub for information on all forms of renewables.

www.worldenergy.org   the World Energy Council site contains information and resources on all aspects of energy.

POINTS FOR DISCUSSION

- How would an increase in our reliance on renewable energy sources, as against fossil fuels, contribute to solving other environmental issues?
- Human society's unsustainable use of fossil fuels is the root cause of virtually all other environmental issues. How far do you agree?
- What energy-saving measures can be introduced in your home and workplace? (Think both short- and long-term.)
- To be truly sustainable, all countries should favour the forms of energy production most suitable to their physical geography. Discuss.
- Should governments do more to encourage the development of renewable energy sources or should they leave it to the power companies?
- Review the arguments for and against the banning of nuclear power.

19 *Mining*

LEARNING OUTCOMES

At the end of this chapter you should:

- Know that many low- and middle-income countries rely heavily upon mineral exports for government revenues.
- Understand the paradox that some countries with abundant mineral resources have less economic growth, weaker democracy, and worse development outcomes, an idea known as the resource curse.
- Know some details about the environmental impacts of mining, including habitat destruction, geomorphological impacts, and pollution.
- Be familiar with some of the many techniques available for reduction and rehabilitation of mining damage.
- Appreciate the challenges involved in resolving transfrontier chloride pollution of the River Rhine.
- Appreciate some of the differences between transnational mining corporations and artisanal/small-scale mining.
- Understand the case of Nauru as an example of how 'weak' sustainability can result in widespread and severe environmental degradation.

KEY CONCEPTS

resource curse, tailings, mountaintop mining, subsidence, crownhole, rock burst, acid mine drainage, industrial barren, biodiversity offsetting, bioleach, corporate social responsibility (CSR), triple bottom-line

We only need to look at the names used to describe key periods in the early development of societies – the Stone Age, the Bronze Age and the Iron Age – to realize the long importance of mining as a human activity. People have been using minerals from the Earth's crust since *Homo habilis* first began to fashion stone tools 2.5 million years ago. Today, we are more dependent than ever before upon the extraction of minerals from the Earth. Virtually every material thing in modern society is either a direct mineral product or the result of processing with the aid of mineral derivatives such as steel, energy or fertilizers.

The naturally occurring elements and compounds mined are usually classified into four groups:

- metals (e.g. aluminium, copper, iron)
- industrial minerals (e.g. lime, soda ash)
- construction materials (e.g. sand, gravel)
- energy minerals (e.g. coal, uranium, oil, natural gas).

In terms of volume extracted, aggregates or construction minerals such as sand and gravel are by far the largest product of the world's mining industry. They are found and extracted in every country and world consumption is estimated to exceed 50 billion tonnes per year by 2020 (Freedonia, 2016). About 7500 tonnes of crushed rock, for example, is used to build 1 km of a single motorway lane. Metallic minerals are mined in smaller quantities. World production of iron ore in 2015 was 3.3 billion tonnes, bauxite 274 million tonnes, and phosphate 223 million tonnes (USGS, 2016).

GLOBAL ECONOMIC ASPECTS OF MINERAL PRODUCTION

The location of mineral concentrations in the Earth's crust is a fact of nature, but there are a host of human and natural drivers that determine if and when a mineral is recognized as a resource, classified as a reserve, and eventually exploited. Such factors include global and local economic and political influences, as well as the local availability of basic requirements for mining such as energy and water. While the use of minerals dates from before the time of *Homo sapiens*, it was the Industrial Revolution that sparked their large-scale exploitation. Between 1750 and 1900, global mineral use increased tenfold as the population doubled (Bosson and Varon, 1977). Between 1900 and 2005, as the global population increased by a factor of four, the global consumption of industrial minerals and metallic ores increased 27-fold and our use of construction minerals rose 34-fold (Krausmann *et al.*, 2009). In the case of copper, Meinert *et al.* (2016) point out that historic mine production recovered more than 658 million tonnes of copper between 1700 and 2015, but that about 26 per cent of this total was produced in just the last ten years of that period. The percentages indicate that global copper production has doubled every 25 years in order to accommodate the accelerating trend in global copper consumption.

This rapid increase in society's use of minerals has periodically sparked concern over their imminent depletion, a sentiment expressed strongly in the 1970s by the Club of Rome (Meadows *et al.*, 1972). Although the amount of minerals in the lithosphere is finite, improvements in technology and changing price/cost relationships, as well as shifting perceptions of the political risks of operating in particular countries, mean that estimates of reserves are constantly being re-evaluated.

Another important factor is the simple fact that large areas of the world's land surface have not been mapped in detail for their minerals, let alone explored, particularly where underlying as opposed to surface geology is concerned. New ores are regularly discovered and new mines opened. These factors combine to balance the fears of those who suggest that the end is nigh for global mineral exploitation. Mudd *et al.* (2013: 1164) emphasize the importance of new discoveries and technologies in keeping pace with demand

and conclude for copper that there are 'substantial resources well within the reach of present technology and economics' and that social and environmental issues will be more important in determining the future of mining.

On the national scale, most countries have adequate reserves of minerals used in the construction industry and these materials (with the exception of cement) are seldom traded internationally. The use of other minerals, particularly metals, is very heavily concentrated in the rich countries, however, and their movement is an important component of international trade. Many low- and middle-income countries rely heavily upon mineral exports as sources of government revenues (Table 19.1), a dependence that has been increasing over the last two decades (ICMM, 2016).

Paradoxically, the populations of some of these countries that rely significantly on mineral resource exports often suffer in both economic and political terms when these resources are extracted and exported, a phenomenon that has been labelled the 'resource curse', although such negative outcomes are by no means automatic (Brunnschweiler and Bulte, 2008).

The sources of minerals for major industrial nations vary. Some industrialized countries have considerable domestic reserves of certain minerals: Australia mined 29 per cent of global bauxite production in 2015, for example, while for other minerals less developed countries are important sources. Generally, industrialized countries import their minerals from proximate areas of the developing world: from Latin America to the USA, Africa to western Europe, and Asia and Oceania to Japan. Most such ores are shipped from their country of origin with minimal processing, since trade barriers often act to prevent developing country producers from adding value to their raw materials.

The demand for minerals inevitably fluctuates for numerous reasons, but there is no doubt that as global population grows and living standards rise, demand will continue. Recycling and substitution will make contributions towards satisfying this demand, but there are limitations to identifying potential substitutes,

Table 19.1 Top ten countries with economies most dependent on the mining of minerals

Country	Main minerals
Congo, DR	Copper, diamonds
Mauritania	Iron ore
Burkina Faso	Gold
Madagascar	Cobalt, ilmenite, nickel, rutile, zirconium
Botswana	Diamonds, copper, nickel
Guyana	Gold, bauxite
Uzbekistan	Gold, uranium
Liberia	Diamonds, iron ore
Kyrgyzstan	Gold
Tajikistan	Antimony, mercury

Note: Ranking according to the Mining Contribution Index (MCI), based on four main quantitative indicators.
Source: after ICMM (2016).

as Nassar (2015) describes for the platinum-group metals. There is also no doubt, therefore, that the mining industry will continue to discover, develop and exploit new resources in the foreseeable future.

ENVIRONMENTAL IMPACTS OF MINING

Concern at the environmental impacts of mining is by no means a recent phenomenon. Georgius Agricola (1556), author of the world's first mining textbook (Figure 19.1), could have been writing today as he outlined the strongest arguments of mining's detractors in sixteenth-century Germany:

> [T]he woods and groves are cut down, for there is need of an endless amount of wood for timbers, machines and the smelting of metals. And when the woods and groves are felled, then are exterminated the beasts and birds, very many of which furnish a pleasant and agreeable food for man. Further, when the ores are washed, the water which has been used poisons the brooks and streams, and either destroys the fish or drives them away.
>
> (Agricola, 1556: 8)

Figure 19.1 Some of the surface and subsurface impacts of medieval mining techniques as depicted in De re metallica, Georgius Agricola's mining textbook published in 1556.

Little appears to have changed in the more than 450 years since Agricola. In recent times, environmentalists have often quoted this passage as evidence of the long-standing environmental impact of mining. Interestingly, the quote is often used out of context, since Agricola in fact dismissed these anti-mining sentiments on the grounds that the environmental cost was far outweighed by the benefits to society:

> If we remove metals from the service of man, all methods of protecting and sustaining health and more carefully preserving the course of life are done away with. If there were no metals, men would pass a horrible and wretched existence in the midst of wild beasts; they would return to the acorns and fruits and berries of the forest.

(ibid.: 14)

There can be no doubting that mining involves an environmental impact, and the destruction of habitats suggested in Agricola's textbook is perhaps the most obvious from among a diverse catalogue of ecological changes that the various types of mining operation can cause (Table 19.2). The question as to whether such impacts should be regarded as worth the benefits derived from the minerals extracted is a complex one, which will be returned to after consideration of the physical impacts involved.

Habitat destruction

Vegetation is stripped and soil and rock moved, both for the extraction process itself and the concomitant building of plant, administration and housing facilities, and the history of mining activity has left

Table 19.2 Environmental problems associated with mining

Problem	Type of mining operation			
	Open pit and quarrying	Opencast (as in coal)	Underground	Dredging (as in tin or gold)
Habitat destruction	X	X	—	X
Dump failure/erosion	X	X	X	—
Subsidence	—	—	X	—
Water pollution	X	X	X	X
Air pollution*	X	X	X	—
Noise	X	X	—	—
Air/blast/ground vibration	X	X	—	—
Dereliction	X	—	X	X

Notes: X = problem present; — = problem unlikely; *can be associated with smelting, which may not be at the site of ore/mineral extraction.
Source: after Whitlow (1990).

considerable areas of disrupted and degraded habitat. The scale of damage is greatest at so-called open-cast or strip mining sites. Several small coral islands in the Pacific Ocean exploited for their phosphate, the fossilized remains of centuries of birds' droppings, are among the worst affected. Opencast phosphate mining on the Pacific island state of Nauru involved scraping off the surface soil to enable removal of the phosphate from between the walls and columns of ancient coral. A severely degraded landscape of highly irregular solution-pitted limestone remains over about 80 per cent of the island's 2100 ha area. Many of the indigenous plants and animals that previously inhabited the mined-out areas have either disappeared or are endangered. Indeed, virtually all of these mined areas, except for narrow corridors either side of gravel roads, are now 'totally unusable for habitation, crops, or anything else that might benefit the people of Nauru' (Gowdy and McDaniel, 1999: 334).

The wide range of activities that drive logging and other forms of forest disturbance in the Congo Basin, where artisanal and small-scale mining represent by far the largest mining sector, is shown in Table 19.3. Hund *et al.* (2013) point out that impacts on the forest vary according to several factors, including the mining techniques used, the previous state of the ecosystem (intact or already disrupted), and whether or not it is a 'rush' situation (when many people suddenly invade a site in the hope of finding precious metals, without any control or accountability).

Disposal of waste rock and/or 'tailings' – the impurities left after a mineral has been extracted from its ore – usually degrades still larger areas of natural ecosystems. This is an increasing problem, both because of the growth in demand for minerals and because as rich ores are mined out, so lower-grade deposits are worked, producing more waste per unit of mineral produced. While four centuries ago the average grade of copper ore mined was about 8 per cent (Bosson and Varon, 1977), the average grade at the turn of the twenty-first century was 0.91 per cent, which means about 990 million tonnes of waste generated for the 9 million tonnes of copper produced.

On the national scale, such mining impacts can cover a significant proportion of a country's land area. In Estonia, where excavation of a 9-billion-tonne oil shale reserve – the world's largest commercially exploited deposit – has continued since 1916, about 1 per cent of the national land area is classed as 'disturbed mined-out land' (Gavrilova *et al.*, 2010). About half of this area is considered unstable and a quarter suffers from subsidence.

To date, however, the largest areas of strip mining in the world are for coal (Figure 19.2). In the USA, which produces about 60 per cent of its coal by this method, it has affected vast areas of the Appalachian Mountains and large areas of the states of Ohio, West Virginia and Kentucky. On average, surface mining for coal consumes 10–30 ha of land for every million tonnes of coal produced in the USA, but in other

---

Table 19.3 Drivers of habitat disturbance by artisanal mining in forests of the Congo Basin

Forest clearance to expose substrate for mining
Timber and branches to build worker camps, reinforce mineshafts, make mining tools and carrying baskets
Firewood for cooking and warmth in camps, to generate heat to crack rocks in underground workings
Bark to make pans for washing minerals
Medicinal plants for use in camps

Source: after Hund *et al.* (2013).

Figure 19.2 Opencast mining of brown coal in Sindh, Pakistan.

parts of the world, local conditions and available technology can result in much higher land consumption-to-production ratios. In southern Russia's Kuznetsk Basin, for example, the figure is about 166 ha per million tonnes (Bond and Piepenburg, 1990).

In areas where smelting takes place near the extraction site, further damage can be caused by cutting trees to fuel smelters. Historically, this has taken a significant toll on forests in many parts of the world. One estimate of the amount of wood converted to charcoal for iron smelting in India during pre-colonial times is 15 million tonnes a year (Jyotishi and Sivramkrishna, 2011). The results of a similarly large demand for hardwood timber to make charcoal for a large iron industry in the US state of Pennsylvania are documented by Linehan (2010). At the end of the nineteenth century, 10 per cent of the state's entire land area was described in a government survey as 'wholly waste or worse' due to deforestation. The knock-on effects included soil erosion and flooding, which had been instrumental in the abandonment of very large areas of farmland, and a serious threat to the state's supply of drinking water.

The environmental impacts of mining are not just limited to terrestrial ecosystems. The mining of coral and sand for road and building materials is a widespread threat to coastlines in many parts of the world and particularly to island reefs in the Pacific Ocean. Apart from the physical destruction of reefs to provide limestone, and the introduction of toxic substances to the marine environment during the mining process, the dredging of sand causes beach erosion and alters water circulation, leading to sedimentation on reefs (see also p. 140).

Marine ecosystems can also be degraded by any form of mining in the coastal zone. A range of effects have been observed on the coral reefs fringing Misima Island in Papua New Guinea following the start of opencast mining for gold in 1989 (Fallon *et al.*, 2002). Increased soil erosion from the mine caused higher sedimentation rates on the reef, resulting in effects ranging from coral mortality due to smothering to contamination from metals contained in the sediment.

When exploitation of polymetallic (manganese) nodules and sea-floor massive sulphides from the deep-sea bed becomes a commercial proposition, although this is likely to be a long time in the future, they too will bring disturbances to the world's oceans. Impacts will occur at different levels in the water column: at the sediment surface due to collector impact, immediately above the sea bed due to the plumes of sediment raised, and at other levels as abraded materials and sediments are discharged (Sharma, 2011). The International Seabed Authority, which has already begun to issue exploration contracts, is in the process of developing a Mining Code for eventual exploitation of deep-sea mineral resources.

Geomorphological impacts

Mining involves the movement of quantities of soil and rock, and thus by definition is a human-induced geomorphological process. The scale of operations has reached the level whereby estimates suggest that as much as a hundred times more material is stripped from the Earth by mining than by all the natural erosion carried out by rivers (Goudie, 2006: 162). The impacts of mining, however, are much more localized than the denudation caused by rivers. The opencast copper mine at Bingham Canyon in Utah, USA, reputed to be the largest human excavation in the world, has involved the removal of more than 3500 million tonnes of material over an area of over 7 km², to a depth of more than 1000 m. Nevertheless, such features can attain significant proportions on the national scale: surficial scars on the landscape of Kuwait, created by the oil-generated construction boom, cover large parts of the national land area, with some individual quarries measuring up to 40 km in length (Figure 19.3). Al-Awadhi (2001) points out that the gravel quarries in northern Kuwait comprise more than 2 per cent of the country's land area and that their excavation has disrupted local relief, eliminated vegetation cover and compacted topsoil.

One major form of surface mining that has a very significant impact on topography is known as 'mountaintop mining'. Under this approach, which is widespread in the Appalachian region of the USA, for example, forests in high-elevation areas are cleared, topsoil is stripped and explosives used to break up rocks to access buried coal (Figure 19.4). Excess rock, so-called minespoil, is then deposited in adjacent valleys, with an overall levelling effect on the landscape (Palmer *et al.*, 2010). New landforms can also be created by the deposition of minespoil, and these can create new geomorphological hazards, such as landslides and subsidence. The death of 150 people at Aberfan, South Wales, in 1966, when waste piles from the Merthyr Vale coal mine suddenly collapsed, graphically illustrated the dangers of unregulated waste tipping. Self-ignition is another potential hazard associated with certain mining wastes. Gavrilova *et al.* (2010) report that nine waste rock heaps deposited by the Estonian oil shale mining industry self-combusted between 1960 and 1980, and the average burning period was about ten years.

Care must be taken with opencast excavations to avoid such problems associated with slope instability, which can occur in both active and abandoned workings. The city centre of Izmir, in Turkey, is peppered with 70 abandoned quarries formerly worked for their building stone, most of it good-quality andesite. The quarries were initially located beyond the urban area but are now within the city centre thanks to Izmir's expansion. Most of these old stone quarries were abandoned without taking any precautions against

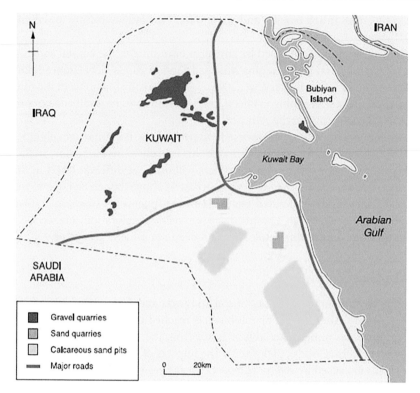

Figure 19.3 Distribution of sand and gravel quarries in Kuwait (after Khalaf, 1989).

Figure 19.4 A large mountaintop coal mining operation in West Virginia, USA.

possible slope failures, and sliding and toppling are now common hazards faced by local residents (Koca and Kincal, 2004).

Surface subsidence can result from underground workings, and is most commonly associated with coal mining where whole seams of material, frequently several metres thick, are removed, resulting in overlying strata collapsing downwards. This may occur rapidly and with little warning or take place gradually over a period of time. Two major types of mining subsidence can be identified (Scott and Statham, 1998):

- *crownholes* – localized crater-like holes that appear at the surface following collapse of strata into a mine
- *general subsidence* – settlement of the ground surface over a wide area resulting from the collapse of part of the mine.

Damage to mining equipment, buildings, communications and agriculture through the disruption of drainage systems are the most common results. Subsidence may be exacerbated by the reactivation of geological faults in the area. Fault reactivation can result in further disruption of the ground surface due to the formation of fault scarps. These features can vary from subtle disturbances on the ground surface to major topographic scarps up to 4 metres high and several kilometres long, as Bell *et al.* (2005) describe in the South Wales coalfield. These authors also cite examples of subsidence associated with the extraction of a range of other minerals, including slate in Germany, ancient chalk mines in southern England, and the extensive extraction of pumice on the flanks of Galeras volcano in Colombia. In the last case, subsidence of abandoned mine workings beneath the city of Pasto has resulted in severe damage to roads and buildings and a number of deaths. Subsidence into old gold mine workings meant that an entire street of houses had to be moved in the town of Malartic in Quebec, Canada (Bétournay, 2009).

Collapse also occurs when water is pumped out of mine workings, lowering local water tables, which results in the drying out and shrinkage of materials such as clays, and through the reduction in underground fluid pressure that results from oil and natural gas extraction. Nagel (2001) notes that marine inundation is a danger caused by subsidence from oil and gas extraction in areas near sea level and quotes examples from oil fields on the shore of Venezuela's Lake Maracaibo and the Groningen gas field in the Netherlands where subsidence of just a few centimetres caused serious concern because large areas of the country are nominally below sea level and protected by dikes. A particularly dramatic example occurred in the Wilmington oil field in southern California where abstraction between 1928 and 1971 caused 9.3 m of subsidence at Long Beach, Los Angeles (Table 19.4). An unexpected degree of sea-floor subsidence beneath the West Ekofisk oil field in the North Sea identified in the mid-1980s resulted in dangers to oil rigs and pipelines.

The stresses in surrounding rocks caused by mine excavations can on occasion exceed the strength of the rocks and cause earthquakes, traditionally known as 'rock bursts' or 'coal bumps'. A country renowned for large mining-induced earthquakes is South Africa, where the problem became apparent in the early twentieth century as mines penetrated to several hundred metres in depth. It is now a major issue facing deep mining operations for gold and platinum, and considerable efforts are made to mitigate the risk, although fatality rates continue to number several tens of deaths per year. A particularly large mining-related seismic event occurred in the Klerksdorp district of South Africa in 2005 where a main shock and aftershocks seriously damaged the nearby town of Stilfontein, injuring 58 people. Two mineworkers in a nearby gold mine lost their lives and thousands of others were evacuated under difficult circumstances (Durrheim, 2010). In China, the problem of mining-induced seismicity is particularly associated with coal mines, an issue that is becoming more serious as mines increase in depth of

Table 19.4 Examples of ground subsidence due to oil and gas abstraction

Location	Area affected (km²)	Maximum subsidence (m)	Period	Rate (mm/year)
Inglewood, USA	–	2.9	1917–63	63
Maracaibo, Venezuela	450	5.0	1929–90	82
West Ekofisk oil field, North Sea	–	3.5	1977–98	167
Wilmington, USA	78	9.3	1928–71	216

Source: after Cooke and Doornkamp (1990); Nagel (2001); Goudie (2006).

extraction and volume of material removed. Li *et al.* (2007) reported that in 2007 some 102 coal mines and 20 other mines reported seismic activity.

Pollution

Pollution from mineral extraction, transportation and processing can also present serious environmental problems, affecting soil, water and air quality. Examples of pollution emanating from the transport of oil are given elsewhere (see Chapters 6 and 7), while here the emphasis is on mine workings and smelting.

Four major pollution problems are commonly associated with mining, all of which can cause serious damage to the environment and, on occasion, to human health:

■ heavy metal pollution
■ acid mine drainage
■ eutrophication
■ deoxygenation.

Many tailings contain metal ores and other contaminants, formerly locked up in solid rock, which can be leached into soils and waterways, and blown into the surrounding atmosphere. Metals are also released into the environment during smelting. Contamination by heavy metals released by these activities has been very extensive from the earliest times. The concentration of copper found in Greenland ice layers began to exceed the natural background concentration about 2500 years ago due to smelting emissions during Roman and medieval times, particularly in Europe and China. Similarly, the lead content of Greenland ice was about four times the background concentration during the period 2500 BP to 1700 BP (Nriagu, 1996). Deposits of such metals in other parts of the environment may continue to act as significant secondary sources of contamination for hundreds of years after mining has ceased. Historic lead and zinc mining in the English Pennines has left a legacy of metals in alluvial deposits that are released under flood conditions, significantly affecting water quality in this part of northern England (Macklin *et al.*, 1997).

Some of the highest heavy metal concentrations are derived from uranium tailings and pumped water from iron mines, but one of the most notorious examples of deleterious effects on human health came from a lead–zinc mine in Toyama prefecture, Japan. Cadmium-polluted water from the Kamioka mine, which was used in rice paddy fields, led to an outbreak of Itai-itai disease, a cadmium-induced bone disorder. Several hundred people are thought to have died from Itai-itai disease among long-term residents of the area since the disease was first recorded in the 1930s (Kaji, 2012).

Even small flows of polluted water can be dangerous to ecosystems and people, but damage can be greatly intensified by large accidental discharges. Some notorious examples have occurred as a result of tailings dam failure (Table 19.5). One such incident, caused by the failure in 2015 of the Fundão iron-ore tailings dam, resulted in one of the worst environmental disasters in Brazilian history. More than 50 million m^3 of sludge buried the houses of Bento Rodrigues, a village in the state of Minas Gerais, leaving 20 people dead and over 600 homeless. Vegetation was directly destroyed by the event in an area of about 17 km^2, half of which was critically endangered Brazilian Atlantic riparian forest (Garcia *et al.*, 2016). The wave of mud reached the Rio Doce, adversely affecting biodiversity along hundreds of kilometres of the river, its riparian lands and the Atlantic coast due to the rapid rise in turbidity (Figure 19.5), the rapid drop in oxygen level and high concentrations of pollutants (arsenic, lead, cadmium, chromium, nickel, selenium and manganese all exceeded legal safety levels). Hundreds of thousands of people were also left without access to clean water.

Acid mine drainage is responsible for water pollution problems in major coal-mining and metal-mining areas around the world. It was originally thought to be associated only with coal mining, but any deposit containing sulphide, and particularly pyrite, can be a source, and it is produced from the working of many other minerals. Reactions with air and water produce sulphuric acid, and pH values of polluted waterways may be as low as 2 or 3. Acid mine drainage is not just a problem of operational workings, but may continue

Table 19.5 Some examples of tailings dam failures

Mine (year)	Main ore mined/ material released	Tailings released (m^3)	Cause	Immediate fatalities
Cerro Negro No. 4, Chile (1985)	Copper	2,000,000	Breach after earthquake	Unknown
Merriespruit, South Africa (1994)	Gold	600,000	Breach after heavy rain/poor maintenance	17
Aznalcóllar, Spain (1998)	Sulphide	1,300,000	Foundation failure/ poor maintenance	0
Baia Mare and Baia Borsa, Romania (2000)	Silver, gold/ cyanide	Two incidents of 100,000	Breach after heavy rain and snowmelt	0
Kingston plant, USA (2008)	Coal fly ash/ radium isotopes/ arsenic/mercury	4,100,000	Retention wall failure	0
Ajkai Timfoldgyar Zrt alumina plant, Hungary (2010)	Aluminium/alkali	6,000,000–7,000,000	Unknown	10
Fundão, Brazil (2015)	Iron	>50,000,000	Unknown	20

Source: updated after Kossoff *et al.* (2014).

Figure 19.5 The heavily polluted flow of the Rio Doce in southern Brazil entering the Atlantic Ocean 25 days after the Fundão tailings dam failed, releasing a catastrophic wave of toxic mud.

as a pollution source for some years after the mine closure (Akcil and Koldas, 2006). Surface tailings can also produce acid effluents; Hägerstrand and Lohm (1990) report that sulphuric acid produced from sulphides in tailings around the Falun copper mine in central Sweden, which has been worked for 1000 years, has seriously damaged several lakes in the region.

When tailings dry out, dust blow can also cause localized pollution. Fine material blown from the quartzite dumps that surround Johannesburg, a city that grew up on the gold-mining activity in South Africa's Witwatersrand, used to stop work at food-processing factories and cause traffic chaos due to decreased visibility on windy days before attempts to stabilize many dumps with vegetation. However, dust blowing from gold-mine tailings remains a common form of air pollution in the city with adverse health implications for nearby residents (Ojelede *et al.*, 2012).

The processing of minerals can also release dangerous compounds into surrounding environments. Areas around metal smelting operations that have lost much vegetation due to the deposition of airborne pollutants are termed 'industrial barrens' by Kozlov and Zvereva (2007) who highlight the extreme and long-lasting effects, typically including large-scale forest dieback and extensive bleak landscapes with little vegetation and highly acidic, metal-contaminated soil and water. Table 19.6 shows some examples.

Table 19.6 Some examples of 'industrial barrens'

Location	Poluter (established–closed)	Primary vegetation	Barren area in ha (estimate year)	Other stressors
Anaconda, USA	Cu smelter (1884–1980)	Forest (subalpine lodgepole pine and Douglas fir)	4,500 (1992)	Unfertile soil; extreme climatic fluctuations
Ashio, Japan	Cu smelter (1877–1988)	Forest (beech and conifer)	1,395 (1893) >2,500 (1970)	Logging; soil erosion
Harjavalta, Finland	Cu–Ni smelter (1945–)	Forest (southern boreal – Scots pine)	<50 (2000)	–
Karabash, Russia	Cu smelter (1907–)	Forest (taiga)	<1,000 (2002)	Soil erosion
Lubenik, Slovakia	Magnesite plant (1958–)	Forest (broadleaved)	200 (2006)	–
Quinteros, Chile	Cu smelter (1964–)	Shrubby grassland	2,000 (2005)	–
Sudbury, Canada	Ni–Cu smelter (1888–)	Forest (mixed boreal conifers)	19,565 (1970)	Logging; fire; pests; soil erosion; frost

Source: after Kozlov and Zvereva (2007).

Pollution by acidification from SO_2 as well as heavy metals has a history of more than 100 years at Sudbury, in Ontario, Canada, site of a major mining operation to exploit nickel–copper ores since the late nineteenth century. The effects have been great on both terrestrial and aquatic ecosystems, with severe damage to spruce and pine trees and some 7000 lakes within a 17,000-km² area in the Sudbury region acidified to a pH of 6.0 or less and with nickel concentrations up to 300 g/l (compared with a global mean for lake waters of 10 g/l). In the 1970s, several efforts to reduce SO_2 and heavy metal emissions were begun, including installation of a very tall (380 m) stack at the largest smelter, closure of another smelter, and technical improvements in the smelting operations. The result has been a reduction in SO_2 and metal emissions by approximately 90 per cent since the 1970s, and many studies (e.g. Tropea *et al.*, 2010) have documented chemical recovery (increased pH and decreased concentrations of sulphur and metals) in terrestrial and aquatic systems (see Chapter 12).

Artisanal and small-scale gold mining, which is practised in more than 70 countries, is thought to be the largest anthropogenic source of mercury in the environment (UNEP, 2013). These miners use mercury for extracting gold particles by amalgamation, but mercury is released into soils and waterways during the amalgam process and into the atmosphere during amalgam roasting. Mercury is an extremely toxic metal that accumulates in the food chain, and can cause birth defects, kidney dysfunction and various neurological problems. Miners and other residents of gold-mining areas can be exposed to mercury via three major routes:

■ Miners can have exposure through their skin when they mix mercury with gold ore.
■ Mercury vapours can be inhaled when amalgams are heated.
■ Methylmercury can be consumed from contaminated fish.

High levels of mercury and associated health impacts have been widely reported from artisanal and small-scale gold-mining communities in many parts of the developing world (Gibb and O'Leary, 2014), and mercury in the environment can be long-lasting. The use of mercury in the 1950s and 1960s at the Discovery mine in the Canadian Northwest Territories resulted in nearby Giauque Lake still being designated a contaminated site by Environment Canada decades later (Meech *et al.*, 1998).

Very long-lasting ecological effects of sulphur dioxide (SO_2) emissions have been reported from the Falun copper mine in Sweden where the first step in copper extraction was to roast the ore in the open air to reduce its high sulphur content. This process oxidized the sulphide, emitting sulphur to the air as SO_2. These emissions, which date from at least the thirteenth century, reached a maximum during the 1600s (Figure 19.6) when Falun produced two-thirds of the world's copper supply. Deposition of SO_2 in the area around Falun exceeded the critical load in this region for centuries and acidification of local soils is still apparent, although it is limited to the most heavily polluted area 12 km to the north-west and to the south-east of the mine. Lakes in the Falun area also became more acidic dating from the seventeenth century, although the pH decrease has been moderate. However, despite 300 years of lowered emissions since the 1600s (the mine was closed in 1993), none of the acidified lakes in Falun shows pH recovery due to a large store of sulphate in the soil, although land-use change in the last 100 years is also likely to have had some effect (Ek *et al.*, 2001).

When elevated concentrations of pollutants enter the food chain, they can remain hazardous to human health sometimes long after their release by mining activities. High accumulations of harmful elements

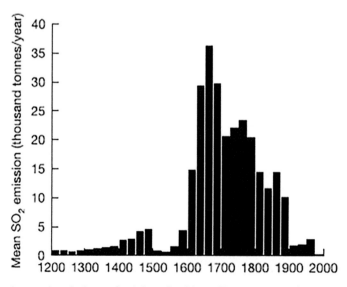

Figure 19.6 Estimated annual emissions of sulphur dioxide as 25-year means from the Falun mine, Sweden, 1200–2000 (Ek et al., 2001).

have been measured in wild-growing mushrooms – very popular among the inhabitants of many Central and East European countries – in numerous areas of contemporary and historical mining and smelting. Harmful levels of cadmium and lead, for example, were determined in wild mushrooms near a historical silver-mining area in South Bohemia, Czech Republic, by Svoboda *et al.* (2006) despite the fact that the mining and processing of silver ores took place mainly during the second half of the sixteenth century. The elevated cadmium and lead concentrations detected by these authors led them to recommend that restrictions should be introduced on the consumption of mushrooms picked in the area.

Disease, poverty and the resource curse

Itinerant miners venturing into little-explored terrain frequently carry with them new diseases and problems associated with alcohol, bringing often disastrous consequences for local populations. Gold prospectors in the Brazilian Amazon, so-called garimpeiros, have spread virulent strains of malaria into reserves occupied by the indigenous Yanomani Indians in northern Amazonas. Local tribes with little resistance to the new strains of parasites have suffered severe outbreaks of malaria (Smith *et al.*, 1991), and met with previously unknown diseases such as measles. Malaria has also been rife more recently in illegal gold-mining camps in forested areas of French Guiana (de Santi *et al.*, 2016).

In general terms, mining is among the world's most hazardous occupations, particularly for artisanal miners. In southern Africa, exposure to dust inside mines causes long-term damage to the lungs, and social conditions outside the mines have historically been major drivers of epidemics, particularly of HIV and tuberculosis (Stuckler *et al.*, 2013). Links between mining and health are by no means confined to the developing countries, however. In a study of the Appalachian coalfields of eastern USA, Hendryx (2010) concluded that residents of mountaintop mining areas experience persistently elevated rates of poverty and mortality and that pollution from mountaintop mining may be a driver of these higher rates. A considerable body of research has argued that this area's dependence on coal mining has contributed to its deep poverty through a variety of mechanisms, including weak local governance and educational attainment, limited economic alternatives, poor health and a degraded environment. This set of drivers is broadly associated with the natural resources curse in the international development literature.

Interestingly, there is also a growing body of research arguing that small-scale mining can help to reduce rural poverty in sub-Saharan Africa. In many parts of the continent, mining has close links with farming, the two sectors often complementing one another over a calendar year. Hilson (2016) offers many examples where money made in the mining sector is used to support food production, including farmers in Zimbabwe using the proceeds from gold mining to purchase fertilizers, and families in Liberia hiring labourers to work artisanal diamond fields, exchanging part of their rice crop for a percentage of the diamond sales.

One possible outcome of the environmental injustice associated with the resource curse (and associated political ramifications) is the outbreak of violent conflict. Table 19.7 shows several examples of violent separatist groups that have made mineral resource exploitation a focus of their independence movement. A classic example in this respect can be cited from the Panguna copper mine on the island of Bougainville in Papua New Guinea (PNG), where discontent over the distribution of mining revenues was a significant factor in precipitating a revolt out of a long-simmering dispute between the islanders and the PNG government (Hyndman, 2001). Discontent over the unequal distribution of the income derived from mining was not helped by the fact that 130,000 tonnes of metal-contaminated tailings were dumped into the Kawerong and Jaba rivers every day, severely depleting biological activity in the waters. After several years of nonviolent disputes over these issues, the Bougainville Revolutionary Army was formed in late 1988 and launched a series of attacks on the mine. By May 1989, they had forced the Panguna mine to close, but the

Table 19.7 Examples of secessionist movements connected to mineral resources

Country	Region	Duration	Mineral resources
Congo, DR	Katanga/Shaba	1960–65	Copper
Indonesia	Papua	1969–present	Copper, gold
Morocco	Western Sahara	1975–88	Phosphates
Myanmar	Hill tribes	1983–95	Tin, gems
Papua New Guinea	Bougainville	1988–97	Copper, gold
Sudan	South Sudan	1983–2005	Oil

Source: after Ross (2008).

conflict continued until 1997 and may have claimed more than 10,000 lives. Bougainville is still the only place in the world where host community violence has resulted in the long-term closure of a large-scale mine (Regan, 2017).

REHABILITATION AND REDUCTION OF MINING DAMAGE

Many techniques are available for the amelioration, curtailment or restoration of damage caused by mining and its associated activities (Bell and Donnelly, 2006), and there is little doubt that for a variety of reasons the mining industry in general has recently adopted a more benevolent approach to its environmental impact than in times past. Site surveys prior to mining, retention and replacement of topsoil after excavation, and careful reseeding with original species, can be employed to return environments to something close to their original states. However, this goal may be particularly difficult on landscapes where fundamental changes have occurred to almost every component, along with the persistence of mining-induced features such as open pits and waste heaps. In such radically disturbed cases, new combinations of physical and biological components may have to be acceptable to all stakeholders (Doley and Audet, 2013).

An alternative approach that has gained in popularity in the past decade or so is 'biodiversity offsetting', which involves trying to compensate for the damage to species and habitats caused by mine development by creating an 'ecologically equivalent' benefit elsewhere. Biodiversity offsets have increasingly become integral to the corporate social responsibility programmes of multinational mining companies, to demonstrate good environmental stewardship, conform to national regulations, and reduce reputational risks (Virah-Sawmy *et al.*, 2014).

Remediation of subsidence problems is inevitably a significant undertaking with considerable economic costs (Nagel, 2001). In the Venezuelan example noted in Table 19.4, more than 150 km of dikes had to be built and repeatedly reinforced to combat the coastal flooding. Remediation efforts to combat sea-floor subsidence beneath the West Ekofisk oil field in the North Sea involved both water injection into the depleted chalk reservoir rocks and a very challenging operation to raise up the oil platforms by extending their legs. The total cost of the remediation programme approached US$1 billion.

For mining operations located near a coast, disposal of tailings underwater can be an attractive alternative to land-based impoundment. Under the right conditions, it can also be less costly and less environmentally damaging. A good example occurred at Island Copper Mine off Vancouver Island in western Canada where more than 400 million tonnes of tailing solids generated over the course of the mine's 25 years of operation were deposited deep on the ocean floor with little environmental impact. The mine's environmental monitoring programme, which continued after closure in 1995, showed that by the year 2000 water quality and plankton biodiversity were indistinguishable from that in control areas (Ellis, 2003). Pipeline transport technology now makes it feasible and cost-effective to implement such underwater tailing placement for a mine up to 200 km from the shoreline.

Waste dumps and tailings on land can be stabilized with vegetation both to ameliorate their visual impact and to reduce the risks of slope failure, or flattened and properly drained as in the case of the waste tips at Aberfan. Acidic minespoil often requires the application of alkaline material before a vegetation cover can be successfully established and one method of so doing uses waste by-products from dry flue gas desulphurization scrubbers fitted to industrial chimneys to remove sulphur gases (Stehouwer *et al.*, 1995). The scrubbers use lime, limestone or dolomite sorbents to react with SO_2, forming anhydrite. The resulting by-products – a mixture of coal ash, anhydrite and residual sorbent – often have pH levels of 12 or greater due to the residual sorbent. The use of this material to help neutralize acidic mining wastes before rehabilitation is a good illustration of how the reappraisal of waste products can create new resources (see p. 421). Disposal of scrubber waste has previously been in landfill sites, but the efforts of many countries to reduce acid rain emissions mean that the volume of these by-products has risen substantially in recent years. Using them on acidic minespoil is therefore a much more sustainable method that helps to resolve two different environmental issues.

The full range of options available to deal with the problems of acid mine drainage is reviewed by Johnson and Hallberg (2005). Source control methods (Table 19.8) are designed to prevent or minimize the generation of acidic waters by excluding either, or both, oxygen and water, which are needed for acid mine drainage to form. This can be achieved by flooding and sealing abandoned deep mines. Alternatively, mine tailings can be stored underwater or sealed on the ground surface in some way, usually with a layer of clay, which can then be covered in topsoil and planted with vegetation. Other ways of neutralizing the acid include the application of some form of surface coating to isolate the relevant chemicals, or the application of biocides to kill the lithotrophic (rock-eating) bacteria that play a pivotal role in generating acid mine drainage.

Table 19.8 Source control methods used to prevent or minimize the generation of acid mine drainage waters

Flooding/sealing of underground workings
Underwater storage of tailings
Land-based storage in sealed waste heaps
Total solidification of tailings
Blending of acid-generating and acid-consuming mineral wastes
Microencapsulation (coatings)
Application of biocides

Source: after Johnson and Hallberg (2005).

Table 19.9 Reuse and recycling options for mine and quarry waste rock and tailings

Waste type	Reuse/recycling options
Waste rock	Landscaping material
	Aggregate in embankment, road, dam and building construction
	Substrate for revegetation at mine sites
	Aggregate for concrete
	Backfill for open voids
Tailings	Reprocessing to extract minerals and metals
	Auxiliary sources to produce building materials (e.g. cement, hollow bricks, concrete, glass)
	Iron-rich tailings mixed with fly ash and sewage sludge to produce lightweight ceramics
	Energy recovery from compost–coal tailings mixtures
	Backfill in underground mines

Source: after Lottermoser (2011).

Preventing the formation of acid mine drainage is not always practical, however, and a number of strategies to minimize the impact of the polluting water on streams and rivers are also available. These techniques are often divided into 'active' and 'passive' processes. The former generally means the continuous application of alkaline materials (e.g. lime, calcium carbonate, magnesium oxide) to neutralize the acidic waters, while the latter refers to the use of natural or constructed wetland ecosystems that have the same effect. The passive approaches have the advantage of requiring relatively little maintenance, and few recurrent costs, when compared to active approaches, although they may be expensive to set up in the first place. However, there are still numerous cases in which comprehensive treatment of abandoned mine waters is simply not achievable for any conceivable level of investment. An example can be cited from the Wheal Jane mine in south-west England where extensive investigations revealed that treatment of the entire flow in the river would still fail to achieve regulatory standards, due to the pervasive presence of vast sources of recontamination downstream (Younger *et al.*, 2005).

Mining wastes themselves are used for a variety of purposes in some countries, particularly in road building and for landfill prior to construction projects (Table 19.9), but transportation costs to areas of high demand often limit their use. Indeed, Lottermoser (2011) concludes that the economic costs associated with many of the proposed reuse and recycling concepts for mine wastes mean that the great majority of wastes generated by mines are still placed into waste storage facilities.

Many of the efforts made to reduce pollution originating in mining wastes have also stemmed from the realization that the pollutants are themselves economically valuable. Part of the reduction in SO_2 emissions from the Sudbury nickel smelters, for example, was achieved in the late 1960s when the gas was converted into sulphuric acid. Similarly, the use of new biotechnologies for recovering wasted resources, such as microbes to 'bioleach' metals from tailings, has also been developed in response to the lower grades of accessible ores and lower operating costs. Microbial mining of copper sulphide ores has been practised on an industrial scale since the late 1950s and, since then, bioleaching has also become common at uranium and gold mines. Another development in mining biotechnology is the use of bacterial cells to detoxify waste cyanide solution from gold-mining operations. The recycling of minerals, particularly metals, also helps to reduce environmental impacts, of course (see Chapter 17).

In recent decades, a growing public concern in the industrialized nations at the environmental cost of mining has led to increasing legislative controls on the activities of mining companies, including the requirement to conduct Environmental Impact Assessments. The generally higher public profile of environmental issues and the pressures brought to bear by environmentalists, particularly in developed countries, have encouraged multinational mining companies to give environmental concerns a higher profile. The formation in 2001 of the International Council on Metals and the Environment (ICMM), a body comprising most of the world's major mining companies, is a clear indication of this trend. The ICMM aims to improve sustainable development in the mining and metals industry.

Increasingly, responsible mine operations have adopted the precautionary principle, with considerable attention to the possible pollution effects, involving regular monitoring of groundwater, air quality and noise, to maintain acceptable standards. Environmentally sound management and clean-ups cost money, however, and legal obligations are often necessary to force appropriate practices. There is little doubt that decades of environmentally unregulated coal mining in the USA have resulted in widespread pollution of water bodies, subsidence effects, scarred landscapes and extensive waste dumps. Rehabilitation of this environmental degradation only became standard practice with the passing of the Surface Mining Control and Reclamation Act in 1977. The Act requires mine operators to reclaim strip mines to the approximate original contours of the landscape and to restore appropriate permanent vegetation. It also imposes a tax on active mines to address the legacy of improperly abandoned mines in coalfield areas. However, complete restoration of pre-mining ecosystem structure, function and services has proved to be very difficult if not impossible to date (Simmons *et al.*, 2008) and the Act only applies to coal mines. No comparable law has been enacted for other types of mine in the USA.

Similar legal obligations are now well established in many developed countries of the global North, although they are often inadequate in countries of the South where the importance of mining to the national economy (Table 19.1) makes the imposition of costly rehabilitation measures unattractive. Legal measures are sometimes developed in response to significant pollution incidents, as was the case in Japan in the 1970s when a law suit filed against the company that owned the Kamioka mine, responsible for the outbreak of Itai-itai disease described above, resulted in a pollution control agreement being applied to the company. Kaji (2012) describes it as a 'rare example' of successful pollution control in Japan, because the subsequent annual water-quality inspections show a reduction of cadmium concentrations in the river to natural levels. Legal instruments also cover international pollution issues (Box 19.1).

Care must be taken, when establishing clean-up targets, to set appropriate standards. Runnels *et al.* (1992) point out that levels of minerals and pH in rivers can be naturally high in areas of undisturbed mineral deposits due to weathering and leaching of metalliferous ores. Ironically, over the course of time, some areas previously damaged by mineral extraction have come to be regarded as sites of beauty and environmental importance with minimal active rehabilitation. Lundholm and Richardson (2010) give numerous examples of abandoned workings becoming important refuges for threatened species, including the colonization of flooded lignite mines in Germany by rare orchid species and records of many nationally

Box 19.1 Resolving chloride pollution in the River Rhine

An interesting example of how legal instruments have been used in attempts to resolve an international pollution issue resulting from mining can be cited from the River Rhine in northern Europe. Massive salinization of the Rhine by chlorides occurred from the 1930s and mining was a major source of this pollution, particularly potash mines in the Alsace region of France and German coal mines further upstream. The high levels of chloride in the Rhine were a particular issue for the Netherlands, where the pollution caused problems for maintaining drinking-water quality and agricultural production.

This classic example of an upstream–downstream conflict of interests over river water quality resulted in negotiations that began in the early 1930s and culminated in the Convention on the Protection of the Rhine against Chlorides. This was signed in 1976 by Switzerland, the former West Germany, France and the Netherlands. One interesting aspect of this Convention was the agreement by countries involved in the dispute to pay for alternative disposal of the chlorides in proportion to the degree to which they were polluted, a clear contradiction of the 'polluter pays' principle.

However, despite more than 40 years of negotiations and a reversal of the polluter pays principle, the Convention still had virtually no effect until 1985 because the French government refused to ratify it until that year. This was because of protests by French farmers who feared that the alternative, local, means of salt storage and disposal would present them with pollution problems. Meanwhile, a group of Dutch horticulturists supported by a local non-governmental organization, the Clean Water Foundation, brought the chlorides case before the District Court in Rotterdam and applied for compensation from the French potash mines.

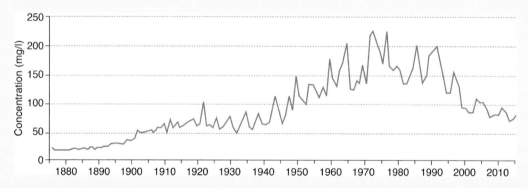

Figure 19.7 Chloride concentrations in the Rhine at Lobith (German–Dutch border), 1875–2015 (after RIWA, 2016).

Legal wrangling over who should pay what continued between France and the Netherlands until 2004 when, coincidentally, the last Alsatian potash mine closed. The remaining stockpiles of chloride have since been gradually discharged into the Rhine. As Figure 19.7 indicates, chloride pollution has been declining in recent decades, but most observers agree that the Convention itself has contributed very little to this trend. The decline of Alsace potash mines and German coal mines has been much more important in falling chloride pollution. The example of chloride pollution in the Rhine shows how difficult it can be to resolve even relatively simple transfrontier environmental issues.

Source: Dieperink (2011).

endangered butterfly species at Czech limestone quarries. In the UK, the country's largest protected wetlands are the Norfolk Broads, an area that is entirely the work of medieval peat cutters (Matless, 2014). Granted National Park status in 1988, the Broads also include two Ramsar sites, 29 Sites of Special Scientific Interest (SSSIs) scheduled under the UK's Wildlife and Countryside Act 1981, and two Special Areas of Conservation (SACs) with special protection under the European Union's Habitats Directive.

In other situations, mining may even be advocated as an option for revitalizing ecosystems damaged by other forms of development. Björk and Digerfeldt (1991) describe how peat extraction in two former wetland areas of Jamaica has allowed their re-establishment. The Black River and Negril wetlands were drained and canalized for the cultivation of rice and cannabis, respectively, but successful revitalization of the wetland habitat after peat cutting on the Negril has led to similar proposals for the Black River.

SUSTAINABLE MINING

The value to society of the products of mining is undeniable, providing as they do the basic building blocks of everyday life for virtually the entire human population. Mining inevitably entails disruption to the natural environment, although given minerals' importance to society and the relatively localized nature of the impacts, the disruption is on a far smaller scale than that of agriculture, for example.

Many positive steps are being taken to limit the environmental impacts of mining. The recycling of metals, for example, means less new damage and often makes economic sense even when the economics only take extraction, processing and distribution into account. Other new technologies for recovering wasted resources, such as the use of microbes to bioleach metals from tailings, have also been developed in response to lower-grade ores. Operational mines, too, are increasingly incorporating environmental rehabilitation into their plans. Although such action often takes a higher priority in countries where legislation demands it, multinationals operating in countries with less-stringent requirements also yield to pressures for more environmentally sustainable and socially responsible practices called for by shareholders and environmental pressure groups. For transnational corporations (TNCs) whose largest stock in trade is their reputation, public feeling over the importance of the environment has undoubtedly influenced practice both directly and indirectly through legislation (Amezaga et al., 2011).

Ultimately, it is for society to decide whether the environmental cost is worth the societal benefits of mining output. Bingham Canyon, which has operated since 1904 and could continue into the 2030s, has long provided about 15 per cent of US copper and made a significant contribution to the economy of the state of Utah. It is for American society to decide whether the environmental cost, in terms of a large hole in the ground and associated disruption to the local environment, is worth this contribution to society. Interestingly, the mine was designated a National Historic Landmark in 1966, indicating its value to the heritage of the USA.

Problems may occur, however, when those who benefit are not fully coincident with those who pay the costs. Generally the greatest costs in environmental and social terms fall on local populations around mine sites, while the benefits, usually measured in economic and political terms, tend to be concentrated at the national and international scales (Figure 19.8). This is an issue of environmental justice. When the costs are particularly high for local communities who see few of the benefits of exploiting minerals in their region, difficulties may arise, as illustrated by the experience of Bougainville (see above). Redressing the imbalances can also act retrospectively: a global precedent was set in 1993 when Australia paid off Nauru to settle the island's claim against the British Phosphate Commission for environmental degradation of two-thirds of its territory.

Figure 19.8 Collahuasi mine in Chile, reputed to be the world's fourth-largest copper mine, which exports most of its production. Situated in a remote part of the Andes, the mine has had little negative impact on a very sparse local population, but elsewhere local people bear the brunt of the costs of mining for benefits that tend to be concentrated at the national and international scales.

The importance of the social dimension to mining, often in remote areas inhabited by marginal communities, has seen a growth in the idea of corporate social responsibility (CSR) among mining TNCs (Yakovleva, 2005). Since the early 2000s, CSR programmes have increasingly been used to try to mitigate conflict with local communities, although critics claim that such programmes serve to legitimize the mining industry rather than to transform it. An example comes from the highlands of Peru where one study concluded that social unrest and resistance remained strong around mining sites in part because the CSR programming of mining TNCs is 'deeply embedded in legitimising this contentious industry' (Gamu and Dauvergne, 2018: 12).

Just how the ideas of sustainability or sustainable development should be interpreted by mine managers, whose primary aim is to run a successful business, remains open to debate. Many agree that a sustainable approach to mining should balance economic, environmental and social considerations, often referred to as the 'triple bottom-line', although Laurence (2011) adds mine safety and efficient use of the mineral resource as equally important factors. If these five issues are addressed appropriately, the life of the mine is optimized, the community benefits are maximized, and the industry itself will have wider community acceptance. In this way, the mining business can make a contribution to poverty reduction and both economic and human development.

Nauru: one type of sustainability

The experience of Nauru offers some interesting further insights into the complicated issue of mining and sustainability. About 100 years of phosphate mining on the island have left 80 per cent of the country in a state of near complete environmental devastation. In other words, most of the national territory is now uninhabitable and effectively useless. Given the exorbitant cost of rehabilitating the land, relocation of the entire population was considered the only option in the 1960s, but the Nauruans refused to leave their homeland (McAdam, 2017).

For much of the island's recent history, its inhabitants have borne a disproportionate burden of the costs of mining, while its benefits have largely been enjoyed elsewhere (on Australian agricultural land), but this imbalance has been redressed considerably in more recent decades. Since independence in 1968, a much greater proportion of mining proceeds has gone to the people of Nauru, and the pay-off in 1993 for previous damage further increased their economic wealth.

Gowdy and McDaniel (1999) point out that Nauru is a clear example of the logical outcome of so-called 'weak' sustainability taken to an extreme. 'Weak' sustainability, as outlined in Chapter 3, implies that development is sustainable as long as its capacity to generate income for future generations is maintained. The citizens of Nauru have lost most of their natural capital. It was transformed into economic capital in the form of the state-owned Nauru Phosphate Royalties Trust, a sovereign wealth fund which at its peak was worth more than US$1 billion. This money, generated by phosphates, gave Nauru a standard of living that was far beyond that on most other Pacific islands, although mismanagement of the fund eventually led to its bankruptcy and it was dissolved in 2014.

The process of colonization transformed Nauruans from a self-sufficient society, living within the constraints of their island's natural resources, into economic beings who are now fully integrated into the world economy. Nauru can no longer support itself with local resources, but imports food and other necessities (even water) from outside. For a period these imports were paid for with income generated by the Nauru Phosphate Royalties Trust, but since its collapse most of the country's economy is financed by foreign aid from Australia, which ironically has instituted a quasi-colonial arrangement. Further, the remaining useful part of the country, a narrow coastal strip, is currently threatened by an unintended consequence of global economic growth: climate change. If predictions of future sea-level rise prove correct, the only part of Nauru left above the water level will be the mined-out central portion of the island.

FURTHER READING

Edwards, D.P., Sloan, S., Weng, L., Dirks, P., Sayer, J. and Laurance, W.F. 2014 Mining and the African environment. *Conservation Letters* 7(3): 302–11. A good overview. Open access: onlinelibrary.wiley.com/doi/10.1111/conl.12076/full.

Himley, M. 2010 Global mining and the uneasy neoliberalization of sustainable development. *Sustainability* 2: 3270–90. Analysis of TNC-led community development at a gold mine in Peru. Open access: www.mdpi.com/2071-1050/2/10/3270.

Lima, A.T., Mitchell, K., O'Connell, D.W., Verhoeven, J. and Van Cappellen, P. 2016 The legacy of surface mining: remediation, restoration, reclamation and rehabilitation. *Environmental Science and Policy* 66: 227–33. An interesting look at terminology and end-goals with examples of peat, coal and oil sands.

Maconachie, R. and Hilson, G. 2012 *Natural resource extraction and development*. London, Routledge. An introduction to key issues in developing countries.

Majer, J.D., Heterick, B., Gohr, T., Hughes, E., Mounsher, L. and Grigg, A. 2013 Is thirty-seven years sufficient for full return of the ant biota following restoration? *Ecological Processes* 2: 19. An interesting case study of rehabilitation at an Australian bauxite mine. Open access: ecologicalprocesses.springeropen.com/articles/10.1186/2192-1709-2-19.

Simate, G.S. and Ndlovu, S. 2014 Acid mine drainage: challenges and opportunities. *Journal of Environmental Chemical Engineering* 2: 1785–803. Reviewing one of the mining industry's more serious environmental problems.

WEBSITES

dev.consejominero.cl   the Consejo Minero (Mining Council) in Chile.

www.chamberofmines.org.za   the South African Chamber of Mines has mining data, educational resources and information on environmental policy.

www.icmm.com   the International Council on Mining and Metals (ICMM) aims to improve sustainable development performance in the mining and metals industry.

www.internationaltin.org   the organization dedicated to supporting the tin industry.

www.miningfacts.org   a site dedicated to the Canadian mining industry's activities and impacts on the environment, communities and economies in Canada and around the world.

www.mpi.org.au   the Mineral Policy Institute, an Australian body with a focus on social and environmental impacts of mining in Australia and the Asia Pacific region.

POINTS FOR DISCUSSION

- Is society eventually going to run out of minerals?
- Is mineral resource wealth an economic blessing or curse?
- Is environmental damage simply an inevitable outcome of society's need for minerals?
- There are many techniques available to rehabilitate and reduce the environmental impact of mining, so why are they not always employed?
- Is 'sustainable mining' an oxymoron?
- Transnational mining companies should not be the target of environmental activists because the resources the companies exploit are for the benefit of everyone. How far do you agree?

20 *Warfare*

Aggression appears to be a fundamental characteristic of human nature, and violence has been used to resolve disputes since prehistoric times. Warfare is no less characteristic of the world today than in the past: since the end of the Second World War, more than 250 wars and violent internal conflicts have raged in more than 80 countries, most of these in the developing world. With the course of history, however, technological developments have greatly enhanced humankind's capacity for destruction and while the devastation to human life and civilization is well appreciated, the ecological impacts of war are less well documented. They are, nonetheless, wide-ranging and not always obvious. In the broad perspective, war and the preparations for war are the antithesis to sustainable development, squandering, as they do, scarce resources and eroding the international confidence necessary to promote development. While competition for the control of resources has long been a reason for conflict, in recent years, resource degradation is increasingly being seen as a cause of war, and one that looks set to become more important with an increasing world population.

COST OF BEING PREPARED

Keeping armies operational and their weaponry updated represents a huge drain on economic, human and natural resources. Globally, military expenditure as a percentage of GDP has been between 2 per cent and 3 per cent since the early 1990s, having reached nearly 4 per cent in 1989. World military expenditure in 2016 exceeded US$1690 billion, accounting for 2.2 per cent of global GDP. The USA remained by far the world's largest military spender in 2016, accounting for 36 per cent of the world total, more than double that of China in second place, and together these two countries accounted for no less than 49 per cent of the global total spend (SIPRI, 2017). The scale of military budgets, some of their contradictions and trade-offs with social and environmental priorities are highlighted by SIPRI (2016) which compared military spending to the costs of achieving the various Sustainable Development Goals (SDGs):

- SDG 4 on education could be comfortably achieved at a cost of well under 10 per cent of one year's global military spending.
- SDGs 1 and 2 (eliminating extreme poverty and hunger) would cost just over 10 per cent of annual global military spending.

Although data on the military's consumption of resources are, like much information on the military, difficult to obtain, an indication of the scale of impact is given in the estimates collated by Renner (1991). The US military consumes 2–3 per cent of total US energy demand, while worldwide nearly 25 per cent of all jet fuel is used for military purposes. An estimate of the military consumption of non-fuel mineral resources is shown in Table 20.1. This estimate suggests that the quantities of aluminium, copper, nickel and platinum used by the military worldwide exceed the entire developing world's demand for these minerals.

Military installations occupy land that cannot be used for productive purposes (Figure 20.1) and an estimated 6 per cent of the world's land surface is used for military training and weapons testing (Lindenmayer et al., 2016), land that is for the most part closed to civilian access. In the Netherlands, 21 per cent of the largest area of continuous Dutch forest (the Veluwe) is in the possession of the military, as is a third of all Dutch heathland (Vertegaal, 1989). In some cases, such land is considered to have high conservation value. The Ministry of Defence in the UK is the custodian of more than 250 Sites of Special Scientific Interest (SSSIs), three of which form part of the Salisbury Plain training area, which contains the largest known expanse of unimproved chalk grassland in north-west Europe. This use of Salisbury Plain, some parts of which have been managed by the military since the late nineteenth century, has protected the grassland from the changes that have been wrought on most other lowland grasslands in Britain – urban development, agricultural improvement and industrial expansion.

However, military training on Salisbury Plain has its impacts too, of course. In a 50-year survey Hirst et al. (2000) found that in the 1990s habitat disturbance was occurring at a greater rate than natural regeneration following increased use of Salisbury Plain after the end of the Cold War, when many training facilities in Europe were closed. The higher levels of disturbance from heavy vehicles represent a significant threat

Table 20.1 Estimated military consumption of selected non-fuel minerals as a proportion of total global consumption in the early 1980s

Mineral	Share (%)
Copper	11.1
Lead	8.1
Aluminium	6.3
Nickel	6.3
Silver	6.0
Zinc	6.0
Fluorspar	6.0
Platinum group	5.7
Iron ore	5.1

Source: after Kim (1984).

Figure 20.1 Hundreds of thousands of gun emplacements like these were built in strategic locations throughout Albania during the second half of the twentieth century to defend the country against invasion.

to the chalk grassland through habitat loss and fragmentation. Elsewhere, military training has probably contributed to high biodiversity by creating disturbance heterogeneity, and the departure of the Soviet Army from Eastern Europe and closure of US Army training bases in Germany may have adversely affected biodiversity in consequence (Warren *et al.*, 2007).

Military activities also generate large quantities of waste material, much of it hazardous. In the USA, the military generates half a billion tonnes of toxic waste per year, more than the top five chemical companies combined (Donohoe, 2003), and more than 1500 military bases with consequent environmental problems have been identified. The largest of such contaminated sites, the Rocky Mountain arsenal in Colorado, was used to dump 125 different dangerous chemicals over a 25-year period. Disposal of such material is no easy matter since its precise effects on the environment are not always well known. The USA is now disposing of many chemical weapons by incineration, but historically some of the methods used for their disposal have been less circumspect. Glasby (1997) notes that large quantities of chemical warfare agents were dumped into the Baltic Sea after the Second World War. These highly toxic chemicals were contained in wooden crates, which have probably disintegrated, allowing the widespread dispersal of chemicals. The long-term effects on the marine environment are not known.

Some of the worst environmental degradation has been associated with the development of nuclear weapons. Testing of over 500 nuclear weapons in the atmosphere has caused the dispersion of radioactive material across the globe, adding a small increment to the background exposure of the world's population. A partial ban on atmospheric testing was signed in 1963, but it was not until 1980 that the last atmospheric test was carried out in north-west China. Underground testing of nuclear weapons, which often causes small to moderate earthquakes, still continues. The former Soviet nuclear test ground near Semipalatinsk in eastern Kazakhstan, which was closed in 2000 after nearly 40 years of testing, covers about 2000 km² of contaminated land. The location of the first nuclear explosion, carried out in July 1945 at Trinity Site in New Mexico, USA, is now considered safe enough to be open to visitors. However, visiting parties are strictly controlled, and children and pregnant women are actively discouraged. Part of the site will remain moderately radioactive for the foreseeable future (Szasz, 1995).

Testing of nuclear weapons was carried out by the US and French governments in the Pacific Ocean. Between 1946 and 1958, the USA conducted 23 nuclear tests at sites in the northern atolls of the Marshall Islands (Figure 20.2). Among the environmental impacts described immediately post-test was a surface sea-water temperature of 55,000°C, the destruction of coral reefs and three entire small islands, as well as extreme waves, increased sediment loading and exposure to high levels of radiation. A survey of coral biodiversity at Bikini Atoll five decades after the end of nuclear testing concluded that although many species have proved to be remarkably resilient to such colossal anthropogenic disturbance, 28 coral species recorded in the atoll prior to the nuclear explosions have been lost (Richards *et al.*, 2008). Bikini islanders and their descendants have lived in exile since they were moved for the first weapons tests in 1946.

Serious contamination from industrial plants that produce weapons has also come to light in recent times. By far the most serious accident involving radioactive wastes in the former Soviet Union, prior to Chernobyl, was an incident that occurred in 1957 at a secret nuclear weapons processing plant in the southern Urals known as the Mayak Production Association, located 80 km north-west of Chelyabinsk. The fact that the accident was not officially admitted to until 1989, when the then Soviet government stated that 2 million curies of radioactive elements had been released from an exploded waste tank and deposited over an area of 90,000 ha, gives some indication of the secrecy surrounding such military operations. More than 12,000 people were evacuated from the area, and parts of the region were still restricted more than 25 years later (Batorshin and Mokrov, 2018).

Figure 20.2 A nuclear weapon test by the US military at Bikini Atoll, Marshall Islands, in 1946.

Long-lasting environmental effects are also attributable to areas where biological weapons have been tested. A classic case is the island of Gruinard off the west coast of Scotland, site of experiments with highly contagious anthrax spores during the Second World War. The island remained uninhabited by government decree until 1988, but even now complete decontamination is difficult to guarantee (Szasz, 1995).

Anthrax was also one of numerous biological agents tested in the open air on Vozrozhdeniye Island, situated towards the western coast of the shrinking Aral Sea, site of a major testing ground for biological weapons during the Soviet era. Others include plague, Q fever, Venezuelan equine encephalitis and smallpox, many being special strains developed by the military to be resistant to conventional forms of treatment (Bozheyeva *et al.*, 1999).

Like Gruinard, Vozrozhdeniye was partly chosen for biological weapons testing because of its geographical isolation. Its insular location prevented the transmission of pathogenic micro-organisms to neighbouring mainland areas by animals or insects. It was also easy to protect: fast patrol boats guarded Vozrozhdeniye against intruders over nearly 40 years of testing. But the island's Soviet base was abandoned (Figure 20.3) in 1992 after the break-up of the USSR, and Vozrozhdeniye became connected to the mainland in 2001 as the Aral Sea continued to shrink. Connection to the mainland eliminated the remaining natural security benefits of the island, hence potentially infected animals living on Vozrozhdeniye now have easy access to the mainland, and vice versa, with unpredictable consequences for the local biosphere. The desiccation of the Aral Sea is well known as the world's worst human-induced environmental disaster (see p. 179), but the eco-tragedy also has another dimension.

In contrast to most of the above examples, positive aspects of incipient military conflict can also arise, as in the case of demilitarized areas maintained as buffer zones between opposing forces. The demilitarized zone separating North and South Korea, for example, has provided sanctuary to endangered and threatened animals and plants for more than 60 years. The 4-kilometre-wide, 250-kilometre-long corridor

Figure 20.3 The author at the defunct Soviet biological weapons testing base, now in Uzbekistan, on Vozrozhdeniye, a former island in the Aral Sea.

extends across the Korean peninsula, traversing a major river delta and old farmlands in the west and rugged mountains in the east. It has remained virtually untouched by human activity since its establishment in 1953, and thus represents a unique, rigidly protected nature reserve (Kim, 1997). The zone is home to 3514 species, equating to 67 per cent of the species diversity of the Korean Peninsula, most of which are endemic to this small plot of land (Healy, 2007). In a similar vein, legal wranglings over the huge clean-up costs at the Rocky Mountain arsenal in Colorado, USA (see above), abated when bald eagles began to nest on the contaminated site, and the area has now been designated a nature reserve (Havlick et al., 2014).

DIRECT WARTIME IMPACTS

The relationship between people and their environment can be changed significantly during wartime. Priorities are altered, and certain resources are used more rapidly than during peacetime in order to fuel the war effort. In the time of Henry VIII, for example, many of England's oak trees were cut down to build warships. In addition to the destruction of agricultural land and woodland during the prolonged trench warfare on the coastal plains of France and Belgium during the First World War, wide areas of forest were felled beyond the battle zones to supply the war effort. In total, Belgium lost most of its remaining forests and France lost about 10 per cent of its forested area at this time (Graves, 1918).

Deliberate destruction and manipulation of the environment have also been used by armies to gain military advantage for centuries. The so-called scorched earth policy, in which vegetation and crops are deliberately destroyed to prevent their use by the enemy, is an age-old military tactic: in biblical times, for example, Samson tied burning torches to the tails of 300 foxes and drove them into the olive groves and grain fields of the Philistines. Similar tactics were used during the war sparked by the Soviet occupation of Afghanistan from 1979 to 1991. Vineyards, orchards, ornamental trees and shrubs were routinely felled for security reasons by the Soviet army and troops loyal to the former communist government of Afghanistan, and vegetation was widely cleared along highways in the provinces of Kabul and Parwan. The destruction of forests, both deliberate and inadvertent (as a result of wildfires set off by bombing), is thought to have reduced the forest cover from 3.4 per cent of the country to 2.6 per cent in just ten years (Formoli, 1995). The Turkish army has used similar tactics more recently, against uprisings in Turkish Kurdistan, burning forests, agricultural fields and villages (de Vos *et al.*, 2008). The wide scope for environmental modification for military purposes in the modern era is indicated in Table 20.2, a list submitted during the negotiations for the 1977 Environmental Modification Convention (see below), although it should be noted that this list is by no means exhaustive in that it does not include the effects of the numerous biological and chemical weapons that have been developed.

Table 20.2 Military techniques for modifying the environment

Technique	Military application	Feasibility
ATMOSPHERE		
Fog/cloud dispersion	Make target areas more visible	Relatively easy for supersaturated fog, difficult for warm fog
Fog/cloud generation	Protect target areas from attack and nuclear flash	Depends on availability of equipment and materials
Hailstone generation	Damage thin-skinned equipment, power/communication wires and antennas	Restricted to hail cloud conditions
Introduction of electric fields	Interrupt communications and remote sensing	Very high energies needed
Generating and directing destructive storms	Damage battlefields, ports, airfields	High energies needed but some success in hurricane dispersal
Rain and snow making	Inhibit mobility, block routes, hinder communications	Very dependent on cloud systems, highly localized, short duration
Change of high atmosphere or ionosphere	Strategic effect on food production and possibly on human survival	Uncertain
OCEANS		
Change of physical, chemical and electrical parameters	Affect acoustic paths, possible effect on potential food supplies	Uncertain, but large energies may be necessary

(continued)

Table 20.2 continued

Technique	Military application	Feasibility
Tsunami generation	Destroy low-lying areas, surface fleets	Uncertain and difficult
LAND		
Burning of vegetation	Generally destructive	Easily started but dependent on flammability of vegetation, and weather conditions
Earthquake/tsunami generation	Damage battlefields, major military bases and strategic facilities	Specific to certain locations and mechanism not fully understood
Generation of avalanches and landslides	Disrupt communications	Only in mountainous areas or in regions of unstable soils
Modification of permafrost areas	Destroy roads, rail foundations, alteration of stream sources	Relatively easy by removal of insulating cover but effect limited to melting season
River diversions	Flooding, river navigation hazards	Large engineering effort necessary

Source: after Goldblat (1975).

A number of these modifications have been employed in conflicts. During the Vietnam War, for example, the US army diverted water from the Mekong delta to drain the Plain of Reeds where an enemy base was located, and a classified rain-making programme was conducted from 1967 to 1972, seeding clouds from the air using silver and lead iodide to increase normal monsoon rainfall over the Ho Chi Minh trail, a vital supply line for North Vietnamese forces (Jasani, 1975). The Vietnam War was also the scene of the most extensive systematic destruction of vegetation ever undertaken in warfare. About 100 million litres of herbicides (including the infamous 'Agent Orange', nicknamed according to the coloured identification band painted on the storage barrels) were sprayed over thousands of square kilometres of the country between 1961 and 1971 in an effort to deprive North Vietnamese forces of the cover provided by dense jungle (Stellman *et al.*, 2003). Most spraying was targeted on forest but cropland was also sprayed as a tactic for decreasing enemy food supplies, and although missions were only conducted during light winds to avoid non-target contamination, drifting sprays actually caused damage over much wider areas (Orians and Pfeiffer, 1970).

About half of Vietnam's mangrove forests were completely destroyed by aerial spraying, an area that has been slow to recover. Revegetation of the sprayed mangroves was essentially complete by the mid-1980s, but not always with the same commercially preferred species. Dioxins in the defoliants used were also toxic to other forms of life. The effects on animals are not well documented but a few studies and much anecdotal evidence suggest decreased abundance of many species, and herbicide-related illness in domestic livestock (Westing, 1984). It is thought that at least 3000 hamlets were sprayed directly with defoliant, affecting between 2 million and 4 million people (Stellman *et al.*, 2003). There is no doubt that the dioxins linked to Agent Orange and other defoliants entered the food chain in Vietnam. Elevated dioxin levels are still clearly detectable in soils, livestock, fish and people decades after Agent Orange was last used in the country (Schecter *et al.*, 2003).

The military experience of defoliant chemical use in Vietnam was subsequently repeated on a smaller scale in El Salvador during the 1980s, where government forces, with US backing, laid waste areas held by or sympathetic to anti-government guerrillas. The term 'ecocide' has been coined for such widespread destruction (Weinberg, 1991). In more recent times, large-scale use of defoliant chemicals has been employed in some countries in the so-called War on Drugs, attempting to destroy coca plantations in particular. In a study of the campaign conducted in Colombia in the early 2000s, Rincón-Ruiz and Kallis (2013) highlight the unintended social and environmental effects of this policy. A critical impact has been to push deforestation into new areas as illegal groups have responded to spraying by quickly relocating, clearing new areas of forest and restarting coca production. These authors view the problem as one of social justice because the local people living in these new areas of deforestation are typically forced out of their homelands or compelled to cultivate coca poppies.

Another military method involving large-scale devastation to the natural environment, sometimes with that very aim in mind, is the destruction of various forms of infrastructure. Dams have also been common targets in this respect. Allied bombing of the Möhne, Eder and Sorpe dams in the Ruhr valley of Germany during the Second World War, for example, aimed to cripple the German economic heartland. Water released from the damaged Möhne and Eder dams killed 1300 people, left 120,000 homeless and ruined 3000 ha of arable land as well as numerous factories and power plants. But the most devastating example of deliberate dam breaching occurred during the Second Sino-Japanese War of 1937–45 when the Chinese dynamited the Huayuankow dike on the Huang Ho River near Chengchow to stop the advance of Japanese forces. The mission was successful in this respect and several thousand Japanese troops were drowned, but the flow of water also ravaged major areas downstream: several million hectares of farmland were inundated, as well as 11 cities and more than 4000 villages. At least several hundred thousand Chinese drowned and several million were left homeless (Bergström, 1990).

Oil fields have also been the target of deliberate damage. Retreating Austrian forces systematically destroyed facilities on Romanian oil fields in 1916–17, as did Iraqi forces before being driven out of Kuwait in 1991. In the latter example, an estimated 7–8 million barrels of oil were discharged, more than twice the size of the world's previously largest spillage caused by a blow-out in the Ixtoc-1 well in the Bay of Campeche off the Mexican coast in 1979. The resulting oil slick contaminated large areas of Kuwait and a 460-kilometre stretch of the northern coast of Saudi Arabia. An estimated 30,000 seabirds died in the immediate aftermath of the spills (Readman *et al.*, 1992), but the long-term effects on the Gulf's ecology are remarkably few. Just three and a half years later, there were no visible signs of immediate or delayed effects on coral reefs in Saudi waters (Vogt, 1995) and rapid recovery has been recorded in most other parts of the marine ecosystem. This finding is consistent with observations made during the Al-Nowruz oil-well spill during the Iran–Iraq War, which suggests that the Gulf's ecology has high resilience to oil pollution (Saenger, 1994).

Fires started in 613 Kuwaiti oil wells caused the burning of 4–8 million barrels a day, and resulted in massive clouds of smoke and gaseous emissions, although fears of effects on the climate were unfounded. Other disturbances to the desert surfaces of Kuwait were caused by the large-scale movements of military vehicles and the building of an earth embankment along the border with Saudi Arabia (Figure 20.4). Experience of similar surface destabilization during the desert campaigns of North Africa in 1941 and 1942 saw wind erosion dramatically increase during the height of the fighting, but deflation subside back to pre-war levels within a few years of the end of the campaign (Oliver, 1945). Large military vehicles also have more long-lasting effects in desert areas, compacting soil, crushing and shearing vegetation, and altering the

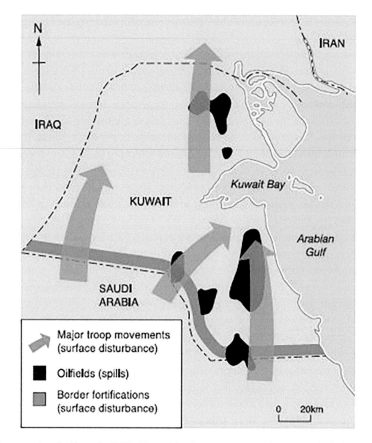

Figure 20.4 Military action in Kuwait 1990–91 and its impacts on surface terrain (after Middleton, 1991).

vertical and horizontal structure of plant communities. Compositional changes to soil and vegetation from a single tank pass during military manoeuvres in the Mojave Desert were found still to be evident 40 years after training had ceased (Prose, 1985).

Numerous impacts on wildlife have also been documented. Some mammal species have been brought close to and beyond extinction due to the direct and indirect effects of war. The last semi-wild Père David's deer was killed by foreign troops during the Boxer Rebellion in China in 1898–1900, although the species still survives in captivity. Similarly, the European bison was pushed close to extinction during the First World War by hunting to supply troops with food; it recovered in numbers during the interwar years but its numbers were again decimated in the Second World War. A proliferation of guns left over from two decades of conflict in Cambodia and a consequent rise in hunting has had a long-lasting impact on wildlife in Cambodia (Box 20.1).

Many Pacific island endemic bird taxa were also lost or pushed towards extinction by warfare and its associated disturbances during the Second World War. On the twin islands of Midway, for example, extinctions of the Laysan finch and the Laysan rail occurred (Fisher and Baldwin, 1946). In Central Africa, warfare has several times in recent years threatened the survival of the 600 mountain gorillas in the Virunga

Box 20.1 Cambodia's declining wildlife: a legacy of armed conflict

Cambodia experienced a protracted recent period of armed conflict, which was particularly intense in the 1970s but persisted into the late 1990s. Declines in the country's abundance of wildlife and its species richness have occurred since 1953, when Cambodia became independent from France, with the sharpest declines coinciding with the worst of the violence in the 1970s. This was a time of profound changes to rural Cambodian lifestyles in which wildlife hunting for income emerged as an important secondary activity after subsistence rice farming, a combination of livelihood strategies that has persisted to the present day. This decline in wildlife was driven by three synergistic social processes associated with armed conflict particularly in the 1970s: the proliferation of guns; the emergence of a trade in wildlife products; and government policies that mandated commercial hunting by local villagers.

Wildlife in Cambodia continues to be threatened by commercial hunting in the post-conflict era. Products for sale internationally include antlers, horns, dried meat and bone (particularly from tigers). Since the end of the conflict, Cambodia has adopted several national policies that should directly or indirectly help to conserve wildlife, including establishment of a protected area system, a national gun confiscation programme, and development of a national biodiversity strategy. However, these post-conflict conservation efforts have met with limited local success. The legacy of war has had additional impacts in this respect: the loss of educated people to run effective administrations and the decline of functional systems of governance.

Source: Loucks *et al.* (2009).

volcanoes, which straddle the borders of Rwanda, Zaire and Uganda. Needless to say, conservation efforts are usually sidelined during times of war, but Hanson *et al.* (2009) have discovered an alarming geographical overlap between armed conflicts and biodiversity hotspots (see p. 384). Their analysis found that over 90 per cent of the major armed conflicts between 1950 and 2000 took place within countries containing biodiversity hotspots, and more than 80 per cent occurred actually within the hotspots, many suffering repeated episodes of violence.

Conversely, however, some species of wild animals may benefit during wartime because of decreased exploitation. Westing (1980) noted the cases of various North Atlantic fisheries during the Second World War and a number of fur-bearing mammals in northern and north-eastern Europe, such as the polar bear and red fox. The Second World War effectively created a large marine protected area in the North Sea where commercial fishing was prevented for six years because trawlers avoided areas threatened by naval or aerial strikes and because of the danger of mines. The reduction in fishing effort allowed commercial fish populations to flourish and populations further proliferated because of changes to the age-structure that resulted in a larger proportion of mature and larger fish (Beare *et al.*, 2010). Large adaptable predators may also benefit in times of warfare. During the Vietnam War, tigers increased in abundance in some areas thanks to the ready availability of food provided by corpses in much the same way that the wolf population in Poland increased during the Second World War (Orians and Pfeiffer, 1970).

Disruption and destruction do not necessarily cease with the end of hostilities. The legacy of conflict can be serious for many aspects of the environment, including wildlife, as Box 20.1 illustrates. Other threats

Table 20.3 Countries worst affected by landmines

Countries with at least 100 km² contaminated	Countries with 20–99 km² contaminated
Afghanistan	Colombia
Angola	Eritrea
Azerbaijan	Ethiopia
Bosnia and Herzegovina	Lebanon
Cambodia	South Sudan
Chad	Sri Lanka
Croatia	Zimbabwe
Iraq	
Thailand	
Turkey	

Source: after ICBL-CMC (2017).

to wildlife, humans and their livestock typically remain from unexploded weaponry because, on average, 10 per cent of all munitions used in any war fail to explode. This very dangerous inheritance inevitably impedes post-war reconstruction and rehabilitation efforts in agriculture, forestry, fishing, mining and other related activities. Landmines are a particular problem; they are present on a huge scale in some countries (Table 20.3) and can take decades to clear. Berhe (2007) points out that there is roughly one mine in the ground for each citizen of Angola, Bosnia and Cambodia and that landmines pose a lose–lose situation because they present a form of land degradation whether left in the ground or detonated.

Nuclear war

There is little direct information on the possible environmental effects of a nuclear war, but some insight can be gained from the detonation of the only two nuclear devices used in warfare to date, in Japan at the end of the Second World War (Table 20.4), although these were relatively small devices by today's standards. The destructive energy released by detonation of a nuclear warhead is divided into three categories: thermal, kinetic (or blast), and radioactive energies; and the most important cause of death and physical destruction stems from the combined effects of blast and thermal energy. The fireball created by the blasts in Japan was intense enough to vaporize humans at the epicentre and burn human skin up to 4 km away. Ionizing radiation comprised about 15 per cent of the explosive yield of the bombs, and among its effects was radiation sickness among many survivors of the blast.

There is a limited amount of information on the ecological effects of nuclear explosions, which has been gleaned from observations made after above-ground tests. Severe damage to vegetation due to the thermal and kinetic impacts caused by detonations decreases more or less geometrically with distance from the blast epicentre. After explosions, plants re-invade damaged areas at a rate of succession that would be

Table 20.4 Destruction caused by the first atom bombs dropped in wartime

	Hiroshima	Nagasaki
Date of detonation	6 August 1945	9 August 1945
Type	Uranium 235	Plutonium
Height of explosion (m)	580	503
Yield (kiloton TNT)	12.5	22
Total area demolished (km²)	13	6.7
Proportion of buildings completely destroyed (%)	67.9	25.3
Proportion of buildings partially destroyed (%)	24.0	10.8
Number of people killed (by 31 December 1945)	90,000–120,000	70,000

Source: Ohkita (1984).

expected after a severe disturbance, as Shields and Wells (1962) documented for a Nevada Desert test site after numerous above-ground detonations (see also the impacts on Bikini Atoll above).

Considerable research has been undertaken into the possible global effects of a major nuclear exchange, much of it focusing on the potential climatic effects. Although results vary with different scenarios, simulations using general circulation models (GCMs) have raised the spectre of a 'nuclear winter' in which the aftermath of a nuclear war would be characterized by darkened skies over large parts of the Earth due to smoke and dust injected into the atmosphere. Such effects could continue for weeks or months, blocking sunlight, reducing temperatures and affecting the global distribution of rainfall. Such climatic effects could have wide-ranging impacts on agriculture and major biomes, with serious implications for food production and distribution far beyond the areas of immediate devastation. It seems that such catastrophic impacts could result from as few as 100 detonations on urban areas (Toon *et al.*, 2007).

INDIRECT WARTIME IMPACTS

The dislocation of economies and societies caused by war can result in many perturbations to the way people interact with their environment. Environmental management, for instance, is often neglected as priorities change. During the 15-year civil war in Lebanon, the country's agricultural resource base was severely damaged because much of the extensive network of stone-wall terraces in upland areas was not properly maintained. Accelerated erosion and landslides have resulted in many such areas, and the cost of repairing the terraces probably far exceeds the budget of the Ministry of Agriculture (Zurayk, 1994). Significant changes in land use also occurred over 26 years of conflict in the Jaffna Peninsula in Sri Lanka during the country's long-running civil war that ended in 2009. A considerable decrease in agricultural land, and concomitant increase in non-agricultural land uses, were caused by a series of drivers (Suthakar and Bui, 2008), including economic blockades which translated into a general lack of fuel for irrigation, ploughing and harvesting as well as the transportation of crops, and the imposition by government troops of high-security zones in 3000 ha of high-potential agricultural lands.

A similar impact on irrigated agriculture has been documented in Iraq by Gibson *et al.* (2017) who assessed the mean cultivated area along the Euphrates River before, during and after the Iraq War that raged for nearly a decade in the early 2000s. Their results, shown in Table 20.5, revealed that the pre-war area of 991 km² had been reduced by almost a half to 538 km² in the late Iraq War period (2007–11) due to the insecurity and political turmoil but also due to the impact of a severe drought from 2007 to 2009 which resulted in abnormally low river levels. In the post-war period assessed (2012–15) the irrigated area had returned to 692 km², which was still not as extensive as during the pre-war or early-war periods.

Such interruptions to conventional agricultural practices, along with forced mass migrations, disruption to transport systems and many other aspects of the economy, can also leave people less able to cope with hazards of the natural environment, increasing their vulnerability to such disasters as famine. On occasion, deliberate destruction of the environment and infrastructure may lead to famine. A recent example is the mass starvation that swept southern Somalia in 1992–93 following a six-month period in which grain stores were plundered, and water pumps and other agricultural infrastructure and machinery were destroyed in the Bay region by forces loyal to former president Siad Barre (Drysdale, 2002). Since then, Somalia has been characterized by sporadic yet persistent violent conflict for more than two decades. Another famine in the country in 2011–12 coincided with widespread drought in East Africa, but conflict remained an important driver of food shortages. Many in Somalia's large population of mobile pastoralists were unable to follow traditional or alternative migratory routes to escape drought due to armed conflict, and insecurity in many parts of the country meant that the delivery of humanitarian assistance was severely restricted (Harris *et al.*, 2013). Other countries linked by conflict, food-insecure populations and climate-related shocks are detailed in Box 13.1 (see p. 328).

Numerous indirect environmental impacts also occur due to the mass movements of refugees away from areas of conflict. During medieval Europe's Hundred Years War, for example, whole regions were depopulated, leaving abandoned farmland to be colonized by brush and woodland. Refugees tend to concentrate in safer zones, in cities, towns or refugee camps in rural areas, exerting intense pressures on local resources. Deforestation for fuel and to build shelters, resultant soil erosion and local pollution, particularly of water sources with associated threats to human health, are impacts commonly associated with large concentrations of refugees. In the Quetta Valley of western Pakistan, for example, the population has grown from 260,000 in 1975 to 1.2 million in 2010, mainly due to the immigration of refugees from war-torn neighbouring Afghanistan. The impact on local topography has been assessed by Khan *et al.* (2013) who attributed ground subsidence rates of up to 100 mm/year to excessive groundwater extraction needed to support the growing population.

Table 20.5 Estimated cultivated area along the Euphrates River in Iraq, 2000–15

Period	Irrigated area (km²)
Pre-Iraq War period (2000–03)	991
Early Iraq War period (2003–06)	835
Late Iraq War period (2007–11)	538
Post-Iraq War period (2012–15)	692

Source: after Gibson *et al.* (2017).

Many examples can also be quoted from sub-Saharan Africa, scene of numerous conflicts in recent decades. In the Darfur region of Sudan, civil war had driven nearly 2 million internally displaced persons (IDPs) to live in rural camps and informal settlements by 2007. UNEP (2007) reported that Darfur refugee camps were associated with deforestation, shortages of fuelwood, overharvesting, food shortages, and falling groundwater levels owing to a proliferation of boreholes in and around camps (Figure 20.5). In north-western Tanzania, the camp of 300,000 refugees from Burundi at Benaco absorbed 410,000 Rwandans fleeing civil war in a month in mid-1994. Suggestions from aid workers to ease the pressure on local forests included the trucking of refugees outside the 4–5-kilometre radius round the camp, which was rapidly denuded in the quest for fuelwood, the marking of trees that would take a long time to regenerate, and the distribution of types of food aid that required less cooking, such as maize powder instead of maize grain (UNHCR, 1994).

The scale of the refugee problem imposed on relatively safe neighbouring countries can reach high levels. Malawi in southern Africa, already one of the region's most densely populated countries, accommodated a million Mozambican refugees during the late 1980s and early 1990s, in addition to its own population of 10 million. Within Mozambique, where conflict was more or less continuous between 1964 and 1992, no fewer than 4.5 million people were estimated to have been displaced within the country by the time a peace accord was signed between the government and the opposition forces of Renamo. Natural resources around

Figure 20.5 Collecting water in a camp for internally displaced people on the outskirts of El Geneina, West Darfur, Sudan.

Figure 20.6 Collecting fuelwood near the village of Maúa in northern Mozambique. The civil war, which lasted from the mid-1970s to 1992, meant a great concentration of dislocated people around relatively safe settlements, dramatically increasing pressure on local resources.

safer villages, towns and cities suffered for years from the increased population pressures (Figure 20.6). Displaced people living in the capital city of Maputo had to travel up to 70 km into the dangerous surrounding countryside in the search for fuelwood, and little forest was left in the early 1990s in the Beira corridor area, which was protected by Zimbabwean troops during the war. On the coast, mangrove forests were severely exploited, presenting a serious threat to prawn and shrimp fisheries, one of Mozambique's major exports, which reproduce in the mangroves.

Cultivated land was not allowed adequate fallow periods, which traditionally last for up to seven years, due to high population densities. Additionally, the plots allocated to displaced people were often too small for a family, leading to further pressure and overutilization (Dejene and Olivares, 1991). The war's impact on wildlife in Mozambique was also severe. At independence in 1975, there were about 3000 elephants in the Gorongosa National Park, a Renamo stronghold, but just over 100 elephants were recorded during an aerial survey in 1994. All the evidence points to Renamo having been heavily involved in the ivory trade, selling tusks to South Africa in return for arms. Hunting by armed forces during the conflict also resulted in sharp declines in other mammal species in Gorongosa and surrounding areas and in the Zambezi Delta (Hatton *et al.*, 2001). In the long term, it seems likely that this decimation of Gorongosa's large mammals – most of them herbivores – has been instrumental in the observed increase in tree cover in the park (Herrero *et al.*, 2017).

LIMITING THE EFFECTS

Numerous treaties, conventions and agreements have been adopted to prevent utter human and environmental devastation in times of war, although they do not specifically cover nuclear weapons, and the effectiveness of such agreements as a deterrent and during the course of war is difficult to evaluate and enforce. Some of the most important relevant conventions are shown in Table 20.6. Other parts of multilateral agreements also have a bearing on the environmental impacts of war, such as the two additional 1977 protocols to the Geneva Conventions of 1949, which bind belligerents to respect and protect the natural environment 'against widespread, long-term and severe damage', and prohibit the destruction of works and installations that harbour dangerous forces, including dams, dikes and nuclear generating stations. Such treaties are only binding on those states that are a party to them, however, and the protocol that bans widespread, long-term and severe environmental damage has not been ratified by major powers such as the USA, France or the UK since these nations are concerned that it could be interpreted as outlawing nuclear weapons. Some treaties meant to exclude any military activities from certain areas have also been adopted. These include the Spitzbergen Treaty (1920), the Åland Convention (1921) and the Antarctic Treaty (1959).

Table 20.6 Major multilateral arms control agreements with a bearing on the environmental effects of warfare, 1970–2017

Agreement	Opened for signature	Entry into force
Treaty prohibiting emplacement of nuclear weapons and other weapons of mass destruction on the sea bed	1971	1972
Bacterial and Toxin Weapons Convention prohibiting development, production and stockpiling of biological and toxin weapons, and/or their destruction	1972	1975
Environmental Modification Convention prohibiting military or other hostile use of environmental modification techniques	1977	1978
Inhumane Weapons Convention prohibiting or restricting use of certain conventional weapons deemed to be excessively injurious or to have indiscriminate effects	1981	1983
Treaty of Rarotonga (South Pacific nuclear-free treaty) 1985		1986
Chemical Weapons Convention prohibiting development, production, stockpiling and use of chemical weapons	1993	1997
Convention on the prohibition of the use, stockpiling, production and transfer of anti-personnel mines and on their destruction	1997	1999
Comprehensive Nuclear-Test-Ban Treaty	1996	Not yet in force
Treaty on the Prohibition of Nuclear Weapons	2017	Not yet in force

Source: UNODA Treaties Database, accessed March 2018.

The widespread ecological effects of the Gulf War prompted renewed calls from conservation agencies for a new Geneva Convention to limit the environmental effects of military conflicts, and the idea of a Green Cross – a new, impartial, international organization, similar to the Red Cross and the Red Crescent, to assess environmental damage resulting from war – has also been suggested.

Although idealists might suggest that effective protection for the environment can only be achieved by global demilitarization, this prospect seems unlikely. A small number of sovereign nations (e.g. Costa Rica, Japan and Panama) are non-militarized in a formal sense by virtue of their national constitutions, although in fact they support more or less potent armed forces. Indeed, the role played by environmental resources in causing conflict can be interpreted as a reason for increasing military might in a world of rising population.

ENVIRONMENTAL CAUSES OF CONFLICT

The environment has often been a cause of political tension and military conflict. Countries have fought for control over raw materials, energy supplies, land, river basins, sea passages and other environmental resources (Figure 20.7). In the Second World War, for example, Japan sought control over oil, minerals

Figure 20.7 This defunct tank near the border between Ethiopia and Eritrea is a relic of Ethiopia's civil war, which ended in 1991. Eritrea's independence in 1993, an outcome of this conflict, meant that Ethiopia became landlocked. This loss of access to the sea was a significant factor in the war between Ethiopia and Eritrea in 1998–2000.

and other resources in China and South-East Asia; rich phosphate deposits have been a root cause of the long-running conflict in the western Sahara; and Israel's lengthy occupation of a large portion of the west bank of the River Jordan has been both for strategic military purposes and to secure access to large groundwater reserves. Similarly, if the Gulf region had not been such a key supplier of crude oil supplies to the west, it seems unlikely that western troops would have occupied Iraq during the Iraq War from 2003 to 2011. Indeed, Downey *et al.* (2010) present evidence to suggest that the base of natural resources – particularly minerals – on which many industrial societies depend is largely constructed through the use and threatened use of armed violence. Their analysis shows that when armed violence is used to protect extraction activities at mines, it is often employed in response to popular protest or rebellion against these activities (see also Table 19.7).

Some observers suggest that environmental change, particularly in the form of degradation and dwindling renewable resources, is set to become an increasingly potent cause of conflict as world population is likely to exceed 9 billion within the next 50 years and global economic output may increase fivefold (Homer-Dixon *et al.*, 1993). The degradation and loss of productive land, fisheries, forests and species diversity, the depletion and scarcity of water resources, and pressures brought about by stratospheric ozone loss and potentially significant climatic change may precipitate civil or international strife. In decades to come, five general types of violent conflict produced by environmental scarcity are envisaged by Homer-Dixon (1999) whose list of potential flashpoints expands in spatial scale from local to possibly global:

- disputes arising directly from local environmental degradation (e.g. due to factory emissions, logging or dam construction)
- ethnic clashes arising from population migration and deepened social cleavages caused by environmental scarcity
- civil strife (e.g. insurgency, banditry and coups d'état) caused by environmental scarcity that affects economic productivity and hence people's livelihoods, the behaviour of elite groups, and the ability of states to meet these changing demands
- inter-state war induced by environmental scarcity (e.g. water)
- North–South conflicts (developed vs. developing worlds) over migration of, adaptation to, and compensation for global environmental problems such as threats to biodiversity, global warming, ozone depletion and decreases in fish stocks.

Such root causes of conflict have been identified from past eras. Pennell (1994), for example, has suggested that environmental degradation and economic isolation, and consequent impoverishment, made piracy an attractive option for inhabitants of the Guelaya Peninsula in north-western Morocco in the mid-nineteenth century. A key aspect of such scenarios is the role played by certain groups that are faced with dwindling finite resources. Such groups may be within a certain country or they may be identifiable on the global scale as countries with unequal access to resources. A proposed chain of factors in this equation, which leads to violence, is suggested in Figure 20.8.

Some researchers have highlighted examples of such events in more recent history. In Central America, Durham (1979) considers that the root causes of the 1969 'Soccer War' between El Salvador and Honduras ran much deeper than the national rivalry over the outcome of a football match that ignited the conflict. He traces the origins of a war in which several thousand people were killed in a few days to changes in agriculture and land distribution in El Salvador that began in the mid-nineteenth century, forcing poor farmers to concentrate in the uplands. Despite their efforts to conserve the land's resources, growing human pressure in a country with annual population growth rates of 3.5 per cent reduced land availability, and resulted

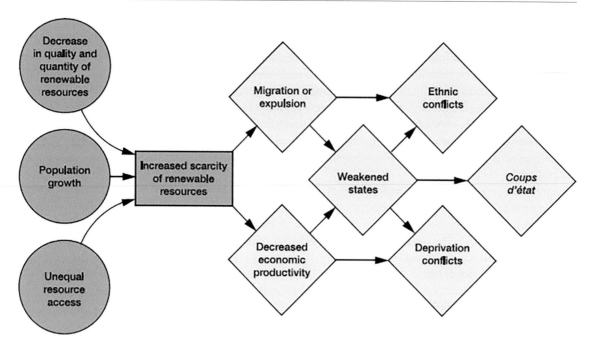

Figure 20.8 Some sources and consequences of renewable resource scarcity (Homer-Dixon *et al.*, 1993).

in deforestation and soil erosion on the steep hillsides. Many farmers consequently moved to Honduras and it was their eventual expulsion that precipitated the war. In the years following the Soccer War, this competition for land was not addressed and *campesino* support for leftist guerrillas provided a powerful contribution to the country's subsequent ten-year civil war.

The importance of population pressure and unequal access to resources has also been emphasized in other domestic settings. In one analysis of the terrible genocidal violence that occurred during 1994 in Rwanda, one of the most densely populated countries in Africa, the intense pressure on land resources was stressed (André and Platteau, 1998). These authors do not suggest that the acute competition for land was a direct cause of Rwanda's 1994 genocide, in which more than 0.5 million people, most of them from the Tutsi ethnic group, were killed. However, they do suggest that Rwanda's progressively increasing scarcity of land, one of the most basic resources in an essentially agricultural nation (Figure 20.9), 'goes a long way towards explaining why the violence spread so quickly and devastatingly throughout the countryside' (ibid.: 38). At its simplest level, this perspective sees the Rwandan genocide as the result of a Malthusian crisis in which the human population had become too large for the country's available resources.

Other research has even suggested that countries whose wealth is largely dependent on the export of primary commodities (including both agricultural produce and mineral resources) are particularly prone to political instability and civil violence (Collier and Hoeffler, 2000). This theory points to numerous civil wars that are at least in part a fight for control of certain primary commodities and/or in which trade in natural resources is a significant source of finance for particular armed groups (recent examples include diamonds in Sierra Leone, drugs in Colombia, and timber in Cambodia). The authors refer to their explanation as a 'simple greed model'. This association between conflict and certain mineral resources in particular has resulted in attempts by many transnational corporations (TNCs) to distance themselves from what have

Figure 20.9 Intensive agricultural land use in north-west Rwanda. The scarcity of land in this very densely populated country probably played some part in the ethnic tensions that led to genocide in 1994.

become known as 'conflict minerals' and 'blood diamonds': primary commodities whose exploitation and trade contribute to human rights violations in the country of extraction and surrounding areas. Most of these efforts examine companies' supply chains and attempt to sever any links between mining or minerals trading and armed conflict or the funding thereof (Hofmann *et al.*, 2018).

Interest in links between climatic change and the risk of conflict has flourished in recent years, with particular attention given to regions regarded as especially vulnerable to climate change impacts, including the Arctic, much of Africa, small islands, and large, densely populated deltas in coastal areas of Africa and Asia. Establishing such connections is by no means straightforward, although the ways in which climate changes may affect resource scarcity have emerged as a key issue (Buhaug *et al.*, 2010). The relatively short, episodic impact of El Niño/Southern Oscillation (ENSO) events has apparently had a destabilizing effect on some modern societies by increasing the probability of new civil conflicts arising, according to Hsiang *et al.* (2011).

To end this section on a more positive note, the conservation value of the Korean demilitarized zone mentioned earlier in this chapter has been put forward as a possible mechanism to help build trust and understanding between rival powers. Kim (1997) suggests that joint maintenance of the ready-made nature reserve, already completely protected and clearly delimited thanks to its origin as a buffer zone, can foster mutual respect between North and South Korea, while helping to preserve the peninsula's biodiversity. Similar arguments for such 'peace parks' are proposed in other borderland areas that have been a focus for international conflict (Ali, 2007). A successful example is the establishment in 2004 of the Condor-Kutuku Peace Park by Ecuador and Peru after a decades-long dispute over access to the Amazon River that culminated in armed clashes with up to 500 casualties in January 1995.

FURTHER READING

Hupy, J.P. and Koehler, T. 2012 Modern warfare as a significant form of zoogeomorphic disturbance upon the landscape. *Geomorphology* 157–58: 169–82. An assessment of how explosive munitions affect topography with examples from France and Vietnam.

Lawrence, M.J., Stemberger, H.L., Zolderdo, A.J., Struthers, D.P. and Cooke, S.J. 2015 The effects of modern war and military activities on biodiversity and the environment. *Environmental Reviews* 23(4): 443–60. A catalogue of the military's environmental impacts. Open access: www.nrcresearchpress.com/doi/full/10.1139/er-2015-0039#.WoGIwufLiY1.

Le Billon, P. 2007 Geographies of war: perspectives on 'resource wars'. *Geography Compass* 1/2: 163–82. A review of three main perspectives on resource wars.

Lockwood, J.A. 2012 Insects as weapons of war, terror, and torture. *Annual Review of Entomology* 57: 205–27. Review of the use of insects as weapons of war.

Rendall, M. 2007 Nuclear weapons and intergenerational exploitation. *Security Studies* 16: 525–54. An argument for how nuclear deterrence benefits us at the expense of future generations.

Silva Arimoro, O.A., Reis Lacerda, A.C., Tomas, W.M., Astete, S., Roig, H.L. and Marinho-Filho, J. 2017 Artillery for conservation: the case of the mammals protected by the Formosa Military Training Area, Brazil. *Tropical Conservation Science* 10: 1940082917727654. Case study assessing the conservation value of a military training zone. Open access: journals.sagepub.com/doi/full/10.1177/1940082917727654.

WEBSITES

www.gcint.org one of Green Cross International's major programmes is concerned with the environmental impact of warfare.

www.icbl.org the International Campaign to Ban Landmines site.

www.opcw.org site of the Chemical Weapons Convention.

www.sipri.org the Stockholm International Peace Research Institute site contains information on military capabilities and expenditure.

www.the-monitor.org the Landmine and Cluster Munition Monitor provides research for the campaign to ban landmines.

www.un.org/disarmament the UN Office for Disarmament Affairs

POINTS FOR DISCUSSION

- Does the human cost of warfare mean that the environmental cost is unimportant?
- Explain how and why warfare is the antithesis to sustainable development.
- Do you think that conflicts over natural resources are set to become more frequent in coming decades?
- Describe and explain the possible consequences of a nuclear winter.
- Should atomic weapons be banned and is it realistic to do so?
- Can national armies be managed sustainably and should different rules apply during conflicts?

21

Natural hazards

Natural hazards such as earthquakes, floods, tropical cyclones and disease epidemics are normal functions of the natural environment. They do, therefore, affect all living organisms, but they are usually only referred to as disasters or catastrophes when they impact human society to cause social disruption, material damage and loss of life. As such, natural hazards should be defined and studied both in terms of the physical processes involved and the human factors affecting the vulnerability of certain groups of people to disasters.

Although some places are more hazardous than others, all locations are at risk – there is always the chance of a disaster – from some natural hazard or other. All places also have some natural advantages, however, and the presence or absence of human activities in any location is the result of weighing up the risks relative to the advantages. In many locations, the physical phenomenon responsible for a hazard also offers some of the advantages: a river, for example, will flood, which may be hazardous, but it is also a source of water and its floodplain is a location with flat land and fertile soils. Hence, the hazard should be seen as an occasionally disadvantageous aspect of a phenomenon that is beneficial to human activity over a different timescale. This is illustrated in Figure 21.1 where the shaded zone represents an acceptable variability of a physical element: variations in its magnitude that society has adapted to and considers a resource. When this variability exceeds a certain threshold, however, the same physical element becomes a hazard for a period of time. In the case of the river, this could mean either a flood or a water shortage. Ironically, beyond the timescale of the hazardous event occurring, the very same hazard may contribute to the resource element of the location: when a flood happens, it presents a hazard to the occupants of the floodplain, but the floodplain is only there because the river floods occasionally.

To complicate matters, however, not all people who choose to live in a particular area enjoy the same degree of choice – a flexibility that may be dictated by the economic means at their disposal, among other things. To complicate matters further, there is a wide range of ways in which people respond to hazards and, again, the rich often have more options in this respect than poorer members of society. The study of natural hazards is very much a geographical subject, therefore, since it includes analysis of both the physical and human environment.

HAZARD CLASSIFICATIONS

Natural hazards can be classified in many different ways. Physical geographical approaches may divide them according to their geological, hydrological, atmospheric and biological origins. They can also be

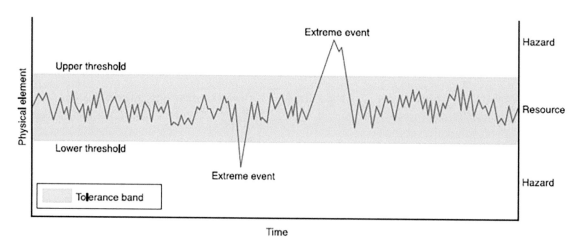

Figure 21.1 A physical element is perceived as a resource by society when its variations in magnitude are within certain limits, but is perceived as a hazard when they exceed certain thresholds (after Hewitt and Burton, 1971).

classified spatially, as certain hazards only occur in certain regions: avalanches occur in snowy, mountainous areas, for example, and most volcanic eruptions and earthquakes occur at tectonic plate margins. Another approach divides hazards into rapid-onset, intensive events (short, sharp shocks such as tornadoes) and slow-onset pervasive events, which often affect larger areas over longer periods of time (such as droughts – see p. 105). Table 21.1 is an attempt to classify hazards in this way, and gives some indication of the frequency and predictability of events in time.

Table 21.1 Natural disasters classified according to frequency of occurrence, duration of impact and length of forewarning

Disaster	Frequency or type of occurrence	Duration of impact	Length of forewarning (if any)
Lightning	Random	Instant	Seconds–hours
Avalanche	Seasonal/diurnal; random	Seconds–minutes	Seconds–hours
Earthquake	Log-normal	Seconds–minutes	Minutes–years
Tornado	Seasonal	Seconds–hours	Minutes
Landslide	Seasonal/irregular	Seconds–decades	Seconds–years
Intense rainstorm	Seasonal/diurnal	Minutes	Seconds–hours
Hail	Seasonal/diurnal	Minutes	Minutes–hours
Tsunami	Log-normal	Minutes–hours	Minutes–hours
Flood	Seasonal; log-normal	Minutes–days	Minutes–days
Subsidence	Sudden or progressive	Minutes–decades	Seconds–years
Windstorm	Seasonal/exponential	Hours	Hours
Frost or ice storm	Seasonal/diurnal	Hours	Hours
Hurricane	Seasonal/irregular	Hours	Hours
Snowstorm	Seasonal	Hours	Hours
Environmental fire	Seasonal; random	Hours–days	Seconds–days
Insect infestation	Seasonal; random	Hours–days	Seconds–days
Fog	Seasonal/diurnal	Hours–days	Minutes–hours
Volcanic eruption	Irregular	Hours–years	Minutes–weeks
Coastal erosion	Seasonal/irregular; exponential	Hours–years	Hours–decades
Soil erosion	Progressive (threshold may be crossed)	Hours–millennia	Hours–decades
Drought	Seasonal/irregular	Days–years	Days–weeks
Crop blight	Seasonal/irregular	Weeks–months	Days–months
Expansive soil	Seasonal/irregular	Months–years	Months–years
Meteorite impact	Irregular	Days–years	Days–months

Source: after Alexander (1993).

Although these classifications provide a useful summary of hazard types, many disasters involve composite hazards. Hence, an earthquake may cause a tsunami wave at sea, landslides or avalanches on slopes, building damage and fires in urban areas, and flooding due to the failure of dams, as well as ground shaking and displacement along faults. Similarly, many natural disasters cause disruption to public hygiene and consequently result in heightened risks of disease transmission.

Other approaches to defining and classifying disasters focus on such factors as the areal extent or the scale of damage, typically incorporating a threshold such as death or injury to a certain number of people, or economic damage above a certain amount. These approaches are most useful when related to the society concerned because a certain level of damage will have a different impact on a society depending upon its vulnerability or strength to cope with the disaster, a recognition used to assess disaster-proneness at the national level (see below). One drawback of these approaches is that they tend to focus on the immediate or short-term impact, but natural disasters can also have impacts that are very long-lasting, as Box 21.1 indicates.

Box 21.1 Long-term impacts of natural disasters

Natural disasters by definition cause substantial damages – including death and destruction – but the deleterious consequences for societies also include impacts on nutrition, education, health and ways of generating income. Some of these impacts can be particularly long-lasting for the most vulnerable groups.

To cope with a disaster people frequently sell assets, such as livestock and land, which could be the basis of their longer-term growth, and many examples show that it takes a significant amount of time for affected households to return to at least their pre-disaster situation. A study of the 1999–2000 drought in South Wollo, Ethiopia, found that families who dropped below the poverty line after the drought took up to ten years to reach the poverty threshold again. Similar long-term consequences of droughts have been identified in Brazil where drops in rural earnings still affected individuals a decade later.

The knock-on effects of natural disasters can have long-term impacts on children in several ways. A short-term lack of adequate food affects a child's physical development, particularly if it occurs at between 12 and 24 months old, the most critical time for child growth. In Zimbabwe, the effects of the 1982–84 droughts were still detectable in children even when they reached late adolescence: stunting by about 2 cm in height. The effects of disasters on malnutrition early in life can also impose a timelagged effect on education: early-life malnourished children are prone to start school later and to perform worse at school.

Education may suffer because of damage to infrastructure. In Mozambique, for instance, the floods in 2000 destroyed more than 500 schools. But a child's education may also suffer because their parents decide to withdraw them from school after a natural disaster, their labour being needed to help the family recover. Children in households most affected by the 2001 earthquakes in El Salvador were almost three times more likely to work after the shock, neglecting their schooling in consequence. Child labour also increased significantly in Nicaragua as a consequence of Hurricane Mitch in 1998. A child that drops out of school is less likely to return and the lack of education will have negative effects on the individual's future productivity and welfare in adulthood.

Source: Baez *et al.* (2010).

There is a case for defining a disaster at any scale because a particular event may impact just one person, or a region, or a country, or even the entire planet. In practice it is generally fair to say that very poor societies commonly suffer the highest numbers of casualties while very rich societies suffer the highest property damage. In a survey of natural disasters from 1977 to 1997, Alexander (1997) found that about 90 per cent of disaster-related deaths occurred in developing countries while some 82 per cent of economic losses were suffered by developed countries. However, poorer countries suffer more in relative terms from both death and destruction. Gaiha and Thapa (2006) concluded that people in low-income countries are four times more likely to die due to natural disasters and the cost per disaster as a share of GDP is considerably higher in developing than in OECD countries.

Another difficulty arises in distinguishing between purely 'natural' events and human-induced events. This chapter is not concerned with such obviously human-induced hazards and catastrophes as industrial accidents or pesticide poisonings. However, in one sense, all 'natural' disasters can be thought of as human-induced since it is the presence of people that defines whether or not an event creates a disaster. Many of the natural physical processes that cause disasters can also be triggered or made worse by human action: while a volcanic eruption is a purely natural process, for example, the failure of a slope, producing a landslide, can be induced by many human activities, such as road-cutting, construction or deforestation. The composite nature of many disasters also blurs the distinction: the major cause of death due to an earthquake is usually crushing beneath buildings, so is the disaster natural or human-induced? Some researchers believe that the difficulty of making this distinction has made the division pointless, and prefer to talk of 'environmental' hazards that refer to a spectrum with purely natural events at one end and distinctly human-induced events at the other (e.g. Smith, 2001). The phrase 'na-tech' (natural-technological) is also used.

DISASTER MANAGEMENT

The steps taken to manage a disaster, whether natural or otherwise, can be taken by governments, particular groups or the whole society, but some can also be taken by individuals. The steps are typically divided into four categories:

- *mitigation*: activities that reduce or eliminate the probability of a disaster (e.g. stringent building codes in earthquake-prone areas, vegetation clearance in areas of high fire risk, hazard control structures such as dams)
- *preparedness*: planning in advance how to respond in case a disaster does occur (e.g. installing early warning systems, developing emergency management plans, training exercises)
- *response*: activities necessary to address the immediate and short-term effects of a disaster, which focus primarily on actions needed to save lives, to protect property and to meet basic human needs (e.g. search and rescue, securing areas prone to looting, damage assessment)
- *recovery*: activities that bring communities back to normal and at least to their pre-disaster state (e.g. clean-up, reconstruction, legal assistance to affected people).

A major international agreement that focuses on disaster management is the Sendai Framework, which has been formally adopted by the UN as the global blueprint over the period 2015–2030 for all efforts to reduce the impacts of hazards on people, communities and societies. The Sendai Framework recognizes that the state has the primary role to reduce disaster risk but that responsibility should be shared with others including local government, the private sector and other stakeholders. The framework has four priorities for action which are shown in Table 21.2.

Management that aims to address the possible impact of future disasters is inevitably related to the information available on past events and the probability of recurrence, the perception of that information, and the awareness of particular opportunities. Some approaches take action to reduce the risk of impact by preventing or modifying events, although some hazards are currently impossible to modify (e.g. volcanic eruptions). Floods, for example, can be prevented over a long period by impounding water behind a dam, and frosts can be prevented by installing a heating system in an orchard. Other approaches aim to modify the loss potential through prediction, either in time (so issuing hazard warnings to enable evacuation) or in space (by producing hazard zone maps that can be used to regulate land use in hazard-prone areas). These examples illustrate the ways in which disaster management can operate over a variety of timescales, from the long-term view of building a dam or regulating land use, to the immediate efforts involved in evacuating a threatened area. Most of these mitigation and preparedness steps are more generally characteristic of the more developed countries, which have sufficient economic resources, infrastructure and trained personnel to organize and carry out such actions. For large natural disasters, the response and recovery phases are typically international operations.

Other forms of management effectively accept the losses imposed by a hazard and cope with these either by planning for them in the form of economic insurance, or by relying on help from a wider community

Table 21.2 The Sendai Framework's four priorities for action

Priority	Detail
Understanding disaster risk	Disaster risk management should be based on an understanding of disaster risk in all its dimensions of vulnerability, capacity, exposure of persons and assets, hazard characteristics and the environment. Such knowledge can be used for risk assessment, prevention, mitigation, preparedness and response.
Strengthening governance to manage disaster risk	Disaster risk governance at the national, regional and global levels is very important for prevention, mitigation, preparedness, response, recovery, and rehabilitation. It fosters collaboration and partnership.
Investing in disaster risk reduction for resilience	Public and private investment in disaster risk prevention and reduction through structural and non-structural measures is essential to enhance the economic, social, health and cultural resilience of persons, communities, countries and their assets, as well as the environment.
Enhancing disaster preparedness for effective response and to 'build back better' in recovery, rehabilitation and reconstruction	The growth of disaster risk means there is a need to strengthen disaster preparedness for response, take action in anticipation of events, and ensure capacities are in place for effective response and recovery at all levels. The recovery, rehabilitation and reconstruction phase is a critical opportunity to build back better, including through integrating disaster risk reduction into development measures.

Source: www.unisdr.org/we/coordinate/sendai-framework.

when they happen, whether from fellow members of a family or from local, national or foreign governments and aid agencies. Economic insurance is again an option usually available only in the more prosperous economies (Linnerooth-Bayer *et al.*, 2011), and in some cases it can actually encourage individuals and societies to persist with activities in hazardous areas.

The adoption of a particular form of management is determined by assessment of the risk, which is dependent on the probability of occurrence of the event, the population exposed, its vulnerability and coping strategies. Each of these elements varies in time and space. Assessment may be made scientifically, by calculating the probability of a particular hazard causing a disaster, based on information of past event frequency and area of impact plus the current vulnerability of people and property. An important measure of frequency is the 'return period' for a particular magnitude of event. Engineers calculate the statistical probability of a certain size of event occurring, and hence speak of the 50-year flood, for example, and design structures to withstand the magnitude of hazard likely to affect the structure within its lifetime. It is not a fail-safe measure, of course, because land-use changes in a catchment can alter its flood response characteristics, and because it is still possible that a bridge built to withstand a flood that happens on average only every 50 years is, in fact, hit by a 100-year flood. Conversely, assessments are often made on the basis of how a particular hazard is perceived. For many reasons, not least the lack of long-term records for many areas, perception is involved in most hazard risk assessments.

HAZARDOUS AREAS AND VULNERABLE POPULATIONS

While people have always been at risk from natural hazards, the perception that the world is becoming a more hazardous place is widespread. Proving this hypothesis is difficult since there is still no comprehensive, long-term worldwide database of hazards or disasters. The rising worldwide trend in the numbers of natural catastrophes, compiled by the insurance industry (Figure 21.2a), is at least partly due to improved reporting of events (Swiss Re, 1998), although there are well-founded fears that certain types of hazard, such as heatwaves, will increase in intensity with global climate change (see Chapter 11). With an increase in world population, however, it is also logical to suppose that the number of people living in high-risk areas is also rising, although the effects of efforts to reduce the risks must also be considered. The resulting cost of damage caused by natural catastrophes has markedly increased on average since the late 1980s (Figure 21.2b), a period characterized by particularly high losses in certain years. A small number of severe events pushed up these losses: Hurricane Andrew in 1992, the winter storms across western Europe in 1999, Hurricanes Katrina, Wilma and Rita in 2005, and Hurricane Ike in 2008. Indeed, it is the growing exposure of people and assets, for example through rapid economic and urban growth in cyclone-prone coastal areas and earthquake-prone cities, that prompts many experts to declare that risk levels for numerous hazards are increasing over time, even assuming constant hazard frequency and severity. Peduzzi *et al.* (2010) were involved in a major UN global assessment on disaster risk reduction and concluded that economic loss risk is increasing faster than mortality risk and that although vulnerability decreases as countries develop, it is generally not enough to compensate for the increase in exposure. On the national scale, a 50-year study in the USA (Gall *et al.*, 2011) found that direct, inflation-adjusted economic losses from natural hazards have steadily increased over the period, as have per capita losses, showing that impacts outpace population growth.

There is no doubt that some countries are perceived to be more prone to disasters than others. The impact of a particular event depends on where the event occurs and how prepared the area is to cope with the disaster. In the case of an earthquake in an urban area, for example, the impact will vary according to such factors as the quality of the buildings, the ability of the society to respond to any destruction, and the

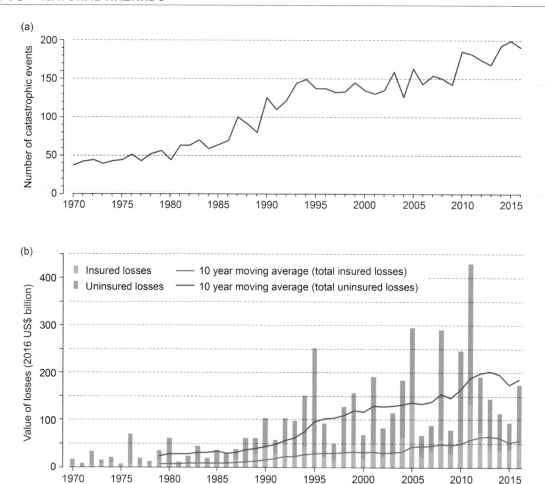

Figure 21.2 Global natural catastrophes, 1970–2016: (a) frequency; (b) losses at 2016 prices (after Swiss Re, 2017).

timing of the event. A significant decline in the number of flood fatalities in Australia since the 1850s has been put down to increased awareness of the flood hazard, better warning systems and the use of structural flood mitigation works (Coates, 1999). The success of preventing flood fatalities in Australia is reflected in the fact that death rates have fallen both in absolute and relative terms (i.e. despite the increase in population). The overall decadal death rate from flooding has fallen from nearly 24 per 100,000 population in the 1800s to 0.04 per 100,000 in the 1990s. An opposite trend has been identified in the USA, however, where Smith and Ward (1998) documented a steady increase in the number of deaths from flooding over the period 1925–94, probably associated with the continuing problem of flash floods. Indeed, Gall *et al.* (2011) found that economic flood losses tripled in the USA over their period of study.

One attempt to rank countries of the world according to their natural hazard risk combines exposure to natural hazards with the vulnerability of a society to produce an overall World Risk Index or WRI (UNU-EHS, 2011). The index consists of indicators in four components:

- *exposure*: to natural hazards such as earthquakes, hurricanes, floods, droughts and sea-level rise
- *susceptibility*: the predisposition of society and ecosystems to suffer harm depending on infrastructure, nutrition, housing and economic conditions
- *coping*: capacities to reduce negative consequences depending on governance, disaster preparedness and early warning, medical services, social and material security
- *adaptation*: capacities for long-term strategies for societal change in relation to future natural events and climate change.

Hence, the WRI takes into account not only the natural hazard but also the social, economic and ecological factors as well as governance aspects characterizing a society. The combination is critical to the scale of a disaster or indeed whether a particular natural hazard event becomes a disaster or not. For example, both Japan and Haiti have a relatively high degree of exposure to natural hazards but very different vulnerabilities. The differences were borne out by the impacts of recent earthquakes in the two countries: about 220,000 people died in the 2010 Haiti earthquake (Figure 21.3) as compared to a death toll of 28,000 people in the Japan earthquake of 2011, despite the fact that the earthquake in Japan had a magnitude that was 100 times stronger.

Figure 21.3 Damage to a street in Port-au-Prince, Haiti after the earthquake of January 2010.

The WRI was not calculated for every country – no data were available in 19 cases – but, of the 173 countries studied, those with the highest natural hazard risk ranked according to the WRI are shown in Table 21.3. Three small island developing states (SIDS) – Vanuatu, Tonga and the Solomon Islands – appear high in the ranking, reflecting their generally great vulnerability to natural hazards, particularly to tropical cyclones. This is partly a function of their small national land area and population, of course, meaning that any calamity that does occur is likely to have an impact that is proportionally greater than a similar event in a physically larger country. Other factors that increase the intrinsic vulnerability of SIDS are examined by Pelling and Uitto (2001) and include relatively small economies often dependent on the export of natural resources, a limited ability to forecast hazards and relatively little insurance cover.

Several other SIDS not listed in the WRI exercise are widely regarded to be similarly disaster-prone. One of these is Dominica, a small island state in the Eastern Caribbean with a population of about 74,000 people occupying a national territory of some 750 km^2 that is vulnerable to a wide range of natural hazards (Figure 21.4). The most common and historically most significant are tropical storms and hurricanes. The strong winds and high seas associated with such events can be particularly damaging because most of the population and infrastructure are located on the island's coast. This coastal concentration of economic activity is a simple reflection of Dominica's generally rugged terrain. The island is almost completely volcanic in origin and has the unenviable distinction of hosting the highest concentration of potentially active volcanoes (nine) in the Lesser Antilles arc and one of the highest worldwide, making it extremely suscep-tible to volcanic hazards. There is also a related tectonic risk of earthquakes. Dominica's steep slopes and humid tropical climate also combine to make landslides a common feature of life and the landscape. Other potential hazards include droughts, storm surges, floods, bush fires and tsunamis (Benson and Clay, 2001).

While all people living in hazard areas are vulnerable, some groups of people are typically more vul-nerable than others. Several examples of poorer sectors of society living in more hazardous urban zones, because they have no other option, are documented in Chapter 10. The impacts of natural disasters often

Table 21.3 The ten countries with the highest natural hazard risk

Rank	Country	World Risk Index (%)
1	Vanuatu	32.00
2	Tonga	29.08
3	Philippines	24.32
4	Solomon Islands	23.51
5	Guatemala	20.88
6	Bangladesh	17.45
7	Timor-Leste	17.45
8	Costa Rica	16.74
9	Cambodia	16.58
10	El Salvador	16.49

Source: UNU-EHS (2011).

Figure 21.4 A public information board in Dominica, a typical small island developing state, indicating the range of hazards faced by the island's inhabitants.

fall disproportionately on the most vulnerable people in society generally: the poor, minorities, children, the elderly and disabled. These groups are often the least prepared for an emergency and have the fewest resources with which to prepare for a hazard. They also tend to live in the highest-risk locations, in sub-standard housing, and lack knowledge or the social and political connections necessary to take advantage of opportunities that would speed their recovery. This vulnerability can be assessed and measured in numerous ways. In the WRI above, vulnerability is measured by combining measures of susceptibility with coping and adaptation capacities. In effect, the physical world acts as a sort of trigger, exposing the vulnerability of societies. Another view of the ways in which vulnerability is created in certain groups is illustrated in Figure 21.5.

EXAMPLES OF NATURAL DISASTERS

The following sections take a closer look at a number of major natural causes of disaster. The selection is based on an analysis of the causes of deaths from all types of disasters over the period 1900–90 compiled by the US Office of Foreign Disaster Assistance (Blaikie *et al.*, 1994). Civil strife and famine were by far the largest causes of death, accounting for nearly 88 per cent of all casualties. Although natural events such as drought and flooding are often implicated as triggers for the onset of famine (see Chapter 5), the most

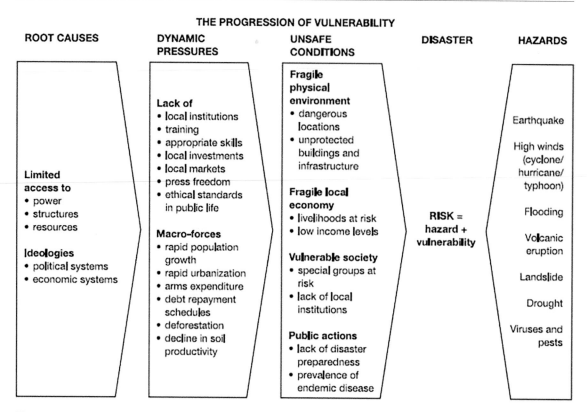

Figure 21.5 Model of pressures that create vulnerability and result in disasters (after Blaikie *et al.*, 1994).

significant direct natural causes of death were earthquakes, volcanoes, tropical cyclones, epidemics and floods.

Earthquakes

Each year, around 1 million earth tremors are recorded over the Earth's surface, but the vast majority of these are so small that they are not felt by people. The handful of major earthquakes that occur every year can cause widespread damage, however, causing some of the world's most devastating natural disasters. China has suffered some of the worst fatalities in major earthquakes: the 1556 event in Shensi caused more than 800,000 deaths, and the Tangshan earthquake of 1976 resulted in over 250,000 fatalities. Tangshan is the most deadly event of modern times (Table 21.4).

Most earthquakes are caused by abrupt tectonic stress release around the margins of the Earth's lithospheric plates, although they also occur at weak spots near the centre of plates. While the locations of plate boundaries and most associated faults are well known, the precise timing and location of individual earthquakes are virtually impossible to predict in any reliable way. A combination of seismic monitoring and observations of natural phenomena, including animal behaviour, were used in the only successful prediction of a large-scale earthquake, which devastated the Chinese city of Haicheng in Liaoning Province

Table 21.4 The most deadly earthquakes, 1970–2016

Date (start)	Country/region	Insured damage (2016 US$ million)	Victims (dead and missing)
28 July 1976	China	–	255,000
12 January 2010	Haiti	110	222,570
26 December 2004	Indian Ocean, Indonesia, Thailand, *et al.* tsunami	2541	220,000
12 May 2008	Sichuan, China	409	87,449
8 October 2005	Pakistan, India, Afghanistan	–	74,310
31 May 1970	Peru	–	66 000
20 June 1990	Iran	211	40,000
26 December 2003	Iran	–	26,271
7 December 1988	Armenia	–	25,000
16 September 1978	Iran	–	25,000

Source: after Swiss Re (2017).

in February 1976. An evacuation of Haicheng and two other cities was ordered 48 hours before the main earthquake, almost certainly saving many thousands of lives (Adams, 1975).

The difficulties surrounding earthquake prediction were compounded in 2012 when six earthquake scientists were convicted of manslaughter for having given inappropriate advice to the public before the L'Aquila earthquake in central Italy in 2009, an event in which 308 people were killed. Although the convictions were subsequently overturned, the incident underlined the mismatch between society's expectations of earthquake scientists and what the scientists are currently able to provide. In an assessment of the incident, Joffe *et al.* (2018) highlight the need for better communication with the public on earthquake risk, with a focus on how scientists portray uncertainty.

The main environmental hazard created by seismic earth movements is ground shaking, which is related to the magnitude of the shock and normally assessed on the Richter scale, a complex logarithmic scale that measures the vibrational energy of a shock. A commonly used alternative measure of earthquake strength, the Mercalli scale, uses the observations of people who experienced the earthquake to estimate its intensity. Local site conditions have an important effect on ground motion, with greater structural damage usually found in areas underlain by unconsolidated material as opposed to rock. Particularly severe damage was caused during the earthquake of 1985 in the central parts of Mexico City, which is built on a dried lake bed. Similarly, a relatively modest-magnitude earthquake, registering 5.4 on the Richter scale, in San Salvador in 1986, caused unusually large destruction since most of the city is built on volcanic ash up to 25 m thick.

Another serious earthquake hazard associated with soft, water-saturated sediments is soil liquefaction, in which intense shaking causes sediments to lose strength temporarily and behave as a fluid. Loss of bearing strength can cause buildings to subside and soils can flow on slopes of greater than 3 degrees. Liquefaction was a major cause of damage to buildings in Christchurch, New Zealand during a series of strong earthquakes in 2010–11 (Green *et al.*, 2014). Mudflows, landslides, and rock and snow avalanches triggered by ground vibrations often play a major role in earthquake disasters, particularly in mountainous areas. In a study of global earthquakes over a 40-year period, earthquake fatalities were found to be dominated by shaking-related causes (i.e. from partial or total building collapse), while secondary effect-induced fatalities were dominated by landslide deaths (Marano *et al.*, 2010).

Tectonic displacement of the sea bed is the main cause of large sea waves, or tsunamis, which can travel thousands of kilometres at velocities greater than 900 km/h and cause great damage when they hit coastlines. Tsunamis are most common in the Pacific where one of the largest and most damaging waves, measuring 24 m in height, drowned 26,000 people in Sanriku, Japan in 1896. The Indian Ocean tsunami of December 2004 caused even greater loss of life – an estimated 220,000 people dead or missing – making it one of the most devastating natural disasters in modern history. A megathrust earthquake off the coast of northern Sumatra was one of the largest ever recorded, registering 9.3 on the Richter scale, and triggered waves up to 30 m high that spread out across the ocean at the speed of a jet aircraft (Lay *et al.*, 2005). The tsunami directly affected about 2 million people. Most of the fatalities and the most intense and widespread destruction occurred in the province of Aceh, Sumatra, the coast nearest to the epicentre, only about 10 minutes away in terms of tsunami propagation time. The next most heavily impacted areas were the coasts of Thailand, India and Sri Lanka, but damage and casualties were also reported from much further afield, including Bangladesh, Seychelles and Somalia. The total economic impact of this international disaster was in billions of US dollars.

Some attempts have been made to regulate the intensity of earthquakes on land by injecting pressurized water along faults to stimulate many small tremors and thus reduce the probability of major movements, but the technique still runs the risk of itself inducing a major earthquake. Hence, responses to the hazards associated with earthquakes focus on reducing risk and coping with the losses. Since the failure of structures is a major cause of injury during seismic shaking, much effort has gone into building location and design. Earthquake-hazard zoning maps have been produced for many countries and some of the approaches used for building in particularly hazardous sites are shown in Figure 21.6. Building codes are effective only if enforced, however, and several recent examples of major damage and loss of life in urban areas have occurred in cities where such codes have not been followed (see p. 244).

Even in places where strict building codes are adequately enforced, major damage and loss of life can still occur, as in March 2011 when one of the world's most powerful earthquakes (9.0 on the Richter scale) off the north-east coast of Japan was followed by one of the world's largest tsunamis and the initiation of the world's second largest nuclear accident (Kingston, 2012). The earthquake directly affected nearly 41 million people with very strong to extreme shaking, reducing entire towns to rubble and killing more than 15,000 people. Relief and recovery were severely hampered by the extensive damage to infrastructure, including the closure of many ports, the major regional airport in Sendai, and numerous roads and rail lines. Damage to three reactors at the Fukushima nuclear power station led to the evacuation of 80,000 people living within 20 km of the plant. The overall cost of the catastrophe is estimated to be between US$200 and 300 billion, making it the costliest natural disaster in history.

Volcanic eruptions

Volcanic eruptions have killed around 250,000 people in the last 400 years due to a range of associated hazards, which include falls of rock and ash, the force of lateral blasts, emission of poisonous gases, debris avalanches, lava flows, mudflows known as 'lahars', and 'pyroclastic flows' made up of suspended rock froth and gases (Chester, 1993), and their environmental effects can extend over very great distances (Table 21.5).

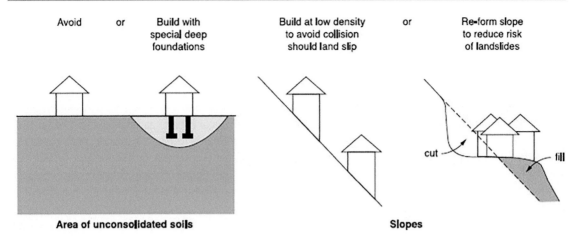

Figure 21.6 Some approaches to building on sites at high risk of disturbance during earthquakes (after Smith, 2001).

Table 21.5 Hazards associated with volcanic eruptions

Distance from volcano	Hazard
Up to 10 km	Lava flows
	Pyroclastic flows and surges
	Earthquakes
	Ground subsidence and cracking
	Landslides
	Jökulhlaups
Up to 100 km	Atmospheric blast
	Lahars
	Debris avalanches
Up to 1000 km	Tephra fall
	Gases
Up to 10,000 km	Tsunami
Planet-wide	Climatic effects

Active volcanoes may also present more persistent hazards to human health through long-term exposure to carbon dioxide, radon and other pollutants (Baxter, 2000). Only about 500 volcanoes are thought to have actually erupted in historical time, although the traditional categories of 'active', 'dormant' and 'extinct' are not infrequently found wanting when supposedly inactive volcanoes erupt, sometimes causing widespread destruction (e.g. Mount Pinatubo, Philippines, 1991). Nonetheless, at least 0.5 billion people worldwide live within potential exposure range of a volcano that has been active within recorded history (Hansell *et al.*, 2006). Like earthquakes, the distribution and activity of volcanoes are controlled by global tectonics.

Solid particles (generally known as 'tephra') and gases ejected from a volcano during an eruption can cause a variety of hazards many hundreds of kilometres from the volcano. Whilst still in the atmosphere, tephra and associated gases emitted by major eruptions can have planet-wide effects on climate, as shown by the eruption in 1815 of the Tambora volcano in Indonesia which ejected 60 to 80 megatons of sulphate particles into the stratosphere. The effects on climate were felt as far apart as China, India and Europe, where the following year became known as the 'year without a summer' (Wirakusumah and Rachmat, 2017). Less spectacular eruptions can also have significant impacts more locally. The ash emitted by the 2010 eruption of the Icelandic volcano Eyjafjallajökull caused severe disruption to civil aviation across much of Europe for just over a week, with profound effects on more than 8 million passengers (Alexander, 2013).

Smaller volcanic particles may be inhaled, with consequent respiratory problems such as asthma and bronchitis, and cause irritation to eyes and skin. Tephra falls can also accumulate on buildings, causing collapse from loading on roofs. Damage to vegetation, including crops, can also occur due to burial or pollution. Tephra may be contaminated by toxic volcanic gases such as fluorine, emitted in large quantities in some eruptions, which can pose a danger to grazing animals on pastures. Fluoride poisoning is a particular risk to livestock in Iceland, causing the deaths of thousands of sheep after the 1970 Hekla eruption. Drinking water may also become contaminated by fluorine from tephra, although fluorine poisoning in human populations appears to be rare (Witham *et al.*, 2005).

Lava flows, one of the most characteristic volcanic hazards, are rarely fast enough to cause death but can completely destroy anything in their path. A distinction can be made between the responses to lava flow hazards typically employed in more developed countries and those more common in less developed countries. Developing countries tend to rely upon hazard warning and evacuation, whilst some wealthier nations have attempted to modify the hazard by bombing to divert flows, erecting barriers to protect inhabited areas, and controlling the flow using water (e.g. in Hawaii, Iceland, Italy and Japan), although these countries also employ warning and evacuation measures. Such a distinction can also be made with regard to the whole suite of volcanic hazards. Chester suggests that 'many deaths are now avoidable often through the application of fairly simple methods of prediction' (1993: 186) since the activity of most volcanoes builds up to a major eruption and the build-up can be detected, given adequate monitoring using seismometers, tiltmeters and gas sample analysers. Nonetheless, the fact remains that a great deal of infrastructure worldwide lies in the path of potential volcanic disasters, not least in major cities located near volcanoes, including Mexico City (Popocatepetl), Bandung (Tangkuban Parahu), Naples (Vesuvius), Quito (Guagua Pichincha), Guatemala City (Pacaya) and Managua (Masaya Nindiri and Apoyeque), all of which lie within 50 km of their respective volcanoes. Urbanization, particularly in developing countries, has led to increasing global exposure to the risks posed to large cities by volcanic eruptions (Heiken, 2013).

To counterbalance these facts, however, it is also worth pointing out that people have lived near volcanoes for a very long time. In many cases this is because of the benefits of the locations concerned. In much of southern Italy, for example, agriculture is difficult due to poor-quality soils, but the area around Mount Vesuvius near the city of Naples has been farmed for centuries because of its rich volcanic soils, despite the risk of eruptions (Figure 21.7). Indeed, Grattan (2006) highlights the fact that several great centres of civilization have thrived despite being close to centres of active volcanism. They include the classical civilizations of the Mediterranean basin and the civilizations of Mexico and Central America. While there is no doubt that volcanic eruptions have resulted in some devastating impacts in these and other areas, the relationship between volcanoes and human society is perhaps rather more complex than at first appears.

Figure 21.7 View of Mount Vesuvius and Naples painted by Mautonstrada in 1836. The Naples area has been inhabited since the Stone Age despite occasional eruptions.

In the longer term, resilient societies recover from volcanic disasters and adapt to the persistent hazard. In such cases volcanic activity can frequently act as a stimulus rather than a brake to cultural development.

In the contemporary era, a sharp contrast can be drawn between the successful prediction and evacuation of the area around Mount Pinatubo on Luzon Island in the Philippines in 1991, and the failure to evacuate in time for the 1985 eruption of Nevado del Rúiz in the Colombian Andes. Seismic and other anomalies in activity were reported from Nevado del Rúiz for almost a year before the eruption, and numerous geologists and seismologists were called in to assess the situation, but poor communication and bureaucratic inefficiency prevented action being taken in time. A hazard map was prepared, the final version of which was due to be presented to the authorities the day after the eruption started on 13 November. The eruption melted part of the summit's snow and ice cap, and a torrent of meltwater and pyroclastic debris flowed from the summit, coalescing in lower channels to entrain vegetation, water and mud to form rapidly flowing lahars. Nearly 70 per cent of the population of Armero (about 20,000 people) were killed in the mudflow and another 1800 died on the volcano's western flanks. A total of 50 schools, two hospitals, 58 industrial plants, roads, bridges and power lines were destroyed or damaged, and 60 per cent of the region's livestock perished. The total cost to the economy of Colombia was estimated at US$7.7 billion, or 20 per cent of the country's GNP in that year. As Voight (1990: 151) concludes: 'The catastrophe was not caused

Table 21.6 Volcanoes having greatest impact on society in the twentieth century

By number killed		By number evacuated/affected (million)	
Pelée (Martinique)	29,000	Pinatubo (Philippines)	1.8
Nevado Del Rúiz (Colombia)	23,000	Guagua Pichincha (Ecuador)	1.2
Santa Maria (Guatemala)	13,750	Agung (Indonesia)	0.3
Kelut (Indonesia)	5500	Mayon (Philippines)	0.2
Lamington (Papua New Guinea)	3000	Arenal (Costa Rica)	0.1

Source: after Witham (2005).

by technological ineffectiveness or defectiveness, not by an overwhelming eruption, or by an improbable run of bad luck, but rather by cumulative human error – by misjudgement, indecision and bureaucratic short-sightedness.' At the time, the Colombian government had other problems on its mind. On 6 November, the president ordered troops to storm Bogota's Palace of Justice, which had been occupied by guerrillas; 100 people were killed, including ten senior judges.

The deficiencies in governance that were so dominant in explaining the Nevado del Rúiz catastrophe stand in stark contrast to the efficient evacuation of the area around Mount Pinatubo in the Philippines in 1991. Hence, these two volcanoes both rank highly among those with the greatest impact on society during the twentieth century, but for different reasons (Table 21.6). In many cases, the volcanoes owe their appearance in Table 21.6 to one large eruption, but not in all. Santa Maria volcano in Guatemala, for instance, had two major eruptions in the twentieth century, resulting in 8750 deaths in 1902 and another 5000 in 1929. Mount Pinatubo affected 1.8 million people in two major events: in 1991 and 1992.

Tropical cyclones

Tropical cyclones – also known as hurricanes in the Atlantic and typhoons in the western Pacific – are among the most destructive of all natural hazards. They are generated over warm tropical oceans and rely on a complex combination of processes to grow, mature and eventually die. Warm ocean water is needed to fuel the tropical cyclone's heat engine, so they only form when the temperature of the top 50 m or so of the ocean surface is at least 26.5°C. The air above has to be marked by a fairly rapid drop in temperature with height, allowing thunderstorms to form initially. It is thunderstorm activity that allows the heat from the ocean waters to be liberated, helping the tropical cyclone to develop. These conditions also need to occur at least 500 km from the Equator. This is because the force exerted by the spinning Earth (the Coriolis force) has to be at a certain strength for tropical cyclone formation, and this strength drops off towards the Equator.

The destruction caused by tropical cyclones occurs in a combination of very strong winds (the threshold wind speed is 33 m/s but winds can exceed 80 m/s), torrential rainfall (which can cause flooding, landslides and mudflows), high waves, and particularly by storm surge, which results from the shearing effect of wind on water to cause local rises in sea level and inundation of coasts where cyclones make landfall. Around 15 per cent of the world's population is considered to be at risk (Smith, 2001), placing tropical cyclones second only to floods according to the number of hazard-related deaths, most from drowning in the storm surge. An indication of the death tolls associated with some of the worst tropical cyclones of recent times is given in Table 21.7, which also shows how costly some of these events were in economic terms in places where insurance was available.

Since attempts to control, alter or destroy tropical cyclones have not been successful, most attention is focused on reducing society's vulnerability through being prepared. Meteorological satellites provide some of the most important instruments for detecting and monitoring tropical cyclones, and early warning systems are the most widely used response. Improvements in forecasting, warnings and evacuation plans have reduced the human death toll in many countries in recent decades, although property damage has risen over the period. In the USA, 70 million people are estimated to be at risk from hurricanes, nearly 10 per cent of them resident in Florida. Several observers have noted the marked difference in impact between developed and developing countries, the former tending to be better organized and able to cope with communicating warnings and facilitating evacuation. However, similarly marked differences have also been observed within countries depending on socio-economic status (see above). Inequalities in impacts from Hurricane Katrina in the US city of New Orleans in 2005 prompted several scholars to call the hurricane an 'unnatural disaster' (Laska and Morrow, 2006), inferring that the associated impacts were more related to underlying social inequalities within the affected population than the hurricane's intensity.

The devastating effects of tropical cyclones on the economies of some small island developing states have been indicated above (Table 21.3). Cyclone Pam's direct impact on Vanuatu in March 2015 was the worst tropical cyclone in the state's recorded history, but Pam (Figure 21.8) was just one element in a multiple-disaster sequence over a two-month period that also included a severe earthquake, a small tsunami and a

Table 21.7 The most deadly tropical cyclones, 1970–2016

Date (start)	Name	Country/region	Insured damage (2016 US$ million)	Victims (dead and missing)
2 May 2008	Tropical cyclone Nargis	Myanmar, Bay of Bengal	–	138,400
29 April 1991	Tropical cyclone Gorky	Bangladesh	4	138,000
29 October 1999	Tropical cyclone 05B	Orissa, India	144	15,000
20 November 1977	Tropical cyclone	Andhra Pradesh, India	–	14,200
22 October 1998	Hurricane Mitch	Central America	589	11,700
25 May 1985	Tropical cyclone	Bangladesh, Bay of Bengal	–	11,100
26 October 1971	Odisha cyclone	India, Bay of Bengal	–	10,800
8 November 2013	Typhoon Haiyan	Philippines, Vietnam, China, Palau	525	8100
5 November 1991	Typhoon Thelma (Uring)	Philippines	–	6300

Source: after Swiss Re (2017).

Figure 21.8 Cyclone Pam shortly before crossing the island of Efate, location of Vanuatu's capital city, Port Vila, on 13 March 2015.

powerful volcanic eruption. The February/March 2015 Vanuatu disaster combination destroyed 70–80 per cent of crops and emergency food supplies in the affected areas and damaged or destroyed 90 per cent of Vanuatu's physical structures. Cyclone Pam brought sustained one-minute wind speeds of 75 m/s and gusts peaking at 89 m/s but, despite its ferocity, the storm caused just 15 fatalities. The country's early warning system and the provision of government evacuation centres have been credited with preventing a far higher death toll (Shultz *et al.*, 2016).

The country probably most frequently and severely affected by tropical cyclones is Bangladesh, which was hit by 61 severe cyclones, 32 of which were accompanied by storm surges, between 1797 and 1997 (Khalil, 1992; Swiss Re, 1998). Some of these events have been characterized by huge fatalities. The cyclone of October 1876, which produced a storm surge 12 m high, resulted in the loss of 400,000 lives; 300,000 people were killed in a 10-metre storm surge in November 1970; and 140,000 people lost their lives in tropical cyclone Gorky in April 1991. Bangladesh's high population density has meant that survivors quickly re-occupy the coastal lowlands after cyclone disasters strike, so it is likely that the most hazardous zones will always be occupied. Currently, protection for coastal inhabitants consists of 123 polders (diked areas), an early warning and evacuation system, and more than 2400 emergency shelters. More effective protection and replanting of threatened areas of mangrove forest in the Sundarbans (see p. 163) and elsewhere along the coast have also been called for, to provide a quasi-natural defence against storm surges. The long-term effectiveness of these measures is questionable, however, in light of some of the predicted changes in the region due to climate and sea-level change. Dasgupta *et al.* (2010) estimate that by 2050 nearly half of the polders would be overtopped during storm surges and another 5500 cyclone shelters (each with a capacity of 1600 people) will be needed. With other protection costs (e.g. strengthening polders, afforestation) the total cost would be US$2.4 billion with an annual recurrent cost of more than US$50 million, but these costs would be far outweighed by the potential damage in the absence of adaptation measures.

The long-term effects of natural disasters on economic development have only recently become a focus of attention for economists and an attempt to relate the impact of tropical cyclones on national economic development has been made by Berlemann and Wenzel (2018). These authors conclude that the reaction of a national economy to a tropical cyclone depends strongly on the country's level of development. Perhaps unsurprisingly, tropical cyclones had a systematically negative long-term effect on the economies of both middle- and low-income countries, but the impact on the richest countries was reversed: growth in high-income countries showed a slightly positive long-term effect from tropical cyclones. This is probably because in most cases economic investment remained stable but fertility rates declined after a tropical cyclone so that a decreasing population enjoyed an increase in per capita income.

Epidemics

Infectious diseases, many of which are preventable and curable, are the greatest cause of morbidity and mortality on the global scale. They account for nearly half of all deaths in developing countries, where children under five are the most susceptible; whereas, in developed countries, diseases of old age and abundance, such as heart disease and cancers, are the most common killers. This clear difference is linked to the fact that far greater numbers of people in developing countries suffer from malnutrition, inadequate water supply and sanitation, poor hygiene practices and overcrowded living conditions. Indeed, the global trend towards urbanization, with many of the fastest urban growth rates being recorded in the poorer countries of the world, has been cited as a major factor in the increased frequency of many infectious diseases (Olshansky *et al.*, 1997).

An infectious disease is caused by a host being invaded by a parasite or other pathogen, which might be a bacterium, virus or worm. The pathogen carries out part of its life cycle inside the host and the disease is often a by-product of these activities. Many pathogens also need a vector, an 'accomplice', to help them spread from host to host, a role often played by an insect. Many of the major diseases of the developing world are associated with water, an association in which three typical situations can be distinguished (Table 21.8), although some diseases may fall under more than one category. Although any disease represents a natural hazard to individuals and groups of people, the emphasis here will be on epidemics, relatively sudden outbreaks of a disease in a certain area. Two diseases associated with water will be examined: cholera and malaria.

Cholera is caused by a bacterium known as Vibrio, which can survive outside the human gut in such environments as water, moist foodstuffs, human faeces and soiled hands, all of which can act as media for its transmission. Cholera is a severely dehydrating illness that can kill a healthy person within 12–24 hours of the onset of diarrhoea and has become endemic in about 50 countries worldwide, killing approximately 100,000 people each year. The disease originated in South Asia, notably in the Ganges–Brahmaputra delta, and appears to spread and recede geographically in cycles. It first reached Europe in the nineteenth century, causing six major pandemics, and realization of the key transmission role played by dirty water was a central driver behind the public health movement in many European countries in the nineteenth century (Figure 21.9). These measures have helped to all but eradicate cholera from most countries in the global North, but a seventh pandemic spread from Indonesia in the early 1960s, reaching Africa and southern Europe within ten years. It reached South America at Lima, Peru, in January 1991 where it is thought to have been brought from Asia in a ship's ballast water. The warming of coastal and inshore waters by an El Niño/Southern Oscillation event at the time may have stimulated the growth of a plankton harbouring the cholera bacterium (Reid, 1995). The bacterium contaminated fish and shellfish, which were consumed by the local human population, and the spread of the disease was facilitated when it entered Lima's

Table 21.8 Classification of diseases associated with water

Role of water	Comments and examples
Waterborne	Diseases arise from presence in water of human or animal faeces or urine infected with disease pathogens, which are transmitted when water is used for drinking or food preparation (e.g. diarrhoea, dysentery, cholera, typhoid, guinea worm)
Water hygiene	Diseases result from inadequate use of water to maintain personal cleanliness (e.g. waterborne diseases as above as well as infestation with lice or mites)
Water habitat	Diseases for which water provides the habitat for vectors. The pathogen may enter the human body from the water directly (e.g. bilharzia is transmitted when a fluke leaves a snail vector and penetrates human skin as people wash or swim), or indirectly via a fly (e.g. trypanosomiasis – sleeping sickness) or mosquito (e.g. malaria)

Source: after GEMS (1988).

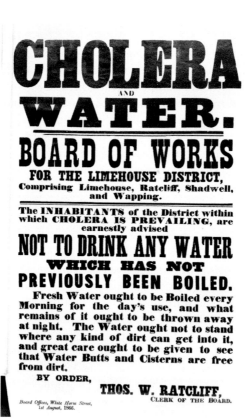

Figure 21.9 Public information broadsheet: 'Cholera and Water', posted in London, 1866.

drinking-water supply via inadequately treated wastewater. From its landfall in Peru, the disease spread throughout much of South America within a year. All previous pandemics had faded within 5–20 years, but the seventh marked its 50th year in 2011, making it the longest cholera pandemic on record, currently infecting 3–5 million people annually (Hu *et al.*, 2016).

Although it may not be possible to prevent totally the emergence of new cholera strains and the cyclic spread of the disease, there is no doubt that the introduction of basic sanitary measures, the provision of safe drinking water and the promotion of less crowded living conditions in developing countries would do a great deal towards limiting the spread of this often fatal disease. It is the lack of these basic provisions that makes cholera a common killer in refugee camps in many tropical areas.

Unlike cholera, the transmission of malaria requires a vector. The disease is caused by a one-cell parasite called plasmodium that is transmitted from one person to another through the bite of a female Anopheles mosquito, which needs blood to nurture its eggs. Transmission of this plasmodium parasite occurs naturally within certain climatic limits: the 15°C July and January isotherms, which are roughly coincident with extremes of latitude at 64° north and 32° south. Hence, malaria is a disease that occurs under natural circumstances across most of the world's land surface. In recent times, however, its geographical distribution has been radically changed by significant successes in malaria eradication.

If we go back roughly 100 years, to the early 1900s, malaria-affected areas stretched as far north as southern Canada, Norway, Sweden, Finland and Russia. In England, the disease was known as 'marsh ague' or 'marsh fever', and it was a common cause of death among many inhabitants in and around wetlands in the nineteenth century (Dobson, 1997). Malaria was still common in parts of Europe, such as the Rhine delta and low-lying parts of Mediterranean countries, until after the Second World War, but western Europe was declared free of the disease in 1975.

Efforts to control malaria have resulted in significant successes and its eradication in numerous areas. In 1900, 140 countries reported cases of the disease (Hay *et al.*, 2004), but by 2010 ongoing transmission was reported from 99 countries and 43 of these registered reductions in the number of malaria cases of more than 50 per cent between 2000 and 2010 (WHO, 2011). In terms of the proportion of the global population at risk from the disease, this reduction in its range equates to a decrease from 77 per cent of the world's population at risk in 1900, to 48 per cent in 2010.

The first attempts to combat malaria were based on the association between the disease and stagnant waters and swamps, recognized by the ancient Greeks and Romans as early as 2600 BP. This awareness led to operations to drain wetland areas with the aim of improving the health of the nearby population. These schemes also had the additional benefit of making more land available for agricultural use, thus increasing food production.

The role of a vector – the mosquito, which breeds in standing water – was realized in the late nineteenth century, enabling more concerted efforts at environmental management. A major triumph in controlling mosquitoes was recorded in the area of the Panama Canal during the early twentieth century. Initial attempts to construct the canal by the French in the late 1800s had failed partly due to the large numbers of workers who died from malaria and yellow fever, another mosquito-borne disease. When the canal was finally completed by the Americans in 1914, it was thanks in no small part to the control of these diseases through mosquito eradication. Swamp areas were drained and any remaining standing or slow-moving bodies of water were covered with a combination of oil and insecticide. The towns of Panama City and Colon were provided with piped running water to do away with the need for the domestic water containers that served as perfect breeding sites for the mosquito, and roads were paved to further eradicate puddles (Figure 21.10).

The first half of the twentieth century saw many similar examples of environmental management in other parts of the world, based on drainage and infrastructural improvements, designed to reduce the occurrence

Figure 21.10 Urban infrastructure here in Casco Viejo, the old town area of Panama City, was significantly improved as part of a successful strategy to combat malaria while constructing the Panama Canal during the early twentieth century. This involved paving the streets and putting the town's sewerage system and water supplies underground.

of malaria. This approach was complemented in the 1950s by campaigns using DDT and other powerful insecticides. DDT kills mosquitoes and was thus widely used to interrupt transmission of the disease. In 1955, the World Health Organization (WHO) adopted its Global Eradication Campaign emphasizing chemical control of mosquitoes by spraying DDT in homes. These strategies helped to eradicate malaria from most temperate and many subtropical areas, and the disease was all but wiped out in Europe and North America (although, in more recent times, increasing international travel to the tropics has resulted in cases among some temperate-zone residents).

Malaria, however, is still endemic in most parts of the developing world and, despite numerous attempts at global eradication since the 1950s, has reached epidemic proportions in a number of countries in association with high population densities, poor sanitation, environmental degradation and civil unrest. Accurate assessment of the malaria burden is difficult because most deaths from the disease occur at home, the clinical features of malaria are very similar to those of many other infectious diseases and good-quality microscopy is available in just a few centres. However, worldwide, malaria is thought to account for at least half a million deaths annually, most in young children, and 90 per cent of these deaths occur in sub-Saharan Africa (WHO, 2011). Resistance of the pathogen to drugs and resistance of mosquitoes to insecticides, as well as the rudimentary level of health services in many areas, combined to increase the malaria problem in sub-Saharan Africa in the late twentieth century. However, since the year 2000, a concerted campaign against the disease in Africa was catalysed by one of the Millennium Development Goals (MDGs), and the incidence of clinical malaria fell by 40 per cent on the continent between 2000 and 2015 (Bhatt *et al.*, 2015).

Insecticide-treated sleeping nets, the most widespread intervention, were by far the largest contributor to this success.

Another influence on the geographical pattern of malaria incidence that has received considerable attention in recent times is climate change, since even subtle changes in temperature can have significant effects on the distribution of mosquitoes. Numerous predictions have been made that in the coming decades tens of millions more cases will occur in regions where the disease is already present, and that transmission will extend to higher latitudes and altitudes. However, as Reiter (2008) points out, the global correlation between malaria and climate has been substantially weakened by a century of eradication efforts (see above). He suggests that an 'obsessive emphasis on "global warming" as a dominant parameter is indefensible' and that the principal determinants of malaria distribution are linked to ecological and societal change, politics and economics. An illustration can be made from the highlands of East Africa, where increases in the incidence of malaria have too hastily been attributed to global warming (Hay *et al.*, 2002). The rise in anti-malarial drug resistance, population migrations, and the breakdown of health services and vector control operations have all contributed to the resurgence of malaria in East African areas where no significant climate change has been detected.

Floods

Floods are one of the most common natural hazards and are experienced in every country. They occur when land not normally covered by water becomes inundated, and they can be classified into coastal floods, most of which are associated with storm surges (see above and Chapter 7), and river channel floods. Atmospheric phenomena are usually the primary cause of river flooding. Heavy rainfall is most commonly responsible, but melting snow and the temporary damming of rivers by floating ice can be important seasonally. The likelihood of these factors leading to a flood is affected by many different characteristics of the drainage basin, such as topography, drainage density and vegetation cover, and human influence often plays a role – especially by modifying land use (including the effects of urbanization) and through conscious attempts to modify the flood hazard.

Techniques for predicting floods are well developed. Most floods have a seasonal element in their occurrence and they are often forecasted using meteorological observations, with the lag time to peak flow of a particular river in response to a rainfall event being calculated using a flood hydrograph. Return periods for particular flood magnitudes can be calculated for engineering purposes, although long periods of historical data are required, which are by no means available for all river basins, and developments can alter a river's flood characteristics. Flood hazard maps are commonly utilized for land-use zoning.

A range of options for flood management is shown in Figure 21.11 with additional information to assist in deciding on which options to adopt given on the figure's axes. The relative costs and benefits of the various options are shown on the vertical axis, so that buying flood insurance for example is considered to have relatively low benefits relative to the costs, while reducing social vulnerability (e.g. through better communications, education, flood awareness-raising) has the highest benefits relative to costs. Another important factor in decision-making is indicated on the horizontal axis: the robustness of adaptation measures to uncertainties about future climate. Robustness is the degree to which benefits brought by a management option vary with assumptions about future climate. 'No regrets' options with high robustness, such as early warning systems, have strong benefits in any climate. Options with lower robustness, such as structural flood defences (e.g. dikes, levées), are more dependent on assumptions about the future climate.

Many of the options for managing the flood hazard shown in Figure 21.11 (see also Table 8.2) have been put into practice on the Mississippi River in the USA, one of the world's largest and most intensively

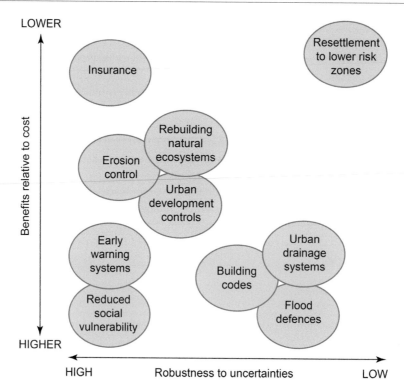

Figure 21.11 Flood management options (after Jha *et al.*, 2012). (Note that the robustness and cost–benefit ratios of measures vary case-by-case, so the bubble locations in the figure are illustrative only.)

managed rivers. Hard engineering structures have been widely employed: more than 4500 km of the Mississippi are lined with embankments or levées designed to confine floodwaters, for example.

Despite numerous attempts at hazard mitigation, the Mississippi still overflows its banks, with devastating consequences. The 1993 flood on the river's upper reaches was a severe example. Heavy rains on already-wet soil enhanced runoff between April and August 1993 and record flooding occurred along many stretches of the Mississippi, Missouri and several tributaries as levées failed in more than 1000 places (Figure 21.12). Floodwaters washed over about 4 million hectares, destroying or seriously damaging more than 40,000 buildings (Williams, 1994), with the period of inundation prolonged in many places by the levées, which prevented the return of water to the channel once the peak had passed. The cost of the damage to insurance companies totalled US$755 million, while total damage was estimated at US$12 billion, and 45 people were killed (Swiss Re, 1994).

One reaction to the 1993 flood by the Federal Emergency Management Agency (FEMA) was a buy-out programme designed to reduce potential future flood damage by removing constructions from the floodplain. Urban growth rates in 1993 flooded areas were significantly smaller between 1990 and 2000, at least partly as a result of this programme, but a study of one stretch of the river by Collenteur *et al.* (2015) found that during the following decade (2000–10), population growth in the 1993 floodplain accelerated rapidly, possibly because of the construction of nine new levées and a resulting false sense of safety. This

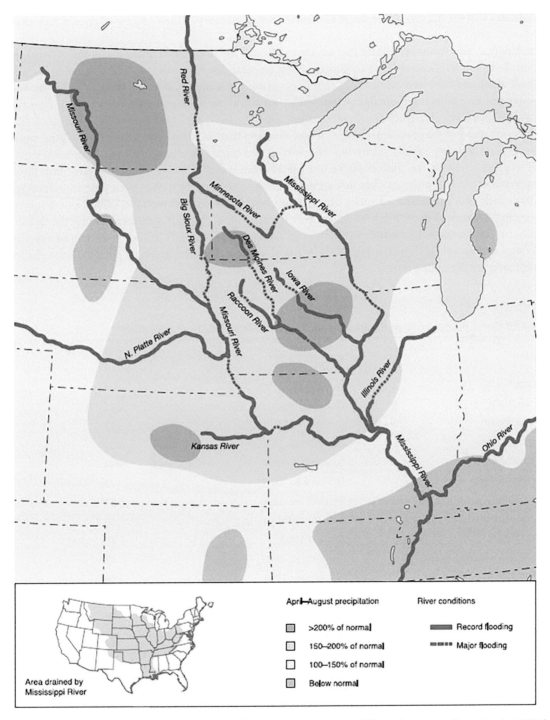

Figure 21.12 Precipitation anomalies and flooding on the upper Mississippi and tributaries in 1993 (after Williams, 1994).

study indicates that the occurrence of the 1993 disaster built a memory of the flood experience, which then decayed over time.

Interestingly, it also seems very likely that efforts to manage the river flooding hazard on the Mississippi have contributed to an increased risk of damage from tropical storms on the Gulf of Mexico coast. The levées built along the river have contributed to the loss of coastal wetlands, starving them of sediment and fresh water, thereby reducing their dampening effect on storm surge levels and possibly enhancing the economic damage from Hurricane Katrina (Stokstad, 2005).

Flash floods stand in sharp contrast to the type of flooding that occurs on the Mississippi. As the name suggests, these floods are rapidly occurring, short-lived and highly localized, making them notoriously difficult to predict. Hence, the time available to issue advance warnings is very limited. Many flash flood warning systems use computer models that simulate the hydrological processes occurring in a drainage basin, but a lack of accurate and adequate observations of rainfall and soil moisture conditions is a serious problem for many of these models (Alfieri and Thielen, 2012). Flash floods rank among the deadliest weather-related hazards in a number of European countries, particularly in the Mediterranean region (Italy, France and Spain), as shown in Table 21.9. Many of the casualties are drivers trapped in their vehicles and swept away by the raging flow.

Table 21.9 Some European flash floods with particularly high numbers of casualties

Location	Date	Deaths
Lynmouth, UK	August 1952	34
Salerno, Italy	October 1954	322
Alicante, Spain	October 1957	77
Barcelona, Spain	September 1962	>800
Lisbon, Portugal	November 1967	464
Piedmont, Italy	November 1968	72
Genoa, Italy	September 1970	37
Valencia, Spain	October 1982	40
Bilbao, Spain	August 1983	45
Vaison-la-Romaine, France	September 1992	47
Piedmont, Italy	November 1994	69
Biescas, Spain	August 1996	87
Sarno, Italy	May 1998	147
Aude, France	November 1999	36
Garde, France	September 2002	25
Draguignan, France	June 2010	25
Madeira, Portugal	February 2010	42
Skopje, Macedonia	August 2016	22

Source: after Barredo (2007); Gaume *et al.* (2009); and media reports.

The inability of structures to prevent river flooding has been highlighted in discussions over a controversial proposal to build large embankments along stretches of the major rivers of the Bangladesh delta. About 80 per cent of the national territory is floodplain, prone both to cyclonic flooding (see above) and floods from the major rivers (Ganges, Brahmaputra and Meghna) which flow through the country. Recent major river floods occurred in 1988 and 1998, inundating more than 60 per cent of the country on each occasion, three times the area normally flooded each year (Mirza *et al.*, 2001). These events had return periods of 100 years or more, and for some their effects justified an engineering solution as part of a comprehensive flood control action plan coordinated by the World Bank.

Not least among the arguments put forward against the embankments – an expensive high-tech, top-down solution – is the fact that they are designed to prevent all floods. Yet Bangladeshi villagers distinguish between beneficial rainy season floods, known as *barsha* in Bengali, and harmful floods of abnormal depth and timing, which are termed *bonna*. *Barsha* floods water the paddy fields and enhance soil fertility both through sediment inputs and through nitrogen-fixing algae, which thrive in the floodwater (Figure 21.13). Fish caught in flooded fields are also the main source of animal protein for rural inhabitants (Boyce, 1990).

Figure 21.13 Not all floods are considered as hazards by rural Bangladeshis, since some enhance crop production. Yields from both the paddy rice being planted here, and the taller jute plants behind, benefit from a period of inundation.

While there is widespread agreement that Bangladesh needs to improve resource management, opponents of the embankment proposal suggest that the scale of natural hazards in Bangladesh is possibly beyond the scope of human control. They argue for a social science approach to flood management, less technological and more ecocentric in its application, aiming to actively involve local inhabitants. New ponds could be constructed for fishing and water storage, and low embankments would check flooding only during the early growth stages of the rice crop. Better preparation for unusually high and rapid flooding would complement these approaches. In addition to these measures, others advocate a more deep-rooted, less technocentric, strategy that seeks to alleviate the factors that cause people to be vulnerable to flooding. Better access to land, replacements for land lost to erosion, and compensation for animals and other assets lost to natural disasters would all protect livelihoods and reduce vulnerability, as would better health care and facilities (Wisner *et al.*, 2004).

Despite the controversies in Bangladesh surrounding the desirability of large-scale physical structures to protect societies from flooding, such projects remain popular methods for flood control. Many of the objections to a permanent structure can be addressed by a movable flood barrier like the Thames Barrier in London, designed to protect the UK's capital from unusually high tides caused by storm surges. These occur when very strong winds and waves associated with an area of low pressure in the North Sea force water towards Britain's east coast. When a storm surge coincides with a high tide, the result is often flooding on a large scale, as in 1953 when parts of the English coast were inundated, from Yorkshire to Dover, drowning more than 300 people.

For most of the time, the Thames Barrier is open, allowing shipping to operate normally, and minimizing interference with the natural flow of the river. An early warning system operates to predict exceptional water levels from surge, tide and river flows several hours in advance, allowing time for the Barrier to be closed. Following completion in 1982, the general trend over its first 35 years or so has been towards more frequent closures, although there is considerable year-to-year variability (Figure 21.14). Most closures are

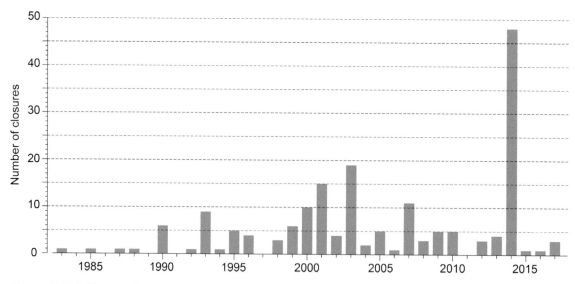

Figure 21.14 The number of times the Thames Barrier has been closed to protect London from flooding, 1983–2017 (data from the UK Environment Agency).

to protect against tidal flooding but on occasion they are to assist in preventing fluvial flooding from upriver by stopping sea water from enhancing already unusually high river levels. In total, nearly 80 per cent of closures have been since 2000. The designers of the Thames Barrier envisaged that the number of closures would increase to around ten per year on average in the first two decades of the twenty-first century because of predicted sea-level rise due to both ground subsidence in south-east England and the effects of global warming.

As sea level continues to rise, the Barrier will have to close more frequently to prevent overtopping of the flood defences upstream of the Barrier. The Thames Barrier was designed to protect London from flooding until at least the year 2030, but the future flood threat to London will be managed using an 'adaptation pathway approach' that is designed to keep options open as circumstances change. The existing system will continue over the next 25 years, followed by a period when existing defences will be enhanced. Beyond 2070, if sea-level rise accelerates, more substantial measures will probably be needed, including the possibility of a new barrier (Penning-Rowsell *et al.*, 2012).

FURTHER READING

Berkes, F. 2007 Understanding uncertainty and reducing vulnerability: lessons from resilience thinking. *Natural Hazards* 41: 283–95. A theoretical paper focusing on how society can cope with hazards.

Dixon, T.H. 2017 *Curbing catastrophe: natural hazards and risk reduction in the modern world.* Cambridge, Cambridge University Press. This book focuses on a few specific disasters to assess how to reduce the risk they pose to society.

Jongman, B., Winsemius, H.C., Aerts, J.C., de Perez, E.C., van Aalst, M.K., Kron, W. and Ward, P.J. 2015 Declining vulnerability to river floods and the global benefits of adaptation. *Proceedings of the National Academy of Sciences* 112(18): E2271–80. An examination of the vulnerability of societies to flooding. Open access: www.pnas.org/content/112/18/E2271.full.

Pierson, T.C., Wood, N.J. and Driedger, C.L. 2014 Reducing risk from lahar hazards: concepts, case studies, and roles for scientists. *Journal of Applied Volcanology* 3(1): 16. An excellent overview of lahar hazards. Open access: appliedvolc.springeropen.com/articles/10.1186/s13617-014-0016-4.

Reiter, P. 2008 Global warming and malaria: knowing the horse before hitching the cart. *Malaria Journal* 7(suppl. 1): S3. A critical review of the potential impact of climate change on malaria. Open access: www.malariajournal.com/content/7/S1/S3.

Smith, K. 2013 *Environmental hazards: assessing risk and reducing disaster*, 6th edn. London, Routledge. All major rapid-onset events of both the natural and human environments – including seismic, mass movement, atmospheric, hydrological and technological hazards – are covered in this book, which emphasizes the physical aspects of hazards and their management.

WEBSITES

www.emdat.be   designed for humanitarian agencies, the Emergency Events Database, EM-DAT, contains data on the occurrence and effects of mass disasters across the world since 1900.

www.eri.u-tokyo.ac.jp   the Earthquake Research Institute at the University of Tokyo.

www.ivhhn.org   the International Volcanic Health Hazard Network facilitates research and disseminates information on the health effects of volcanic emissions.

www.preventionweb.net   a huge information hub on disaster reduction.

www.swissre.com   includes annual 'sigma' reports on hazards and catastrophes compiled by the insurance industry.

www.unisdr.org   the UN Office for Disaster Risk Reduction helps societies to become resilient to the effects of natural and technological hazards and disasters.

POINTS FOR DISCUSSION

- Can natural hazards ever be construed as good for human society?
- Since the precise timing and location of earthquakes are virtually impossible to predict, why do people live in zones that are renowned for their earthquake activity?
- Discuss the proposal that economic development is the best way of reducing the risks posed by natural hazards.
- Are some countries naturally more hazardous than others?
- Is vulnerability to natural hazards simply an environmental justice issue?
- If floods are the most common natural hazard, why aren't we better at coping with them?

22 *Conclusions*

Environmental issues are not new phenomena. People have always interacted with the natural world and these interactions have always thrown up challenges to human societies. Mistakes have been made and solutions often found to problems encountered or created. However, as human society has become more complex, as populations have increased, so the range and scale of environmental issues have multiplied. Although there are still many sizeable portions of the Earth that show little obvious evidence of human impact (e.g. many hyper-arid deserts, the deep oceans, parts of the polar regions and some of the tropical rain forests), there is nowhere that is not affected to some extent by changes in the chemical makeup of the atmosphere and associated changes in climate and pollution levels. Likewise, the nature of relatively unaffected parts of the planet also has an impact on those regions that are directly used by human populations, through their effects on global climate and biogeochemical cycles.

It is the global nature of so many environmental issues in the Anthropocene that has thrust them to prominence in our minds. Environmental scientists have recognized two types of global environmental change induced by human society (Table 22.1): systemic change through a direct impact on globally functioning systems (e.g. emissions of greenhouse gases that affect global climate) and cumulative changes that attain global significance through their worldwide distribution (e.g. species loss) or because of their effects on a large proportion of a global resource (e.g. soil degradation on prime agricultural land). The impacts of both are recognized as being of global proportions, but the appropriate responses to systemic and cumulative change vary. Addressing a systemic global environmental issue like global climate change requires a truly global strategy: all countries must take action together to combat the potential dangers. Addressing a cumulative global environmental issue can also benefit from a worldwide approach, through the exchange of data and comparison of strategies, for example. However, many cumulative global environmental issues need to be tackled at the local level because the precise causes of problems are often different in different places.

Several common themes emerge from the preceding chapters on individual issues. Numerous issues have arisen as a direct result of deliberate attempts to manipulate the natural environment to our advantage (e.g. widespread soil erosion on agricultural land, the Aral Sea tragedy, the impacts of big dams). In these cases, the issues stem from our managing environments to maximize the production of one ecosystem service which results in unexpected and unwanted declines in the provision of other ecosystem services. Other issues have arisen as unexpected, indirect repercussions of activities conducted to improve human societies (e.g. contamination by fertilizer residues, global warming, acid rain). The complexities of many environmental issues are also clear. Most are related to some form of environmental change. Sometimes this change is rapid, at other times it is more gradual, but always these changes are occurring in complex

Table 22.1 Types of human-induced global environmental change

Type	Characteristic	Examples
Systemic	Direct impact on globally functioning system	Industrial and land-use emissions of greenhouse gases Land cover changes in albedo
Cumulative	Impact due to worldwide distribution of change	Biodiversity loss Groundwater depletion
	Impact due to scale of change (proportion of global resource)	Deforestation Soil degradation

Source: after Turner *et al.* (1990).

systems that involve both people and nature. Cause and effect are not always obvious (e.g. the many factors affecting forest decline: see Table 12.3), and different human activities can sometimes combine to give a synergistic effect (see Table 11.7). Impacts on the natural environment are also not always immediate or expected, due to the operation of thresholds, feedbacks and timelags. Many environmental issues overlap and interact. Deforestation in the tropics is a major cause of biodiversity loss, and global warming has been seen to have all sorts of potential knock-on effects for many other issues.

A full understanding of many issues is hampered by our imperfect knowledge of the physical environment. There are still many 'blank spaces on the map', both literally and metaphorically (Figure 22.1), because we remain a long way off knowing all the basic information about this planet and indeed its peoples. The collection and analysis of data, both the measurement and monitoring of contemporary processes and the study of palaeoenvironmental reconstructions using proxy indicators, continue to be important tasks for all of those disciplines involved in the study of environmental issues. A scientific approach to these tasks will increasingly be complemented by other perspectives, such as traditional ecological knowledge (see Box 11.1).

As our knowledge and understanding improve, so our perspective on certain issues can change. Reappraisal of the Sahelian fuelwood crisis, widely feared in the 1970s but now thought to be much less of a

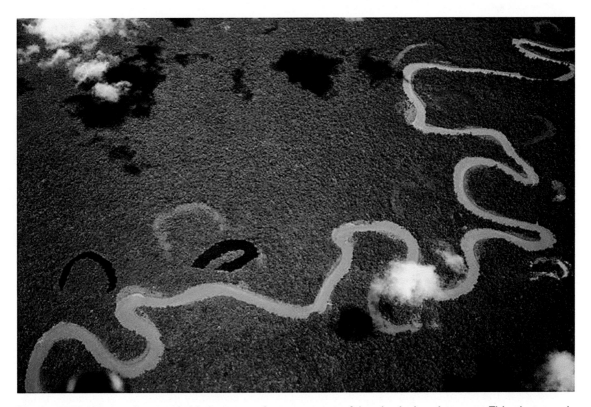

Figure 22.1 We remain remarkably ignorant of many aspects of the physical environment. This photograph is taken over tropical rain forest – a biome about which we know little – in New Guinea, a large island that remains relatively unexplored by scientists.

concern, is a case in point here. We can only base our approach to managing environmental issues on existing knowledge, while continuing to conduct research to improve our understanding. But in the meantime we must be aware of uncertainties, hence the emergence of important tenets such as the precautionary principle and 'no regrets' approaches.

Some elements of human society emerge as common to several issues. Ownership, or a lack of ownership, is a pertinent issue in many atmospheric and marine environmental problems. Economic factors are also important. Poverty can drive people to degrade their resources, or force individuals to live in marginal and hazardous areas, and the poor tend to be harder hit by disasters, an issue of environmental justice. Societies and governments may not tackle environmental problems because they lack the necessary economic resources even though workable solutions to their problems may be well known. This is not to say, however, that poor people necessarily cause environmental problems, nor that richer sectors of society do not cause them. Greater economic prosperity certainly tends to be equated with greater environmental impact: the ecological footprint of richer societies and countries is much larger than that of poorer groups (Figure 22.2). Indeed, where systemic global environmental change issues, such as global climate change, are concerned, responsibility lies primarily with richer people in richer countries. This is clearly a case

Figure 22.2 The city of Los Angeles, USA, shrouded in smog caused by its own waste products. Urban areas, particularly those in rich countries of the North, represent a huge concentration of resources brought from all over the world. A key question that underlies many environmental issues is whether everyone on this planet can consume resources at the rates enjoyed by people who live in countries such as the USA.

of global-scale and intergenerational environmental injustice because the impacts of climate change are disproportionately felt by certain poorer countries and regions, and by future generations. That said, sometimes environmental injustices are indiscriminate: affecting people over large areas and across the social spectrum. Fallout from the accident at the Chernobyl nuclear power plant is one example (see Chapter 18).

Political factors also play a key role. International political forces have worked together to develop many agreements and conventions designed to address environmental issues (see below). Political and economic power often go hand in hand. Many wars and other conflicts arise due to competition over resources, and inadequate enforcement of conservation measures may result from political priorities that lie elsewhere or through the weakening of civil authority. Political corruption and illegal logging have featured prominently in the tropical deforestation process (see Chapter 4).

Perceptions and priorities are important too. The way we interact with the environment can change because our view of certain aspects of nature is subject to change. The motivation behind assessing the global area of forests, for example, has altered over the past 100 years or so. Initially driven by interest in timber resources, it is now more closely related to global environmental concerns such as biodiversity loss and climate change. This shift reflects a more general move in our view of the natural environment as a whole: from nature as resource to nature as life-sustaining global ecosystem that also harbours resources. Perceptions of a particular landscape may vary with social and technological changes within a particular culture. Individuals within that culture may also see the same landscape in a different way, perhaps depending on socio-economic and political inequalities. Furthermore, different cultures may see different resources and/or challenges in the same places. The landscape of Greenland (Figure 22.3) was viewed as

Figure 22.3 The Inuit population has successfully hunted wildlife here in northern Greenland for about 4500 years but for Norse settlers who arrived in medieval times the terrain was seen as very harsh.

very marginal in late medieval times by Norse settlers wanting to plough the land and herd animals, but their Inuit contemporaries considered the region to be exceptionally rich because they obtained their food resources by hunting.

Our approaches to explaining change in the environment are also subject to shifting perceptions and priorities, themselves in part reflected in shifts in intellectual fashion. A case in point is the intense current interest in the importance of climate as a driver of change in both the physical and human environment. The interest stems from concerns about anthropogenic impacts on global warming, which most agree is real enough, but one result has been an intellectual climate that is increasingly sympathetic to the idea of cultural catastrophes triggered by environmental change. Palaeoenvironmental researchers have produced numerous studies showing apparent correlations between climatic change and cultural collapse, while a few have used similar lines of evidence to argue that climatic change also appears to have played a significant role in the emergence of civilizations (see p. 266). This fashion for climatic explanations of change marks the return to prominence of a form of environmental determinism.

However, there is a risk that we are putting too much emphasis on climate as a driver of change. There is the possibility that such associations may be purely coincidental or of marginal importance. Indeed, we should also examine the numerous instances where cultures have survived rapid environmental change unperturbed or collapsed without any environmental forcing. A wider assessment of historical collapses of society has identified five groups of interacting factors – which may drive the change unilaterally or in combinations – as being especially important (Diamond, 2005): the damage that people have inflicted on their environment; climatic change; changes in friendly trading partners; the deleterious effects of enemies; and the society's political, economic and social responses to these shifts.

Examples in history, prehistory, and indeed in the contemporary era, will always be subject to new interpretations and fresh analyses. The supposedly textbook example cited in Chapter 3 of a culture that doomed itself by destroying its own habitat, that of Rapa Nui or Easter Island in the Pacific, has also been interpreted in several different ways (Table 22.2). One alternative explanation cites evidence for abrupt climatic and environmental change in the Pacific around 1300 CE, ultimately driven by an increase in El Niño events. Another prefers a synergy of impacts, including the devastating ecological effects on the island's trees of rats inadvertently introduced by the original settlers, and the later impacts of European contact on demography.

Another message from the preceding chapters of this book is that supposedly advanced societies can learn useful lessons about environmental management from indigenous cultures, which often remain 'closer to nature' than urban, industrialized groups. Long experience of coping with the environmental dynamism of drylands, for example, has led to numerous strategies for managing risk and variability in Sahelian Africa (see Table 5.7). However, again, this is not necessarily to say that societies operating with rudimentary levels of technology are always better at avoiding environmental problems. Examples to the contrary have also been highlighted in this book: accelerated soil erosion rates in central Mexico were as great before the arrival of the plough brought by Spanish invaders, as after, and evidence suggests that at least half of Hawaiian bird species became extinct in the pre-European period.

These examples also serve to re-emphasize the fact that people have been leaving a significant mark on nature for a long time – in some cases a very long time. The first urban cultures began to develop around 5000 years ago (see Chapter 10); we think the domestication of plants and animals began some 10,000 years BP (see Chapter 13), and many believe overhunting by our Stone Age ancestors was at least partly responsible for the disappearance of large mammals such as the mammoth even before that (see Chapter 15). The starting date for the Anthropocene, the era when human action has emerged as a critical force in global biophysical systems, is debatable (see Chapter 3), but certainly the importance of human

Table 22.2 Some suggested explanations for Rapa Nui's environmental catastrophe and population collapse

Main driver of change	Causal chain	Reference
Ecocide	Deforestation leads to accelerated soil erosion, falling crop yields and food shortages resulting in intertribal warfare, social disintegration and population collapse	Bahn and Flenly (1992)
Environmental change	Climatic deterioration leads to resource depletion, increased competition and food shortages resulting in intertribal warfare, social disintegration and population collapse	Nunn (2000)
Biological invasion and genocide	Rats responsible for forest decline. European contact brings new diseases and slave-trading resulting in social disintegration and population collapse	Hunt (2007)

activity has become heightened since the Great Acceleration of the mid-twentieth century (Steffen *et al.*, 2015). This is also the time period during which human society has spread novel materials such as concrete and plastics worldwide. Arguably, the most significant global impact occurred as a result of the fallout from nuclear weapons testing (see Chapter 20), but the development of concrete has enabled us to build dams larger than ever before (see Chapter 9), enhancing our ability to transform the natural environment on a huge scale (Figure 22.4). Plastics are used for many purposes but their largest market is for packaging and such single-use plastics (e.g. plastic bags) have become a very significant source of pollution both on land and at sea (see Chapter 6).

Nevertheless, and despite these obvious recent examples of large-scale human impact, long histories of profound human influence may make the traditional division between the impact of nature on society, on the one hand, and human impact on the environment, on the other, inappropriate. The division will undoubtedly continue because human actions represent the one set of driving forces for change that we can control in a predictable way. However, in some cases it may be that human activities and environmental change are better viewed together, as a co-evolutionary and adaptive process.

This book has also highlighted many areas where environmental issues have been tackled successfully. A general trend towards improving urban air quality has been documented with higher development levels, and there are many examples of waste disposal problems being ameliorated by viewing society's waste products as resources waiting to be used. Sometimes, however, environmental recovery may be slow because as one pressure recedes another takes its place. Examples can be cited from efforts to remediate the impacts of acidification. Methods for rehabilitating and protecting the environment are constantly being developed, and the fact that many issues are interrelated creates synergies, so that efforts to tackle one issue often have beneficial effects for others. As Figure 22.5 shows, policies that bring about reductions in air pollutants produce numerous benefits to many aspects of the physical and human environments.

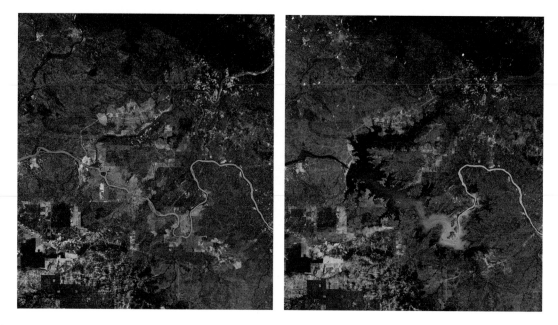

Figure 22.4 Satellite images of the Lower Sesan II Dam in Cambodia before (2017 left) and after (2018 right) its completion resulted in the permanent inundation of 75 square kilometres of territory along two tributaries of the Mekong River. The concrete dam is 75 metres high and 8 kilometres long.

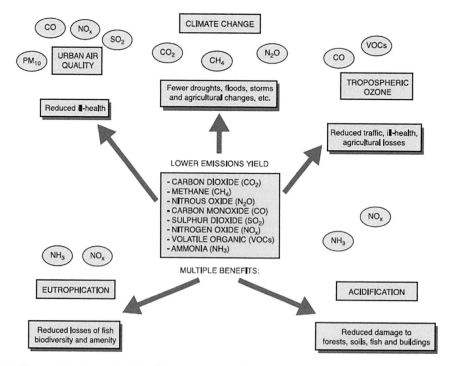

Figure 22.5 The multipollutant/multi-effect approach (after EEA, 1997).

Of course the very rise in interest in sustainable development – 'the most recent big idea to emerge in the history of the Anthropocene' (Clark *et al.*, 2005: 2) – can be seen as an appropriate response by society to the scale and critical nature of many environmental issues. This book has been organized on an issue-by-issue basis, but the remainder of this chapter is devoted to a more regional approach, looking at the ways in which particular societies are reacting to the many environmental issues that affect them.

REGIONAL PERSPECTIVES

A distinction can be made between three global regions that should have different priorities in their contributions to global environmental sustainability (Goodland *et al.*, 1993a). The North must concentrate on reducing its long history of environmental damage due to overconsumption and affluence; the main priority in the South should be to stabilize population growth; and in the countries of the East, the former communist bloc, modernization of wasteful and polluting technology should be the central priority (Figure 22.6). To promote these strategies, it is in the self-interest of the North to accelerate the transfer of technology to the East and South.

Figure 22.6 Much of the industry in the former communist countries of the East was developed with pollution control as a relatively low priority, as at this iron and steel complex at Elbasani, Albania. The transfer of improved, cleaner technologies from the North is ultimately in the interest of all concerned.

Measures of population density and relative affluence (which also has a major bearing on technological capacity) have been used to classify some major world regions, as shown in Figure 22.7. More detailed design and implementation of sustainable development strategies for these regions will vary according to their conditions and characteristics. Low-income regions with low population densities such as Amazonia and Malaya–Borneo include many of the world's remaining settlement frontiers where large-scale clearance for agriculture, grazing and timber has begun only recently, after a longer history of resource use in shifting cultivation, small-scale plantation and mining sites. The widespread poverty of farmers engaged in land clearance, and the poorly developed state of local institutions of governance, which might guide sustainable development, mean that appropriate management in such areas will be particularly difficult to attain.

Regions of low income and high population density, such as on the Ganges–Brahmaputra floodplains of the Indian subcontinent and the Huang–Huai–Hai plains of China, by contrast, have long histories of human agricultural endeavour. This agricultural character has been augmented in recent decades by rapid industrial development and urbanization, bringing the types of pollution problems that have been faced by industrial European countries within the past 100 years. The introduction of employment opportunities that can relieve pressure on agricultural land, but that do not simultaneously exacerbate urbanization and industrial pollution problems, is the critical management challenge.

Other developing world areas, such as the Basin of Mexico, with an agricultural history over hundreds of years, have experienced very rapid population growth, speedy industrialization and a burgeoning consumer culture in recent decades. The grave problems of pollution and congestion are typical, if extreme, for primate cities of the developing world. Development in such cases can only be made sustainable, and problems eased, by a concerted national effort at decentralization.

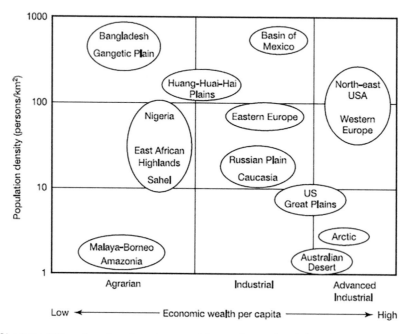

Figure 22.7 Characteristics of regional environmental transformation (after Clark, 1989; Kates *et al.*, 1990).

Areas that bear the more direct mark of environmental impact from the relatively wealthy countries include such frontier regions as the Arctic and the Australian desert – environments perceived as 'harsh' and 'sensitive' by the residents of more temperate climes. Exploitation, primarily for mineral resources, in these areas of low population density involves sophisticated technology and large economic investments. Although our knowledge of such environments remains poor, and hence the consequences of inappropriate actions can be great, the involvement of relatively few powerful corporate and governmental institutions means that the potential for introducing sustainable development practices is relatively good.

The greatest potential and the largest responsibilities for appropriate change lie in the densely populated wealthy industrialized regions of the world. Such regions have perpetrated a disproportionately large environmental impact, both locally and at the continental and global levels, through such pathways as emissions of greenhouse gases, their typically large ecological footprints effected through the processes of globalization, and the export of environmentally damaging practices to other parts of the world. In recent times these areas have also achieved some successes in improving environmental quality locally. Some of the leads these countries have taken are considered in more detail in the following section.

Most of the initiatives described in this section will be taken at the national and regional level, and will contribute to both systemic and cumulative global environmental change issues. However, the mitigation of and adaptation to these environmental issues may also be affected by one or several major regional phase shifts precipitated by the crossing of tipping points associated with global climate change. Some of these have been described in Chapter 11 and are depicted in Figure 22.8. Should any of these regional thresholds be breached, whether it be this century or some time in the more distant future, the pathways towards sustainability are likely to require some equally dramatic redirection.

NATIONAL EFFORTS

While some of the technologies and efforts to curb environmental problems have been introduced in the preceding chapters, numerous instances of environmental improvements being hampered by economic and political forces have also been noted. Hence, it is appropriate to look at some of the ways in which governments and other institutions have worked to promote sustainable practices. The rise in interest in environmental issues and sustainable development has inspired numerous efforts to gather information on national environmental situations and to develop national strategies for sustainable development. The suggestion that all countries should prepare such a strategy was made at UNCED, repeated at Rio+20, and embodied in the UN's 2030 Agenda for Sustainable Development, which launched the Sustainable Development Goals (SDGs) in 2016. Guidelines from the UN and the OECD define sustainable development strategies as cyclical strategic processes that are supposed to combine formal planning and incremental learning. Some countries have made good progress in this regard but by no means all. For those strategies that are in place, the use of economic, social and environmental indicators to measure changes and the efficacy of policy instruments is a key feature.

Probably the first formalized national strategy for achieving sustainable development was launched in the Netherlands. The Dutch National Environmental Policy Plan, first introduced in 1989 and subsequently updated, provides the framework for policies to drive the country towards sustainability. Since 1970, Dutch environmental policy has evolved through three distinct phases (Table 22.3) and has expanded from a focus on protecting human health and ecosystem integrity to ensuring a sustainable supply of natural resources at national and international levels. This expansion has been achieved by changing the style of policies, while management of the physical environment has evolved from a focus on eliminating pollution after it occurred to avoiding it altogether. At the beginning of the twenty-first century, Dutch environmental

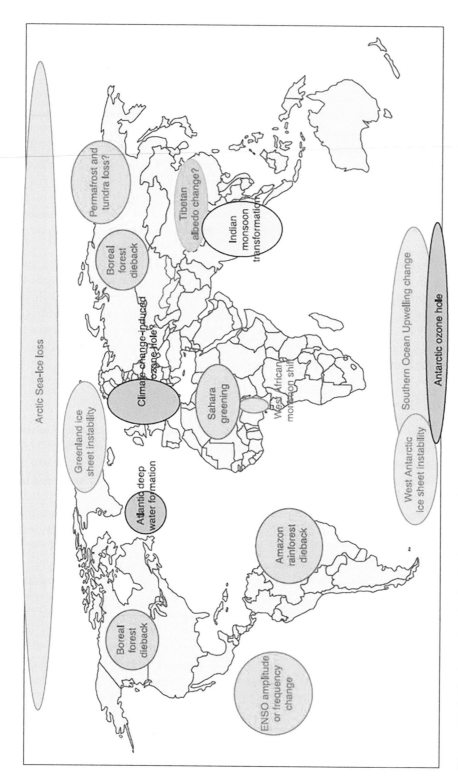

Figure 22.8 Tipping points in the Earth system that could be breached by changes wrought by global warming (after Kemp, 2005; and Lenton et al., 2008).

Table 22.3 Evolution of environmental policies in the Netherlands since 1970

Period	Policy focus	Style of decision processes	Instruments
Delimiting the ecological arena (1970–83): improving national conditions for human health	Cleaning up national stocks of air and water; starting soil clean-up operations	Top-down national legislation and quality standards; no stakeholder involvement	National and European laws; regulatory licensing enforced by local authorities; add-on technology
Encouraging pollution prevention (1984–89): protecting national health conditions for people and ecosystems	Pollution prevention to preserve national stocks of air, water, soil and biodiversity; initial attention to acidification and ozone depletion	Start of stakeholder involvement in design and implementation of programmes; allowing room for technology choices and flexible timing	Emission reduction targets; environmental care; EIA; financial incentives; legal liability; stricter enforcement
Enhancing eco-efficiency (1990–99): building on previous actions to include international responsibility for global air quality	More attention to transnational issues of acidification, global warming and ozone depletion	Greater autonomy granted to local authorities and private enterprise to set objectives; more flexible design and implementation	Targets; target groups; negotiated agreements; economic, technological, fiscal and social instruments
Super-optimization for sustainable development (2000–30): opportunities for multisectoral benefits	More attention to limiting national use and improving management of global stocks of biodiversity, energy and minerals	Provoking integration of ecological, economic and social interests nationally and internationally; new processes to develop joint objectives	Target global resources; incentives for producers and consumers; breakthrough technologies; get prices right; quality consumption

Source: after Keijzers (2000).

policy began shifting from pollution prevention to the sustainable management of resource stocks (Smith and Kern, 2009).

The first period, from 1970 to 1983, focused on cleaning up national stocks of air and water in the Netherlands primarily by a 'command and control' approach of setting national quality standards enforced by environmental laws. This legislation and the policies it contained were based on a number of general principles, which continue to underpin subsequent development of policy. They are:

- *the polluter pays principle* (polluters are liable for the costs of clean-up and the prevention of pollution)
- *stand still principle* (polluted areas should not be polluted further and clean areas must remain clean)

- *principle of isolation and control* (pollution should be controlled at its source, not exported elsewhere through the air or water)
- *principle of priority for pollution abatement at source* (as opposed to end-of-pipe solutions)
- *the use of 'best technical means' technology* (to eliminate emissions when serious risks to public health are at stake regardless of economic costs, and the use of 'best practical means' technology when health effects are limited, so enabling industry to consider economic costs)
- *the principle of avoiding unnecessary pollution.*

This approach proved to be effective at dealing with the most obvious pollution problems. Air and water quality improved thanks to significant reductions in emissions of pollutants including heavy metals, SO_2, and phosphates from detergents. However, the excessive use of fertilizers and pesticides in agriculture, and emissions from the transport sector, continued to cause serious pollution.

The second period, 1984 to 1989, saw a shift in emphasis from reacting to pollution problems to actively trying to prevent them from becoming an issue. This shift was accompanied by a change in policy, away from the command and control approach of environmental laws, to a structure of economic incentives, social institutions and self-regulation. Policies aimed to encourage pollution prevention by starting to involve all stakeholders, including business, government and local communities, in the design and implementation of programmes. This focus on preventing pollution at source spawned new instruments such as environmental care programmes for businesses and environmental impact assessments. The process enabled a shift in perception from seeing environmental standards as being in conflict with economic interests (because achieving the standards cost money) towards viewing a clean environment as a necessary precondition for a sound economy. The results from this period were also encouraging as chemical industries developed substitutes for phosphate detergents, agreements were reached to reduce packaging materials and the production of mercury batteries was stopped. Industry and energy suppliers made further drastic reductions in emissions of heavy metals and SO_2, and pollution from agriculture began to decline.

The third period in the evolution of Dutch environmental policy, from 1990 to 1999, saw the introduction of the National Environmental Policy Plans (NEPPs): NEPP1 in 1989, followed by NEPP2 in 1993 and NEPP3 in 1998. These NEPPs have continued the policy of emissions reductions, which aims to safeguard the quality of air, water and soil by setting targets for pollution prevention. The focus has been on eight themes: dispersal of toxic substances, waste disposal, disturbance by noise and external safety from hazards, eutrophication, acidification, climate change (global warming and ozone depletion), groundwater depletion, and squandering of resources. By this time, a clear premise of the Dutch environmental management philosophy had emerged: that a high-quality environment cannot be achieved through conventional pollution-control measures alone. A mixture of new, cleaner technologies and structural changes in patterns of production and consumption is also needed. The NEPPs have emphasized the need to improve eco-efficiency (the level of emissions and resources used per unit of production), and further encouraged the integration of ecological and economic concerns (Figure 22.9).

The style of policies to implement the NEPPs has also continued to develop away from environmental legislation and regulations towards more open decision-making with more flexibility and autonomy for local authorities and enterprises to define their own ways of achieving targets set. When launching NEPP3, the Dutch government announced further significant achievements since NEPP1: energy efficiency in industry had increased by more than 10 per cent, the sale of CFCs ceased in 1995, the proportion of waste reused increased from 61 per cent to 72 per cent in the period 1990–96, and acidification had decreased by 50 per cent since 1985.

Figure 22.9 Windmills at Kinderdjik in the Netherlands, built in the eighteenth century to drain water from an area which is below sea level. They are symbolic of the Dutch domination over nature (more than half the national land area is made up of reclaimed wetlands) that is taking on a progressively more sustainable approach.

Since the introduction of NEPP1 in 1989, the Netherlands has succeeded in achieving economic growth with a reduction in environmental pressure in many areas. However, policies to curb energy use, to reduce emissions of CO_2 and NO_x, and to minimize resource use were still far from sustainable levels as described in NEPP1. Key areas for attention in NEPP4, launched in 2002, include managing CO_2 emissions, controlling future infrastructure development, minimizing resource use and reducing the burden on biodiversity.

Nonetheless, the Netherlands, like all countries, does not operate in a vacuum. The country is affected by pollutants that arrive from outside its borders, such as water pollution in the River Rhine (see Box 19.1). The low-lying country is also particularly vulnerable to any rise in sea level that may be consequent upon human-induced climatic change. Dutch exports produced under a sustainable regime may be less competitively priced than equivalent products derived from unsustainable methods elsewhere, and Dutch imports of resources from other countries may effectively contribute to environmentally damaging practices elsewhere. Various attempts have been made to measure the Netherlands' ecological footprint, which has a considerable overseas element. Rood *et al.* (2004) concluded that the size of this footprint increased by approximately 40 per cent over the 40-year period 1960–2000, and that trend is confirmed in data from

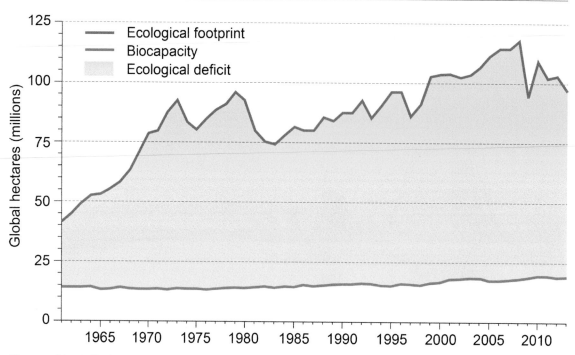

Figure 22.10 Netherlands' ecological footprint vs. biocapacity, 1961–2013 (Global Footprint Network National Footprint Accounts, 2017 Edition, accessed February 2018 from data.footprintnetwork.org).

the Global Footprint Network from 1960 to 2013 (Figure 22.10). While the NEPP is undoubtedly a very significant step in the right direction at the national level, its measures can only work effectively if paralleled by similar initiatives elsewhere.

In a global analysis, Szigeti *et al.* (2017) concluded that the average ecological footprint intensity (calculated as a country's ecological footprint divided by its GDP) had improved significantly in the first decade of the twenty-first century and that some countries had managed to reduce their ecological footprints overall. This study also found that a handful of countries (Norway, New Zealand, Japan, UK) had managed a significant reduction in their ecological footprint in absolute terms while maintaining significant GDP growth, in effect decoupling economic growth from environmental impact.

GLOBAL DIMENSIONS

National efforts towards sustainable development are certainly important, but globalization has increased the interconnectedness of peoples and countries. Although most pressure upon ecosystems occurs within national borders, there is also very significant displacement of pressure on ecosystems through international trade. In general terms, high-income countries require more biologically productive land per capita than low-income countries, and this need is fulfilled through global trade links (Weinzettel *et al.*, 2013). In other words, affluence drives the global displacement of land use. In an analysis of these global supply chains, Yu *et al.* (2013) showed that nearly half of cropland in Brazil and 88 per cent of Argentinean cropland is used for consumption purposes outside their national territories, in these cases mainly in EU countries and China. Other environmental issues have more direct global dimensions. Many issues occur across international boundaries and all countries contribute to, and are impacted by, systemic global environmental change issues.

International attempts to promote conservation and sustainable development through effective environmental governance have been mentioned already in this and other chapters. These include the establishment of the UN Environment Programme, the recommendations of organizations such as the World Commission on Environment and Development, and many global international agreements, such as the conventions on biological diversity and climate change negotiated at UNCED (the Rio Earth Summit of 1992). There are also many other regional agreements and conventions. Since the turn of the current century, global efforts towards addressing a range of environmental and development issues have focused on first the Millennium Development Goals (MDGs) and then their successors, the Sustainable Development Goals (SDGs).

Table 22.4 shows the major global conventions that address environmental issues. Such conventions typically arise as concern at the environmental effects of certain activities reaches a threshold at which global action is considered necessary. They reflect the fact that a sufficient number of countries recognize a common interest in trying to combat certain issues, including those rooted in both systemic and cumulative global environmental change. The international community can act quickly when issues are perceived to be particularly acute. The Chernobyl accident in April 1986 precipitated two conventions on nuclear accident notification and assistance in the same year.

The years indicated in Table 22.4 are when the conventions were first agreed. A country becomes a signatory to a treaty when someone given authority by a national government signs it. Usually a signatory has no duty to perform the obligations under the treaty until the treaty comes into force in that country, which means the national government has to ratify the treaty, or otherwise adopt its provisions in national law, and the treaty itself is ratified by a prescribed number of countries. This can take a long time. The UN Convention on the Law of the Sea, for example, was first signed in 1982, but did not come into force until 1994. Countries can still join conventions after the first signing session. The Convention on Wetlands of International Importance Especially as Waterfowl Habitats (the Ramsar Convention) was first signed by 18 countries in 1971, but by early 2018 the original 18 had been joined by 151 other contracting parties.

These conventions represent a significant step forward at the international level, but international agreements and conventions still have their weaknesses. Enforcement is a particular difficulty. While non-compliance with an agreement can lay a country open to criticism, compliance can rarely be enforced by law. Part of this problem is related to the difficulties of translating agreements in principle into national laws. Another weakness is structural. Any agreement requires compromise between the parties, and international agreements, often involving hundreds of nation-states, can result in provisions that are imprecise and not very constraining.

Many of the subjects of the global environmental conventions shown in Table 22.4 are, of course, interlinked. For example, the links between three major global change issues discussed in detail at the Earth Summit – biodiversity, climate change and desertification/land degradation – are numerous. Land degradation can often result in a loss of biodiversity which can in turn have impacts on regional and global climate through changes in surface energy budgets and the carbon cycle. Changes in rainfall patterns and amounts, and the consequent production of biomass, will affect land degradation. Hence, any attempt to deal with one of these environmental issues will inevitably have impacts on others. So a scheme financed under the UN Convention to Combat Desertification to promote sustainable land use, for instance, may well also enhance biodiversity.

Table 22.4 Major global environmental conventions

Topic	Convention	Agreed	Main aims
Antarctica	Antarctic Treaty and Convention	Washington, DC, 1959	Ensure Antarctica is used for peaceful purposes, such as scientific research
Wetlands	Convention on Wetlands of International Importance Especially as Waterfowl Habitats	Ramsar, 1971	Stem progressive encroachment on and loss of wetlands
World heritage	Convention Concerning the Protection of the World Cultural and Natural Heritage	Paris, 1972	Establish a system of collective protection of cultural and natural heritage sites of outstanding universal value
Ocean dumping	Convention on the Prevention of Marine Pollution by Dumping of Wastes and Other Matter	London, Mexico City, Moscow and Washington, DC, 1972	Control marine pollution by prohibiting the dumping of certain materials and regulating ocean disposal of others, encouragement of regional agreements and establishment of mechanisms to assess liability and settle disputes
Biological and toxin weapons	Convention on the Prohibition of the Development, Production, and Stockpiling of Bacteriological (Biological) and Toxin Weapons, and on their Destruction	London, Moscow and Washington, DC, 1972	Prohibit acquisition and retention of biological agents and toxins not justified for peaceful purposes, and the means of delivering them for hostile purposes
Endangered species	Convention on International Trade in Endangered Species of Wild Fauna and Flora	Washington, DC, 1973	Protect endangered species from overexploitation by controlling trade in live or dead individuals and in their parts through a system of permits
Ship pollution	Protocol of 1978 Relating to the International Convention for the Prevention of Pollution from Ships, 1973	London, 1978	Modify the 1973 convention to eliminate international pollution by oil and other harmful substances and to minimize accidental discharge of such substances
Migratory species	Convention on the Conservation of Migratory Species of Wild Animals	Bonn, 1979	Promote international agreements to protect wild animals that migrate across international borders
Southern Ocean	Convention on the Conservation of Antarctic Marine Living Resources	Canberra, 1980	Safeguard the environment and protect marine ecosystems surrounding Antarctica

Topic	Convention	Place, Year	Objective
Law of the sea	UN Convention on the Law of the Sea	Montego Bay, 1982	Establish a comprehensive legal regime for the seas and oceans, including environmental standards and rules to control marine pollution
Ozone layer	Vienna Convention for the Protection of the Ozone Layer	Vienna, 1985	Protect human health and the environment by promoting research, monitoring of the ozone layer and control measures against activities that produce harmful effects
Nuclear accident notification	Convention on Early Notification of a Nuclear Accident	Vienna, 1986	Provide relevant information about nuclear accidents to minimize transboundary radiological consequences
Nuclear accident assistance	Convention on Assistance in the Case of a Nuclear Accident or Radiological Emergency	Vienna, 1986	Facilitate prompt provision of assistance following a nuclear accident or radiological emergency
CFC control	Protocol on Substances that Deplete the Ozone Layer	Montreal, 1987	Require developed states to cut consumption of CFCs and halons with allowances for increases in consumption by developing countries
Hazardous waste movement	Basel Convention on the Control of Transboundary Movements of Hazardous Wastes and their Disposal	Basel, 1989	Establish obligations to reduce transboundary movements of wastes, minimize generation of hazardous wastes and ensure their environmentally sound management
Biodiversity	UN Convention on Biological Diversity	Nairobi, 1992	Protect biological resources, and regulate biotechnology firms' access to and ownership of genetic material, and compensation to developing countries for extraction of their genetic materials
Climate change	Convention on Climate Change	New York, 1993	Stabilize atmospheric concentrations of greenhouse gases
Desertification	UN Convention to Combat Desertification	Paris, 1994	Promote sustainable development of drylands
Persistent Organic Pollutants (POPs)	Convention on Persistent Organic Pollutants (POPs)	Stockholm, 2001	International control of 12 chemicals

Many of these international conventions have also been associated with new global bodies established to assess the issues scientifically and to offer advice to policy-makers. A good example is the Intergovernmental Panel on Climate Change (IPCC) which has made tremendous advances in our understanding of climate change since its establishment in 1988. A similar initiative took place between 2001 and 2005 in the form of the Millennium Ecosystem Assessment, which aimed to provide a scientific appraisal of the condition and trends in the world's ecosystems and the services they provide, as well as the scientific basis for action to conserve and use them sustainably. The evolution of the IPCC reflects the fact that global environmental governance describes world politics that are no longer confined to the governments of nation-states, but are characterized by the increasing participation and relevance of other actors (Biermann and Pattberg, 2008). The IPCC was initiated not by governments but by international organizations (the World Meteorological Organization and the UN Environment Programme or UNEP). It is composed of private actors – experts, scientists, and their autonomous professional organizations – which are nonetheless engaged in a constant dialogue with governments and their representatives. Other non-state actors include environmentalist non-governmental organizations (NGOs), transnational corporations, cities and provinces, as well as new agencies set up by governments, including intergovernmental organizations and international courts.

These and other initiatives have all made a positive contribution, but they have done little to address one of the major underlying difficulties in achieving global sustainability: poverty and the unequal distribution of resources. Inequalities exist at all levels (see Chapter 2), but on the global scale an obvious imbalance is evident between North and South. Some minor, if significant, attempts have been made to address both debt problems and environmental problems in developing countries by converting part of the external debt of a country into a domestic obligation to support a specific programme. Some examples of 'debt-for-nature swaps' are detailed in Table 22.5. An international conservation group, or in some cases a national government, purchases part of a debtor country's foreign debt on the secondary market, at a fraction of the theoretical face value of the debt. The low cost of the original debt on the secondary market is a reflection of the fact that creditors have a low expectation of the debt itself being repaid. Once the debt has been acquired, the 'investing' conservation organization agrees a rate of exchange with the debtor country so that the debt is converted into local currency, which is then used to finance the intended aims.

Supporters of debt-for-nature swaps point out that they are 'win-win-win' transactions, agreements in which all parties involved stand to gain. Conservation organizations increase the spending power of their money, through the difference between the secondary market cost and the face value of the debt purchased, and are able to exert influence on conservation policies in developing countries. The debtor gains by reducing its foreign debt, reducing the government's need to raise foreign currency to service its debts, and also wins finance to support conservation programmes, which the government has some control over. For some debtor countries, however, such agreements may be seen as an infringement upon national sovereignty, passing some control and influence over national resources to foreign interests. Stimulated by the apparent success of environmental organizations in mobilizing extra resources through debt-for-nature swaps, a large number of NGOs and other non-profit organizations became actively involved in debt-for-development swaps under various guises (e.g. debt-for-education swaps, debt-for-health swaps).

Despite the various types of debt swap, part of the packages of development aid that flow from North to South, the problems of indebtedness and unequal trading relations remain essentially unchanged. Redistribution from rich to poor countries on any significant scale still appears to be politically impossible, yet it is the resolution or otherwise of this issue, perhaps more than any other, that will dictate the pace of global change towards sustainability.

Table 22.5 Examples of established debt-for-nature agreements

Debtor	Date	Debt face value (US$ million)	Cost (US$ million)	Investor(s)	Terms and aims
Poland	1990	0.05	0.01	WWF-International and WWF-Sweden	Clean-up of River Vistula and development of Biebrza National Park
Peru	2002	14.3	6.61	WWF-US, The Nature Conservancy, Conservation International and US government	Build long-term forest conservation and sustainable forestry initiatives in 10 areas of tropical forest
Seychelles	2016	21.6	15.2	The Nature Conservancy and several private groups	Marine conservation, management of fisheries and tourism

Having drawn attention to these international dimensions to global environmental issues, it is appropriate also to highlight drawbacks that some identify with the underlying belief that these problems are solvable through globally coordinated action. Adger *et al.* (2001) continue the theme of synergies between issues in their analysis of the major discourses associated with four global environmental issues: deforestation, desertification, biodiversity use and climate change. They find striking similarities (Table 22.6) in the approaches to these issues at the global level. These approaches reflect a technocentric worldview rooted in the belief that blueprints based on external policy interventions can solve global environmental dilemmas. By contrast, each issue also has a more populist discourse, often focused through NGOs, that portrays local actors as victims of external interventions bringing about degradation and exploitation. Bridging the gap between global environmental management and policy-making institutions, on the one hand, and individual resource users, on the ground, on the other, is essential if these global environmental issues are to be successfully managed in a sustainable way.

THE SUSTAINABLE FUTURE

It is important to realize that sustainable development does not mean no human impact on the environment. Such a situation is impossible to achieve so long as there are people on the planet. The ideal scenario to strive for is one in which all environmental impacts can be undertaken consciously, in the full knowledge of the costs and consequences, even though this situation is a long way off, not least because we still have much to learn about the operation of nature. And even when and if such a status quo is attained, accidents will still happen. Some 17 years before the supertanker *Exxon Valdez* released 36,000 tonnes of oil into Prince William Sound in Alaska in March 1989, causing widespread environmental damage, the author of a report on human activities in the region declared that: 'the volume of tanker traffic in this area [Prince William Sound at Valdez] makes occasional massive oil spills a distinct likelihood' (Price, 1972: 77).

Table 22.6 Two approaches to global environmental issues

Issue	Global environmental management discourse	Populist discourse
Deforestation	Neo-Malthusian discourse on increasing population and agricultural conversion in developing countries with slash-and-burn farmers portrayed as the primary villains	Populist discourse sees deforestation as a significant issue caused by the marginalization of rural poor and external forces of globalization (e.g. Northern consumption of timber products)
Desertification	Neo-Malthusian discourse suggests that local resource users in drylands degrade the ecosystems on which they depend. Only international action and strict regulation can prevent further desertification	Populist discourse accepts evidence that desertification is important but sees it is as the inevitable consequence of historic marginalization of pastoralists and smallholders in both colonial and post-colonial periods
Biodiversity use	Bioprospecting discourse promotes sustainable use of biodiversity as the solution to an impending extinction crisis. This solution can be promoted though international cooperation and institutions	Biopiracy discourse portrays an extinction crisis promoted by the institutions and interests of capitalism, threatening both cultural and biological diversity
Climate change	Managerial discourse on the compelling science of climate change requiring new markets for carbon and global institutions	Profligacy discourse also accepts climate change as a major problem and as the key symptom in the global overconsumption crisis espoused by capitalism

Source: after Adger *et al.* (2001).

Despite stringent safety procedures, which can reduce the chances of catastrophe, the best-laid human plans can malfunction for unexpected reasons, among them human error. Hence, there will always be a need for continual vigilance and the existence of contingency plans for when the unexpected does occur.

However, one of the central themes of this book has been the need to change the ways in which socio-economic systems work in order to reduce environmental impacts that feed back on the operation of society. The emphasis on economic growth, which relies on increasing the amount of resources channelled through society, must be replaced by an emphasis on sustainable development – the qualitative improvement in human welfare. There is no doubt that great modifications to society are necessary to achieve a globally sustainable future. Some suggest that the scale of alteration is comparable to only two other changes in the history of humankind: the agricultural revolution of the late Neolithic period and the Industrial Revolution of the eighteenth and nineteenth centuries. However, while these previous revolutions were gradual and largely unconscious, the sustainable revolution will have to be a conscious one. It needs to be constantly flexible, guided by scientific and other forms of information and interpretation of events, steered by governmental and private policy, and sensitive to cultural norms.

Figure 22.11 Mowing the lawn in front of the Indian parliament building in New Delhi: an image that symbolizes the fact that the world's developing countries have much to offer richer nations in terms of appropriate use of nature and technology.

A basic prerequisite for this revolution is a fundamental revision of the way in which we view human society and the operation of nature. The need to evaluate natural systems properly, in economic and other terms, has been mentioned, as has the need to realize that maintaining a sustainable environment is a global challenge that must be based on a long-term perspective, and is required for the entire world's population, not just a select and privileged few. If a large section of the members of our species are poor, we cannot hope to live in a peaceful and sustainable world, and if the transfer of technology and wealth from rich to poor is not forthcoming on a much greater scale than at present, then efforts by less developed nations to improve their conditions along the same lines as the rich nations have followed will result in increasing worldwide ecological damage. At the same time, however, many lessons can be learnt by the richer countries – which are often over-reliant upon polluting and unnecessary technology – from more appropriate innovations in developing countries (Figure 22.11).

Another fundamental change that many have to make is the need to return to the realization that humans are part of nature and not separate from it. Industrial and urban societies have benefited greatly from their histories of economic development, but one of the casualties has been this basic truism. Traditional societies have maintained their environments through spiritual connections with plants and animals and other aspects of the natural world, which, although not lost by all (Figure 22.12), have been widely replaced by a notion of human domination of nature. This notion must change. In simple and not overly dramatic terms,

Figure 22.12 Natural materials on sale for medicinal, magical and religious uses in Agadez, Niger. The contrast between this scene and its equivalent in urban industrialized societies illustrates the real and perceived gap that has grown between many people and nature. Sustainable development requires a return to the realization that we are a part of, and not apart from, nature.

the continuance of the human race depends upon our ability to abstain from destroying the natural systems that regenerate our planet.

These changes in values will also have to be encouraged, cajoled and in some cases enforced. Existing powerful institutional forces thus need to change their direction too, and be supplemented with new ones. Money, trade and national defence are the concerns of today's most influential institutions, but the priorities of governments change and the existence of organizations such as the World Bank and transnational corporations has a relatively short history. Our current global institutions and structures of governance were built with the nation-state at their cores, mainly designed to protect national interests, with relatively little thought given to the concept of global responsibility. The nation-state has certainly helped to improve the well-being of many individuals, but arguably at the cost of reduced global resilience. Although some successes have been achieved in tackling common aims, in many cases aided by the increasing prominence of NGOs, they have too frequently been limited by the narrow interests of individual nation-states. Difficult though it may sound, our common future relies upon reducing this sense of national importance (Figure 22.13), to create more truly global coalitions designed to manage truly global issues. As realization of the importance of environmental concerns continues to grow, historical experience suggests that such

Figure 22.13 Flags on sale at Independence Day celebrations in Carmelo, Uruguay. Successful management of global environmental issues will often mean diminishing the interests of individual nation-states in favour of the global common good.

power bases should be able to adjust accordingly. The alternative, which also has historical precedents, is institutional collapse.

Perhaps, above all, it is important to realize that a sustainable future lies in our hands. The need to alter values, beliefs and behaviour should by now be clear. Anyone who has read this book as far as this point should be well aware of that.

FURTHER READING

Dauvergne, P. 2016 *Environmentalism of the rich*. Cambridge, MA, MIT Press. A strong argument in favour of change to a global political economy based on growth and consumption.

Dearing, J.A. 2006 Climate–human–environment interactions: resolving our past. *Climate of the Past* 2: 187–203. A cogent review of how lessons from the past can inform current views of environmental issues. Open access: www.clim-past.net/2/187/2006/cp-2-187-2006.pdf.

Folke, C., Biggs, R., Norström, A.V., Reyers, B. and Rockström, J. 2016 Social-ecological resilience and biosphere-based sustainability science. *Ecology and Society* 21(3). An argument for integrating society and ecology in achieving sustainability. Open access: www.jstor.org/stable/26269981.

Martin, J.L., Maris, V. and Simberloff, D.S. 2016 The need to respect nature and its limits challenges society and conservation science. *Proceedings of the National Academy of Sciences* 201525003. Arguing for profound shifts in societal values away from economic growth. Open access: www.pnas.org/content/113/22/6105.full.

Rockstrom, J. and 29 others 2009 Planetary boundaries: exploring the safe operating space for humanity. *Ecology and Society* 14(2): 32. An approach to global sustainability in which the authors define planetary boundaries within which they think humanity can operate safely. *Open access: www.ecologyandsociety.org/vol14/iss2/art32/.*

Wysocki, J. 2012 The environment has no standing in environmental governance. *Organization and Environment* 25: 25–38. A thoughtful assessment of the division between people and nature.

WEBSITES

For ease of reference, sites are shown in tables. Tables 22.7 and 22.8 give selections of national and international sites that contain information on environmental programmes, data and state-of-the-environment reports. Table 22.9 is a selection of sites with large directories and news on all types of environmental issues.

Table 22.7 National environment websites (English versions available unless stated otherwise)

Country	Agency	Site address
Angola	Ministry of Environment	www.minamb.gov.ao
Argentina	Environment Ministry (in Spanish)	www.ambiente.gob.ar
Australia	Department of Environment and Energy	www.environment.gov.au
Austria	Federal Environment Agency	www.umweltbundesamt.at
Bangladesh	Ministry of Environment and Forests	www.moef.gov.bd
Brazil	Environment Ministry (in Portuguese)	www.ibama.gov.br
Brunei	Department of Environment, Parks and Recreation	www.env.gov.bn
Canada	Environment and Climate Change Canada	www.ec.gc.ca
Chile	Environment Ministry (in Spanish)	www.mma.gob.cl
China	Ministry of Environmental Protection	www.sepa.gov.cn
Costa Rica	Ministry of Environment and Energy (in Spanish)	www.minae.go.cr
Denmark	Ministry of Environment and Food	www.mim.dk
Egypt	Ministry of Environment	www.eeaa.gov.eg
Eire	Environmental Protection Agency	www.epa.ie
Estonia	Environment Ministry	www.envir.ee
Finland	Environmental Administration	www.environment.fi
France	Environment Ministry (in French)	www.ecologique-solidaire.gouv.fr
Germany	Federal Ministry of Environment, Nature Conservation and Nuclear Safety	www.bmub.bund.de
Ghana	Environment Ministry	mesti.gov.gh
Iceland	Ministry for the Environment and Natural Resources	www.environment.is

Country	Agency	Site address
India	Ministry of Environment, Forests and Climate Change	envfor.nic.in
Israel	Environmental Protection Ministry	www.sviva.gov.il
Italy	Environment Ministry (in Italian)	www.minambiente.it
Jamaica	National Environment and Planning Agency	nepa.gov.jm
Japan	Environment Ministry	www.env.go.jp
Kenya	Ministry of Environment and Forestry	www.environment.go.ke
Kuwait	Environment Public Authority	epa.org.kw
Lebanon	Environment Ministry	www.moe.gov.lb
Malaysia	Department of Environment	www.doe.gov.my
Mauritius	Environment Ministry	environment.govmu.org
Mexico	Environment Ministry (in Spanish)	www.gob.mx/semarnat
Namibia	Ministry of Environment and Tourism	www.met.gov.na
Netherlands	Environmental Assessment Agency	www.pbl.nl
New Zealand	Environment Ministry	www.mfe.govt.nz
Nicaragua	Ministry of Environment and Natural Resources (in Spanish)	www.marena.gob.ni
Nigeria	Environment Ministry	http://environment.gov.ng
Norway	Environment Agency	www.miljodirektoratet.no
Pakistan	Environmental Protection Agency	environment.gov.pk
Panama	National Association for Conservation of Nature (in Spanish)	www.ancon.org
Poland	Environment Ministry (in Polish)	www.mos.gov.pl
Qatar	Environment Ministry	www.mme.gov.qa
Romania	Ministry of Environment and Forests	www.mmediu.ro
Russia	Ministry of Natural Resources and Environment	www.mnr.gov.ru
Seychelles	Ministry of Environment, Energy and Climate Change	meecc.seydevplus.com
Singapore	Ministry of Environment and Water	www.mewr.gov.sg
Slovakia	Environment Ministry	www.minzp.sk
South Africa	Environmental Affairs	www.environment.gov.za
South Korea	Environment Ministry	www.me.go.kr
Spain	Environment Ministry	www.mapama.gob.es
Sri Lanka	Central Environmental Authority	www.cea.lk
Sweden	Environmental Protection Agency	www.naturvardsverket.se
Switzerland	Federal Office for Environment	www.bafu.admin.ch

(continued)

Table 22.7 continued

Country	Agency	Site address
Thailand	Thailand Environment Institute	www.tei.or.th
Tunisia	Ministry of Environment and Sustainable Development	www.environnement.gov.tn
Turkey	Environment and Forestry Ministry	www.ormansu.gov.tr
Uganda	National Environment Management Authority	nema.go.ug
UK	Environment Agency	www.environment-agency.gov.uk
Ukraine	Ministry of Ecology and Natural Resources	www.menr.gov.ua
USA	Environmental Protection Agency	www.epa.gov
Vietnam	Ministry of Natural Resources and Environment	www.monre.gov.vn
Zambia	Environmental Management Agency	www.zema.org.zm

Table 22.8 International environment websites

Region	Agency	Site address
Arctic	AMAP: the Arctic Monitoring and Assessment Programme	www.amap.no
Asia (S and E)	AECEN: Asian Environmental Compliance and Enforcement Network	www.aecen.org
Central Europe	Central European Environmental Data Request Facility	www.cedar.at
Europe	European Environment Agency	www.eea.europa.eu
Global	Millennium Ecosystem Assessment	www.millenniumassessment.org
Global	UN Environment	www.unenvironment.org
Global	UNEP's Global Resource Information Database	www.grid.unep.ch
Global	World Resources Institute	www.wri.org
Global	Food and Agriculture Organization	www.fao.org
Global	Global Environment Facility	www.thegef.org
Global	Global Green Growth Institute	gggi.org
Global	Global Forest Coalition	globalforestcoalition.org
Global	Conservation International	www.conservation.org
Global	Communal information database	ecoport.org
Global	International Union for Conservation of Nature	www.iucn.org
Global	World Wildlife Fund	www.worldwildlife.org
Middle East	Environmental Portal of Gulf Cooperation Council	www.gcceportal.com
Pacific	Pacific Regional Environment Programme	www.sprep.org

Table 22.9 Directories and news sites on all types of environmental issues

Topics	Site address
EnviroLink	www.envirolink.org
Environmental behaviour change organization	www.planetark.org
Environment and science news	www.sciencedaily.com
Environment news	www.theguardian.com/uk/environment
Environment news	www.edie.net

POINTS FOR DISCUSSION

- Will any environmental issues become more critical in coming decades? Conversely, will any become less so?
- Using specific examples, assess the differences between 'systemic' and 'cumulative' global environmental changes.
- Should efforts towards global environmental management take more notice of individual resource users?
- Do you think that the best approach to resolving environmental issues is technocentric or ecocentric, or should the approach vary depending on the issue?
- Given that so many environmental issues occur across international boundaries, are there any solely national issues?
- What do you consider to be the most urgent environmental issue facing your country? Is enough being done to resolve it (by government, civil society, business)?

Glossary

abiotic: non-biological (see also **biotic**)

afforestation: the planting of a forest in an area where no forest originally existed

albedo: the reflectivity of a surface to sunlight. High albedo means that the majority of the incoming radiation is reflected (e.g. snow); low albedo means that the majority of the incoming radiation is absorbed (e.g. water)

algae: photosynthetic, mostly aquatic plant group, including single-cell **phytoplankton** and multicellular forms such as kelp

anadromous: fish that migrate from the sea into fresh water for breeding (e.g. salmon)

anaerobic: the term applied to organisms that live, or processes that occur, in the absence of gaseous or dissolved oxygen. Anaerobic bacteria, for example, obtain oxygen by breaking down such things as vegetable matter

Anthropocene: a term used to describe the most recent planetary era characterized by the emergence of human action as a critical force in a range of biophysical systems

anthropogenic: produced as the result of human action

aquaculture: the farming of fish and other aquatic organisms such as seaweeds, frogs and turtles, either in artificial environments such as tanks, or in natural or semi-natural ecosystems such as ponds, rivers or enclosed areas of the sea

atmosphere: the gaseous layer surrounding a planet. The Earth's atmosphere is divided into four parts distinguished according to the rate of change of temperature with height: the **troposphere**, **stratosphere**, mesosphere and thermosphere

bacteria: single-celled microscopic organisms that lack chlorophyll and an obvious nucleus. Bacteria play a significant role in the decomposition of organic matter

benthic: applied to organisms living close to the bottom of a lake, river or sea

bioaccumulation: the build-up in concentration of a toxic substance in the body of an individual organism over time that occurs when an organism is unable to excrete the substance or break it down within the body. (Confusingly, some authors also use the term for the increasing concentration of a substance up a food chain, a process better referred to as biomagnification)

biochemical oxygen demand (BOD): the amount of oxygen used in biological/chemical processes that decompose organic material in waste effluent. BOD is often used as an indicator of the polluting capacity of wastes because the amount of dissolved oxygen in water has a profound effect on the plants and animals living in it

biodegradation: the decomposition or breakdown of organic matter by micro-organisms, particularly by oxygen-using or 'aerobic' bacteria

biodiversity: a term that refers to the number, variety and variability of living organisms. It is commonly defined in terms of genes, species and ecosystems, corresponding to three fundamental levels of biological organization

biogeochemical cycle: the chemical interactions that take place among the atmosphere, biosphere, hydrosphere, and lithosphere. Examples include the carbon, nitrogen and phosphorus cycles, and the hydrological cycle

biological classification: the arrangement of organisms into a hierarchy of groups that reflects evolutionary relationships. Usually the smallest group, or taxon (plural: taxa), is the species, although **species** may be subdivided into subspecies and varieties. Similar species are grouped into **genera**, genera into **families**,

families into orders, orders into classes, classes into phyla (singular: phylum) for animals – the equivalent for plants is divisions – and these into kingdoms. The formal Latin name given to an organism has two parts: the European crested newt, for example, is known as *Triturus cristatus*, indicating that it is the species *cristatus* in the genus *Triturus*

biomass: the total mass of living organisms, usually expressed in dry weight per unit area

biome: ecosystem of a large geographical area with characteristic plants and typical climate (e.g. **tundra**, **savanna**)

biosphere: the part of the Earth and its atmosphere in which organisms live

biotechnology: the deliberate manipulation of living organisms, their cells or sub-cellular components to change their character for the benefit of human society. Examples include traditional applications such as the making of bread, cheese and wine, and more modern ones such as cell culture, genetically modified foods or cloning plants and animals

biotic: living or biological in origin, as opposed to abiotic

boreal forest: coniferous forest occurring in the low Arctic

business as usual: a situation in which individuals, industries and societies continue to act as they have in the past, with no changes made to deal with some existing or anticipated problem, such as global warming

carbon cycle: an example of a **biogeochemical cycle**: the complex series of reactions by which carbon passes through the Earth's atmosphere, biosphere, hydrosphere, and lithosphere. For example, plants remove carbon in the form of carbon dioxide from the atmosphere and use it to produce carbohydrates in living organisms (**photosynthesis**). When those organisms die, the carbon is returned to the Earth as carbon dioxide, as fossil fuels (during decay), or as inorganic compounds such as calcium carbonate (limestone)

carbon sequestration: a natural or artificial process whereby carbon dioxide in the atmosphere is removed and stored over the long term in solid or liquid form

carcinogen: a substance with the potential to cause cancer

carrying capacity: the maximum number of a species that can be sustained in an area. If the carrying capacity is exceeded, **degradation** of the habitat will occur

cations: atoms that have lost one or more electrons and are thus left with a net positive charge

cetacean: a whale, dolphin or porpoise. The cetaceans are an order of aquatic mammals

CFCs (chlorofluorocarbons): organic compounds that include atoms of carbon, chlorine and fluorine. CFCs have been created for use in industrial processes and manufacturing, and have become a cause for environmental concern due to their inadvertent release into the atmosphere. Breakdown of CFCs in the stratosphere releases chlorine, which is thought to contribute to **ozone** depletion

clathrates: sea-floor crystals of ice, also known as gas hydrates, that enclose gases such as methane

cogeneration: the process by which two different and useful forms of energy are produced at the same time (e.g. when boiling water to generate electricity the leftover steam can be sold for heating or industrial processes)

coliform organisms: a group of bacteria abundant in the intestines of humans and other warm-blooded animals. Their concentration in water is used as an indicator of water quality

commodity: any useful material or product that might be traded

contamination: the occurrence of a relatively large amount of a toxic substance, relative to the normal ambient condition. Contamination does not necessarily imply ecological damage, but further contamination may result in **pollution**

cost–benefit analysis: an economic evaluation in which both the costs and consequences of different actions are expressed in monetary units. Many items and services can be assigned a market value, but a monetary value can be derived for other things by assessing the willingness to accept or pay compensation for gains or losses

critical load: a damage threshold for pollutant deposition

cryosphere: the portion of the Earth's surface that is frozen throughout the year

curie: the former unit of measurement for radioactivity, now replaced by the becquerel (1 curie = 3.7×10^{10} becquerels)

defoliation: the removal of plant foliage by mechanical means, chemical spray or excessive eating by herbivores

deforestation: the removal of forests from an area of land. People commonly cut down trees to use as fuel or for building materials, or to clear land for agricultural, developmental or other purposes

degradation: a reduction in quality or decline in usefulness of an environmental resource. Human causes of degradation occur when a resource is used in an unsustainable manner

denitrification: the breakdown of **nitrates** by soil bacteria resulting in the release of nitrogen

desert: a biome of a predominantly dry climate with relatively little plant life. The term is often also used in a looser sense to indicate a relative lack of fresh water and/or organic life

dioxins: a family of chlorinated organic compounds that are highly toxic. They are by-products of the manufacture of certain classes of herbicides, disinfectants and bleaches, and can also be formed when chlorine compounds are burned at low temperatures in waste incinerators

DNA (deoxyribonucleic acid): the principal material of inheritance, found in chromosomes

domestication: alterations to wild plants and animals under human control whereby the genetic makeup of the species changes to make them more suited to the needs of people and/or to an environment that is deliberately managed for agriculture

drought: a temporary dry period when precipitation is significantly below normal recorded levels

ecocide: an intentional, anti-environmental action carried out over a large area (e.g. as a military strategy)

ecological footprint: the area of land and sea required to produce the resources a human population consumes and to assimilate its wastes, given prevailing technology. An ecological footprint can be calculated for any group, such as a country or city, or an individual. Since some societies draw resources from – and/or pollute – places far away, their ecological footprint includes these areas wherever they happen to be

ecosystem: a community of organisms and the **abiotic** environment they inhabit and interact with. Ecosystems are recognized at many different geographical scales from a pond, for example, up to the scale of a **biome**

ecosystem services: the benefits people acquire from ecosystems. These include actual products, such as food, and less immediate benefits, such as those gained from the regulation of ecosystem processes (e.g. climate, water). Other benefits people derive from ecosystems include services in spiritual, recreational and cultural terms, and supporting services that are necessary for the provision of all other ecosystem services (e.g. production of biomass and atmospheric oxygen, soil formation and nutrient cycling)

El Niño/Southern Oscillation (ENSO) event: a distinct warming of the surface sea water in the south-eastern tropical Pacific (El Niño) that occurs with a rapid change in atmospheric pressure over coastal South America and Indonesia (the Southern Oscillation). In the eastern Pacific, ENSO events are often associated with an abrupt decrease in fisheries, caused by a decline in coastal upwelling, and an increase in rainfall. ENSO events also affect climate variations far from the Pacific Ocean through **teleconnections**

endemism: an endemic species is one found only in a particular geographical region, due to such factors as isolation, particular soils or climate

environmental determinism: the idea that the physical environment predisposes societies towards particular development trajectories

Environmental Impact Assessment (EIA): an interdisciplinary process by which the possible environmental consequences of proposed actions are predicted and considered. An EIA may include assessments of ecological, anthropological, sociological, geomorphological and other environmental effects

environmental justice: an area of research concerned with the even distribution of environmental risk irrespective of race, income, class or any other differentiating feature of socio-economic status

environmental monitoring: repeated measurements of environmental parameters, including inorganic, ecological, social and economic factors, with a view to documenting and often predicting important changes

epidemic: an unusual outbreak of a disease affecting a number of people in a relatively short time

epiphyte: a plant that grows on the outside of another plant, using it for support and not as a source of nutrients (e.g. lichen on trees)

erodibility: the susceptibility of soil to erosion

erosivity: the power of an agent of erosion (water, wind, ice, gravity) to move soil material

eutrophication: the addition of mineral **nutrients** to an ecosystem, so raising primary productivity. The process occurs naturally, particularly in saltwater and freshwater bodies over long periods, but can also be human-induced, in which case it is sometimes referred to as 'cultural eutrophication'. Cultural eutrophication occurs by the, often inadvertent, addition of nutrients in sewage and detergents, and from runoff contaminated with fertilizers

evapotranspiration: combined term for water lost as a vapour from soil or open water (evaporation) and water lost from the surface of a plant, mainly through stomata (transpiration)

extinction: an event resulting in the complete loss of all surviving individuals of a species or other taxon

extirpation: the localized loss of a species or other taxon from an area, with the taxon still surviving elsewhere

family: a group of similar **genera** (see **biological classification**)

feedback: the process by which a system responds to a disturbance that makes it deviate from its initial state. When the feedback has the net effect of amplifying a perturbation, thus reinforcing or accentuating the change, it is known as 'positive feedback'. A 'negative feedback' is one that has the net effect of dampening a perturbation, hence acting to stabilize the system

fossil fuel: a mined fuel comprising the remains of plant or animal life from a previous geological period primarily composed of hydrocarbons (e.g. coal, oil, natural gas)

genera (singular: genus): groups of similar species (see **biological classification**)

general circulation model (GCM): a computer program that uses fundamental laws of physics and chemistry to analyse the interaction of temperature, pressure, solar radiation and other climatic factors in order to predict climatic conditions for the past, present or future

glacial maximum: the maximum extent of glaciation at the height of an **Ice Age**

GNI (gross national income): a measure of the market value of an economy's production over a given period (usually a year), that is now more commonly used than GNP, an identical measure. GNI is adjusted to give GDP (gross domestic product) by removing the value of profits from overseas investments and those profits from the economy that go to foreign investors

GNP (gross national product): see **GNI**

greenhouse effect: the process by which atmospheric gases trap long-wave radiation emitted from the surface of a planet to warm its atmosphere

groundwater: water found beneath the ground surface, but usually excluding soil moisture

heavy metals: a broad group of metals of atomic weight higher than that of sodium and having a specific gravity in excess of 5.0. Those recognized as being of environmental concern, when present at certain concentrations that are harmful to living organisms, include copper, cadmium, mercury, tin, lead, antimony, vanadium, chromium, molybdenum, manganese, cobalt and nickel

Holocene: the latest epoch of the geological timescale which has to date lasted about 11,500 years (10,000 radiocarbon years). Together with the Pleistocene epoch, it completes the **Quaternary**

humus: decomposed organic matter in soils

hydrosphere: the water present at or near the Earth's surface as a liquid, solid or gas, including oceans, lakes, rivers, streams, groundwater and water vapour in the atmosphere

ice age (glacial period, glacial epoch): recurring periods in Earth's history when the climate was colder and glaciers expanded to cover larger areas of the Earth's surface

interglaciation (interglacial period): the time between glaciations when Earth's climate is warmer and the ice sheets have retreated from large areas of the continents. The climate and extent of ice today are considered an interglacial period

jökulhlaup: meltwater flood resulting from a volcanic eruption under a glacier

K-strategy: a species lifestyle in which the reproduction rate is relatively low and individuals relatively long-living, which allows a species to persist in a particular place for a long period (cf. *r*-strategy)

keystone species: a species whose impact on its community is disproportionately large relative to its abundance. Hence, removal of a keystone species from its **ecosystem** can lead to a series of extinctions in that ecosystem

lahar: a type of mudflow composed of volcanic rock fragments and water flowing down the slopes of a volcano

leachate: solution formed when water percolates through a permeable medium

leaching: removal of soil materials in solution

lichen: a symbiotic association of **algae** and fungus: the fungus provides protection and moisture while the algae provide food for the fungus

lithosphere: the Earth's crust and the upper part of the mantle, consisting of rocks, soil and **sediments**

macrobenthic: large **benthic** organisms

mammal: a member of a class of animals that are warm-blooded, have vertebrae, secrete milk to their young through mammary glands, and have hair on their bodies (e.g. humans, whales, tigers, bats)

mass extinction: the extinction of an unusually large number of taxa in a relatively short period of time

maximum sustainable yield: the largest quantity of a renewable resource (e.g. fish, fruit, timber) that can be extracted from nature without impairing nature's ability to produce a similar yield at a later date

Mesolithic (Middle Stone Age): a transitional period in the development of human societies between the **Palaeolithic** and **Neolithic** periods. The Mesolithic occupies different timespans in different places

methylation: the addition of a methyl group (CH_3) to an element or compound, usually through the activity of **anaerobic** bacteria. The process of methylation has important implications when it involves **heavy metals** since it releases them from materials, so enabling them to enter the food chain

monoculture: the cultivation of a single-species crop

morbidity rate: the number of people in a population with physical injury and disease in a given time interval

mortality rate: the number of deaths in a population in a given time interval

mycorrhizae: structures formed by an association of certain fungi and the roots of some plants

natural archives: accumulations of sediments or biological remains that can be examined and interpreted as an archive of environmental change (e.g. ice cores, tree rings). Natural archives provide a form of **proxy evidence**

Neolithic (New or Late Stone Age): a level of human development characterized by the use of polished stone tools and the beginnings of settled agriculture. The Neolithic occupies different timespans in different places

net primary production: the amount of organic material produced by living organisms from inorganic sources minus that used in respiration

NIMBY syndrome: an individual or collective reaction to new developments that involve environmental change; stands for 'Not In My Back Yard'

nitrate: a chemical compound containing nitrogen, an essential nutrient for all living things. Nitrates are formed naturally in the soil by micro-organisms and are also produced industrially for use as fertilizers

non-point source: a term applied to a source of pollution meaning a dispersed source, such as a field, as opposed to a 'point source' such as an industrial waste pipe

nutrient: an element or compound needed by a living organism. A distinction is often made between 'macronutrients', which are needed in relatively large quantities (e.g. carbon, hydrogen, oxygen, nitrogen, potassium and phosphorus), and 'micronutrients' needed in smaller amounts (e.g. iron, manganese, zinc, copper)

onchocerciasis: a tropical disease caused by the worm *Onchocerca volvulus* and transmitted by a biting fly that breeds in fast-flowing streams. The worm's larvae can enter the eye, causing blindness, hence the disease's common name 'river blindness'

order: see **biological classification**

organic matter: living and dead **biomass**

organochlorines: organic compounds containing chlorine, which tend to **bioaccumulate** and are often long-lasting in the environment. They include several types of pesticides (e.g. aldrin, dieldrin, DDT) and PCBs

overexploitation: the unsustainable use of a renewable natural resource

oxisol: a relatively infertile loamy and clayey soil found only in the tropics

ozone: a chemical compound made up of three oxygen atoms. It is a natural constituent of the atmosphere with highest concentrations in the stratosphere (the 'ozone layer'), where it plays an important role in screening the Earth's surface from ultraviolet solar radiation. Elevated concentrations of ozone also sometimes occur in the troposphere above urban areas due to photochemical reactions between oxides of nitrogen and hydrocarbons emitted by motor vehicles in bright sunlight. In this situation ozone is considered a pollutant, a common constituent of some smogs

Palaeolithic (Early Stone Age): the stage in the development of a human society when people obtain food by hunting, fishing and gathering wild plants, as opposed to engaging in settled agriculture. The Palaeolithic occupies different timespans in different places

palaeontology: the study of fossils

pandemic: a disease prevalent over a whole country or the world. Commonly used to describe global **epidemics**

PCBs (polychlorinated biphenyls): a group of chlorinated hydrocarbons (see **organochlorines**) used mainly in high-voltage transformers. There is evidence that these compounds are highly persistent and toxic when released into the environment

pelagic: organisms that inhabit the open water at sea or in a lake

permafrost: permanently frozen ground where temperatures below 0°C have persisted for at least two consecutive winters and the intervening summer. Permafrost covers about 26 per cent of the Earth's land surface

pest: an organism considered to be undesirable to human populations or their activities

pesticide: a chemical or a microbial pathogen that is toxic to pests. The term is used to group insecticides, herbicides and fungicides

pheromones: chemicals used by an animal as a social cue

photosynthesis: a set of coordinated biochemical reactions that occur in green plants, blue-green **algae** and **phytoplankton**. They are powered by sunlight absorbed by chlorophyll and other pigments, in which organic compounds are synthesized from carbon dioxide and water, with the release of oxygen as a waste product

phytoplankton: see **plankton**

phytotoxic: poisonous to green plants

plankton: animals and plants, many of them microscopically small, that float or swim in fresh or salt water. The animals (zooplankton) are mainly protozoa, small crustacea and larval stages of molluscs. The plants (**phytoplankton**) are almost all algae

podsolic soils (podsols): a soil type formed at an advanced stage of leaching, which has left an acid humus layer

pollution: the occurrence of toxic substances or energy in a quantity that exceeds the tolerance of an ecological community or species, resulting in damage (cf. **contamination**)

POPs (persistent organic pollutants): a wide range of environmentally hazardous chemicals that tend to be long-lived and accumulate up the food chain. POPs include PCBs, polychlorinated dioxins and pesticides such as aldrin, chlordane, DDT and dieldrin

primary/secondary forest: primary forest is that which has not been disturbed by human action, while secondary forest is a forest ecosystem that has regrown following human disturbance. In practice, it can be difficult to distinguish in the field between the two types of forest, particularly where secondary forest is more than around 60 years old

proxy evidence: data or records relating to an environmental variable that were derived indirectly or reconstructed from **natural archives** as opposed to data that is directly observed or measured

Quaternary: the most recent period of the geological timescale, continuing to the present day and comprising the Pleistocene and **Holocene** epochs

r-strategy: a species lifestyle in which the reproduction rate is relatively high and individuals relatively short-living, which means that a species is a good colonizer but tends to persist in a particular place for a short period (cf. **K-strategy**)

radiative forcing: a measure used by the Intergovernmental Panel on Climate Change (IPCC) to mean an external perturbation in the energy budget of the Earth's climate that may lead to climatic changes. A positive forcing (more incoming energy) tends to warm the system, while a negative forcing (more outgoing energy) tends to cool it

radionuclide: an unstable nucleus of an atom that undergoes spontaneous radioactive decay, thereby emitting radiation and eventually changing from one element into another

rain forest: a mature forest with large biomass in parts of the temperate or tropical zone where precipitation totals are high

rangeland: any extensive area of native herbaceous or shrubby vegetation that is grazed by wild or domesticated animals (e.g. **steppe**, **savanna**, **tundra**)

raptor: a bird of prey that hunts by day (e.g. eagle, goshawk, kite, osprey, peregrine falcon, vulture)

regime shift: a rapid change between two alternate stable environmental states that occurs when inter-relationships between key variables in an ecosystem change fundamentally

renewable energy: natural sources of energy, the supply of which is effectively replaced as they are used. They include the sun, wind, water (rivers, tides and waves), biomass and geothermal sources

resilience: the ability of a system to recover from a disturbance

resistance: the ability of a system to tolerate a disturbance without experiencing a significant change of state

return period: the amount of time that passes on average between consecutive events of similar magnitude in a particular location

sacred grove: an area of habitat – usually a forest or woodland – that is set aside for religious or cultural purposes and governed by traditional rules that often prohibit the felling of trees and the killing of animals, but do allow the collection of firewood, fodder and medicinal plants

savanna: a tropical or subtropical grassland with scattered trees or shrubs

schistosomiasis: any one of a group of diseases caused by infestation by blood flukes of humans and some other mammals. The diseases are common in low latitudes and are also known as bilharzia

sediment: unconsolidated particles, usually produced by the breakdown of rocks, ranging from clay-size to boulders that may be carried by natural agents (wind, water, ice) and eventually deposited to form sedimentary deposits. Organisms and chemical precipitation can also produce sediment. In arid regions, evaporation can result in the deposition of salts, another type of sediment

smog: an atmospheric condition of poor visibility and high concentration of air pollutants. The word is derived from 'smoke' and 'fog'

soil: unconsolidated mineral and/or organic matter found at the surface of the Earth that is capable of supporting plant growth

species: groups of living individuals that look alike and can interbreed, but cannot interbreed with other species (see **biological classification**)

steppe: a **temperate** grassland **biome** with no trees in Eurasia. Similar environments in other parts of the world have different names: prairie (North America), pampa (South America), veldt (South Africa)

Stone Age: the period in the development of human societies during which tools and implements were made of stone, bone or wood, but no metals were used. The dates of the period vary widely from place to place, and the term is sometimes used to describe present-day pre-agricultural peoples, although they may use imported metal implements. See also **Palaeolithic, Mesolithic** and **Neolithic**

stratosphere: the layer of the atmosphere above the **troposphere**, which extends on average between 10 km and 50 km above the Earth's surface

succession: the natural development of vegetation over time involving the progressive replacement of earlier plant communities with others towards a so-called 'climax' in which vegetation is supposedly in equilibrium with the existing environmental conditions. A distinction is made between 'primary' succession, which takes place on land with no previous vegetation, and 'secondary' succession, which occurs on land where vegetation has been partially or wholly destroyed

symbiotic: situation in which two species live together and both of them benefit from the relationship

synergy: the combined effect of two or more environmental influences that exceeds the sum of their individual effects

taxon: see **biological classification**

technology: a term used to describe the total system of means by which people interact with their environment and each other. It includes tools, information and knowledge, and the organization of resources for productive activity

tectonic: relating to the forces and movements of the Earth and its crust. Earthquakes, volcanoes, and mountain building are related to tectonic activity

teleconnection: a link between environmental events in time and space, especially connections between climatic variations occurring in locations that are geographically remote from one another. Teleconnections take place through the dynamism of the Earth-atmosphere system, with events in one part of the system causing change in another part of the system after a **timelag**. Some of the most distinct teleconnections recognized are associated with **El Niño/Southern Oscillation (ENSO) events**

temperate: mild or moderate. The middle latitudes, between polar and tropical climate regimes, are generally referred to as having a temperate climate

tephra: the general term for solid particles ejected from a volcano during an eruption. Such particles range in size from ash (<2 mm) to boulders >1 m in diameter

terrestrial: land above sea level. Terrestrial flora and fauna live within land-based environments, not within the aquatic realm

threshold: the level of magnitude of a system process at which sudden or rapid change occurs

timelag: an interval of time between two related phenomena (e.g. a cause and its effect)

topography: surface relief of the land

total economic value (TEV): an approach used to compare the benefits and costs associated with ecosystems and measure the ecosystem goods and services used by people. The classification typically divides TEV into two primary categories – use values and non-use values – some of which can be further subdivided

toxic: poisonous

traditional ecological knowledge (TEK): knowledge, innovations and practices of indigenous and local communities developed from experience gained over the centuries and adapted to the local culture and environment. TEK is mainly of a practical nature, particularly in such fields as agriculture, fisheries, health, horticulture, and forestry

transnational corporations (TNCs): large business organizations whose activities span international boundaries

troposphere: the lowest layer of the Earth's atmosphere, which varies in depth between 10 and 12 km over the poles to 17 km over the Equator. It is the layer in which most weather features occur

trypanosomiasis: a group of debilitating, long-lasting diseases caused by infestation with microscopic single-celled organisms. In Africa, sleeping sickness among humans and ngana in cattle, both transmitted by tsetse flies, are common forms of trypanosomiasis. A common form in the Americas is the incurable Chagas' disease

tsunami: a sea wave caused by a submarine earthquake, underwater volcanic explosion or massive slide of sea-bed sediments

tundra: treeless vegetation that occurs at high latitude or high altitude

ultisol: a relatively infertile loamy and clayey tropical soil

vector: an organism that conveys a parasite from one host to another (e.g. the *Anopheles* mosquito, which transmits malaria)

volatile organic compounds (VOCs): the term used for a class of atmospheric pollutants, also known as hydrocarbons, which can produce ozone in combination with NO_x and sunlight. There are many hundreds of VOCs (e.g. methane, propane, toluene, ethanol) emitted into the atmosphere from both natural and human sources

volatilization: the collective term for evaporation and sublimation

wetland: any habitat that is regularly saturated by water at or near the surface (e.g. bogs, fens, marshes, salt marshes, swamps)

wise use (of an ecosystem): sustainable utilization for the benefit of humankind in a way that is compatible with maintenance of the ecosystem's natural properties

zooplankton: see **plankton**

Bibliography

Aakvik, A. and Tjøtta, S. 2011 Do collective actions clear common air? The effect of international environmental protocols on sulphur emissions. *European Journal of Political Economy* 27: 343–51.

Abd-El Monsef, H., Smith, S.E. and Darwish, K. 2015 Impacts of the Aswan high dam after 50 years. *Water Resources Management* 29: 1873–85.

Abdullah, M.M., Feagin, R.A., Musawi, L., Whisenant, S. and Popescu, S. 2016 The use of remote sensing to develop a site history for restoration planning in an arid landscape. *Restoration Ecology* 24(1): 91–9.

Abidin, H.Z., Andreas, H., Gumilar, I. and Brinkman, J.J. 2015 Study on the risk and impacts of land subsidence in Jakarta. *Proceedings of the International Association of Hydrological Sciences* 372: 115–20.

Abraham, K.F., Jefferies, R.L. and Alisauskas, R.T. 2005 The dynamics of landscape change and snow geese in mid-continent North America. *Global Change Biology* 11: 841–55.

Ackerman, F. and Heinzerling, L. 2004 *Priceless: on knowing the price of everything and the value of nothing*. New York, The New Press.

Adams, J. 1993 The emperor's old clothes: the curious comeback of cost–benefit analysis. *Environmental Values* 2: 247–60.

Adams, R. 1975 The Haicheng earthquake of 4 February 1975: the first successfully predicted major earthquake. *Earthquake Engineering and Structural Dynamics* 4: 423–37.

Adger, W.N., Benjaminsen, T.A., Brown, K. and Svarstad, H. 2001 Advancing a political ecology of global environmental discourses. *Development and Change* 32: 681–715.

Ågren, C. 1993 SO_2 emissions: the historical trend. *Acid News* 5: 1–4.

Agricola, G. 1556 *De re metallica*. Trans. Hoover, H.C. and Hoover, L.H. 1950. New York, Dover Publications.

Ahmed, M.K., Baki, M.A., Kundu, G.K., Islam, M.S., Islam, M.M. and Hossain, M.M. 2016 Human health risks from heavy metals in fish of Buriganga river, Bangladesh. *SpringerPlus* 5: 1697.

Ahn, C., Chang, E. and Kang, H. 2011 Epiphytic macrolichens in Seoul: 35 years after the first lichen study in Korea. *Journal of Ecology and Environment* 34(4): 381–91.

Ainley, D.G. and Pauly, D. 2014 Fishing down the food web of the Antarctic continental shelf and slope. *Polar Record* 50(1): 92–107.

Akcil, A. and Koldas, S. 2006 Acid Mine Drainage (AMD): causes, treatment and case studies. *Journal of Cleaner Production* 14: 1139–45.

Akiner, S., Cooke, R.U. and French, R.A. 1992 Salt damage to Islamic monuments in Uzbekistan. *Geographical Journal* 158: 257–72.

Al Naber, M. and Molle, F. 2017 Controlling groundwater over abstraction: state policies vs. local practices in Jordan highlands. *Water Policy* 19(4): 692–708.

Al-Awadhi, J.M. 2001 Impact of gravel quarrying on the desert environment of Kuwait. *Environmental Geology* 41: 365–71.

Alexander, D. 1993 *Natural disasters*. London, University College London Press.

Alexander, D. 1997 The study of natural disasters, 1977–1997: some reflections on a changing field of knowledge. *Disasters* 21: 284–304.

Alexander, D. 2013 Volcanic ash in the atmosphere and risks for civil aviation: a study in European crisis management. *International Journal of Disaster Risk Science* 4(1): 9–19.

Alexander, P., Brown, C., Arneth, A., Finnigan, J., Moran, D. and Rounsevell, M.D. 2017 Losses, inefficiencies and waste in the global food system. *Agricultural Systems* 153: 190–200.

Alfieri, L. and Thielen, J. 2012 A European precipitation index for extreme rain-storm and flash flood early warning. *Meteorological Applications* doi: 10.1002/met.1328.

Alheit, J. and Bakun, A. 2010 Population synchronies within and between ocean basins: apparent teleconnections and implications as to physical–biological linkage mechanisms. *Journal of Marine Systems* 79(3): 267–85.

Ali, S.H. 2007 *Peace parks: conservation and conflict resolution*. Cambridge, MA, MIT Press.

Allen-Diaz, B., Chapin, F.S., Diaz, S., Howden, M., Puigdefabregas, J. and Stafford Smith, M. 1996 Rangelands in a changing climate: impacts, adaptations and mitigation. In Watson, W.T., Zinyowera, M.C., Moss, R.H. and Dokken, D.J. (eds) *Climate change 1995 – impacts, adaptation and mitigation*. Cambridge University Press, Cambridge: 131–58.

Alley, R.B., Clark, P.U., Huybrechts, P. and Joughin, I. 2005 Ice-sheet and sea-level changes. *Science* 310: 456–60.

Alström, K. and Åkerman, A.B. 1992 Contemporary soil erosion rates on arable land in southern Sweden. *Geografiska Annaler* 74(A): 101–8.

Ambraseys, N. and Bilham, R. 2011 Corruption kills. *Nature* 469(7329): 153–5.

Amezaga, J.M., Rotting, T.S., Younger, P.L., Nairn, R.W., Noles, A. and Quintanilla, J. 2011 A rich vein? Mining and the pursuit of sustainability. *Environmental Science and Technology* 45: 21–6.

Andersen, A., Simcox, D.J., Thomas, J.A. and Nash, D.R. 2014 Assessing reintroduction schemes by comparing genetic diversity of reintroduced and source populations: a case study of the globally threatened large blue butterfly (Maculinea arion). *Biological Conservation* 175: 34–41.

Andersen, F.M. and Larsen, H.V. 2012 FRIDA: a model for the generation and handling of solid waste in Denmark. *Resources, Conservation and Recycling* 65: 27–56.

Andersen, M.S. 1998 Assessing the effectiveness of Denmark's waste tax. *Environment* 40(4): 10–15, 38–41.

André, C. and Platteau, J.-P. 1998 Land relations under unbearable stress: Rwanda caught in the Malthusian trap. *Journal of Economic Behavior and Organization* 34: 1–47.

Andrews, A. 2009 Beyond the ban: can the Basel Convention adequately safeguard the interests of the world's poor in the international trade of hazardous waste? *Law, Environment and Development Journal* 5(2): 167–84.

Angold, P.G. 1997 The impact of a road upon adjacent heathland vegetation: effects on plant species composition. *Journal of Applied Ecology* 34: 409–17.

Anneville, O., Molinero, J.C., Souissi, S., Balvay, G. and Gerdeaux, D. 2007 Long-term changes in the copepod community of Lake Geneva. *Journal of Plankton Research* 29: 49–59.

Anneville, O., Vogel, C., Lobry, J. and Guillard, J. 2017 Fish communities in the Anthropocene: detecting drivers of changes in the deep peri-alpine Lake Geneva. *Inland Waters* 7: 65–76.

Anyamba, A., Small, J.L., Tucker, C.J. and Pak, E.W. 2014 Thirty-two years of Sahelian zone growing season non-stationary NDVI3g patterns and trends. *Remote Sensing* 6(4): 3101–22.

Arita, I. and Francis, D. 2014 Is it time to destroy the smallpox virus? *Science* 345(6200): 1010.

Arnell, N.W. and Gosling, S.N. 2016 The impacts of climate change on river flood risk at the global scale. *Climatic Change* 134(3): 387–401.

Arrossi, S. 1996 Inequality and health in metropolitan Buenos Aires. *Environment and Urbanization* 10: 167–86.

Arvizu, D., Balaya, P., Cabeza, L., Hollands, T., Jager-Waldau, A., Kondo, M., Konseibo, C., Meleshko, V., Stein, W., Tamaura, Y., Xu, H. and Zilles, R. 2011 Direct solar energy. In O. Edenhofer, *et al.* (eds) *IPCC Special Report on Renewable Energy Sources and Climate Change Mitigation*. Cambridge, Cambridge University Press.

Asher, J., Warren, M., Fox, R., Harding, P., Jeffcoate, G. and Jeffcoate, S. 2001 *The millennium atlas of butterflies in Britain and Ireland*. Oxford, Oxford University Press.

Ashraf, M., Athar, H.R., Harris, P.J.C. and Kwon, T.R. 2008 Some prospective strategies for improving crop salt tolerance. *Advances in Agronomy* 97: 45–110.

Asner, G.P., Knapp, D.E., Broadbent, E.N., Oliveira, P.J.C., Keller, M. and Silva, J.N. 2005 Selective logging in the Brazilian Amazon. *Science* 310: 480–2.

Astrup, T.F., Tonini, D., Turconi, R. and Boldrin, A. 2015 Life cycle assessment of thermal waste-to-energy technologies: review and recommendations. *Waste Management* 37: 104–15.

Athens, J.S. 2009 Rattus exulans and the catastrophic disappearance of Hawai'i's native lowland forest. *Biological Invasions* 11: 1489–501.

Avissar, R. and Werth, D. 2005 Global hydroclimatological teleconnections resulting from tropical deforestation. *Journal of Hydrometeorology* 6: 134–45.

Awiti, A.O. 2011 Biological diversity and resilience: lessons from the recovery of cichlid species in Lake Victoria. *Ecology and Society* 16(1): 9.

Azevedo-Santos, V.M., Vitule, J.R., Pelicice, F.M., García-Berthou, E. and Simberloff, D. 2016 Nonnative fish to control Aedes mosquitoes: a controversial, harmful tool. *BioScience* 67: 84–90.

Baer, K.E. and Pringle, C.M. 2000 Special problems of urban river conservation: the encroaching megalopolis. In Boon, P.J., Davies, B.R. and Petts, G.E. (eds) *Global perspectives on river conservation: science, policy and practice.* Chichester, Wiley: 385–402.

Baez, J., de la Fuente, A. and Santos, I. 2010 *Do natural disasters affect human capital? An assessment based on existing empirical evidence.* IZA Discussion Paper No. 5164.

Bahaj, A.S. 2011 Generating electricity from the oceans. *Renewable and Sustainable Energy Reviews* 15: 3399–416.

Bahn, P. and Flenly, V. 1992 *Easter Island.* London, Thames & Hudson.

Bahr, K.D., Jokiel, P.L. and Rodgers, K.S. 2015 The 2014 coral bleaching and freshwater flood events in Kneohe Bay, Hawai'i. *PeerJ* 3: e1136.

Bakir, F., Damlaji, S., Amin-Zaki, L., Murtadha, M., Khalidi, A., Al-Rawi, N., Tikriti, S., Dhakir, H., Clarkson, T., Smith, J. and Doherty, R. 1973 Methylmercury poisoning in Iraq. *Science* 181: 230–41.

Bakir, H.A. 2001 Sustainable wastewater management for small communities in the Middle East and North Africa. *Journal of Environmental Management* 61: 319–28.

Bakker, M.M., Govers, G., Kosmas, C., Vanacker, V., Van Oost, K. and Rounsevell, M.D.A. 2005 Soil erosion as a driver of land-use change. *Agriculture, Ecosystems and Environment* 105: 467–81.

Banerjee, O., Bark, R., Connor, J. and Crossman, N.D. 2013 An ecosystem services approach to estimating economic losses associated with drought. *Ecological Economics* 91: 19–27.

Banister, D. 2008 The Big Smoke: congestion charging and the environment. In Richardson, H.W. and Bae, C.C. (eds) *Road congestion pricing in Europe: implications for the United States.* Cheltenham, Edward Elgar: 176–94.

Banister, D., Anderton, K., Bonilla, D., Givoni, M. and Schwanen, T. 2011 Transportation and the environment. *Annual Review of Environment and Resources* 36: 247–70.

Banks, A.N., Crick, H.Q.P., Coombes, R., Benn, S., Ratcliffe, D.A. and Humpreys, E.M. 2010 The breeding status of Peregrine Falcons Falco peregrines in the UK and Isle of Man in 2002. *Bird Study* 57: 421–36.

Barber, C.P., Cochrane, M.A., Souza, C.M. and Laurance, W.F. 2014 Roads, deforestation, and the mitigating effect of protected areas in the Amazon. *Biological Conservation* 177: 203–9.

Barbier, E.B. 2012 Scarcity, frontiers and development. *The Geographical Journal* 178: 110–22.

Barragán, F., Moreno, C.E., Escobar, F., Halffter, G. and Navarrete, D. 2011 Negative impacts of human land use on dung beetle functional diversity. *PLoS ONE* 6(3): e17976. doi:10.1371.

Barratt, B.I.P., Moran, V.C., Bigler, F. and van Lenteren, J.C. 2018 The status of biological control and recommendations for improving uptake for the future. *BioControl* 63: 155–67.

Barredo, J.I. 2007 Major flood disasters in Europe: 1950–2005. *Natural Hazards* 42: 125–48.

Barrios, E., Delve, R.J., Bekunda, M., Mowo, J., Agunda, J., Ramisch, J., Trejo, M.T. and Thomas, R.J. 2006 Indicators of soil quality: a South–South development of a methodological guide for linking local and technical knowledge. *Geoderma* 135: 248–59.

Barrow, C.J. 1981 Health and resettlement consequences and opportunities created as a result of river impoundment in developing countries. *Water Supply and Management* 5: 135–50.

Barrow, C.J. 1991 *Land degradation.* Cambridge, Cambridge University Press.

Bartone, C. 1990 Economic and policy issues in resource recovery from municipal solid wastes. *Resources, Conservation and Recycling* 4: 7–23.

Basagaoglu, H., Mariño, M.A. and Botzan, T. M. 1999 Land subsidence in the Los Banos-Kettleman City area, California: past and future occurrence. *Physical Geography* 20: 67–82.

Bastos, P. 2017 Drought impacts and cost analysis for Northeast Brazil. In De Nys, E., Engle, N.E. and Magalhães, A.R. (eds) *Drought in Brazil: Proactive management and policy.* Boca Raton, CRC Press: 119–42.

Bates, T.S., Lamb, B.K., Guenther, A., Dignon, J. and Stoiber, R.E. 1992 Sulphur emissions to the atmosphere from natural sources. *Journal of Atmospheric Chemistry* 14: 315–37.

Batorshin, G.S. and Mokrov, Y.G. 2018 Experience and the results of emergency management of the 1957 accident at the Mayak Production Association. *Journal of Radiological Protection* 38(1): R1.

Battisti, D.S. and Naylor, R.L. 2009 Historical warnings of future food insecurity with unprecedented seasonal heat. *Science* 323: 240–4.

Baxter, P.J. 2000 Impacts of eruptions on human health. In Sigurdsson, H., Houghton, B.F., McNutt, S.R., Rymer, H. and Stix, J. (eds) *Encyclopedia of volcanoes.* San Diego, Academic Press: 1035–43.

Bayer, P., Rybach, L., Blum, P. and Brauchler, R. 2013 Review on life cycle environmental effects of geothermal power generation. *Renewable and Sustainable Energy Reviews* 26: 446–63.

Bayfield, N.G., Barker, D.H. and Yah, K.C. 1992 Erosion of road cuttings and the use of bioengineering to improve slope stability in Peninsular Malaysia. *Singapore Journal of Tropical Geography* 13: 75–89.

Beare, D., Hölker, F., Engelhard, G.H., McKenzie, E. and Reid, D.G. 2010 An unintended experiment in fisheries science: a marine area protected by war results in Mexican waves in fish numbers-at-age. *Naturwissenschaften* 97(9): 797–808.

Beaumont, P., Blake, G.H. and Wagstaff, J.M. 1988 *The Middle East: a geographical study*, 2nd edn. London, David Fulton.

Beck, T. and Ghosh, M.G. 2000 Common property resources and the poor: findings from West Bengal. *Economic and Political Weekly* 35: 147–53.

Been, V. 1994 Locally undesirable land uses in minority neighbourhoods: disproportionate siting or market dynamics? *The Yale Law Journal* 103: 1383–422.

Beerkens, R.G.C. and van Limpt, J. 2001 Energy efficiency benchmarking of glass furnaces. Paper presented at the 62nd Conference on Glass Problems. University of Illinois at Urbana-Champaign, Illinois, USA.

Behnke, R. and Mortimore, M. 2016 Introduction: the end of desertification? In Behnke, R. and Mortimore, M. (eds) *The end of desertification? Disputing environmental change in the drylands*. Berlin, Springer: 1–34.

Behnke, R.H. and Scoones, I. 1993 Rethinking range ecology: implications for range management in Africa. In Behnke, R.H., Scoones, I. and Kerven, C. (eds) *Range ecology at disequilibrium*. London, Overseas Development Institute: 1–30.

Belcher, B. and Schreckenberg, K. 2007 Commercialisation of non-timber forest products: a reality check. *Development Policy Review* 25: 355–77.

Bell, F.G. and Donnelly, L.J. 2006 *Mining and its impact on the environment*. London, Taylor & Francis.

Bell, F.G., Donnelly, L.J., Genske, D.D. and Ojeda, J. 2005 Unusual cases of mining subsidence from Great Britain, Germany and Colombia. *Environmental Geology* 47: 620–31.

Bell, M.L., Davis, D.L. and Fletcher, T. 2004 A retrospective assessment of mortality from the London smog episode of 1952: the role of influenza and pollution. *Environmental Health Perspectives* 112: 6–8.

Bennett, E.M., Peterson, G.D. and Gordon, L.J. 2009 Understanding relationships among multiple ecosystem services. *Ecology Letters* 12: 1–11.

Benson, C. and Clay, E.J. 2001 Dominica: natural disasters and economic development in a small island state. Washington, DC, World Bank, Disaster Risk Management Working Paper, Series No. 2.

Benstead, J.P., Stiassny, M.L.J., Loiselle, P.V., Riseng, K.J. and Raminosoa, N. 2000 River conservation in Madagascar. In Boon, P.J., Davies, B.R. and Petts, G.E. (eds) *Global perspectives on river conservation: science, policy and practice*. Chichester, Wiley: 205–31.

Bentley, N. 1998 An overview of the exploitation, trade and management of corals in Indonesia. *TRAFFIC Bulletin* 17: 67–78.

Beresford, N.A., Fesenko, S., Konoplev, A., Skuterud, L., Smith, J.T. and Voigt, G. 2016 Thirty years after the Chernobyl accident: what lessons have we learnt? *Journal of Environmental Radioactivity* 157: 77–89.

Bergström, M. 1990 The release in war of dangerous forces from hydrological facilities. In Westing, A.H. (ed.) *Environmental hazards of war*. London, Sage: 38–47.

Berhe, A.A. 2007 The contribution of landmines to land degradation. *Land Degradation and Development* 18: 1–15.

Beringer, T., Lucht, W. and Schaphoff, S. 2011 Bioenergy production potential of global biomass plantations under environmental and agricultural constraints. *GCB Bioenergy* 3: 299–312.

Berkes, F. (ed.) 1989 *Common property resources: ecology and community-based sustainable development*. London, Belhaven.

Berkes, F. 2012 *Sacred ecology: traditional ecological knowledge and resource management*, 3rd edn. London, Routledge.

Berlemann, M. and Wenzel, D. 2018 Hurricanes, economic growth and transmission channels: empirical evidence for countries on differing levels of development. *World Development* 105: 231–47.

Berry, L., Olson, J. and Campbell, D. 2003 *Assessing the extent, cost and impact of land degradation at the national level: findings and lessons learned from seven pilot case studies*. Rome, Global Mechanism of the UNCCD.

Bertani, R. 2016 Geothermal power generation in the world 2010–2014 update report. *Geothermics* 60: 31–43.

Bétournay, M.C. 2009 Abandoned metal mine stability risk evaluation. *Risk Analysis* 29: 1355–1370.

Bett, B., Said, M.Y., Sang, R., Bukachi, S., Wanyoike, S., Kifugo, S.C., Otieno, F., Ontiri, E., Njeru, I., Lindahl, J. and Grace, D. 2017 Effects of flood irrigation on the risk of selected zoonotic pathogens in an arid and semi-arid area in the eastern Kenya. *PloS ONE* 12(5): e0172626.

Betzold, C. 2015 Adapting to climate change in small island developing states. *Climatic Change* 133(3): 481–9.

Bhatt, S., Weiss, D.J., Cameron, E., Bisanzio, D., Mappin, B., Dalrymple, U., Battle, K.E., Moyes, C.L., Henry, A., Eckhoff, P.A. and Wenger, E.A. 2015 The effect of malaria control on Plasmodium falciparum in Africa between 2000 and 2015. *Nature* 526(7572): 207.

Biazin, B., Sterk, G., Temesgen, M., Abdulkedir, A. and Stroosnijder, L. 2012 Rainwater harvesting and management in rainfed agricultural systems in sub-Saharan Africa: a review. *Physics and Chemistry of the Earth* 47–8: 139–51.

Biesbroek, G.R., Swart, R.J., Carter, T.R., Cowan, C., Henrichs, T., Mela, H., Morecroft, M.D. and Rey, D. 2010 Europe adapts to climate change: comparing national adaptation strategies. *Global Environmental Change* 20: 440–50.

Bigazzi, A.Y. and Rouleau, M. 2017 Can traffic management strategies improve urban air quality? A review of the evidence. *Journal of Transport and Health* 7B: 111–24.

Bilodeau, F., Therrien, J. and Schetagne, R. 2017 Intensity and duration of effects of impoundment on mercury levels in fishes of hydroelectric reservoirs in northern Québec (Canada). *Inland Waters* 7(4): 493–503.

Binet, T., Failler, P., Chavance, P.N. and Mayif, M.A. 2013 First international payment for marine ecosystem services: the case of the Banc d'Arguin National Park, Mauritania. *Global Environmental Change* 23: 1434–43.

Bird, E.F.C. 1985 *Coastline changes: a global review*. Chichester, Wiley.

Björk, S. and Digerfeldt, G. 1991 Development and degradation, redevelopment and preservation of Jamaican wetlands. *Ambio* 20: 276–84.

Blaikie, P. 1989 Explanation and policy in land degradation and rehabilitation for developing countries. *Land Degradation and Rehabilitation* 1: 23–37.

Blaikie, P., Cannon, T., Davis, I. and Wisner, B. 1994 *At risk: natural hazards, people's vulnerability and disasters*. London, Routledge.

Blair, J. (ed.) 2007 *Waterways and canal-building in Medieval England*. Oxford, Oxford University Press.

Blinn, D.W. and Poff, N.L. 2003 The Colorado River system. In Benke, A.C. and Cushing, C.E. (eds) *Rivers of North America*. New York, Academic Press: 483–538.

Blockstein, D.E., 2002 Passenger pigeon (Ectopistes migratorius). In Poole, A. and Gill, F. (eds) *The Birds of North America*. No. 611. Philadelphia, PA.

Blöschl, G., Hall, J., Parajka, J., Perdigão, R.A., Merz, B., Arheimer, B., Aronica, G.T., Bilibashi, A., Bonacci, O., Borga, M. and anjevac, I. 2017 Changing climate shifts timing of European floods. *Science* 357(6351): 588–590.

Blumenthal, D.M., Kray, J.A., Ortmans, W., Ziska, L.H. and Pendall, E. 2016 Cheatgrass is favored by warming but not CO2 enrichment in a semiarid grassland. *Global Change Biology* 22(9): 3026–38.

Boakes, E.H., Mace, G.M., McGowan, P.J.K. and Fuller, R.A. 2010 Extreme contagion in global habitat clearance. *Proceedings of the Royal Society of London B* 277 (1684): 1081–5.

Boardman, J. 2006 Soil erosion science: reflections on the limitations of current approaches. *Catena* 68(2–3): 73–86.

Bob, M., Rahman, N., Elamin, A. and Taher, S. 2016 Rising groundwater levels problem in urban areas: a case study from the Central Area of Madinah City, Saudi Arabia. *Arabian Journal for Science and Engineering* 41: 1461–72.

Bodkin, J.L., Ballachey, B.E., Coletti, H.A., Esslinger, G.G., Kloecker, K.A., Rice, S.D., Reed, J.A. and Monson, D.H. 2012 Long-term effects of the 'Exxon Valdez' oil spill: sea otter foraging in the intertidal as a pathway of exposure to lingering oil. *Marine Ecology Progress Series* 447: 273–87.

Boesch, D.F. 2006 Scientific requirements for ecosystem-based management in the restoration of Chesapeake Bay and coastal Louisiana. *Ecological Engineering* 26: 6–26.

Boët, P., Belliard, J., Berrebi-dit-Thomas, R. and Tales, E. 1999 Multiple human impacts by the City of Paris on fish communities in the Seine river basin, France. *Hydrobiologia* 410: 59–68.

Bond, A.L., Hobson, K.A. and Branfireun, B.A. 2015 Rapidly increasing methyl mercury in endangered ivory gull (Pagophila eburnea) feathers over a 130 year record. *Proceedings of the Royal Society of London B: Biological Sciences* 282(1805): 20150032.

Bond, A.R. and Piepenburg, K. 1990 Land reclamation after surface mining in the USSR: economic, political, and legal issues. *Soviet Geography* 31: 332–65.

Boroffka, N., Oberhänsli, H., Sorrel, P., Demory, F., Reinhardt, C., Wünnemann, B., Alimov, K., Baratov, S., Rakhimov, K., Saparov, N., Shirinov, T., Krivonogov, S.K. and Röhl, U. 2006 Archaeology and climate: settlement and lake-level changes at the Aral Sea. *Geoarchaeology* 21: 721–34.

Borre, L., Barker, D.R. and Duker, L.E. 2001 Institutional arrangements for managing the great lakes of the world: results of a workshop on implementing the watershed approach. *Lakes and Reservoirs: Research and Management* 6: 199–209.

Borrelle, S.B., Rochman, C.M., Liboiron, M., Bond, A.L., Lusher, A., Bradshaw, H. and Provencher, J.F. 2017 Opinion: Why we need an international agreement on marine plastic pollution. *Proceedings of the National Academy of Sciences* 114(38): 9994–7.

Borysova, O., Kondakov, A., Paleari, S., Rautalahti-Miettinen, E., Stolberg, F. and Daler, D. 2005 *Eutrophication in the Black Sea region: impact assessment and causal chain analysis*. Kalmar, Sweden, University of Kalmar.

Boserüp, E. 1965 *The conditions of agricultural growth: the economics of agrarian change under population pressure*. London, Allen & Unwin.

Bosson, R. and Varon, B. 1977 *The mining industry and the developing countries*. New York, Oxford University Press.

Bostock, J., McAndrew B., Richards, R., Jauncey, K., Telfer, T., Lorenzen, K., Little, D., Ross, L., Handisyde, N., Gatward, I. and Corner, R. 2010 Aquaculture: global status and trends. *Philosophical Transactions of Royal Society B* 365: 2897–912.

Boulton, C.A., Booth, B.B. and Good, P. 2017 Exploring uncertainty of Amazon dieback in a perturbed parameter Earth system ensemble. *Global Change Biology* 23(12).

Boutron, C.F., Görlach, U., Candelone, J.-P., Bolshov, M.A. and Delmas, R.J. 1991 Decrease in anthropogenic lead, cadmium and zinc in Greenland snows since the late 1960s. *Nature* 353: 153–6.

Bowen, B.B. and Benison, K.C. 2009 Geochemical characteristics of naturally acid and alkaline saline lakes in southern Western Australia. *Applied Geochemistry* 24: 268–84.

Boyce, J.K. 1990 Birth of a megaproject: political economy of flood control in Bangladesh. *Environmental Management* 14: 419–28.

Boyle, J. 1998 Cultural influences on implementing environmental impact assessment: insights from Thailand, Indonesia, and Malaysia. *Environmental Impact Assessment Review* 18: 95–116.

Bozheyeva, G., Kunakbayev, Y. and Yeleukenov, D. 1999 *Former Soviet biological weapons facilities in Kazakhstan: past, present, and future*. Monterey Institute of International Studies, Center for Nonproliferation Studies Occasional Papers 1.

BP 2017 Statistical review of world energy 2017. London, BP.

Bradshaw, C.J.A., Sodhi, N.S., Peh, K.S.-H. and Brook, B.W. 2007 Global evidence that deforestation amplifies flood risk and severity in the developing world. *Global Change Biology* 13: 2379–395.

Brandt, M., Mbow, C., Diouf, A.A., Verger, A., Samimi, C. and Fensholt, R. 2015 Ground and satellite-based evidence of the biophysical mechanisms behind the greening Sahel. *Global Change Biology* 21(4): 1610–20.

Brasier, C.M. and Kirk, S.A. 2001 Designation of the EAN and NAN races of *Ophiostoma novo-ulmi* as subspecies. *Mycological Research* 105: 547–54.

Braun, S. and Fluckiger, W. 1984 Increased population of the aphis *Aphis pomi* at a motorway. Part 2 the effect of drought and deicing salt. *Environmental Pollution (series A)* 36: 261–70.

Bravo, A.H., Soto, A.R., Sosa, E.R., Sanchez, A.P., Alarcón, J.A.L., Kahl, J. and Ruíz, B.J. 2006 Effect of acid rain on building material of the El Tajin archaeological zone in Veracruz, Mexico. *Environmental Pollution* 144: 655–60.

Brodie, J. and Waterhouse, J. 2012 A critical review of environmental management of the 'not so Great' Barrier Reef. *Estuarine, Coastal and Shelf Science* 104–5: 1–22.

Bromet, E.J., Havenaar, J.M. and Guey, L.T. 2011 A 25 year retrospective review of the psychological consequences of the Chernobyl accident. *Clinical Oncology* 23: 297–305.

Brönmark, C. and Hansson, L. 2002 Environmental issues in lakes and ponds: current state and perspectives. *Environmental Conservation* 29: 290–307.

Brook, B.W., Sodhi, N.S. and Ng, P.K.L. 2003 Catastrophic extinctions follow deforestation in Singapore. *Nature* 424: 420–3.

Brook, B.W., Sodhi, N.S. and Bradshaw, C.J.A. 2008 Synergies among extinction drivers under global change. *Trends in Ecology and Evolution* 23: 453–60.

Brookes, A. 1985 River channelization: traditional engineering methods, physical consequences, and alternative practices. *Progress in Physical Geography* 9: 44–73.

Brooks, N. 2006 Cultural responses to aridity in the Middle Holocene and increased social complexity. *Quaternary International* 151: 29–49.

Bruinsma, J. 2003 *World agriculture: towards 2015/2030: an FAO perspective*. London, Earthscan.

Brunnschweiler, C.N. and Bulte, E.H. 2008 The resource curse revisited and revised: a tale of paradoxes and red herrings. *Journal of Environmental Economics and Management* 55: 248–64.

Buhaug, H., Gleditsch, N.P. and Theisen, O.M. 2010 Implications of climate change for armed conflict. In Mearns, R. and Norton, A. (eds) *Social dimensions of climate change: equity and vulnerability in a warming world*. Washington, DC, World Bank: 75–102.

Burke, L., Reytar, K., Spalding, M. and Perry, A. 2011 *Reefs at risk revisited*. Washington, DC, World Resources Institute.

Burkhardt, U. and Kärcher, B. 2011 Global radiative forcing from contrail cirrus. *Nature Climate Change* 1: 54–8.

Burney, D.A., Burney, L.P., Godfrey, L.R., Jungers, W.L., Goodman, S.M., Wright, H.T. and Jull, A.J. T. 2004 A chronology for late prehistoric Madagascar. *Journal of Human Evolution* 47: 25–63.

Burt, T. 1994 Long-term study of the natural environment: perceptive science or mindless monitoring. *Progress in Physical Geography* 18: 475–96.

Buschiazzo, D.E., Aimar, S.B. and Garcia Queijeiro, J.M. 1999 Long-term maize, sorghum and millet monoculture effects on an Argentina typic ustipsamment. *Arid Soil Research and Rehabilitation* 13: 1–15.

Butler-Stroud, C. 2016 What drives Japanese whaling policy? *Frontiers in Marine Science* 3: 102.

Caissie, D. 2006 The thermal regime of rivers: a review. *Freshwater Biology* 51: 1389–1406.

Canfield, D.E., Glazer, A.N. and Falkowski, P.G. 2010 The evolution and future of earth's nitrogen cycle. *Science* 330(6001): 192–6.

Capra, F. 1982 *The turning point: science, society and the rising culture*. New York, Simon & Schuster.

Carbognin, L., Teatini, P., Tomasin, A. and Tosi, L. 2010 Global change and relative sea level rise at Venice: what impact in term of flooding? *Climate Dynamics* 35: 1039–47.

Carney, J., Gillespie, T.W. and Rosomoff, R. 2014 Assessing forest change in a priority West African mangrove ecosystem: 1986–2010. *Geoforum* 53: 126–35.

Carpenter, K.E. and 38 others 2008 One-third of reef-building corals face elevated extinction risk from climate change and local impacts. *Science* 321: 560–3.

Carson, R. 1962 *Silent spring*. Boston, Houghton Mifflin.

Carter, F.W. 1993 Czechoslovakia. In Carter, F.W. and Turnock, D. (eds) *Environmental problems in Eastern Europe*. London, Routledge: 63–88.

Cascão, A.E. 2009 Changing power relations in the Nile river basin: unilateralism vs. cooperation? *Water Alternatives* 2: 245–68.

Cashore, B., Leipold, S., Cerutti, P.O., Bueno, G., Carodenuto, S., Xiaoqian, C., de Jong, W., Denvir, A., Hansen, C., Humphreys, D. and McGinley, K. 2016 Global governance approaches to addressing illegal logging: uptake and lessons learnt (No. IUFRO World Series no. 35). International Union of Forest Research Organizations (IUFRO), Vienna, Austria.

Caujapé-Castells, J., Tye, A., Crawford, D.J., Santos-Guerra, A., Sakai, A., Beaver, K., Lobin, W., Vincent Florens, F.B., Moura, M., Jardim, R., Gómes, I. and Kueffer, C. 2010 Conservation of oceanic island floras: present and future global challenges. *Perspectives in Plant Ecology, Evolution and Systematics* 12: 107–29.

Caviedes, C.N. and Fik, T.J. 1992 The Peru–Chile eastern Pacific fisheries and climatic oscillation. In Glantz, M.H. (ed.) *Climatic variability, climate change and fisheries*. Cambridge, Cambridge University Press: 355–75.

Cederholm, C.J., Kunze, M.D., Murota, T. and Sibatani, A. 1999 Pacific salmon carcasses: essential contributions of nutrients and energy for aquatic and terrestrial ecosystems. *Fisheries* 24(10): 6–15.

Cesaroni, G., Forastiere, F., Stafoggia, M., Andersen, Z.J., Badaloni, C., Beelen, R., Caracciolo, B., de Faire, U., Erbel, R., Eriksen, K.T. and Fratiglioni, L. 2014 Long term exposure to ambient air pollution and incidence of acute coronary events: prospective cohort study and meta-analysis in 11 European cohorts from the ESCAPE Project. *British Medical Journal* 348: f7412.

Chambers, J.Q., Negron-Juarez, R.I., Marra, D.M., Di Vittorio, A., Tews, J., Roberts, D., Ribeiro, G.H., Trumbore, S.E. and Higuchi, N. 2013 The steady-state mosaic of disturbance and succession across an old-growth Central Amazon forest landscape. *Proceedings of the National Academy of Sciences* 110: 3949–54.

Chan, D.S.W. 2017 Asymmetric bargaining between Myanmar and China in the Myitsone Dam controversy: social opposition akin to David's stone against Goliath. *The Pacific Review* 30(5): 674–91.

Chao, B.F. 1995 Anthropological impact on global geodynamics due to water impoundment in major reservoirs. *Geophysical Research Letters* 22: 3533–6.

Chapman, D. (ed.) 1996 *Water quality assessments*, 2nd edn. London, E. & F.N. Spon.

Chapman, N. and Hooper, A. 2012 The disposal of radioactive wastes underground. *Proceedings of the Geologists' Association* 123: 46–63.

Chaudhuri, S. and Ale, S. 2014 Long term (1960–2010) trends in groundwater contamination and salinization in the Ogallala aquifer in Texas. *Journal of Hydrology* 513: 376–90.

Chavez, F.P., Ryan, J., Lluch-Cota, S.E. and Ñiquen, M. 2003 From anchovies to sardines and back: multidecadal change in the Pacific Ocean. *Science* 299(5604): 217–21.

Chazdon, R.L., Peres, C.A., Dent, D., Sheil, D., Lugo, A.E., Lamb, D., Stork, N.E. and Miller, S.E. 2009 Where are the wild things? Assessing the potential for species conservation in tropical secondary forests. *Conservation Biology* 23: 1406–7.

Chen, J.L., Pekker, T., Wilson, C.R., Tapley, B.D., Kostianoy, A.G., Cretaux, J.F. and Safarov, E.S. 2017a Long-term Caspian Sea level change. *Geophysical Research Letters* 44(13): 6993–7001.

Chen, X., Zhang, X., Church, J.A., Watson, C.S., King, M.A., Monselesan, D., Legresy, B. and Harig, C. 2017b The increasing rate of global mean sea-level rise during 1993–2014. *Nature Climate Change* 7(7): 492.

Chendev, Y.G., Sauer, T.J., Ramirez, G.H. and Burras, C.L. 2015 History of east European chernozem soil degradation; protection and restoration by tree windbreaks in the Russian steppe. *Sustainability* 7: 705–24.

Chernobyl Forum 2005 *Chernobyl's legacy: health, environmental and socio-economic impacts.* Vienna, International Atomic Energy Agency.

Chester, D. 1993 *Volcanoes and society.* London, Edward Arnold.

Chester, M.V. and Horvath, A. 2009 Environmental assessment of passenger transportation should include infrastructure and supply chains. *Environmental Research Letters* 4: 024008.

Chittum, A. and Østergaard, P.A. 2014 How Danish communal heat planning empowers municipalities and benefits individual consumers. *Energy Policy* 74: 465–74.

Choobari, O.A., Zawar-Reza, P. and Sturman, A. 2014 The global distribution of mineral dust and its impacts on the climate system: a review. *Atmospheric Research* 138: 152–65.

Chorus, I. and Bartram, J. 1999 *Toxic cyanobacteria in water: a guide to their public health consequences, monitoring and management.* World Health Organization, London, E. & F.N. Spon.

Choun, H.F. 1936 Dust storms in southwestern plains area. *Monthly Weather Review* 64: 195–9.

Christensen, V., Coll, M., Piroddi, C., Steenbeek, J., Buszowski, J. and Pauly, D. 2014 A century of fish biomass decline in the ocean. *Marine Ecology Progress Series* 512: 155–66.

Cincotta, R.P., Wisnewski, J. and Engelman, R. 2000 Human population in the biodiversity hotspots. *Nature* 404: 990–2.

Ciudad de México 2010 *Calidad del aire en la Ciudad de México.* Mexico, DF, Ciudad de México.

Clair, T.A. and Hindar, A. 2005 Liming for the mitigation of acid rain effects in freshwaters: a review of recent results. *Environmental Reviews* 13: 91–128.

Clapham, P.J. 2016 Managing leviathan: conservation challenges for the great whales in a post-whaling world. *Oceanography* 29(3): 214–25.

Clark, M. and Tilman, D. 2017 Comparative analysis of environmental impacts of agricultural production systems, agricultural input efficiency, and food choice. *Environmental Research Letters* 12: 111002.

Clark, R.B. 1992 *Marine pollution*, 3rd edn. Oxford, Clarendon Press.

Clark, W.C. 1989 Managing planet Earth. *Scientific American* 261(3): 19–26.

Clark, W.C., Crutzen, P.J. and Schellnhuber, H.J. 2005 Science for global sustainability: toward a new paradigm. CID Working Paper No. 120. Cambridge, MA, Science, Environment and Development Group, Center for International Development, Harvard University.

Clayton, K. 1991 Scaling environmental problems. *Geography* 76: 2–15.

Clements, F.E. 1916 *Plant succession: an analysis of the development of vegetation.* Publication 242. Washington, DC, Carnegie Institute.

Cline-Cole, R. and Maconachie, R. 2016 Wood energy interventions and development in Kano, Nigeria: a longitudinal, 'situated' perspective. *Land Use Policy* 52: 163–73.

Coates, L. 1999 Flood fatalities in Australia, 1788–1996. *Australian Geographer* 30: 391–408.

Cohen, M.J., Brown, M.T. and Shepherd, K.D. 2006 Estimating the environmental costs of soil erosion at multiple scales in Kenya using energy synthesis. *Agriculture, Ecosystems and Environment* 114: 249–69.

Colaizzi, P.D., Gowda, P.H., Marek, T.H. and Porter, D.O. 2009 Irrigation in the Texas High Plains: a brief history and potential reductions in demand. *Irrigation and Drainage* 58: 257–74.

Colby, M.E. 1989 The evolution of paradigms of environmental management in development. World Bank Strategic Planning and Review Discussion Paper. Washington, DC, World Bank.

Colinvaux, P. 1993 *Ecology 2.* New York, Wiley.

Collar, N.J., Crosby, M.J. and Stattersfield, A.J. 1994 *Birds to watch 2: the world list of threatened birds.* Cambridge, BirdLife International.

Collenteur, R.A., de Moel, H., Jongman, B. and Di Baldassarre, G. 2015 The failed-levee effect: do societies learn from flood disasters? *Natural Hazards* 76(1): 373–88.

Collier, P. and Hoeffler, A. 2000 Greed and grievance in civil war. World Bank, Policy Research Working Paper 2355.

Collins, C.O. and Scott, S.L. 1993 Air pollution in the Valley of Mexico. *Geographical Review* 83: 119–33.

Connell, J.H. 1978 Diversity in tropical rain forests and coral reefs. *Science* 199: 1302–10.

Connelly, N.A., O'Neill, C.R., Knuth, B.A. and Brown, T.L. 2007 Economic impacts of zebra mussels on drinking water treatment and electric power generation facilities. *Environmental Management* 40: 105–12.

Cook, B.I., Miller, R.L. and Seager, R. 2009 Amplification of the North American 'Dust Bowl' drought through human-induced land degradation. *Proceedings of the National Academy of Sciences* 106: 4997–5001.

Cooke, R.U. and Doornkamp, J.C. 1990 *Geomorphology in environmental management*, 2nd edn. Oxford, Clarendon Press.

Coote, T. and Loève, E. 2003 From 61 species to five: endemic tree snails of the Society Islands fall prey to an ill-judged biological control programme. *Oryx* 37: 91–6.

Coppus, R., Imeson, A.C. and Sevink, J. 2003 Identification, distribution and characteristics of erosion sensitive areas in three different central Andean ecosystems. *Catena* 51: 315–28.

Corlett, R.T. 1992 The ecological transformation of Singapore, 1819–1900. *Journal of Biogeography* 19: 411–20.

Costa, M.H., Botta, A. and Cardille, J.A. 2003 Effects of large-scale changes in land cover on the discharge of the Tocantins River, Southeastern Amazonia. *Journal of Hydrology* 283: 206–17.

Costanza, R., Hart, M., Posner, S. and Talberth, J. 2009 Beyond GDP: the need for new measures of progress. Boston University Pardee Paper 4.

Costello, M.J., May, R.M. and Stork, N.E. 2013 Can we name Earth's species before they go extinct? *Science* 339(6118): 413–16.

Cózar, A., Echevarría, F., González-Gordillo, J.I., Irigoien, X., Úbeda, B., Hernández-León, S., Palma, Á.T., Navarro, S., García-de-Lomas, J., Ruiz, A. and Fernández-de-Puelles, M.L. 2014 Plastic debris in the open ocean. *Proceedings of the National Academy of Sciences* 111(28): 10239–44.

CPRC (Chronic Poverty Research Centre) 2004 *The chronic poverty report 2004–05*. Manchester, CPRC.

Crighton, E.J., Barwin, L., Small, I. and Upshur, R. 2011 What have we learned? A review of the literature on children's health and the environment in the Aral Sea area. *International Journal of Public Health* 56: 125–38.

Critchley, W.R.S., Reij, C. and Willcocks, T.J. 1994 Indigenous soil and water conservation: a review of the state of knowledge and prospects for building on traditions. *Land Degradation and Rehabilitation* 5: 293–314.

Crocker, R.L. and Major, J. 1955 Soil development in relation to vegetation and surface age at Glacier Bay, Alaska. *Journal of Ecology* 43: 427–48.

Croitoru, L. 2010 Deforestation and forest degradation: the case of the Islamic Republic of Iran. In Croitoru, L. and Sarraf, M. (eds) *The cost of environmental degradation: case studies from the Middle East and North Africa*. Washington, DC, World Bank: 53–74.

Crutzen, P. 2006 Albedo enhancement by stratospheric sulfur injections: a contribution to resolve a policy dilemma? *Climatic Change* 77: 211–20.

Crutzen, P.J. and Stoermer, E.F. 2000 The 'anthropocene'. *IGBP Newsletter* 41: 17–18.

Cuellar, A. and Webber, M. 2010 Wasted food, wasted energy: the embedded energy in food waste in the United States. *Environmental Science and Technology* 44: 6464–9.

Daggupati, P., Srinivasan, R., Dile, Y.T. and Verma, D. 2017 Reconstructing the historical water regime of the contributing basins to the Hawizeh marsh: implications of water control structures. *Science of the Total Environment* 580: 832–45.

Dahl, T.E. 2000 *Status and trends of wetlands in the conterminous United States 1986–1997*. Onalaska, WI, US Fish and Wildlife Service.

Dahl, T.E. 2006 *Status and trends of wetlands in the conterminous United States 1998–2004*. Onalaska, WI, US Fish and Wildlife Service.

Dai, A. 2010 Drought under global warming: a review. *WIREs Climate Change* 2: 45–65.

Daly, H.E. 1993 The perils of free trade. *Scientific American* 269(5): 24–9.

Daniels, A.E., Bagstad, K., Esposito, V., Moulaert, A. and Rodriguez, C.M. 2010 Understanding the impacts of Costa Rica's PES: are we asking the right questions? *Ecological Economics* 69: 2116–26.

Dasgupta, S., Huq, M., Khan, Z.H., Ahmed, M.M.Z., Nandan Mukherjee, N., Khan, M.F. and Pandey, K. 2010 Vulnerability of Bangladesh to cyclones in a changing climate: potential damages and adaptation cost. Washington, DC, World Bank Policy Research Working Paper 5280.

Daskalov, G.M. 2002 Overfishing drives a trophic cascade in the Black Sea. *Marine Ecology Progress Series* 225: 53–63.

Davidson, N.C. 2014 How much wetland has the world lost? Long-term and recent trends in global wetland area. *Marine and Freshwater Research* 65: 934–41.

Davies, A. 2009 Environmentalism. In Kitchin, R. and Thrift, N. (eds) *International encyclopedia of human geography*. Amsterdam, Elsevier: 65–72.

Davies, B.R., Boon, P.J. and Petts, G.E. 2000 River conservation: a global imperative. In Boon, P.J., Davies, B.R. and Petts, G.E. (eds) *Global perspectives on river conservation: science, policy and practice*. Chichester, Wiley: xi–xvi.

Davis, D.K. 2004 Desert 'wastes' of the Maghreb: desertification narratives in French colonial environmental history of North Africa. *Cultural Geographies* 11: 359–87.

Davis, L.W. 2017 Saturday driving restrictions fail to improve air quality in Mexico City. *Scientific Reports* 7: 41652.

Davis, M.B. 1976 Erosion rates and land use history in southern Michigan. *Environmental Conservation* 3: 139–48.

Davis, T.J. (ed.) 1993 *Towards the wise use of wetlands*. Gland, Switzerland, Wise Use Project, Ramsar Convention Bureau.

Dawson, C.J. and Hilton, J. 2011 Fertiliser availability in a resource-limited world: production and recycling of nitrogen and phosphorus. *Food Policy* 36: S14–S22.

De Almeida, A.T., Patrão, C., Fonseca, P., Araújo, R., Nunes, U., Rochas, C., Bulgakova, J., Rivière, P., Da Silva, D., Rahbar, A. and Fjordbak, T. 2011 *Standby and off-mode energy losses in new appliances measured in shops*. Coimbra, University of Coimbra.

de Andrade, R.M.T. and Miccolis, A. 2011 Policies and institutional and legal frameworks in the expansion of Brazilian biofuels. Working Paper 71. CIFOR, Bogor, Indonesia.

de Bauer, M. and Hernández Tejeda, T. 2007 A review of ozone-induced effects on the forests of central Mexico. *Environmental Pollution* 147: 446–53.

de Brito, F.C. and Joshi, R.C. 2016 The golden apple snail Pomacea canaliculata: a review on invasion, dispersion and control. *Outlooks on Pest Management* 27(4): 157–63.

de Jong, J. and Wiggens, A.J. 1983 Polders and their environment in the Netherlands. In *Polders of the world, an international symposium: final report*. Wageningen, International Institute for Land Reclamation and Improvement: 221–41.

de Mora, S.J. and Turner, T. 2004 The Caspian Sea: a microcosm for environmental science and international cooperation. *Marine Pollution Bulletin* 48: 26–9.

de Santi, V.P., Dia, A., Adde, A., Hyvert, G., Galant, J., Mazevet, M., Nguyen, C., Vezenegho, S.B., Dusfour, I., Girod, R. and Briolant, S. 2016 Malaria in French Guiana linked to illegal gold mining. *Emerging Infectious Diseases* 22(2): 344.

de Souza Jr, A.B. 2000 Emergency planning for hazardous industrial areas: a Brazilian case study. *Risk Analysis* 20: 483–94.

De Vente, J. and Poesen, J. 2005 Predicting soil erosion and sediment yield at the basin scale: scale issues and semi-quantitative models. *Earth-Science Reviews* 71: 95–125.

de Vivero, J.L.S. and Mateos, J.C.R. 2015 Marine governance in the Mediterranean Sea. In Gilek, M. and Kern, K. (eds) *Governing Europe's marine environment: Europeanization of regional seas or regionalization of EU policies?* Farnham, Ashgate: 203–24.

de Vos, H., Jongerden, J. and van Etten, J. 2008 Images of war: using satellite images for human rights monitoring in Turkish Kurdistan. *Disasters* 32(3): 449–66.

DEA (Danish Energy Authority) 2017 *Danish experiences from offshore wind*. Copenhagen, DEA.

Dean, J.L. and Sholley, M.G. 2006 Groundwater basin recovery in urban areas and implications for engineering projects. Engineering Geology for Tomorrow's Cities, Theme 2.

Dearing, J.A. 2006 Climate-human-environment interactions: resolving our past. *Climate of the Past* 2: 187–203.

De'ath, G., Fabricius, K.E., Sweatman, H. and Puotinen, M. 2012 The 27-year decline of coral cover on the Great Barrier Reef and its causes. *Proceedings of the National Academy of Sciences* 109(44): 17995–9.

Décamps, H. and Fortuné, M. 1991 Long-term ecological research and fluvial landscapes. In Risser, P.G. (ed.) *Long-term ecological research*. SCOPE Report 47. Chichester, Wiley: 135–51.

Defeo, O., McLachlan, A., Schoeman, D.S., Schlacher, T.A., Dugan, J., Jones, A., Lastra, M. and Scapini, F. 2009 Threats to sandy beach ecosystems: a review. *Estuarine, Coastal and Shelf Science* 81: 1–12.

Degu, A.M., Hossain, F., Niyogi, D., Pielke Sr, R., Shepherd, J.M., Voisin, N. and Chronis, T. 2011 The influence of large dams on surrounding climate and precipitation patterns. *Geophysical Research Letters* 38: L04405, doi:10.1029/2010GL046482.

Deichmann, U. and Eklundh, L. 1991 *Global digital datasets for land degradation studies: a GIS approach*. UNEP/GEMS GRID Case Study Series 4. Nairobi, UNEP.

Dejene, A. and Olivares, J. 1991 *Integrating environmental issues into a strategy for sustainable agricultural development: the case of Mozambique*. World Bank Technical Paper 146.

Delgado, C.L. 2003 Rising consumption of meat and milk in developing countries has created a new food revolution. *The Journal of Nutrition* 133: 3907S–10S.

Demarest, A., Rice, P.M. and Rice, D.S. (eds) 2004 *The terminal classic in the Maya lowlands: collapse, transition, and transformation*. Boulder, CO, University of Colorado.

Demaria, F. and Schindler, S. 2016 Contesting urban metabolism: struggles over waste to energy in Delhi, India. *Antipode* 48(2): 293–313.

deMenocal, P.B. 2001 Cultural responses to climate change during the late Holocene. *Science* 292: 667–73.

Desbiens, C. 2004 Producing North and South: a political geography of hydro development in Québec. *The Canadian Geographer* 48: 101–18.

Dessens, O., Köhler, M.O., Rogers, H.L., Jones, R.L. and Pyle, J.A. 2014 Aviation and climate change. *Transport Policy* 34: 14–20.

Dessler, A.E. 2010 A determination of the cloud feedback from climate variations over the past decade. *Science* 330: 1523–7.

Devine-Wright, P. (ed.) 2014 *Renewable energy and the public: from NIMBY to participation.* Abingdon, Routledge.

D'Haen, K., Verstraeten, G. and Degryse, P. 2012 Fingerprinting historical fluvial sediment fluxes. *Progress in Physical Geography* 36(2): 154–86.

Dhakal, S. 2004 *Urban energy use and greenhouse gas emissions in Asian mega-cities: policies for a sustainable future.* Kitakyushu, Japan, Institute for Global Environmental Strategies.

Dhara, V.R. and Dhara, R. 2002 The Union Carbide disaster in Bhopal: a review of health effects. *Archives of Environmental Health* 57: 391–404.

Diamond, J. 2005 *Collapse: how societies choose to fail or survive.* New York, Viking Penguin.

Diarra, D.C. and Akuffo, F.O. 2002 Solar photovoltaic in Mali: potential and constraints. *Energy Conversion and Management* 43: 151–63.

Dias, S.M. 2016 Waste pickers and cities. *Environment and Urbanization* 28: 375–90.

Diaz, R.J. and Rosenberg, R. 2008 Spreading dead zones and consequences for marine ecosystems. *Science* 321: 926–9.

Dickey-Collas, M. 2016 North Sea herring: longer term perspective on management science behind the boom, collapse and recovery of the North Sea herring fishery. In Edwards, C.T.T. and Dankel, D.J. (eds) *Management science in fisheries: an introduction to simulation-based methods.* Abingdon, Routledge: 365–408.

Dickey-Collas, M., Nash, R.D.M., Brunel, T., Damme, C.J.G. van, Marshall, C.T., Payne, M.R., Corten, A., Geffen, A.J., Peck, M.A., Hatfield, E.M.C., Hintzen, N.T., Enberg, K., Kell, L.T. and Simmonds, E.J. 2010 Lessons learned from stock collapse and recovery of North Sea herring: a review. *ICES Journal of Marine Science* 67: 1875–86.

Dickinson, G., Murphy, K. and Springuel, I. 1994 The implications of the altered water regime for the ecology and sustainable development of Wadi Allaqi, Egypt. In Millington, A.C. and Pye, K. (eds) *Environmental change in drylands: biogeographical and geomorphological perspectives.* Chichester, Wiley: 379–91.

Dieperink, C. 2011 International water negotiations under asymmetry: lessons from the Rhine chlorides dispute settlement (1931–2004). *International Environmental Agreements* 11: 139–57.

Dijkema, G.P.J., Reuter, M.A. and Verhoef, E.V. 2000 A new paradigm for waste management. *Waste Management* 20: 633–8.

Dinerstein, E., Loucks, C., Wikramanayake, E., Ginsberg, J., Sanderson, E., Seidensticker, J., Forrest, J., Bryja, G., Heydlauff, A., Klenzendorf, S., Leimgruber, P., Mills, J., O'Brien, T.G., Shrestha, M., Simons, R. and Songer, M. 2007 The fate of wild tigers. *BioScience* 57: 508–14.

Dixon, J.A., Talbot, L.M. and Le Moigne, J.-M. 1989 *Dams and the environment.* World Bank Technical Paper 110.

DME (Danish Ministry of Energy) 1990 *Energy 2000: a plan of action for sustainable development.* Copenhagen, DME.

Dobson, M. 1997 *Contours of death and disease in early modern England.* Cambridge, Cambridge University Press.

Doley, D. and Audet, P. 2013 Adopting novel ecosystems as suitable rehabilitation alternatives for former mine sites. *Ecological Processes* 2: 22.

Dolk, H., Vrijheid, M., Armstrong, B., Abramsky, L., Banchi, F., Garne, E., Nelen, V., Robert, E., Scott, J.E.S., Stone, D. and Tenconi, R. 1998 Risk of congenital abnormalities near hazardous waste landfill sites in Europe: the EUROHAZCON study. *Lancet* 352: 423–7.

Doney, S.C., Mahowald, N., Lima, I., Feely, R.A., Mackenzie, F.T., Lamarque, J.-F. and Rasch, P.J. 2007 Impact of anthropogenic atmospheric nitrogen and sulfur deposition on ocean acidification and the inorganic carbon system. *Proceedings of the National Academy of Sciences* 104: 14580–5.

Donner, S.D., Rickbeil, G.J. and Heron, S.F. 2017 A new, high-resolution global mass coral bleaching database. *PloS ONE* 12(4): e0175490.

Donohoe, M. 2003 Causes and health consequences of environmental degradation and social injustice. *Social Science and Medicine* 56: 573–87.

Doolette, J.B. and Smyle, J.W. 1990 Soil and moisture conservation technologies: review of literature. In Doolette, J.B. and Magrath, W.B. (eds) *Watershed development in Asia: strategies and technologies*. World Bank Technical Paper 127.

Döös, B.R. 1994 Why is environmental protection so slow? *Global Environmental Change* 4: 179–84.

Dottridge, J. and Abu Jaber, N. 1999 Groundwater resources and quality in northeastern Jordan: safe yield and sustainability. *Applied Geography* 19: 313–23.

Douglas, I. 2008 Environmental change in peri-urban areas and human and ecosystem health. *Geography Compass* 2: 1095–137.

Dovers, S.R. and Handmer, J.W. 1993 Contradictions in sustainability. *Environmental Conservation* 20: 217–22.

Dovers, S.R. and Handmer, J.W. 1995 Ignorance, the precautionary principle, and sustainability. *Ambio* 24: 92–7.

Downey, L., Bonds, E. and Clark, K. 2010 Natural resource extraction, armed violence, and environmental degradation. *Organization and Environment* 23(4): 417–45.

Downs, P.W. and Gregory, K.J. 2004 *River channel management: towards sustainable catchment hydrosystems*. London, Arnold.

Driscoll, C.T., Lawrence, G.B., Bulger, A.J., Butler, T.J., Cronan, C.S., Eager, C., Lambert, K.F., Likens, G.E., Stoddard, J.L. and Weathers, K.C. 2001 Acidic deposition in the Northeastern United States: sources and inputs, ecosystem effects, and management strategies. *BioScience* 51: 180–98.

Drysdale, J. 2002 *Whatever happened to Somalia?* 2nd edn. London, Haan.

Du Pisani, J.A. 2006 Sustainable development: historical roots of the concept. *Environmental Sciences* 3: 83–96.

Duan, L., Yu, Q., Zhang, Q., Wang, Z., Pan, Y., Larssen, T., Tang, J. and Mulder, J. 2016 Acid deposition in Asia: emissions, deposition, and ecosystem effects. *Atmospheric Environment* 146: 55–69.

Dudley, N. 1986 Acid rain and British pollution control policy. In Goldsmith, E. and Hildyard, N. (eds) *Green Britain or industrial wasteland?* Cambridge, Polity Press: 95–107.

Dudley, N., Groves, C., Redford, K.H. and Stolton, S. 2014 Where now for protected areas? Setting the stage for the 2014 World Parks Congress. *Oryx* 48(4): 496–503.

Duran-Alvarez, J.C. and Jimenez-Cisneros, B. 2016 Reuse of municipal wastewater for irrigated agriculture: a review. In Goyal, M. (ed.) *Wastewater management for irrigation: principles and practices*. Oakville, CRC Press: 147–226.

Durham, W.H. 1979 *Scarcity and survival in Central America: ecological origins of the Soccer War*. Stanford, CA, Stanford University Press.

Durrheim, R.J. 2010 Mitigating the risk of rockbursts in the deep hard rock mines of South Africa: 100 years of research. In Brune, J. (ed.) *Extracting the science: a century of mining research*. Littleton, CO, Society for Mining, Metallurgy, and Exploration: 156–71.

D'Yakanov, K.N. and Reteyum, A.Y. 1965 The local climate of the Rybinsk reservoir. *Soviet Geography* 6: 40–53.

Earthquest 1991 *Science capsule*, vol. 5(1). Washington, DC, Office for Interdisciplinary Earth Studies.

Easterling, D.R. and Wehner, M.F. 2009 Is the climate warming or cooling? *Geophysical Research Letters* 36: L08706, doi:10.1029/2009GL037810.

Easterling, W.E., Aggarwal, P.K., Batima, P., Brander, K.M., Erda, L., Howden, S.M., Kirilenko, A., Morton, J., Soussana, J.-F., Schmidhuber, J. and Tubiello, F.N. 2007 Food, fibre and forest products. In Parry, M.L., Canziani, O.F., Palutikof, J.P., van der Linden, P.J. and Hanson, C.E. (eds) *Climate change 2007: impacts, adaptation and vulnerability*. Contribution of Working Group II to the Fourth Assessment Report of the Intergovernmental Panel on Climate Change. Cambridge, Cambridge University Press: 273–313.

Edwards, D.P., Tobias, J.A., Sheil, D., Meijaard, E. and Laurance, W.F. 2014 Maintaining ecosystem function and services in logged tropical forests. *Trends in Ecology and Evolution* 29(9): 511–20.

EEA (European Environment Agency) 1997 *Air pollution in Europe*. Copenhagen, EEA Environmental Monograph 4.

EEA 2005 *The European environment: state and outlook 2005*. Copenhagen, EEA.

EEA 2012 *European waters – assessment of status and pressures*. EEA Report No 8/2012. Copenhagen, EEA.

EEA 2015 *The European environment: state and outlook 2015*. Copenhagen, EEA.

EEA 2017 *Managing exposure to noise in Europe*. Copenhagen, EEA.

Ehrlich, P. and Ehrlich, A.H. 1981 *Extinction: the causes and consequences of the disappearance of species*. New York, Random House.

Ehrlich, P.R. and Ehrlich, A.H. 2002 Population, development, and human natures. *Environment and Development Economics* 7: 158–70.

Eigaard, O.R., Bastardie, F., Breen, M., Dinesen, G.E., Hintzen, N.T., Laffargue, P., Mortensen, L.O., Nielsen, J.R., Nilsson, H.C., O'Neill, F.G. and Polet, H. 2015 Estimating seabed pressure from demersal trawls, seines, and dredges based on gear design and dimensions. *ICES Journal of Marine Science* 73(suppl. 1): i27–i43.

Ek, A.S., Löfgren, S., Bergholm, J. and Qvarfort, U. 2001 Environmental effects of one thousand years of copper production at Falun, central Sweden. *Ambio* 30: 96–103.

El-Gohary, M.A. 2016 A holistic approach to the assessment of the groundwater destructive effects on stone decay in Edfu temple using AAS, SEM-EDX and XRD. *Environmental Earth Sciences* 75: 1.

Elkington, J. and Burke, T. 1987 *The green capitalists*. London, Victor Gollancz.

Ellis, D.V. 2003 The concept of 'sustainable ecological succession', and its value in assessing the recovery of sediment seabed biodiversity from environmental impact. *Marine Pollution Bulletin* 46: 39–41.

Ellsworth, W.L. 2013 Injection-induced earthquakes. *Science* 341(6142): 1225942.

Elsom, D. 1996 *Smog alert: managing urban air quality*. London, Earthscan.

Elvingson, P. 1993 Younger stands now affected. *Acid News* 5: 8–9.

Elvingson, P. 1997 Still many trees damaged. *Acid News* 4–5: 14–15.

EMEP 2006 Transboundary air pollution by main pollutants (S, N, O_3) and PM: Sweden. Norwegian Meteorological Institute, MSC-W Data Note 1/2006.

Engler, R. and Guisan, A. 2009 MIGCLIM: Predicting plant distribution and dispersal in a changing climate. *Diversity and Distributions* 15: 590–601.

Environmental Services in the Developing World 2011 Framing pan-tropical analysis and comparison. ICRAF Working Paper No. 32. Nairobi, World Agroforestry Centre.

ESA 2000 Acid deposition, Ecological Society of America. www.esa.org/education/edupdfs/aciddeposition.pdf.

Eskeland, G. and Harrison, A. 2003 Moving to greener pastures? Multinationals and the pollution-haven hypothesis. *Journal of Development Economics* 70: 1–23.

Estrada, A., Garber, P.A., Rylands, A.B., Roos, C., Fernandez-Duque, E., Di Fiore, A., Nekaris, K.A.I., Nijman, V., Heymann, E.W., Lambert, J.E. and Rovero, F. 2017 Impending extinction crisis of the world's primates: why primates matter. *Science Advances* 3(1): p.e1600946.

Eurostat 1997 *Environmental statistics 1996*. Luxembourg, European Commission.

Evans, E. and Wilcox, A.C. 2014 Fine sediment infiltration dynamics in a gravel bed river following a sediment pulse. *River Research and Applications* 30(3): 372–84.

Evans, R. 2005 Monitoring water erosion in lowland England and Wales: a personal view of its history and outcomes. *Catena* 64(2–3): 142–61.

Evrard, O., Heitz, C., Liegeois, M., Boardman, J., Vandaele, K., Auzet, A.-V. and van Wesemael, B. 2010 A comparison of management approaches to control muddy floods in central Belgium, northern France and southern England. *Land Degradation and Development* 21: 322–35.

Ezeh, A., Oyebode, O., Satterthwaite, D., Chen, Y.F., Ndugwa, R., Sartori, J., Mberu, B., Melendez-Torres, G.J., Haregu, T., Watson, S.I. and Caiaffa, W. 2017 The history, geography, and sociology of slums and the health problems of people who live in slums. *Lancet* 389(10068): 547–58.

Fahmi, W. and Sutton, K. 2010 Cairo's contested garbage: sustainable solid waste management and the Zabaleen's right to the city. *Sustainability* 2: 1765–83.

Fallon, S.J., White, J.C. and McCulloch, M.T. 2002 *Porites* corals as recorders of mining and environmental impacts: Misima Island, Papua New Guinea. *Geochimica et Cosmochimica Acta* 66: 45–62.

FAO (Food and Agriculture Organization) 1993 *The state of food and agriculture*. Rome, FAO.

FAO 1995 *Forest resources assessment 1990: global synthesis*. FAO Forestry Paper 124.

FAO 2000 *The challenges of sustainable forestry development in Africa*. Rome, FAO.

FAO 2007 *The world's mangroves, 1980–2005*. FAO Forestry Paper 153. Rome, FAO.

FAO 2010 *Global forest resources assessment, 2010*. Rome, FAO.

FAO 2012 *The state of world fisheries and aquaculture 2012*. Rome, FAO.

FAO 2015 *Global forest resources assessment, 2015*. Rome, FAO.

FAO 2016a *State of the world's forests 2016*. Rome, FAO.

FAO 2016b *The state of world fisheries and aquaculture 2016*. Rome, FAO.

FAO, IFAD, UNICEF, WFP and WHO 2017 *The state of food security and nutrition in the world. Building resilience for peace and food security*. Rome, FAO.

Farfán, M.A., Duarte, J., Real, R., Muñoz, A.R., Fa, J.E. and Vargas, J.M. 2017 Differential recovery of habitat use by birds after wind farm installation: a multi-year comparison. *Environmental Impact Assessment Review* 64: 8–15.

Fearnside, P.M. 1990 Environmental destruction in the Brazilian Amazon. In Goodman, D. and Hall, A. (eds) *The future of Amazonia: destruction or sustainable development?* Basingstoke, Macmillan: 179–225.

Fearnside, P.M. 2005 Deforestation in Brazilian Amazonia: history, rates, and consequences. *Conservation Biology* 19: 680–8.

Fehlenberg, V., Baumann, M., Gasparri, N.I., Piquer-Rodriguez, M., Gavier-Pizarro, G. and Kuemmerle, T. 2017 The role of soybean production as an underlying driver of deforestation in the South American Chaco. *Global Environmental Change* 45: 24–34.

Feng, K., Siu, Y.L., Guan, D. and Hubacek, K. 2012 Assessing regional virtual water flows and water footprints in the Yellow River Basin, China: a consumption-based approach. *Applied Geography* 32: 691–701.

Fergutz, O., Dias, S. and Mitlin, D. 2011 Developing urban waste management in Brazil with waste picker organizations. *Environment and Urbanization* 23: 597–608.

Fernando, C.H. 1991 Impact of fish introductions in tropical Asia and America. *Canadian Journal of Fisheries and Aquatic Sciences* 48: 24–32.

Feshbach, M. and Friendy, A. 1992 *Ecocide in the USSR: health and nature under siege*. New York, Basic Books.

Field, R.D., van der Werf, G.R., Fanin, T., Fetzer, E.J., Fuller, R., Jethva, H., Levy, R., Livesey, N.J., Luo, M., Torres, O. and Worden, H.M. 2016 Indonesian fire activity and smoke pollution in 2015 show persistent nonlinear sensitivity to El Niño-induced drought. *Proceedings of the National Academy of Sciences* 113: 9204–9.

Filatova, T., Mulder, J.P. and van der Veen, A. 2011 Coastal risk management: how to motivate individual economic decisions to lower flood risk? *Ocean and Coastal Management* 54(2): 164–72.

Fimbel, R.A., Gramal, A. and Robinson, J.G. 2001 Logging and wildlife in the tropics. In Fimbel, R.A., Grajal, A. and Robinson, J.G. (eds) *The cutting edge: conserving wildlife in logged tropical forest*. New York, Columbia University Press: 667–95.

Fioletov, V.E., McLinden, C.A., Krotkov, N., Li, C., Joiner, J., Theys, N., Carn, S. and Moran, M.D. 2016 A global catalogue of large SO2 sources and emissions derived from the Ozone Monitoring Instrument. *Atmospheric Chemistry and Physics* 16(18): 11497.

Fisher, H.I. and Baldwin, P.H. 1946 War and the birds of Midway Atoll. *Condor* 48: 3–15.

Fisher, R., O'Leary, R.A., Low-Choy, S., Mengersen, K., Knowlton, N., Brainard, R.E. and Caley, M.J. 2015 Species richness on coral reefs and the pursuit of convergent global estimates. *Current Biology* 25(4): 500–5.

Floehr, T., Xiao, H., Scholz-Starke, B., Wu, L., Hou, J., Yin, D., Zhang, X., Ji, R., Yuan, X., Ottermanns, R. and Roß-Nickoll, M. 2013 Solution by dilution? A review on the pollution status of the Yangtze River. *Environmental Science and Pollution Research* 20: 6934–71.

Foale, S., Adhuri, D., Aliño, P., Allison, E.H., Andrew, N., Cohen, P., Evans, L., Fabinyi, M., Fidelman, P., Gregory, C. and Stacey, N. 2013 Food security and the Coral Triangle initiative. *Marine Policy* 38: 174–83.

Foley, J.A., Coe, M.T., Scheffer, M. and Wang, G.L. 2003 Regime shifts in the Sahara and Sahel: interactions between ecological and climatic systems in northern Africa. *Ecosystems* 6: 524–39.

Foley, M.M., Bellmore, J.R., O'Connor, J.E., Duda, J.J., East, A.E., Grant, G.E., Anderson, C.W., Bountry, J.A., Collins, M.J., Connolly, P.J. and Craig, L.S. 2017 Dam removal – listening in. *Water Resources Research* 53: 5229–46.

Folke, C. and Jansson, A.M. 1992 The emergence of an ecological economics paradigm: examples from fisheries and aquaculture. In Svedin, U. and Aniansson, B. (eds) *Society and the environment: a Swedish perspective*. Dordrecht, Kluwer: 69–87.

Forestry Commission 2003 *National inventory of woodland and trees*. Edinburgh, Forestry Commission.

Forman, R.T.T., Sperling, D., Bissonette, J.A., Clevenger, A.P., Cutshall, C.D., Dale, V.H., Fahrig, L., France, R., Goldman, C.R., Heanue, K., Jones, J.A., Swanson, F.J., Turrentine, T. and Winter, T.C. 2003 *Road ecology*. Washington, DC, Island Press.

Formoli, T.A. 1995 Impacts of the Afghan–Soviet war on Afghanistan's environment. *Environmental Conservation* 22: 66–9.

Forseth, I. and Innis, A. 2004 Kudzu (Pueraria montana): history, physiology, and ecology combine to make a major ecosystem threat. *Critical Reviews in Plant Science* 23: 401–13.

Forster, P., Ramaswamy, V., Artaxo, P., Berntsen, T., Betts, R., Fahey, D.W., Haywood, J., Lean, J., Lowe, D.C., Myhre, G., Nganga, J., Prinn, R., Raga, G., Schulz, M. and Van Dorland, R. 2007 Changes in atmospheric constituents and in radiative forcing. In Solomon, S., Qin, D., Manning, M., Chen, Z., Marquis, M., Averyt, K.B., Tignor, M. and Miller, H.L. (eds) *Climate change 2007: the physical science basis*. Contribution of Working Group I to the Fourth Assessment Report of the Intergovernmental Panel on Climate Change. Cambridge and New York, Cambridge University Press.

Foster, M.S. and Schiel, D.R. 2010 Loss of predators and the collapse of southern California kelp forests (?): alternatives, explanations and generalizations. *Journal of Experimental Marine Biology and Ecology* 393(1): 59–70.

Fowler, D., Coyle, M., Skiba, U., Sutton, M.A., Cape, J.N., Reis, S., Sheppard, L.J., Jenkins, A., Grizzetti, B., Galloway, J.N. and Vitousek, P. 2013 The global nitrogen cycle in the twenty-first century. *Philosophical Transactions of the Royal Society B* 368(1621): 20130164.

Frank, K.T., Petrie, B., Fisher, J.A.D. and Leggett, W.C. 2011 Transient cynamics of an altered large marine ecosystem. *Nature* 477: 86–9.

Franklin, M. and Fruin, S. 2017 The role of traffic noise on the association between air pollution and children's lung function. *Environmental Research* 157: 153–9.

Franzén, L.G. and Cropp, R.A. 2007 The peatland/ice age hypothesis revised, adding a possible glacial pulse trigger. *Geografiska Annaler: Series A, Physical Geography* 89(4): 301–30.

Freedonia 2016 *World construction aggregates*. Industry Study No. 3389. Cleveland, The Freedonia Group.

Freer-Smith, P.H. 1998 Do pollutant-related forest declines threaten the sustainability of forests? *Ambio* 27: 123–31.

French, H.M. 2007 *The periglacial environment*, 3rd edn. Chichester, Wiley.

Fruergaard, T., Christensen, T.H. and Astrup, T. 2010 Energy recovery from waste incineration: assessing the importance of district heating networks. *Waste Management* 30: 1264–72.

FSIN (Food Security Information Network) 2017 *Global report on food crises 2017*. Rome, FAO.

Fujita, M.S. and Tuttle, M.D. 1991 Flying foxes (*Chiroptera: Pteropodidae*): threatened animals of key ecological and economic importance. *Conservation Biology* 5: 455–63.

Furuno, K., Kagawa, A., Kazaoka, O., Kusuda, T. and Nirei, H. 2015 Groundwater management based on monitoring of land subsidence and groundwater levels in the Kanto Groundwater Basin, Central Japan. *Proceedings of the International Association of Hydrological Sciences* 372: 53–7.

Gabbay, S. 1998 *The environment in Israel*. Jerusalem, Ministry of the Environment.

Gaiha, R. and Thapa, G. 2006 Natural disasters, vulnerability and mortalities: a cross-country analysis. International Fund for Agricultural Development Working Paper.

Gaillard, J.C. 2009 From marginality to further marginalization: experiences from the victims of the July 2000 Payatas trashslide in the Philippines. *Jàmbá: Journal of Disaster Risk Studies* 2(3): 197–215.

Gall, M., Borden, K.A., Emrich, C.T. and Cutter, S.L. 2011 The unsustainable trend of natural hazard losses in the United States. *Sustainability* 3: 2157–81.

Gamu, J.K. and Dauvergne, P. 2018 The slow violence of corporate social responsibility: the case of mining in Peru. *Third World Quarterly* February: 1–17.

Gandy, M. 1996 Crumbling land: the postmodernity debate and the analysis of environmental problems. *Progress in Human Geography* 20: 23–40.

Garcia, L.C., Ribeiro, D.B., Oliveira Roque, F., Ochoa-Quintero, J.M. and Laurance, W.F. 2016 Brazil's worst mining disaster: corporations must be compelled to pay the actual environmental costs. *Ecological Applications* 27: 5–9.

Garonna, I., Jong, R., Wit, A.J., Mücher, C.A., Schmid, B. and Schaepman, M.E. 2014 Strong contribution of autumn phenology to changes in satellite-derived growing season length estimates across Europe (1982–2011). *Global Change Biology* 20(11): 3457–70.

Gaume, E. and 24 others 2009 A compilation of data on European flash floods. *Journal of Hydrology* 367: 70–8.

Gavrilova, O., Vilu, R. and Vallner, L. 2010 A life cycle environmental impact assessment of oil shale produced and consumed in Estonia. *Resources, Conservation and Recycling* 55: 232–45.

Geertz, C. 1963 *Agricultural involution: the process of change in Indonesia*. Berkeley, CA, University of California Press.

Gellis, A.C., Webb, R.M.T., McIntyre, S.C. and Wolfe, W.J. 2006 Land-use effects on erosion, sediment yields, and reservoir sedimentation: a case study in the Lago Loíza Basin, Puerto Rico. *Physical Geography* 27: 39–69.

GEMS (Global Environment Monitoring System) 1988 *Assessment of freshwater quality*. Nairobi, UNEP/WHO.

Gensburg, L.J., Pantea, C., Fitzgerald, E., Stark, A. and Kim, N. 2009 Mortality among former Love Canal residents. *Environmental Health Perspectives* 117: 209–16.

George, D.J. 1992 Rising groundwater: a problem of development in some urban areas of the Middle East. In McCall, G.J.H., Laming, D.J.C. and Scott, S.C. (eds) *Geohazards: natural and man-made hazards*. London, Chapman & Hall: 171–82.

Gerber, R., Smit, N.J., Van Vuren, J.H., Nakayama, S.M., Yohannes, Y.B., Ikenaka, Y., Ishizuka, M. and Wepener, V. 2016 Bioaccumulation and human health risk assessment of DDT and other organochlorine pesticides in an apex aquatic predator from a premier conservation area. *Science of the Total Environment* 550: 522–33.

Gerhard, J. and Haynie, F.H. 1974 *Air pollution effects on catastrophic failure of metals*. EPA-650/3-74-009. Research Triangle Park, North Carolina.

GESAMP (Joint Group of Experts on the Scientific Aspects of Marine Environmental Protection) 1990 *The state of the marine environment*. Oxford, Blackwell Scientific.

GESAMP 2009 *Pollution in the open oceans: a review of assessments and related studies*. GESAMP Report Study 79.

Gharehchahi, E., Mahvi, A.H., Amini, H., Nabizadeh, R., Akhlaghi, A.A., Shamsipour, M. and Yunesian, M. 2013 Health impact assessment of air pollution in Shiraz, Iran: a two-part study. *Journal of Environmental Health Science and Engineering* 11: 11.

Ghassemi, F., Jakeman, A.J. and Nix, H.A. 1995 *Salinisation of land and water resources: human causes, extent, management and case studies*. Sydney, University of New South Wales Press.

Gianessi, L., Rury, K. and Rinkus, A. 2009 An evaluation of pesticide use reduction policies in Scandinavia. *Outlooks on Pest Management* 20(6): 268–74.

Gibb, H. and O'Leary, K.G. 2014 Mercury exposure and health impacts among individuals in the artisanal and small-scale gold mining community: a comprehensive review. *Environmental Health Perspectives* 122(7): 667.

Gibbon, D., Lake, A. and Stocking, M. 1995 Sustainable development: a challenge for agriculture. In Morse, S. and Stocking, M. (eds) *People and environment*. London, UCL Press: 31–68.

Gibbs, H.K., Munger, J., L'Roe, J., Barreto, P., Pereira, R., Christie, M., Amaral, T. and Walker, N.F. 2016 Did ranchers and slaughterhouses respond to zero-deforestation agreements in the Brazilian Amazon? *Conservation Letters* 9: 32–42.

Gibson, G.R., Taylor, N.L., Lamo, N.C. and Lackey, J.K. 2017 Effects of recent instability on cultivated area along the Euphrates River in Iraq. *The Professional Geographer* 69(2): 163–76.

Giebułtowicz, J. and Nał-cz-Jawecki, G. 2014 Occurrence of antidepressant residues in the sewage-impacted Vistula and Utrata rivers and in tap water in Warsaw (Poland). *Ecotoxicology and Environmental Safety* 104: 103–9.

Gieré, R., LaFree, S.T., Carleton, L.E. and Tishmack, J.K. 2004 *Environmental impact of energy recovery from waste tyres*. Geological Society, London, Special Publications 236: 475–98.

Gignoux, C.R., Henn, B.R. and Mountain, J.L. 2011 Rapid, global demographic expansions after the origins of agriculture. *Proceedings of the National Academy of Sciences* 108: 6044–9.

Gilland, B. 1993 Cereals, nitrogen and population: an assessment of the global trends. *Endeavour* 17: 84–7.

Gilland, B. 2002 World population and food supply: can food production keep pace with population growth in the next half-century? *Food Policy* 27: 47–63.

Gilliom, R.J. 2007 Pesticides in U.S. streams and groundwater. *Environmental Science and Technology* 41: 3407.

Gillis, A.M. 1992 Keeping aliens out of paradise. *Bioscience* 42: 482–5.

Gillson, L. and Hoffman, M.T. 2007 Rangeland ecology in a changing world. *Science* 315: 53–4.

Gilman, E.L., Ellison, J., Duke, N.C. and Field, C. 2008 Threats to mangroves from climate change and adaptation options: a review. *Aquatic Botany* 89: 237–50.

Giordano, M.A. and Wolf, A.T. 2003 Sharing waters: post-Rio international water management. *Natural Resources Forum* 27: 163–71.

Giri, C., Ochieng, E., Tieszen, L., Zhu, Z., Singh, A., Loveland, T., Masek, J. and Duke, N. 2011 Status and distribution of mangrove forests of the world using Earth observation satellite data. *Global Ecology and Biogeography* 20: 154–9.

Glantz, M.H. and Orlovsky, N. 1983 Desertification: a review of the concept. *Desertification Control Bulletin* 9: 15–22.

Glantz, M.H., Rubinstein, A.Z. and Zonn, I. 1993 Tragedy in the Aral Sea basin: looking back to plan ahead. *Global Environmental Change* 3: 174–98.

Glasby, G.P. 1997 Disposal of chemical weapons in the Baltic Sea. *Science of the Total Environment* 206: 267–73.

Gleick, P.H. (ed.) 1993 *Water in crisis: a guide to the world's fresh water resources*. New York, Oxford University Press.

Godefroid, S. and 19 others 2011 How successful are plant species reintroductions? *Biological Conservation* 144: 672–82.

Godfray, C., Beddington, J.R., Crute, I.R., Haddad, L., Lawrence, D., Muir, J.F., Pretty, J., Robinson, S., Thomas, S.M. and Toulmin, C. 2010 Food security: the challenge of feeding 9 billion people. *Science* 327: 812–18.

Goldblat, J. 1975 The prohibition of environmental warfare. *Ambio* 4: 186–90.

Goldemberg, J., Johansson, T.B., Reddy, A.K.N. and Williams, R.H. 1988 *Energy for a sustainable world*. New Delhi, Wiley.

Goldschmidt, T. 1996 *Darwin's dreampond: drama on Lake Victoria*. Cambridge, MA, MIT Press.

Goldsmith, E. and Hildyard, N. 1984 *The social and environmental effects of large dams*, vol. 1. Wadebridge, Cornwall, Wadebridge Ecological Centre.

Goodland, R. 1990 The World Bank's new environmental policy for dams and reservoirs. *Water Resources Development* 6: 226–39.

Goodland, R.J.A., Daly, H.E. and Serafy, S. El 1993a The urgent need for rapid transformation to global environmental sustainability. *Environmental Conservation* 20: 297–309.

Goodland, R.J.A., Juras, A. and Pachauri, R. 1993b Can hydro-reservoirs in tropical moist forests be environmentally sustainable? *Environmental Conservation* 20: 122–30.

Goossens, D. and Buck, B. 2011 Effects of wind erosion, off-road vehicular activity, atmospheric conditions and the proximity of a metropolitan area on PM10 characteristics in a recreational site. *Atmospheric Environment* 45: 94–107.

Gordon, L.J., Peterson, G.D. and Bennett, E.M. 2008 Agricultural modifications of hydrological flows create ecological surprises. *Trends in Ecology and Evolution* 23: 211–19.

Gorham, E. 1958 The influence and importance of daily weather conditions in the supply of chloride, sulphate and other ions to freshwaters from atmospheric precipitation. *Philosophical Transactions of the Royal Society, London* 241B: 147–78.

Goudie, A.S. 2006 *The human impact on the natural environment*, 6th edn. Oxford, Blackwell.

Gowdy, J. and McDaniel, C. 1999 The physical destruction of Nauru: an example of weak sustainability. *Land Economics* 75: 333–8.

Grace, J. 2004 Understanding and managing the global carbon cycle. *Journal of Ecology* 92: 189–202.

Grainger, A. 1993a *Controlling tropical deforestation*. London, Earthscan.

Grainger, A. 1993b Rates of deforestation in the humid tropics: estimates and measurements. *Geographical Journal* 159: 33–44.

Granéli, E. and Haraldson, C. 1993 Can increased leaching of trace metals from acidified areas influence phytoplankton growth in coastal waters? *Ambio* 22: 308–11.

Grassini, P., Eskridge, K.M. and Cassman, K.G. 2013 Distinguishing between yield advances and yield plateaus in historical crop production trends. *Nature Communications* 4: 2918.

Grattan, J. 2006 Aspects of Armageddon: an exploration of the role of volcanic eruptions in human history and civilization. *Quaternary International* 151: 10–18.

Graveland, J., van der Wal, R., van Balen, J.H. and van Noordwijk, A.J. 1994 Poor reproduction in forest passerines from decline of snail abundance on acidified soils. *Nature* 368: 446–8.

Graves, H.S. 1918 Effect of the war on forests of France. *American Forestry* 24: 707–17.

Great Barrier Reef Marine Park Authority 2001 *Water quality: a threat to the Great Barrier Reef*. Townsville, Queensland, Great Barrier Reef Marine Park Authority.

Green, R.A., Cubrinovski, M., Cox, B., Wood, C., Wotherspoon, L., Bradley, B. and Maurer, B. 2014 Select liquefaction case histories from the 2010–2011 Canterbury earthquake sequence. *Earthquake Spectra* 30(1): 131–53.

Greene, O.J., Holt, S.E. and Wilkinson, A. 2004 *Ammunition stocks: promoting safe and secure storage and disposal*. Bradford, Department of Peace Studies, University of Bradford.

Gregory, C., Brierley, G. and Le Heron, R. 2011 Governance spaces for sustainable river management. *Geography Compass* 5/4: 182–99.

Gregory, M.R. 2009 Environmental implications of plastic debris in marine settings: entanglement, ingestion, smothering, hangers-on, hitch-hiking and alien invasions. *Philosophical Transactions of the Royal Society B* 364: 2013–25.

Grieve, I.C. 2001 Human impacts on soil properties and their implications for the sensitivity of soil systems in Scotland. *Catena* 42: 361–74.

Grigg, D.B. 1992 *The transformation of agriculture in the West*. Oxford, Blackwell.

Grigg, D.B. 1993 The role of livestock products in world food consumption. *Scottish Geographical Magazine* 109: 66–74.

Gröger, J.P., Kruse, G.H. and Rohlf, N. 2009 Slave to the rhythm: how large-scale climate cycles trigger herring (Clupea harengus) regeneration in the North Sea. *ICES Journal of Marine Science* 67: 454–65.

Groombridge, B. and Jenkins, M.D. 2000 *Global biodiversity: Earth's living resources in the 21st century*. Cambridge, WCMC.

Gross, M. 2016 A planet with two billion cars. *Current Biology* 26(8): R307–10.

Grossman, L.S. 1992 Pesticides, caution, and experimentation in Saint Vincent, Eastern Caribbean. *Human Ecology* 20: 315–36.

Grove, R. 1990 The origins of environmentalism. *Nature* 345: 11–14.

Grube, A., Donaldson, D., Kiely, T. and Wu, L. 2011 *Pesticides industry sales and usage: 2006 and 2007 market estimates*. Washington, DC, US Environmental Protection Agency.

Gunn, J.M., Kielstra, B.W. and Szkokan-Emilson, E. 2016 Catchment liming creates recolonization opportunity for sensitive invertebrates in a smelter impacted landscape. *Journal of Limnology* 75(s2).

Guo, J.H., Liu, X.J., Zhang, Y., Shen, J.L., Han, W.X., Zhang, W.F., Christie, P., Goulding, K., Vitousek, P. and Zhang, F.S. 2010 Significant acidification in major Chinese croplands. *Science* 327: 1008–10.

Gupta, H.K. 2002 A review of recent studies of triggered earthquakes by artificial water reservoirs with special emphasis on earthquakes in Koyna, India. *Earth-Science Reviews* 58: 279–310.

Gurjar, B.R., Ravindra, K. and Nagpure, A.S. 2016 Air pollution trends over Indian megacities and their local-to-global implications. *Atmospheric Environment* 142: 475–95.

Guy, C.S., Treanor, H.B., Kappenman, K.M., Scholl, E.A., Ilgen, J.E. and Webb, M.A. 2015 Broadening the regulated-river management paradigm: a case study of the forgotten dead zone hindering pallid sturgeon recovery. *Fisheries* 40: 6–14.

Ha-Duong, M. and Journé, V. 2014 Calculating nuclear accident probabilities from empirical frequencies. *Environment Systems and Decisions* 34(2): 249–58.

Hagerman, S.M. and Pelai, R. 2016 'As far as possible and as appropriate': implementing the Aichi Biodiversity Targets. *Conservation Letters* 9(6): 469–78.

Hägerstrand, T. and Lohm, U. 1990 Sweden. In Turner II, B.L., Clark, W.C., Kates, R.W., Richards, J.F., Mathews, J.T. and Meyer, W.B. (eds) *The Earth as transformed by human action*. Cambridge, Cambridge University Press: 605–22.

Hahs, A.K., McDonnell, M.J., McCarthy, M.A., Vesk, P.A., Corlett, R.T., et al. (2009) A global synthesis of plant extinction rates in urban areas. *Ecology Letters* 12: 1165–73.

Hairat, M.K. and Ghosh, S. 2017 100GW solar power in India by 2022 – a critical review. *Renewable and Sustainable Energy Reviews* 73: 1041–50.

Hall, D.R. 2002 Albania. In Carter, F.W. and Turnock, D. (eds) *Environmental problems of East Central Europe*, 2nd edn. London, Routledge: 251–82.

Hall, K., Guo, J., Dore, M. and Chow, C. 2009 The progressive increase of food waste in America and its environmental impact. *PLoS ONE* 4(11): e7940.

Halleaux, D.G. and Rennó, N.O. 2014 Aerosols–climate interactions at the Owens 'Dry' Lake, California. *Aeolian Research* 15: 91–100.

Hamilton, S.E. and Casey, D. 2016 Creation of a high spatio-temporal resolution global database of continuous mangrove forest cover for the 21st century (CGMFC-21). *Global Ecology and Biogeography* 25: 729–38.

Hammar, L., Perry, D. and Gullström, M. 2016 Offshore wind power for marine conservation. *Open Journal of Marine Science* 6: 66–78.

Hanan, N.P., Prevost, Y., Diouf, A. and Diallo, O. 1991 Assessment of desertification around deep wells in the Sahel using satellite imagery. *Journal of Applied Ecology* 28: 173–86.

Hand, J.L., Gill, T.E. and Schichtel, B.A. 2017 Spatial and seasonal variability in fine mineral dust and coarse aerosol mass at remote sites across the United States. *Journal of Geophysical Research: Atmospheres* 122: 3080–97.

Hanger, S., Komendantova, N., Schinke, B., Zejli, D., Ihlal, A. and Patt, A. 2016 Community acceptance of large-scale solar energy installations in developing countries: evidence from Morocco. *Energy Research and Social Science* 14: 80–9.

Hansell, A.L., Horwell, C.J. and Oppenheimer, C. 2006 The health hazards of volcanoes and geothermal areas. *Occupational and Environmental Medicine* 63: 149–56.

Hansen, L.T., Breneman, V.E., Davison, C.W. and Dicken, C.W. 2002 The cost of soil erosion to downstream navigation. *Journal of Soil and Water Conservation* 57: 205–12.

Hanson, S., Nicholls, R., Ranger, N., Hallegatte, S., Corfee-Morlot, J. and Herweijer, C. 2011 A global ranking of port cities with high exposure to climate extremes. *Climatic Change* 104: 89–111.

Hanson, T., Brooks, T.M., Da Fonseca, G.A.B., Hoffmann, M., Lamoreux, J.F., Machlis, G., Mittermeier, C.G., Mittermeier, R.A. and Pilgrim, J.D. 2009 Warfare in biodiversity hotspots. *Conservation Biology* 23: 578–87.

Hardin, G. 1968 The tragedy of the commons. *Science* 162: 1243–8.

Hardoon, D., Fuentes-Nieva, R. and Ayele, S. 2016 *An economy for the 1%: how privilege and power in the economy drive extreme inequality and how this can be stopped*. Oxford, Oxfam.

Harris, K., Keen, D. and Mitchell, T. 2013 *When disaster and conflicts collide: improving links between disaster resilience and conflict prevention*. London, ODI.

Harrison, R.D. 2011 Emptying the forest: hunting and the extirpation of wildlife from tropical nature reserves. *Bioscience* 61: 919–24.

Harrison, R.D., Sreekar, R., Brodie, J.F., Brook, S., Luskin, M., O'Kelly, H., Rao, M., Scheffers, B. and Velho, N. 2016 Impacts of hunting on tropical forests in Southeast Asia. *Conservation Biology* 30: 972–81.

Harrison, R.M. and Hester, R.E. (eds) 2017 *Environmental impacts of road vehicles: past, present and future*. London, Royal Society of Chemistry.

Hatton, J., Couto, M. and Oglethorpe, J. 2001 *Biodiversity and war: a case study of Mozambique*. Washington, DC, Biodiversity Support Program.

Havlick, D.G., Hourdequin, M. and John, M. 2014 Examining restoration goals at a former military site: Rocky Mountain Arsenal, Colorado. *Nature and Culture* 9(3): 288.

Hay, S.I., Cox, J., Rogers, D.J., Randolph, S.E., Stern, D.I., Shanks, G.D., Myers, M.F. and Snow, R.W. 2002 Climate change and the resurgence of malaria in the East African highlands. *Nature* 415: 905–9.

Hay, S.I., Guerra, C.A., Tatem, A.J., Noor, A.M. and Snow, R.W. 2004 The global distribution and population at risk of malaria: past, present, and future. *Lancet Infectious Diseases* 4: 327–36.

Hayhow, D.B., Ausden, M.A., Bradbury, R.B., Burnell, D., Copeland, A.I., Crick, H.Q.P., Eaton, M.A., Frost, T., Grice, P.V., Hall, C., Harris, S.J., Morecroft, M.D., Noble, D.G., Pearce-Higgins, J.W., Watts, O. and Williams, J.M. 2017 *The state of the UK's birds 2017*. Sandy, Bedfordshire, RSPB, BTO, WWT, DAERA, JNCC, NE and NRW.

Hays, J.D., Imbrie, J. and Shackleton, N.J. 1976 Variations in the earth's orbit: pacemaker of the ice ages. *Science* 235: 1156–67.

Heacock, M., Kelly, C.B., Asante, K.A., Birnbaum, L.S., Bergman, Å.L., Bruné, M.N., Buka, I., Carpenter, D.O., Chen, A., Huo, X. and Kamel, M. 2016 E-waste and harm to vulnerable populations: a growing global problem. *Environmental Health Perspectives* 124(5): 550.

Healy, H. 2007 Korean demilitarized zone: peace and nature park. *International Journal of World Peace* 24(4): 61–83.

Heath, J., Pollard, E. and Thomas, J.A. 1984 *Atlas of butterflies in Britain and Ireland*. Harmondsworth, Viking.

Hedley, P.J., Bird, M.I. and Robinson, R.A.J. 2010 Evolution of the Irrawaddy delta region since 1850. *The Geographical Journal* 176: 138–49.

Heiken, G. 2013 *Dangerous neighbors: volcanoes and cities*. Cambridge, Cambridge University Press.

Held, I.M. and Soden, B.J. 2006 Robust responses of the hydrological cycle to global warming. *Journal of Climate* 19: 5686–99.

Hellden, U. 1988 Desertification monitoring: is the desert encroaching? *Desertification Control Bulletin* 17: 8–12.

Henderson, J.V., Storeygard, A. and Deichmann, U. 2017 Has climate change driven urbanization in Africa? *Journal of Development Economics* 124: 60–82.

Henderson, S., Dawson, T.P. and Whittaker, R.J. 2006 Progress in invasive plants research. *Progress in Physical Geography* 30: 25–46.

Hendryx, M. 2010 Poverty and mortality disparities in Central Appalachia: mountaintop mining and environmental justice. *Journal of Health Disparities Research and Practice* 4(3): Article 6.

Herdt, R.W. 2006 Biotechnology in agriculture. *Annual Review of Environment and Resources* 31: 265–95.

Hernandez, R.R., Easter, S.B., Murphy-Mariscal, M.L., Maestre, F.T., Tavassoli, M., Allen, E.B., Barrows, C.W., Belnap, J., Ochoa-Hueso, R., Ravi, S. and Allen, M.F. 2014 Environmental impacts of utility-scale solar energy. *Renewable and Sustainable Energy Reviews* 29: 766–79.

Herrero, H.V., Southworth, J., Bunting, E. and Child, B. 2017 Using repeat photography to observe vegetation change over time in Gorongosa National Park. *African Studies Quarterly* 17(2): 65.

Herrero, M., Wirsenius, S., Henderson, B., Rigolot, C., Thornton, P., Havlík, P., De Boer, I. and Gerber, P.J. 2015 Livestock and the environment: what have we learned in the past decade? *Annual Review of Environment and Resources* 40: 177–202.

Herrmann, D.L., Schwarz, K., Shuster, W.D., Berland, A., Chaffin, B.C., Garmestani, A.S. and Hopton, M.E. 2016 Ecology for the shrinking city. *BioScience* 66: 965–73.

Hertel, T.W., Ramankutty, N. and Baldos, U.L.C. 2014 Global market integration increases likelihood that a future African Green Revolution could increase crop land use and CO_2 emissions. *Proceedings of the National Academy of Sciences* 111(38): 13799–804.

Hewitt, K. and Burton, I. 1971 *The hazardousness of a place: a regional ecology of damaging events*. Toronto, University of Toronto Geography Department.

Higham, J.E., Bejder, L., Allen, S.J., Corkeron, P.J. and Lusseau, D. 2016 Managing whale-watching as a non-lethal consumptive activity. *Journal of Sustainable Tourism* 24: 73–90.

Hilson, G. 2016 Farming, small-scale mining and rural livelihoods in Sub-Saharan Africa: a critical overview. *The Extractive Industries and Society* 3: 547–63.

Hintz, W.D., Mattes, B.M., Schuler, M.S., Jones, D.K., Stoler, A.B., Lind, L. and Relyea, R.A. 2017 Salinization triggers a trophic cascade in experimental freshwater communities with varying food-chain length. *Ecological Applications* 27(3): 833–44.

Hinzman, L.D. and 34 others 2005 Evidence and implications of recent climate change in northern Alaska and other Arctic regions. *Climatic Change* 72: 251–98.

Hirst, R.A., Pywell, R.F. and Putwain, P.D. 2000 Assessing habitat disturbance using an historical perspective: the case of Salisbury Plain military training area. *Journal of Environmental Management* 60: 181–93.

Hodgson, D.A. and Johnston, N.M. 1997 Inferring seal populations from lake sediments. *Nature* 387: 30–1.

Hodgson, J.A., Moilanen, A., Bourn, N.A., Bulman, C.R. and Thomas, C.D. 2009 Managing successional species: modelling the dependence of heath fritillary populations on the spatial distribution of woodland management. *Biological Conservation* 142(11): 2743–51.

Hofmann, H., Schleper, M.C. and Blome, C. 2018 Conflict minerals and supply chain due diligence: an exploratory study of multi-tier supply chains. *Journal of Business Ethics* 147(1): 115–41.

Hogland, W., Marques, M. and Nimmermark, S. 2004 Landfill mining and waste characterization: a strategy for remediation of contaminated areas. *Journal of Material Cycles and Waste Management* 6(2): 119–24.

Holdgate, M.W. 1991 Conservation in a world context. In Spellerberg, I.F., Goldsmith, F.B. and Morris, M.G. (eds) *The scientific management of temperate communities for conservation*. Oxford, Blackwell Scientific: 1–26.

Hole, D.G., Perkins, A.J., Wilson, J.D., Alexander, I.H., Grice, P.V. and Evans, A.D. 2005 Does organic farming benefit biodiversity? *Biological Conservation* 122: 113–30.

Holland, G.J. and Webster, P.J. 2007 Heightened tropical cyclone activity in the North Atlantic: natural variability or climate trend? *Philosophical Transactions of the Royal Society A*. doi:10.1098/rsta.2007.2083.

Holling, C.S. 1995 What barriers? What bridges? In Gunderson, L.H., Holling, C.S. and Light, S.S. (eds) *Barriers and bridges to the renewal of ecosystems and institutions*. New York, Columbia University Press: 10–20.

Holmes, J., Lowe, J., Wolff, E. and Srokosz, M. 2011 Rapid climate change: lessons from the recent geological past. *Global and Planetary Change* 79(3–4): 157–62.

Homer-Dixon, T.F. 1999 *Environment, scarcity and violence*. Princeton, NJ, Princeton University Press.

Homer-Dixon, T.F., Boutwell, J.H. and Rathjens, G.W. 1993 Environmental change and violent conflict. *Scientific American* 268(2): 16–23.

Hosonuma, N., Herold, M., De Sy, V., De Fries, R.S., Brockhaus, M., Verchot, L., Angelsen, A. and Romijn, E. 2012 An assessment of deforestation and forest degradation drivers in developing countries. *Environmental Research Letters* 7(4): 044009.

Houghton, J.T., Jenkins, G.J. and Ephraums, J.J. (eds) 1990 *Climate change: the IPCC scientific assessment*. Cambridge, Cambridge University Press.

Houghton, J.T., Ding, Y., Griggs, D.J., Noguer, M., van der Linden, P.J. and Xiaosu, D. (eds) 2001 *Climate change 2001: the scientific basis*. Cambridge, Cambridge University Press.

Howden, N.J.K., Burt, T.P., Worrall, F., Whelan, M.J. and Bieroza, M. 2010 Nitrate concentrations and fluxes in the River Thames over 140 years (1868–2008): are increases irreversible? *Hydrological Processes* 24: 2657–62.

Howells, G. 1990 *Acid rain and acid waters*. London, Ellis Horwood.

Howey, D.A. 2012 Policy: a challenging future for cars. *Nature Climate Change* 2: 28–9.

Howitt, R., Medellín-Azuara, J., MacEwan, D., Lund, J.R. and Sumner, D. 2015 *Economic analysis of the 2014 drought for California agriculture*. University of California, Davis, CA, Center for Watershed Sciences.

Hsiang, S.M., Meng, K.C. and Cane, M.A. 2011 Civil conflicts are associated with the global climate. *Nature* 476: 438–41.

Hu, D., Liu, B., Feng, L., Ding, P., Guo, X., Wang, M., Cao, B., Reeves, P.R. and Wang, L. 2016 Origins of the current seventh cholera pandemic. *Proceedings of the National Academy of Sciences* 113(48): E7730–9.

Hu, Y., Nie, Y., Wei, W., Ma, T., Horn, R.V., Zheng, X., Swaisgood, R.R., Zhou, Z., Zhou, W., Yan, L. and Zhang, Z. 2017 Inbreeding and inbreeding avoidance in wild giant pandas. *Molecular Ecology* 26(20): 5793–806.

Hudson, N.W. 1991 *A study of the reasons for success or failure of soil conservation projects*. FAO Soils Bulletin 64.

Hughes, R.N., Hughes, D.J. and Smith, I.P. 2014 Limits to understanding and managing outbreaks of crown-of-thorns starfish (Acanthaster spp.). *Oceanography and Marine Biology: An Annual Review* 52: 133–200.

Huho, J.M., Ngaira, J.K.W. and Ogindo, H.O. 2011 Living with drought: the case of the Maasai pastoralists of northern Kenya. *Educational Research* 2: 779–89.

Hulme, M. 2001 Climatic perspectives on Sahelian desiccation: 1973–1998. *Global Environmental Change* 11: 19–29.

Hulme, M. 2014 *Can science fix climate change? A case against climate engineering*. Chichester, John Wiley & Sons.

Humborg, C., Ittekkot, V., Cociasu, A. and Bodungen, B.V. 1997 Effect of Danube River dam on Black Sea biogeochemistry and ecosystem structure. *Nature* 386: 385–8.

Hund, K., Megevand, C., Gomes, E.P., Miranda, M. and Reed, E. 2013 *Deforestation trends in the Congo Basin: mining*. Washington, DC, World Bank.

Hunt, T.L. 2007 Rethinking Easter Island's ecological catastrophe. *Journal of Archaeological Science* 34: 485–502.

Hunt, T.L. and Lipo, C.P. 2006 Late colonization of Easter Island. *Science* 311: 1603–6.

Huntingdon, E. 1907 *The pulse of Asia: a journey in Central Asia illustrating the geographic basis of history.* Boston, Houghton Mifflin.

Hurley, R.R., Woodward, J.C. and Rothwell, J.J. 2018 Microplastic contamination of river beds significantly reduced by catchment-wide flooding. *Nature Geoscience.* doi: 10.1038/s41561-018-0080-1.

Hurni, H. 1993 Land degradation, famine, and land resource scenarios in Ethiopia. In Pimental, D. (ed.) *World soil erosion and conservation.* Cambridge, Cambridge University Press: 27–61.

Hvelplund, F., Østergaard, P.A. and Meyer, N.I. 2017 Incentives and barriers for wind power expansion and system integration in Denmark. *Energy Policy* 107: 573–84.

Hyndman, D. 2001 Digging the mines in Melanesia. *Cultural Survival Quarterly* 15(2): 32–9.

ICBL-CMC (International Campaign to Ban Landmines – Cluster Munition Coalition) 2017 *Landmine monitor 2017.* Geneva, ICBL-CMC.

ICMM (International Council on Mining and Metals) 2016 *Role of mining in national economies*, 3rd edn. London, ICMM.

IEA (International Energy Agency) 2011 *25 energy efficiency policy recommendations – 2011 update.* Paris, IEA.

IEA (International Energy Agency) 2011 *Key world energy statistics.* Paris, IEA.

IFC (International Finance Corporation) 2002 *Treasure or trouble? Mining in developing countries.* Washington, DC, IFC.

IGES (Institute for Global Environmental Strategies) 2006 *Sustainable groundwater management in Asian cities.* Kanagawa, Japan, IGES.

IGN 1992 Mali: transect methodology to assess ecosystem change. In *UNEP world atlas of desertification.* London, Edward Arnold: 62–5.

Inkpen, R.J., Viles, H.A., Moses, C., Baily, B., Collier, P., Trudgill, S.T. and Cooke, R.U. 2012 Thirty years of erosion and declining atmospheric pollution at St Paul's Cathedral, London. *Atmospheric Environment* 62: 521–9.

Intergovernmental Panel on Climate Change (IPCC) 1992 *Climate change 1992: the supplementary report to the IPCC scientific assessment.* Report by Working Group I. Cambridge, Cambridge University Press.

IPCC 2012 *Managing the risks of extreme events and disasters to advance climate change adaptation: a special report of Working Groups I and II of the Intergovernmental Panel on Climate Change*, ed. C.B. Field, *et al.* Cambridge, Cambridge University Press.

IPCC 2014 *Climate change 2014: synthesis report.* Contribution of Working Groups I, II and III to the Fifth Assessment Report of the Intergovernmental Panel on Climate Change [Core Writing Team, R.K. Pachauri and L.A. Meyer (eds)]. Geneva, IPCC.

Islam, M.Z., Ismail, K. and Boug, A. 2011 Restoration of the endangered Arabian Oryx *Oryx leucoryx*, Pallas 1766 in Saudi Arabia: lessons learnt from the twenty years of re-introduction in arid fenced and unfenced protected areas. *Zoology in the Middle East* 54(suppl. 3): 125–40.

Issanova, G., Abuduwaili, J., Galayeva, O., Semenov, O. and Bazarbayeva, T. 2015 Aeolian transportation of sand and dust in the Aral Sea region. *International Journal of Environmental Science and Technology* 12: 3213–224.

Ito, K., Uchiyama, Y., Kurokami, N., Sugano, K. and Nakanishi, Y. 2011 Soil acidification and decline of trees in forests within the precincts of shrines in Kyoto (Japan). *Water, Air, and Soil Pollution* 214(1–4): 197–204.

Ito, T.Y., Lhagvasuren, B., Tsunekawa, A. and Shinoda, M. 2017 Habitat fragmentation by railways as a barrier to great migrations of ungulates in Mongolia. In Borda-de-Água, L., Barrientos, R., Beja, P. and Pereira, H. (eds) *Railway ecology.* Cham, Springer: 229–46.

IUCN (International Union for Conservation of Nature and Natural Resources) 1988 *Coral reefs of the world.* Vol. 3: *Central and Western Pacific.* Cambridge, IUCN.

IUCN/UNEP/WWF 1980 *World conservation strategy: living resource conservation for sustainable development.* Gland, Switzerland, IUCN.

Jabareen, Y. 2008 A new conceptual framework for sustainable development. *Environment, Development and Sustainability* 10: 179–92.

Jablonski, D. 2004 Extinction: past and present. *Nature* 427: 589.

Jackson, J., Kirby, M., Berger, W., Bjorndal, K., Dotsford, L., Bourque, B., Brabury, R., Cooke, R., Erlandson, J., Estes, J., Hughes, T., Kidwell, S., Lange, C., Lenihan, H., Pandolfi, J., Peterson, C., Steneck, R., Tigner, M. and Warner, R. 2001 Historical overfishing and the recent collapse of coastal ecosystems. *Science* 293: 629–38.

Jackson, T. 2017 *Prosperity without growth – foundations for the economy of tomorrow*, 2nd edn. London: Routledge.

James, G.K., Adegoke, J.O., Saba, E., Nwilo, P. and Akinyede, J. 2007 Satellite-based assessment of the extent and changes in the mangrove ecosystem of the Niger Delta. *Marine Geodesy* 30: 249–67.

Jamieson, A.J., Malkocs, T., Piertney, S.B., Fujii, T. and Zhang, Z. 2017 Bioaccumulation of persistent organic pollutants in the deepest ocean fauna. *Nature Ecology and Evolution* 1: 0051.

Jantan, I., Bukhari, S.N.A., Mohamed, M.A.S., Wai, L.K. and Mesaik, M.A. 2015 The evolving role of natural products from the tropical rainforests as a replenishable source of new drug leads. In Vallisuta, O. and Olimat, S. (eds) *Drug discovery and development: from molecules to medicine*. Intech: ch. 1.

Janzen, J. 1994 Somalia. In Glantz, M.H. (ed.) *Drought follows the plow*. Cambridge, Cambridge University Press: 45–57.

Jasani, B. 1975 Environmental modification – new weapons of war? *Ambio* 4: 191–8.

Jepson, P., Harvie, J.K., Mackinnon, K. and Monk, K.A. 2001 The end for Indonesia's lowland forests? *Science* 292: 859–61.

Jernelöv, A. 2010 The threats from oil spills: now, then, and in the future. *Ambio* 39: 353–66.

Jha, A.K., Bloch, R. and Lamond, J. 2012 *Cities and flooding: a guide to integrated urban flood risk management for the 21st century*. Washington, DC, World Bank.

Jickells, T.D., Carpenter, R. and Liss, P.S. 1990 Marine environment. In Turner II, B.L., Clark, W.C., Kates, R.W., Richards, J.F., Mathews, J.T. and Meyer, W.B. (eds) *The Earth as transformed by human action*. Cambridge, Cambridge University Press: 313–34.

Jodha, N.S. 1992 Common property resources. Washington, DC, World Bank Discussion Paper.

Joffe, H., Rossetto, T., Bradley, C. and O'Connor, C. 2018 Stigma in science: the case of earthquake prediction. *Disasters* 42(1): 81–100.

Johnson, D.B. and Hallberg, K.B. 2005 Acid mine drainage remediation options: a review. *Science of the Total Environment* 338: 3–14.

Jokiel, P.L. and Brown, E.K. 2004 Global warming, regional trends and inshore environmental conditions influence coral bleaching in Hawaii. *Global Change Biology* 10: 1627–41.

Joly, F., Saïdi, S., Begz, T. and Feh, C. 2012 Key resource areas of an arid grazing system of the Mongolian Gobi. *Mongolian Journal of Biological Sciences* 10(1–2): 13–24.

Jones, P.D., Briffa, K.R., Barnett, T.P. and Tett, S.F.B. 1998 High-resolution palaeoclimatic records for the last millennium: interpretation, integration and comparison with General Circulation Model control-run temperatures. *The Holocene* 8: 455–71.

Jones, P.D., Osborn, T.J. and Briffa, K.R. 2001 The evolution of climate over the last Millennium. *Science* 292: 622–7.

Jones, T.G., Ratsimba, H.R., Ravaoarinorotsihoarana, L., Cripps, G. and Bey, A. 2014 Ecological variability and carbon stock estimates of mangrove ecosystems in northwestern Madagascar. *Forests* 5(1): 177–205

Jørgensen, D. and Renöfält, B.M. 2012 Damned if you do, dammed if you don't: debates on dam removal in the Swedish media. *Ecology and Society* 18(1): 18.

Jutila, E. 1992 Restoration of salmonid rivers in Finland. In Boon, P.J., Calow, P. and Petts, G.E. (eds) *River conservation and management*. Chichester, Wiley: 353–62.

Jyotishi, A. and Sivramkrishna, S. 2011 The forest and the trees: delineating the protected area debate in India. Working Paper No.110, Amrita School of Business, Tamil Nadu, India.

Kaczensky, P., Walzer, C., Ganbataar, O., Enkhsaikhan, N., Altansukh, N. and Stauffer, C. 2011 Re-introduction of the 'extinct in the wild' Przewalski's horse to the Mongolian Gobi. In Soorae, P.S. (ed.) *Global re-introduction perspectives, 2011: more case studies from around the globe*. Gland, Switzerland, IUCN/SSC Re-introduction Specialist Group: 199–204.

Kafumbata, D., Jamu, D. and Chiotha, S. 2014 Riparian ecosystem resilience and livelihood strategies under test: lessons from Lake Chilwa in Malawi and other lakes in Africa. *Philosophical Transactions of the Royal Society B: Biological Sciences* 369(1639): 20130052.

Kaiser, J. 1996 Acid rain's dirty business: stealing minerals from soil. *Science* 272: 198.

Kaji, M. 2012 Role of experts and public participation in pollution control: the case of Itai-itai disease in Japan. *Ethics in Science and Environmental Politics* 12(2): 99–111.

Kambezidis, H.D. and Kalliampakos, G. 2012 Mapping atmospheric corrosion on materials of archaeological importance in Athens. *Water, Air, and Soil Pollution* 223: 2169–80.

Kang, S.M., Polvani, L.M., Fyfe, J.C. and Sigmond, M. 2011 Impact of polar ozone depletion on subtropical precipitation. *Science* 332: 951–4.

Kärcher, B. 2016 The importance of contrail ice formation for mitigating the climate impact of aviation. *Journal of Geophysical Research: Atmospheres* 121(7): 3497–505.

Kashulina, G., Reimann, C., Finne, T.E., Halleraker, J.H., Äyräs, M. and Chekushin, V.A. 1997 The state of the ecosystems in the central Barents Region: scale, factors and mechanism of disturbance. *Science of the Total Environment* 206: 203–25.

Kaygusuz, K. 2010 Sustainable energy, environmental and agricultural policies in Turkey. *Energy Conversion and Management* 51: 1075–84.

Keddy, P.A., Fraser, L.H., Solomeshch, A.I., Junk, W.J., Campbell, D.R., Arroyo, M.K. and Alho, C.J.R. 2009 Wet and wonderful: the world's largest wetlands are conservation priorities. *BioScience* 59: 39–51.

Keijzers, G. 2000 The evolution of Dutch environmental policy: the changing ecological arena from 1970–2000 and beyond. *Journal of Cleaner Production* 8: 179–200.

Keiser, J., de Castro, M.C., Maltese, M.F., Bos, R., Tanner, M., Singer, B.H. and Utzinger, J. 2005 Effect of irrigation and large dams on the burden of malaria on a global and regional scale. *American Journal of Tropical Medicine and Hygiene* 72: 392–406.

Keller, R.P., Drake, J.M., Drew, M. and Lodge, D.M. 2011 Linking environmental conditions and ship movements to estimate invasive species transport across the global shipping network. *Diversity and Distributions* 17: 93–102.

Keller, W., Heneberry, J., Gunn, J.M., Snucins, E., Morgan, G. and Leduc, J. 2004 *Recovery of acid and metal-damaged lakes near Sudbury Ontario: trends and status*. Cooperative Freshwater Ecology Unit, Laurentian University, Sudbury, Ontario.

Kelly, G. 2011 History and potential of renewable energy development in New Zealand. *Renewable and Sustainable Energy Reviews* 15: 2501–9.

Kelman, I. and West, J.J. 2009 Climate change and small island developing states: a critical review. *Ecological and Environmental Anthropology* 5(1): unpaginated.

Kemp, M. 2005 Science in culture: inventing an icon. *Nature* 437: 1238.

Kennedy, C.A., Stewart, I., Facchini, A., Cersosimo, I., Mele, R., Chen, B., Uda, M., Kansal, A., Chiu, A., Kim, K.G. and Dubeux, C. 2015 Energy and material flows of megacities. *Proceedings of the National Academy of Sciences* 112: 5985–90.

Kesavachandran, C.N., Fareed, M., Pathak, M.K., Bihari, V., Mathur, N. and Srivastava, A.K. 2009 Adverse health effects of pesticides in agrarian populations of developing countries. *Reviews of Environmental Contamination and Toxicology* 200: 33–52.

Khalaf, F.I. 1989 Desertification and aeolian processes in Kuwait. *Journal of Arid Environments* 12: 125–45.

Khalil, G.M. 1992 Cyclones and storm surges in Bangladesh: some mitigative measures. *Natural Hazards* 6: 11–24.

Khan, A.S., Khan, S.D. and Kakar, D.M. 2013 Land subsidence and declining water resources in Quetta Valley, Pakistan. *Environmental Earth Sciences* 70(6): 2719–27.

Khan, A.U. 1994 History of decline and present status of natural tropical thorn forest in Punjab. *Biological Conservation* 67: 205–10.

Khodorevskaya, R.P., Ruban, G.I. and Pavlov, D.S. 2009 *Behaviour, migrations, distribution and stocks of sturgeons in the Volga-Caspian basin*. Norderstedt, Germany, Books on Demand GmbH.

Kibert, N.C. 2004 Extended producer responsibility: a tool for achieving sustainable development. *Journal of Land Use and Environmental Law* 19: 503–10.

Kilbane Gockel, C. and Gray, L.C. 2011 Debt-for-nature swaps in action: two case studies in Peru. *Ecology and Society* 16(3): 13.

Kilburn, C.R. and Petley, D.N. 2003 Forecasting giant, catastrophic slope collapse: lessons from Vajont, Northern Italy. *Geomorphology* 54: 21–32.

Kim, K.C. 1997 Preserving biodiversity in Korea's demilitarized zone. *Science* 278: 242.

Kim, S.S. 1984 *The quest for a just world*. Boulder, CO, Westview Press.

Kingston, J. (ed.) 2012 *Natural disaster and nuclear crisis in Japan: response and recovery after Japan's 3/11*. London, Routledge.

Kirchherr, J., Pomun, T. and Walton, M.J. 2016 Mapping the social impacts of 'Damocles Projects': The case of Thailand's (as yet unbuilt) Kaeng Suea Ten Dam. *Journal of International Development*.

Kiss, A. 1985 The protection of the Rhine against pollution. *Natural Resources Journal* 25: 613–32.

Klimont, Z., Smith, S.J. and Cofala, J. 2013 The last decade of global anthropogenic sulfur dioxide: 2000–2011 emissions. *Environmental Research Letters* 8(1): 014003.

Knowler, D. and Bradshaw, B. 2007 Farmers' adoption of conservation agriculture: a review and synthesis of recent research. *Food Policy* 32: 25–48.

Knox, P., Agnew, J. and McCarthy, L. 2014 *The geography of the world economy*, 6th edn. London, Hodder Education.

Koca, M.Y. and Kincal, C. 2004 Abandoned stone quarries in and around the Izmir city centre and their geo-environmental impacts. *Engineering Geology* 75: 49–67.

Koh, L.P. and Wilcove, D.S. 2008 Is oil palm agriculture really destroying tropical biodiversity? *Conservation Letters* 1: 60–4.

Koike, K. 1985 Japan. In Bird, E.C.F. and Schwartz, M.L. (eds) *The world's coastline*. Stroudsburg, PA, Van Nostrand Reinhold: 843–55.

Kondolf, G.M., Gao, Y., Annandale, G.W., Morris, G.L., Jiang, E., Zhang, J., Cao, Y., Carling, P., Fu, K., Guo, Q. and Hotchkiss, R. 2014 Sustainable sediment management in reservoirs and regulated rivers: experiences from five continents. *Earth's Future* 2(5): 256–80.

Koné, M., Coulibaly, L., Kouadio, Y.L., Neuba, D.F. and Malan, D.F. 2016 Multitemporal monitoring of the forest cover in Côte d'Ivoire from the 1960s to the 2000s, using Landsat satellite images. Geoscience and Remote Sensing Symposium (IGARSS), 2016 IEEE International: 1325–8.

Kossoff, D., Dubbin, W.E., Alfredsson, M., Edwards, S.J., Macklin, M.G. and Hudson-Edwards, K.A. 2014 Mine tailings dams: characteristics, failure, environmental impacts, and remediation. *Applied Geochemistry* 51: 229–45.

Kovacs, K.M., Lydersen, C., Overland, J.E. and Moore, S.E. 2010 Impacts of changing sea-ice conditions on Arctic marine mammals. *Marine Biodiversity*. doi: 10.1007/s12526-010-0061-0.

Kozlov, M.V. and Zvereva, E.L. 2007 Industrial barrens: extreme habitats created by non-ferrous metallurgy. *Reviews in Environmental Science and Biotechnology* 6: 231–59.

Krausmann, F., Gingrich, S., Eisenmenger, N., Erb, K.-H., Haberl, H. and Fischer-Kowalski, M. 2009 Growth in global materials use, GDP and population during the 20th century. *Ecological Economics* 68: 2696–705.

KREM (Korean Republic Environment Ministry) *Green Korea 2006*. Gwacheon, KREM.

KREM 2012 *Environmental statistics yearbook 2011*. Gwacheon, KREM.

Kroon, F.J., Thorburn, P., Schaffelke, B. and Whitten, S. 2016 Towards protecting the Great Barrier Reef from land based pollution. *Global Change Biology* 22: 1985–2002.

Kroonenberg, S.B., Badyukovab, E.N., Stormsa, J.E.A., Ignatovb, E.I. and Kasimov, N.S. 2000 A full sea-level cycle in 65 years: barrier dynamics along Caspian shores. *Sedimentary Geology* 134: 257–74.

Kühl, A., Mysterud, A., Grachev, I.A., Bekenov, A.B., Ubushaev, B.S., Lushchekina, A.A. and Milner-Gulland, E.J. 2009 Monitoring population productivity in the saiga antelope. *Animal Conservation* 12: 355–63.

Kühni, C. and Spycher, B. 2014 Nuclear power plants and childhood leukaemia: lessons from the past and future directions. *Swiss Medical Weekly* 144: w13912.

Kumar, A., Schei, T., Ahenkorah, A., Caceres Rodriguez, R., Devernay, J.-M., Freitas, M., Hall, D., Killingtveit, A. and Liu, Z. 2011 Hydropower. In O. Edenhofer, *et al.* (eds) *IPCC Special Report on Renewable Energy Sources and Climate Change Mitigation*. Cambridge, Cambridge University Press.

La Sorte, F.A., Aronson, M.F., Williams, N.S., Celesti Grapow, L., Cilliers, S., Clarkson, B.D. and Winter, M. 2014 Beta diversity of urban floras among European and non European cities. *Global Ecology and Biogeography* 23: 769–79.

Lahlou, A. 1996 Environmental and socio-economic impacts of erosion and sedimentation in North Africa. In Walling, D.E. and Webb, B.W. (eds) *Erosion and sediment yield: global and regional perspectives*. Wallingford, International Association of Hydrological Sciences Publication No. 236: 491–500.

Lahnsteiner, J., van Rensburg, P. and Esterhuizen, J. 2018 Direct potable reuse – a feasible water management option. *Journal of Water Reuse and Desalination* 8: 14–28.

Laidler, G.J. 2006 Inuit and scientific perspectives on the relationship between sea ice and climate change: the ideal complement? *Climatic Change* 78: 407–44.

Laidre, K.L., Stern, H., Kovacs, K.M., Lowry, L., Moore, S.E., Regehr, E.V., Ferguson, S.H., Wiig, Ø., Boveng, P., Angliss, R.P. and Born, E.W. 2015 Arctic marine mammal population status, sea ice habitat loss, and conservation recommendations for the 21st century. *Conservation Biology* 29(3): 724–37.

Lajewski, C.K., Mullins, H.T., Patterson, W.P. and Callinan, C.W. 2003 Historic calcite record from the Finger Lakes, New York: impact of acid rain on a buffered terrain. *Geological Society of America Bulletin* 115: 373–84.

Lal, R. and Stewart, B.A. 1990 Need for action: research and development priorities. *Advances in Soil Science* 11: 331–6.

Lal, R., Fifield, L.K., Tims, S.G. and Wasson, R.J. 2017 239 Pu fallout across continental Australia: implications on 239 Pu use as a soil tracer. *Journal of Environmental Radioactivity* 178: 394–403.

Lambin, E.F., Geist, H.J., and Lepers, E. 2003 Dynamics of land-use and land-cover change in tropical regions. *Annual Review of Environment and Resources* 28: 205–41.

Lamprey, H.F. 1975 *Report on the desert encroachment reconnaissance in northern Sudan, October 21–November 10, 1975*. Khartoum, National Council for Research, Ministry of Agriculture, Food and Resources.

Langmead, O., McQuatters-Gollop, A., Mee, L.D., Friedrich, J., Gilbert, A.J., Gomoiu, M., Jackson, E.L., Knudsen, S., Minicheva, G. and Todorova, V. 2010 Recovery or decline of the northwestern Black Sea: a societal choice revealed by socio-ecological modelling. *Ecological Modelling* 220: 2927–39.

Larsen, I.J., MacDonald, L.H., Brown, E., Rough, D., Welsh, M.J., Pietraszek, J.H., Libohova, Z., de Dios Benavides-Solorio, J. and Schaffrath, K. 2009 Causes of post-fire runoff and erosion: water repellency, cover, or soil sealing? *Soil Science Society of America Journal* 73(4): 1393–1407.

Larssen, T., Lydersen, E., Tang, D., He, Y., Gao, J., Liu, H., Duan, L., Seip, H.M., Vogt, R.D., Mulder, J., Shao, M., Wang, Y., Shang, H., Zhang, X., Solberg, S., Aas, W., Okland, T., Eilertsen, O., Angell, V., Li, Q., Zhao, D., Xiang, R., Xiao, J. and Luo, J. 2006 Acid rain in China. *Environmental Science and Technology* 40: 418–25.

Laska, S. and Morrow, B.H. 2006 Social vulnerabilities and Hurricane Katrina: an unnatural disaster in New Orleans. *Marine Technology Society Journal* 40(4): 16–26.

Laurance, W.F., Delamônica, P., Laurance, S.G., Vasconcelos, H.L. and Lovejoy, T.E. 2000 Rainforest fragmentation kills big trees. *Nature* 404: 836.

Laurance, W.F., Goosem, M. and Laurance, S.G.W. 2009 Impacts of roads and linear clearings on tropical forests. *Trends in Ecology and Evolution* 24: 659–69.

Laurence, D. 2011 Establishing a sustainable mining operation: an overview. *Journal of Cleaner Production* 19: 278–84.

Laurian, L. 2008 Environmental injustice in France. *Journal of Environmental Planning and Management* 51: 55–79.

Lavigne, F., Wassmer, P., Gomez, C., Davies, T.A., Hadmoko, D.S., Iskandarsyah, T.Y.W., Gaillard, J.C., Fort, M., Texier, P., Heng, M.B. and Pratomo, I. 2014 The 21 February 2005, catastrophic waste avalanche at Leuwigajah dumpsite, Bandung, Indonesia. *Geoenvironmental Disasters* 1: 10.

Lavorgna, A. 2014 Wildlife trafficking in the Internet age. *Crime Science* 3(1): 5.

Lavorgna, A., Rutherford, C., Vaglica, V., Smith, M.J. and Sajeva, M. 2017 CITES, wild plants, and opportunities for crime. *European Journal on Criminal Policy and Research*: 1–20.

Lawrence, G.B. 2002 Persistent episodic acidification of streams linked to acid rain effects on soil. *Atmospheric Environment* 36: 1589–98.

Lay, T., Kanamori, H., Ammon, C.J., Nettles, M., Ward, S.N., Aster, R., Beck, S.L., Bilek, S.L., Brudzinski, M.R., Butler, R., DeShon, H.R., Ekstrom, G., Satake, K. and Sipkin, S. 2005 The great Sumatra-Andaman earthquake of 26 December 2004. *Science* 308: 1127–32.

Layrisse, M. 1992 The 'holocaust' of the Amerindians. *Interciencia* 17: 274.

Leach, G. and Mearns, R. 1988 *Beyond the fuelwood crisis: people, land and trees in Africa*. London, Earthscan.

Leblois, A., Damette, O. and Wolfersberger, J. 2017 What has driven deforestation in developing countries since the 2000s? Evidence from new remote-sensing data. *World Development* 92: 82–102.

Ledec, G., Quintero, J.D. and Mejia, M.C. 1997 Good dams and bad dams: environmental and social criteria for choosing hydroelectric project sites. Washington, DC, World Bank Sustainable Dissemination Note 1.

Lee, D.S., Fahey, D.W., Forster, P.M., Newton, P.J., Wit, R.C.N., Lim, L.L., Owen, B. and Sausen, R. 2009 Aviation and global climate change in the 21st century. *Atmospheric Environment* 43: 3520–37.

Lee, J.Y. and Anderson, C.D. 2013 The restored Cheonggyecheon and the quality of life in Seoul. *Journal of Urban Technology* 20(4): 3–22.

Lee-Smith, D. 2010 Cities feeding people: an update on urban agriculture in equatorial Africa. *Environment and Urbanization* 22: 483–99.

Lees, A.C. and Pimm, S.L. 2015 Species, extinct before we know them? *Current Biology* 25(5): R177–80.

Lees, A.C., Peres, C.A., Fearnside, P.M., Schneider, M. and Zuanon, J.A. 2016 Hydropower and the future of Amazonian biodiversity. *Biodiversity and Conservation* 25(3): 451–66.

Lejeune, Q., Davin, E.L., Guillod, B.P. and Seneviratne, S.I. 2015 Influence of Amazonian deforestation on the future evolution of regional surface fluxes, circulation, surface temperature and precipitation. *Climate Dynamics* 44: 2769–86.

Lelek, A. 1989 The Rhine river and some of its tributaries under human impact in the last two centuries. *Canadian Special Publication of Fisheries and Aquatic Science* 106: 469–87.

Lenton, T.M., Held, H., Kriegler, E., Hall, J.W., Lucht, W., Rahmstorf, S. and Schellnhuber, H.J. 2008 Tipping elements in the Earth's climate system. *Proceedings of the National Academy of Sciences* 105: 1786–93.

Leone, S. 2017 Deadly landslide an avoidable tragedy. *Green Left Weekly* 1150: 15.

Lepers, E. 2003 *Synthesis of the main areas of land-cover and land-use change.* Millennium Ecosystem Assessment, Final Report. New York, Island Press.

Leprun, J.-C., da Silveira, C.O. and Sobral Filho, R.M. 1986 Efficacité des pratiques culturales antiérosives testées sous différents climats brésiliens. *Cahiers ORSTOM Série Pédologie* 22: 223–33.

Letcher, R.J., Morris, A.D., Dyck, M., Sverko, E., Reiner, E.J., Blair, D.A.D., Chu, S.G. and Shen, L. 2018 Legacy and new halogenated persistent organic pollutants in polar bears from a contamination hotspot in the Arctic, Hudson Bay Canada. *Science of the Total Environment* 610: 121–36.

Leuven, R.S., van der Velde, G., Baijens, I., Snijders, J., van der Zwart, C., Lenders, H.R. and bij de Vaate, A. 2009 The river Rhine: a global highway for dispersal of aquatic invasive species. *Biological Invasions* 11: 1989.

Levins, R., Awerbuch, T., Brinkman, U., Eckardt, I., Epstein, P., Makhoul, N., Albuquerque de Possas, C., Puccia, C., Spielman, A. and Wilson, M.E. 1994 The emergence of new diseases. *American Scientist* 82: 52–60.

Lewis, L.A. and Berry, L. 1988 *African environments and resources.* London, Allen & Unwin.

Lewis, S.L. and Maslin, M.A. 2015 Defining the Anthropocene. *Nature* 519(7542): 171–80.

Li, C., McLinden, C., Fioletov, V., Krotkov, N., Carn, S., Joiner, J., Streets, D., He, H., Ren, X., Li, Z. and Dickerson, R.R. 2017 India is overtaking China as the world's largest emitter of anthropogenic sulfur dioxide. *Scientific Reports* 7(1): 14304.

Li, T., Cai, M.F. and Cai, M. 2007 A review of mining-induced seismicity in China. *International Journal of Rock Mechanics and Mining Sciences* 44(8): 1149–71.

Li, X., Mitra, C., Marzen, L. and Yang, Q. 2016 Spatial and temporal patterns of wetland cover changes in East Kolkata Wetlands, India from 1972 to 2011. *International Journal of Applied Geospatial Research* 7(2): 1–13.

Lichtenstein, G. 2010 Vicuña conservation and poverty alleviation? Andean communities and international fibre markets. *International Journal of the Commons* 4: 100–21.

Likens, G.E. 2010 The role of science in decision making: does evidence-based science drive environmental policy? *Frontiers in Ecology and the Environment* 8: e1–9.

Lima, A.C., Sayanda, D., Agostinho, C.S., Machado, A.L., Soares, A.M. and Monaghan, K.A. 2018 Using a trait based approach to measure the impact of dam closure in fish communities of a Neotropical river. *Ecology of Freshwater Fish* 27: 408–20.

Lin, J.C. 1996 Coastal modification due to human influence in south-western Taiwan. *Quaternary Science Reviews* 15: 895–900.

Lindenmayer, D.B., MacGregor, C., Wood, J., Westgate, M.J., Ikin, K., Foster, C., Ford, F. and Zentelis, R. 2016 Bombs, fire and biodiversity: vertebrate fauna occurrence in areas subject to military training. *Biological Conservation* 204: 276–83.

Linderholm, H.W. 2006 Growing season changes in the last century. *Agricultural and Forest Meteorology* 137: 1–14.

Linehan, P. 2010 Saving Penn's woods: deforestation and reforestation in Pennsylvania. *Pennsylvania Legacies* 10: 20–5.

Linnerooth-Bayer, J., Mechler, R. and Hochrainer-Stigler, S. 2011 Insurance against losses from natural disasters in developing countries: evidence, gaps and the way forward. *Journal of Integrated Disaster Risk Management.* doi: 10.5595/idrim.2011.0013.

Linnhoff, S., Volovich, E., Martin, H.M. and Smith, L.M. 2017 An examination of millennials' attitudes toward genetically modified organism (GMO) foods: is it Franken-food or super-food? *International Journal of Agricultural Resources, Governance and Ecology* 13(4): 371–90.

Liu, C.W., Lin, K.H. and Kuo, Y.M. 2003 Application of factor analysis in the assessment of groundwater quality in a blackfoot disease area in Taiwan. *Science of the Total Environment* 313: 77–89.

Liu, J., Linderman, M., Ouyang, Z., An, L., Yang, J. and Zhang, H. 2001 Ecological degradation in protected areas: the case of Wolong Nature Reserve for giant pandas. *Science* 292: 98–101.

Liverman, D.M. 2009 Conventions of climate change: constructions of danger and the dispossession of the atmosphere. *Journal of Historical Geography* 35: 279–96.

Lloyd, G.O. and Butlin, R.N. 1992 Corrosion. In Radojevic, M. and Harrison, R.M. (eds) *Atmospheric acidity: sources, consequences and abatement.* London, Elsevier Applied Science: 405–34.

Lockeretz, W. 1978 The lessons of the Dust Bowl. *American Scientist* 66: 560–9.

Löfstedt, R. 1998 Sweden's biomass controversy: a case study of communicating policy issues. *Environment* 40(4): 16–20, 42–5.

Loft, L., Pham, T.T., Wong, G., Brockhaus, M., Le, D.N., Tjajadi, J.S. and Luttrell, C. 2016 Risks to REDD+: potential pitfalls for policy design and implementation. *Environmental Conservation*, 12pp.

Logan, C.A. 2010 A review of ocean acidification and America's response. *BioScience* 60: 819–28.

Lonergan, S.C. 1993 Impoverishment, population, and environmental degradation: the case for equity. *Environmental Conservation* 20: 328–34.

López-Feldman, A. and Wilen, J.E. 2008 Poverty and spatial dimensions of non-timber forest extraction. *Environment and Development Economics* 13: 621–42.

Lorenzen, E.D. and 55 others 2011 Species-specific responses of Late Quaternary megafauna to climate and humans. *Nature* 479: 359–63.

Lorius, C., Jouzel, J., Raynaud, D., Hansen, J. and Le Treut, H. 1990 The ice-core record: climate sensitivity and future greenhouse warming. *Nature* 347: 139–45.

Lottermoser, B.G. 2011 Recycling, reuse and rehabilitation of mine wastes. *Elements* 7: 405–10.

Lottermoser, B.G. and Morteani, G. 1993 Sewage sludge: toxic substances, fertilizers, or secondary metal resources? *Episodes* 16: 329–33.

Loucks, C., Mascia, M.B., Maxwell, A., Huy, K., Duong, K., Chea, N., Long, B., Cox, N. and Seng, T. 2009 Wildlife decline in Cambodia, 1953–2005: exploring the legacy of armed conflict. *Conservation Letters* 2: 82–92.

Loucks, C.J., Lü, Z., Dinerstein, E., Wang, H., Olson, D.M., Zhu, C. and Wang, D. 2001 Giant pandas in a changing landscape. *Science* 294(5546): 1465.

Lovei, M. 1998 Phasing out lead from gasoline: worldwide experience and policy implications. World Bank Technical Paper No. 397, Pollution Management Series.

Lovelock, J.E. 1989 *The ages of Gaia*. Oxford, Oxford University Press.

Lovelock, J.E. 2006 *The revenge of Gaia: why the earth is fighting back – and how we can still save humanity*. London, Allen Lane.

Lu, Z. and Streets, D.G. 2012 Increase in NO_x emissions from Indian thermal power plants during 1996–2010: unit-based inventories and multisatellite observations. *Environmental Science and Technology* 46: 7463–70.

Lucht, W., Schaphoff, S., Erbrecht, T., Heyder, U., and Cramer, W. 2006 Terrestrial vegetation redistribution and carbon balance under climate change. *Carbon Balance and Management* 1: 6.

Lücking, R. and Matzer, M. 2001 High foliicolous lichen alpha-diversity on individual leaves in Costa Rica and Amazonian Ecuador. *Biodiversity and Conservation* 10: 2139–52.

Lund, H.G. 2007 Accounting for the world's rangelands. *Rangelands* 29: 3–10.

Lundholm, J.T. and Richardson, P.J. 2010 Habitat analogues for reconciliation ecology in urban and industrial environments. *Journal of Applied Ecology* 47: 966–75.

Mackinnon, D. and Cumbers, A. 2007 *An introduction to economic geography: globalization, uneven development and place*. Harlow, Pearson Education.

Macklin, M.G., Hudson-Edwards, K.A. and Dawson, E.J. 1997 The significance of pollution from historic metal mining in the Pennine orefields on river sediment contaminant fluxes to the North Sea. *Science of the Total Environment* 194–5: 391–7.

Magrin, G.O., Travasso, M.I. and Rodríguez, G.R. 2005 Changes in climate and crop production during the 20th century in Argentina. *Climatic Change* 72: 229–49.

Mahboob, S., Alkkahem Al-Balwai, H.F., Al-Misned, F., Al-Ghanim, K.A. and Ahmad, Z. 2014 A study on the accumulation of nine heavy metals in some important fish species from a natural reservoir in Riyadh, Saudi Arabia. *Toxicological and Environmental Chemistry* 96: 783–98.

Mahmood, R., Foster, S.A., Keeling, T., Hubbard, K.G., Carlson, C. and Leepe, R. 2006 Impacts of irrigation on 20th century temperature in the northern Great Plains. *Global and Planetary Change* 54: 1–18.

Mahmoudpour, M., Khamehchiyan, M., Nikudel, M.R. and Ghassemi, M.R. 2016 Numerical simulation and prediction of regional land subsidence caused by groundwater exploitation in the southwest plain of Tehran, Iran. *Engineering Geology* 201: 6–28.

Maitland, P.S. 1991 Conservation of fish species. In Spellerberg, I.F., Goldsmith, F.B. and Morris, M.G. (eds) *The scientific management of temperate communities for conservation*. Oxford, Blackwell Scientific: 129–48.

Maki, A.W. 1991 The Exxon oil spill: initial environmental impact assessment. *Environmental Science and Technology* 25: 24–9.

Malthus, T.R. 1798 *An essay on the principle of population*. London, Johnson.

Mani, M. and Wheeler, D. 1998 In search of pollution havens? Dirty industry in the world economy, 1960 to 1995. *The Journal of Environment and Development* 7: 215–47.

Manney, G.L. and 28 others 2011 Unprecedented Arctic ozone loss in 2011. *Nature* 478: 469–75.

Mantyka-Pringle, C., Martin, T.G. and Rhodes, J.R. 2012 Interactions between climate and habitat loss effects on biodiversity: a systematic review and meta-analysis. *Global Change Biology.* doi: 10.1111/j. 1365-2486.2011.02593.x.

Maragos, J.E. 1993 Impact of coastal construction on coral reefs in the US-affiliated Pacific Islands. *Coastal Management* 21: 235–69.

Marano, K.D., Wald, D.J. and Allen, T.I. 2010 Global earthquake casualties due to secondary effects: a quantitative analysis for improving rapid loss analyses. *Natural Hazards* 52: 319–28.

MARD (Ministry of Agriculture and Rural Development Viet Nam) 2015 Forest Sector Development Report Year 2014. Hanoi, Vietnam.

Marlier, M.E., Jina, A.S., Kinney, P.L. and DeFries, R.S. 2016 Extreme air pollution in global megacities. *Current Climate Change Reports* 2: 15–27.

Marques, M.C.M., Burslem, D.F., Britez, R.M. and Silva, S.M. 2009 Dynamics and diversity of flooded and unflooded forests in a Brazilian Atlantic rain forest: a 16-year study. *Plant Ecology and Diversity* 2: 57–64.

Marsh, G.P. 1874 *The Earth as modified by human actions*. New York, Sampson Low.

Marshall, B.E. and Junor, F.J.R. 1981 The decline of *Salvinia molesta* on Lake Kariba. *Hydrobiologia* 83: 477–84.

Marszal, A.J., Heiselberg, P., Bourrelle, J.S., Musall, E., Voss, K., Sartori, I. and Napolitano, A. 2011 Zero Energy Building: a review of definitions and calculation methodologies. *Energy and Buildings* 43: 971–9.

Martiello, M.A. and Giacchi, M.V. 2010 High temperatures and health outcomes: a review of the literature. *Scandinavian Journal of Public Health* 38: 826–37.

Martinez, J., Dabert, P., Barrington, S. and Burton, C. 2009 Livestock waste treatment systems for environmental quality, food safety, and sustainability. *Bioresource Technology* 100: 5527–36.

Martínez, M.L., Intralawan, A., Vázquez, G., Pérez-Maqueo, O., Sutton, P. and Landgrave, R. 2007 The coasts of our world: ecological, economic and social importance. *Ecological Economics* 63: 254–72.

Martinot, E. 2016 Grid integration of renewable energy: flexibility, innovation, and experience. *Annual Review of Environment and Resources* 41: 223–51.

Mashreky, S.R., Bari, S., Sen, S.L., Rahman, A., Khan, T.F. and Rahman, F. 2010 Managing burn patients in a fire disaster: experience from a burn unit in Bangladesh. *Indian Journal of Plastic Surgery: Official Publication of the Association of Plastic Surgeons of India* 43(suppl.): S131.

Maslin, M., Malhi, Y., Phillips, O. and Cowling, S. 2005 New views on an old forest: assessing the longevity, resilience and future of the Amazon rainforest. *Transactions of the Institute of British Geographers* 30: 477–99.

Massard-Guilbaud, G. and Rodger, R. (eds) 2011 *Environmental and social justice in the city: historical perspectives*. Isle of Harris, White Horse Press.

Mather, A.S. 1990 *Global forest resources*. London, Belhaven.

Mather, A.S. 2005 Assessing the world's forests. *Global Environmental Change* Part A, 15: 267–80.

Mathiesen, B.V., Lund, H. and Karlsson, K. 2011 100% renewable energy systems, climate mitigation and economic growth. *Applied Energy* 88: 488–501.

Matless, D. 2014 *In the nature of landscape: cultural geography on the Norfolk Broads*. Chichester, Wiley.

May, R.M. 1978 Human reproduction reconsidered. *Nature* 272: 491–5.

Maynard, R. 2004 Key airborne pollutants: the impact on health. *Science of the Total Environment* 334: 9–13.

McAdam, J. 2017 The high price of resettlement: the proposed environmental relocation of Nauru to Australia. *Australian Geographer* 48: 7–16.

McCall, G.J.H. 1998 Geohazards and the urban environment. In Maund, J.G. and Eddlestone, M. (eds) *Geohazards in engineering geology*. London, Geological Society: 309–18.

McCauley, J.F., Breed, C.S., Grolier, M.J. and Mackinnon, D.J. 1981 The US dust storm of February 1977. In Péwé, T.L. (ed.) *Desert dust: origins, characteristics and effects on man*. Geological Society of America Special Paper 186: 123–47.

McCully, P. 1996 *Silenced rivers: the ecology and politics of large dams*. London, Zed Books.

McFarlane, G.A., King, J.R. and Beamish, R.J. 2000 Have there been recent changes in climate? Ask the fish. *Progress in Oceanography* 47: 147–69.

McGarr, A., Simpson, D., Seeber, L. and Lee, W.H.K. 2002 Case histories of induced and triggered seismicity. *International Geophysics Series* 81(A): 647–64.

McGranahan, G. and Mitlin, D. 2016. Learning from sustained success: how community-driven initiatives to improve urban sanitation can meet the challenges. *World Development* 87: 307–17.

McKinney, M.L. 2006 Urbanization as a major cause of biotic homogenization. *Biological Conservation* 127: 247–60.

McLaughlin, C. and Krantzberg, G. 2011 An appraisal of policy implementation deficits in the Great Lakes. *Journal of Great Lakes Research* 37: 390–6.

McNeely, J. 1994 Lessons from the past: forests and biodiversity. *Biodiversity and Conservation* 3: 3–20.

McTainsh, G. and Strong, C. 2007 The role of aeolian dust in ecosystems. *Geomorphology* 89: 39–54.

Meadows, D.H., Meadows, D.L., Randers, J. and Behrens III, W.W. 1972 *The limits to growth: a report to the Club of Rome's project on the predicament of mankind*. New York, Potomac Associates.

Meadows, D.H., Randers, J. and Meadows, D.L. 2004 *The limits to growth: the 30-year update*. London, Earthscan.

Meadows, M.E. and Hoffman, M.T. 2002 The nature, extent and causes of land degradation in South Africa: legacy of the past, lessons for the future. *Area* 34: 428–37.

Meadows, P.S. and Campbell, J.I. 1988 *An introduction to marine science*. London, Blackie & Son.

Mearns, R. and Norton, A. 2010 *Social dimensions of climate change: equity and vulnerability in a warming world*. Washington, DC, World Bank.

Meech, J.A., Veiga, M.M. and Tromans, D. 1998 Reactivity of mercury from gold mining activities in darkwater ecosystems. *Ambio* 27: 92–8.

Meehan, T.D., Werling, B.P., Landis, D.A. and Gratton, C. 2011 Agricultural landscape simplification and insecticide use in the Midwestern United States. *Proceedings of the National Academy of Sciences* 108: 11500–5.

Meinert, L.D., Robinson, G.R. and Nassar, N.T. 2016 Mineral resources: reserves, peak production and the future. *Resources* 5(1): 14.

Mekonnen, M.M. and Hoekstra, A.Y. 2012 The blue water footprint of electricity from hydropower. *Hydrology and Earth System Sciences* 16: 179–87.

Melse, R.W., Ogink, N.W. and Rulkens, W.H. 2009 Overview of European and Netherlands' regulations on airborne emissions from intensive livestock production with a focus on the application of air scrubbers. *Biosystems Engineering* 104: 289–98.

Menzel, A. and Fabian, P. 1999 Growing season extended in Europe. *Nature* 397: 659.

Mercer, J. 2010 Disaster risk reduction or climate change adaptation: are we reinventing the wheel? *Journal of International Development* 22: 247–64.

Metcalfe, S and Derwent, R.G. 2005 *Atmospheric pollution and environmental change*. London, Hodder Arnold.

Meybeck, M. 2002 Riverine quality at the Anthropocene: propositions for global space and time analysis, illustrated by the Seine River. *Aquatic Sciences* 64: 376–93.

Meybeck, M., Chapman, D. and Helmer, R. 1989 *Global freshwater quality: a first assessment*. Oxford, Blackwell.

Meyfroidt, P. and Lambin, E.F. 2009 Forest transition in Vietnam and displacement of deforestation abroad. *Proceedings of the National Academy of Sciences* 106: 16139–44.

Meyfroidt, P. and Lambin, E.F. 2011 Global forest transition: prospects for an end to deforestation. *Annual Reviews of Environment and Resources* 36: 343–71.

Michalopoulos, A., Landeweerd, L., Van der Werf-Kulichova, Z., Puylaert, P.G.B. and Osseweijer, P. 2011 Contrasts and synergies in different biofuel reports. *Interface Focus* 2: 248–54.

Micheli, F., Halpern, B.S., Walbridge, S., Ciriaco, S., Ferretti, F., Fraschetti, S., Lewison, R., Nykjaer, L. and Rosenberg, A.A. 2013 Cumulative human impacts on Mediterranean and Black Sea marine ecosystems: assessing current pressures and opportunities. *PloS ONE* 8(12): e79889.

Micklin, P. 2010 The past, present, and future Aral Sea. *Lakes and Reservoirs: Research and Management* 15: 193–213.

Middleton, N.J. 1985 Effect of drought on dust production in the Sahel. *Nature* 316: 431–4.

Middleton, N.J. 1991 *Desertification*. Oxford, Oxford University Press.

Middleton, N.J. 2002 The Aral Sea. In Shahgedanova, M. (ed.) *The physical geography of Northern Eurasia*. Oxford, Oxford University Press: 497–510.

Middleton, N.J. 2017 Desert dust hazards: a global review. *Aeolian Research* 24: 53–63.

Middleton, N.J. and Thomas, D.S.G. 1997 *World atlas of desertification*, 2nd edn. London, Arnold.

Middleton, N.J. and van Lynden, G.W.J. 2000 Secondary salinization in South and Southeast Asia. *Progress in Environmental Science* 2: 1–19.

Middleton, N.J., Stringer, L., Goudie, A. and Thomas, D. 2011 *The forgotten billion: MDG achievement in the drylands*. New York, UNDP-UNCCD.

Millennium Ecosystem Assessment 2005 *Ecosystems and human well-being: biodiversity synthesis*. Washington, DC, World Resources Institute.

Miller, G., Mangan, J., Pollard, D., Thompson, S., Felzer, B. and Magee, J. 2005 Sensitivity of the Australian Monsoon to insolation and vegetation: implications for human impact on continental moisture balance. *Geology* 33: 65–8.

Miller, L.M., Gans, F. and Kleidon, A. 2011 Estimating maximum global land surface wind power extractability and associated climatic consequences. *Earth System Dynamics* 2(1): 1–12.

Mirza, M.M.Q. (ed.) 2004 *The Ganges water diversion: environmental effects and implications.* Heidelberg, Springer.

Mirza, M.M.Q., Warrick, R.A., Ericksen, N.J. and Kenny, G.J. 2001 Are floods getting worse in the Ganges, Brahmaputra and Meghna basins? *Environmental Hazards* 3: 37–48.

Moffatt, I. 1999 Edinburgh: a sustainable city? *International Journal of Sustainable Development and World Ecology* 6: 135–48.

MOHA (Ministry of Home Affairs, Maldives) 2001 *First national communication of the Republic of Maldives to the UN Framework Convention on Climate Change.* Ministry of Home Affairs, Housing and Environment, Malé, Republic of Maldives.

Mol, J.H., de Mérona, B., Ouboter, P.E. and Sahdew, S. 2007 The fish fauna of Brokopondo Reservoir, Suriname, during 40 years of impoundment. *Neotropical Ichthyology* 5: 351–68.

Molina, M.J. and Molina, L.T. 2004 Megacities and atmospheric pollution. *Journal of the Air and Waste Management Association* 54: 644–80.

Molina, M.J. and Rowland, F.S. 1974 Stratospheric sink chlorofluoromethanes: chlorine atom catalyzed destruction of ozone. *Nature* 249: 810–14.

Moller, H., Berkes, F., Lyver, P.O. and Kislalioglu, M. 2004 Combining science and traditional ecological knowledge: monitoring populations for co-management. *Ecology and Society* 9(3): 2.

Mooers, H.D., Cota-Guertin, A.R., Regal, R.R., Sames, A.R., Dekan, A.J. and Henkels, L.M. 2016 A 120-year record of the spatial and temporal distribution of gravestone decay and acid deposition. *Atmospheric Environment* 127: 139–54.

Moore, N.W., Hooper, M.D. and Davis, B.N.K. 1967 Hedges, I. Introduction and reconnaissance studies. *Journal of Applied Ecology* 4: 201–20.

Moriarty, P. and Honnery, D. 2009 Hydrogen's role in an uncertain energy future. *International Journal of Hydrogen Energy* 34: 31–9.

Morley, S.A. and Karr, J.R. 2002 Assessing and restoring the health of urban streams in the Puget Sound Basin. *Conservation Biology* 16: 1498–509.

Mortimore, M. 2010 Adapting to drought in the Sahel: lessons for climate change. *Wiley Interdisciplinary Reviews: Climate Change* 1: 134–43.

Mortimore, M. and Turner, B. 2005 Does the Sahelian smallholder's management of woodland, farm trees, rangeland support the hypothesis of human-induced desertification? *Journal of Arid Environments* 63: 567–95.

Mortimore, M. with contributions from Anderson, S., Cotula, L., Davies, J., Faccer, K., Hesse, C., Morton, J., Nyangena, W., Skinner, J. and Wolfangel, C. 2009 *Dryland opportunities: a new paradigm for people, ecosystems and development.* Gland, Switzerland, IUCN; London, IIED; and Nairobi, Kenya, UNDP/DDC.

Motz, H. and Geiseler, J. 2001 Products of steel slags an opportunity to save natural resources. *Waste Management* 21: 285–93.

Mudd, G.M., Weng, Z. and Jowitt, S.M. 2013 A detailed assessment of global Cu resource trends and endowments. *Economic Geology* 108: 1163–83.

Murphy, E.J., Thorpe, S.E., Tarling, G.A., Watkins, J.L., Fielding, S. and Underwood, P. 2017 Restricted regions of enhanced growth of Antarctic krill in the circumpolar Southern Ocean. *Scientific Reports* 7(1): 6963.

Murray, I. 1994 Time and tide rip into the frontier of old England. *Times*, 23 March, p. 7.

Mussells, O., Dawson, J. and Howell, S. 2017 Navigating pressured ice: risks and hazards for winter resource-based shipping in the Canadian Arctic. *Ocean and Coastal Management* 137: 57–67.

Myers, N. 1979 *The sinking ark: a new look at the problem of disappearing species.* Oxford, Pergamon.

Myers, N., Mittermeier, R.A., Mittermeier, C.G., Da Fonseca, G.A.B. and Kent, J. 2000 Biodiversity hotspots for conservation priorities. *Nature* 403: 853–8.

Mylona, S. 1993 *Trends of sulphur dioxide emissions, air concentrations and depositions of sulphur in Europe since 1880.* EMEP/MSC-W Report 2/93. Oslo, EMEP.

Nagel, N.B. 2001 Compaction and subsidence issues within the petroleum industry: from Wilmington to Ekofisk and beyond. *Physics and Chemistry of the Earth, Part A: Solid Earth and Geodesy* 26(1–2): 3–14.

Nakahara, O., Takahashi, M., Sase, H., Yamada, T., Matsuda, K., Ohizumi, T., Fukuhara, H., Inoue, T., Takahashi, A., Kobayashi, H. and Hatano, R. 2010 Soil and stream water acidification in a forested catchment in central Japan. *Biogeochemistry* 97(2–3): 141–58.

Nakajima, H., Kaneko, H. and Tsuchida, M. 2012 The management of land subsidence and groundwater conservation in Tokyo. *Journal of Groundwater Hydrology* 52: 35–47.

Nandeesha, M.C. 2002 Sewage fed aquaculture systems of Kolkata: a century-old innovation of farmers. *Aquaculture Asia* 7: 28–32.

Naser, H.A. 2013 Assessment and management of heavy metal pollution in the marine environment of the Arabian Gulf: a review. *Marine Pollution Bulletin* 72: 6–13.

Nassar, N.T. 2015 Limitations to elemental substitution as exemplified by the platinum-group metals. *Green Chemistry* 17: 2226–35.

Naylor, R.L., Goldburg, R.J., Primavera, J.H., Kautsky, N., Beveridge, M.C.M., Clay, J., Folke, C., Lubchenco, J., Mooney, H. and Troell, M. 2000 Effect of aquaculture on world fish supplies. *Nature* 405: 1017–24.

Neary, B.P., Dillon, P.J., Munro, J.R., and Clark, B.J. 1990 *The acidification of Ontario lakes: an assessment of their sensitivity and current status with respect to biological damage.* Technical Report, Dorset, ON, Ontario Ministry of Environment.

Nellemann, C. and INTERPOL Environmental Crime Programme 2012 *Green carbon, black trade: illegal logging, tax fraud, and laundering in the world's tropical forests: a rapid response assessment.* UNEP, GRID-Arendal.

Nellemann, C., Vistnes, I., Jordhoy, P., Strand, O. and Newton, A. 2003 Progressive impact of piecemeal infrastructure development on wild reindeer. *Biological Conservation* 113: 307–17.

Nelson, F.E., Anisimov, O.A. and Shiklomanov, N.I. 2002 Climate change and hazard zonation in the circum-Arctic permafrost regions. *Natural Hazards* 26: 203–25.

NEPA (National Environmental Protection Agency) 1997 *1996 report on the state of the environment.* Beijing, NEPA.

Nepstad, D., Schwartzman, S., Bamberger, B., Santilli, M., Ray, D., Schlesinger, P., Lefebvre, P., Alencar, A., Prinz, E., Fiske, G. and Rolla, A. 2006 Inhibition of Amazon deforestation and fire by parks and indigenous lands. *Conservation Biology* 20: 65–73.

Nepstad, D., McGrath, D., Stickler, C., Alencar, A., Azevedo, A., Swette, B., Bezerra, T., DiGiano, M., Shimada, J., da Motta, R.S. and Armijo, E. 2014 Slowing Amazon deforestation through public policy and interventions in beef and soy supply chains. *Science* 344(6188): 1118–23.

Neumann, A.C. and Macintyre, I. 1985 Reef response to sea level rise: keep-up, catch-up or give-up. *Proceedings of the 5th International Coral Reef Congress, Tahiti* 3: 105–10.

Neumann, K., Stehfest, E., Verburg, P.H., Siebert, S., Müller, C. and Veldkamp, T. 2011 Exploring global irrigation patterns: a multilevel modelling approach. *Agricultural Systems* 104: 703–13.

Neumayer, E. 2010 *Weak versus strong sustainability: exploring the limits of two opposing paradigms*, 3rd edn. Cheltenham, Edward Elgar Publishing.

Neves, K. 2010 Cashing in on cetourism: a critical ecological engagement with dominant E-NGO discourses on whaling, cetacean conservation, and whale-watching. *Antipode* 42: 719–41.

Newson, M.D. 2002 Geomorphological concepts and tools for sustainable river ecosystem management. *Aquatic Conservation: Marine and Freshwater Ecosystems* 12: 365–79.

Nichol, J.E. 1989 Ecology of fuelwood production in Kano region, northern Nigeria. *Journal of Arid Environments* 16: 347–60.

Nichols, C.A., Vandewalle, M.E. and Alexander, K.A. 2017 Emerging threats to dryland forest resources: elephants and fire are only part of the story. *Forestry: An International Journal of Forest Research*: 1–12

Nilsson, C., Reidy, C.A., Dynesius, M. and Revenga, C. 2005 Fragmentation and flow regulation of the world's large river systems. *Science* 308: 405–8.

Nishida, K., Nagayoshi, Y., Ota, H. and Nagasawa, H. 2001 Melting and stone production using MSW incinerated ash. *Waste Management* 21: 443–9.

Nixon, S.W. 1993 Nutrients and coastal waters: too much of a good thing? *Oceanus* 36(2): 38–47.

Njiru, M., Waithaka, E., Muchiri, M., van Knaap, M. and Cowx, I.G. 2005 Exotic introductions to the fishery of Lake Victoria: what are the management options? *Lakes and Reservoirs: Research and Management* 10: 147–55.

Nortcliff, S. and Gregory, P.J. 1992 Factors affecting losses of soil and agricultural land in tropical countries. In McCall, G.J.H., Laming, D.J.C. and Scott, S.C. (eds) *Geohazards: natural and man-made hazards.* London, Chapman & Hall: 183–90.

Norton, D.A. 1991 *Trilepidea adamsii*: an obituary for a species. *Conservation Biology* 5: 52–7.

Novara, A., Gristina, L., Saladino, S.S., Santoro, A. and Cerdà, A. 2011 Soil erosion assessment on tillage and alternative soil managements in a Sicilian vineyard. *Soil and Tillage Research* 117: 140–7.

Nriagu, J. 1996 A history of global metal pollution. *Science* 272: 223–4.

Nriagu, J., Blankson, M.L. and Ocran, K. 1996 Childhood lead poisoning in Africa: a growing public health problem. *Science of the Total Environment* 181: 93–100.

Nunn, P.D. 1990 Recent environmental changes on Pacific islands. *Geographical Journal* 156: 125–40.

Nunn, P.D. 2000 Environmental catastrophe in the Pacific Islands around AD 1300. *Geoarchaeology* 15: 715–40.

Nunn, P.D. 2003 Nature-society interactions in the Pacific Islands. *Geografiska Annaler* 85B: 219–29.

Nyland, K.E., Shiklomanov, N.I. and Streletskiy, D.A. 2017 Climatic- and anthropogenic-induced land cover change around Norilsk, Russia. *Polar Geography* 40(4): 257–72.

Obeng, L. 1978 Environmental impacts of four African impoundments. In Gunnerson, C.G. and Kalbermatten, J.M. (eds) *Environmental impacts of international civil engineering projects and practices.* New York, American Society of Civil Engineers.

O'Connor, S., Campbell, R., Cortez, H. and Knowles, T. 2009 *Whale watching worldwide: tourism numbers, expenditures and expanding economic benefits.* A special report from the International Fund for Animal Welfare, Yarmouth, MA, USA, prepared by Economists at Large.

O'Connor, T.G., Puttick, J.R. and Hoffman, M.T. 2014 Bush encroachment in southern Africa: changes and causes. *African Journal of Range and Forage Science* 31(2): 67–88.

OECD (Organisation for Economic Co-operation and Development) 1991a *The state of the environment.* Paris, OECD.

OECD 1991b *Environmental indicators: a preliminary set.* Paris, OECD.

OECD 2001 *Extended producer responsibility: a guidance manual for governments.* Paris, OECD.

OECD 2005 *Environmental data compendium 2004.* Paris, OECD.

OECD 2017 Waste water treatment (indicator). doi: 10.1787/ef27a39d-en, accessed July 2017.

O'Hara, S.L. and Metcalfe, S.E. 2004 Late Holocene environmental change in west central Mexico: evidence from the basins of Patzcuaro and Zacapu. In Redman, C.L. James, S.R., Fish, P.R. and Rogers, J.D. (eds) *The archaeology of global change: the impact of humans on their environment.* Washington, DC, Smithsonian Books: 95–112.

Ohkita, T. 1984 Health effects on individuals and health services of the Hiroshima and Nagasaki bombs. In WHO, *Effects of nuclear war on health and health service.* Geneva, WHO.

Ojelede, M.E., Annegarn, H.J. and Kneen, M.A. 2012 Evaluation of aeolian emissions from gold mine tailings on the Witwatersrand. *Aeolian Research* 3: 477–86.

Oke, T.R., Mills, G., Christen, A. and Voogt, J.A. 2017 *Urban climates.* Cambridge, Cambridge University Press.

Oki, T. and Kanae, S. 2006 Global hydrological cycles and world water resources. *Science* 313(5790): 1068–72.

Olden, J.D. 2006 Biotic homogenization: a new research agenda for conservation biogeography. *Journal of Biogeography* 33: 2027–39.

Oliver, F.W. 1945 Dust storms in Egypt and their relation to the war period, as noted in Maryut, 1939–45. *Geographical Journal* 106: 26–49.

Olshansky, S.J., Carnes, B., Rogers, R.G. and Smith, L. 1997 Infectious diseases: new and ancient threats to world health. *Population Bulletin* 52(2).

Olson, S.L. and James, H.F. 1984 The role of Polynesians in the extinction of the avifauna of the Hawaiian Islands. In Martin, P.S and Klein, R.G. (eds) *Quaternary extinctions: a prehistoric revolution.* Tucson, AZ, University of Arizona Press: 768–80.

Olsson, L. 1993 On the causes of famine: drought, desertification and market failure in the Sudan. *Ambio* 22: 395–403.

Onodera, S. 2011 Subsurface pollution in Asian megacities. In Taniguchi, M. (ed.) *Groundwater and subsurface environments: human impacts in Asian coastal cities.* Tokyo, Springer: 159–84.

O'Reilly, C.M., Sharma, S., Gray, D.K., Hampton, S.E., Read, J.S., Rowley, R.J., Schneider, P., Lenters, J.D., McIntyre, P.B., Kraemer, B.M. and Weyhenmeyer, G.A. 2015 Rapid and highly variable warming of lake surface waters around the globe. *Geophysical Research Letters* 42(24).

Orians, G.H. and Pfeiffer, E.W. 1970 Ecological effects of the war in Vietnam. *Science* 168: 544–54.

O'Riordan, T. and Turner, R.K. 1983 *An annotated reader in environmental planning and management.* Oxford, Pergamon Press.

Ortiz, I. and Cummins, M. 2011 *Global inequality: beyond the bottom billion: a rapid review of income distribution in 141 countries.* New York, UNICEF.

Ortiz, N., Pires, M.A.F. and Bressiani, J.C. 2001 Use of steel converter slag as nickel adsorber to wastewater treatment. *Waste Management* 21: 631–5.

OSPAR Commission 2000 *Quality status report 2000.* London, OSPAR Commission.

Ostling, J.L., Butler, D.R. and Dixon, R.W. 2009 The biogeomorphology of mangroves and their role in natural hazards mitigation. *Geography Compass* 3: 1607–24.

Ostro, B. 1994 Estimating the health effects of air pollutants: a method with an application to Jakarta. World Bank, Policy Research Working Paper 1301.

Othman, A., Sultan, M., Becker, R., Alsefry, S., Alharbi, T., Gebremichael, E., Alharbi, H. and Abdelmohsen, K. 2018 Use of geophysical and remote sensing data for assessment of aquifer depletion and related land deformation. *Surveys in Geophysics.* doi: org/10.1007/s10712-017-9458-7.

Otto, F.E. 2017 Attribution of weather and climate events. *Annual Review of Environment and Resources* 42: 627–46.

Ovando-Shelley, E., Ossa, A. and Romo, M.P. 2007 The sinking of Mexico City: its effects on soil properties and seismic response. *Soil Dynamics and Earthquake Engineering* 27: 333–43.

Oyama, M.D. and Nobre, C.A. 2003 A new climate-vegetation equilibrium state for Tropical South America. *Geophysical Research Letters* 30: 2199. doi:10.1029/2003GL018600.

Özerdem, A. and Barakat, S. 2000 After the Marmara earthquake: lessons for avoiding short cuts to disasters. *Third World Quarterly* 21: 425–39.

Ozyavas, A., Khan, S.D. and Casey, J.F. 2010 A possible connection of Caspian Sea level fluctuations with meteorological factors and seismicity. *Earth and Planetary Science Letters* 299(1–2): 150–8.

Pabian, S.E. and Brittingham, M.C. 2007 Terrestrial liming benefits birds in an acidified forest in the northeast. *Ecological Applications* 17: 2184–94.

Pall, P., Aina, T., Stone, D.A., Stott, P.A., Nozawa, T., Hilberts, A.G.J., Lohmann, D. and Allen, M.R. 2011 Anthropogenic greenhouse gas contribution to flood risk in England and Wales in autumn 2000. *Nature* 470: 382–5.

Palmer, M.A., Bernhardt, E.S., Schlesinger, W.H., Eshleman, K., Foufoula-Georgiou, N.E., Hendryx, M.S., Lemly, A.D., Likens, G.E., Loucks, O.L., Power, M.E., White, P.S. and Wilcock, P.R. 2010 Mountaintop mining consequences. *Science* 327(5962): 148–9.

Paoletti, E., Schaub, M., Matyssek, R., Wieser, G., Augustaitis, A., Bastrup-Birk, A.M., Bytnerowicz, A., Günthardt-Goerg, M.S., Muller-Starck, G. and Serengil, Y. 2010 Advances of air pollution science: from forest decline to multiple-stress effects on forest ecosystem services. *Environmental Pollution* 158: 1986–9.

Park, C.E., Jeong, S.J., Joshi, M., Osborn, T.J., Ho, C.H., Piao, S., Chen, D., Liu, J., Yang, H., Park, H. and Kim, B.M. 2018 Keeping global warming within 1.5° C constrains emergence of aridification. *Nature Climate Change* 8: 70–4.

Partecke, J. and Gwinner, E. 2007 Increased sedentariness in European blackbirds following urbanization: a consequence of local adaptation? *Ecology* 88: 882–90.

Parton, W.J., Gutmann, M.P. and Ojima, D. 2007 Long-term trends in population, farm income, and crop production in the Great Plains. *BioScience* 57: 737–47.

Pasternak, G.B., Brush, G.S. and Hilgartner, W.B. 2001 Impact of historic land-use change on sediment delivery to a Chesapeake Bay subestuarine delta. *Earth Surface Processes and Landforms* 26: 409–27.

Pastor, M., Sadd, J.L. and Morello-Frosch, R. 2002 Who's minding the kids? Pollution, public schools, and environmental justice in Los Angeles. *Social Science Quarterly* 83: 263–80.

Patrick, S.T., Timberlid, J.A. and Stevenson, A.C. 1990 The significance of land-use and land management change in the acidification of lakes in Scotland and Norway: an assessment utilizing documentary sources and pollen analysis. *Philosophical Transactions of the Royal Society, London* 327(1240): 363–7.

Pauly, D., Christensen, V., Dalsgaard, J., Froese, R. and Torres, F. 1998 Fishing down marine food webs. *Science* 279: 860–3.

Pearce, D. 1993 *Economic values and the natural world*. London, Earthscan.

Pearce, J. and Kingham, S. 2008 Environmental inequalities in New Zealand: a national study of air pollution and environmental justice. *Geoforum* 39: 980–93.

Pederson, N., Hessl, A.E., Baatarbileg, N., Anchukaitis, K.J. and Di Cosmo, N. 2014 Pluvials, droughts, the Mongol Empire, and modern Mongolia. *Proceedings of the National Academy of Sciences* 111(12): 4375–9.

Peduzzi, P. and 14 others 2010 *The Global Risk Analysis for the 2009 Global Assessment Report on Disaster Risk Reduction*. Extended summary for the International Disaster and Risk Conference IDRC, Davos 2010, 30 May–3 June 2010, on-line conference proceedings.

Pejchar, L. and Mooney, H.A. 2009 Invasive species, ecosystem services and human well-being. *Trends in Ecology and Evolution* 24: 497–504.

Pelicice, F.M., Pompeu, P.S. and Agostinho, A.A. 2015 Large reservoirs as ecological barriers to downstream movements of Neotropical migratory fish. *Fish and Fisheries* 16: 697–715.

Pelletier, N., Audsley, E., Brodt, S., Garnett, T., Henriksson, P., Kendall, A., Kramer, J., Murphy, D., Nemecek, T. and Troell, M. 2011 Energy intensity of agriculture and food systems. *Annual Review of Environment and Resources* 36: 223–46.

Pelling, M. and Uitto, J.I. 2001 Small island developing states: natural disaster vulnerability and global change. *Environmental Hazards* 3: 49–62.

Pennell, C.R. 1994 The geography of piracy: northern Morocco in the mid-nineteenth century. *Journal of Historical Geography* 20: 272–82.

Penning-Rowsell, E.C., Haigh, N., Lavery, S. and McFadden, L. 2012 A threatened world city: the benefits of protecting London from the sea. *Natural Hazards* 66(3): 1383–404.

Pérez-Lombard, L., Ortiz, J. and Pout, C. 2008 A review on buildings energy consumption information. *Energy and Buildings* 40: 394–8.

Perkins, J.S. and Thomas, D.S.G. 1993 Environmental responses and sensitivity to permanent cattle ranching, semi-arid western central Botswana. In Thomas, D.S.G. and Allison, R.J. (eds) *Landscape sensitivity*. Chichester, Wiley: 273–86.

Pernetta, J.C. 1992 Impacts of climate change and sea-level rise on small island states: national and international responses. *Global Environmental Change* 2: 19–31.

Pescott, O.L., Simkin, J.M., August, T.A., Randle, Z., Dore, A.J. and Botham, M.S. 2015 Air pollution and its effects on lichens, bryophytes, and lichen-feeding Lepidoptera: review and evidence from biological records. *Biological Journal of the Linnean Society* 115(3): 611–35.

Peters, R.L. and Lovejoy, T.E. 1990 Terrestrial fauna. In Turner II, B.L., Clark, W.C., Kates, R.W., Richards, J.F., Mathews, J.T. and Meyer, W.B. (eds) *The Earth as transformed by human action*. Cambridge, Cambridge University Press: 353–69.

Petersen, J.K. and Malm, T. 2006 Offshore windmill farms: threats to or possibilities for the marine environment. *Ambio* 35: 75–80.

Pfaff, A., Broad, K. and Glantz, M. 1999 Who benefits from climate forecasts? *Nature* 397: 645–6.

Phantumvanit, D. and Liengcharernsit, W. 1989 Coming to terms with Bangkok's environmental problems. *Environment and Urbanization* 1: 31–9.

Pimental, D. 1984 Energy flows in food systems. In Pimental, D. and Hall, C. (eds) *Food and energy resources*. New York, Academic Press: 1–23.

Pimental, D. 1991 Diversification of biological control strategies in agriculture. *Crop Protection* 10: 243–53.

Pimentel, D. 2005 Environmental and economic costs of the application of pesticides primarily in the United States. *Environment, Development and Sustainability* 7: 229–52.

Pimentel, D., Gardner, J., Bonnifield, A., Garcia, X., Grufferman, J., Horan, C., Schlenker, J. and Walling, E. 2009 Energy efficiency and conservation for individual Americans. *Environment, Development and Sustainability* 11: 523–46.

Pimentel, D., Lach, L., Zuniga, R. and Morrison, D. 2014 Environmental and economic costs associated with non-indigenous species in the United States. In Pimentel, D. (ed.) *Biological invasions: economic and environmental costs of alien plant, animal, and microbe species*. Boca Raton, CRC Press: 285–306.

Pimm, S.L., Jenkins, C.N., Abell, R., Brooks, T.M., Gittleman, J.L., Joppa, L.N., Raven, P.H., Roberts, C.M. and Sexton, J.O. 2014 The biodiversity of species and their rates of extinction, distribution, and protection. *Science* 344(6187): 1246752.

Plucknett, D.L. and Smith, N.J.H. 1986 Sustaining agricultural yields. *BioScience* 36: 40–5.

Poff, N.L. and Zimmerman, J.K.H. 2010 Ecological responses to altered flow regimes: a literature review to inform the science and management of environmental flows. *Freshwater Biology* 55: 194–205.

Pomeroy, D., Tushabe, H. and Loh, J. 2017 The state of Uganda's biodiversity 2017. Kampala, National Biodiversity Data Bank, Makerere University.

Pompa, S., Ehrlich, P.R. and Ceballos, G. 2011 Global distribution and conservation of marine mammals. *Proceedings of the National Academy of Sciences* 108: 13600.

Postel, S.L., Daily, G.C. and Ehrlich, P.R. 1996 Human appropriation of renewable fresh water. *Science* 271: 785–8.

Posthumus, H., Deeks, L.K., Rickson, R.J. and Quinton, J.N. 2015 Costs and benefits of erosion control measures in the UK. *Soil Use and Management* 31(S1): 16–33.

Pounds, J.A., Bustamante, M.R., Coloma, L.A., Consuegra, J.A., Fogden, M.P., Foster, P.N., La Marca, E., Masters, K.L., Merino-Viteri, A., Puschendorf, R., *et al.* 2006 Widespread amphibian extinctions from epidemic disease driven by global warming. *Nature* 439: 161–7.

Preston, S.H. and Van de Walle, E. 1978 Urban French mortality in the nineteenth century. *Population Studies* 32(2): 275–97.

Price, L.W. 1972 *The periglacial environment, permafrost, and man*. Commission on Geographical Resources Paper 14. Washington, DC, Association of American Geographers.

Prince, S.D. 2016 Where does desertification occur? Mapping dryland degradation at regional to global scales. In Behnke, R. and Mortimore, M. (eds) *The end of desertification? Disputing environmental change in the drylands*. Berlin, Springer: 225–63.

Pringle, C.M., Freeman, M.C. and Freeman, B.J. 2000 Regional effects of hydrologic alterations on riverine macrobiota in the new world: tropical–temperate comparisons. *BioScience* 50: 807–23.

Progiou, G. and Ziomas, I.C. 2011 Road traffic emissions impact on air quality of the Greater Athens Area based on a 20 year emissions inventory. *Science of the Total Environment* 410–11: 1–7.

Prosdocimi, M., Cerdà, A. and Tarolli, P. 2016 Soil water erosion on Mediterranean vineyards: a review. *Catena* 141: 1–21.

Prose, D.V. 1985 Persisting effects of armoured military manoeuvres on some soils of the Mojave Desert. *Environmental Geology and Water Sciences* 7: 163–70.

Pryde, P.R. 1972 *Conservation in the Soviet Union*. Cambridge, Cambridge University Press.

Pryde, P.R. 1991 *Environmental management in the Soviet Union*. Cambridge, Cambridge University Press.

Psenner, R. 1999 Living in a dusty world: airborne dust as a key factor for alpine lakes. *Water, Air, and Soil Pollution* 112: 217–27.

Purvis, O.W. 2010 Lichens and industrial pollution. In Batty, L.C. and Hallberg, K.B. (eds) *Ecology of industrial pollution*. Cambridge, Cambridge University Press: 41–69.

Putz, F.E., Zuidema, P.A., Synnott, T., Peña-Claros, M., Pinard, M.A., Sheil, D., Vanclay, J.K., Sist, P., Gourlet-Fleury, S., Griscom, B. and Palmer, J. 2012 Sustaining conservation values in selectively logged tropical forests: the attained and the attainable. *Conservation Letters* 5(4): 296–303.

Qadir, M., Quillérou, E., Nangia, V., Murtaza, G., Singh, M., Thomas, R.J., Drechsel, P. and Noble, A.D. 2014 Economics of salt-induced land degradation and restoration. *Natural Resources Forum* 38(4): 282–95.

Qi, S., Leipe, T., Rueckert, P., Di, Z. and Harff, J. 2010 Geochemical sources, deposition and enrichment of heavy metals in short sediment cores from the Pearl River Estuary, Southern China. *Journal of Marine Systems* 82: S28.

Quah, E. and Boon, T.L. 2003 The economic cost of particulate air pollution on health in Singapore. *Journal of Asian Economics* 14: 73–90.

Qureshi, A.S., McCornick, P.G., Qadir, M. and Aslam Z. 2008 Managing salinity and waterlogging in the Indus Basin of Pakistan. *Agricultural Water Management* 95: 1–10.

Qureshi, A.S., McCornick, P.G., Sarwar, A. and Sharma, B.R. 2010 Challenges and prospects for sustainable groundwater management in the Indus Basin, Pakistan. *Water Resources Management* 24: 1551–69.

Rackham, O. 1986 *The history of the countryside*. London, Dent.

Ragnarsson, A. 2015. Geothermal development in Iceland 2010–2014. In Proceedings of the World Geothermal Congress 2015, Melbourne, Australia, 19–24 April.

Raine, A.F., Gauci, M. and Barbara, N. 2016 Illegal bird hunting in the Maltese Islands: an international perspective. *Oryx* 50(4): 597–605.

Ramírez, F., Afán, I., Davis, L.S. and Chiaradia, A. 2017 Climate impacts on global hot spots of marine biodiversity. *Science Advances* 3(2): e1601198.

Ramos, M.C. and Martínez-Casasnovas, J.A. 2006 Trends in precipitation concentration and extremes in the Mediterranean Penedès-Anoia region, NE Spain. *Climatic Change* 74: 457–74.

Ramsar Convention Secretariat 2010 *Wise use of wetlands: concepts and approaches for the wise use of wetlands*, 4th edn, vol. 1. Gland, Switzerland, Ramsar Convention Secretariat.

Raper, G.P., Speed, R., Simons, J., Kendle, A., Blake, A., Ryder, A., Smith, R., Stainer, G. and Bourke, L. 2014 Groundwater trend analysis for south-west Western Australia 2007–12. *Resource Management Technical Report* 388.

Rashad, S.M. and Ismail, M.A. 2000 Environmental impact assessment of hydro-power in Egypt. *Applied Energy* 65: 285–302.

Raskin, P.D. 1995 Methods for estimating the population contribution to environmental change. *Ecological Economics* 15: 225–33.

Rasmussen, J.L., Regie, H.A., Sparks, R.E. and Taylor, W.W. 2011 Dividing the waters: the case for hydrologic separation of the North American Great Lakes and Mississippi River Basins. *Journal of Great Lakes Research* 37: 588–92.

Ray, D.K., Welch, R.M., Lawton, R.O. and Nair, U.S. 2006 Dry season clouds and rainfall in northern Central America: implications for the Mesoamerican Biological Corridor. *Global and Planetary Change* 54: 150–62.

Raynolds, M.K., Walker, D.A., Ambrosius, K.J., Brown, J., Everett, K.R., Kanevskiy, M., Kofinas, G.P., Romanovsky, V.E., Shur, Y. and Webber, P.J. 2014 Cumulative geoecological effects of 62 years of infrastructure and climate change in ice-rich permafrost landscapes, Prudhoe Bay Oilfield, Alaska. *Global Change Biology* 20(4): 1211–24.

Readman, J.W., Fowler, S.W., Villeneuve, J.-P., Cattini, C., Oregioni, B. and Mee, L.D. 1992 Oil and combustion product contamination of the Gulf marine environment following the war. *Nature* 358: 662–5.

Rees, W.E. 2003 Understanding urban ecosystems: an ecological economics perspective. In Berkowitz, A.R., Nilon, C.H. and Hollweg, K.S. (eds) *Understanding urban ecosystems: a new frontier for science and education*. New York, Springer-Verlag: 115–36.

Regan, A.J. 2017 Bougainville: origins of the conflict, and debating the future of large-scale mining. In Filer, C. and le Meur, P.-Y. (eds) *Large-scale mines and local-level politics*. Acton, ANU Press: 353–414.

Rehfuess, E.A., Puzzolo, E., Stanistreet, D., Pope, D. and Bruce, N.G. 2014 Enablers and barriers to large-scale uptake of improved solid fuel stoves: a systematic review. *Environmental Health Perspectives* 122(2): 120.

Reid, W.V. 1995 Biodiversity and health: prescription for progress. *Environment* 37: 36.

Reij, C., Tappan, G. and Smale, M. 2009 Agroenvironmental transformation in the Sahel: another kind of 'Green Revolution'. IFPRI Discussion Paper 00914.

Reijnen, R. and Foppen, R. 1994 The effects of car traffic on breeding bird populations in woodland. 1 Evidence of reduced habitat quality for willow warblers breeding close to a highway. *Journal of Applied Ecology* 31: 85–94.

Reiter, P. 2008 Global warming and malaria: knowing the horse before hitching the cart. *Malaria Journal* 7(suppl. 1): S3.

REN21 2017 *Renewables 2017 global status report*. Paris, REN21.

Renberg, I., Korsman, T. and Birks, H.J.B. 1993 Prehistoric increases in the pH of acid-sensitive Swedish lakes caused by land-use change. *Nature* 362: 824–7.

Renberg, I., Bigler, C., Bindler, R., Norberg, M., Rydberg, J. and Segerstrom, U. 2009 Environmental history: a piece in the puzzle for establishing plans for environmental management. *Journal of Environmental Management* 90: 2794–800.

Renn, O. and Marshall, J.P. 2016 Coal, nuclear and renewable energy policies in Germany: From the 1950s to the 'Energiewende'. *Energy Policy* 99: 224–32.

Renner, M. 1991 Assessing the military's war on the environment. In *State of the world 1991*. New York, W.W. Norton.

Renwick, W.H., Smith, S.V., Bartley, J.D. and Buddemeier, R.W. 2005 The role of impoundments in the sediment budget of the conterminous United States. *Geomorphology* 71: 99–111.

Restrepo, J.D. and Escobar, H.A. 2016 Sediment load trends in the Magdalena River basin (1980–2010): anthropogenic and climate-induced causes. *Geomorphology*.

Reynolds, J.F. and 16 others 2007 Global desertification: building a science for dryland development. *Science* 316: 847–51.

Ribbing, C. 2007 Environmentally friendly use of non-coal ashes in Sweden. *Waste Management* 27: 1428–35.

Ribbink, A.J. and Roberts, M. 2006 African Coelacanth Ecosystem Programme: an overview of the conference contributions. *South African Journal of Science* 102: 409–15.

Ribot, J.C. 1999 A history of fear: imagining deforestation in the West African dryland forests. *Global Ecology and Biogeography* 8(3–4): 291–300.

Ricciardi, A. 2006 Patterns of invasion in the Laurentian Great Lakes in relation to changes in vector activity. *Diversity and Distributions* 12: 425–33.

Richards, Z.T., Beger, M., Pinca, S. and Wallace, C.C. 2008 Bikini Atoll coral biodiversity resilience five decades after nuclear testing. *Marine Pollution Bulletin* 56(3): 503–15.

Richardson, C.J. and Hussain, N.A. 2006 Restoring the Garden of Eden: an ecological assessment of the marshes of Iraq. *Bioscience* 56: 477–89.

Richardson, M.L. 1993 The assessment of hazards and risks to the environment caused by war damage to industrial installations in Croatia. Paper presented at the International Conference on the Effects of War on the Environment, Zagreb, 15–17 April.

Richter, B.D. and Thomas, G.A. 2007 Restoring environmental flows by modifying dam operations. *Ecology and Society* 12: 12.

Richter, E.D. 2002 Acute pesticide poisonings. In D. Pimentel (ed.) *Encyclopedia of pest management*. Boca Raton, FL, CRC Press: 3–6.

Riitters, K.H. and Wickham, J.D. 2003 How far to the nearest road? *Frontiers in Ecology and the Environment* 1: 125–9.

Rincón-Ruiz, A. and Kallis, G. 2013 Caught in the middle, Colombia's war on drugs and its effects on forest and people. *Geoforum* 46: 60–78.

Ringrose, S. and Matheson, W. 1992 The use of Landsat MSS imagery to determine the areal extent of woody vegetation cover change in the west-central Sahel. *Global Ecology and Biogeography Letters* 2: 16–25.

RIWA 2016 *Jaarrapport 2015*. Nieuwegein, RIWA.

Roberts, C.M., McClean, C.J., Veron, J.E.N., Hawkins, J.P., Allen, G.R., McAllister, D.E., Mittermeier, C.G., Schueler, F.W., Spalding, M., Wells, F., Vynne, C. and Werner, T.B. 2002 Marine biodiversity hotspots and conservation priorities for tropical reefs. *Science* 295: 1280–4.

Roberts, J.M. and Cairns, S.D. 2014 Cold-water corals in a changing ocean. *Current Opinion in Environmental Sustainability* 7: 118–26.

Roberts, L. 1988 Conservationists in Panda-monium. *Science* 241: 529–31.

Robinson, R.A. and Sutherland, W.J. 2002 Post-war changes in arable farming and biodiversity in Great Britain. *Journal of Applied Ecology* 39: 157–76.

Rodolfo, K.S. and Siringan, F.P. 2006 Global sea-level rise is recognised, but flooding from anthropogenic land subsidence is ignored around northern Manila Bay, Philippines. *Disasters* 30: 118–39.

Rodrigues, R.R., Gandolfi, S., Nave, A.G., Aronson, J., Barreto, T.E., Vidal, C.Y. and Brancalion, P.H.S. 2011 Large-scale ecological restoration of high-diversity tropical forests in SE Brazil. *Forest Ecology and Management* 261: 1605–13.

Rodriguez, C., Buynder, P.V., Lugg, R., Blair, P., Devine, B., Cook, A. and Weinstein, P. 2009 Indirect potable reuse: a sustainable water supply alternative. *International Journal of Environmental Research and Public Health* 6: 1174–209.

Rodway-Dyer, S.J. and Walling, D.E. 2010 The use of 137Cs to establish longer-term soil erosion rates on footpaths in the UK. *Journal of Environmental Management* 91: 1952–62.

Roem, W.J. and Berendse, F. 2000 Soil acidity and nutrient supply ratio as possible factors determining changes in plant species diversity in grassland and heathland communities. *Biological Conservation* 92: 151–61.

Rogers, H.S., Buhle, E.R., HilleRisLambers, J., Fricke, E.C., Miller, R.H. and Tewksbury, J.J. 2017 Effects of an invasive predator cascade to plants via mutualism disruption. *Nature Communications* 8: 14557.

Rood, G.A., Wilting, H.C., Nagelhout, D., ten Brink, B.J.E., Leewis, R.J. and Nijdam, D.S. 2004 *Spoorzoeken naar de invloed van Nederlanders op de mondiale biodiversiteit: Model voor een ecologische voetafdruk* [Tracking the effects of inhabitants on biodiversity in the Netherlands and abroad: an ecological footprint model]. RIVM Report 500013005. Bilthoven, RIVM.

Ross, M.L. 2008 Mineral wealth, conflict, and equitable development. In Bebbington, A.J., Dani, A.A., de Haan, A. and Walton, M. (eds) *Institutional pathways to equity: addressing inequality traps.* Washington, DC, World Bank: 193–215.

Roy, P., Nei, D., Orikasa, T., Xu, Q., Okadome, H., Nakamura, N. and Shiina, T. 2009 A review of life cycle assessment (LCA) on some food products. *Journal of Food Engineering* 90: 1–10.

Ruddiman, W.F. 2014 *Earth's climate: past and future*, 3rd edn. New York, Freeman and Company.

Rudel, T.K. 2007 Changing agents of deforestation: from state-initiated to enterprise driven processes, 1970–2000. *Land Use Policy* 24: 35–41.

Rudel, T.K., Defries, R., Asner, G.P. and Laurance, W.F. 2009 Changing drivers of deforestation and new opportunities for conservation. *Conservation Biology* 23: 1396–405.

Runnels, D.D., Shepherd, T.A. and Angino, E.E. 1992 Metals in water. *Environmental Science and Technology* 26: 2316–23.

Ruslandi, O.V. and Putz, F.E. 2011 Over-estimating the costs of conservation in Southeast Asia. *Frontiers in Ecology and the Environment* 9: 542–4.

Russell, J.C. and Blackburn, T.M. 2017 The rise of invasive species denialism. *Trends in Ecology and Evolution* 32: 3–6.

Sachdev, H.S., Akella, A.K. and Kumar, N. 2015 Analysis and evaluation of small hydropower plants: a bibliographical survey. *Renewable and Sustainable Energy Reviews* 51: 1013–22.

Sadhwani, J.J., Veza, J.M. and Santana, C. 2005 Case studies on environmental impact of seawater desalination. *Desalination* 185: 1–8.

Saenger, P. 1994 Cleaning up the Arabian Gulf: aftermath of an oil spill. *Search* 25: 19–22.

Sahin, V. and Hall, M.J. 1996 The effects of afforestation and deforestation on water yields. *Journal of Hydrology* 178: 293–309.

Saitanis, C., Karandinos, M.G., Riga-Karandinos, A.N., Lorenzini, G. and Vlassi, A. 2003 Photochemical air pollutant levels and ozone phytotoxicity in the region of Mesogia Attica, Greece. *International Journal of Environment and Pollution* 19: 197–208.

Sakai, S.I., Yoshida, H., Hiratsuka, J., Vandecasteele, C., Kohlmeyer, R., Rotter, V.S., Passarini, F., Santini, A., Peeler, M., Li, J. and Oh, G.J. 2014 An international comparative study of end-of-life vehicle (ELV) recycling systems. *Journal of Material Cycles and Waste Management* 16: 1–20.

Samoli, E., Nastos, P.T., Paliatsos, A.G., Katsouyanni, K. and Priftis, K.N. 2011 Acute effects of air pollution on pediatric asthma exacerbation: evidence of association and effect modification. *Environmental Research* 111: 418–24.

Sánchez-Bayo, F., Goulson, D., Pennacchio, F., Nazzi, F., Goka, K. and Desneux, N. 2016 Are bee diseases linked to pesticides? A brief review. *Environment International* 89: 7–11.

Sanders, T.G.M., Michel, A.K. and Ferretti, M. 2016 *30 years of monitoring the effects of long-range transboundary air pollution on forests in Europe and beyond.* Eberswalde, UNECE/ICP Forests.

Sapountzaki, K. and Chalkias, C. 2005 Coping with chronic and extreme risks in contemporary Athens: confrontation or resilience? *Sustainable Development* 13: 115–28.

Sapozhnikova, S.A. 1973 *Map diagram of the number of days with dust storms in the hot zone of the USSR and adjacent territories.* Report HT-23-0027. Charlottesville, VA, US Army Foreign and Technology Center.

Saraiva, P.P., Ribeiro, L.A., Camara, I.P. and da Silva, T.L. 2018 How technologies contribute to urban sustainability: the case of Curitiba – Brazil. In Azeiteiro, U.M., Akerman, M., Leal Filho, W., Setti, A.F. and Brandli, L.L. (eds) *Lifelong learning and education in healthy and sustainable cities.* Cham, Springer: 507–20.

Sarver, T., Al-Qaraghuli, A. and Kazmerski, L.L. 2013 A comprehensive review of the impact of dust on the use of solar energy: history, investigations, results, literature, and mitigation approaches. *Renewable and Sustainable Energy Reviews* 22: 698–733.

Sassen, N., DeMott, P.J. and Prospero, J.M. 2003 Saharan dust storms and indirect effects on clouds: CRYSTAL-FACE results. *Geophysical Research Letters* 30: article 1633.

Sather, J.M. and Smith, R.D. 1984 *An overview of major wetland functions and values.* Washington, DC, Fish and Wildlife Service, FWS/OBS-84/18.

Sathirathai, S. and Barbier, E.B. 2001 Valuing mangrove conservation in southern Thailand. *Contemporary Economic Policy* 19: 109–22.

Satterthwaite, D. 2016 Missing the Millennium Development Goal targets for water and sanitation in urban areas. *Environment and Urbanization* 28: 99–118.

Savchenko, V.K. 1991 The Chernobyl catastrophe and the biosphere. *Nature and Resources* 27(1): 37–46.

Sayer, J.A., Harcourt, C.S. and Collins, N.M. (eds) 1992 *The conservation atlas of tropical forests: Africa.* London, Macmillan/IUCN.

Sayre, N.F. 2008 The genesis, history, and limits of carrying capacity. *Annals of the Association of American Geographers* 98: 120–34.

Schecter, A., Quynh, H.T., Pavuk, M., Papke, O., Malisch, R. and Constable, J.D. 2003 Food as a source of dioxin exposure in the residents of Bien Hoa City, Vietnam. *Journal of Occupational and Environmental Medicine* 45: 781–8.

Scheierling, S.M., Bartone, C.R., Mara, D.D. and Drechsel, P. 2011 Towards an agenda for improving wastewater use in agriculture. *Water International* 36: 420–40.

Schiettecatte, W., Ouessar, M., Gabriels, D., Tanghe, S., Heirman, S. and Abdelli, F. 2005 Impact of water harvesting techniques on soil and water conservation: a case study on a micro catchment in southeastern Tunisia. *Journal of Arid Environments* 61: 297–313.

Schindler, D.W., Curtis, P.J., Parker, B.R. and Stainton, M.P. 1996 Consequences of climate warming and lake acidification for UV-B penetration in North American boreal lakes. *Nature* 379: 705–8.

Schlesinger, W.H. 1991 *Biogeochemistry: an analysis of global change.* San Diego, Academic Press.

Schlömer, S., Bruckner, T., Fulton, L., Hertwich, E., McKinnon, A., Perczyk, D., Roy, J., Schaeffer, R., Sims, R., Smith, P. and Wiser, R. 2014 Annex III: Technology-specific cost and performance parameters. In Edenhofer, O., Pichs-Madruga, R., Sokona, Y., Farahani, E., Kadner, S., Seyboth, K., Adler, A., Baum, I., Brunner, S., Eickemeier, P., Kriemann, B., Savolainen, J., Schlömer, S., von Stechow, C., Zwickel, T. and Minx, J.C. (eds) *Climate change 2014: mitigation of climate change.* Contribution of Working Group III to the Fifth Assessment Report of the Intergovernmental Panel on Climate Change. Cambridge, Cambridge University Press.

Schneider, P. and Hook, S.J. 2010 Space observations of inland water bodies show rapid surface warming since 1985. *Geophysical Research Letters* 37(22): L22405.

Schneider, S.H. 1989 *Global warming: are we entering the greenhouse century?* San Francisco, Sierra Club Books.

Schrank, W.E. 2005 The Newfoundland fishery: ten years after the moratorium. *Marine Policy* 29: 407–20.

Schreiber, M.A., Ñiquen, M. and Bouchon, M. 2011 Coping strategies to deal with environmental variability and extreme climatic events in the Peruvian anchovy fishery. *Sustainability* 3: 823–46.

Schwartz, J. 1994 Low level lead exposure and children's IQ: a meta-analysis and search for a threshold. *Environmental Research* 65: 42–55.

Schwarz, H. 2004 *Urban renewal, municipal revitalization: the case of Curitiba, Brazil*. Alexandria, VA, Hugh Schwarz.

Schweitzer, M.D., Calzadilla, A.S., Salamo, O., Sharifi, A., Kumar, N., Holt, G., Campos, M. and Mirsaeidi, M. 2018 Lung health in era of climate change and dust storms. *Environmental Research* 163: 36–42.

Scodanibbio, L. and Mañez, G. 2005 The World Commission on Dams: a fundamental step towards integrated water resources management and poverty reduction? A pilot case in the Lower Zambezi, Mozambique. *Physics and Chemistry of the Earth* 30: 976–83.

Scott, C.A. and Shah, T. 2004 Groundwater overdraft reduction through agricultural energy policy: insights from India and Mexico. *Water Resource Development* 20: 149–64.

Scott, M.J. and Statham, I. 1998 Development advice: mining subsidence. In Maund, J.G. and Eddlestone, M. *Geohazards in engineering geology*. London, Geological Society: 391–400.

Secretariat, Convention on Biological Diversity 2010 *Global biodiversity outlook 3*. Montréal, Convention on Biological Diversity.

Seitzinger, S.P., Mayorga, E., Bouwman, A.F., Kroeze, C., Beusen, A.H.W., *et al.* 2010 Global river nutrient export: a scenario analysis of past and future trends. *Global Biogeochemical Cycles* 24: GB0A08.

Sen, A.K. 1981 *Poverty and famines: an essay on entitlement and deprivation*. Oxford, Clarendon Press.

Servheen, C., Herrero, S. and Peyton, B. 1999 *Bears: status survey and conservation action plan*. Gland, Switzerland, IUCN.

Seufert, V., Ramankutty, N. and Foley, J.A. 2012 Comparing the yields of organic and conventional agriculture. *Nature* 485: 229–32.

Sevaldrud, I.H., Muniz, I.P. and Kalvenes, S. 1980 Loss of fish populations in southern Norway: dynamics and magnitude of the problem. In Drablos, D. and Tollan, A. (eds) *Ecological impact of acid precipitation*. Norway, SNSF-Project: 350–1.

Shah, P. and Baylis, K. 2015 Evaluating heterogeneous conservation effects of forest protection in Indonesia. *PloS ONE* 10(6): e0124872.

Shahgedanova, M. 2002 Air pollution. In Shahgedanova, M. (ed.) *The physical geography of Northern Eurasia*. Oxford, Oxford University Press: 476–96.

Shahriary, E., Palmer, M.W., Tongway, D.J., Azarnivand, H., Jafari, M. and Saravi, M.M. 2012 Plant species composition and soil characteristics around Iranian piospheres. *Journal of Arid Environments* 82: 106–14.

Shang, J. and Wilson, J.P. 2009. Watershed urbanization and changing flood behavior across the Los Angeles metropolitan region. *Natural Hazards* 48: 41–57.

Sharma, R. 2011 Deep-sea mining: economic, technical, technological, and environmental considerations for sustainable development. *Marine Technology Society Journal* 45(5): 28–41.

Sharma, V.P. 1996 Re-emergence of malaria in India. *Indian Journal of Medical Research* 103: 26–45.

Sheeline, L. 1993 Pacific fruit bats in trade: are CITES controls working? *Traffic USA* 12(1): 1–4.

Shields, L.M. and Wells, P.V. 1962 Effects of nuclear testing on desert vegetation. *Science* 135: 38–40.

Shultz, J.M., Cohen, M.A., Hermosilla, S., Espinel, Z. and McLean, A. 2016 Disaster risk reduction and sustainable development for small island developing states. *Disaster Health* 3(1): 32–44.

Sibley, C.G. and Monroe, B.L. 1990 *Distribution and taxonomy of birds of the world*. New Haven, CT, Yale University Press.

Sidle, R.C., Ziegler, A.D., Negishi, J.N., Rahim Nik, A., Siew, R. and Turkelboom, F. 2006 Erosion processes in steep terrain: truths, myths, and uncertainties related to forest management in Southeast Asia. *Forest Ecology and Management* 224: 199–225.

Siebe, C. and Cifuentes, E. 1995 Environmental impact of wastewater irrigation in Central Mexico: an overview. *International Journal of Environmental Health Research* 5: 161–73.

Siegert, F., Ruecker, G., Hinrichs, A. and Hoffmann, A.A. 2001 Increased damage from fires in logged forests during droughts caused by El Niño. *Nature* 414: 437–40.

Simatele, D. and Etambakonga, C.L. 2015 Scavenging for solid waste in Kinshasa: a livelihood strategy for the urban poor in the Democratic Republic of Congo. *Habitat International* 49: 266–74.

Simberloff, D. 2009 We can eliminate invasions or live with them: successful management projects. *Biological Invasions* 11: 149–57.

Simmons, J.A., Currie, W.S., Eshleman, K.N., Kuers, K., Monteleone, S., Negley, T.L., Pohlad, B.R. and Thomas, C.L. 2008 Forest to reclaimed mine land use change leads to altered ecosystem structure and function. *Ecological Applications* 18: 104–18.

Simon, J.L. 1981 *The ultimate resource*. Oxford, Martin Robertson.

Sinden, G. 2007 Characteristics of the UK wind resource: long-term patterns and relationship to electricity demand. *Energy Policy* 35: 112–27.

SIPRI (Stockholm International Peace Research Institute) 2016 *SIPRI Yearbook 2016: Armaments, disarmament and international security*. Oxford, Oxford University Press.

SIPRI 2017 *SIPRI Yearbook 2017: Armaments, disarmament and international security*. Oxford, Oxford University Press.

Sivan, O., Yechieli, Y., Herut, B. and Lazar, B. 2005 Geochemical evolution and timescale of seawater intrusion into the coastal aquifer of Israel. *Geochimica et Cosmochimica Acta* 69(3): 579–92.

Skaf, R. 1988 A story of a disaster: why locust plagues are still possible. *Disasters* 12: 122–7.

Skinner, I., van Essen, H., Smokers, R. and Hill, N. 2010 *Towards the decarbonisation of EU's transport sector by 2050*. Final report produced under the contract ENV.C.3/SER/2008/0053 between European Commission Directorate-General Environment and AEA Technology plc.

Slaymaker, O. and French, H.M. 1993 Cold environments and global change. In French, H.M. and Slaymaker, O. (eds) *Canada's cold environments*. Montreal, McGill-Queen's University Press: 313–34.

Slingerland, M. and Masdewel, M. 1996 Mulching on the central plateau of Burkina Faso. In Reij, C., Scoones, I. and Toulmin, C. *Sustaining the soil: indigenous soil and water conservation in Africa*. London, Earthscan: 85–9.

Smink, C.K. 2007 Vehicle recycling regulations: lessons from Denmark. *Journal of Cleaner Production* 15: 1135–46.

Smith, A. and Kern, F. 2009 The transitions storyline in Dutch environmental policy. *Environmental Politics* 18(1): 78–98.

Smith, B. 1997 Water: a critical resource. In King, R., Proudfoot, L. and Smith, B. (eds) *The Mediterranean: environment and society*. London, Arnold: 227–51.

Smith, K. 2001 *Environmental hazards: assessing risk and reducing disaster*, 3rd edn. London, Routledge.

Smith, K. and Ward, R. 1998 *Floods: physical processes and human impacts*. Chichester, Wiley.

Smith, N.J.H., Alvim, P., Homma, A., Falesi, I. and Serráo, A. 1991 Environmental impacts of resource exploitation in Amazonia. *Global Environmental Change* 1: 313–20.

Smith, R.A. 1852 On the air and rain of Manchester. *Memoirs and Proceedings of the Manchester Literary and Philosophical Society* 2: 207–17.

Smith, S.J., van Aardenne, J., Klimont, Z., Andres, R., Volke, A. and Delgado Arias, S. 2010 Anthropogenic sulfur dioxide emissions: 1850–2005. *Atmospheric Chemistry and Physics* 10: 16111–51.

Smith, V.H. and Schindler, D.W. 2009 Eutrophication science: where do we go from here? *Trends in Ecology and Evolution* 24(4): 201–7.

Smyth, C.G. and Royle, S.A. 2000 Urban landslide hazards: incidence and causative factors in Niterói, Rio de Janeiro State, Brazil. *Applied Geography* 20: 95–118.

Sneddon, C. and Fox, C. 2006 Rethinking transboundary waters: a critical hydropolitics of the Mekong basin. *Political Geography* 25: 181–202.

Sobels, J., Curtis, A. and Lockie, S. 2001 The role of Landcare group networks in rural Australia: exploring the contribution of social capital. *Journal of Rural Studies* 17(3): 265–76.

Sokolow, S.H., Jones, I.J., Jocque, M., La, D., Cords, O., Knight, A., Lund, A., Wood, C.L., Lafferty, K.D., Hoover, C.M. and Collender, P.A. 2017 Nearly 400 million people are at higher risk of schistosomiasis because dams block the migration of snail-eating river prawns. *Philosophical Transactions of the Royal Society B* 372(1722): 20160127.

Solomon, S., Plattner, G.K., Knutti, R. and Friedlingstein, P. 2009 Irreversible climate change due to carbon dioxide emissions. *Proceedings of the National Academy of Sciences* 106(6): 1704–9.

Solomon, S., Ivy, D.J., Kinnison, D., Mills, M.J., Neely, R.R. and Schmidt, A. 2016 Emergence of healing in the Antarctic ozone layer. *Science* 353(6296): 269–74.

Sørensen, M., Hvidberg, M., Andersen, Z.J., Nordsborg, R.B., Lillelund, K.G., Jakobsen, J., Tjønneland, A., Overvad, K. and Raaschou-Nielsen, O. 2011 Road traffic noise and stroke: a prospective cohort study. *European Heart Journal* 32: 737–44.

Sousa, P.M., Trigo, R.M., Aizpurua, P., Nieto, R., Gimeno, L. and Garcia-Herrera, R. 2011 Trends and extremes of drought indices throughout the 20th century in the Mediterranean. *Natural Hazards and Earth System Sciences* 11: 33–51.

Spalding, M., Burke, L., Wood, S.A., Ashpole, J., Hutchison, J. and zu Ermgassen, P. 2017 Mapping the global value and distribution of coral reef tourism. *Marine Policy* 82: 104–13.

Spalding, M.D., Meliane, I., Milam, A., Fitzgerald, C. and Hale, L.Z. 2013 Protecting marine spaces: global targets and changing approaches. *Ocean Yearbook Online* 27(1): 213–48.

Spencer, P.D. and Collie, J.S. 1997 Patterns of population variability in marine fish stocks. *Fisheries Oceanography* 6(3): 188–204.

Staley, J.T., Bullock, J.M., Baldock, K.C., Redhead, J.W., Hooftman, D.A., Button, N. and Pywell, R.F. 2013 Changes in hedgerow floral diversity over 70 years in an English rural landscape, and the impacts of management. *Biological Conservation* 167: 97–105.

Stanley, D.J. 1996 Nile delta: extreme case of sediment entrapment on a delta plain and consequent coastal land loss. *Marine Geology* 129: 189–95.

Steadman, D.W. 2006 *Extinction and biogeography of tropical Pacific birds.* Chicago, University of Chicago Press.

Steffen, W., Grinevald, J., Crutzen, P. and McNeill, J. 2011 The Anthropocene: conceptual and historical perspectives. *Philosophical Transactions of the Royal Society A* 369: 842–67.

Steffen, W., Broadgate, W., Deutsch, L., Gaffney, O. and Ludwig, C. 2015 The trajectory of the Anthropocene: the great acceleration. *The Anthropocene Review* 2: 81–98.

Stehouwer, R.C., Sutton, P. and Dick, W.A. 1995 Minespoil amendment with dry flue desulfurization by-products: plant growth. *Journal of Environmental Quality* 24: 861–9.

Steinhauser, G., Brandl, A. and Johnson, T.E. 2014 Comparison of the Chernobyl and Fukushima nuclear accidents: a review of the environmental impacts. *Science of the Total Environment* 470: 800–17.

Steinmann, P., Keiser, J., Bos, R., Tanner, M. and Utzinger, J. 2006 Schistosomiasis and water resources development: systematic review, meta-analysis, and estimates of people at risk. *Lancet Infectious Diseases* 6: 411–25.

Stellman, J.M., Stellman, S.D., Christian, R., Weber, T. and Tomasallo, C. 2003 The extent and patterns of usage of Agent Orange and other herbicides in Vietnam. *Nature* 422: 681–7.

Stiassny, M.L.J. and Raminosoa, N. 1994 The fishes of the inland waters of Madagascar. *Annales du Musée Royal de l'Afrique Centrale Zoologie* 275: 133–49.

Stiles, D. 2009 CITES: approved ivory sales and elephant poaching. *Pachyderm* 45: 150–3.

Stirling, I. and Derocher, A.E. 2012 Effects of climate warming on polar bears: a review of the evidence. *Global Change Biology* 18: 2694–706.

Stivari, S.M.S., de Oliveira, A.U.P. and Soares, J. 2005 On the climate impact of the local circulation in the Itaipu Lake area. *Climatic Change* 72: 103–21.

Stokstad, E. 2005 Louisiana's wetlands struggle for survival. *Science* 310: 1264–6.

Stonich, S.C. and DeWalt, B.R. 1996 The political ecology of deforestation in Honduras. In Sponsel, L.E., Headland, T.N. and Bailey, R.C. (eds) *Tropical deforestation: the human dimension*. New York, Columbia University Press: 187–215.

Straus, J. 2008 How to break the deadlock preventing a fair and rational use of biodiversity. *The Journal of World Intellectual Property* 11: 229–95.

Stuber, N., Forster, P., Rädel, G. and Shine, K. 2006 The importance of the diurnal and annual cycle of air traffic for contrail radiative forcing. *Nature* 441: 864–7.

Stuckler, D., Steele, S., Lurie, M. and Basu, S. 2013 Introduction: 'dying for gold': the effects of mineral mining on HIV, tuberculosis, silicosis, and occupational diseases in southern Africa. *International Journal of Health Services* 43: 639–49.

Suarez, R.K. and Sajise, P.E. 2010 Deforestation, swidden agriculture and Philippine biodiversity. *Philippine Science Letters* 3: 91–9.

Suau-Sanchez, P., Pallares-Barbera, M. and Paül, V. 2011 Incorporating annoyance in airport environmental policy: noise, societal response and community participation. *Journal of Transport Geography* 19: 275–84.

Sullivan, S. 1999 The impacts of people and livestock on topographically diverse open wood- and shrub-lands in arid north-west Namibia. *Global Ecology and Biogeography* 8: 257–77.

Sultan, B., Labadi, K., Guégan, J. and Janicot, S. 2005 Climate drives the meningitis epidemics onset in West Africa. *PLoS Med* 2(1): e6.

Suthakar, K. and Bui, E.N. 2008 Land use/cover changes in the war-ravaged Jaffna Peninsula, Sri Lanka, 1984–early 2004. *Singapore Journal of Tropical Geography* 29: 205–20.

Svidén, O. 1993 Clean fuel and engine systems for twenty-first-century road vehicles. In Giannopoulos, G. and Gillespie, A. (eds) *Transport and communications innovation in Europe*. London, Belhaven: 122–48.

Svoboda, L., Havlíčková, B. and Kalač, P. 2006 Contents of cadmium, mercury and lead in edible mushrooms growing in a historical silver-mining area. *Food Chemistry* 96: 580–5.

Swallow, B., Kallesoe, M., Iftikhar, U., van Noordwijk, M., Bracer, C., Scherr, S., Raju, K.V., Poats, S., Duraiappah, A., Ochieng, B., Mallee, H. and Rumley, R. 2007 Compensation and rewards for environmental services in the developing world: framing pan-tropical analysis and comparison. ICRAF Working Paper No. 32. Nairobi, World Agroforestry Centre.

Swinnerton, C.J. 1984 Protection of groundwater in relation to waste disposal in Wessex Water Authority. *Quarterly Journal of Engineering Geology* 17: 3–8.

Swiss Re 1994 Natural catastrophes and major losses in 1993: insured damage drops significantly. *sigma* 2/94. Zurich, Swiss Reinsurance Company.

Swiss Re 1998 Natural catastrophes and major losses in 1997: exceptionally few high losses. *sigma* 3/98. Zurich, Swiss Reinsurance Company.

Swiss Re 2017 Natural catastrophes and man-made disasters in 2016: a year of widespread damages, *sigma* 2/17. Zurich, Swiss Reinsurance Company.

Syvitski, J.P.M., Vörösmarty, C.J., Kettner, A.J. and Green, P. 2005 Impact of humans on the flux of terrestrial sediment to the global coastal ocean. *Science* 308: 376–80.

Syvitski, J.P.M., Kettner, A.J., Overeem, I., Hutton, E.W.H., Hannon, M.T., Brakenridge, G.R., Day, J., Vörosmarty, C., Saito, Y., Giosan, L. and Nicholls, R.J. 2009 Sinking deltas due to human activities. *Nature Geoscience* 2: 681.

Szasz, F.M. 1995 The impact of World War II on the land: Gruinard Island, Scotland and Trinity Site, New Mexico as case studies. *Environmental History Review* 19(4): 15–30.

Szigeti, C., Toth, G. and Szabo, D.R. 2017 Decoupling – shifts in ecological footprint intensity of nations in the last decade. *Ecological Indicators* 72: 111–17.

Tacon, A.G.J. and Metian, M. 2008 Global overview on the use of fish meal and fish oil in industrially compounded aquafeeds: trends and future prospects. *Aquaculture* 285: 146–58.

Tajrishy, M. and Abrishamchi, A. 2005 Integrated approach to water and wastewater management for Tehran, Iran. In National Research Council, *Water conservation, reuse, and recycling: proceedings of an Iranian-American workshop*. Washington, DC, The National Academies Press: 217–30.

Tamminen, P. and Derome, J. 2005 Temporal trends in chemical parameters of upland forest soils in southern Finland. *Silva Fennica* 39: 313–30.

Taranger, G.L., Karlsen, Ø., Bannister, R.J., Glover, K.A., Husa, V., Karlsbakk, E., Kvamme, B.O., Boxaspen, K.K., Bjørn, P.A., Finstad, B. and Madhun, A.S. 2014 Risk assessment of the environmental impact of Norwegian Atlantic salmon farming. *ICES Journal of Marine Science* 72(3): 997–1021.

Tawfik, R. 2016 The Grand Ethiopian Renaissance Dam: a benefit-sharing project in the Eastern Nile? *Water International* 41(4): 574–92.

Taylor, M. 2007 Voluntary travel behavior change programs in Australia: the carrot rather than the stick in travel demand management. *International Journal of Sustainable Transportation* 1: 173–92.

Teatini, P., Ferronato, M., Gambolati, G., Bertoni, W. and Gonella, M. 2005 A century of land subsidence in Ravenna, Italy. *Environmental Geology* 47: 831–46.

Tegegne, Y.T., Lindner, M., Fobissie, K. and Kanninen, M. 2016 Evolution of drivers of deforestation and forest degradation in the Congo Basin forests: exploring possible policy options to address forest loss. *Land Use Policy* 51: 312–24.

Temple, S.A. 1977 Plant–animal mutualism: coevolution with dodo leads to near extinction of plant. *Science* 197: 885–6.

Thanh, N.C. and Tam, D.M. 1990 Water systems and the environment. In Thanh, N.C. and Biswas, A.K. (eds) *Environmentally-sound water management*. Delhi, Oxford University Press: 1–29.

Themelis, N.J. and Ulloa, P.A. 2007 Methane generation in landfills. *Renewable Energy* 32: 1243–7.

Thiebault, T., Chassiot, L., Fougère, L., Destandau, E., Simonneau, A., Van Beek, P., Souhaut, M. and Chapron, E. 2017 Record of pharmaceutical products in river sediments: a powerful tool to assess the environmental impact of urban management? *Anthropocene* 18: 47–56.

Thitamadee, S., Prachumwat, A., Srisala, J., Jaroenlak, P., Salachan, P.V., Sritunyalucksana, K., Flegel, T.W. and Itsathitphaisarn, O. 2016 Review of current disease threats for cultivated penaeid shrimp in Asia. *Aquaculture* 452: 69–87.

Thomas, D.S.G. and Middleton, N.J. 1994 *Desertification: exploding the myth*. Chichester, Wiley.

Thomas, D.S.G., Knight, M. and Wiggs, G.F.S. 2005 Remobilization of southern African desert dune systems by twenty-first century global warming. *Nature* 435: 1218–21.

Thomas, J.A., Simcox, D.J. and Clarke, R.T. 2009 Successful conservation of a threatened Maculinea butterfly. *Science* 325 (5936): 80–3.

Thompson, D.R., Becker, P.H. and Furness, R.W. 1993 Long-term changes in mercury concentrations in herring gulls *Larus argentatus* and common terns *Sterna hirundo* from the German North Sea coast. *Journal of Applied Ecology* 30: 316–20.

Thrupp, L.A. 1991 Sterilization of workers from pesticide exposure: causes and consequences of DBCP-induced damage in Costa Rica and beyond. *International Journal of Health Services* 21: 731–9.

Tiffen, M., Mortimore, M. and Gichuki, F. 1994 *More people, less erosion: environmental recovery in Kenya.* Chichester, Wiley.

Tilman, D., Reich, P.B. and Isbell, F. 2012 Biodiversity impacts ecosystem productivity as much as resources, disturbance, or herbivory. *Proceedings of the National Academy of Sciences* 109(26): 10394–7.

Tipping, E., Bass, J.A.B., Hardie, D., Haworth, E.Y., Hurley, M.A. and Wills, G. 2002 Biological responses to the reversal of acidification in surface waters of the English Lake District. *Environmental Pollution* 116: 137–46.

Tolba, M.K. 1990 Building an environmental institutional framework for the future. *Environmental Conservation* 17: 105–10.

Tolba, M.K. 1992 *Saving our planet.* London, Chapman & Hall.

Tolba, M.K. and El-Kholy, O.A. (eds) 1992 *The world environment 1972–1992.* London, Chapman & Hall.

Toledo, V.M., Batis, A.I., Becerra, R., Martinez, E. and Ramos, C.H. 1995 La selva util: etnobotánica cuantitativa de los grupos indígenas del trópico húmedo de México. *Interciencia* 20: 177–87.

Tolouie, E., West, J.R. and Billam, J. 1993 Sedimentation and desiltation in the Sefid-Rud reservoir, Iran. In McManus, J. and Duck, R.W. (eds) *Geomorphology and sedimentology of lakes and reservoirs.* Chichester, Wiley: 125–38.

Toon, O.B., Robock, A., Turco, R.P., Bardeen, C., Oman, L., Stenchikov, G.L. 2007 Consequences of regional-scale nuclear conflicts. *Science* 315: 1224–5.

Trenberth, K.E., Jones, P.D., Ambenje, P., Bojariu, R., Easterling, D., KleinTank, A., Parker, D., Rahimzadeh, F., Renwick, J.A., Rusticucci, M., Soden, B. and Zhai, P. 2007 Observations: surface and atmospheric climate change. In Solomon, S., Qin, D., Manning, M., Chen, Z., Marquis, M., Averyt, K.B., Tignor, M. and Miller, H.L. (eds) *Climate change 2007: the physical science basis.* Contribution of Working Group I to the Fourth Assessment Report of the Intergovernmental Panel on Climate Change. Cambridge and New York, Cambridge University Press.

Trevino, H.S., Skibiel, A.L., Karels, T.J. and Dobson, F.S. 2007 Threats to avifauna on oceanic islands. *Conservation Biology* 21: 125–32.

Trigo, E., Cap, E., Malach, V. and Villarreal, F. 2009 The case of zero-tillage technology in Argentina. IFPRI Discussion Paper 00915. Washington, DC, International Food Policy Research Institute (IFPRI).

Triplett, G.B. and Dick, W.A. 2008 No-tillage crop production: a revolution in agriculture! *Agronomy Journal* 100: S-153–65.

Trivelpiece, W.Z., Hinke, J.T., Miller, A.K., Reiss, C.S., Trivelpiece, S.G. and Watters, G.M. 2011 Variability in krill biomass links harvesting and climate warming to penguin population changes in Antarctica. *Proceedings of the National Academy of Sciences* 108(18): 7625–8.

Tropea, A.E., Paterson, A.M., Keller, W. and Smol, J.P. 2010 Sudbury sediments revisited: evaluating limnological recovery in a multiple-stressor environment. *Water, Air, and Soil Pollution* 210: 317–33.

Tucker, C.J. and Nicholson, S.E. 1999 Variations in the size of the Sahara desert from 1980 to 1997. *Ambio* 28: 587–91.

Turner, A. 2010 Marine pollution from antifouling paint particles. *Marine Pollution Bulletin* 60: 159–71.

Turner II, B.L., Kasperson, R.E., Meyer, W.B., Dow, K.M., Golding, D., Kasperson, J.X., Mitchell, R.C. and Ratick, S.J. 1990 Two types of global environmental change: definitional and spatial-scale issues in their human dimensions. *Global Environmental Change* 1: 14–22.

Turner, S.W., Ng, J.Y. and Galelli, S. 2017 Examining global electricity supply vulnerability to climate change using a high-fidelity hydropower dam model. *Science of the Total Environment* 590: 663–75.

Tyrrell, T. 2011 Anthropogenic modification of the oceans. *Philosophical Transactions of the Royal Society A* 369: 887–908.

UN Habitat 2010 *Solid waste management in the world's cities.* London, UN Human Settlement Programme.

UNCTAD 2015 *State of commodity dependence 2014.* Geneva, UNCTAD.

UNDESA (UN Department of Economic and Social Affairs/Population Division) 2004 *World urbanization prospects: the 2003 revision.* New York, UN.

UNDESA 2006 *World urbanization prospects: the 2005 revision.* New York, UN.

UNDESA 2010 *World urbanisation prospects, the 2009 revision.* New York, UN.

UNDESA 2015 *World population prospects: the 2015 revision.* New York, UN.

UNDP and WHO 2009 *The energy access situation in developing countries: a review focusing on least developed countries and SSA.* Sustainable Energy Programme Environment and Energy Group Report. New York, UN.

UNDP/UNICEF 2002 *The human consequences of the Chernobyl nuclear accident: a strategy for recovery.* Vienna, IAEA.

UNECLA (UN Economic Commission for Latin America and the Caribbean) 1990 *The water resources of Latin America and the Caribbean: planning, hazards, and pollution.* Santiago, UNECLA.

UNECLA 1991 *Sustainable development: changing production patterns, social equity and the environment.* Santiago, UNECLA.

UNEP 1992a *World atlas of desertification.* Sevenoaks, Edward Arnold.

UNEP 1992b *The Aral Sea: diagnostic study for the development of an action plan for the conservation of the Aral Sea.* Nairobi, UNEP.

UNEP 1995 *Global biodiversity assessment.* Cambridge, Cambridge University Press.

UNEP 2001 *The Mesopotamian Marshlands: demise of an ecosystem.* Early Warning and Assessment Technical Report, UNEP/DEWA/TR.01-3 Rev. 1. Nairobi, United Nations Environment Programme.

UNEP 2007 *Sudan: post-conflict environmental assessment.* Geneva, UNEP.

UNEP 2011 *Towards a green economy: pathways to sustainable development and poverty eradication.* www.unep.org/greeneconomy.

UNEP 2012 *UNEP Yearbook 2012.* Nairobi, UNEP.

UNEP 2013 *Global mercury assessment 2013: sources, emissions, releases and environmental transport.* Geneva, UNEP Chemicals Branch.

UNEP IE/PAC 1993 *Cleaner production worldwide.* Paris, IE/PAC.

UNEP/WHO 1988 *Assessment of freshwater quality.* Nairobi, UNEP.

UNEP/WHO 1992 *Urban air pollution in megacities of the world.* Oxford, Blackwell.

UNESCO 1978 *World water balance and water resources of the Earth.* Paris, UNESCO.

UN-HABITAT 2006 *State of the world's cities report 2006/7.* London, Earthscan.

UNHCR (UN High Commissioner for Refugees) 1994 Environmental issues in Benaco refugee camp. Press release, 21 June.

UNICEF and WHO 2012 *Progress on drinking water and sanitation: 2012 update.* New York, UN.

UNICEF and WHO 2015 *Progress on sanitation and drinking water: 2015 update and MDG assessment.* Geneva, UN.

United Nations 1989 *Prospects of world urbanization.* Population Studies 112.

United Nations 2001 *Johannesburg summit 2002: world summit on sustainable development.* New York, UN.

United Nations 2016 *The first global integrated marine assessment.* World Ocean Assessment. New York, UN.

UNU-EHS (UN University Institute for Environment and Human Security) 2011 *World risk report 2011.* Brussels, UNU-EHS.

UNU-IHDP and UNEP 2012 *Inclusive wealth report 2012: measuring progress toward sustainability.* Cambridge, Cambridge University Press.

USEPA (US Environmental Protection Agency) 1990 *National water quality inventory. 1988 Report to Congress, Office of Water.* EPA 440-4-90-003. Washington, DC, USEPA.

USEPA 2016 *2015 program progress – cross-state air pollution rule and acid rain program.* Washington, DC, USEPA.

USGS (US Geological Survey)
2016 *Mineral commodity summaries 2016.* Reston, VA, USGS.

Valencia, R., Balslev, H. and Paz y Miño, G. 1994 High tree alpha-diversity in Amazonian Ecuador. *Biodiversity and Conservation* 3: 21–8.

Valiela, I. and Martinetto, P. 2007 Changes in bird abundance in Eastern North America: urban sprawl and global footprint? *BioScience* 57: 360–70.

Valiela, I., Bowen, J. and York, J. 2001 Mangrove forests: One of the world's threatened major tropical environments. *Bioscience* 51: 807–15.

Van, T.T., Wilson, N., Thanh-Tung, H., Quisthoudt, K., Quang-Minh, V., Xuan-Tuan, L., Dahdouh-Guebas, F. and Koedam, N. 2015 Changes in mangrove vegetation area and character in a war and land use change affected region of Vietnam (Mui Ca Mau) over six decades. *Acta Oecologica* 63: 71–81.

Van Beers, D., Corder, G., Bossilkov, A. and Van Berkel, R. 2007 Industrial symbiosis in the Australian minerals industry: the cases of Kwinana and Gladstone. *Journal of Industrial Ecology* 11: 55–72.

Van den Hoek, R.E., Brugnach, M., Mulder, J.P.M. and Hoekstra, A.Y. 2014 Analysing the cascades of uncertainty in flood defence projects: how 'not knowing enough' is related to 'knowing differently'. *Global Environmental Change* 24: 373–88.

van der Wulp, S.A., Dsikowitzky, L., Hesse, K.J. and Schwarzbauer, J. 2016 Master Plan Jakarta, Indonesia: the giant seawall and the need for structural treatment of municipal waste water. *Marine Pollution Bulletin* 110: 686–93.

Van Looy, K., Tormos, T. and Souchon, Y. 2014 Disentangling dam impacts in river networks. *Ecological Indicators* 37: 10–20.

Van Lynden, G.W.J. and Oldeman, L.R. 1997 *The assessment of the status of human-induced soil degradation in South and Southeast Asia.* Nairobi, UNEP, and Wageningen, ISRIC.

van Uhm, D. and Siegel, D. 2016 The illegal trade in black caviar. *Trends in Organized Crime* 19(1): 67–87.

Vázquez-Suñé, E., Sanchez-Vila, X. and Carrera, J. 2005 Introductory review of specific factors influencing urban groundwater, an emerging branch of hydrogeology, with reference to Barcelona, Spain. *Hydrogeology Journal* 13: 522–33.

VEPA (Viet Nam Environment Protection Agency) 2005 *Overview of wetlands status in Viet Nam following 15 years of Ramsar Convention implementation.* Hanoi, VEPA.

Verheijen, F.G.A., Jones, R.J.A., Rickson, R.J. and Smith, C.J. 2009 Tolerable versus actual soil erosion rates in Europe. *Earth-Science Reviews* 94: 23–38.

Vertegaal, P.J.M. 1989 Environmental impact of Dutch military activities. *Environmental Conservation* 16: 54–64.

Viet Nam 1985 *Viet Nam: national conservation strategy.* Ho Chi Minh City, Committee for Rational Utilisation of National Resources and Environmental Protection Programme 52–02.

Virah-Sawmy, M., Ebeling, J. and Taplin, R. 2014 Mining and biodiversity offsets: a transparent and science-based approach to measure 'no-net-loss'. *Journal of Environmental Management* 143: 61–70.

Vogt, H.P. 1995 Coral reefs in Saudi Arabia: 3.5 years after the Gulf War oil spill. *Coral Reefs* 14: 271–3.

Voight, B. 1990 The 1985 Nevado del Rúiz volcano catastrophe: anatomy and retrospection. *Journal of Volcanology and Geothermal Research* 42: 151–88.

von Glasow, R., Jickells, T.D., Baklanov, A., Carmichael, G.R., Church, T.M., Gallardo, L., Hughes, C., Kanakidou, M., Liss, P.S., Mee, L. and Raine, R. 2013 Megacities and large urban agglomerations in the coastal zone: interactions between atmosphere, land, and marine ecosystems. *Ambio* 42: 13–28.

Wada, Y., Flörke, M., Hanasaki, N., Eisner, S., Fischer, G., Tramberend, S., Satoh, Y., Van Vliet, M., Yillia, P., Ringler, C. and Wiberg, D. 2016 Modeling global water use for the 21st century: Water Futures and Solutions (WFaS) initiative and its approaches. *Geoscientific Model Development* 9: 175–222.

Waitz, I.A., Lukachko, S.P. and Lee, J.J. 2005 Military aviation and the environment: historical trends and comparison to civil aviation. *Journal of Aircraft* 42: 329–39.

Wali, A. 1988 *Kilowatts and crisis: a study of development and social change in Panama.* Boulder, CO, Westview.

Waller, N.L., Gynther, I.C., Freeman, A.B., Lavery, T.H. and Leung, L.K.P. 2017 The Bramble Cay melomys Melomys rubicola (Rodentia: Muridae): a first mammalian extinction caused by human-induced climate change? *Wildlife Research* 44(1): 9–21.

Walling, D.E. and Fang, D. 2003 Recent trends in the suspended sediment loads of the world's rivers. *Global and Planetary Change* 39: 111–26.

Walls, M. 2006 Extended producer responsibility and product design: economic theory and selected case studies. Discussion Paper. RFF DP 06-08. Washington, DC, Resources for the Future.

Walsh, C.J., Roy, A.H., Feminella, J.W., Cottingham, P.D., Groffan, P.M. and Morgan, R.P. 2005 The urban stream syndrome: current knowledge and the search for a cure. *Journal of the North American Benthological Society* 24: 706–23.

Walsh, K.J., McBride, J.L., Klotzbach, P.J., Balachandran, S., Camargo, S.J., Holland, G., Knutson, T.R., Kossin, J.P., Lee, T.C., Sobel, A. and Sugi, M. 2016 Tropical cyclones and climate change. *Wiley Interdisciplinary Reviews: Climate Change* 7(1): 65–89.

Walsh, P.D., Henschel, P., Abernethy, K.A., Tutin, C.E.G., Telfer, P. and Lahm, S.A. 2004 Logging speeds little red fire ant invasion of Africa. *Biotropica* 36: 637–41.

Walter, K., Gunkel, G. and Gamboa, N. 2012 An assessment of sediment reuse for sediment management of Gallito Ciego Reservoir, Peru. *Lakes and Reservoirs: Research and Management* 17(4): 301–14.

Walther, G.-R., Post, E., Convey, P., Menzel, A., Parmesan, C., Beebee, T.J.C., Fromentin, J.-M., Hoegh-Guldberg, O. and Bairlein, F. 2002 Ecological responses to recent climate change. *Nature* 416: 389–95.

Wang, W.C., Yung, Y.L., Lacis, A.A., Mo, T. and Hansen, J.E. 1976 Greenhouse effect due to manmade perturbations of other gases. *Science* 194: 685–90.

Wang, Z. and Hu, C. 2009 Strategies for managing reservoir sedimentation. *International Journal of Sediment Research* 24: 369–84.

Warren, A. and Agnew, C. 1988 *An assessment of desertification and land degradation in arid and semi-arid areas.* Drylands Paper 2. London, International Institute for Environment and Development.

Warren, A., Osbahr, H., Batterbury, S. and Chappell, A. 2003 Indigenous views of soil erosion at Fandou Béri, southwestern Niger. *Geoderma* 111: 439–56.

Warren, S.D., Holbrook, S.W., Dale, D.A., Whelan, N.L., Elyn, M., Grimm, W. and Jentsch, A. 2007 Biodiversity and the heterogeneous disturbance regime on military training lands. *Restoration Ecology* 15: 606–12.

Washington, W. and Parkinson, C.L. 2005 *An introduction to three-dimensional climate modeling*, 2nd edn. Mill Valley, CA, University Science Books.

Watt, W.D., Scott, C.D. and White, W.J. 1983 Evidence of acidification of some Nova Scotia rivers and its impact on Atlantic salmon, *Salmo salar. Canadian Journal of Fisheries and Aquatic Science* 40: 462–73.

Watts, N., Adger, W.N., Agnolucci, P., Blackstock, J., Byass, P., Cai, W., Chaytor, S., Colbourn, T., Collins, M., Cooper, A. and Cox, P.M. 2015 Health and climate change: policy responses to protect public health. *Lancet* 386(10006): 1861–914.

WCED (World Commission on Environment and Development) 1987 *Our common future.* Oxford, Oxford University Press.

WCED 1992 *Our common future reconvened.* London, WCED.

Webb, E.L., Jachowski, N.R., Phelps, J., Friess, D.A., Than, M.M. and Ziegler, A.D. 2014 Deforestation in the Ayeyarwady Delta and the conservation implications of an internationally-engaged Myanmar. *Global Environmental Change* 24: 321–33.

Weber, P. 1994 Safeguarding oceans. In Brown, L.R. (ed.) *State of the world 1994.* New York, W.W. Norton: 41–60.

WEC (World Energy Council) 2016 *Survey of energy resources 2016.* London, WEC.

WEC 2017 *World energy resources: bioenergy 2016.* London, WEC.

Wehrli, B. 2011 Climate science: renewable but not carbon-free. *Nature Geoscience* 4: 585–6.

Wei, W., Chen, D., Wang, L., Daryanto, S., Chen, L., Yu, Y., Lu, Y., Sun, G. and Feng, T. 2016 Global synthesis of the classifications, distributions, benefits and issues of terracing. *Earth-Science Reviews* 159: 388–403.

Weinberg, B. 1991 *War on the land: ecology and politics in Central America.* London, Zed Books.

Weinzettel, J., Hertwich, E.G., Peters, G.P., Steen-Olsen, K. and Galli, A. 2013 Affluence drives the global displacement of land use. *Global Environmental Change* 23: 433–8.

Weiss, H., Courty, M.-A., Wetterstrom, W., Guichard, F., Senior, L., Meadow, R. and Curnow, A. 1993 The genesis and collapse of third millennium north Mesopotamian civilisation. *Science* 261: 995–1004.

Wellner, F.-W. and Kürsten, M. 1992 International perspective on mineral resources. *Episodes* 15: 182–94.

Westing, A.H. 1980 *Warfare in a fragile world: military impact on the human environment.* London, Taylor & Francis.

Westing, A.H. 1994 Environmental security for the Horn of Africa: an overview. In Polunin, N. and Burnett, J. (eds) *Surviving with the biosphere.* Edinburgh, Edinburgh University Press: 354–7.

Westlake, K. 1997 Sustainable landfill – possibility or pipe-dream? *Waste Management and Research* 15: 453–61.

Wheatley, S., Sovacool, B.K. and Sornette, D. 2016 Reassessing the safety of nuclear power. *Energy Research and Social Science* 15: 96–100.

White, L.J. 1967 The historical roots of our ecological crisis. *Science* 155: 1203–7.

White, P. 2005 War and food security in Eritrea and Ethiopia, 1998–2000. *Disasters* 29: 92–113.

Whitlow, R.J. 1990 Mining and its environmental impacts in Zimbabwe. *Geographical Journal of Zimbabwe* 21: 50–80.

Whittaker, R.H. and Likens, G.E. 1973 Carbon in the biota. In Woodwell, G.M. and Pecan, E.V. (eds) *Carbon and the biosphere.* Washington, DC, US Department of Commerce: 281–302.

WHO (World Health Organization) 2011 *World malaria report 2011.* Geneva, World Health Organization.

Wiles, G.J., Bart, J., Beck, R.E. and Aguon, C.F. 2003 Impacts of the brown tree snake: patterns of decline and species persistence in Guam's avifauna. *Conservation Biology* 17: 1350–60.

Wilkening, K.E. 2004 *Acid rain science and politics in Japan: a history of knowledge and action toward sustainability.* Cambridge, MA, MIT Press.

Wilkinson, B.H. 2005 Humans as geologic agents: a deep-time perspective. *Geology* 33: 161–4.

Wilkinson, C. (ed.) 2004 *Status of coral reefs of the world: 2004.* Townsville, Queensland, Australian Institute of Marine Science.

Williams, A.E., Duthie, H.C. and Hecky, R.E. 2005 Water hyacinth in Lake Victoria: why did it vanish so quickly and will it return. *Aquatic Botany* 81: 300–14.

Williams, D.D., Williams, N.E. and Cao, Y. 2000 Road salt contamination of groundwater in a major metropolitan area and development of a biological index to monitor its impact. *Water Research* 34: 127–38.

Williams, F., Eschen, R., Harris, A., Djeddour, D., Pratt, C., Shaw, R.S., Varia, S., Lamontagne-Godwin, J., Thomas, S.E. and Murphy, S.T. 2010 *The economic cost of invasive non-native species on Great Britain.* Wallingford, CABI Project No. VM10066.

Williams, J. 1994 The great flood. *Weatherwise* 47: 18–22.

Williams, J.G., Armstrong, G., Katopodis, C., Larinier, M. and Travade, F. 2012 Thinking like a fish: a key ingredient for development of effective fish passage facilities at river obstructions. *River Research and Applications* 28(4): 407–17.

Williams, M.A.J., Dunkerley, D.L., De Deckker, P., Kershaw, A.P. and Stokes, T. 1993 *Quaternary environments.* London, Edward Arnold.

Willis, K.J. and Whittaker, R.J. 2002 Species diversity: scale matters. *Science* 295: 1245–8.

Willis, K.J., Gillson, L. and Brncic, T.M. 2004 How 'virgin' is virgin rainforest? *Science* 304: 402–3.

Willson, B. 1902 *Lost England: the story of our submerged coasts.* London, George Newnes.

Wilson, D.C., Velis, C. and Cheeseman, C. 2006 Role of informal sector recycling in waste management in developing countries. *Habitat International* 30: 797–808.

Wilson, E.O. 1989 Threats to biodiversity. *Scientific American* 261(3): 60–6.

Wilson, J.S. 1858 The general and gradual desiccation of the Earth and atmosphere. *Report of the Proceedings of the British Association for the Advancement of Science*: 155–6.

Wingeyer, A.B., Amado, T.J., Pérez-Bidegain, M., Studdert, G.A., Varela, C.H.P., Garcia, F.O. and Karlen, D.L. 2015 Soil quality impacts of current South American agricultural practices. *Sustainability* 7: 2213–42.

Wirakusumah, A.D. and Rachmat, H. 2017 Impact of the 1815 Tambora Eruption to global climate change. *IOP Conference Series: Earth and Environmental Science* 71(1): 012007.

Wirawan, N. 1993 The hazard of fire. In Brookfield, H. and Byron, Y. (eds) *South-East Asia's environmental future.* Tokyo, UN University Press: 242–60.

Wiser, R., Yang, Z., Hand, M., Hohmeyer, O., Infield, D., Jensen, P.H., Nikolaev, V., O'Malley, P., Sinden, G. and Zervos, A. 2011 Wind energy. In Edenhofer, O. *et al.* (eds) *IPCC special report on renewable energy sources and climate change mitigation.* Cambridge, Cambridge University Press.

Wisner, B., Blaikie, P., Cannon, T. and Davis, I. 2004 *At risk: natural hazards, people's vulnerability and disasters*, 2nd edn. London, Routledge.

Witham, C.S. 2005 Volcanic disasters and incidents: a new database. *Journal of Volcanology and Geothermal Research* 148: 191–233.

Witham, C.S., Oppenheimer, C. and Horwell, C.J. 2005 Volcanic ash-leachates: a review and recommendations for sampling methods. *Journal of Volcanology and Geothermal Research* 141: 299–326.

Wohl, E. 2006 Human impacts to mountain streams. *Geomorphology* 79: 217–48.

Wolf, A.T. 1998 Conflict and cooperation along international waterways. *Water Policy* 1: 251–65.

Wolf, A.T., Yoffe, S.B. and Giordano, M. 2003 International waters: identifying basins at risk. *Water Policy* 5: 29–60.

Wood, L.B. 1982 *The restoration of the tidal Thames.* London, Hilger.

Woodwell, G.M., Wurster, C.F. and Isaacson, P.A. 1967 DDT residues in an east coast estuary: a case of biological concentration of a persistent insecticide. *Science* 156: 821–4.

World Bank 2000 *A review of the World Bank's 1991 forest strategy and its implementation.* Washington, DC, World Bank.

World Commission on Dams 2000 *Dams and development: a new framework for decision-making.* London, Earthscan.

Worster, D. 1979 *Dust bowl.* New York, Oxford University Press.

Wright, R.F. and Hauhs, M. 1991 Reversibility of acidification: soils and surface waters. *Proceedings of the Royal Society of Edinburgh* 97B: 169–91.

Wright, R.F., Couture, R.M., Christiansen, A.B., Guerrero, J.L., Kaste, Ø. and Barlaup, B.T. 2017 Effects of multiple stresses hydropower, acid deposition and climate change on water chemistry and salmon populations in the River Otra, Norway. *Science of the Total Environment* 574: 128–38.

WWAP (United Nations World Water Assessment Programme) 2017 *The United Nations world water development report 2017. Wastewater: the untapped resource.* Paris, UNESCO.

WWF 2013 *WWF guide to building REDD+ strategies: a toolkit for REDD+ practitioners around the globe.* Washington DC, WWF-FCI.

Xu, J. 1993 A study of long-term environmental effects of river regulation on the Yellow River of China in historical perspective. *Geografiska Annaler* 75: 61–72.

Xu, J., Grumbine, R.E., Shrestha, A., Eriksson, M., Yang, X., Wang, Y. and Wilkes, A. 2009 The melting Himalayas: cascading effects of climate change on water, biodiversity, and livelihoods. *Conservation Biology* 23: 520–30.

Xu, X., Tan, Y. and Yang, G. 2013 Environmental impact assessments of the Three Gorges Project in China: issues and interventions. *Earth-Science Reviews* 124: 115–25.

Yakovleva, N. 2005 *Corporate social responsibility in the mining industries.* Aldershot, Ashgate Publishing.

Yalden, D. 1999 *The history of British mammals.* London, T. & A.D. Poyser.

Yan, N.D., Keller, W., Scully, N.M., Lean, D.R.S. and Dillon, P.J. 1996 Increased UV-B penetration in a lake owing to drought-induced acidification. *Nature* 381: 141–3.

Yang, S.L., Milliman, J.D., Li, P. and Xu, K. 2011 50,000 dams later: erosion of the Yangtze River and its delta. *Global and Planetary Change* 75: 14–20.

Yang, Y., Ji, C., Ma, W., Wang, S., Wang, S., Han, W., Mohammat, A., Robinson, D. and Smith, P. 2012 Significant soil acidification across northern China's grasslands during 1980s–2000s. *Global Change Biology* 18: 2292–300.

Yatsukhno, V., 2012. Rural development in the Belarusian Polesie area. The Baltic University Programme, Uppsala University.

Yhdego, M. 1991 Scavenging solid wastes in Dar es Salaam, Tanzania. *Waste Management and Research* 9: 259–65.

Young, A. and Mitchell, N. 1994 Microclimate and vegetation edge effects in a fragmented podocarp-broadleaf forest in New Zealand. *Biological Conservation* 67: 63–72.

Young, J.E. 1992 *Mining the Earth.* Worldwatch Paper 109. Washington, DC, Worldwatch Institute.

Younger, P.L., Coulton, R.H. and Froggatt, E.C. 2005 The contribution of science to risk-based decision-making: lessons from the development of full-scale treatment measures for acidic mine waters at Wheal Jane, UK. *Science of the Total Environment* 338: 137–54.

Yu, Y., Feng, K. and Hubacek, K. 2013 Tele-connecting local consumption to global land use. *Global Environmental Change* 23: 1178–86.

Yu, Z., Beilman, D.W., Frolking, S., MacDonald, G.M., Roulet, N.T., Camill, P. and Charman, D.J. 2011 Peatlands and their role in the global carbon cycle. *Eos, Transactions American Geophysical Union* 92(12): 97–8.

Yusuf, M. 2010 Ethical issues in the use of the terminator seed technology. *African Journal of Biotechnology* 9: 8901–4.

Zacarias-Farah, A. and Geyer-Allely, E. 2003 Household consumption patterns in OECD countries: trends and figures. *Journal of Cleaner Production* 11: 819–27.

Zarfl, C., Lumsdon, A.E., Berlekamp, J., Tydecks, L. and Tockner, K. 2015 A global boom in hydropower dam construction. *Aquatic Sciences* 77: 161–70.

Zavada, M.S., Wang, Y., Rambolamanana, G., Raveloson, A. and Razanatsoa, H. 2009 The significance of human induced and natural erosion features (lavakas) on the central highlands of Madagascar. *Madagascar Conservation and Development* 4(2): 120–7.

Zedler, J.B. and Kercher, S. 2005 Wetland resources: status, trends, ecosystem services, and restorability. *Annual Review of Environment and Resources* 30: 39–74.

Zekri, S. 2008 Using economic incentives and regulations to reduce seawater intrusion in the Batinah coastal area of Oman. *Agricultural Water Management* 95: 243–52.

Zektser, I.S. 2000 *Groundwater and the environment: applications for the global community.* Boca Raton, FL, Lewis Publishers.

Zhan, X., Li, M., Zhang, Z., Goossens, B., Chen, Y., Wang, H., Bruford, M.W. and Wei, F. 2006 Molecular censusing doubles giant panda population estimate in a key nature reserve. *Current Biology* 16: R451–2.

Zhao, B. 2016 Facts and lessons related to the explosion accident in Tianjin Port, China. *Natural Hazards* 84: 707–13.

Zhao, G., Mu, X., Wen, Z., Wang, F. and Gao, P. 2013 Soil erosion, conservation, and eco-environment changes in the loess plateau of China. *Land Degradation and Development* 24: 499–510.

Zhao, W., Jiao, E., Wang, G. and Meng, X. 1992 Analysis on the variation of sediment yield in Sanchuanhe River basin in 1980s. *International Journal of Sedimentary Research* 7: 1–19.

Zhao, Y., Duan, L., Xing, J., Larssen, T., Nielsen, C.P. and Hao, J. 2009 Soil acidification in China: is controlling SO_2 emissions enough? *Environmental Science and Technology* 43: 8021–6.

Zhu, Q., de Vries, W., Liu, X., Hao, T., Zeng, M., Shen, J. and Zhang, F. 2018 Enhanced acidification in Chinese croplands as derived from element budgets in the period 1980–2010. *Science of the Total Environment* 618: 1497–505.

Zika, M. and Erb, K.H. 2009 The global loss of net primary production resulting from human-induced soil degradation in drylands. *Ecological Economics* 69: 310–18.

Zingoni, E., Love, D., Magadza, C., Moyce, W. and Musiwa, K. 2005 Effects of a semi-formal urban settlement on groundwater quality: Epworth (Zimbabwe): case study and groundwater quality zoning. *Physics and Chemistry of the Earth A/B/C* 30(11): 680–8.

Zítková, J., Hegrová, J. and And-l, P. 2018 Bioindication of road salting impact on Norway spruce (Picea abies). *Transportation Research Part D: Transport and Environment* 59: 58–67.

Zonn, I.S. 1995 Desertification in Russia: problems and solutions (an example in the Republic of Kalmykia-Khalmg Tangch). *Environmental Monitoring and Assessment* 37: 347–63.

Zurayk, R.A. 1994 Rehabilitating the ancient terraced lands of Lebanon. *Journal of Soil and Water Conservation* 49: 106–12.

Index

Page numbers in **bold** indicate tables and in *italic* indicate figures.